Second Edition

# Handbook of
# ALTERNATIVE FUEL TECHNOLOGIES

# GREEN CHEMISTRY AND CHEMICAL ENGINEERING

## Series Editor: Sunggyu Lee
### Ohio University, Athens, Ohio, USA

**Second Edition**

# Handbook of

# ALTERNATIVE FUEL TECHNOLOGIES

**Edited by**
**Sunggyu Lee**
**James G. Speight**
**Sudarshan K. Loyalka**

CRC Press
Taylor & Francis Group
Boca Raton London New York

CRC Press is an imprint of the
Taylor & Francis Group, an **informa** business

CRC Press
Taylor & Francis Group
6000 Broken Sound Parkway NW, Suite 300
Boca Raton, FL 33487-2742

© 2015 by Taylor & Francis Group, LLC
CRC Press is an imprint of Taylor & Francis Group, an Informa business

No claim to original U.S. Government works

Printed on acid-free paper
Version Date: 20140117

International Standard Book Number-13: 978-1-4665-9456-2 (Hardback)

### Library of Congress Cataloging-in-Publication Data

Handbook of alternative fuel technologies / editors, Sunggyu Lee, James G. Speight, and
   Sudarshan K. Loyalka. -- Second edition.
       pages cm -- (Green chemistry and chemical engineering)
    Summary: "This book focuses on the different intermediates and raw material options
that can generate energy output and products equivalent to conventional petroleum sources.
It presents short-term options for clean alternative energy sources that complement the
development of long-term sustainable energy infrastructures. Detailing the chemical
processes for each technology, the text also assesses the environmental impact, benefits,
and performance of the various processes and fuel products. It also summarizes processing
and transportation issues, safety concerns, regulations, and other practical considerations
associated with alternative fuels"-- Provided by publisher.
    Includes bibliographical references and index.
    ISBN 978-1-4665-9456-2 (hardback)
    1. Fuel--Handbooks, manuals, etc. 2. Fuel switching--Handbooks, manuals, etc. 3. Power
resources--Handbooks, manuals, etc. I. Lee, Sunggyu. II. Speight, James G. III. Loyalka, S. K.

TP318.H28285 2014
662.6028'6--dc23                                                                2014000192

**Visit the Taylor & Francis Web site at**
**http://www.taylorandfrancis.com**

**and the CRC Press Web site at**
**http://www.crcpress.com**

# Contents

# Series Statement

The subjects and disciplines of chemistry and chemical engineering have encountered a new landmark in the way of thinking about, developing, and designing chemical products and processes. This revolutionary philosophy, termed *green chemistry and chemical engineering*, focuses on the designs of products and processes that are conducive to reducing or eliminating the use and/or generation of hazardous substances. In dealing with hazardous or potentially hazardous substances, there may be some overlaps and interrelationships between environmental chemistry and green chemistry. While environmental chemistry is the chemistry of the natural environment and the pollutant chemicals in nature, green chemistry proactively aims to reduce and prevent pollution at its very source. In essence, the philosophy of green chemistry and of chemical engineering tend to focus more on industrial application and practice rather than academic principles and phenomenological science. However, as both a chemistry and chemical engineering philosophy, green chemistry and chemical engineering derive from and build upon organic chemistry, inorganic chemistry, polymer chemistry, fuel chemistry, biochemistry, analytical chemistry, physical chemistry, environmental chemistry, thermodynamics, chemical reaction engineering, transport phenomena, chemical process design, separation technology, automatic process control, and more. In short, green chemistry and chemical engineering are the rigorous use of chemistry and chemical engineering for pollution prevention and environmental protection.

The Pollution Prevention Act of 1990 in the United States established a national policy to prevent or reduce pollution at its source whenever feasible. Adhering to the spirit of this policy, the Environmental Protection Agency (EPA) launched its Green Chemistry Program in order to promote innovative chemical technologies that reduce or eliminate the use or generation of hazardous substances in the design, manufacture, and use of chemical products. The global efforts in green chemistry and chemical engineering have recently gained a substantial amount of support from the international communities of science, engineering, academia, industry, and government in all phases and aspects.

Some of the successful examples and key technological developments include the use of supercritical carbon dioxide as a green solvent in separation technologies; application of supercritical water oxidation for destruction of harmful substances; process integration with carbon dioxide sequestration steps; solvent-free synthesis of chemicals and polymeric materials; exploitation of biologically degradable materials; use of aqueous hydrogen peroxide for efficient oxidation; development of hydrogen proton exchange membrane (PEM) fuel cells for a variety of power generation needs; advanced biofuel productions; devulcanization of spent tire rubber; avoidance of the use of chemicals and processes causing generation of volatile organic compounds (VOCs); replacement of traditional petrochemical processes by microorganism-based bioengineering processes; replacement of chlorofluorocarbons (CFCs) with nonhazardous alternatives; advances in design of energy-efficient processes;

use of clean, alternative, and renewable energy sources in manufacturing; and much more. This list, even though it is only a partial compilation, is undoubtedly growing exponentially.

This series on Green Chemistry and Chemical Engineering by CRC Press/Taylor & Francis is designed to meet the new challenges of the twenty-first century in the chemistry and chemical engineering disciplines by publishing books and monographs based upon cutting-edge research and development to the effect of reducing adverse impacts upon the environment by chemical enterprise. In achieving this, the series will detail the development of alternative sustainable technologies, which will minimize the hazard and maximize the efficiency of any chemical choice. The series aims at delivering to the readers in academia and industry an authoritative information source in the field of green chemistry and chemical engineering. The publisher and its series editor are fully aware of the rapidly evolving nature of the subject and its long-lasting impact upon the quality of human life in both the present and the future. As such, the team is committed to making this series the most comprehensive and accurate literary source in the field of green chemistry and chemical engineering.

**Sunggyu Lee**

# Preface

Energy has always been the foremost important resource upon which humans have relied for survival and productive activities. Industrialization and technological advancement of modern society have also been possible through the effective use of energy. There is a strong correlation between the index for quality of life and energy consumption. Heightened economic strength of a country, technological prosperity of a society, higher production output of an industry, improved finances of a household, and increased activities of an individual are also realized by effective utilization of energy.

A number of important factors have historically dominated the trend, market, and type of energy utilization. These factors are (1) resource availability, (2) convenience of energy utilization, (3) efficiency of conversion, (4) technological feasibility, (5) portability and ease of transportation, (6) sustainability, (7) renewability, (8) cost and affordability, (9) regional strength, (10) safety and health effects, and (11) environmental acceptance and impact. The technological success and prosperity of petrochemical industries in the twentieth and early twenty-first centuries can largely be attributed to the vast utilization of fossil fuels, especially petroleum, as well as technological breakthroughs and innovations by process industries. Industry and consumers have seen and come to expect a wide array of new and improved polymeric materials and other chemical and petrochemical products. However, the fossil fuel resources upon which industry is heavily dependent are limited in available quantities and are expected to be close to depletion in the near future. The unprecedented popularity and successful utilization of petroleum resources observed in the twentieth century may decline in the twenty-first century due to a lack of resource availability, thus making prospects for future sustainability seem grim. Public appetites for convenient and alternative fuel sources and superior high-performance materials are, however, growing. Therefore, additional and alternative sources for fuels and petrochemical feedstocks not only should be developed further but are also needed for immediate commercial exploitation. Use of alternative fuels is no longer a matter for the future; it is a realistic issue of the present.

Additional and alternative sources for intermediate and final products, whether fuels or petrochemicals, directly contribute to the conservation of petroleum resources of the world by providing additional raw material options for generating the same products for consumers. Examples may include biogas for methanol; grain ethanol; lignocellulosic alcohol; biodiesel from crop oil and algae; BTX (benzene, toluene, and xylenes) and valuable petrochemicals from coal; biogas or bioliquid from agricultural wastes and plant biomass; hydrogen as transportation fuel and for fuel cells; biohydrogen from a variety of biological sources; jet fuel from shale oil or crop oil; Fischer–Tropsch fuel from coal or biomass; bisphenols from agricultural sources; liquid transportation fuels from a natural gas source by catalysis; dimethyl ether (DME) for internal combustion engines; target olefins via conversion of synthesis gas; use of coal-derived acetylene for petroleum-derived ethylene as a building block chemical; liquid fuels from spent tires or mixed wastes; natural gas from shale gas, tight gas, and natural gas hydrates; polymer synthesis using wet gas from oil shale; solar liquid fuels; and so forth.

If usable energy or deliverable power is the final product to be desired, alternate sources for energy may strongly and directly affect the lifestyle of consumers, as well as their energy consumption patterns. A good example can be found in electric cars that are powered by powerful rechargeable batteries. These powerful batteries serve no use for conventional gasoline motors, while in turn, premium gasoline is not needed in these electric cars. Another good example is the solar house whose climate control inside the house is provided solely by solar energy. Other examples include liquefied petroleum gas (LPG) vehicles, DME buses, hybrid cars, E85 vehicles, hydrogen vehicles, fuel cell vehicles, solar-powered equipment and vehicles, wind energy–powered equipment, geothermal heating and cooling, and so forth.

During the past several decades, there has been a considerable increase in research and development (R&D) in areas of environmentally acceptable alternative fuels. Synthetic fuels were of prime interest in the 1970s, due to a sudden shortage of petroleum supply kindled by an oil embargo in 1973, as well as public concerns of dwindling petroleum reserves. While synfuels seemed to be a most promising solution to the conservation of petroleum resources (or, at least, frugal use of the resources) and the development of additional sources for conventional liquid fuels, some of the focus has been shifted toward environmental acceptance of the fuel and the long-term sustainability of world prosperity in the last decade of the twentieth century. Efforts have been made to reduce emissions of air pollutants associated with combustion processes whose sources include electric power generation and vehicular transportation. Air pollutants that have been targeted for minimization or elimination include $SO_x$, $NO_x$, $CO_x$, VOCs, particulate matter (PM), mercury, and selenium. These efforts have significantly contributed to the enhancement of air quality and associated technologies.

Concerns of global warming, via greenhouse gases (GHGs), have further intensified the issue of environmental acceptance of fuel consumption. Combustion of fossil fuels inevitably generates carbon dioxide due to an oxidation reaction of hydrocarbons and carbonaceous materials. Carbon dioxide is known as a major GHG, whose emission needs to be significantly reduced. Therefore, new developments in alternative fuels and energy have focused more on nonfossil sources or on mitigation and fixation of carbon dioxide in fossil fuel utilization. Renewable energy sources are certainly very promising due to their long-term sustainability and environmental friendliness. Of particular interest are solar (solar thermal and photovoltaic), wind, hydropower, tidal, and geothermal energies, in addition to biomass (wood, wood waste, plant/crop-based renewables, agricultural wastes, food wastes, and algae) and biofuels including bioethanol, biohydrogen, bio-oil, and biodiesel. It should be noted that hydropower is also regarded as a "conventional" energy source, as it has provided a significant amount of electrical energy for over a century. Government mandates, tax incentives, and stricter enforcement of environmental regulations are pushing environmentally friendly alternative fuels into the marketplace at an unprecedented rate.

The number of alternative-fueled vehicles in use in the world is expected to increase sharply. These alternative-fueled vehicles are powered by LPG, liquefied natural gas (LNG), compressed natural gas (CNG), ethanol 85% (E85), methanol

85% (M85), electricity, neat methanol (M100), ethanol 95% (E95), DME, Fischer–
Tropsch fuel, and hydrogen, among which hydrogen presently accounts for very
little but is considered quite promising by many. It should be noted that this list of
alternative fuels in vehicles represents only the successful results of the past developments and does not include recent advances and breakthroughs in the field. R&D
efforts in alternative-fueled vehicles and utilization of renewable energy sources have
intensified in the past few years. Alternative-fueled vehicles and emission-free cars
are expected to gain more popularity due in part to enforcement of stricter emission
standards, the unmistakable fate of depletion for conventional transportation fuels,
and numerous tax incentives for such vehicles. This intensified interest is coupled
with the record-high prices of gasoline and petroleum-based products experienced
all over the world. Perhaps the key difference between the 1973 oil embargo era and
the present is that this time around, efforts are likely to firmly latch on to the roster
of ongoing priorities most exigent to mankind.

Energy from wastes cannot be neglected as a valuable energy source. If effectively harnessed, energy from wastes, including municipal solid waste (MSW), agricultural refuse, plastic wastes and spent tires, and mixed wastes can be employed to
alleviate the current burden for energy generation from fossil fuel sources. Moreover,
energy generation from wastes bears extra significance in reducing the volume of
wastes, thus saving landfill space and utilizing resources that would otherwise be of
no value. Environmental aspects involving waste energy generation are to be fully
addressed in commercial exploitation.

A great number of research articles, patents, reference books, textbooks, monographs, government reports, and industry brochures are published and referenced
everyday. However, these literary sources are not only widely scattered and massive
in volume, but they are also lacking in scientific consistency and technological comprehensiveness. Further, most of the published articles focus on the justification and
potential availability of alternative fuel sources rather than environmental and technical readiness of the fuel as a principal energy source for the future postpetroleum era.

This handbook aims to present comprehensive information regarding the science and technology of alternative fuels and their processing technologies. Special
emphasis has been placed on environmental and socioeconomic issues associated
with the use of alternative energy sources, such as sustainability, applicable technologies, mode of utilization, and impacts on society.

Chapter 1 focuses on the current concerns in the area of consumption of conventional energy sources and highlights the importance of further development and
utilization of alternative, renewable, and clean energy sources. This chapter presents
past statistics as well as future predictions for each of the major conventional and
alternative energy sources of the world.

Chapter 2 deals with the science and technology of coal gasification to produce
synthesis gas. Synthesis gas is a crucially important petrochemical feedstock and
also serves as an intermediate for other valuable alternative fuels such as methanol,
DME, ethanol, gasoline, diesel, jet fuel, and hydrogen. Since the technology developed for gasification of coal has been widely modified and applied to processing of
other fuel sources such as oil shale, biomass, and mixed wastes, details of various

gasifiers and gasification processes of technological significance are presented in this chapter.

Chapter 3 covers the science and technology of coal liquefaction for production of clean liquid fuels. All aspects of pyrolysis, direct liquefaction, indirect liquefaction, and coal–oil coprocessing liquefaction are addressed in detail. This chapter has significant relevance to the development and production of alternative transportation fuels that can replace or supplement the conventional transportation fuels. The scientific and technological concepts developed for coal liquefaction serve as invaluable foundations for other kinds of fuel processing.

Chapter 4 deals with the science and technology of coal slurry fuels. Major topics in this chapter include slurry properties and characterization, hydrodynamics, slurry types, transportation, and environmental issues. The chapter also explains the important factors that are essential in developing successful slurry fuels that involve nontraditional carbon sources such as petroleum coke.

Chapter 5 discusses the liquid fuels obtained from natural gas. Special emphasis is also placed upon Fischer–Tropsch synthesis, whose chemistry, catalysis, and commercial processes are detailed.

Chapter 6 presents the science and technology of petroleum resids. Properties and characterization of resids as well as process conversion of resids are detailed in this chapter.

Chapter 7 describes the occurrence, production, and properties of oil sand bitumen and the methods used to convert the bitumen to synthetic crude oil. Properties of synthetic crude oil are also discussed.

Chapter 8 explores the science and technology of oil shale utilization. In particular, occurrence, extraction, and properties of oil shale kerogen are discussed. A variety of oil shale retorting processes as well as shale oil upgrading processes are described. This chapter is focused on the conventional aspects of oil shale development.

Chapter 9 discusses modern efforts of shale gas and shale fuel based on horizontal drilling and hydraulic fracture. The chapter discusses the shale fuel recovery technology currently being utilized, economic and environmental prospects, technology development directions, details of fracking chemicals being used, and more.

Chapter 10 focuses on the synthesis of methanol from synthesis gas. Chemical reaction mechanisms, synthesis catalysis, and industrial process technologies of methanol synthesis are described. The methanol economy is also explained.

Chapter 11 deals with the production of fuel ethanol from corn. The chapter elucidates the chemistry, fermentation technology, and unit operations involved in the production process. Both wet milling and dry milling processes are detailed. Moreover, the chapter discusses the environmental benefits of the use of ethanol as internal combustion fuel or as oxygenated additives.

Chapter 12 discusses the detailed process steps and technological issues that are involved in the conversion of lignocellulosic materials into fuel ethanol. The issues of sustainability and renewability are also addressed.

Chapter 13 deals with biodiesel production via transesterification of triglycerides obtained from crop oil. Details of process chemistry, biodiesel properties, source-specific natures, and merits of biodiesel utilization are explained.

Chapter 14 discusses various aspects of algae fuel technology in terms of microalgae growth, algae harvesting, algae oil extraction, oil upgrading, and more. The chapter also discusses the advantages of algae fuel as an alternative fuel source.

Chapter 15 focuses on the thermochemical conversion of biomass. The chapter covers biomass gasification, fast pyrolysis of biomass, production of biochar, and more. Details of the technology developed and designs of reactor systems are also discussed.

Chapter 16 focuses on energy generation from waste resources. Particular emphasis is placed on beneficial utilization of MWSs, mixed wastes, polymeric waste, and scrap tires.

Chapter 17 describes the occurrence, renewability, and environmentally beneficial utilization of geothermal energy. Geothermal power plants, district heating, and geothermal heat pumps are also discussed.

Chapter 18 deals with the science and technology of nuclear energy. The chapter describes nuclear reactor physics, nuclear fuel cycles, types of reactors, and electricity generation from nuclear reactors. Public concerns of safety and health are also discussed.

Chapter 19 presents the basic concepts of fuel cells and describes a number of different types of fuel cells and their characteristics. Hydrogen production and storage are also discussed in this chapter.

This book is unique in its nature, scope, perspectives, and completeness. Detailed description and assessment of available and feasible technologies, environmental health and safety issues, governmental regulations, issues and agendas for R&D, and alternative energy networks for production, distribution, and consumption are covered throughout the book. For R&D scientists and engineers, this handbook serves as a single-volume comprehensive reference that will provide necessary information regarding chemistry, technology, and alternative routes as well as scientific foundations for further enhancements and breakthroughs.

This book can also be used as a textbook for a three-credit-hour course entitled "Alternative Fuels," "Renewable Energy," "Fuel Engineering," or "Fuel Processing." The total number of chapters is slightly more than the total number of weeks in a typical college semester. However, depending upon the specific nature and aims of the course, certain chapters can be combined and omitted in the actual course curriculum. This book may also be adopted as a reference book for a more general subject on fuel science and engineering, energy and environment, energy and environmental policy, and others. Professors and students may find this book a vital source for their design or term projects for a number of other core courses.

All chapters are carefully authored for scientific accuracy, style consistency, notational and unit consistency, and cross-reference convenience so that readers will enjoy the consistency and comprehensiveness of this book.

Finally, the authors are deeply indebted to their former graduate students, colleagues, and family members for their assistance, encouragement, and helpful comments.

# Editors

**Dr. Sunggyu Lee** is the Russ Ohio Research Scholar in Coal Syngas Utilization and professor of chemical and biomolecular engineering at Ohio University in Athens, Ohio. Prior to his current position, he also taught at the University of Akron, University of Missouri, and the Missouri University of Science and Technology. His specialty areas include alternative fuels and renewable energy, advanced and functional materials, as well as chemical reaction and process engineering. Dr. Lee has 34 U.S. patents and over 180 international patents in addition to more than 500 archival publications. He is the editor of the *Encyclopedia of Chemical Processing* published by Taylor & Francis. He earned his PhD from Case Western Reserve University and his BS and MS from Seoul National University.

**Dr. James G. Speight,** C CChem, FRSC, FCIC, FACS, earned his BSc and PhD from the University of Manchester, England—he also holds a DSC in The Geological Sciences (VINIGRI, St. Petersburg, Russia) and a PhD in Petroleum Engineering (Dubna International University, Moscow, Russia). Dr. Speight is the author of more than 50 books in petroleum science, petroleum engineering, and environmental sciences. Formerly the CEO of the Western Research Institute (now an independent consultant), he has served as adjunct professor in the Department of Chemical and Fuels Engineering at the University of Utah and in the Departments of Chemistry and Chemical and Petroleum Engineering at the University of Wyoming. In addition, he has also been a visiting professor in chemical engineering at the following universities: University of Missouri-Columbia, Technical University of Denmark, and University of Trinidad and Tobago.

Dr. Speight was elected to the Russian Academy of Sciences in 1996 and awarded the Gold Medal of Honor that same year for outstanding contributions to the field of petroleum sciences. He has also received the Scientists Without Borders Medal of Honor of the Russian Academy of Sciences. In 2001, the Academy also awarded Dr. Speight the Einstein Medal for outstanding contributions and service in the field of geological sciences.

**Dr. Sudarshan Loyalka** was educated in Pilani, India (BE 1964) and Stanford University (MS 1965, PhD 1967). He is currently a Curators' Professor of Nuclear Engineering in the Nuclear Science and Engineering Institute at the University of Missouri in Columbia. His educational and research interests are in transport theory, aerosol mechanics, the kinetic theory of gases, and neutron reactor physics and safety. He is a fellow of both the American Physical Society (since 1982, for elucidating the role of gas–surface interactions on molecular transport) and the American Nuclear Society (since 1985, for contributions to reactor physics and safety and aerosol mechanics). He is a recipient of the David Sinclair Award (1995) of the American Association for Aerosol Research and the Glenn Murphy Award (1998) of the American Society for Engineering Education. Professor Loyalka's research has been supported by the NSF, EPA, NASA, DOD, NIH, DOE, NRC, and the industry. He has also consulted with national laboratories and several companies in both the nuclear (fission product transport and heat transfer) and environmental (molecular and aerosol transport) areas. He is a coauthor/coeditor of five books, including *Aerosol Science: Theory and Practice with Special Applications to the Nuclear Industry*, by M. M. R. Williams and S. K. Loyalka, Pergamon/Elsevier (1991). Professor Loyalka is a member of the editorial advisory boards of both *Progress in Nuclear Energy* and the *Annals of Nuclear Energy*. He is a registered professional nuclear engineer in Missouri.

# Contributors

**Aaron Gonzales**
Sustainable Energy and Advanced
    Materials (SEAM) Lab
Ohio University
Athens, Ohio

**Mihaela F. Ion**
Nuclear Science and Engineering
University of Missouri–Columbia
Columbia, Missouri

**H. Bryan Lanterman**
DRS Technologies, Inc.
Alexandria, Virginia

**Sunggyu Lee**
Chemical and Biological Engineering
University of Missouri–Rolla
Rolla, Missouri

**Sudarshan K. Loyalka**
Nuclear Science and Engineering
University of Missouri–Columbia
Columbia, Missouri

**James G. Speight**
CD&W Inc.
Laramie, Wyoming

**Ryan Tschannen**
Sustainable Energy and Advanced
    Materials (SEAM) Lab
Ohio University
Athens, Ohio

**Amber Tupper**
Sustainable Energy and Advanced
    Materials (SEAM) Lab
Ohio University
Athens, Ohio

**Maxwell Tobias Tupper**
Sustainable Energy and Advanced
    Materials (SEAM) Lab
Ohio University
Athens, Ohio

**Barbara Wheelden**
Sustainable Energy and Advanced
    Materials (SEAM) Lab
Ohio University
Athens, Ohio

# 1 Global and U.S. Energy Overview

*Sunggyu Lee and Barbara Wheelden*

## CONTENTS

## 1.1   WORLD ENERGY CONSUMPTION

History shows that world energy consumption has been steadily increasing and also that a huge increase in the energy consumption has taken place in the last two centuries. The reasons for this steady increase include enhancements in quality of human life, population increase, industrialization, rapid economic growth of developing countries, increased transportation of people and goods, and so forth. There are many different types of fuel available worldwide, the demand for which strongly depends on application and use, geographical location and regional resources, cost and affordability, "cleanness" and environmental impact factors, safety of generation and utilization, socioeconomic and technoeconomic factors, global and regional politics, new energy technologies, and so forth. The energy utilization cycle consists of three phases: generation, distribution, and consumption, all of which must be closely balanced for an ideal energy infrastructure. Any bottlenecking or shortage would immediately affect the entire cycle as a limiting factor. If there is a decrease in production of a certain type of fuel, the distribution and consumption of this specific fuel would also decrease accordingly. As propagating consequences, fuel switching from this type to another, if possible, as well as forced conservation of more affected fuel types become inevitable. Further, based on the supply-and-demand principle, the consumer price of this fuel type would undoubtedly rise. Even a breakdown in the transportation system of a certain fuel type would affect the consumer market directly, and consequences such as fuel shortages and price hikes would be realized at least for a limited time in the affected region. Similar shortage and instability situations can also be caused by unexpected shutdowns of major production facilities, natural disasters, infrastructure overloads, national and international politics, and more.

Figure 1.1 summarizes world energy consumption, both history and future projections, for each of the principal fuel types from 1990 to 2040 [1]. The analysis and projections reported by the Energy Information Administration (EIA), U.S. Department of Energy, are regarded as the most reliable data source. As shown in Figure 1.1, all types of energy sources recorded steady increases for the reported period between 1990 and 2010. Based on the trend over the past 20 years, the following observations can be made:

1. All types of energy have seen steady increases.
2. Coal energy consumption recorded the steepest growth of all types for the past 20 years. In particular, its consumption increase in the first decade of the twenty-first century is quite remarkable and easily outpaced all other types. This sharp increase in coal consumption may be attributable to several reasons, including (1) comparatively much lower energy costs than other conventional fossil fuels, in particular, natural gas and petroleum; (2) coal being regionally more reliable and, cost-wise, a more stable fuel source; and (3) technological advances achieved during prior decades in clean coal utilization and emission control.
3. The consumption of natural gas and liquids steadily increased in a nearly monotonous trend. The increased consumption of petroleum resources was mainly due to the rapid consumption growth of the emerging economy of China, India, and other rapidly industrializing nations.

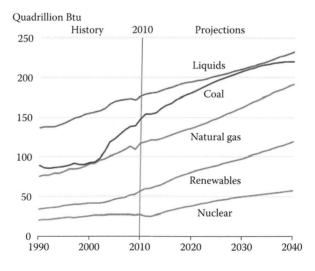

**FIGURE 1.1** World energy consumption by fuel type, 1990–2040. (From EIA Annual Energy Outlook 2013.)

4. The consumption growth in nuclear energy has been slow for the past 20-year period, while new orders for construction of nuclear plants in developing nations has been sharply increasing in recent years.
5. The consumption of renewable energy increased at a fast pace, especially in more recent years, due to advances in new technologies. Since the statistical data for liquid biofuels, which are also renewable, are assigned to that of liquids, the total renewable energy consumption is higher than that reported in Figure 1.1.

Figure 1.1 also shows the EIA's future consumption projections for all energy types based on available information. The International Energy Outlook 2013 (IEO2013) projects that the world's total energy consumption will grow by 56% for the period from 2010 to 2040, from 524 quadrillion to 820 quadrillion Btu. One quadrillion Btu is $10^{15}$ Btu. This rate of growth is, approximately, at an annual rate of 1.5% growth. IEO2013 also projects that the Organization for Economic Co-operation and Development (OECD) nations' consumption growth will be only 17%, while the non-OECD nations' consumption growth will be 90%. The following observations regarding the future projections can be made:

1. Fossil fuels will continue to supply a great portion, approximately 80%, of the world energy through 2040.
2. Renewable energy will be the single fastest-growing energy source in the world, at an annual rate of approximately 2.5% per year. This optimism is based on the recent technological advances in the field as well as the global commitments in energy sustainability.
3. The average annual growth rate of 1.5% is lower than the generally expected economic growth of the world for the period. Considering that the energy consumption has somewhat outpaced that of the economic growth for the past 100 years, this may seem to be low at first glance. However, this projection by the IEO2013 includes the global efforts and achievements in energy efficiency increase in all aspects of energy consumption.
4. Projected growth in natural gas consumption is faster than that for liquids, which may reflect optimism developed in recent years with shale gas development in the U.S. Since shale fuel is distributed abundantly throughout the world, its commercial development would increase the availability of affordable natural gas. Further, this is also attributable to stronger demands for natural gas in industrial and residential heating, increased installations of natural gas-based electric power plants, and new discoveries of large natural gas deposits, both conventional and unconventional.
5. A sharp increase in nuclear energy consumption is also projected for the period, which reflects the growing interest in and large orders for new power plants in non-OECD countries.

## 1.2   U.S. ENERGY CONSUMPTION

Figure 1.2 shows the total U.S. energy consumption in quadrillion Btu for the period from 1950 to 2011 [2]. Based on the data, it is noted that U.S. total energy

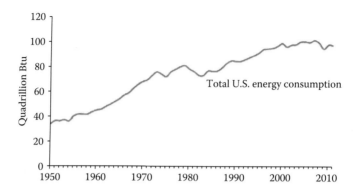

**FIGURE 1.2**   Total U.S. energy consumption, 1950–2011. (From EIA Annual Energy Review 2011.)

consumption has tripled over the past 50 years, that is, from 1950 to 2000. Over the first 25 years of this period, the increase was about 2.4 times, whereas it was about 1.3 times over the following 25 years. A slowdown of the pace of U.S. energy consumption was noticed immediately after the oil crisis of 1973. Many factors may have contributed to this: to name a few, increase in energy conversion efficiency, energy conservation across the board, energy-efficient products, and even climates becoming milder due to global warming. However, if we consider separately the period from 1973 to 1988, for which total U.S. energy consumption was fairly stable and did not change much, the recent rate of increase for the period from 1988 to 2000 was as steep as that for the initial 25 years, that is, from 1950 to 1975. The period from 1973 to 1988 also coincides with the years when energy process development efforts in the U.S. were very active and public awareness of energy conservation was quite strong. During the 1990s, energy prices were stable, and research and development in energy technology took a backseat, partly due to the lack of immediate market competitiveness of alternative fuels. This was also the time when energy consumption sharply increased again in the United States as there was little fear of an imminent global energy crisis in the consumers' minds. Another decreasing trend was noticed in 2007 and 2008, when the market prices for petroleum and natural gas were both high. Around this time, the U.S. domestic production of both crude oil and natural gas started to increase sharply, mainly due to the so-called "shale fuel boom."

Figure 1.3 shows the U.S. energy consumption data by source from 1950 to 2011 [3]. The data are reported in quadrillion Btu. Based on the past 60 years of data, the following may be summarized:

1. The consumption of petroleum has fluctuated significantly in response to the market conditions.
2. In the initial 30 years, petroleum consumption had been growing at a steady rate, whereas the consumption growth for the next 30 years has been more

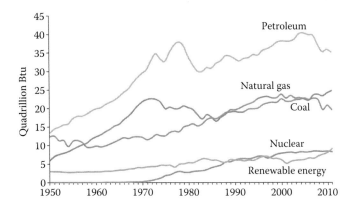

**FIGURE 1.3**   U.S. energy consumption by fuel type, 1950–2011. (From EIA Annual Energy Review 2011.)

fluctuating and has slowed down. The total consumption figure for 2010 was even lower than the maximum annual consumption recorded in 1977. This is attributable to a number of factors including (1) increased use of bio-based liquid fuels such as ethanol and biodiesel, (2) improved vehicular fuel efficiency, and so forth.

3. Natural gas consumption has very steadily increased in the recent 50 years. In 2007, the natural gas consumption in the United States exceeded that of coal. This was mainly due to the increasing trend of switching from coal to natural gas for electric power generation in the United States. The switch to natural gas was partly due to the fast-falling price of natural gas in the early 2010s.

4. Also seen from the past statistics, the consumption of renewable energy sources has sharply increased. This increase was mainly due to the technological advances as well as improved long-term economics in comparison to other nonrenewable energy sources.

## 1.3   U.S. ENERGY PRODUCTION

Figure 1.4 shows the U.S. energy production by energy source for the period from 1950 to 2011. The following can be summarized:

1. Coal has been the most produced source in the United States since 1968. Coal has been considered the most reliable energy source in the United States in all aspects of production, consumption, and distribution.

2. Domestic crude oil production has been steadily declining since 1980 due to depletion of conventional oil wells. The trend has been quickly reversed since 2008, mainly due to crude oil production from oil shale formations.

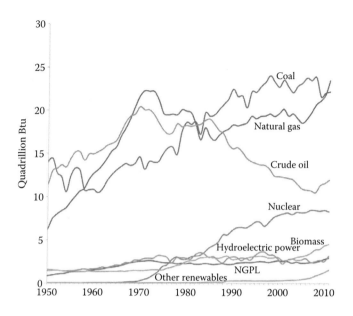

**FIGURE 1.4**    U.S. energy production by source, 1950–2011. (From EIA Annual Energy Review 2011.)

3. Nuclear electric power generation in the United States has also been steadily increasing since 1970.
4. Biomass energy has seen an accelerated production growth since 2000, which is mainly due to increased biofuel production. On this chart, bioethanol and biodiesel are included in biomass energy.

Figure 1.5 shows the annual data for U.S. total energy consumption, production, imports, and exports for the period from 1950 to 2011 [2]. Energy imports by the United States steadily increased from 1983 till 2007. The trend of increase of energy imports has changed to a decreasing trend, while energy exports by the United States have been sharply increasing since 2007. Energy production in the United States has also been sharply increasing since 2007, which also explains the transition in the export–import trend. The difference between the energy consumption in the United States and the domestic energy production is the net annual deficit amount of energy for the United States, which has to be covered by the net international trade, that is, imports minus exports. The difference between the energy consumption and production in the United States has been decreasing since roughly 2007, and this trend provides a renewed optimism in energy independence for the United States in the foreseeable future. Needless to say, rapid development and market adoption of alternative fuels and renewable energy will be a major contributor for the energy independence and sustainability.

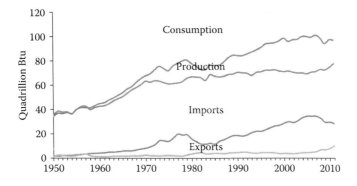

**FIGURE 1.5**    U.S. primary energy overview, 1950–2011. (From EIA Annual Energy Review 2011.)

## 1.4  PETROLEUM

Worldwide petroleum consumption data are summarized in Table 1.1 for the period from 2008 to 2012 [4].

For this period, world petroleum consumption has grown at an average rate of 1.36% a year. The petroleum consumption in North America and Europe decreased for the period, while the petroleum consumptions in Asia and Oceania, the Middle East, Central and South America, Eurasia, and Africa have seen increases. In fact, the total worldwide consumption growth of 4.71 million barrels a day for the period is outnumbered by the consumption growth in the Asia and Oceania region of 4.857 million barrels a day for the same period. The largest increase in petroleum consumption was realized in China, due to its rapid economic growth.

**TABLE 1.1**
**Total Petroleum Consumption by Region in Thousand Barrels per Day**

|  | 2008 | 2009 | 2010 | 2011 | 2012 |
|---|---|---|---|---|---|
| North America | 23,893 | 23,014 | 23,534 | 23,337 | 22,995 |
| Central and South America | 6014 | 6106 | 6331 | 6571 | 6765 |
| Europe | 16,152 | 15,375 | 15,337 | 14,961 | 14,423 |
| Eurasia | 4156 | 4133 | 4160 | 4366 | 4529 |
| Middle East | 6500 | 6752 | 6991 | 7538 | 7637 |
| Africa | 3141 | 3260 | 3374 | 3297 | 3360 |
| Asia and Oceania | 24,710 | 26,172 | 27,719 | 28,593 | 29,567 |
| World | 84,565 | 84,813 | 87,446 | 88,662 | 89,275 |

*Source:*   U.S. Energy Information Administration (EIA). International energy statistics. 2013. Available:
http://www.eia.gov/countries/data.cfm.

Estimation of the years for which petroleum can be supplied and consumed at the current consumption rate has often been made by professionals and policymakers, but the numbers have been inconsistent and fluctuating from year to year. This uncertainty comes from the difficulty of estimating the future recoverable amount of petroleum from all the proved and unproved reserves. The Society of Petroleum Engineers (SPE) and the World Petroleum Council (WPC) have developed and approved several definitions of petroleum reserve-related terms to facilitate consistency among professionals using these terms [5]:

> *Proved reserves* are those quantities of petroleum that, by analysis of geological and engineering data, can be estimated with reasonable certainty to be commercially recoverable, from a given date forward, from known reservoirs and under current economic conditions, operating methods, and government regulations. Proved reserves can be further categorized as *developed* or *undeveloped.*
>
> *Unproved reserves* are based on geologic and engineering data similar to that used in estimates of proved reserves, but technical, contractual, economic, or regulatory uncertainties preclude such reserves being classified as proved. Unproved reserves may be further classified as *probable reserves* and *possible reserves.*
>
> *Probable reserves* are those unproved reserves that an analysis of geological and engineering data suggests are more likely to be recoverable than not. If a probabilistic interpretation is to be given, there should be at least a 50% probability that the quantities actually recovered will equal or exceed the sum of estimated proved plus probable reserves.
>
> *Possible reserves* are those unproved reserves that an analysis of geological and engineering data suggests are less likely to be recoverable than probable reserves. If a probabilistic interpretation is to be given, there should be at least a 10% probability that the quantities actually recovered will equal or exceed the sum of estimated proved plus probable plus possible reserves.

According to the *Oil and Gas Journal* [6], the world proven petroleum reserve as of January 1, 2012, is estimated to be 1.523 trillion barrels. If we use the world petroleum consumption rate of 2012 as a fixed rate, the worldwide petroleum reserve would be able to sustain the current level of consumption for an additional 47.6 years. The very same number calculated using the 2004 available data was 43.4 years [7]. This seemingly controversial increase in the estimated number of years for petroleum reserve is mainly due to new discoveries of new oil fields and advances in oil production and recovery technologies.

Table 1.2 shows the top 10 nations with major petroleum reserves. The top nations' reserves account for about 80% of the world's total proven oil reserve estimate. The U.S. ranks fourteenth with 26.8 billion barrels.

Currently, transportation, fuel, and petrochemical industries depend very heavily upon petroleum-based feedstocks. Therefore, alternative hydrocarbon feedstocks replacing petroleum-based feedstocks for petrochemicals and supplementing

**TABLE 1.2**

**Top 10 Nations with Proven Oil Reserves (2012)**

| Country | Proven Oil Reserve |
|---|---|
| Venezuela | 296.5 |
| Saudi Arabia | 265.4 |
| Canada | 175 |
| Iran | 151.2 |
| Iraq | 143.1 |
| Kuwait | 101.5 |
| United Arab Emirates | 136.7 |
| Russia | 74.2 |
| Kazakhstan | 49 |
| Libya | 47 |

*Source:* Organization of the Petroleum Exporting Countries (OPEC). *OPEC share of world crude oil reserves 2012.* Available: http://www.opec.org/opec_web/en/data_graphs/330.htm. Units: billion barrels.

petroleum-derived products must be developed and utilized more. Necessary infrastructure also needs to be developed and changed to make a transition from the current petroleum economy.

## 1.5 NATURAL GAS

In recent years, natural gas has gained popularity among many industrial sectors. It burns cleaner than coal or petroleum, thus providing environmental benefits. It is distributed mainly via pipelines and in a liquid phase (called *liquefied natural gas* [LNG]) transported across oceans by tankers.

Table 1.3 shows the worldwide consumption of dry natural gas by regions for the period from 2008 to 2012 [4].

As shown, the worldwide consumption of natural gas steadily increased at a pace of 2.3% per year, which is significantly faster than that for petroleum. According to the U.S. EIA, the world's proven natural gas reserve as of January 1, 2013, is 6845 trillion cubic feet. Assuming that the current level of natural gas consumption for the world is maintained, the reserve would be enough to last for another 57.7 years, provided factors such as increased yearly consumption, discovery of new deposits, and advances in technology, such as utilization of natural gas hydrates, are not taken into account. Even though this rough estimate may look somewhat better than that of petroleum, the fate of natural gas is more or less the same as that of petroleum.

Table 1.4 shows the worldwide distribution of natural gas for regions with major reserves [4]. Projection of the world reserves of natural gas has generally increased, at least by numbers, due to new discoveries of major natural gas fields, whose estimated reserves offset more than the annual consumption.

**TABLE 1.3**
**Dry Natural Gas Consumption (Billion Cubic Feet)**

|                          | 2008      | 2009      | 2010      | 2011      | 2012     |
|--------------------------|-----------|-----------|-----------|-----------|----------|
| North America            | 28,416.9  | 28,107.6  | 29,286.1  | 29,912.8  | NA       |
| United States            | 23,277.0  | 22,910.0  | 24,087.0  | 24,385.0  | 25,502.0 |
| Central and South America| 4655.3    | 4383.5    | 5106.4    | 5193.2    | NA       |
| Europe                   | 20,556.2  | 19,405.9  | 20,377.7  | 18,909.0  | NA       |
| Eurasia                  | 22,948.8  | 19,105.8  | 21,113.0  | 24,816.6  | NA       |
| Middle East              | 11,664.2  | 12,468.6  | 13,277.5  | 14,096.5  | NA       |
| Africa                   | 3543.2    | 3384.4    | 3557.9    | 3780.9    | NA       |
| Asia and Oceania         | 17,447.1  | 18,407.1  | 20,602.5  | 22,000.8  | NA       |
| World                    | 109,231.6 | 105,262.8 | 113,321.1 | 118,709.7 | NA       |

*Note:* The U.S. consumption data have already been included in the North America total.

**TABLE 1.4**
**Proven Reserves of Natural Gas (Trillion Cubic Feet)**

|                          | 2009    | 2010    | 2011    | 2012    | 2013    |
|--------------------------|---------|---------|---------|---------|---------|
| North America            | 315.7   | 347.2   | 378.5   | 412.4   | NA      |
| United States            | 244.7   | 272.5   | 304.6   | 334.1   | NA      |
| Central and South America| 266.5   | 266.8   | 268.5   | 270.0   | 268.9   |
| Europe                   | 169.1   | 166.3   | 153.8   | 146.9   | 145.5   |
| Eurasia                  | 1993.8  | 2164.8  | 2164.8  | 2164.8  | 2177.8  |
| Middle East              | 2591.7  | 2658.3  | 2686.4  | 2800.0  | 2823.2  |
| Africa                   | 495.1   | 496.2   | 518.6   | 545.7   | 514.8   |
| Asia and Oceania         | 430.5   | 538.7   | 537.6   | 504.8   | 504.4   |
| World                    | 6262.4  | 6638.2  | 6708.2  | 6844.6  | NA      |

*Source:* U.S. Energy Information Administration (EIA). International energy statistics. 2013. Available: http://www.eia.gov/countries/data.cfm.
*Note:* The U.S. data are included in the North American total.

Russia has about 1688 trillion cubic feet of proven natural gas reserve, which accounts for 24.7% of the world natural gas reserves, whereas the combined total for the Middle East accounts for 40.9% (at 2800 trillion cubic feet). In terms of natural gas consumption based on the 2011 statistics, the United States accounts for 20.5%, whereas Asia and Oceania account for 18.5%. This is quite different from the consumption pattern for petroleum, which is the globally preferred transportation fuel. Energy provided by natural gas can be obtained by other sources or replaced by other types of energy depending upon the region's infrastructure, available energy resources, and supply-and-demand dynamics.

As shown in Figure 1.3, consumption of natural gas in the United States has increased while consumption of coal has decreased in recent years. Natural gas was the second-most-used energy source in the United States in 2011. The major consumers of natural gas are public utilities, manufacturers, residential consumers (heating homes and cooking), and commercial users, mainly for heating buildings, as shown in Table 1.5.

Natural gas helps manufacture a wide variety of goods and chemicals including plastics, fertilizers, photographic films, inks, synthetic rubber, fibers, detergents, glues, methanol, ethers, insect repellents, and much more [9]. It is used in electric power generation as it burns cleaner and more efficiently than coal and has less emission-related problems than other popular fossil fuels. However, natural gas has only a limited market share as a transportation fuel, even though it can be used in regular internal combustion engines. This is mainly due to its low energy density per volume unless it is compressed under very high pressure. Over half of U.S. homes use natural gas as the main heating fuel. Any major disruption in the natural gas supply would bring out quite grave consequences in the nation's energy management, at least for the short term and/or for a certain affected region, as natural gas is heavily utilized by both electric power-generating utilities and residential homes. However, the regional energy dependence problem has been somewhat mitigated by deregulation of utilities, which altered the business practices of electric utilities and natural gas industry. Deregulation allows customers to purchase their natural gas from suppliers other than their local utility, thus providing choices for consumers and eventually resulting in better value for them.

Natural gas is distributed primarily via pipelines. In the United States, more than 1 million miles of underground pipelines are connected between natural gas fields and major cities. This gas can be liquefied by cooling to −260°F (−162°C) and is much more condensed in volume (615 times) when compared to natural gas at room temperature, making it easier to store or transport. LNG in special tanks can be transported by trucks or by ships as LNG has the fluidity and volume compactness of other liquid fuels. In this regard, it has some of the necessary qualities for a transportation fuel. As a result, a large number of LNG storage facilities are currently being operated in the world, and the number is rapidly increasing.

Like all other fossil fuels, natural gas also generates carbon dioxide (a major greenhouse gas [GHG]) upon combustion. Also, natural gas by itself is a GHG.

**TABLE 1.5**
**End Use of Natural Gas in the United States (2011)**

| End Use | Percentage |
|---|---|
| Electric power generation | 31.1 |
| Industrial | 27.4 |
| Residential | 20.1 |
| Commercial | 13.0 |
| Oil and gas operations | 5.4 |
| Pipeline fuel | 2.8 |
| Vehicular fuel | 0.1 |

*Source:* U.S. Energy Information Administration (EIA). International energy statistics. 2013. Available: http://www.eia.gov/countries/data.cfm.

Therefore, in all phases of generation, storage, and transportation, preventive measures must be undertaken to ensure that accidental release of natural gas does not occur due to any leakage.

## 1.6 COAL

Coal is primarily consumed in electric power generation and in industrial sectors. In 2007, coal consumption accounted for 28% of the total energy consumption in the world. About 65% of coal consumption was used for electric power generation, 31% for industrial consumers such as steel manufacturers and steam generators, and much of the remaining 4% for consumers in residential and commercial sectors. Coal was once an important transportation fuel for powering steam engines; however, coal nowadays is rarely used in transportation. Table 1.6 shows worldwide consumption of coal by regions and countries for the period from 2008 to 2012 [4].

Table 1.7 shows the world's top 10 countries with major coal production in 2012 [10]. China is the world's largest coal producer with 2.83 billion tons, which is about 36% of the total world coal production.

According to the EIA projection, the world coal consumption is expected to grow at an average rate of 1.3% per year, from 147 quadrillion Btu in 2010 to 180 quadrillion Btu in 2040. In the near term, coal consumption is expected to expand significantly in China, India, and non-OPEC nations. In the United States, coal use in electric power generation as well as industrial sectors has slowed due to environmental policies and regulations that encourage the use of cleaner energy sources as well as to the development of shale gas, which has made natural gas economically competitive [1]. However, the use of coal in the United States is still expected to be steady in future years as the most dependable fossil fuel source.

## TABLE 1.6
## Worldwide Total Coal Consumption (in 1000 Short Tons)

|  | 2008 | 2009 | 2010 | 2011 | 2012 |
|---|---|---|---|---|---|
| North America | 1,200,108 | 1,069,811 | 1,124,933 | 1,071,818 | NA |
| United States | 1,120,548 | 997,478 | 1,051,307 | 1,003,066 | 890,483 |
| Central and South America | 44,791 | 39,342 | 46,971 | 49,568 | NA |
| Europe | 1,034,173 | 959,646 | 966,325 | 1,003,697 | NA |
| Eurasia | 421,200 | 395,151 | 412,682 | 428,365 | NA |
| Middle East | 17,159 | 15,233 | 16,316 | 16,811 | NA |
| Africa | 230,678 | 224,394 | 222,543 | 214,972 | NA |
| Asia and Oceania | 4,351,152 | 4,768,246 | 4,960,749 | 5,338,369 | NA |
| World | 7,299,261 | 7,471,823 | 7,750,518 | 8,123,601 | NA |

*Source:* U.S. Energy Information Administration (EIA). International energy statistics. 2013. Available: http://www.eia.gov/countries/data.cfm.

*Note:* The U.S. data are included in the North America total.

**TABLE 1.7**

**Top 10 Coal Producing Countries (2012) (in Million Short Tons)**

| Country | Production |
|---|---|
| China | 2831 |
| United States | 849 |
| India | 509 |
| Indonesia | 373 |
| South Africa | 250 |
| Australia | 199 |
| Russia | 178 |
| Kazakhstan | 98 |
| Colombia | 80 |
| Poland | 65 |

*Source:* World Coal Association. *Coal statistics.* Available: http://www.worldcoal. org/resources/coal-statistics/.

Total recoverable reserves of coal around the world are estimated at 948 billion tons as of 2008 [4], which would be enough to last approximately 130 years if maintained at the 2008 consumption level of 7.299 billion tons. The reserve amount was recently adjusted downward after applying more restrictive criteria, that is, safe and economical recoverability. Even though coal deposits are distributed widely throughout the world, about 57% of the world's recoverable coal reserves are located in three countries: United States (27%), Russia (17%), and China (13%). After these three, six countries account for 33% of the total reserves: India, Australia, South Africa, Ukraine, Kazakhstan, and Yugoslavia. Coal is also very unequally and unevenly distributed, just as are other fossil fuels such as petroleum and natural gas.

Coal has been studied extensively for conversion into gaseous and liquid fuels, as well as hydrocarbon feedstocks. Largely thanks to its relative abundance and stable fuel price on the market, coal has been a focal target for synthetic conversion into other forms of fuel, that is, synfuels. Research and development (R&D) work has seen research ups and downs due to external factors, including the comparative fossil fuel market, as well as the international energy outlook of the era. Coal can be gasified, liquefied, pyrolyzed, and coprocessed with other fuels including oil, biomass, scrap tires, and municipal solid wastes (MSWs). Secondary conversion of coal-derived gas and liquids can generate a wide array of petrochemical products, as well as alternative fuels.

In 2012, coal was the second-largest leading source of carbon dioxide emissions from the consumption and flaring of fossil fuels, accounting for 31% of the total. The leading primary source of carbon dioxide emission was from the consumption of petroleum, accounting for 43% of the total. In third place was natural gas at 26% [11].

## 1.7 NUCLEAR ENERGY

Table 1.8 shows worldwide nuclear electric power generation data, and Figure 1.6 shows the projected increase in nuclear electricity generation by country, from 2010 to 2040. Approximately 31.8% of the world nuclear electric power generation in 2011 was in the United States, followed by France (17.7%), Russia (6.7%), South Korea (6.0%), Germany (4.2%), and Japan (3.9%). The market share of world electric power

**TABLE 1.8**
**Nuclear Electricity Net Generation (Billion Kilowatt-Hours)**

|  | 2008 | 2009 | 2010 | 2011 | 2012 |
|---|---|---|---|---|---|
| North America | 904.775 | 894.418 | 898.679 | 886.177 | 868.727 |
| United States | 806.208 | 798.855 | 806.968 | 790.204 | 769.331 |
| Central and South America | 20.9335 | 20.06259 | 20.4886 | 20.7683 | 21.173 |
| Europe | 907.281 | 865.29173 | 892.804 | 885.394 | 859.652 |
| Eurasia | 250.214 | 246.402 | 248.331 | 248.911 | 252.519 |
| Middle East | 0 | 0 | 0 | 0.09798 | 1.328 |
| Africa | 11.317 | 13.004 | 12.806 | 12.099 | 12.398 |
| Asia and Oceania | 508.12545 | 529.234 | 547.108 | 453.769 | 327.734 |
| World | 2602.64595 | 2568.41231 | 2620.2166 | 2507.21628 | 2343.531 |

*Source:* U.S. Energy Information Administration (EIA). International energy statistics. 2013. Available: http://www.eia.gov/countries/data.cfm.

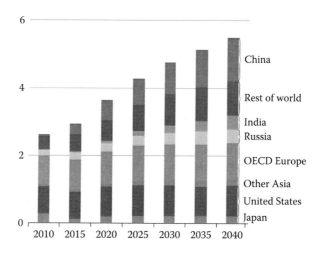

**FIGURE 1.6** World net electricity generation from nuclear power by region, 2010–2040 (trillion kilowatt-hours). (From U.S. Energy Information Administration (EIA). International energy outlook 2013. Available: http://www.eia.gov/forecasts/ieo/more_highlights.cfm.)

generation in 2011 by nuclear energy is 11.7%, while that for the United States is 19.0%. Dependence of electric power generation on nuclear energy in 2011 was by far the heaviest for France at 79.4% [12].

According to the IEO2013, electric power generation using nuclear power plants from around the world is projected to increase from 2620 billion kilowatt-hours in 2010 to 4255 billion kilowatt-hours in 2025 and 5492 billion kilowatt-hours in 2040 [1]. The outlook for nuclear energy in general improved substantially over recent years, due to a number of reasons that include

1. Higher fossil fuel prices
2. Higher capacity utilization rates reported for many existing nuclear facilities
3. Expectation that most existing plants in the mature markets and transitional economies will be granted extensions in operating lives
4. Enforcement of Kyoto protocol
5. Anticipation for hydrogen economy and need for cost-effective electrical energy

However, predicting the trend for nuclear energy is still very difficult, owing to considerable uncertainties originating from political and socioeconomic factors.

## 1.8   RENEWABLE ENERGY

All fossil fuels are nonrenewable, and as such, they will eventually be depleted. As they are based on finite resources and their distributions are heavily localized in certain areas of the world, they will become expensive. Further, energy generation from fossil fuels requires combustion, thus damaging the environment with pollutants and GHG emission. In order to sustain the future of the world with a clean environment and nondepletive energy resources, renewable energy is the obvious choice. Renewable energy sources include solar energy, wind energy, geothermal energy, biomass, and hydrogen. Most renewable energy, except for geothermal energy, comes directly or indirectly from the sun. Benefits of renewable energy are numerous, and they include

1. Environmental cleanness without pollutant emission
2. Nondepletive nature
3. Availability throughout the world
4. No cause for global warming
5. Waste reduction
6. Stabilization of energy costs
7. Creation of jobs

As can be seen in Figure 1.3, the share of renewable energy is still very minute, about 6% of the total energy consumption. With increases in the prices of petroleum and natural gas, as was experienced in 2005 and 2006, the relative competitiveness of renewable and alternative fuels is drastically improving. Further, technological advances in the alternative renewable energy areas, as well as public awareness backed by strong governmental support and incentives, make the outlook of alternative and renewable energy very promising (Table 1.9).

Worldwide generation of geothermal, solar, wind, wood, and waste electric power increased at an average annual rate of 11.7% from 2001 to 2011. In 2011, the United States led the world with 205 billion kilowatt-hours, followed by China with 110 billion, Germany with 109 billion, Spain with 56.71 billion, and Italy with 40 billion. These five countries accounted for about 56% of the world's geothermal, solar, wind, wood, and waste electric power generation in 2011 [4].

Ethanol from corn has been increasingly used as gasoline-blending fuel. One new brand is E85, which contains 85% ethanol and 15% gasoline. Many gas stations in the United States have started to stock E85 fuels regularly, and many automakers are offering multiple lines of automobiles that can be operated on either conventional gasoline or E85. Phasedown of methyl tertiary–butyl ether (MTBE), once the most popular oxygenated blend fuel, in many U.S. states also accelerated the use of ethanol as an oxygenated gasoline-blend fuel. Public awareness of clean-burning and energy-efficient hydrogen has also propelled unprecedented interest in hydrogen technology and fuel cell R&D. Many experts predict the future to be a hydrogen economy. For the hydrogen economy to be realized, a long list of technological advances must be accomplished, which includes technologies for inexpensive generation, safe distribution and storage, safe and efficient materials for hydrogen handling, hydrogen internal combustion engines, hydrogen fuel cells, loss prevention, and so forth.

Energy generation utilizing biomass and MSWs are also promising in regions where landfill spaces are very limited. Technological advances in the field have made this option both efficient and environmentally safe.

## TABLE 1.9
## Total Renewable Electricity Net Generation (Billion Kilowatt-Hours)

|  | 2008 | 2009 | 2010 | 2011 | 2012 |
|---|---|---|---|---|---|
| North America | 822.78967 | 845.63804 | 853.97435 | 967.90898 | NA |
| United States | 392.73567 | 429.65204 | 440.23135 | 524.46198 | 507.78804 |
| Central and South America | 702.42315 | 725.30892 | 736.09638 | NA | NA |
| Europe | 816.82089 | 847.67309 | 947.62524 | NA | NA |
| Eurasia | 238.61203 | 245.526 | 246.87 | NA | NA |
| Middle East | 11.90615 | 13.22073 | 17.94634 | NA | NA |
| Africa | 97.64847 | 102.10877 | 110.9038 | NA | NA |
| Asia and Oceania | 1039.45939 | 1088.78896 | 1253.78324 | NA | NA |
| World | 3729.65975 | 3868.26451 | 4167.19935 | NA | NA |

*Source:* U.S. Energy Information Administration (EIA). International energy statistics. 2013. Available: http://www.eia.gov/countries/data.cfm.

# REFERENCES

1. U.S. Energy Information Administration (EIA). International energy outlook. Available at http://www.eia.gov/forecasts/ieo/more_highlights.cfm, 2013.
2. U.S. Energy Information Administration (EIA). Annual energy review: Primary energy overview. Available at http://www.eia.gov/totalenergy/data/annual/pdf/sec1_4.pdf, 2013.
3. U.S. Energy Information Administration (EIA). Annual energy review: Primary energy consumption by source, September 27, 2013.
4. U.S. Energy Information Administration (EIA). International energy statistics. Available at http://www.eia.gov/countries/data.cfm, 2013.
5. Society of Petroleum Engineers. Petroleum reserves and resources definitions. Available at http://www.spe.org/industry/reserves.php, 2013.
6. D. Rachovich. World's top 23 proven oil reserve holders. Available at http://petroleuminsights.blogspot.com/2012/01/worlds-top-23-proven-oil-reserves.html#.UlsqDdLOnZc, January 1, 2012.
7. S. Lee, J. G. Speight and S. K. Loyalka. *Handbook of Alternative Fuel Technologies.* Boca Raton, FL: CRC Press, 2007.
8. Organization of the Petroleum Exporting Countries (OPEC). OPEC share of world crude oil reserves. Available at http://www.opec.org/opec_web/en/data_graphs/330.htm, 2012.
9. S. Lee. *Methane and its Derivatives.* New York: Marcel Dekker, 1997.
10. World Coal Association. Coal statistics. Available at http://www.worldcoal.org/resources/coal-statistics/, 2013.
11. U.S. Energy Information Administration (EIA). Total energy monthly energy review—environment, EIA, September 2013.
12. International Energy Agency. Key world energy statistics, OECD/IEA, Paris, France, 2013.

# 2 Gasification of Coal

*Sunggyu Lee*

## CONTENTS

## 2.1  BACKGROUND

Conversion of solid coal by any combination of chemical, physical, and thermal treatments to produce a mixture of combustible gases is termed *coal gasification*, even though a large number of chemical reactions, which may not be typically classified as coal gasification reactions, are also involved. Although the product gases of coal gasification mainly involve combustible chemical species, the purpose of gasification is not limited to generation of gaseous fuel only, since the product gas can be easily processed to generate other valuable chemical and petrochemical feedstock. Commercial gasification of coal generally entails the controlled partial oxidation (POX) of the coal to convert it into desired gaseous products. The coal can be heated either directly by thermal energy generated by combustion or indirectly by another heat source. A gasifying medium is typically passed over (or through) the heated coal particles or beds to facilitate intimate molecular contact between the gaseous reactants and coal chemicals for efficient gasification reaction. The gaseous reactants react with carbonaceous matters of coal (i.e., coal hydrocarbons) and also with other primary decomposition products and process intermediates of coal to produce gaseous products. These gaseous products are termed as raw coal syngas, raw gasifier effluent, or raw product gas. Not all the gaseous products generated by such processes are desirable from the standpoints of fuel quality, further petrochemical processing, and environmental safety. Therefore, coal gasification is always performed in connection with downstream processes not only for final applications but also for gas-cleaning purposes. The primary emphases of coal gasification may be on electric power generation via *integrated gasification combined cycle* (IGCC) types, on syngas production for pipeline applications, on *hydrogen production*, on processing of mixed feedstocks of biomass and coal, or on synthesis of *liquid fuels* and petrochemicals as alternative sources for raw materials. With the advent of a hydrogen economy, the role of coal gasification in generation of hydrogen may become even more important [1].

Conversion of coal from its solid form to a gaseous fuel (or gaseous chemical) is widely practiced today. During earlier years (1920–1940), coal gasification was being employed to produce *manufactured gas* in hundreds of plants worldwide, and such plants were called *manufactured gas plants* (MGPs) [2,3]. This technology became obsolete in the post-World War II era due to the abundant supply of petroleum and natural gas at affordable prices. With the advent of the *oil embargo* in the early 1970s and subsequent escalation of petroleum prices, as well as globally experienced shortages of natural gas and petroleum supply, the interest in coal gasification as well as its further commercial exploitation was revived. Surging interest in fuel cell technology also prompted keen interest in coal gasification as a potential means of obtaining reliable and inexpensive hydrogen sources. Many major activities in research, development, and demonstration of coal gasification have recently resulted in significant improvements in conventional technology and thus made coal gasification more competitive in modern fuel and petrochemical markets. However, rapid commercial development and exploitation of shale gas in the United States and other regions of the world has recently contributed to the stabilization of natural gas supply and has impacted on the commercial interest in coal gasification for syngas production in those nations [4].

The concept of electric power generation based on coal gasification received its biggest boost in the 1990s when the U.S. Department of Energy's Clean Coal Technology Program provided federal cost sharing for the first true commercial-scale IGCC plants in the United States. Tampa Electric Company's Polk Power Station near Mulberry, Florida, is the nation's first "greenfield" (built as a brand new plant, not a retrofit) commercial IGCC power station [1]. The plant, dedicated in 1997, is capable of producing 313 MWe (megawatts of electricity) and removing more than 98% of sulfur in coal that is converted into commercial products. On the other hand, the Wabash River Coal Gasification Repowering Project was the first full-size commercial gasification combined cycle plant built in the United States, located in West Terre Haute, Indiana. The plant started full operations in November 1995. It is capable of producing 292 MWe and is still one of the largest single-train IGCCs operating commercially [1]. There are several other IGCC plants currently operating and noteworthy in the world, and they include Willem-Alexander IGCC Plant (253 MWe, commercial production started in 1998), Buggenum, The Netherlands; Puertollano IGCC Plant (335 MWe, production started in 1999) in Puertollano, Spain; and JGC IGCC Plant (342 MWe, built in 2003) in Kanagawa, Japan.

The power generation based on the IGCC technology provides numerous benefits besides the enhanced energy efficiency of the plant, and they include (a) lower emissions of sulfur, particulates, and mercury; (b) providing technoeconomic solutions for carbon dioxide capture and sequestration (i.e., so-called "capture ready"); and (c) applicability to high-sulfur coals, biomass, and heavy petroleum resids. However, a major hurdle in the commercial integration and implementation of the IGCC technology is its high capital cost, which is estimated to be 1.2 to 2.8 times higher than a conventional clean coal facility. However, the IGCC technology may become more attractive if we include the cost factors of the carbon capture and sequestration, which is an important feature in the next-generation IGCC technology.

Coal gasification includes a series of reaction steps that convert coal containing C, H, and O, as well as impurities such as S and N, into *synthesis gas* (SG) and other forms of hydrocarbons. This conversion is generally accomplished by introducing a gasifying agent (typically air or oxygen, usually co-fed with steam) into a reactor vessel containing coal feedstock where the temperature, pressure, and flow pattern (moving bed, fluidized bed, or entrained flow) are controlled. The proportions of the resultant product gases ($CO$, $CO_2$, $CH_4$, $H_2$, $H_2O$, $N_2$, $H_2S$, $SO_2$, etc.) depend on the type of coal and its composition, the gasifying agent (or gasifying medium), and the thermodynamics and chemistry of the gasification reactions as controlled by the process operating parameters.

Coal gasification technology can be utilized in the following energy systems of potential importance:

1. Production of fuel for use in electric power generation units
2. Manufacturing synthetic or substitute natural gas (SNG) for use as pipeline gas supplies
3. Producing hydrogen for ammonia production and for fuel cell applications

4. Production of syngas for use as a chemical feedstock
5. Gasification of combined or mixed feedstocks of biomass and coal
6. Generation of fuel gas (low-Btu or medium-Btu gas) for industrial purposes

Coal has been recognized as the largest recoverable fossil fuel resource in the United States as well as in the world. Due to the enhanced recoverability of shale gas and natural gas hydrates, however, this statement may have to be modified. syngas production serves as the starting point for production of a variety of chemicals. The success of the Tennessee Eastman Corp. in producing acetic anhydride from coal shows the great potential of using coal as petrochemical feedstock [5]. A principal concern in coal gasification via POX routes is its generation of carbon dioxide, a major greenhouse gas (GHG), which has to be either converted in the downstream processing or sequestered as much as possible. Another major concern for such a technology involves the contaminants in coal. Coal contains appreciable amounts of sulfur, which is of principal concern to the downstream processes because many catalysts that might be used in the production of chemicals are highly susceptible to *sulfur poisoning.* Coals also contain nonnegligible amounts of alkali metal compounds that contribute to the fouling and *corrosion* of the reactor vessels in the form of slag. Further, coal also contains a number of trace elements that may also affect downstream processes and potentially create environmental and safety risks. If coal gasification is to be adopted to produce certain target chemicals, the choice of the specific gasification technology becomes very critical because a different process will produce a different quality (or composition) of syngas as well as alter the economics of production.

*Syngas* is a very important starting material for both fuels and petrochemicals. It is also called *syn gas* or *syngas,* of which the latter is universally more accepted. It can be obtained from various sources including petroleum, natural gas, coal, biomass, biogas, and even municipal solid wastes (MSWs). Syngas is conveniently classified, based on its principal composition, as (1) $H_2$-rich gas, (2) CO-rich gas, (3) $CO_2$-rich gas, (4) $CH_4$-rich gas, etc. Alternately, syngas may be classified based on its heating value as (1) high-BTU gas, (2) medium-BTU gas, and (3) low-BTU gas. Principal fuels and chemicals directly made from syngas include hydrogen, carbon monoxide, methane, ammonia, methanol, dimethylether (DME), gasoline, diesel fuel, jet fuel, ethylene, propylene, isobutylene, mixture of $C_2–C_4$ olefins, $C_1–C_5$ alcohols, ethanol, ethylene glycol, phosgene, etc. [6]. Secondary fuels and chemicals synthesized via methanol routes include formaldehyde, acetic acid, gasoline, diesel fuel, methyl formate, methyl acetate, acetaldehyde, acetic anhydride, vinyl acetate, DME, dimethyl carbonate (DMC), ethylene, propylene, isobutylene, ethanol, $C_1–C_5$ alcohols, propionic acid, methyl *tert*-butyl ether (MTBE), ethyl *tert*-butyl ether (ETBE), *tert*-amyl methyl ether (TAME), benzene, toluene, xylenes, ethyl acetate, a methylating agent, etc. Some of the chemicals are listed in both primary and secondary lists, since these chemicals can be produced by direct routes as well as via indirect synthesis routes. For example, the synthesis route leading to such chemicals as DME, diesel, propylene, and acetic acid via methanol as an intermediate may be called *indirect synthesis.*

## 2.2  SYNGAS CLASSIFICATION BASED ON ITS HEATING VALUE

Depending on the heating values of the resultant syngas produced by gasification processes, product gases are typically classified as three types of gas mixtures [7]:

1. *Low-Btu gas* consisting of a mixture of carbon monoxide, hydrogen, and some other gases with a heating value typically lower than 300 Btu/scf. This type of gas usually contains large amounts of nitrogen and carbon dioxide.
2. *Medium-Btu gas* consisting of a mixture of methane, carbon monoxide, hydrogen, and various other gases with a heating value in the range of 300–700 Btu/scf.
3. *High-Btu gas* consisting predominantly of methane with a heating value of approximately 1000 Btu/scf. It is also referred to as *SNG*. This type of gas contains a high amount of methane. In comparison, the gross heating value or higher heating value (HHV) of pure methane is 1011 Btu/scf. As shown, the heating value of syngas is most frequently expressed in Btu/scf (British thermal unit per standard cubic foot).

Coal gasification involves the reactions of coal carbon (precisely speaking, macromolecular coal hydrocarbons) and other pyrolysis products with oxygen, hydrogen, and steam to provide fuel gases, that is, syngas.

### 2.2.1  LOW-BTU GAS

For production of low-Btu gases, air is typically used as a combusting (or gasifying) agent. As air, instead of pure oxygen, is used, the product gas inevitably contains a large concentration of noncombustible or undesirable constituents such as nitrogen or nitrogen-containing compounds. Therefore, it results in a low heating value of 150–300 Btu/scf. Sometimes, this type of gasification of coal may be carried out in air-blown gasifiers or *in situ*, that is, underground, where mining of coal by other techniques is not deemed economically favorable. For such *in situ* gasification, low-Btu gas may be a desired product. Low-Btu gas contains five principal components with around 50% v/v nitrogen, some quantities of hydrogen and carbon monoxide (combustible), carbon dioxide, and some traces of methane. The presence of such high contents of nitrogen classifies the product gas as low Btu. The other two noncombustible components ($CO_2$ and $H_2O$) further lower the heating value of the product gas. The presence of these components limits the applicability of low-Btu gas to chemical synthesis. The two major combustible components are hydrogen and carbon monoxide; their ratio varies depending on the gasification conditions employed. One of the most undesirable components is hydrogen sulfide ($H_2S$), which occurs in a ratio proportional to the sulfur content of the original coal. It must be removed by gas-cleaning procedures before product gas can be used for other useful purposes such as further processing and upgrading. There are efficient processes developed and practically used for recovering useful sulfur by-products, viz., elemental sulfur by modified Claus plant and sulfuric acid via wet sulfuric acid (WSA) process.

## 2.2.2 Medium-Btu Gas

In the production of medium-Btu gas, pure oxygen rather than air is used as a combusting agent, which results in an appreciable increase in the heating value, by about 300–400 Btu/scf. The product gas predominantly contains carbon monoxide and hydrogen with some methane and carbon dioxide. This type of syngas is primarily used in the *synthesis of methanol*, higher hydrocarbons via *Fischer–Tropsch synthesis* (FTS), and a variety of other chemicals. It can also be used directly as a fuel to generate steam or to drive a gas turbine. The *$H_2$-to-CO ratio* in medium-Btu gas varies from about 2:3 (CO-rich gas) to more than 3:1 ($H_2$-rich gas). The increased heating value is attributed to higher contents of methane and hydrogen as well as to lower concentration of carbon dioxide, in addition to the absence of nitrogen in the gasifying agent. Most of oxygen-blown gasifiers generate medium-BTU syngas as their product gas.

## 2.2.3 High-Btu Gas

High-Btu gas consists mainly of pure methane (>95%), and, as such, its heating value is around 900–1000 Btu/scf, that is, close to the heating value of pure methane. It is compatible with natural gas and can be used as a synthetic or substitute natural gas (SNG). This type of syngas is usually produced by catalytic reaction of carbon monoxide and hydrogen, which is called the *methanation reaction*. The feed syngas to the methanation reactor usually contains carbon dioxide and methane in small amounts. Further, steam is usually present in the gas or added to the feed to alleviate carbon fouling in the gasifier, which adversely affects the catalytic effectiveness. Therefore, the pertinent chemical reactions in the methanation system include

$$3H_2 + CO = CH_4 + H_2O$$

$$2H_2 + 2CO = CH_4 + CO_2$$

$$4H_2 + CO_2 = CH_4 + 2H_2O$$

$$2CO = C + CO_2$$

$$CO + H_2O = CO_2 + H_2$$

Among these, the most dominant chemical reaction leading to methane is the first one. One can notice that the stoichiometric reaction for methanation as stated is the reverse reaction of the steam reformation reaction of methane (MSR). The methanation reaction is thermodynamically favored at low temperatures ($T < 650°C$), while the steam reformation of methane (the reverse reaction of methanation) is thermodynamically favored at high temperatures ($T > 650°C$). Even though the thermodynamic equilibrium clearly suggests that a lower temperature reaction be employed for methanation, its kinetic reaction rate at low temperatures is dismally slow in the absence of a suitable catalyst. Therefore, methanation is carried out over a catalyst with a syngas mixture of $H_2$ and CO, and the preferred $H_2$-to-CO ratio of the feed syngas is around 3:1, that is, close to the stoichiometric ratio. The large amount of $H_2O$ produced is removed by condensation and recirculated as process water or steam. During this process, most of the exothermic heat due to the methanation reaction is also recovered through a variety of energy integration processes. Whereas all the reactions listed above are quite strongly exothermic except the forward water gas shift (WGS) reaction, which is mildly exothermic, the heat release depends largely on the amount of CO present in the feed syngas. For each 1% of CO in the feed syngas, an adiabatic reaction will experience a 60°C temperature rise, which may be termed as *adiabatic temperature rise*. Many different types of reactor designs have been proposed and evaluated in order to handle and control the adiabatic reactor temperature rise.

A variety of metals exhibit catalytic effects on the methanation reaction. In the order of catalytic activity of methanation reaction, Ru > Ni > Co > Fe > Mo. Nickel is by far the most commonly used catalyst in commercial processes because of its relatively low cost and also of reasonably high catalytic activity. Nearly all the commercially available catalysts used for this process are, however, very susceptible to sulfur poisoning, and efforts must be taken to remove all hydrogen sulfide ($H_2S$) before the catalytic reaction starts. It is necessary to reduce the sulfur concentration in the feed gas to lower than 0.5 ppm in order to maintain adequate catalyst activity for a long period of time. Therefore, the objective of the catalyst development has been aimed at enhancing the *sulfur tolerance* of the catalyst and/or at developing sulfur-resistant catalyst formulation. Since nickel is an effective catalyst for methanation of syngas, handling of syngas in a high-nickel alloy reactor (such as Hastelloy®, Inconel®, and Haynes® alloys) potentially experiences and/or exhibits a monolithic catalytic reaction behavior, that is, reaction catalyzed by reactor wall materials. However, generalization of this statement is premature, since it still depends on a variety of factors.

Some of the noteworthy commercial methanation processes include Comflux, HICOM, direct methanation, and KBR's Coal-to-SNG process [8]. Comflux is a Ni-based, pressurized fluidized bed (PFB) process converting CO-rich gases into SNG in a single stage, where both methanation and WGS reaction take place simultaneously. The HICOM process developed by British Gas Corporation is a fixed bed process, which involves a series of methanation stages using relatively low $H_2$-to-CO ratio syngas. Direct methanation is a process developed by the Gas Research Institute (GRI, now Gas Technology Institute [GTI]), which methanates equimolar mixtures of $H_2$ and CO, producing $CO_2$ rather than $H_2O$ (steam) in addition to methane:

$$2H_2 + 2CO = CH_4 + CO_2$$

The catalyst developed is claimed to be unaffected by sulfur poisoning, and, as such, the process can be used to treat the raw, quenched gas from a coal gasifier with no or little pretreatment [9]. KBR's coal-to-SNG process is an integrated process between the KBR's transport reactor integrated gasifier (TRIG) technology and the conventional methanation process, in which low rank coal such as lignite is gasified via an advanced gasification process in a pressurized circulating bed gasifier (CFB), and the resultant syngas is subsequently methanated into pipeline quality SNG.

The Great Plains Synfuels Plant near Beulah, North Dakota, operated by the Dakota Gasification Company injects approximately 4.1 million m³/day of SNG into the natural gas grid [10]. The SNG is produced by sequential steps of gasification of lignite coal, gas cleaning, WGS conversion, and methanation.

## 2.3   COAL GASIFICATION REACTIONS

In the chemistry of coal gasification, four principal gasification reactions are crucial:

1. Steam gasification
2. Carbon dioxide gasification or Boudouard reaction
3. Hydrogasification
4. Partial oxidation (POX) reaction

In most gasifiers, several of these reactions, along with the WGS reaction, occur simultaneously. Coal is a macromolecular, hydrogen-deficient, carbonaceous matter and possesses very complex molecular structures. Due to this structural complexity, it is impossible to represent coal chemistry in a few stoichiometric equations. Therefore, investigators examine the equivalent chemical reactions involving carbon and then make necessary interpretations for and connections with coal. Table 2.1 shows the *equilibrium constants* ($K_p$) for these reactions, that is, carbon gasification

**TABLE 2.1**

**Equilibrium Constants for Carbon Gasification Reactions**

| $T$ (K) | $1/T$ | I | II | III | IV | V | VI |
|---|---|---|---|---|---|---|---|
| | | | | $Log_{10}$ $K_p$ | | | |
| 300 | 0.003333 | 23.93 | 68.67 | −15.86 | −20.81 | 4.95 | 8.82 |
| 400 | 0.0025 | 19.13 | 51.54 | −10.11 | −13.28 | 3.17 | 5.49 |
| 500 | 0.002 | 16.26 | 41.26 | −6.63 | −8.74 | 2.11 | 3.43 |
| 600 | 0.001667 | 14.34 | 34.40 | −4.29 | −5.72 | 1.43 | 2.00 |
| 700 | 0.001429 | 12.96 | 29.50 | −2.62 | −3.58 | 0.96 | 0.95 |
| 800 | 0.00125 | 11.93 | 25.83 | −1.36 | −1.97 | 0.61 | 0.15 |
| 900 | 0.001111 | 11.13 | 22.97 | −0.37 | −0.71 | 0.34 | −0.49 |
| 1000 | 0.001 | 10.48 | 20.68 | 0.42 | 0.28 | 0.14 | −1.01 |
| 1100 | 0.000909 | 9.94 | 18.80 | 1.06 | 1.08 | −0.02 | −1.43 |
| 1200 | 0.000833 | 9.50 | 17.24 | 1.60 | 1.76 | −0.16 | −1.79 |
| 1300 | 0.000769 | 9.12 | 15.92 | 2.06 | 2.32 | −0.26 | −2.10 |
| 1400 | 0.000714 | 8.79 | 14.78 | 2.44 | 2.80 | −0.36 | −2.36 |

*Note:* Reaction I: $C + 1/2\ O_2 = CO$; reaction II: $C + O_2 = CO_2$; reaction III: $C + H_2O = CO + H_2$; reaction IV: $C + CO_2 = 2\ CO$; reaction V: $CO + H_2O = CO_2 + H_2$; reaction VI: $C + 2\ H_2 = CH_4$.

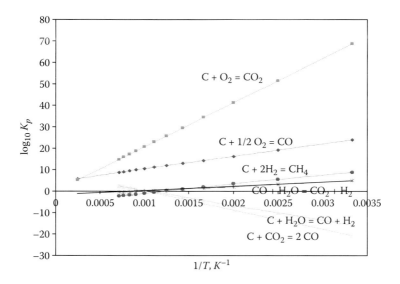

**FIGURE 2.1**    Equilibrium constant ($K_p$) for gasification reactions.

reactions, as functions of temperature. The same data are plotted in Figure 2.1, as $\log_{10} K_p$ vs. $1/T$. From the figure, the following are evident and significant:

1. The plots of $\log_{10} K_p$ vs. $1/T$ are nearly linear for all reactions, as expected from the principles of equilibrium thermodynamics.
2. The exothermicity of reaction is on the same order as the slope of the plot of $\log_{10} K_p$ vs. $1/T$ for each reaction. The steeper the slope, the more severe the exothermicity.
3. By using the criterion of $K_p > 1$ (i.e., $\log_{10} K_p > 0$), it is found that hydrogasification is thermodynamically favored at lower temperatures ($T < 825$ K), whereas $CO_2$ and steam gasification reactions are thermodynamically favored at higher temperatures ($T > 960$ K and $T > 945$ K, respectively).
4. The equilibrium constant for the WGS reaction is the weakest function of the temperature among all the compared reactions, as clearly evidenced by its relatively flat functionality on the plot. This also means that the equilibrium of this reaction can be reversed relatively easily by changing the imposed operating conditions and/or gas compositions. In other words, both forward and reverse reactions can play significant roles in the process chemistry, depending upon the prevailing conditions.
5. The thermodynamic data presented by both Table 2.1 and Figure 2.1 are for carbon gasification reactions, not for coal gasification reactions. While the data provide useful and insightful information as well as close approximations to the coal gasification reactions, the chemical and structural differences between carbon and coal must be clearly recognized in using the information.

## 2.3.1 STEAM GASIFICATION

Steam gasification of coal is very much similar to the steam reformation of natural gas or hydrocarbons. If we consider steam gasification of carbon, its stoichiometric equation can be written as

$$C(s) + H_2O(g) = CO(g) + H_2(g) \quad \Delta H^\circ_{298} = 131.3 \text{ kJ/mol}$$

The steam gasification reaction is endothermic, that is, requiring heat input for the reaction to proceed in its forward direction. As mentioned earlier, the forward reaction is thermodynamically favored at high temperatures, $T > 670°C$. However, the reaction rate for $T < 825°C$ is slow and impractical. Usually, an excess amount of steam is also needed to promote the reaction.

However, excess steam used in this reaction hurts the thermal efficiency of the process and requires a major effort of recovery or recycle. This type of process would require more energy to heat the reactor to the gasification temperatures due to the high latent heat of water. Therefore, this reaction is typically combined with other gasification reactions (such as POX) in practical applications. The $H_2$-to-CO ratio of the product syngas depends on the synthesis chemistry as well as process engineering. Two reaction mechanisms [11,12] have received most attention among scientists for the carbon-steam reactions over a wide range of practical gasification conditions.

*Mechanism A* [11]

$$C_f + H_2O = C(H_2O)_A$$

$$C(H_2O)_A \rightarrow CO + H_2$$

$$C_f + H_2 = C(H_2)_B$$

In the given equations, $C_f$ denotes free carbon sites that are not occupied, that is, free sites; $C(H_2O)_A$ and $C(H_2)_B$ denote chemisorbed species in which $H_2O$ and $H_2$ are adsorbed onto the carbon site, that is, occupied sites; " = " means that the specific mechanistic reaction is reversible; and "→" means that the reaction is predominantly irreversible. In mechanism A, the overall gasification rate is inhibited by hydrogen adsorption on the free sites, as represented by the last equation, thus reducing the availability of the unoccupied active sites for steam adsorption. Therefore, this mechanism may be referred to as *inhibition by hydrogen adsorption*.

*Mechanism B* [12]

$$C_f + H_2O = C(O)_A + H_2$$

$$C(O)_A \rightarrow CO$$

On the other hand, in mechanism B, the gasification rate is affected by competitive reaction of chemisorbed oxygen with hydrogen, as represented by the reverse reaction of the first equation, thus limiting the conversion of chemisorbed oxygen into carbon monoxide (the second reaction equation). Therefore, this mechanism may be referred to as *inhibition by oxygen exchange*.

Both mechanisms are still capable of producing the rate expression for steam gasification of carbon in the form of [13]

$$r = \frac{k_1 p_{H_2O}}{1 + k_2 p_{H_2} + k_3 p_{H_2O}}$$

which was found to correlate with the experimental data quite well. This type of rate expression can be readily derived by taking pseudo-steady state approximation on the adsorbed species of the mechanism as well as the balance on active sites that makes the total number of active sites equal to the sum of free and occupied sites.

It has to be clearly noted here that the mechanistic chemistry discussed in this section is based on the reaction between carbon and gaseous reactants, not for reactions between coal and gaseous reactants. Even though carbon is the dominant atomic species present in coal, its reactivity is quite different from that of coal or coal hydrocarbons. In general, coal is more reactive than pure carbon, for a number of reasons, including the presence of various reactive organic functional groups and the availability of catalytic activity via naturally occurring mineral ingredients. It may now be easy to understand why anthracite, which has the highest carbon content among all ranks of coal, is most difficult to gasify or liquefy. Furthermore, alkali metal salts are known to catalyze the steam gasification reaction of carbonaceous materials, including coals. The order of catalytic activity of alkali metals on coal gasification reaction is Cs > Rb > K > Na > Li. Coal has a variety of mineral matters naturally occurring, while pure carbon does not have any. In the case of catalytic steam gasification of coal, carbon deposition reaction may affect the catalysts' life by fouling the catalyst active sites. This carbon deposition reaction is more likely to take place whenever the steam concentration is lacking.

### 2.3.2 CARBON DIOXIDE GASIFICATION

The reaction of coal with $CO_2$ may be approximated or simplified as the reaction of carbon with carbon dioxide for modeling purposes. Carbon dioxide reacts with carbon to produce carbon monoxide, and this reaction is called *Boudouard reaction*. This reaction is also endothermic in nature, similar to the steam gasification reaction.

$$C(s) + CO_2(g) = 2CO(g) \quad \Delta H^{\circ}_{298} = 172.5 \text{ kJ/mol}$$

The endothermicity of this reaction is more severe than that of the steam gasification reaction, thus implying that this reaction has to overcome a higher energy barrier than the steam gasification. In other words, this reaction is the least energy-efficient reaction of the four principal gasification reactions. The reverse reaction is a carbon deposition reaction that is a major culprit of carbon fouling on many surfaces, such as process catalyst deactivation by fouling. This gasification reaction is thermodynamically favored, that is, $K_p$ becomes greater than 1, at high temperatures ($T > 690°C$), which is also somewhat similar to the steam gasification. The reaction, if carried out alone, requires high temperature (for fast reaction) and high pressure (for higher reactant concentrations and for high gas throughput) for significant conversion. However, this reaction in practical gasification applications is almost never attempted as a single independent chemical reaction, because of a variety of factors including slow kinetic rate, low thermal efficiency, unimpressive process economics, etc. However, this reaction has found added significances in the global efforts of mitigating the carbon dioxide emission as one of the principal reactions that can directly utilize the carbon dioxide in a beneficial way or that can be used to convert carbon dioxide into more reactive carbon monoxide. In other words, this reaction provides one of the practical pathways reducing carbon dioxide to carbon monoxide via a redox reaction in which carbon (or coal) functions as a reducing agent. This reaction also provides a justification for a practical process route known as "dry reformation" in which typical reforming agent of steam is replaced by carbon dioxide, analogously to the interchangeability established between steam gasification and carbon dioxide gasification.

There is general agreement among scientists that experimental data on the rate of carbon gasification by $CO_2$ fit an empirical equation of the form [13]

$$r = \frac{k_1 p_{CO_2}}{1 + k_2 p_{CO} + k_3 p_{CO_2}}$$

where $p_{CO}$ and $p_{CO_2}$ are partial pressures of CO and $CO_2$ in the reactor. This rate equation is shown to be consistent with at least two mechanisms whereby carbon monoxide retards the overall gasification reaction [13].

*Mechanism A*

$$C_f + CO_2 \rightarrow C(O)_A + CO$$

$$C(O)_A \rightarrow CO$$

$$CO + C_f = C(CO)_B$$

*Mechanism B*

$$C_f + CO_2 = C(O)_A + CO$$

$$C(O)_A \rightarrow CO$$

In both mechanisms, carbon monoxide retards the overall reaction rate. The retardation is via carbon monoxide adsorption to the free sites in the case of mechanism A, whereas it is via reaction of chemisorbed oxygen with gaseous carbon monoxide to produce gaseous carbon dioxide in mechanism B.

As mentioned earlier when discussing steam gasification, the $CO_2$ gasification rate of coal is different from that of the carbon–$CO_2$ rate for the very same reason. Generally, the carbon–$CO_2$ reaction follows a global reaction order on the $CO_2$ partial pressure that is around one or lower, that is, $0.5 < n < 1$, whereas the coal–$CO_2$ reaction follows a global reaction order on the $CO_2$ partial pressure that is one or higher, that is, $1 < n < 2$. The observed higher reaction order for the coal reaction may be attributable to the high reactivity of coal for the multiple reasons described earlier. It should also be noted that many investigators have successfully used the global first-order reaction rate expressions to fit their experimental data on carbon dioxide gasification reaction kinetics [14].

### 2.3.3  HYDROGASIFICATION

Direct addition of hydrogen to coal under high pressure forms methane. This reaction is called *hydrogasification* and may be written as

$$Coal + H_2 = CH_4 + Carbonaceous\ matter$$

or

$$C(s) + 2H_2(g) = CH_4(g) \quad \Delta H^\circ_{298} = -74.8\ kJ/mol$$

Of the two equations, the latter represents the hydrogasification of carbon. The hydrogasification reaction is exothermic and is thermodynamically favored at low temperatures ($T < 550°C$), unlike both steam and $CO_2$ gasification reactions. However, at low temperatures, the reaction rate is inevitably too slow, that is, kinetically undesirable. Therefore, high temperature, preferably $T > 900°C$, is always required for kinetic reasons, which in turn requires high pressure of hydrogen, in order to maintain $K_p > 1$, thus forcing the reaction to proceed in the net forward direction. The carbon conversion of the process is usually low, and the rate is slower than other gasification reactions when carried out noncatalytically. This reaction can be catalyzed by $K_2CO_3$, nickel, iron chlorides, iron sulfates, etc. However, use of catalyst in coal gasification suffers from serious economic constraints because of the low raw material value, as well as difficulty in recovering and reusing the catalyst. Therefore, catalytic coal gasification on an industrial scale has not been practiced much.

The hydrogasification can be utilized in the production of SNG from coal and other hydrocarbon feedstocks. The hydrogasification process can be modified or enhanced by addition of steam in the feed gas, thus creating a process called *steam hydrogasification*. With steam input, additional hydrogen can be generated from the gasifier. While steam addition requires more energy input to bring the gasifier to the gasification temperature, the effluent steam could be used, in an integral and beneficial manner, in the processes such as steam methane reformation (SMR) to produce hydrogen, some of which can also be used in the steam hydrogasification reactor. The hydrogasification process does not require an oxygen plant, which can be viewed as a financial benefit in terms of the process economics.

### 2.3.4  PARTIAL OXIDATION

Combustion of coal involves reaction with oxygen, which may be supplied as pure oxygen or as air, and forms carbon monoxide and carbon dioxide. Principal chemical reactions between carbon and oxygen involve

$$C(s) + O_2(g) = CO_2(g) \quad \Delta H^\circ_{298} = -393.5 \text{ kJ/mol}$$

$$C(s) + 1/2\, O_2(g) = CO(g) \quad \Delta H^\circ_{298} = -111.4 \text{ kJ/mol}$$

If sufficient air or oxygen is supplied to coal, combustion proceeds sequentially through vapor-phase oxidation and ignition of volatile matter to eventual ignition of the residual char. Certainly, it is not desirable to carry out the combustion reaction to a great extent or to completion, because it is a wasteful use of carbonaceous resources. Therefore, POX in coal gasification serves for two principal purposes, viz., (1) oxidatively cleavaging the C–C and C–H bonds in macromolecular coal structures and facilitating chemical reactions to produce syngas and (2) generating exothermic heat of reaction that is needed to sustain the gasification reaction.

Even though the combustion or oxidation reactions of carbon may be expressed in terms of simple stoichiometric reaction equations, POX involves a complex reaction mechanism that determines how fast and efficiently combustion progresses. The reaction pathway is further complicated because of the presence of both gas-phase homogeneous reactions and heterogeneous reactions between gaseous and solid reactants. The early controversy involving the carbon oxidation reaction centered on whether carbon dioxide is a primary product of the heterogeneous reaction of carbon with oxygen or a secondary product resulting from the gas-phase homogeneous oxidation of carbon monoxide [13]. Oxidation of carbon involves at least the following four carbon–oxygen interactions, of which only two are stoichiometrically independent:

$$C(s) + 1/2\, O_2(g) = CO(g)$$

$$CO(g) + 1/2\, O_2(g) = CO_2(g)$$

$$C(s) + CO_2(g) = 2\, CO(g)$$

$$C(s) + O_2(g) = CO_2(g)$$

Of the four reactions, only the second reaction takes place homogeneously in the gas phase, while other three reactions are taking place heterogeneously as gas–solid reactions.

Based on a great deal of research work, including isotope labeling studies, it is generally agreed concerning the carbon–oxygen reaction [13] that

1. $CO_2$, as well as CO, is a primary product of heterogeneous carbon oxidation.
2. The ratio of the primary products, CO to $CO_2$, is generally found to increase sharply with increasing temperature.

3. There is disagreement in that the magnitude of the ratio of the primary products is a sole function of temperature and independent of the type of carbon reacted.

Further details on the carbon oxidation can be found from a classical work done by Walker et al. [13].

Combustion or oxidation of coal is much more complex in its nature than oxidation of carbon. Coal is not a pure chemical species; rather, it is a multifunctional, multispecies, heterogeneous macromolecule that occurs in a highly porous form (typical porosity of 0.3–0.5) with a very large available internal surface area (typically in the range of 250–700 $m^2$/g). The internal surface area of coal is usually expressed in terms of specific surface area, which is an intensive property that is a measure of the internal surface area available per unit mass. Therefore, coal combustion involves a very complex system of chemical reactions that occur both simultaneously and sequentially. Certain coal mineral matters also have catalyzing effects on the combustion reaction. Further, the reaction phenomenon is further complicated by transport processes of simultaneous heat and mass transfer. The overall rate of coal oxidation, both complete and partial, is affected by a number of factors and operating parameters, including the reaction temperature, $O_2$ partial pressure, coal porosity and its distribution, coal particle size, types of coal, types and contents of specific mineral matter, heat and mass transfer conditions in the reactor, and more.

Kyotani et al. [15] determined the reaction rate of combustion for five different types of coals in a very wide temperature range between 500°C and 1500°C to examine the effects of coal rank (i.e., carbon content) and catalysis by coal mineral matter. Based on their experimental results, the combustion rates were correlated with various char characteristics. It was found that in a region where chemical reaction rate is controlling the overall rate, that is, typically in a low-temperature region where the kinetic rate is much slower than the diffusional rate of gaseous reactant through the pores, the catalytic effect of mineral matter is a determining factor for coal reactivity. It was also found that for high-temperature regions where the external mass transfer rate controls the overall rate, the reactivity of coal decreased with increasing coal rank. The external mass transfer in this context means the mass transfer of gaseous species between the bulk gas phase and the exterior surface of coal particle. When the external mass transfer rate is limiting (or controlling) the overall rate of reaction, the mechanistic rate of external mass transfer is the slowest of all mechanistic rates, including the surface reaction rate and the pore diffusional rate of the reactant and the product. Such a controlling regime is experienced typically at a very high-temperature operation, as the intrinsic kinetic rate is far more strongly affected by the temperature than the external mass transfer rate is.

## 2.3.5 WGS REACTION

Even though the WGS reaction is not classified as one of the principal gasification reactions, it cannot be omitted in the analysis of chemical reaction systems that involve syngas. Among all reactions involving syngas, this reaction equilibrium is least sensitive to the temperature variation. In other words, its equilibrium constant is least strongly dependent on the temperature. Therefore, this reaction equilibrium can be reversed in a variety of practical process conditions over a wide range of temperatures. WGS reaction can be effected catalytically or noncatalytically, homogeneously or heterogeneously, in a separate stage or as a parallel reaction in the main stage, etc. WGS reaction in its forward direction is mildly exothermic as

$$CO(g) + H_2O(g) = CO_2(g) + H_2(g) \quad \Delta H^\circ_{298} = -41.2 \text{ kJ/mol}$$

Even though all the participating chemical species are in the form of a gas, scientists believe that this reaction occurring in a coal gasifier predominantly takes place at the heterogeneous surfaces of coal and also that the reaction is catalyzed by carbon surfaces. As the WGS reaction is catalyzed by many heterogeneous surfaces and the reaction can also take place homogeneously as well as heterogeneously, a generalized understanding of the WGS reaction in coal conversion has been very difficult to achieve. Even the kinetic rate information reported in the literature may not be immediately useful or applicable to a practical reactor situation.

Once condensable compounds from the gasifier effluent stream are removed, a raw syngas product is obtained. A syngas product from a gasifier also contains a variety of gaseous species besides carbon monoxide and hydrogen. Typically, they include carbon dioxide and methane. If air is used as a gasifying medium, nitrogen and oxygen are also present. Depending on the objective of the ensuing process, the composition of syngas may need to be preferentially readjusted. If the objective of the gasification were to obtain a high yield of methane, it would be preferred to have the molar ratio of hydrogen to carbon monoxide at 3:1, based on the following methanation reaction stoichiometry:

$$CO(g) + 3H_2(g) = CH_4(g) + H_2O(g)$$

If the objective of generating syngas is the synthesis of methanol via vapor-phase low-pressure process [16], the stoichiometrically consistent ratio between hydrogen and carbon monoxide would be 2:1 to 2.5:1. It should be noted that a small amount, 2–4 mole%, of carbon dioxide must be present in the syngas composition for the catalytic stability of the synthesis reaction [17]

$$CO(g) + 2H_2(g) = CH_3OH(g)$$

$$CO_2(g) + 3H_2(g) = CH_3OH(g) + H_2O(g)$$

In such cases, the stoichiometrically consistent syngas mixture, or close to it, is often referred to as *balanced syngas*, whereas a syngas composition that is substantially deviated from the principal reaction's stoichiometry is called *unbalanced syngas*. Frequently in the field of methanol synthesis, CO-rich syngas is simply mentioned as unbalanced syngas.

If the objective of syngas production is to obtain a high yield of hydrogen, it would be advantageous to increase the ratio of $H_2$ to CO by further converting CO (and $H_2O$) into $H_2$ (and $CO_2$) via WGS reaction. However, if the final gaseous product is to be used in fuel cell applications, carbon monoxide and carbon dioxide must be removed to ultralow levels by a process such as acid gas removal, pressure swing adsorption (PSA), membrane separation, or other adsorption processes. In particular, for proton exchange membrane (PEM) fuel cell operation, impurities such as carbon monoxide, carbon dioxide, hydrogen sulfide, and other sulfur species must be thoroughly removed from the hydrogen gas.

The WGS reaction is one of the major reactions taking place in the steam gasification process, where both water and carbon monoxide are present in ample amounts, thereby creating an environment conducive for WGS conversion:

$$CO(g) + H_2O(g) = CO_2(g) + H_2(g)$$

Even though all four chemical species involved in the WGS reaction are gaseous compounds at the reaction stage of most gas processing, scientists believe that the WGS reaction, in the case of steam gasification of coal, predominantly takes place heterogeneously, that is, on the solid surface of coal. This statement does neither mean nor imply that the homogeneous gas-phase WGS reaction does not take place in the system. WGS reaction is catalyzed by a variety of metallic catalysts or metallic ingredients on the coal surface. For example, Cu/ZnO and Cu/ZnO/Al$_2$O$_3$ catalysts have been frequently utilized as industrial WGS catalysts. These catalysts are operated at relatively low temperatures, $T < 300°C$, and therefore known as low temperature shift (LTS) catalysts. If the product syngas from a gasifier needs to be reconditioned by the WGS reaction, this reaction can be catalyzed by a variety of metallic catalysts. Choice of specific kinds of catalysts has always depended on the desired outcome, the prevailing temperature conditions, composition of gas mixture, and process economics. Many investigators have studied the WGS reaction over a variety of catalysts including iron, copper, zinc, nickel, chromium, and molybdenum. Significant efforts have been made in developing a robust catalyst system that has superior sulfur tolerance and wider applicable temperature range.

Reverse WGS (RWGS) reaction is also of significance, since the reaction can provide a direct route for reducing "chemically inactive" carbon dioxide into "more reactive" carbon monoxide. The reaction could serve as an important precursor step for reactive conversion of carbon dioxide into hydrocarbons. The thermodynamic equilibrium constant $K_p$ for the RWGS reaction becomes greater than 1 above 825°C. The energy required for the RWGS reaction to proceed is not insignificant, and a highly effective and robust catalyst system needs to be developed for enhanced process economics.

## 2.4 SYNGAS GENERATION VIA COAL GASIFICATION

### 2.4.1 CLASSIFICATION OF GASIFICATION PROCESSES

In the earlier section, different types of syngas (SG) were classified. Similarly, there are a large number of widely varying gasification processes. The gasification processes can be classified basically in two general ways: (1) by the Btu content of the product gas [18] and (2) by the type of the reactor hardware configuration, as well as by whether the reactor system is operated under pressure or not.

The following processes for conversion of coal to gases are grouped according to *the heating value of the product gas.*

*Medium- or High-Btu Gas Gasification Processes with Options for Low-Btu Gas Generation*

1. Lurgi nonslagging gasifier
2. British Gas/Lurgi (BGL) slagging gasifier
3. Winkler gasifier
4. Synthane gasifier
5. KBR coal-to-SNG process
6. Atgas molten iron coal gasifier

*Low- or Medium-Btu Gas Gasification Processes*

1. Koppers–Totzek gasifier
2. Texaco gasifier
3. Shell gasifier
4. Kellogg's molten salt gasifier
5. $CO_2$-acceptor gasification process
6. KBR gasification process

*Low-Btu Gas Only Gasification Process*

1. U-gas process
2. Underground *in situ* gasification process

Based on the reactor configuration, as well as by the method of contacting between gaseous and solid streams, gasification processes can also be categorized into the following four types [7]:

1. *Fixed or moving bed*: In the fixed bed reactor, coal is supported by a grate and the gasifying media (steam, air, or oxygen) pass upward through the supported bed, whereby the product gases exit from the top of the reactor. Only noncaking coals can be used in the fixed bed reactor. On the other hand, in the moving bed reactor, coal and gaseous streams move countercurrently, that is, coal moves downward by gravity while gas passes upward through the coal bed. The temperature at the bottom of the reactor is higher than that at the top. Because of the lower temperature at the top for coal devolatilization, relatively large amounts of liquid hydrocarbons are also produced in this type of gasifier. In both types of reactor, the residence time of the coal is much longer than that in a suspension type reactor, thus providing ample contact time between reactants. This type of gasifier permits the use of fairly large particle sizes of coal. Ash is removed from the bottom of the reactor as dry ash or slag. Lurgi and Wellman–Galusha gasifiers are examples of this type of a reactor. It should be clearly understood that a moving bed reactor is classified as a

kind of fixed bed reactor, because solids in the bed stay together regardless of the movement of the hardware that supports the coal bed.

2. *Fluidized bed*: It uses finely pulverized coal particles, typically smaller than 6 mm in linear dimensions. As a general scheme, the gas (or gasifying medium) flows upward through the bed and fluidizes the coal particles. The linear superficial velocity of fluidization gas widely varies depending upon designs, but typically ranging between 30 and 100 times of the minimum fluidization velocity. Owing to the ascent of particles and fluidizing gas, a larger coal surface area per particle is made available, which positively promotes the gas–solid chemical reaction, which in turn results in enhancement in carbon conversion. This type of a reactor allows intimate contact between gas and solid coal fines, at the same time providing relatively longer residence times than an entrained flow reactor, but still significantly shorter residence times than those for fixed bed reactors. Either dry ash is removed continuously from the bed or the gasifier is operated at such a high temperature that it can be removed as agglomerates. Such beds, however, have limited ability to handle caking coals, owing to operational complications in fluidization characteristics. Winkler and Synthane processes use this type of reactor.

   Two modified variations from the traditional fluidized bed reactors are bubbling fluidized bed (BFB) and circulating fluidized bed (CFB) reactors. A CFB gasifier uses a cyclone for returning the solid material from the gasifier effluent stream back to the lower part of the gasifier. This type of a gasifier provides excellent emission control. A transport reactor, a modified version of CFB, operates at considerably higher circulation rates and riser densities than conventional CFBs, achieving higher throughput, improved reactant mixing, and higher mass and heat transfer rates. KBR gasification process uses this type of a reactor system. KBR transport gasifier is also known as transport integrated gasification (TRIG™).

3. *Entrained bed*: This type of a reactor is also referred to as an *entrained flow reactor*, because there is no definable bed of solids. This reactor system uses finely pulverized coal particles blown into the gas stream before entry into the reactor, with combustion and gasification occurring inside the coal particles suspended in the gas phase. Because of the entrainment requirement, high space velocity of gas stream and fine powdery coal particles are very essential to the operation of this type of process. Because of the very short residence time (i.e., very high space velocity) in the reactor, a very high gasification temperature is required to achieve good conversion in such a short period of reaction time. This can also be assisted by using excess oxygen or air, which provides the exothermic heat of combustion upon reacting with coal. This bed configuration is typically capable of handling both caking and noncaking coals without much operational difficulty. The particle size used in an entrained flow reactor is typically even finer than that employed in a fluidized bed, while the gasification temperature achieved in an entrained flow mode is higher than that for a fluidized bed. Examples of commercial gasifiers that use this type of reactor include the Koppers–Totzek gasifier, Texaco gasifier, and Shell gasifier.

4. *Molten salt bath reactor*: In this reactor, coal is fed along with steam or oxygen in the molten bath of salt or metal operated at 1000°C–1400°C. The gasification temperature is dictated by the choice of molten salt material used. The choice of salt medium should be based on (1) its melting point that needs to be lower than the gasification temperature, thereby ensuring the molten state at gasification conditions; (2) its boiling point of the salt that needs to be significantly higher than the gasification temperature, thereby ensuring low volatility of the chosen medium at the gasification temperature; (3) its chemical inertness with coal ingredients, thereby assuring the regenerability/reusability of the medium; and (4) low material cost of the salt medium. Ash and sulfur are removed from the reactor as slag. This type of a reactor is used in Kellogg and Atgas molten salt gasification processes [19].

### 2.4.2 HISTORICAL BACKGROUND OF COAL GASIFICATION AND ITS COMMERCIALIZATION

It was known as early as the seventeenth century that gas could be produced by simply heating the coal, that is, pyrolysis of coal in modern terms. Around 1750, in England, coal was subjected to pyrolysis, more precisely devolatilization and thermal decomposition, to form gases that were used for lighting [20]. With the invention of the Bunsen gas burner (at atmospheric pressure), the potential of heating was opened to gas combustion. In 1873, cyclic carbureted water gas process was developed by Thaddeus S. C. Lowe for gas production. In this process, water gas ($H_2$ + CO) was produced by reacting hot coke (i.e., smokeless char) with steam via a simplified reaction of $C + H_2O = CO + H_2$. Heat for the reaction (endothermic, 131 kJ/mol) was supplied by combustion energy by introducing air intermittently to burn a portion of the coke. The development of coal-to-gas processes was a major breakthrough in Europe during those days, because coal was the principal fuel available besides wood. By the early 1920s, there were at least five *Winkler fluid-bed processes* being operated, all of which were air-blown, producing 10 million scf/h of producer gas. Some of them were later converted to use oxygen instead of air in order to produce nitrogen-free syngas, which contained a higher heating value per gas volume.

The *Lurgi gasification process* was developed to manufacture town gas by complete gasification of brown coal in Germany. In 1936, the first commercial plant based on this process went operational. It produced 1 million scf/day of town gas from low-rank lignite coal. By 1966, there were at least ten Lurgi plants at a number of places in Europe and Asia producing syngas. British Gas and Lurgi developed in collaboration an improved version of a slagging gasifier, known as a *BGL gasifier*, based on Lurgi's original nonslagging gasifier. The technology was demonstrated by British Gas and Lurgi from 1974 to 1991 on a wide variety of coal feedstocks.

In 1942, Heinrich Koppers in Germany developed the *Koppers–Totzek (K–T) suspension gasification process* based on the pilot plant work initiated four years earlier. The first industrial plant was built in France around 1949, which produced 5.5 million scf/day of syngas that was later used to produce ammonia and methanol. By the early 1970s, there were at least 20 K–T plants built all over the world. All of them used oxygen as the primary gasification medium, thus producing nitrogen-free syngas.

Winkler, Lurgi, and Koppers–Totzek processes all employed steam and oxygen (or air) to carry out gasification. Most of these developments were originated and perfected in Europe. However, very little development of these processes had taken place in the United States until the energy crisis of the 1970s, mainly because of the discovery of natural gas as a convenient fuel source and also because of the relatively stable supply of liquid petroleum until then. After the oil embargo of 1973, very active research and development efforts were conducted for cleaner use of coal resources in coal gasification, coal liquefaction, clean coal technology, IGCC, etc. Since then, most coal power plants have significantly upgraded their quality of operation in terms of energy efficiency, by-products, emission control, and profitability.

In the twenty-first century, the utilization of coal resources, including both coal gasification and coal power generation, is facing new challenges that are coming from the major GHG regulations as well as the market competition originating from shale gas. New technologies developed for combined horizontal drilling and hydraulic fracture brought about a *shale gas boom* in regions of the United States, which has drastically increased the gas availability in the nation and significantly lowered the market price of natural gas (Please refer to Chapter 9 for more details). The low or stable natural gas price on the marketplace, in turn, affected the economic prospect of coal gasification somewhat less attractive in comparison to the steam reforming of natural gas in some regions of the world, if the syngas generation is its primary objective of coal gasification. Further, the dependence on coal combustion for electric power generation has also been declining in the United States by a new trend of switching over to the natural gas power plants.

### 2.4.3 GENERAL ASPECTS OF GASIFICATION

The kinetic rates and extents of conversion for various gasification reactions are typically functions of temperature, pressure, gas composition, and the nature of the coal being gasified. Strictly from the thermodynamic equilibrium standpoints, some gasification reactions such as carbon–hydrogen reaction producing methane is favored to proceed in the forward direction at high pressures and lower temperatures ($T < 825$ K), whereas low pressures and high temperatures ($T > 950$ K) favor the conversion to syngas from carbon via steam or carbon dioxide gasification reaction. The kinetic rate of reaction is intrinsically higher at higher temperatures, whereas the thermodynamic equilibrium of the reaction dictates the net reaction to proceed in either forward or reverse direction at these temperatures depending on the specific type of gasification reaction and the prevailing concentrations of the chemical species involved. The effect of pressure on the rate also depends on the specific reaction both kinetically and thermodynamically.

Supply and recovery of heat is a key element in the gasification process as well as developing the process flowsheets from the standpoints of thermal efficiency, process economics, plant design, and operability. Partial oxidation of char with steam and oxygen leads to generation of heat and syngas. Another way to produce a hot gas stream is via the cyclic reduction and oxidation of iron ore. The type of coal being gasified is also important to the gasification and downstream operations. Only suspension-type gasifiers such as entrained flow reactor can handle any type of coal, but if caking coals are to be used in fixed or fluidized bed, special measures must be taken so that coal does not agglomerate (or cake) during gasification. If such agglomeration does happen, it would adversely affect the operability of the gasification process. In addition to this, the chemical composition, the volatile matter (VM) content, and the moisture content of coal also play important roles in the coal processing during gasification. The S and N contents of coal seriously affect the quality of the product gas, as well as the downstream gas-cleaning requirements. The sulfur content of coal typically comes from three different sources (forms) of coal sulfur, namely, pyritic sulfur, organic sulfur, and sulfatic sulfur. The first two are more dominant sulfur forms in freshly mined coals, whereas weathered or oxidized coals have more sulfatic forms than fresh coals [21]. Typical breakdown of sulfur forms in coal includes 40%–70% in pyritic, 30%–60% in organic, and 0%–2% in sulfatic form, when total sulfur in freshly mined coal is considered 100%. Sulfurous gas species can be hydrogen sulfide ($H_2S$), carbonyl sulfide (COS), sulfur dioxide ($SO_2$), or mercaptans (R-SH), depending on the nature of the reactive environment and the type of coal.

### 2.4.4 GASIFICATION PROCESSES

In this section, gasifiers of commercial significance are explained in detail with the information of gasifier design features, process flowsheets, associated reaction chemistry, operating conditions, process merits and demerits, thermal efficiency, process integration options and practices, product gas compositions, and more.

#### 2.4.4.1 Lurgi Gasification Process

The Lurgi gasification process is one of the several processes for which commercial technology has been fully developed [22].

Since its development in Germany before World War II, this process has been used in a large number of commercial plants throughout the world. This process produces low- to medium-Btu gas as product gas. It may be classified as a fixed bed process in which the reactor configuration is similar to that of a typical fixed bed reactor. The older version of Lurgi process is *dry ash gasification* process that differs significantly from the more recently developed *slagging gasification process*. The latter is capable of gasifying caking coals, while the former is not.

The *dry ash Lurgi gasifier* is a pressurized vertical reactor that accepts crushed noncaking coals only [23]. The coal feed is supported at the base of the reactor by a revolving grate through which the steam and oxygen mixture is introduced and the ash removed. This process takes place at around 24 to 31 atm and in the temperature range of 620°C to 760°C. The residence time in the reactor is about 1 h. Steam introduced from the bottom of the reactor provides the necessary hydrogen source, and the heat is supplied by the combustion of a portion of the char. The product gas from a high-pressure reactor has a relatively high methane content compared to a non-pressurized gasifier. The high methane content of the product gas is a result of the relatively low gasification temperature as well as the high partial pressure of steam in the reactor. If oxygen is used as an injecting (and gasifying) medium, the exiting gas has a heating value of approximately 450 Btu/scf (a medium-Btu gas). The crude gas leaving the gasifier contains a substantial amount of condensable products including tar, oil, phenol, etc., which are separated in a devolatilizer, where gas is cleaned to remove unsaturated hydrocarbons and naphtha. The gas is then subjected to methanation ($CO + 3H_2 = CH_4 + H_2O$) to produce a high-Btu gas (pipeline quality).

Recent modification of the Lurgi process called *slagging Lurgi gasifier* has been developed to process caking coals [7]. Therefore, the operating temperature of this gasifier is kept higher, and the injection ratio of steam is reduced to 1–1.5 mol/mol of oxygen. These two factors cause the ash to melt easily, and, therefore, the molten ash is removed as a slag. Coal is fed to the gasifier through a lock hopper system and distributor. It is gasified with steam and oxygen injected into the gasifier near the bottom. The upward movement of hot product gases provides convective heat transfer and makes the preheating and devolatilization of coal easier. Both volatile matter liberated from coal and devolatilized char react with gasifying media, that is, steam and oxygen. The molten slag formed during the process passes through the slag tap hole. It is then quenched with water and removed through a slag lock hopper. The amount of unreacted steam passing through the system has to be minimized in this process for high energy efficiency. Also, the high operating temperature and fast removal of product gases lead to higher output rates in a slagging Lurgi gasifier than a conventional dry ash Lurgi system.

The conventional Lurgi gasification is widely recognized for its role as the gasifier technology for South Africa's Sasol I complex. A typical product composition for oxygen-blown operation is given in Table 2.2. As can be seen, the $H_2$-to-CO ratio is higher than 2:1. It is also noted that a relatively large amount of $CO_2$ is present. Therefore, this gasifier is well suited for production of feed syngas for methanol synthesis or hydrocarbon fuel synthesis using Fischer-Tropsch.

The new version of a slagging gasifier was developed by British Gas in collaboration with Lurgi, and the gasifier is referred to as the *BGL (British Gas/Lurgi) gasifier*. The gasifier is based on injection of oxygen and steam, and therefore, POX is a predominant reaction. Due to the POX reaction of coal, the resultant syngas is intrinsically a CO-rich syngas. According to the GL Noble Denton [24], licensor

---

**TABLE 2.2**
**Typical Lurgi Gas Products**

| Species | Mole Percentage |
|---|---|
| CO | 16.9 |
| $H_2$ | 39.4 |
| $CH_4$ | 9.0 |
| $C_2H_6$ | 0.7 |
| $C_2H_4$ | 0.1 |
| $CO_2$ | 31.5 |
| $H_2S + COS$ | 0.8 |
| $N_2 + Ar$ | 1.6 |

*Source:* Lloyd, W.G., *The Emerging Synthetic Fuel Industry*, Thumann, A., Ed., Atlanta, GA: Fairmont Press, 1981, pp. 19–58. With permission.

---

of BGL gasifier technology, typical clean gas composition of the BGL coal gasifier is CO-rich syngas whose $CO/H_2$ ratio is approaching 2:1. The BGL gasifier is well suited for production of substitute natural gas (SNG) and fuel gas.

### 2.4.4.1.1 Lurgi Dry-Ash Gasifier

In this gasifier, coarse coal particles sized between 4 cm and 4 mesh (~4.75 mm) react with steam and oxygen in a slowly moving bed. Because of this mechanical movement of the bed, this type of a gasifier is often called a moving bed gasifier. The process is operated semicontinuously. A schematic of a Lurgi pressure gasifier is shown in Figure 2.2 [25]. The gasifier is equipped with the following hardware parts [26]:

1. An automated *coal lock chamber* for feeding coal from a coal bin to the pressurized reactor. This device is often called a *coal lock hopper*.
2. A *coal distributor* through which coal is uniformly distributed into the moving bed.
3. A *revolving grate* through which the steam and oxygen are introduced into the reacting zone (coal bed) and the ash is removed.
4. An *ash lock chamber* for discharging the ash from the pressurized reactor into an ash bin, where the ash is cooled by water quenching.
5. A *gas scrubber* in which the hot gas is quenched and washed before it passes through a waste heat boiler.

The gasifier shell is water-cooled and steam is produced from the water jacket. A motor-driven distributor is located at the top of the coal bed, which evenly distributes the feed coal coming from the coal lock hopper. The grate at the bottom of the reactor is also driven by a motor to discharge the coal ash into the ash lock hopper. The section between

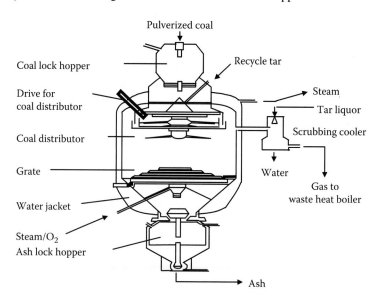

**FIGURE 2.2**  Lurgi nonslagging pressure gasifier.

the inlet and outlet grates has several distinct zones. The topmost zone preheats the feed coal by contacting with the hot crude product gas that is ready to leave the reactor. As the coal gets heated, devolatilization and gasification reactions proceed at temperatures ranging from 620°C to 760°C. Devolatilization of coal is accompanied by gasification of the resulting char. The interaction between devolatilization and gasification is a determining factor in the kinetics of the process, as well as of the product compositions.

The bottom of the bed is the combustion zone, where coal hydrocarbons react with oxygen to yield mainly carbon oxides. The exothermic heat generated by this combustion reaction provides the heat for gasification and devolatilization, both of which are endothermic reactions. By utilizing the exothermic heat of combustion in the gasification and devolatilization, energy integration within the gasifier is accomplished. More than 80% of the coal fed is gasified, with the remainder being burned in the combustion zone. The portion of feed coal burned for *in situ* heat generation may be called *sacrificial coal*. The temperature of the combustion zone must be selected in such a way that it is below the ash fusion point but high enough to ensure complete gasification of coal in subsequent zones. This temperature is also determined by the steam-to-oxygen ratio employed in operation.

The material and energy balance of the Lurgi gasifier is determined by the following process variables:

1. Pressure, temperature, and steam-to-oxygen ratio.
2. The nature of coal: The type of coal determines the nature of gasification and devolatilization reaction. Lignite is the most reactive coal, for which reaction proceeds at 650°C. On the other hand, coke is the least reactive, for which minimum temperature required for chemical reaction is around 840°C. Therefore, more coal is gasified per unit mole of oxygen for lignite compared to other types (ranks) of coal. The higher the coal rank (i.e., the carbon content of coal), the lower the coal reactivity.
3. The ash fusion point of the coal, which limits the maximum operable temperature in the combustion zone, which in turn determines the steam-to-oxygen ratio.
4. Both the amount and chemical composition of VM of the coal, which influence the quality and quantity of tar and oils produced.

The Lurgi gasifier has relatively high thermal efficiency because of its medium-pressure operation and the countercurrent gas–solid flow. At the same time, it consumes a lot of steam, and the concentration of carbon dioxide in the crude product gas is high, as shown in Table 2.2. Also, the crude gas leaving the gasifier contains a substantial amount of carbonization products such as tar, oil, naphtha, ammonia, etc. These carbonization products are results of devolatilization, pyrolytic reactions, and secondary chemical reactions involving intermediates. This crude product gas is passed through a scrubber, where it is washed and cooled down by a waste heat boiler.

*2.4.4.1.2   Slagging Lurgi Gasifier*
This gasifier is an improved version of the Lurgi dry-ash gasifier. A schematic [25] of the slagging Lurgi gasifier is shown in Figure 2.3. The temperature of the combustion

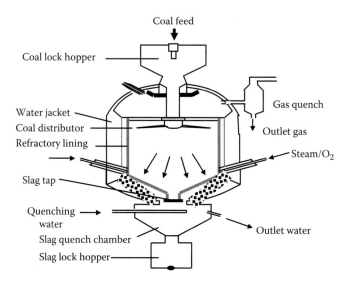

**FIGURE 2.3** Schematic of slagging Lurgi gasifier.

zone is kept higher than the ash fusion point. This is achieved by using a smaller amount of steam than dry-ash Lurgi gasifier, thus lowering the steam/oxygen ratio. The ash is removed from the bottom as slag, not as dry ash. Therefore, the process can handle *caking coals*, unlike the conventional dry-ash gasifier. The main advantage of this gasifier over the conventional dry-ash gasifier is that the range of acceptable feedstock coal is wide, the yield of carbon monoxide and hydrogen is high, and the coal throughput also increases many times. The steam consumption is also minimized [27].

### 2.4.4.1.3 BGL Gasifier

The BGL gasifier [24] is based on a moving bed reactor for gasification, which receives lump coal whose maximum particle size is as large as 5 cm. This coarse or lump coal is introduced into the top of the pressurized gasifier through a lock hopper. The gasifying medium of oxygen and steam is injected from the lower part of the gasifier. As coal moves down the gasifier against the flow of the gasifying medium, coal undergoes different stages of physical and chemical changes, viz., drying, devolatilization, pyrolysis, gasification, and combustion. At the base of the gasifier, where coal combustion takes place between the remaining coal char and oxygen, exothermic heat is generated and the temperature in the core of combustion zone reaches over 2000°C. This high thermal energy is sufficient for driving and sustaining coal gasification operation as well as melting the ash into molten slag. The liquid slag is drained from the bottom, collected, quenched, and made as nonleachable granular glassy slag. The product gas is a CO-rich syngas whose composition in volume% is shown in Table 2.3.

The process has several noteworthy features and they include the following:

1. The BGL gasifier demands low oxygen consumption.
2. The gas temperature at the top of the coal bed is low, typically around 450°C. Therefore, there is no need for an expensive heat recovery effort.

**TABLE 2.3**

**Typical Clean Gas Compositions BGL Gasifier**

| Species | Volume Percentage |
| --- | --- |
| $H_2$ | 30.8 |
| CO | 57.2 |
| $CO_2$ | 4.9 |
| $CH_4$ | 6.2 |
| Hydrocarbons | 0.4 |
| Noncombustibles | 0.5 |

*Source:* A. Williams and M. Olschar, BGL gasification—
project progress and technology developments. Available at http://www.icheme.org/gasification2010/pdfs/
bglgasificationandrewwilliams.pdf, 2010.

3. The slag formed is a dense solid material, which encapsulates the trace elements, which facilitates an easy removal as well as its environmentally safe
utilization as a building material.
4. No fly ash is generated nor emitted from the process.
5. The gasifier wall of the high-temperature region of the gasifier, which is the
base of the gasifier, is protected by a layer of solid slag.
6. The amount of wastewater generated by the process is low and reused on
site after appropriate treatment.
7. Due to the low effluent gas temperature, the thermal efficiency of the gasifier is high.
8. The carbon conversion from the gasifier is high and therefore the carbon
gasification efficiency is also high.

### 2.4.4.2 Koppers–Totzek Gasification Process

This gasification process uses entrained flow technology, in which finely pulverized coal is fed into the reactor with steam and oxygen [28,29]. The process operates at atmospheric pressure. As with all entrained flow reactors, the space time
in the reactor is very short. Therefore, the particle size used has to be very fine to
facilitate efficient entrainment in the reactor and also to eliminate potential pore
diffusional limitation inside the coal particles. The short space time in the reactor
also requires a very fast gasification reaction rate in order to achieve high conversion and yield. Such a high reaction rate is only achievable at very high reactor temperatures. The Koppers–Totzek gasifier typically operates at a temperature of about
1400°C–1500°C and *atmospheric pressure*. Due to the high reactor temperature
requirement, the reactor hardware needs to be carefully designed. The gasifier itself
is a cylindrical, refractory-lined coal burner with at least two burner heads through
which coal, oxygen, and steam are charged. The burner heads are spaced either 180°
(with the two-headed design) or 90° apart (with the four-headed arrangements) and
are designed such that steam covers the flame and prevents the reactor refractory

walls from becoming excessively hot. At this high temperature, the reaction rate of gasification is extremely high, that is, by orders of magnitude higher than that at a temperature in a typical fixed bed reactor. About 90% of carbonaceous matter is gasified in a single pass, depending on the type of coal gasified. Lignite is the most reactive coal, for which reactivity approaches nearly 100% [7].

In contrast to moving bed or fluidized bed reactors, this gasifier has very few limitations on the nature of feed coal in terms of caking behavior and mineral matter (ash) properties. Because of very high operating temperatures, the ash agglomerates and drops out of the combustion zone as molten slag and subsequently gets removed from the bottom of the reactor. The hot effluent gases are quenched and cleaned. This gas product contains no tar, ammonia, or condensable hydrocarbons and is predominantly syngas. It has a heating value of about 280 Btu/scf and can be further upgraded by reacting with steam to form additional hydrogen and carbon dioxide via WGS reaction.

### 2.4.4.2.1  Koppers–Totzek Gasifier

This gasifier is one of the most significant entrained bed (flow) gasifiers in commercial operation today. It accepts almost any type of coal, including highly caking coal, without any major operational restrictions. It has the highest operating temperature (around 1400°C–1500°C) of all the conventional gasifiers. Depending upon the arrangement of the burner heads, there are two versions in terms of process equipment design, a two-headed and a four-headed burner type. A schematic of a Koppers–Totzek two-headed gasifier [30] is shown in Figure 2.4. The original version designed in 1948 in Germany was two-headed, with the burner heads mounted at the ends, that is, 180° apart. The gasifier of this type is ellipsoidal in shape and horizontally situated. Each head contains two burners. The shell of the

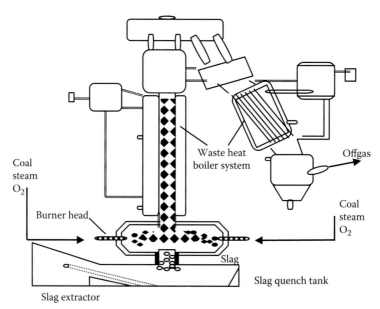

**FIGURE 2.4**  Schematic of Koppers–Totzek gasifier (two-headed burner design).

gasifier is water-jacketed and has an inner refractory lining. An advanced design of four-headed gasifiers began in India around 1970. In this design, burner heads are spaced 90° apart, instead of 180° as in two-headed designs. All the burner heads are installed horizontally. The capacity of a four-headed burner gasifier is larger than its two-headed counterpart and is generally more efficient [31].

### 2.4.4.2.2  Features of the Koppers–Totzek Process

The Koppers–Totzek gasification process has been very successfully operated commercially, and some of the process features are summarized as follows:

1. *High capacity*: These process units are designed for coal feed rates up to 800 tons per day, or for syngas production rate of about 42 million scf/day of 300-Btu/scf gas.
2. *Versatility*: The process is capable of handling a variety of feedstocks, including all ranks of solid fuels, liquid hydrocarbons, and pumpable slurries containing carbonaceous materials. Even feedstocks containing high sulfur and ash contents can be readily used in this process. Therefore, this process is not limited only to coal. This versatility also allows for handling of mixed feedstocks such as coal–biomass mixture, mixed coal–wastes, etc.
3. *Flexibility*: The changeover from solid fuel feed to liquid fuels involves only a change in the burner heads. Multiple feed burners permit wide variations in turndown ratio (which is defined as the numeric ratio between the highest and the lowest effective system capacity). This process is capable of instantaneous shutdown with full production resumable in a remarkably short time period, only 30 min.
4. *Simplicity of construction*: There is no complicated mechanical equipment or pressure-scaling device required. The only moving parts in the gasifiers are the moving screw feeders for solids or pumps for liquid feedstocks.
5. *Ease of operation*: Control of the gasifiers is achieved primarily by maintaining carbon dioxide concentration in the clean gas at a reasonably constant value. Slag fluidity at high process temperatures may be visually monitored. Gasifiers display good dynamic responses. The process can be integrated with a variety of carbon dioxide sequestration process concepts.
6. *Low maintenance*: Simplicity of design and a minimum number of moving parts require little maintenance between the scheduled annual maintenance events.
7. *Safety and efficiency*: The process has a track record of over 50 years of safe operation. The overall thermal efficiency of the gasifier is 85% to 90%. The time on stream (TOS) or availability is better than 95%.

### 2.4.4.2.3  Process Description of Koppers–Totzek Gasification

The Koppers–Totzek gasification process, whose flow schematic is shown in Figure 2.5, employs POX of pulverized coal in suspension with oxygen and steam. The gasifier is a refractory-lined steel shell encased with a steam jacket for producing low-pressure process steam (LP steam) as an energy recovery scheme. A typical two-headed gasifier is capable of handling 400 tons per day of coal. Coal, oxygen,

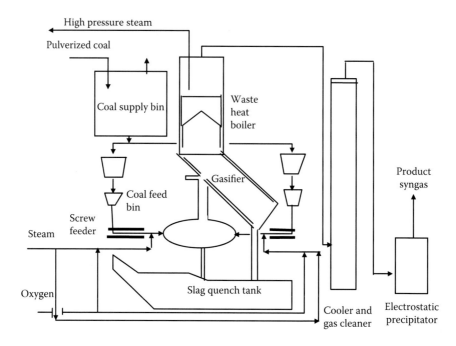

**FIGURE 2.5**  Schematic of the Koppers–Totzek gasification process.

and steam are brought together in opposing gasifier burner heads spaced 180° apart (in the two-headed case). In the case of four-headed gasifiers, these burners are 90° apart. A typical four-head design can handle up to 850 tons of coal per day. Exothermic reactions due to coal combustion produce a flame temperature of approximately 1930°C, which is lowered by heat exchange with a steam jacket. Gasification of coal is nearly complete and instantaneous. The carbon conversion depends on the reactivity of coal, approaching 100% for lignites. The lower the rank of coal, the higher the conversion.

Gaseous and vapor-phase hydrocarbons evolving from coal at moderate temperature are passed through a zone of very high temperature, in which they decompose so rapidly that there is no coagulation of coal particles during the plastic stage that typically happens around 375°C–400°C for bituminous coal. Thus, any coal can be gasified irrespective of its caking property, ash content, or ash fusion temperature. As a result of the endothermic reactions occurring in the gasifier between carbon and steam and radiational heat transfer to the refractory walls, the reactor temperature decreases from 1930°C (flame temperature) to around 1500°C. At these conditions, only gaseous products are produced without tars, condensable hydrocarbons, or phenols formed. Typical compositions of Koppers–Totzek gaseous products are shown in Table 2.4.

Ash in the coal feed becomes molten in the high-temperature zone of the gasifier. Approximately 50% of the coal ash drops out as slag into a slag quench tank below the gasifier. The remaining ash is carried out of the gasifier as fine fly ash. The

**TABLE 2.4**

**Typical Raw Product Gas Compositions of Koppers–Totzek Gasifier (Oxygen-Blown Type)**

| Component | Percentage |
|---|---|
| CO | 52.5 |
| $H_2$ | 36.0 |
| $CO_2$ | 10.0 |
| $H_2S$ + COS | 0.4 |
| $N_2$ + Ar | 1.1 |

*Source:* Lloyd, W.G., *The Emerging Synthetic Fuel Industry*, Thumann, A., Ed., Atlanta, GA: Fairmont Press, 1981, pp. 19–58.

*Note:* Average heating value = 286 Btu/scf; all percentages are in volume percent.

gasifier outlet is equipped with water sprayers to drop the gas temperature below the ash fusion temperature. This cooling prevents slag particles from adhering to the tubes of the waste heat boiler, which is mounted above the gasifier.

The raw gas from the gasifier passes through the waste heat boiler, where high-pressure steam (HP steam) up to 100 atm is produced via waste heat recovery. After leaving the waste heat boiler, the gas at 175°C–180°C is cleaned and cooled in a highly efficient scrubbing system, which reduces the entrained solids to 0.002–0.005 grains/scf or less and further lowers the temperature from 175°C to 35°C. If the gas coming out of the Koppers–Totzek process is to be compressed to high pressures for chemical synthesis, electrostatic precipitators (ESPs) are used for further cleaning. Several gasifiers can share common cleaning and cooling equipment, thus reducing the capital cost of the overall system.

The cool, cleaned gas leaving the gas cleaning system still contains sulfur compounds that must be removed to meet the final gas specifications. The type of the desulfurization system chosen depends on the end uses and the pressure of the product gas. For low pressure and low-Btu gas applications, there are a number of chemically reactive processes, such as *amine and carbonate processes*. Amine process is also known as "amine gas treating" or "gas sweetening" and is very popularly used in petroleum refinery, petrochemical industries, and natural gas processing plants. A number of different amines can be used and the list includes diethanolamine (DEA), monoethanolamine (MEA), methyldiethanolamine (MDEA), and others. At higher pressures (usually 400–1000 psia), physical absorption processes such as *Rectisol process* can be used. *Rectisol* is the trade name for a process that is based on an acid gas removal process that uses cold methanol (about –40°C) as a solvent to separate acid gases such as hydrogen sulfide and carbon dioxide from gas streams. Since Rectisol is a physical solvent, there is no chemical reaction between the solvent and acid gases. The choice of the process also depends on the desired purity of the product gas and its selectivity with respect to the concentration of carbon dioxide and sulfides. Advances in gas cleaning have been quite significant in recent years, owing to more stringent environmental regulations [32] as well as demands arisen from highly specialized and more sophisticated end uses of product syngas.

### 2.4.4.3 Shell Gasification Process

The Shell coal gasification process was developed by Royal Dutch and Shell group in the early 1970s. It uses a pressurized, slagging entrained flow reactor for gasifying dry pulverized coal [33]. Similar to the Koppers–Totzek process, it has the potential to gasify widely different ranks of coals, including low-rank lignites with high moisture content. Unlike other gasifying processes, it uses *pure oxygen* as the gasifying medium, for gasification via POX. Shell Global Solutions licenses two versions of gasification technologies, that is, one for liquid feedstock applications and the other for coal and petroleum coke. A schematic of the Shell coal gasification process is given in Figure 2.6. The process has the following features [34]:

1. Almost 100% conversion of a wide variety of coals, including high-sulfur coals, lignites, and coal fines
2. High thermal efficiency in the range of 75% to 80%
3. Efficient heat recovery through production of high-pressure superheated steam
4. Production of clean gas without any significant amount of by-products
5. High throughput
6. Environmental compatibility

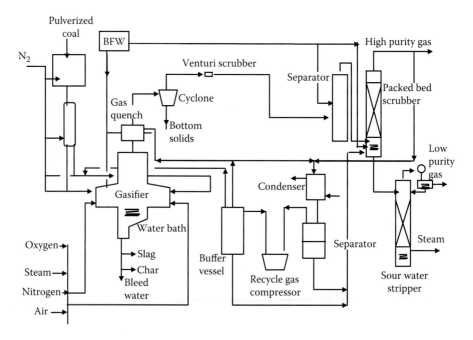

**FIGURE 2.6** Schematic of Shell gasification process.

Coal before feeding to the gasifier vessel is crushed and ground to less than 90 µm size (or about 170 mesh). This pulverized and dried coal is fed through diametrically opposite diffuser guns into the reaction chamber [35]. The coal is then reacted with the pure oxygen and steam, where flame temperature reaches as high as 1800°C–2000°C. A typical operating pressure is around 30 atm, and therefore, this is a pressurized gasification process. Raw product gas typically consists of mainly carbon monoxide (62%–63%) and hydrogen (28%), with some quantities of carbon dioxide. A water-filled bottom compartment is provided in which molten ash is collected. Some amount of ash is entrained with the syngas as dry fly ash, which is then recycled along with the unconverted carbon. A quench section is provided at the reactor outlet to lower the gas temperature. Removal of particulate matter from the raw product gas is integrated with the overall process. This particulate removal system typically consists of *cyclones and scrubbers*. The main advantage of this section is elimination of solid-containing wastewater, thus eliminating the need for filtration. Since the Shell gasification process uses pure oxygen and a high pressure, the physical size of the gasifier is substantially smaller than an atmospheric pressure gasifier.

### 2.4.4.4  Texaco Gasification Process

The Texaco gasification process also uses entrained flow technology for gasification of coal, similar to the Koppers–Totzek and Shell gasification processes. It gasifies coal under relatively high pressure by injection of oxygen (or air) and steam with concurrent gas/solid flow. Fluidized coal is mixed with either oil or water to make it into *pumpable slurry*. This slurry is pumped under pressure into a vertical gasifier, which is basically a pressure vessel lined inside with refractory walls. The slurry reacts with either air or oxygen at high temperature. The product gas contains primarily carbon monoxide, carbon dioxide, and hydrogen with some quantity of methane. Because of high temperature, oil or tar is not produced. This process is basically used to manufacture *CO-rich syngas* [7]. A schematic of the Texaco gasification process is shown in Figure 2.7.

This gasifier has evolved from the commercially proven Texaco POX process [23] used to gasify crude oil and hydrocarbons. Its main feature is the *use of coal slurry feed*, which simplifies the coal-feeding system and operability of the gasifier. The gasifier is a simple, vertical, cylindrical pressure vessel with refractory linings in the upper POX chamber. It is also provided with a slag quench zone at the bottom, where the resultant gases and molten slag are cooled down. In the latter operation, large amounts of high-pressure steam (HP-steam) can be obtained, which boosts the thermal efficiency of the process. Another important factor that affects the gasifier thermal efficiency is the water content of the coal slurry. This water content should be minimized because a large amount of oxygen must be used to supply the heat required to vaporize the slurry water, which would in turn hurt the thermal efficiency of the process. This gasifier favors high-energy dense coals so that the water-to-energy ratio in the feed is small. Therefore, high-rank Eastern U.S. bituminous coals are preferable to low-rank lignites for this gasifier. The gasifier operates at around 1100°C–1370°C and a pressure of 20–85 atm. This gasification is also a pressurized gasification technology.

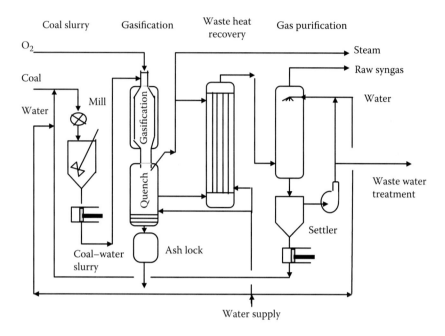

**FIGURE 2.7**   Schematic of Texaco gasification process.

The product gases and molten slag produced in the reaction zone pass downward through a water spray chamber and a slag quench bath, where the cooled gas and slag are then removed for further treatment. The gas, after being separated from slag and cooled, is treated to remove carbon fines and ash. These fines are then recycled to the slurry preparation system, while the cooled gas is treated for acid gas removal and elemental sulfur is recovered from the hydrogen sulfide ($H_2S$)-rich stream by a process like Claus desulfurization process of Claus plant.

### 2.4.4.5   *In Situ* Gasification Process

*In situ* gasification, or underground gasification, is a technology for recovering the energy content of coal deposits that cannot be exploited either economically or technically by conventional mining (or *ex situ*) processes. Coal reserves that are suitable for *in situ* gasification have low heating values, thin seam thickness, great depth, severe weather conditions above ground, high ash or excessive moisture content, large seam dip angle, or undesirable overburden properties. A considerable amount of investigation has been performed on *underground coal gasification (UCG)* in the former USSR and in Australia, but it is only in recent years that the concept has been revived in Europe and North America as a means of fuel gas production. In addition to its potential for recovering deep, low-rank coal reserves, the UCG process in general may offer some advantages with respect to its resource recovery, minimal environmental impact, operational safety, process efficiency, and economic potential. The aim of *in situ* gasification of coal is to convert coal hydrocarbons into combustible gases by combustion of coal seam in the presence of air, oxygen, or steam.

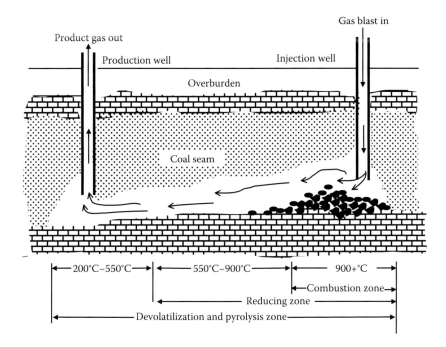

**FIGURE 2.8**    Schematic of *in situ* underground gasification process.

The basic concepts of UCG may be illustrated by Figure 2.8 [36]. The basic principles of *in situ* gasification are still very similar to those involved in the aboveground (*ex situ*) gasification of coal. Combustion process itself could be handled in either *forward* or *reverse* mode. Forward combustion involves movement of the combustion front and injected air in the same direction, whereas in reverse combustion, the combustion front moves in the opposite direction to the injected air. The process involves drilling and subsequent linking of the two boreholes to enable gas flow between the two. Combustion is initiated at the bottom of one borehole called *injection well* and is maintained by the continuous injection of air. Advances in horizontal drilling technologies, as widely practiced in the shale gas field, could be beneficially adopted in the underground gasification technology development.

As illustrated in Figure 2.8, in the initial reaction zone, carbon dioxide is generated by reaction of oxygen (air) with the coal ($C + O_2 = CO_2$), which further reacts with coal to produce carbon monoxide by the Boudouard reaction ($CO_2 + C = 2CO$) in the reduction zone. Further, at such high temperatures, the moisture present in the seam may also react with carbon to form carbon monoxide and hydrogen via the steam gasification reaction ($C + H_2O = CO + H_2$). In addition to all these basic gasification reactions, coal decomposes in the pyrolysis zone owing to high temperatures to produce hydrocarbons and tars, which also contribute to the product gas mixture. The heating value from the air-blown *in situ* gasifier is roughly about 100 Btu/scf, only about one-tenth of the natural gas heating value. The low heating value of the product gas makes it uneconomical for long distance transportation, making it necessary to use the product gas on site. Furthermore, the efficiency of coal utilization and/or energy recovery from the underground coal resource is relatively low. An extensive discussion on *in situ* gasification can be found in references by Thompson [37] and by Gregg and Edgar [38]. A noteworthy R&D effort in UCG has also been conducted by the Commonwealth Scientific and Industrial Research Organization (CSIRO), Australia. CSIRO researchers have developed a model to assist with the implementation of this technology [39]. A number of other trials and trial schemes were evaluated in Europe, China, India, South Africa, and the United States.

### 2.4.4.5.1 Potential Possibility of Using Microbial Processes for In Situ Gasification

Jüntgen [40] in his review article has explored the possibilities of using microbiological techniques for *in situ* conversion of coal into methane. Microorganisms have been found that grow on coal as a sole carbon source. Both forms of sulfur, namely, organic and inorganic (pyritic and sulfatic), are claimed to be removable by biochemical techniques, and microorganisms are able to grow, in principle, in narrow pore structures of solids. The conversion of large-molecular-weight aromatics, including polynuclear aromatics (PNAs), is also potentially feasible. An important precursor of developing such process techniques for *in situ* coal conversion in deep seams is the knowledge of coal properties, both physical and chemical, under the prevailing conditions. The two most important coal properties, which dictate the *in situ* processes, are the *permeability* of coal seam, including the overburden and the *rank* of coal. For microbial conversion of coal, microporosity also becomes an important parameter. The permeability of coal seam in great depths is usually quite low due to the rock compaction effect mainly caused by high overburden pressure. However, accessibility is very important for performing *in situ* processes. There are several ways to increase the permeability of the coal seams at great depths [40]. Some of these ideas are very similar to those used in *in situ* oil shale retorting and shale gas recovery.

The main advantage of using microbiological techniques is that the reaction takes place at ambient temperatures. Progress made in developing these types of processes is quite notable. A remarkable effect of such reactions in coal is that the microorganisms can penetrate into fine pores of the coal matrix and can also create new pores if substances contained in the coal matrix are converted into gaseous compounds.

However, the most difficult and complex problem associated with microorganism-based reactions is the transition from solely oxidative processes to methane-forming reactions. There are at least three reaction steps involved: (1) the aerobic degradation of coal to biomass and high-molecular-weight products; (2) an anaerobic reaction leading to the formation of acetate, hydrogen, and carbon monoxide; and (3) the conversion of these products to methane using methanogenic bacteria. Methanogenic bacteria belong to a group of primitive microorganisms, the *Archaea*. They give off methane gas as a by-product of their metabolism and are common in sewage treatment plants and hot springs, where the temperature is warm and oxygen is absent. Advantages of these processes over other coal conversion processes are lower conversion temperature and more valuable products [40]. However, an intensive investigation must be conducted to adapt reaction conditions and product yields to conditions prevailing in coal seams at great depth, where transport processes play a significant role in the overall reaction.

### 2.4.4.5.2  Underground Gasification System

The underground gasification system involves three distinct sets of operations: pregasification, gasification, and further processing and utilization. Pregasification operations provide access to the coal deposit and prepare it for gasification. Connection between the inlet and outlet through the coal seam is achieved via shafts and boreholes. Linking can be achieved through several means, such as pneumatic, hydraulic, or electric linking, and using explosives, etc. Sometimes, partial linking may also be accomplished by taking advantage of the natural permeability of the coal seam. Among all the linking methods, only directionally drilled boreholes provide positive connections between inlet and outlet sections, and all other methods permit a certain degree of uncertainty to play a role in the system. A schematic view of a *linked-vertical-well underground gasification* plant operated near Moscow [36] is shown in Figure 2.9.

The gasification operations that allow reliable production of low-Btu gas consist of input of gasifying agents such as air or oxygen and steam (or alternating air and steam), followed by ignition. Ignition can be managed either by electrical means or by burning solid fuels. Ignition results in contact between gasifying agents and coal organics at the flame front. The flame front may advance in the direction of gas flow (*forward burning*) or in the direction opposite to the gas flow (*backward burning*). During these operations, the major technical difficulties and challenges are in the area of process control. Owing to the unique nature of underground gasification, there inherently exist problems of controllability and observability.

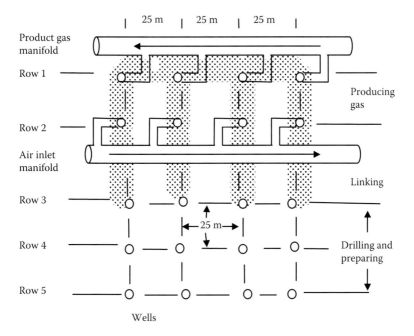

**FIGURE 2.9**  Plane view of linked-vertical-well underground gasification plant operated near Moscow.

The next, and most important, operation is the utilization of the product gas, and it requires a coupling between the gas source and the energy demand. The product gas can be either used as an energy source to produce electricity on site or upgraded to a high-Btu pipeline-quality gas for transmission. In some other applications, it could be utilized near the deposit as a hydrogen source, as a reducing agent, or as a basic raw material for manufacture of other chemicals. With realization of the hydrogen economy, the product gas may have good potential as a hydrogen source. Generally speaking, there are no major technical problems involved with the utilization of product gas, apart from potential environmental concerns.

### 2.4.4.5.3  Methods for Underground Gasification

There are two principal methods that have been tried successfully, *shaft methods* and *shaftless methods* (and combinations of the two) [38,39,41]. Selection of a specific method to be adopted depends on such parameters as the natural permeability of the coal seam, the geochemistry of the coal deposit, the seam thickness, depth, width and inclination, closeness to metropolitan developments, and the amount of mining desired. Shaft methods involve driving of shafts and drilling of other large-diameter openings that require underground labor, whereas shaftless methods use boreholes for gaining access to the coal seam and do not require labor to work underground.

#### 2.4.4.5.3.1  Shaft Methods

1. *Chamber or warehouse method*: This method requires the preparation of underground galleries and the isolation of coal panels with brick wall. The blast of air for gasification is applied from the gallery at the previously ignited face of one side of the panel, and the gas produced is removed through the gallery at the opposite side of the panel. This method relies on the natural permeability of the coal seam for airflow through the system. Gasification and combustion rates are usually low, and the product gas may have variable composition from time to time. To enhance the effectiveness, coal seams are precharged with dynamites to rubblize them in advance of the reaction zone by a series of controlled explosions.
2. *Borehole producer method*: This method typically requires the development of parallel underground galleries and are located about 500 ft. apart within the coal bed. From these galleries, about 4-in.-diameter boreholes are drilled about 15 ft. apart from one gallery to the opposite one. Electric ignition of the coal in each borehole can be achieved by remote control. This method was originally designed to gasify substantially flat-lying seams. Variations of this technique utilize hydraulic and electric linking as alternatives to the use of boreholes.
3. *Stream method*: This method can be applied to steeply pitched coal beds. Inclined galleries following the dip of the coal seam are constructed parallel to each other and are connected at the bottom by a horizontal gallery or "fire-drift." A fire in the horizontal gallery initiates the gasification, which proceeds upward with air coming down one inclined gallery and gas leaving through the other. One obvious advantage of the stream method is that

ash and roof material drop down, tend to fill void space, and do not tend to choke off the combustion zone at the burning coal front. However, this method is structurally less suitable for horizontal coal seams because of roof collapse problems.

*2.4.4.5.3.2  Shaftless Methods*   In shaftless methods, all development, including gasification, is carried out through a borehole or a series of boreholes drilled from the surface into the coal seam. A general approach has been to make the coal bed more permeable between the inlet and outlet boreholes by a chosen linking method, ignite the coal seam, and then gasify it by passing air and other gasifying agents from the inlet borehole to the outlet borehole.

*2.4.4.5.3.3  Percolation or Filtration Methods*   This is the most direct approach to accomplish shaftless gasification of a coal seam using multiple boreholes. The distance required between boreholes depends on the seam permeability. Lower-rank coals such as lignites have a considerable natural permeability and, as such, can be gasified without open linking. However, higher-rank coals such as anthracites are far less permeable, and it becomes necessary to connect boreholes by some efficient linking techniques that will increase the permeability and fracture of the coal seam so that an increased rate of gas flow can be attained. Air or air/steam is blown through one borehole, and product gas is removed from another borehole. Either forward or reverse combustion can be permitted by this method. As the burn-off (a combination of combustion and gasification) progresses, the permeability of the seam also increases, and compressed air blown through the seam helps enlarge cracks or openings in the seam. When the combustion of a zone nears completion, the process is transferred to the next pair of boreholes and continues. In this operation, coal ash and residues should be structurally strong enough to prevent roof collapse.

### 2.4.4.5.4   Potential Problem Areas with In Situ Gasification

There are several issues why the *in situ* gasification processes may not be able to produce a high-quality and constant quantity of product gas, recover a high percentage of coal energy in the ground, and control the groundwater contamination. Potential problem areas in commercial exploitation of this technology are discussed in the following text.

*2.4.4.5.4.1  Combustion Control*   Combustion control is essential for controlling the product gas quality as well as the extent of coal conversion. The reactive contacting between the coal and the gasifying agent should be such that the coal is completely *in situ* gasified, all oxygen in the inlet gas is consumed, and the production of fully combusted carbon dioxide and water is minimized. In a typical *in situ* coal gasification process, as the processing time goes by, the heating value of the product gas decreases. This may be attributable to increasingly poor contact of gas with the coalface, because of large void volumes and from roof collapse. The problem of efficient contacting needs to be solved satisfactorily in this process.

*2.4.4.5.4.2 Roof Structure Control*    After the coal is burned off, a substantial roof area is left unsupported. Uncontrolled roof collapse causes nontrivial problems in the combustion control and also seriously hinders successful operation of the overall gasification process. Further, it potentially results in the leakage of reactant gases, seepage of groundwater into the coal seam, loss of product gas, and surface subsidence above the coal deposit.

*2.4.4.5.4.3 Permeability, Linking, and Fracturing*    An underground coal bed usually does not have a sufficiently high permeability to permit the passage of oxidizing gases through it without a serious pressure drop. Also, intentional linking methods such as pneumatic, hydraulic, and electric, as well as fracturing with explosives, do not result in a uniform increase in permeability throughout the coal bed. They also tend to disrupt the surrounding strata and worsen the leakage problems. Therefore, the use of boreholes is proved to provide a more predictable method of linking and is a preferred technique.

*2.4.4.5.4.4 Leakage Control*    This is one of the most important problems because the loss of substantial amount of product gas can adversely affect the recovered amount of the product gas as well as the gasification economics. Further, the inlet reactant gases should not be wasted. Influx of water can also affect the control of the process. Leakage varies from site to site and also depends on a number of factors including geological conditions, depth of coal seam, types of boreholes and their seals, and permeability of coal bed.

Based on the above considerations, it is imperative that *in situ* gasification never be attempted in a severely fractured area, in shallow seams, or in coal seams adjoining porous sedimentary layers. It is also essential to prevent roof collapse and to properly seal inlet and outlet boreholes after operation. Similarly, leakage control is very important in underground operation of shale gas exploitation.

### 2.4.4.5.5   Monitoring of Underground Processes

Proper monitoring of the underground processes is a necessary component of successful operation and design of an underground gasification system. *A priori* knowledge of all the parameters affecting the gasification is required so that adequate process control philosophy can be adopted and implemented for controlling the operation. These factors include the location, shape, and temperature distribution of the combustion front, the extent and nature of collapsed roof debris, the permeability of coal seam and debris, the leakage of reactant and product gases, the seepage of groundwater, and the composition and yield of the product gases.

### 2.4.4.5.6   Criteria for an Ideal Underground Gasification System

The following are the criteria for successful operation of an ideal UCG system:

1. The process must be operable on a large scale.
2. The process must ensure that no big deposits of coal are left ungasified or partially gasified.
3. The process must be controllable so that desired levels, in terms of quality and quantity, of product gases are consistently produced.

4. The mechanical features must ensure that they should be able to control undesirable phenomena such as groundwater inflow and leakage (as out-flow) of reactants and products.
5. The process should require little or no underground labor, either during operation or even during the installation of the facilities.

### 2.4.4.6 Production of Coal Bed Methane

Many coal beds contain vast amounts of methane that can be extracted by drilling appropriate wells. This useful fuel resource is called *coal bed methane*. Researchers have shown that a good fraction of methane in the coal beds is produced by naturally occurring microorganisms that feed on coal. Researchers also found specific ways to stimulate the microbes to produce more methane from coal, and if the conditions are managed properly, the process and phenomena can provide a valuable avenue for converting coal into methane. This is especially true for the economically unfavor-able or inaccessible coal mines.

Luca Technologies Inc. based in Golden, Colorado, said laboratory evidence indi-cates that anaerobic microbes are turning the coal in the Powder River Basin of northeastern Wyoming into methane. The company is using this approach of stimu-lating the microbes to increase production from coal beds with existing methane wells. On the other hand, Next Fuel, based in Sheridan, Wyoming, showed that a similar technology could be used to produce methane from coal beds that do not have preexisting methane in them, thereby raising the possibility that potentially large amounts of methane could be produced from coal mines that are currently too expensive to operate or not profitable to produce coal due to a variety of reasons. This could be very valuable for the regions and nations where the natural gas price is quite high but abundant coal deposits are available (MIT Technology Review 2012).

### 2.4.4.7 Winkler Gasification Process

This is the oldest commercial process employing fluidized bed technology [42]. The process was developed by Rheinbrau AG (now RWE) in Germany in 1926. The initial process used lignites. Later in the 1970s, enhancements were made for the process employing higher gasification temperatures, thereby improving the carbon conver-sion as well as upgrading the product syngas quality. This high temperature version of Winkler process is called the "High-Temperature Winkler Process," "HT Winkler Process," or "HTW." Recent developments of this technology were on increasing the gas throughput and meeting the gas turbine inlet pressures for IGCC applications by increasing the system operating pressures, which in turn alleviates the need for an intermediate compression stage. There are more than 15 plants in operation today all over the world with the largest having an output of 1.1 million scf/day.

In the classical version of the Winkler process, pulverized coal is dried and fed into a fluidized bed reactor by means of a *variable speed screw feeder*. The gasifier operates at atmospheric pressure and a temperature of 815°C–1000°C. Coal par-ticles react with oxygen and steam to produce offgas rich in carbon monoxide and hydrogen. The relatively high operating temperature leaves very little tar and liquid hydrocarbons in the product gas stream. The gas stream that may carry up to 70% of the generated ash is cleaned by water scrubbers, cyclones, and ESPs. Unreacted char carried over by the fluidizing gas stream is further converted by secondary steam and oxygen in the space above the fluidized bed. As a result, the maximum temperature occurs above the fluidized bed. To prevent ash fines from melting at a high temperature and forming deposits in the exit duct, gas is cooled by a radiant boiler before it leaves the gasifier. Raw hot gas leaving the gasifier is passed through a waste heat recovery section. The gas is then compressed and goes through WGS reaction. The product gas has a heating value of about 275 Btu/scf. The thermal efficiency of the process runs approximately 75%.

#### 2.4.4.7.1  Process Description

In the early 1920s, Winkler, an employee of Davy Power Gas Inc., conceived the idea of using a fluidized bed for gasifying the coal. The first commercial unit was built in 1926. Since then, more than 30 producers and 15 installations have put this process into operation for coal gasification.

In earlier facilities, dryers were used, prior to the introduction of coal into the gas generator, to reduce the coal moisture to less than 8%. It was later realized that as long as the feed coal could be sized, stored, and transported without plugging, dryers could be omitted. Without dryers, moisture in the coal is vaporized in the generator with the heat provided by using additional oxygen for combustion reaction. Drying the coal in the generator also offers an additional advantage, that is, elimination of an effluent stream, the dryer stack, which would require further treatment of particulate and sulfur removal.

#### 2.4.4.7.2  Gasifier (Gas Generator)

A schematic of a Winkler fluidized bed gasifier [36] is shown in Figure 2.10. Pulverized coal is fed to the gasifier through variable-speed feeding screws. These

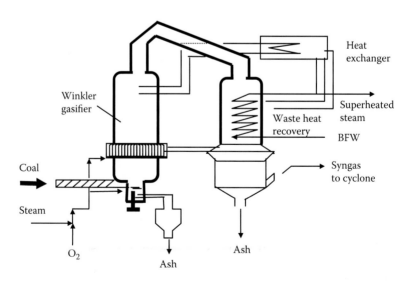

**FIGURE 2.10**  Schematic of Winkler gasification process.

screw-type feeders not only control the coal feed rate but also serve to seal the gasifier by preventing steam from wetting the coal and blocking the pathway by agglomeration. Alternatively, the coal feed can be discharged into the gasifier using a gravity feed pipe in a modern version of the process. A high-velocity gas stream flows upward from the bottom of the gasifier. This gas stream fluidizes the bed of coal, as well as intimately mixes the reactants, thus bringing them into close contact. Fluidization helps the gas-to-solid mass transfer. This also helps in attaining an isothermal condition between the solid and the gas stream, which permits the reactions to reach equilibrium in the shortest possible time. Gasification chemistry in the Winkler gasifier is based on a combination of combustion reaction and WGS reaction.

$$C + O_2 = CO_2$$

$$C + 1/2\, O_2 = CO$$

$$C + H_2O = H_2 + CO$$

$$CO + H_2O = CO_2 + H_2$$

In the preceding reactions, carbon was used instead of coal only for illustrative purposes. Therefore, the actual reactions in the gasifier are much more complex. Owing to the relatively high temperatures of the process, nearly all the tars and heavy hydrocarbons are reacted [43].

As a result of the fluidization, the ash particles get segregated depending upon particle size and specific gravity. About 30% of the ash leaves through the bottom, whereas 70% is carried overhead. Ash is removed from the gasifier bottom by use of an ash discharge system, whose design typically includes an ash screw mover, lock hopper, and discharge bin. The lighter particles carried upward along with the produced gas are further gasified in the space above the bed. Therefore, the quantity of gasifying medium injected into this bed must be adjusted proportionally to the amount of unreacted carbon being carried over. If it is too little, ungasified carbon gets carried out of the generator, resulting in a slightly lower thermal efficiency, and if it is too much, product gas is unnecessarily consumed by combustion. The maximum temperature in the generator occurs in the space above the fluidized bed because of this secondary (further) gasification. Very fine particulate ash and unconverted char particles are *entrained* in the raw product syngas and removed in cyclone and cooled.

A radiant boiler installed immediately above the bed cools the hot product gas down to 150°C–205°C before it leaves the generator. This helps prevent the fly ash from getting sintered on the refractory walls of the exit duct. The sensible heat recovered by the radiant boiler generates superheated steam and is used to preheat the boiler feed water (BFW), as an energy integration scheme. The typical gas composition from a Winkler gasifier is shown in Table 2.5. As can be seen from the data, the product gas is rich in carbon monoxide, making the resultant gas a CO-rich syngas, which is typical for an air or oxygen blown gasifier.

**TABLE 2.5**
**Typical Winkler Gas Products**

| Component | O$_2$-Blown (%) | Air-Blown (%) |
|---|---|---|
| CO | 48.2 | 22.0 |
| H$_2$ | 35.3 | 14.0 |
| CH$_4$ | 1.8 | 1.0 |
| CO$_2$ | 13.8 | 7.0 |
| N$_2$ + Ar | 0.9 | 56.0 |

*Source:* Lloyd, W.G., *The Emerging Synthetic Fuel Industry*, Thumann, A., Ed., Atlanta, GA: Fairmont Press, 1981, pp. 19–58.

*Note:* Heating value, Btu/scf: O$_2$-blown = 288; air-blown = 126.

### 2.4.4.7.3  Features of Winkler Process

The following are the chief characteristics of the Winkler process:

1. A variety of coal feedstocks of widely different ranks, ranging from lignite to coke, can be gasified. Petrologically younger and low-rank lignite is more reactive than older and higher-rank counterparts of bituminous coal and anthracite. With more reactive coal, the required gasification temperature decreases, whereas the overall gasification efficiency increases. For less reactive coals, however, the energy losses through unburned solids inevitably increase.

2. Coal with high ash content can be gasified without difficulty. Although high-ash-content coals result in increased residues and incombustible materials, usually they are less expensive; thus, sources of feed coal can be greatly expanded. Winkler gasifier is not sensitive to variations in the ash content during operation.

3. Winkler gasifier can also gasify liquid fuels in conjunction with coal gasification. The addition of supplementary liquid feeds results in an increase in production and heating value of the product gas, thereby boosting the process economics favorably.

4. Winkler gasification is very flexible in terms of the capacity and turndown ratio. It is limited at the lower end by the minimum flow required for fluidization and at the upper end by the minimum residence time required for complete combustion of residues. A turndown ratio, which is also known as rangeability, is a ratio of the maximum controllable throughput to the minimum controllable throughput.

5. Shutdown can be very easily facilitated by stopping the flows of oxygen, coal, and steam, and can be achieved within minutes. Even for hard coals (with low permeability), which are more difficult to ignite, the heat loss during shutdown may be reduced by brief injection of air into the fuel bed.

6. Maintenance of the gas generator is straightforward, because it consists only of a brick-lined reactor with removable injection nozzle for the gasification medium.

From a more recent study, the *high-temperature Winkler (HTW) process* was chosen to be well suited for gasification of the lignite found in the Rhine area of Germany. The suitability was based on its temperature for gasification and the fluidized bed reactor configuration [44]. The study also discusses the selection criteria of gasification processes. Rhinebraun AG has operated a demonstration plant of HTW process at Berrenrath, Germany, since 1986 [45]. A variety of feedstocks other than coal, namely, plastic wastes, household refuse, and sewage sludge, were successfully processed [45].

The modern version of the HT Winkler process is based on a circulating fluidized bed (CFB) reactor and operates under high pressures in either air of oxygen blown modes. As such, the gasifier can be classified as a dry-feed, pressured, dry-ash, and fluidized bed technology.

A recent effort termed as *N3T* (next thermal treatment technology) is also based on the HTW process in conjunction with a short kiln, which is a result of successful demonstrations of cogeneration of municipal wastes and dried lignites, development and utilization of waste pretreatment technology, prevention of dioxin and furan formation via reductive environment, etc. One of its principal aims, besides energy production, is to reduce the amount of wastes that have to be disposed of by landfilling. Krupp Udhe's PreCon® process is a modular technology for waste gasification that also includes HT Winkler gasification.

A 20 t/day municipal solid waste (MSW) gasification plant based on the HTW process was built and successfully operated at Niihama, Sikuku, Japan, by Sumitomo Heavy Industries, Ltd. (SHI). This process is often referred to as "N3T-HTW" process or "N3T-HTW for MSW."

### 2.4.4.8   KBR Gasification Process

The KBR's transport gasifier, which is also known as TRIG, was developed particularly for low-rank coals such as lignite and subbituminous coals. TRIG stands for "transport integrated gasification" and is an advanced gasification technology whose principal gasification reaction is via POX in a cost-efficient air-blown system. The transport reactor is a variation of a CFB reactor, wherein low-rank coal is gasified with blown air that is partially extracted from the gas turbine. The gasifier reacts and converts coal, air, and steam into low-Btu gas at about 25 bar and 1000°C. Limestone is also fed to the reactor at a predetermined rate, which captures most of the sulfur in coal during gasification. After solids and fines are removed from the effluent gas stream using disengage and cyclone, the raw syngas is cooled to about 350°C–375°C in a heat exchanger. The entrained char is removed using an iron aluminide ($Fe_3AL$) hot gas filter. The unconverted carbon, unreacted limestone, and reacted limestone and coal ash are recovered, and after adding water for dust control, the mixture is sent for landfill. The carbon conversion typically exceeds 97%. Natural gas burner heats the gasifier before the introduction of solids into the gasifier during start-up operation. The raw gas, excluding nitrogen, is classifiable as a CO-rich syngas. An example of design syngas composition is given in Table 2.6 [46].

**TABLE 2.6**

**Design Syngas Composition—Power Systems Development Facility (Wilsonville, Alabama)**

| Species | Percentage |
|---|---|
| $CH_4$ | 2.10 |
| HCN | 0.02 |
| CO | 19.95 |
| $H_2O$ | 4.71 |
| $CO_2$ | 7.46 |
| $NH_3$ | 0.15 |
| $H_2$ | 10.32 |
| $N_2$ | 55.28 |
| $H_2S$ | 0.01 |

*Source:* R. Leonard et al., The PSDF—commercial readiness for coal power—revisited. Available at http://www.netl. doe.gov/technologies/coalpower/gasification/projects/ adv-gas/25140/Revised%20paper-randall.pdf, 2012.

### 2.4.4.9 Wellman–Galusha Gasification Process

This process has been in commercial use for more than 40 years. It is capable of producing low-Btu gas: to be specific, using air (as a gasifying medium) for production of fuel gas or using oxygen (as a gasifying medium) for syngas. There are two types of gasifiers for this process, namely, *the standard type without agitator* and *the modified type with agitator.* The rated capacity of the agitated type is about 25% more than that of a standard type gasifier of the same size. The agitated type can handle volatile caking bituminous coals, whereas the nonagitated type would have technical difficulties with this type of coal [7]. A schematic of a *Wellman–Galusha agitated gasifier* [25] is shown in Figure 2.11.

This gasifier can be classified under the categories of a fixed bed or moving bed type reactor. The gasifier shell is water-jacketed, and hence, the inner wall of the reactor vessel does not require a refractory lining. The upper part of the gasifier operates at about 540°C–650°C and at atmospheric pressure. Pulverized coal is fed to the gasifier from the top through a lock hopper and vertical feed pipes, whereas steam and oxygen are injected at the bottom of the bed through tuyeres (nozzles through which the gasifying medium is blown into). The fuel valves are operated to maintain constant flow of coal to the gasifier, which also helps in stabilizing the bed, thus maintaining the quality of the product gas. The injected air or oxygen passes over the water jacket and generates the steam required for the process. A rotating grate is located at the bottom of the gasifier to remove ash from the bed uniformly. An air–steam mixture is introduced underneath the grate and is evenly distributed through the grate into the bed. This gasifying medium passes through the ash, combustion, and gasifying zones in this specific order while undergoing a variety of chemical reactions. The product gas contains hydrogen, carbon monoxide, carbon

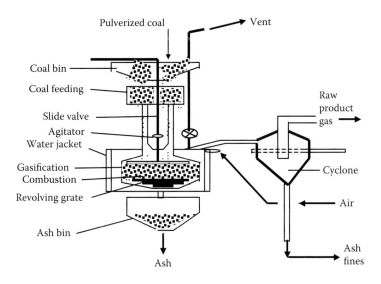

**FIGURE 2.11** Schematic of agitated Wellman–Galusha gasifier.

dioxide, and nitrogen (if air is used as an injecting medium), which, being hot, dries and preheats the incoming coal before leaving the gasifier. The typical product composition of an air-blown Wellman–Galusha gasifier is presented in Table 2.7.

The product gas is passed through a cyclone separator, where char particles and fine particulate ash are removed. It is then cooled and scrubbed in a direct-contact countercurrent water cooler and treated for sulfur removal. If air is used as an oxidant as illustrated in Table 2.7, low-Btu gas is obtained owing to the presence of a large amount of nitrogen; if oxygen is used, then medium-Btu gas would be produced.

Unlike the standard Wellman–Galusha gasifier, the agitated version is equipped with a slowly revolving horizontal arm that spirals vertically below the surface of the

**TABLE 2.7**
**Typical Wellman–Galusha Products (Air-Blown)**

| Component | Percentage |
|-----------|------------|
| CO | 28.6 |
| $H_2$ | 15.0 |
| $CH_4$ | 2.7 |
| $N_2$ | 50.3 |
| $CO_2$ | 3.4 |

*Source:* Lloyd, W.G., *The Emerging Synthetic Fuel Industry,* Thumann, A., Ed., Atlanta, GA: Fairmont Press, 1981, pp. 19–58.

*Note:* Heating value (dry) = 168 Btu/scf.

coal bed to minimize channeling. This arm also helps in providing a uniform bed for gasification.

Principal advantages of the process include simple design, small-scale operability, very high carbon utilization, and adaptation to biomass conversion. One of the drawbacks of the process is the generation of tar, which is mainly attributable to its low gasification temperature.

### 2.4.4.10 U-Gas Process

The process was developed by the Institute of Gas Technology (IGT), Des Plaines, Illinois, to produce gaseous products from coal in an efficient and environmentally acceptable manner. The combination of the IGT and the Gas Research Institute (GRI) formed the Gas Technology Institute (GTI) in 2000. Since then, the process is referred to as "GTI's U-gas Process." The product gas may be used to produce low-Btu gas, medium-Btu gas, and SNG for use as fuels, or as chemical feedstocks for ammonia, methanol, hydrogen, oxo-chemicals, etc., or for electricity generation via an *IGCC*. Based on extensive research and pilot plant testing, it has been established that the process is capable of handling large volumes of gas throughput, achieving a high conversion of coal to gas without producing tar or oil, and causing minimum damage to the environment.

The U-gas process is based on a single-stage, fluidized bed gasifier, as shown in Figure 2.12. The gasifier accomplishes four principal functions in a single stage, namely, (1) decaking coal, (2) devolatilizing coal, (3) gasifying coal, and (4) agglomerating and

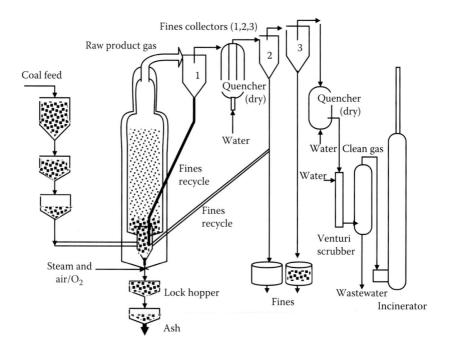

**FIGURE 2.12**   Schematic of U-gas process.

separating ash from char. Coal of about 0.25 in. (or about 6 mm) diameter is dried and pneumatically injected into the gasifier through a lock hopper system. In the fluidized bed reactor, coal reacts with steam and oxygen at a temperature of 950°C–1100°C. The temperature of the bed is determined based on the type of coal feed and is controlled to prevent slagging conditions of ash. The pressure may be flexible, typically ranging from 50 to 350 psi, and is largely determined based on the ultimate use of the final product gas. Oxygen may be substituted with air. If an air-blown system is used, the product gas would be a low-Btu gas due to its high abundance of nitrogen content. In the gasifier, coal is rapidly gasified producing $H_2$, CO, $CO_2$, and small amounts of $CH_4$. The fluidized bed is always maintained under reducing atmosphere, and, as such, all sulfur species present in coal is converted into $H_2S$. Simultaneously with gasification, the ash is agglomerated into spherical particles that grow in size and are separated from the bed into water-filled ash hoppers, from which they are withdrawn as slurry. A portion of fluidizing gas enters the gasifier section through an inclined grid, whereas most of the remaining entering gas flows upward at a high velocity through the ash-agglomerating zone and forms a relatively hot zone within the bed. This step essentially accomplishes some heat exchange between the ash and input gas.

Coal fines elutriated from the bed are collected by two external cyclones. Fines from the first cyclone are returned to the bed, whereas those from the second cyclone are sent to the ash-agglomerating zone. Raw product gas is virtually free of tar and oils, thus simplifying the ensuing energy recovery and gas purification steps. The pilot plant operated by the IGT has a gasifier made of a mild-steel, refractory-lined vessel with an I.D. of 3 ft. and a height of about 30 ft.

An IGCC process based on the GTI's U-gas process was developed by Tampella Power Company, Finland, which later became Carbona Inc. The choice of the IGT process is based on its excellent carbon conversion, as well as its versatility with a wide range of coals and peat. Enviropower Inc. originally licensed the U-gas technology and developed it as Enviropower gasification technology. Later, Enviropower's gasification business was taken over by Carbona Inc. Carbona has developed the technology applicable to biomass gasification and is developing a pressurized fluidized bed gasification plant for the 55 MW cogeneration project with Ignifluid Boilers India Ltd. (IBIL), Chennai, India. The plant is designed for multifuel operation, including biomass [47]. In 2005, the GTI designed and operated a 4–8 MW pilot plant for gasification of coal and biomass using an air/oxygen blown bubbling fluidized bed (BFB) reactor.

### 2.4.4.11  Catalytic Coal Gasification

In recent years, the study of catalytic gasification has received attention because it requires less thermal energy input but yields higher carbon conversion. However, the challenges are in the operational cost of catalytic gasification system as well as the management of the catalyst in the gasifier system, in both retention and activity of the catalyst. Studies on the catalysis of coal gasification have twofold objectives: (1) to understand the kinetics of coal gasification that involves active mineral matter and (2) to design possible processes using these catalysts. The use of efficient catalysts lowers the gasification temperature, which favors product composition under equilibrium conditions as well as high thermal efficiency. However, under normal conditions, a catalytic process cannot compete with a noncatalytic one unless the catalyst is quite inexpensive, easily retainable and/or recoverable, or highly active at low temperatures. In practice, recovery and reuse of a catalyst in the process is undesirable and unattractive in coal gasification because of the expensive separation efforts and the relatively low cost of coal (raw material) and coal syngas (product). Research on catalysis covers mainly three subjects: basic chemistry, application-related problems, and process engineering. Jüntgen [48] published an extensive review article on catalytic gasification. Nishiyama [49] also published a review article, which features some possibilities for a well-defined catalytic research effort. The article contains the following observations of significance:

1. Salts of alkali and alkaline earth metals as well as transition metals are active catalysts for gasification.
2. The activity of a particular catalyst depends on the gasifying agent as well as the gasifying conditions.
3. The main mechanism of catalysis using alkali and alkaline earth metal salts in steam and carbon dioxide gasification involves the transfer of oxygen from the catalyst to carbon through the formation and decomposition of the C–O complex, that is, C(O).

The mechanism of hydrogasification reactions catalyzed by iron or nickel is still not very clear. But a possible explanation is that the active catalyst appears to be in the metallic state, and there are two main steps for the mechanism. These are hydrogen dissociation and carbon activation [50–54]. For the latter case, carbon dissolution into and diffusion through a catalyst particle seems logical. Gasification proceeds in two stages, each of which has a different temperature range and thermal behavior, so that a single mechanism cannot explain the entire reaction. Thus, the catalyst is still assumed to activate the hydrogen.

Calcium as a catalyst has also been studied by several investigators [55–62]. This catalyst has a very high activity in the initial period when it is well dispersed in the other promoter catalyst, but with increasing conversion, the activity gradually drops. The chemical state and dispersion are studied by chemisorption of carbon dioxide, x-ray diffraction (XRD), and some other analytical techniques. They confirmed the existence of two or more states of calcium compounds, as well as the formation of a surface oxygen complex.

Compared to other heterogeneous catalytic reaction systems, the catalysis in gasification is complex because the catalyst is very short-lived and effective only while in contact with the substrate, which itself changes during the course. Therefore, the definition of the activity for such systems is not very straightforward. For an alkali metal catalyst, the rate increases owing to the change in the catalyst dispersion and also to the increase in the ratio of catalyst/carbon in the later stage of gasification. Other possible explanations for the rate increase could be the change in the surface area by pore opening and the change in the chemical state of the catalyst. At the same time, there are some changes that deactivate the catalyst, for example, agglomeration of catalyst particles, coking, and chemical reaction with sulfur or other trace elements. Coking causes fouling on the catalyst surface, and exposure to high-temperature environment causes sintering of the catalyst, whereas reaction with sulfur poisons the catalytic activity. Although all these three modes of catalyst deactivation are expected to take place in typical gasifiers, remedies are neither easily implementable nor straightforward, unless a robust and highly resilient catalytic system is developed.

The activity of the catalyst also depends on the nature of the substrate and gasifying conditions. The main properties of the substrate related to the activity are (1) reactivity of the carbonaceous constituents, (2) catalytic effect of minerals, and (3) effect of minerals on the activity of the added catalyst. The following general trends have been observed in reference to the factors affecting the activity of the catalysts:

1. Nickel catalysts are more effective toward lower-rank coals because they can be more easily dispersed into the coal matrix owing to higher permeability of the coal, whereas the efficiency of a potassium catalyst is independent of the coal rank. In any case, the coal rank alone, as given by the carbon content, cannot predict the catalyst activity.
2. The internal surface area of coal char relates to the overall activity of the catalyst. It can be related to the number of active sites in cases when the amount of the catalyst is large enough to cover the available surface area. For an immobile catalyst, the conversion is almost proportional to the initial surface area.
3. Pretreatment of coal before the catalytic reaction often helps in achieving higher reaction rates. Although the pretreatment of coal may not be directly applicable as a practical process, a suitable selection of coal types or processing methods could enhance the activity of catalysts.
4. The effect of coal mineral matter on the catalyst effectiveness is twofold. Some minerals such as alkali and alkaline-earth metals catalyze the reaction, whereas others such as silica and alumina interact with the catalyst and deactivate it. In general, demineralization results in enhancement of activity for potassium catalysts but only slightly so for calcium and nickel catalysts.

The method of catalyst loading is also important for catalytic activity management. The catalyst should be loaded in such a way that a definite contact between both solid coal and gaseous reactants is ensured. It was observed that when the catalyst was loaded from an aqueous solution, a hydrophobic carbon surface resulted in finer dispersion of the catalyst when compared to a hydrophilic surface.

The most common and effective catalysts for steam gasification are oxides and chlorides of alkali and alkaline-earth metals, separately or in combination [63]. Xiang et al. studied the catalytic effects of the Na–Ca composite on the reaction rate, methane conversion, steam decomposition, and product gas composition, at reaction temperatures of 700°C–900°C and pressures from 0.1 to 5.1 MPa. A kinetic expression was derived with the reaction rate constants and the activation energy determined at elevated pressures. Alkali metal chlorides such as NaCl and KCl are very inexpensive and hence preferred as catalyst raw materials for catalytic gasification. However, their activities are quite low compared to the corresponding carbonates because of the strong affinity between alkali metal ion and chloride ion.

Takarada et al. [64] have attempted to make Cl-free catalysts from NaCl and KCl by an ion exchange technique. The authors ion-exchanged alkali metals to brown coal from an aqueous solution of alkali chloride using ammonia as a pH-adjusting agent. Cl ions from alkali chloride were completely removed by water washing. This Cl-free catalyst markedly promoted the steam gasification of brown coal. This catalyst was found to be catalytically as active as alkali carbonate in steam gasification. During gasification, the chemical form of active species was found to be in the carbonate form and was easily recovered. Sometimes, an effective way of preparing the catalyst is physically mixing K-exchanged coal with the higher-rank coals [65]. This direct contact between K-exchanged and higher-rank coal resulted in enhancement of the gasification rate. Potassium was found to be a highly suitable catalyst for catalytic gasification by the physical mixing method. Weeda et al. [66] studied the high-temperature gasification of coal under product-inhibited conditions whereby they used potassium carbonate as a catalyst to enhance the reactivity. They performed temperature-programmed experiments to comparatively characterize the gasification behavior of different samples. However, the physical mixing method is likely to be neither practical nor economical for large-scale applications. Chin et al. [67] have recovered the catalysts used in the form of a fertilizer of economic significance. They used a combination of catalysts consisting of potassium carbonate and magnesium nitrate in the steam gasification of brown coal. The catalysts along with coal ash were recovered as potassium silicate complex fertilizer.

In addition to the commonly used catalysts such as alkali and alkaline-earth metals for catalytic gasification, some less-known compounds made of rare earth metals as well as molybdenum oxide ($MoO_2$) have been successfully tried for steam and carbon dioxide gasification of coal [68–70]. Some of the rare earth compounds used were $La(NO_3)_3$, $Ce(NO_3)_3$, and $Sm(NO_3)_3$. The catalytic activity of these compounds decreased with increasing burn-off (i.e., carbon conversion) of the coal. To alleviate this problem, coloading with a small amount of Na or Ca was attempted, and the loading of rare earth complexes was done by the ion exchange method.

Coal gasification technology could benefit from the development of suitable and effective catalysts that will help catalyze steam decomposition and carbon/steam reaction. Batelle Science and Technology International [71] has developed a process in which calcium oxide was used to catalyze the hydrogasification reaction. It was also shown that a reasonably good correlation exists between the calcium content and the reactivity of coal chars with carbon dioxide. Other alkali metal compounds, notably chlorides and carbonates of sodium and potassium, can also enhance the gasification rate by as much as 35%–60%. In addition to the oxides of calcium, iron, and magnesium, zinc oxides are also found to substantially accelerate gasification rates by 20%–30%.

Some speculative mechanisms have been proposed by Muralidhara and Sears [71] as to the role of calcium oxide in enhancing the reaction rate. For instance, coal organic matter may function as a donor of hydrogen, which then may be abstracted by calcium oxide by a given mechanism as described in Scheme 1. Scheme 2 explains the mechanism of generating oxygen-adsorbed CaO sites and subsequent desorption of nascent oxygen, which in turn reacts with organic carbon of coal to form carbon monoxide. Scheme 3 explains direct interaction between CaO and coal organics, which results in liberation of carbon monoxide. The scheme further explains an oxygen exchange mechanism that brings the reactive intermediates back to CaO.

*Scheme 1*:

$$Organic \rightarrow Organic * + H_2$$

$$CaO + 2H_2 \rightarrow CaH_2 + H_2O$$

$$Organic* + CO_2 \rightarrow 2CO$$

$$CO_2 + CaH_2 \rightarrow CaO + CO + H_2$$

*Scheme 2*:

$$CaO + CO_2 \rightarrow CaO(O) + CO$$

$$CaO(O) \rightarrow CaO + (O)$$

$$C + (O) \rightarrow CO$$

*Scheme 3*:

$$CaO + 2C \rightarrow CaC_x + CO$$

$$CaC_x + Organic\ (Oxygen) \rightarrow CaO + Organic*$$

Exxon (currently ExxonMobil) has reported that impregnation of 10%–20% of potassium carbonate lowers the optimum temperature and pressure for steam gasification of bituminous coals, from 980°C to 760°C and from 68 to 34 atm, respectively [72]. In their commercial-scale plant design, the preferred form of make-up catalyst was identified as potassium hydroxide. This catalyst aids the overall process in several ways. First, it increases the rate of gasification, thereby allowing a lower gasification temperature. Second, it prevents swelling and agglomeration when handling caking coals, which is another benefit of a lower gasification temperature. Most importantly, it promotes the methanation reaction because it is thermodynamically more favored at a lower temperature and also under high pressure. Therefore, in this process, the production of methane is thermodynamically and kinetically favored in comparison to syngas. A catalyst recovery unit is provided after the gasification stage to recover the used catalyst.

## 2.4.4.12 Molten Media Gasification Process

Generally speaking, molten media may mean one of the following: molten salt, molten metal, or molten slag. When salts of alkali metals and iron are used as a medium to carry out the coal gasification, it is referred to as *molten media gasification*. The molten medium not only catalyzes the gasification reaction but also supplies the necessary heat and serves as a heat exchange medium [7,73]. There have been several distinct commercial processes developed over the years:

1. Kellogg–Pullman molten salt process
2. Atgas molten iron gasification process
3. Rockwell molten salt gasification
4. Rummel–Otto molten salt gasification

Schematics of a Rockwell molten salt gasifier and a Rummel–Otto single-shaft gasifier are shown in Figures 2.13 [25] and 2.14 [36], respectively.

### 2.4.4.12.1 Kellogg Molten Salt Process

In this process, gasification of coal is carried out in a bath of molten sodium carbonate ($Na_2CO_3$) through which steam is passed [74]. The molten salt produced by this process offers the following advantages:

1. The steam–coal reaction, being basic in nature, is strongly catalyzed by sodium carbonate, resulting in complete gasification at a relatively low temperature.
2. Molten salt disperses coal and steam throughout the reactor, thereby permitting direct gasification of caking coals without carbonization.
3. A salt bath can be used to supply heat to the coal undergoing gasification.
4. Owing to the uniform temperature throughout the medium, the product gas obtained is free of tars and tar acids.

Crushed coal is picked up from lock hoppers by a stream of preheated oxygen and steam and carried into the gasifier. In addition, sodium carbonate recycled from the ash rejection system is also metered into the transport gas stream and the combined coal, salt, and carrier are admitted to the gasifier. The main portion of the preheated oxygen and steam is admitted into the bottom of the reactor for passage through the salt bath to support the gasification reactions. Along with the usual gasification reactions, sulfur entering with the coal accumulates as sodium sulfide ($Na_2S$) to

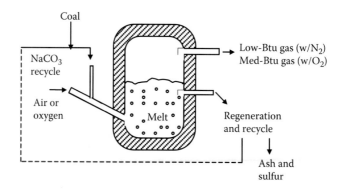

**FIGURE 2.13** Schematic of Rockwell molten salt gasifier.

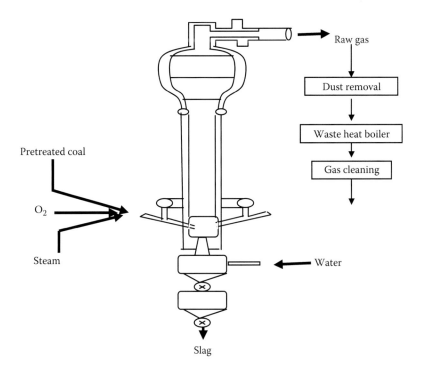

**FIGURE 2.14**   Rummel–Otto single-shaft gasifier.

an equilibrium level. At this level, it leaves the reactor according to the following reaction:

$$Na_2CO_3 + H_2S \rightarrow Na_2S + CO_2 + H_2O$$

Ash accumulates in the melt and leaves along with the bleed stream of salt, where it is rejected and sodium carbonate is recycled. The bleed stream of salt is quenched in water to dissolve sodium carbonate ($Na_2CO_3$) and permit rejection of coal ash by filtration. The dilute solution of sodium carbonate is further carbonated for precipitation and recovery of sodium bicarbonate ($NaHCO_3$). The filtrate is recycled to quench the molten salt stream leaving the reactor. The sodium bicarbonate filtrate cake is dried and heated to regenerate to sodium carbonate for recycle to the gasifier. The gas stream leaving the gasifier is processed to recover the entrained salt and the heat, and is further processed for conversion to the desired product gas such as syngas, pipeline gas, or SNG.

### 2.4.4.12.2   Atgas Molten Iron Coal Gasification

This process is based on the molten iron gasification concept in which coal is injected with steam or air into a molten iron bath. Steam dissociation and thermal cracking of coal volatile matter generate hydrogen and carbon monoxide, that is, principal ingredients of syngas. The coal sulfur is captured by the iron and transferred to a lime slag from which elemental sulfur can be recovered as a by-product. The coal dissolved in the iron is removed by oxidation to carbon monoxide with oxygen or air injected near the molten iron surface. The Atgas process uses coal, steam, or oxygen to yield product gases with heating values of about 900 Btu/scf.

The Atgas molten iron process has several inherent advantages over the gas–solid contact gasification in either fixed or fluidized bed reactors [75]:

1. Gasification is carried out at low pressures; hence, the mechanical difficulty of coal feeding in a pressurized zone is eliminated.
2. Caking properties, ash fusion temperatures, and generation of coal fines are not problematic.
3. The sulfur content of coal does not cause any environmental problem as it is retained in the system and recovered as elemental sulfur from the slag. Elemental sulfur by-product helps the overall process economics.
4. The system is very flexible with regard to the physical and chemical properties of the feed coal. Relatively coarse size particles can be handled without any special pretreatment.
5. Formation of tar is suppressed owing to very high-temperature operation.
6. The product gas is essentially free of sulfur compounds.
7. Shutdown and start-up procedures are greatly simplified compared to fixed bed or fluidized bed reactors.

Coal and limestone are injected into the molten iron through tubes using steam as a carrier gas. The coal goes through devolatilization with some thermal decomposition of the volatile constituents, leaving the fixed carbon and sulfur to dissolve in iron whereupon carbon is oxidized to carbon monoxide. The sulfur, in both organic and pyritic forms ($FeS_2$), migrates from the molten iron to the slag layer where it reacts with lime to produce calcium sulfide (CaS).

The product gas, which leaves the gasifier at approximately 1425°C, is cooled, compressed, and fed to a shift converter (WGS reactor) in which a portion of carbon monoxide is reacted with steam via WGS reaction to attain a CO-to-$H_2$ ratio of 1:3. The carbon dioxide produced is removed from the product gas, and the gas is cooled again. It then enters a methanator in which carbon monoxide and hydrogen react to form methane via $CO + 3H_2 = CH_4 + H_2O$. Excess water is removed from the methane-rich product. The final gaseous product has a heating value around 900 Btu/scf, i.e., high-Btu gas.

### 2.4.4.13 Plasma Gasification

Plasma gasification is a nonincineration thermal process that uses extremely high temperatures in an oxygen-free or oxygen-deprived environment to completely decompose input material into very simple molecules. The extreme heat, aided by the absence of an oxidizing agent such as oxygen, decomposes the input material into basic molecular structure species. The plasma gasification or plasma pyrolysis process was originally developed for treatment of waste materials. However, the process can be very effectively applied to coal gasification or oil shale pyrolysis, capitalizing on its high thermal efficiency, as long as the input energy for plasma generation can be obtained effectively via energy integration or some other inexpensive source of energy. When the plasma gasification is applied to carbonaceous materials such as coal and oil shale kerogen, by-products are normally a combustible gas and an inert slag. Product gas can be cleaned by conventional technologies, including cyclone, scrubbers, and ESPs. Cyclone/scrubber effluents can normally be recycled for further processing.

Plasma is often mentioned as the fourth state. Electricity is fed to a plasma torch that has two electrodes, creating an arc through which inert gas is passed. The inert gas heats the process gas to a very high temperature, as high as 25,000°F. The temperature at a location several feet away from the torch can be as high as 5000°F–8000°F, at which temperature the carbonaceous materials are completely destroyed and broken down into their elemental forms. Furthermore, there is no tar or furan involved or produced in this process. Ash or mineral matter would become completely molten and flow out of the bottom of the reactor. Therefore, the plasma reactor is not specific to any particular kind of coal for gasification. Figure 2.15 illustrates how the plasma torch operates [76].

When applied to waste materials such as MSW, plasma gasification possesses unique advantages for the protection of air, soil, and water resources through extremely low limits of air emissions and leachate toxicity. Because the process is not based on combustion of carbonaceous matters, generation of *greenhouse chemicals*, in particular carbon dioxide, is far less than from any other conventional gasification technology. Furthermore, air emissions are typically orders of magnitude below the current regulations. The slag is monolithic and the leachate levels are orders of magnitude lower than the current EP-toxicity standard, which is one of the four criteria for hazardous waste classification [77]. Slag weight and volume reduction ratios are typically very large; for example, in the case of biomedical wastes, they are 9:1 and 400:1, respectively. Even though the data for a variety of coals are not readily available in the literature, both the mass reduction ratio and the volume reduction ratio for coals are

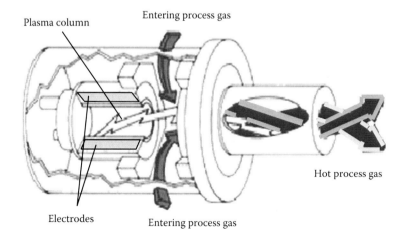

**FIGURE 2.15** Plasma torch. (From Recovered Energy, Inc. Web site, http:www.recovered energy.com/d_plasma.html, 2004. With permission.)

believed to be significantly higher than those for nonplasma gasification technology, thus substantially reducing the burden of waste and spent ash disposal problem.

Activities in Canada and Norway are noteworthy in the technology development of plasma gasification. Resorption Canada Limited (RCL) [78] is a private Canadian entity that was federally incorporated to develop and market industrial processes based on plasma arc technology. They have amassed extensive operating experience in this technology, covering a wide variety of input materials including environmental, biomedical, and energy-related materials and resources.

## 2.5 MATHEMATICAL MODELING OF COAL GASIFIERS

As research and development continues on new and efficient coal gasification concepts, mathematical modeling provides insight into their operation and commercial potential. The influence of design variables and processing conditions on the gasifier performance must be *a priori* determined before any commercial processes are designed. Such models are then used as tools for design modifications, scaling, and optimization.

Coal gasification is performed in different types of reactors in which, depending on the type of gas–solid contact, the bed can be moving, fluidized, entrained, or made up of molten salts.

Different approaches have been used to model various types of reactors. There are mainly two kinds of models. The first kind is the thermodynamic or equilibrium model, which is easier to formulate; but it generates only certain restrictive information such as offgas compositions in a limiting case. The other type of model is the kinetic model, which predicts kinetic behavior inside the reactor. The time-dependent behavior of the process can be either steady state or dynamic in nature. Adanez and Labiano [79] have developed a mathematical model of an atmospheric moving bed countercurrent coal gasifier and studied the effect of operating conditions on the gas yield and composition, process efficiency, and longitudinal temperature profiles. The model was developed for adiabatic reactors. It assumes that the gasifier consists of four zones with different physical and chemical processes taking place. They are the zones for (1) coal preheating and drying, (2) pyrolysis, (3) gasification, and (4) combustion, followed by the ash layer, which acts as a preheater of the reacting (i.e., entering) gases. In reality, however, there is no physical distinction between the zones, and the reactions occurring in each zone vary considerably. The model uses the *unreacted shrinking core model* to define the reaction rate of the coal particles [80]. The unreacted shrinking core model assumes that the dimension (as often represented by the particle size) of unreacted core (of the remaining coal particle) is progressively shrinking as the coal gets reacted. The most critical parameter in the operation of these moving bed gasifiers with dry ash extraction is the longitudinal temperature profile, because the temperature inside the reactor must not exceed the ash-softening (or ash-oozing) point at any time, in order to avoid ash fusion or oozing. The model also takes into account the effect of coal reactivity, particle size, and steam/oxygen ratio. To partially check the validity of the model, predicted data on the basis of the model were compared to real data on the product gas composition for various coals, and good agreement was attained. The authors have concluded that the reactivity of the coals and the emissivity of the ash layer must be known accurately, as they have a strong influence on the temperature profiles, the maximum temperature in the reactor, and its capacity for processing coal.

Lim et al. [81] have developed a mathematical model of a spouted bed gasifier based on simplified first-order reaction kinetics for the gasification reactions. The spouted bed gasifier has been under development in Canada and Japan [82,83]. The spout is treated as a plug flow reactor (PFR) of a fixed diameter with cross-flow into the annulus. The annulus is treated as a series of steam tubes, each being a PFR with no axial dispersion. The model calculates the composition profile of various product gases in the spout as a function of the height, radial composition profiles, and average compositions in the annulus at different heights, average compositions exiting the spout and annulus, and flow rates and linear velocities in the spout and annulus. The model has been further developed as a two-region model including an enthalpy balance [84].

Monazam et al. [85] have developed a similar model for simulating the performance of a cross-flow coal gasifier. Gasification in a cross-flow gasifier is analogous to the batch gasification in a combustion pot. Therefore, the model equations for kinetics as well as mass and energy balances formulated were based on a batch process. In the cross-flow coal gasifier concept, operating temperatures are much higher than 1000°C, and, as such, the diffusion through the gas film and ash layer is a critical factor. The model also assumes shrinking unreacted core model for kinetic formulations. Simulation results of the model were compared to the experimental data obtained in batch and countercurrent gasification experiments, and good agreement was attained. It was also concluded that the performance of the gasifier depends on the gas–solid heat transfer coefficient, whereas the particle size and the bed voidage had a significant effect on the time required for complete gasification.

Watkinson et al. [86] have developed a mathematical model to predict the gas composition and yield from coal gasifiers. Gas composition depends on the contacting pattern of blast and fuel, temperature and pressure of the operation, composition of the blast, and form of fuel feeding. The authors have presented a calculation method, and the predicted data have been compared to the operating data from nine different types of commercial and pilot-scale gasifiers, including Texaco, Koppers–Totzek, and Shell, Winkler-fluidized bed, and Lurgi dry ash as well as Lurgi slagging moving bed gasifier. The model consists of elemental mass balances for C, H, O, N, and S, chemical equilibria for four key chemical reactions, and an optional energy balance. The four key reactions were POX, steam gasification, Boudouard reaction, and WGS reaction. Predictions were most accurate for entrained flow systems, less accurate for fluidized bed gasifiers, and uncertain for moving bed reactors. This was due to the lower temperatures and uncertain volatile yields in the latter ones resulting in deviation between the calculated and experimentally reported values.

Lee et al. [14] developed a single-particle model to interpret kinetic data of coal char gasification with $H_2$, $CO_2$, and $H_2O$. Their model yields asymptotic analytical solutions taking into account all the major physical factors that affect and contribute to the overall gasification rate. Some of the factors taken into account involved changing magnitudes of internal surface area, porosity, activation energy, and effective diffusivity as functions of conversion (or burnoff). Their model closely describes the characterizing shape of the conversion vs. time curves as determined by $CO_2$ gasification studies. The curve shape under certain restrictions leads to a "universal curve" of conversion vs. an appropriate dimensionless time. The model developed is mathematically very simple, and all the parameters in the model equation have physical significance. Therefore, the model is applicable to a wide variety of coals having different physicochemical and petrological properties. The number of adjustable parameters in this model is only two. Their model predictions were compared against experimental data obtained using a novel thermobalance reactor, and excellent agreement was attained [14].

Gururajan et al. [87], in their review, critically examined many of the mathematical models developed for fluidized bed reactors. The review is primarily concerned with the modeling of BFB coal gasifiers. They also discuss the rate processes occurring in a fluidized bed reactor and compare some of the reported models in the literature with their presentation.

When a coal particle is fed into a gasifier, it undergoes several physicochemical transformations, which include (1) drying, (2) devolatilization, and (3) gasification of the residual char in different gaseous atmospheres. These heterogeneous reaction-transport phenomena are accompanied by a number of supplementary reactions that are homogeneous in nature. Detailed kinetic studies are an important prerequisite for the development of a mathematical model. Mathematical models for a BFB coal gasifier can be broadly classified into two kinds, that is, thermodynamic (or equilibrium) and kinetic (or rate) models. Thermodynamic models predict the equilibrium compositions and temperature of the product gas based on a given set of steam/oxygen feed ratios, the operating pressure, and the desired carbon conversion. These models are independent of the type of the gasifier and based on the assumption of complete oxygen consumption. Therefore, they cannot be used to investigate the influence of operating parameters on the gasifier performance. The kinetic model, on the other hand, predicts the composition and temperature profiles inside the gasifier for a given set of operating conditions and reactor configurations and hence can be used to evaluate the performance of the gasifier. They are developed by combining a suitable hydrodynamic model for the fluidized bed with appropriate kinetic schemes for the reactive processes occurring inside the gasifier. Various rate models may be classified into four groups on the basis of hydrodynamic models used [87]:

1. Simplified flow models
2. Davidson–Harrison type models
3. Kunii–Levenspiel type models
4. Kato–Wen type models

The same review [87] also examined and compared the different types of models. Although many investigators have compared their model predictions with experimental data, a detailed evaluation of the influence of model assumptions on its predictions has not been reported. Although efforts have been made to compare the predictions of different models, an attempt to evaluate the model with experimental data from different sources has not been made.

Gururajan et al. [87] in their review article have developed a model of their own for a bottom feeding BFB coal gasifier based on the following assumptions:

1. The bubble phase is in plug flow and does not contain any particles, whereas the emulsion phase is completely mixed and contains the particles in fluidized conditions.
2. Excess gas generated in the emulsion phase passes into the bubble phase. The rate of this excess per unit bed volume is constant.
3. The coal particles in the feed are spherical, homogeneous, and uniform in size.
4. Only WGS reaction occurs in the homogeneous gas phase.
5. Resistances by external mass transfer and intraparticle diffusion are assumed to be negligible in the char gasification reactions.
6. Entrainment, abrasion, agglomeration, or fragmentation of the bed particles is assumed to be negligible.
7. The gasifier is at a steady state and is isothermal.

All the model equations are derived on the basis of the preceding assumptions. The model predictions were compared with the experimental data from three pilot-scale gasifiers reported in the literature [87]. They concluded that the predictions were more sensitive to the assumptions regarding the combustion/decomposition of the volatiles and the products of char combustion than to the rate of char gasification. Hence, in pilot-scale gasifiers, owing to the short residence time of coal particles, the carbon conversion and the product gas yields are mainly determined by the fast-rate coal devolatilization, volatile combustion/decomposition and char combustion, and also slow-rate char gasification reactions. This explains why models based on finite-rate char gasification reactions are able to fit the same pilot-scale gasification data.

A better understanding of coal devolatilization, decomposition of the volatiles, and char combustion under conditions prevailing in a fluidized bed coal gasifier is very important for the development of a model with good predictive capability. There is a strong need to investigate the kinetics of gasification of coal and char in syngas atmospheres and to obtain experimental data for the same coal and char in a pilot-scale plant.

It is well known that there are many physical changes occurring when the coal char particles are gasified. There have been many attempts to unify these dynamic changes through various normalizing parameters such as half-life, coal rank, reactivity, or surface area. According to the study by Raghunathan and Yang [88], the experimental char conversion vs. time data from different experiments can be unified into a single curve where time is considered to be normalized time, $t/t_{1/2}$, $t_{1/2}$ being the half-life of the char–gas reaction. This unification curve with only one parameter is then fitted into the rate models commonly used, for example, the *grain model* and the *random pore model*. With the aid of reported correlations for unification curves, a master curve is derived to approximate the conversion–time data for most of the gasification systems. Also, as the half-life (more precisely, half-conversion time) is simply related to the average reactivity, it can be generally used as a reactivity index for characterizing various char–gas reactions. Further, conversions up to 70% can be predicted with reasonable accuracy over a wide range of temperatures.

A great deal of effort has been devoted to mathematically model a variety of gasifiers and reaction conditions in order to obtain design- and performance-related information. Numerous simplified models and asymptotic solutions have been obtained for coal gasification reactors along with a large database of digital simulation of such systems.

## 2.6 FUTURE OF COAL GASIFICATION

The roles of coal gasification have been changing constantly based on the societal demands of the era. We observed in the past century that the principal roles and foci of coal-derived syngas shifted from domestic heating fuel, to feedstock for Fischer–Tropsch (F-T), to petrochemical feedstocks, to starting materials for alternative fuels, to IGCC, and to hydrogen sources. With the advent of hydrogen economy, coal gasification has again taken center stage as a means for producing hydrogen for fuel cell applications [1]. Further, coal gasification technology can also be easily applied to biomass and solid waste gasification with modifications. Unlike coal, biomass is not only renewable but also available inexpensively, often free of charge. Coal can also be coprocessed together with a variety of other materials, including petroleum products, scrap tires, biomass, municipal wastes, sewage sludge, etc. With advances in flue gas desulfurization (FGD), coal gasification can be more widely utilized in process industries. In electric power generation, IGCC has contributed tremendously to the improvement of power generation efficiency, thus keeping the cost of electric power competitive against all other forms of energy. Keen interest in methanol and dimethylether (DME) is rekindled due to the ever-rising cost of conventional clean liquid fuel. In order to use coal gasification technology in hydrogen production, the steam gasification process, which is essentially very similar to the hydrocarbon reformation process, needs to be refined further. Therefore, more advances are expected in the areas of product gas cleaning, separation and purification, feedstock flexibility, and integrated or combined process concepts.

## REFERENCES

1. United States Department of Energy Office of Fossil Energy, Gasification Technology R&D. Available at http://energy.gov/fe/science-innovation/clean-coal-research/gasification, 2013.
2. U.S. Energy Information Administration (EIA), Electricity market module. U.S. Department of Energy. Available at eia.doe.gov, 2013.
3. M. W. Melaina, "Market transformation lessons for hydrogen from the early history of the manufactured gas industry," in *Hydrogen Energy and Vehicle Systems*, S. E. Grasman, Ed. Boca Raton, FL: CRC Press, 2013, pp. 123–158.
4. I. Howard-Smith and G. J. Werner, *Coal Conversion Technology*. Park Ridge, NJ: Noyes Data Corporation, 1976.
5. J. Wilson, J. Halow and M. R. Ghate, "Gasification—key to chemicals from coal," *CHEMTECH*, pp. 123–128, 1988.
6. S. Lee, *Methane and its Derivatives*. New York, NY: Marcel Dekker, 1997.
7. J. G. Speight, *"Gasification," The Chemistry and Technology of Coal*, 3rd. Ed., Boca Raton, FL: CRC Press, Chapter 2, pp. 37–58, 2012.

8. Dakota Gasification Company, SNG pipeline. Available at http://www.dakotagas.com/Gas_Pipeline/Syngas_Pipeline/index.html (accessed July 2013).

9. A. L. Lee, "Evaluation of coal conversion catalysts, final report submitted to the U.S. Department of Energy," Gas Research Institute, Chicago, IL, Tech. Rep. GRI-87-0005, 1987.

10. S. Ariyapadi, P. Shires, M. Bhargava and D. Ebbern, "KBR's transport gasifier (TRIG™)—an advanced gasification technology for SNG production from low-rank coals," in *Twenty-Fifth Annual International Pittsburgh Coal Conference*, Pittsburgh, PA, September 29–October 2, 2008.

11. J. Gadsby, C. N. Hinshelwood and K. W. Sykes, "The kinetics of the reactions of the steam-carbon system," *Proc. R. Soc. A*, vol. 187, pp. 129–151, 1946.

12. H. F. Johnstone, C. Y. Chen and D. S. Scott, "Kinetics of the steam-carbon reaction in porous graphite tubes," *Ind. Eng. Chem.*, vol. 44, pp. 1564–1569, 1952.

13. P. L. Walker, F. Rusinko and L. G. Austin, "Gas reactions in carbon," in *Advances in Catalysis*, D. D. Ely, P. W. Selwood and P. B. Weisz, Eds. New York, NY: Academic Press, 1959, pp. 133–221.

14. S. Lee, J. C. Angus, R. V. Edwards and N. C. Gardner, "Noncatalytic coal char gasification," *AIChE J.*, vol. 30, pp. 583–593, 1984.

15. T. Kyotani, K. Kubota, J. Cao, H. Yamashita and A. Tomita, "Combustion and $CO_2$ gasification of coals in a wide temperature range," *Fuel Process Technol.*, vol. 36, pp. 209–217, 12, 1993.

16. S. Lee, *Methanol Synthesis Technology*. Boca Raton, FL: CRC Press, 1990.

17. S. Lee, V. Parameswaran, C. J. Kulik and I. Wender, "The roles of carbon dioxide in methanol synthesis," *Fuel Sci. Technol. Int.*, vol. 7, pp. 1021–1057, 1989.

18. W. W. Bodle and K. C. Vyas, "Clean Fuels from Coal," in *IGT Symposium Papers*, Chicago, IL: Institute of Gas Technology, 1973, pp. 49–91.

19. H. M. Braunstein and H. A. Pfuderer, "Environmental health and control aspects of coal conversion: An information overview," Oak Ridge National Laboratory, Oak Ridge, TN, Report ORNL-EIS-94, 1977.

20. K. C. Vyas and W. W. Bodle, "Technical historical background and principles of modern technology," in *Clean Fuels from Coal Symposium II Papers*, Chicago, IL: Institute of Gas Technology, 1975, pp. 53–84.

21. S. Lee, *Alternative Fuels*. Philadelphia, PA: Taylor & Francis, 1996.

22. P. F. H. Rudolph, "The Lurgi process—the route to SNG from coal," in *Fourth Synthetic Pipeline Gas Symposium*, Chicago, IL, October 1972, p. 52.

23. Kuo, J. C. W., "Gasification and Indirect Liquefaction." *The Science and Technology of Coal and Coal Utilization*, B. R. Cooper and W. A. Ellingson, Eds. New York, NY: Plenum Press, 1984, pp. 163–230.

24. A. Williams and M. Olschar, BGL gasification—project progress and technology developments, 2010. Available at http://www1.icheme.org/gasification2010/pdfs/bglgasificationandrewwilliams.pdf (accessed October 2013).

25. W. G. Lloyd, "Synfuels technology update," in *The Emerging Synthetic Fuel Industry*, A. Thumann, Ed. Atlanta, GA: Fairmont Press, 1981, pp. 19–58.

26. J. M. Moe, "SNG from coal via the Lurgi gasification process," in *Clean Fuels from Coal Symposium Papers*, Chicago, IL: Institute of Gas Technology, 1973, pp. 91–110.

27. B. C. Johnson, "The grand forks slagging gasifier," in *Coal Processing Technology, Vol. IV, A CEP Technical Manual*, AIChE, Ed. New York, NY: AIChE, 1978, pp. 94–98.

28. H. J. Michels and H. F. Leonard, "Coal Gasification," *Chem. Eng. Prog.*, vol. 74, p. 85, 1978.

29. M. J. Van der Bergt, "Clean Syngas from Coal," *Hydrocarbon Process.*, vol. 58, p. 161, 1979.

30. D. A. Sams and F. Shadman, "Mechanism of potassium-catalyzed carbon/$CO_2$ reaction," *AIChE J.*, vol. 32, pp. 1132–1137, 1986.

31. J. Farnsworth, H. F. Leonard and M. Mitsak, "Production of gas from coal by the Koppers-Totzek process," in *Clean Fuels from Coal Symposium Papers*, Chicago: Institute of Gas Technology, 1973, pp. 143–163.
32. V. A. S. Singh, "Improve NGL recovery," *Hydrocarbon Process.*, vol. 80, pp. 41–65, 2001.
33. *EPRI J.*, "EPRI evaluates potential utility application of coal gasification systems," pp. 41–44, 1983.
34. R. V. London, "European EOR," *Oil Gas J.*, vol. 83, p. 51, 1985.
35. Anonymous. "Shell gasification process," *Hydrocarbon Process.*, vol. 63, p. 96, 1984.
36. R. F. Probstein and R. E. Hicks, *Synthetic Fuels*. New York: McGraw-Hill, 1982.
37. P. N. Thompson, "Gasifying coal underground," *Endeavour*, vol. 2, pp. 93–97, 1978.
38. D. W. Gregg and T. F. Edgar, "Underground coal gasification," *AIChE J.*, vol. 24, pp. 753–781, 1978.
39. CSIRO, Commonwealth Scientific and Industrial Research Organisation. Available at www.csiro.au, 2013.
40. H. Jüntgen, "Research for future in situ conversion of coal," *Fuel*, vol. 66, pp. 443–453, 4, 1987.
41. R. M. Nadkarni, "Underground gasification of coal," in *Clean Fuels from Coal Symposium Papers*, Chicago, IL: Institute of Gas Technology, 1973, pp. 611–638.
42. W. W. Odell, "Gasification of solid fuels in Germany by the Lurgi, Winkler, and Leuna slagging-type gas-producer processes," Bureau of Mines, Washington, DC, Tech. Rep. BM-IC-7415, 1947.
43. I. N. Banchik, "The Winkler process for the production of low Btu gas from coal," in *Clean Fuels from Coal Symposium II Papers*, Chicago, IL: IIT Research Institute, 1975, pp. 359–374.
44. H. Teggers and L. Schrader, "The high-temperature Winkler process—a way for the generation of synthesis gas from lignite," *Energiewirtschaftliche Tagesfragen*, vol. 31, pp. 397–399, 1981.
45. W. Adlhoch, H. Sato, J. Wolff and K. Radtke, "High Temperature Winkler gasification of municipal solid waste," in *2000 Gasification Technologies Conference*, San Francisco, CA, 2000, pp. 1–15.
46. R. Leonard, T. Pinkston, L. Rogers, R. Rush and J. Wheeldon, THE PSDF—commercial readiness for coal power—revisited. 2012. Available at http://www.netl.doe.gov/technologies/coalpower/gasification/projects/adv-gas/25140/Revised%20paper-randall.pdf.
47. B. Skrifvars and P. Kilpinen, "Biomass combustion technology in Finland," *IFRF (Int. Flame Res. Found.), Ind. Combust. Mag., Tech.*, vol. 2, 1999. Available at http://www.magazine.ifrf.net/9903biomass1/turku/.
48. H. Jüntgen, "Application of catalysts to coal gasification processes. Incentives and perspectives," *Fuel*, vol. 62, pp. 234–238, 2, 1983.
49. Y. Nishiyama, "Catalytic gasification of coals—features and possibilities," *Fuel Process Technol.*, vol. 29, pp. 31–42, 11, 1991.
50. K. Asami and Y. Ohtsuka, "Highly active iron catalysts from ferric chloride for the steam gasification of brown coal," *Ind. Eng. Chem. Res.*, vol. 32, pp. 1631–1636, 1993.
51. H. Yamashita, S. Yoshida and A. Tomita, "Local structures of metals dispersed on coal. 2. Ultrafine FeOOH as active iron species for steam gasification of brown coal," *Energy Fuels*, vol. 5, pp. 52–57, 1991.
52. S. Matsumoto, "Catalyzed hydrogasification of Yallourn char in the presence of supported hydrogenation nickel catalyst," *Energy Fuels*, vol. 5, pp. 60–63, 1991.
53. R. C. Srivastava, S. K. Srivastava and S. K. Rao, "Low temperature nickel-catalysed gasification of Indian coals. 1," *Fuel*, vol. 67, pp. 1205–1207, 9, 1988.

54. T. Haga and Y. Nishiyama, "Influence of structural parameters on coal char gasification: 2. Ni-catalysed steam gasification," *Fuel*, vol. 67, pp. 748–752, 6, 1988.

55. Y. Ohtsuka and K. Asami, "Steam gasification of high sulfur coals with calcium hydroxide," in *1989 International Conference on Coal Science*, Tokyo, Japan, 1989, pp. 353–356.

56. C. Salinas-Martínez de Lecea, M. Almela-Alarcón and A. Linares-Solano, "Calcium-catalysed carbon gasification in $CO_2$ and steam," *Fuel*, vol. 69, pp. 21–27, 1, 1990.

57. J. P. Joly, D. Cazorla-Amoros, H. Charcosset, A. Linares-Solano, N. R. Marcilio, A. Martinez-Alonso and C. S. de Lecea, "The state of calcium as a char gasification catalyst—a temperature-programmed reaction study," *Fuel*, vol. 69, pp. 878–884, 7, 1990.

58. H.-J. Mühlen, "Finely dispersed calcium in hard and brown coals: Its influence on pressure and burn-off dependencies of steam and carbon dioxide gasification," *Fuel Process Technol.*, vol. 24, pp. 291–297, 1, 1990.

59. Y. A. Levendis, S. W. Nam, M. Lowenberg, R. C. Flagan and G. R. Gavalas, "Catalysis of the combustion of synthetic char particles by various forms of calcium additives," *Energy Fuels*, vol. 3, pp. 28–37, 1989.

60. Z. G. Zhang, T. Kyotani and A. Tomita, "Dynamic behavior of surface oxygen complexes during oxygen-chemisorption and subsequent temperature-programmed desorption of calcium-loaded coal chars," *Energy Fuels*, vol. 3, pp. 566–571, 1989.

61. T. Haga, M. Sato, Y. Nishiyama, P. K. Agarwal and J. B. Agnew, "Influence of structural parameters of coal char on potassium- and calcium-catalyzed steam gasifications," *Energy Fuels*, vol. 5, pp. 317–322, 1991.

62. P. Pereira, G. A. Somorjai and H. Heinemann, "Catalytic steam gasification of coals," *Energy Fuels*, vol. 6, pp. 407–410, 1992.

63. R. Xiang-Quan, W. You-Qing, L. Zuo-Liang and L. Shu-Fen, "Effects of catalysis on gasification of Datong coal char," *Fuel*, vol. 66, pp. 568–571, 4, 1987.

64. T. Takarada, T. Nabatame, Y. Ohtsuka and A. Tomita, "Steam gasification of brown coal using sodium chloride and potassium chloride catalysts," *Ind. Eng. Chem. Res.*, vol. 28, pp. 505–510, 1989.

65. T. Takarada, M. Ogiwara and K. Kato, "Catalytic steam gasification of coal by physical mixing of K-exchanged brown coal," *J. Chem. Eng. Jpn.*, vol. 25, pp. 44–48, 1992.

66. M. Weeda, P. J. J. Tromp, B. van der Linden and J. A. Moulijn, "High temperature gasification of coal under severely product inhibited conditions: The potential of catalysis," *Fuel*, vol. 69, pp. 846–850, 7, 1990.

67. G. Chin, L. Guifen and D. Qushi, "New approach for gasification of coal char," *Fuel*, vol. 66, pp. 859–863, 6, 1987.

68. F. Carrasco-Marín, J. Rivera-Utrilla, E. Utrera-Hidalgo and C. Moreno-Castilla, "$MoO_2$ as catalyst in the $CO_2$ gasification of activated carbons and chars," *Fuel*, vol. 70, pp. 13–16, 1, 1991.

69. A. López-Peinado, F. Carrasco-Marín, J. Rivera-Utrilla and C. Moreno-Castilla, "Steam gasification of a lignite char catalysed by metals from chromium to zinc," *Fuel*, vol. 71, pp. 105–108, 1, 1992.

70. T. Suzuki, S. Nakajima and Y. Watanabe, "Catalytic activity of rare earth compounds for the steam and carbon dioxide gasification of coal," *Energy Fuels*, vol. 2, pp. 848–853, 1988.

71. H. S. Muralidhara and J. T. Sears, "Effect of calcium on gasification," in *Coal Processing Technology Vol. IV, A CEP Technical Manual*, AIChE, Ed. New York, NY: AIChE, 1979, pp. 22–25.

72. J. E. Gallagher and H. A. Marshall, "SNG from coal by catalytic gasification," in *Coal Processing Technology, Vol. IV, A CEP Technical Manual*, AIChE, Ed. New York, NY: AIChE, 1979, pp. 199–204.

73. A. E. Cover and W. C. Schreiner, "Molten salt gasification of coal," *Clean Fuels from Coal Symposium Papers*, Chicago, IL: Institute of Gas Technology, September 10–14, 1973, pp. 273–279.
74. A. E. Cover, W. C. Schreiner and G. T. Skaperdas, "Kellogg's coal gasification process," *Chem. Eng. Prog.*, vol. 69, pp. 31–36, 1973.
75. P. LaRosa and R. J. McGarvey, "Fuel gas from molten iron coal gasification," *Clean Fuels from Coal Symposium Papers*, Chicago, IL: Institute of Gas Technology, 1973, pp. 285–300.
76. Recovered Energy Inc, Recovered Energy, Inc. Presents the Recovered Energy System. Available at http://recoveredenergy.com/, 2013.
77. J. G. Speight and S. Lee, *Environmental Technology Handbook*. Philadelphia, PA: Taylor & Francis, 2000.
78. Resorption Canada Ltd. (RCL), Plasco Energy Group. Changing the way communities deal with waste. Available at www.plascoenergygroup.com, 2013.
79. J. Adanez and F. G. Labiano, "Modeling of moving-bed coal gasifiers," *Ind. Eng. Chem. Res.*, vol. 29, pp. 2079–2088, 1990.
80. O. Levenspiel, *Chemical Reaction Engineering*. New York, NY: John Wiley & Sons, 2008.
81. C. J. Lim, J. P. Lucas, M. Haji-Sulaiman and A. P. Watkinson, "A mathematical model of a spouted bed gasifier," *Can. J. Chem. Eng.*, vol. 69, pp. 596–606, 1991.
82. S. K. Foong, G. Cheng and A. P. Watkinson, "Spouted bed gasification of Western Canadian coals," *Can. J. Chem. Eng.*, vol. 59, pp. 625–630, 1981.
83. A. P. Watkinson, G. Cheng and C. J. Lim, "Oxygen-steam gasification of coals in a spouted bed," *Can. J. Chem. Eng.*, vol. 65, pp. 791–798, 1987.
84. J. P. Lucas, C. J. Lim and A. P. Watkinson, "A nonisothermal model of a spouted bed gasifier," *Fuel*, vol. 77, pp. 683–694, 5, 1998.
85. E. R. Monazam, E. K. Johnson and J. W. Zondlo, "Modeling and simulation of a cross-flow coal gasifier," *Fuel Sci. Technol. Int.*, vol. 10, pp. 51–73, 1992.
86. A. P. Watkinson, J. P. Lucas and C. J. Lim, "A prediction of performance of commercial coal gasifiers," *Fuel*, vol. 70, pp. 519–527, 4, 1991.
87. V. S. Gururajan, P. K. Agarwal and J. B. Agnew, "Mathematical modeling of fluidized bed coal gasifiers," *Trans. Inst. ChemE.*, vol. 70, pp. 211–238, Part A, 1992.
88. K. Raghunathan and R. Y. K. Yang, "Unification of coal gasification data and its applications," *Ind. Eng. Chem. Res.*, vol. 28, pp. 518–523, 1989.

# 3 Clean Liquid Fuels from Coal

*Sunggyu Lee*

## CONTENTS

## 3.1 BACKGROUND

There are three principal routes by which liquid fuels can be produced from solid coal: coal pyrolysis, direct liquefaction, and indirect liquefaction. Liquid-like fuels can also be obtained via coal slurry technology.

Even though coal has a reasonably high heating value of approximately 8000–14,000 Btu/lb, its solid state is one of the main reasons it is inconvenient to handle as a consumer fuel. To make this solid fuel more user-friendly, research has been ongoing to convert it into pipeline-quality gaseous fuel or clean liquid fuel. During World War II, production of approximately 100,000 barrels per day of liquid fuel from coal was reported for the German war effort. The German liquefaction process used a high-temperature and high-pressure technology, and the product liquid fuels were of environmentally poor quality by modern environmental standards as cleaning and refining was minimal.

The current process objectives of coal liquefaction are mainly focused on easing the severity of operating conditions, minimizing the hydrogen requirement, and making the liquid product more environmentally acceptable. Due to the recent trend toward higher and fluctuating petroleum prices in the world market, the relative process economics of coal liquefaction are changing much more favorably. Considering the vast amount of coal reserves throughout the world, and the global distribution of major deposits, this alternative is even more attractive and also very practical.

There are inherent technological advantages with coal liquefaction, as coal liquefaction can produce clean liquid fuels that can be sold as *transportation fuels*. It has long been believed that if the crude oil price stays at a level higher than about $35 per barrel for a sustainably long period, production of gasoline and diesel by liquefaction of coal would become economically competitive. Such a claim was made when the crude oil price was substantially lower than $35 per barrel. The crude oil price has been sharply rising in the twenty-first century. The Brent crude spot oil price hit a record high mark of $145 per barrel in July 2008, soon after it sharply fell to a low of $40 per barrel, and since then rose again and fluctuated in wide ranges, hovering at $100 per barrel. Financial experts and oil analysts are predicting that this high crude oil price is here to stay, rather than a temporary phenomenon. Even after considering the changes in various economic factors involving energy industries as well as development of emerging energy and fuel technologies, production of transportation fuels or fuel oils via coal liquefaction is certainly an outstanding option for the future. Further, the products of coal liquefaction can be refined and formulated to possess the properties of conventional *transportation fuels*, requiring neither major infrastructural changes in distribution nor lifestyle changes for consumers.

## 3.2 COAL PYROLYSIS FOR LIQUID FUEL

Pyrolysis of coal yields condensable tar, oil, and water vapor and noncondensable gases, through a process called *destructive distillation*, which involves cleavages of C–C bonds in coal macromolecular structure. In the process of coal pyrolysis, the C–C bond cleavage reactions are largely responsible for the molecular weight reduction in coal hydrocarbons that ultimately converts solid fuel into liquid or gas.

The solid residue of coal pyrolysis that is left behind is called *char*. Therefore, char contains substantially smaller amounts of volatile hydrocarbons than coal and is lower in H content as well. As implied, the ratio of H/C is an effective indicator of the nature of coal product. For instance, formation of coal char from coal is in the direction of the H/C ratio decreasing, whereas conversion of coal into coal liquid is in the direction of the H/C ratio increasing. The condensed pyrolysis product must be further hydrogenated to remove sulfur and nitrogen species as well as to improve the liquid fuel quality. Nitrogen and sulfur species not only generate air pollutants (in forms of $NO_x$ and $SO_x$) when combusted but also poison and deactivate the upgrading catalyst in the downstream processing of liquid product. As the term implies, pyrolysis involves thermal decomposition reactions that induce mainly the cleavage of C–C bonds and partial breakdown of C–S and C–H bonds inside the macromolecular structure of coal, thus producing lower-molecular-weight products such as liquid hydrocarbons.

A number of coal pyrolysis processes are commercially available. Table 3.1 lists various coal pyrolysis processes and their process operating conditions and yields [1]. The factors affecting the process efficiency and product yield include the coal rank, coal particle size, reactor type, process mechanics, hydrogen partial pressure, reactor pressure, processing temperature, coal residence time, and so forth. A quick glance at Table 3.1 shows that higher liquid yields are obtained with shorter residence times and also that a hydrogen atmosphere helps the liquid product yield [2]. A shorter residence time, if properly managed, does not allow a sufficient reaction time to thermally crack the liquid hydrocarbons further (to gaseous hydrocarbons), thereby leaving more liquid hydrocarbons in the product stream. Adding hydrogen to coal hydrocarbons improves the H/C ratio enough to increase the fluidity and to produce a liquid fuel. High tar content in supercritically extracted coal products may be attributable to the supercritical solvent's excellent low-temperature solubility that extracts and dissolves the tar *as is*. In this special case of low-temperature operation, extraction of high-molecular-weight hydrocarbons is taking place mainly due to the superior solvent properties of supercritical fluids, rather than the C–C bond breakage reaction of coal pyrolysis.

### 3.2.1 CHAR OIL ENERGY DEVELOPMENT PROCESS

The char oil energy development (COED) process was originally developed by the FMC Corporation, Princeton, NJ [3]. The process has been improved, and the improved version has become the COED/COal GASification (COGAS) process. The process is based on a fluidized-bed technology that is carried out in four successive fluidized-bed pyrolysis stages at progressively higher temperatures. Figure 3.1 shows a schematic of the COED/COGAS process [1]. The optimal temperatures for the four stages vary, depending on the properties of the feed coal. The temperatures of the stages are selected to be just below the maximum temperature to which the particular feed coal can be heated without agglomerating and plugging the fluid bed [1]. Typical operating temperatures are 315°C–345°C, 425°C–455°C, 540°C, and 870°C in the first, second, third, and fourth stages, respectively [2]. Heat for the process is provided by combusting a portion of the product char with a steam–oxygen mixture

**TABLE 3.1**

**Comparative Summary of Pyrolysis and Hydropyrolysis Processes**

| Process | Developer | Reactor Type | Reaction Temperature (°C) | Reaction Pressure (psi) | Coal Residence Time | Yield (%) | | |
|---|---|---|---|---|---|---|---|---|
| | | | | | | Char | Oil | Gas |
| Lurgi–Ruhrgas | Lurgi–Ruhrgas | Mechanical mixer | 450–600 | 15 | 20 s | 45–55 | 15–25 | 30 |
| COED | FMC Corp. | Multiple fluidized bed | 290–815 | 20–25 | 1–4 h | 60.7 | 20.1 | 15.1 |
| Occidental coal pyrolysis | Occidental | Entrained flow | 580 | 15 | 2 s | 56.7 | 35.0 | 6.6 |
| TOSCOAL | TOSCO | Kiln-type retort vessel | 425–540 | 15 | 5 min | 80–90 | 5–10 | 5–10 |
| Clean coke | U.S. Steel Corp. | Fluidized bed | 650–750 | 100–150 | 50 min | 66.4 | 13.9 | 14.6 |
| Union carbide process | Union Carbide | Fluidized bed | 565 | 1000 | 5–11 min | 38.4 | 29.0 | 16.2 |

*Note:* The TOSCOAL process is to designate the process which involves the application of a successful oil shale retorting technology developed by TOSCO (The Oil Shale Corporation) to the low-temperature pyrolysis of coal. COED, char oil energy development.

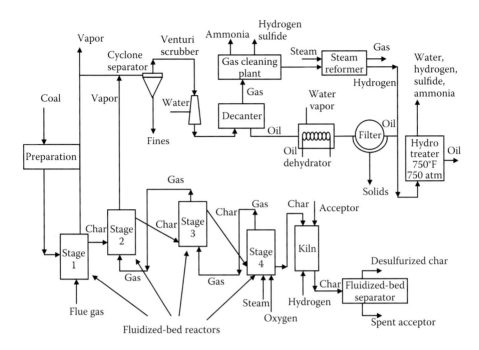

**FIGURE 3.1**  A schematic of the COED/COGAS process. (From Speight, J.G., *The Chemistry and Technology of Coal*, Marcel Dekker, New York, 1983. With permission.)

in the fourth stage. Hot gases flow countercurrently to the char movement and provide the hot fluidizing medium for pyrolysis stages. The gases leaving both the first and second stages are passed to cyclones that remove the fines, but the vapors leaving the cyclones are quenched in a Venturi scrubber to condense the oil, and the gases and oil are separated in a decanter. The gas is desulfurized and then steam-reformed to produce hydrogen and fuel gas. The oil from the decanter is dehydrated, filtered, and hydrotreated to remove nitrogen, sulfur, and oxygen to form a heavy synthetic crude oil of approximately 25° API.

The API gravity is frequently correlated with the specific gravity of oil at 60°F using the following formula:

$$°API \text{ gravity} = (141.5/SG \text{ at } 60°F) - 131.5$$

where SG at 60°F is the specific gravity of oil at 60°F. For example, if the specific gravity of an oil at 60°F is 1.0, the °API gravity would be 10, that is, 10°API gravity. As shown, a lower value of the API gravity means it is a heavier oil.

The properties of synthetic crude oils from coal by the COED process are shown in Table 3.2 [1].

The char is desulfurized in a shift kiln, where it is treated with hydrogen to produce hydrogen sulfide ($H_2S$), which is subsequently absorbed by an acceptor such as dolomite ($CaCO_3 \cdot MgCO_3$) or limestone ($CaCO_3$) [1]. The COGAS process involves

**TABLE 3.2**

**Synthetic Crude Oil Products from Coal by COED Process**

| Properties | Coal | |
|---|---|---|
| | Illinois No. 6 | Utah King |
| Analysis of hydrocarbon types (vol%) | | |
| Paraffins | 10.4 | 23.7 |
| Olefins | – | – |
| Naphthenes | 41.4 | 42.2 |
| Aromatics | 48.2 | 34.1 |
| °API gravity | 28.6 | 28.5 |
| ASTM distillation (°F) | | |
| Initial boiling point (IBP) | 108 | 260 |
| 50% distilled | 465 | 562 |
| End point | 746 | 868 |
| Fractionation yields (wt%) | | |
| IBP–180°F | 2.5 | |
| 180°F–390°F | 30.2 | 5 |
| 390°F–525°F | 26.7 | 35 |
| 390°F–650°F | 51.0 | 65 |
| 650°F–EP | 16.3 | 30 |
| 390°F–EP | 67.3 | 95 |

*Source:* Speight, J.G., *The Chemistry and Technology of Coal*, Marcel Dekker, New York, 1983. With permission.

the gasification of the COED char to produce a synthesis gas ($CO + H_2$). This COED/COGAS process is significant from both liquefaction and gasification standpoints.

## 3.2.2 TOSCOAL Process

A schematic of the TOSCOAL process is shown in Figure 3.2 [1]. In this process, crushed coal is fed to a rotating drum, which contains preheated ceramic balls at temperatures between 425°C and 540°C. The hydrocarbons, water vapor, and gases are drawn off, and the residual char is separated from the ceramic balls in a revolving drum, which has holes in it. The ceramic balls are reheated in a separate furnace by burning some of the product gas [2]. The TOSCOAL process is analogous to the TOSCO process for producing overhead oil from oil shale [4]. In this process analogy, the char replaces the spent shale, whereas the raw coal replaces raw oil shale [1,4]. It is noted that TOSCO is an acronym of The Oil Shale Corporation.

Table 3.3 shows the properties of liquids produced by the TOSCOAL process from Wyodak coal (as mined). As shown, the recovery efficiency is nearly 100%. It should be noted that the recovery efficiency is from the total material balance concept, not the conversion efficiency toward coal liquid or coal gas. As such, the high recovery rate of water comes from the nature of the feed coal. Wyodak coal is

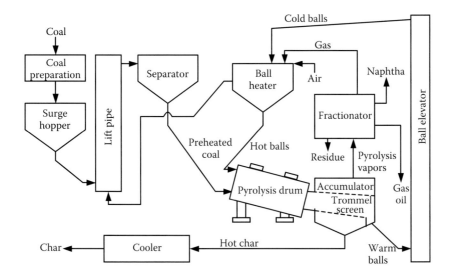

**FIGURE 3.2** A schematic of the TOSCOAL process.

**TABLE 3.3**

**Liquid Products from Wyodak Coal (as Mined) via TOSCOAL Process**

| Temperature (°C) | 425 | 480 | 520 |
|---|---|---|---|
| Yield (wt%) | | | |
| Gas ($\leq C_3$) | 6.0 | 7.8 | 6.3 |
| Oil ($\geq C_4$) | 5.7 | 7.2 | 9.3 |
| Char | 52.5 | 50.6 | 48.4 |
| Water | 35.1 | 35.1 | 35.1 |
| Recovery percentage | 99.3 | 100.7 | 99.1 |

*Source:* Carlson, F.B. et al., Reprints of Clean Fuels from Coal II Symposium, Institute of Gas Technology, Chicago, IL, 1975, p. 504.

the coal from the Tongue River Member of the Fort Union Formation, Powder River Basin, WY. It is subbituminous in rank and has a typical heating value of 8200–8300 Btu/lb. It contains, on average, 5%–6% ash and less than 0.5% sulfur. It is not a coincidence that the processing temperature for the TOSCOAL process is similar to that for the TOSCO oil shale process [4].

## 3.2.3 LURGI–RUHRGAS PROCESS

The process is named after its developers, Lurgi Gesellschaft für Warmetechnik G.m.b.H. and Ruhrgas AG. The Lurgi–Ruhrgas (L–R) process was developed as

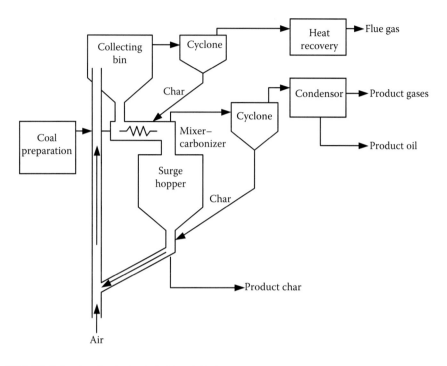

**FIGURE 3.3**    A schematic of the Lurgi–Ruhrgas process.

a low-pressure process for liquid production from lower rank coals. This process developed in Europe is currently in commercial use. A schematic of this process is given in Figure 3.3 [1].

In the L–R process, crushed coal is fed into a mixer and heated rapidly to 450°C–600°C by direct contact with hot recirculating char particles that have been previously heated in a partial oxidation process in an entrained-flow reactor [1]. A cyclone removes the fines from the product gases, and the liquid products are collected by a series of condensers. The liquid products are hydrotreated to yield upgraded products. The high gas yield is due to the relatively long residence time, and the gaseous products include both primary and secondary products. The term *secondary product* is used here to clearly indicate that the product is formed not directly from the coal but, rather, by thermal decomposition of other primary products that are derived directly from coal. Therefore, the reactions involved in the secondary product formation include gas–phase, gas–liquid, and gas–solid reactions.

### 3.2.4   OCCIDENTAL FLASH PYROLYSIS PROCESS

A schematic of the occidental flash pyrolysis process is given in Figure 3.4 [1]. In this process, hot recycle char provides the heat for the flash pyrolysis of pulverized coal in an entrained-flow reactor at a temperature not exceeding 760°C. The process operates with a short residence time, thereby increasing the coal throughput and also increasing the production of liquid products while minimizing the production of gaseous products. At a

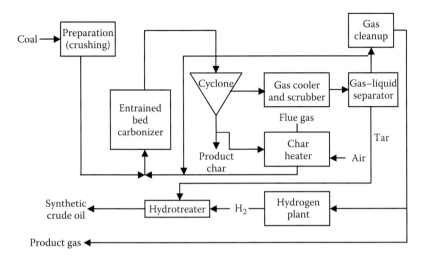

**FIGURE 3.4** A schematic of the occidental flash pyrolysis process. (From Speight, J.G., *The Chemistry and Technology of Coal* (Rev. Ed.), Marcel Dekker, New York, 1994.)

short residence time at a high temperature, pyrolytic decomposition of coal hydrocarbons into liquid-range hydrocarbons actively takes place, but their further conversion into gaseous hydrocarbons is less appreciable. Like other processes, cyclones remove fine char particles from the pyrolysis overhead before quenching in the two-stage collector system [1]. The first stage consists of quenching at approximately 99°C to remove the majority of heavier hydrocarbons, whereas the second stage is for quenching at approximately 25°C to cause water and light oils (i.e., lower-molecular-weight hydrocarbons) to be removed.

### 3.2.5 CLEAN COKE PROCESS

The clean coke process was originally developed by the Unites States Steel Corporation as a conceptual process converting nonmetallurgical-grade coal to metallurgical coke [5]. A schematic of the clean coke process is shown in Figure 3.5 [1]. The process involves feeding oxidized clean coal into a fluidized-bed reactor at temperatures up to 800°C where the coal reacts to produce tar, gas, and low-sulfur char. Alternatively, the coal can be processed by noncatalytic hydrogenation at 455°C–480°C and pressures of up to 340 bars of hydrogen. With direct hydrogenation of coal, the process accomplishes the addition of hydrogen to the coal hydrocarbons while cracking the high-molecular-weight hydrocarbons into lower-molecular-weight hydrocarbons, thus increasing the H/C ratio of the fuel to a level of the liquid hydrocarbon fuel. The liquid products from both the carbonization and hydrogenation stages are combined for further processing to yield synthetic liquid fuels.

### 3.2.6 COALCON PROCESS

The Coalcon process was developed based on Union Carbide's hydrocarbonization studies. A schematic of the Coalcon process is shown in Figure 3.6 [1,6]. This process is based on a dry noncatalytic fluidized bed of coal particles suspended in

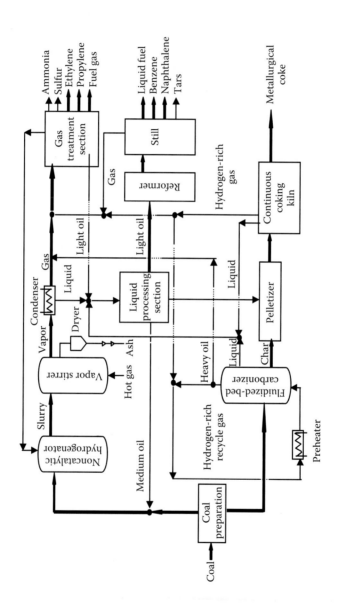

**FIGURE 3.5**  A schematic of the clean coke process. (From Speight, J.G., *The Chemistry and Technology of Coal* (Rev. Ed). Marcel Dekker, New York, 1994.)

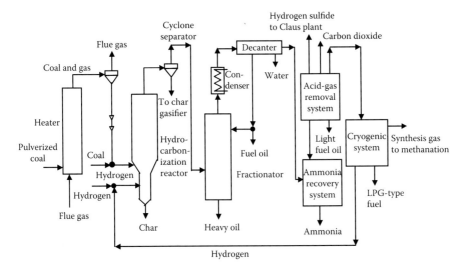

**FIGURE 3.6** A schematic of the Coalcon process. (From Speight, J.G., *The Chemistry and Technology of Coal* (Rev. Ed.), Marcel Dekker, New York, 1994.)

hydrogen gas. It is an intermediate hydrogen pressure process utilizing finely pulverized low-rank coal as feedstock. Hot, oxygen-free flue gas is used to heat the coal to approximately 325°C and also to carry the coal to a feed hopper. A fractionator is employed to subdivide the overhead stream into four streams, namely, (1) gases ($H_2$, CO, $CO_2$, and $CH_4$), (2) light oil, (3) heavy oil, and (4) water. Most of the char is removed from the bottom of the reactor, quenched with water, and cooled [1]. The char can then be used as a feed to a Koppers–Totzek gasifier and reacted with oxygen and steam to produce hydrogen for the process [1].

## 3.3   DIRECT LIQUEFACTION OF COAL

Direct liquefaction of coal is defined to mean *hydroliquefaction*, to distinguish it from pyrolysis, coprocessing, and indirect liquefaction. Hydroliquefaction more specifically means that the liquefaction process is carried out under a hydrogen environment. Direct liquefaction may be categorized into a single-stage or a two-stage process. In two-stage processes, the coal is first hydrogenated in a liquid-phase stage, transforming it into a deashed, liquid product; then, in a second vapor-phase hydrogenation stage, the liquid products are catalytically converted to clean, light distillate fuels. Direct liquefaction has a relatively long history, and various processes have been successfully operated on large scales. Large-scale operations of coal liquefaction, in turn, contributed tremendously to the advances of chemical process industries in all aspects of machinery, design, separation, and knowledge. Recent processes that are ready for demonstration or full commercialization include H-Coal, Solvent Refined Coal (SRC-I, SRC-II), Exxon Donor Solvent (EDS), Integrated Two-Stage Liquefaction (ITSL), Close-Coupled Integrated Two-Stage Liquefaction (CC-ITSL), and Catalytic Two-Stage Liquefaction (CTSL). In this section, several significant direct coal liquefaction processes are reviewed.

### 3.3.1 Bergius-IG Hydroliquefaction Process

The Bergius process was operated very successfully in Germany before and during World War II and was a two-stage process [2]. Even though the process is currently not in use, it has contributed immensely to the development of catalytic coal liquefaction technology. The process involves the catalytic conversion of coal (slurried with heavy oil) in the presence of hydrogen and an iron oxide catalyst, at 450°C–500°C and 200–690 bars (197–681 atm, or 20–69 MPa). The products were usually separated into three principal fractions of light oils, middle distillates, and residuum. Middle distillates, or mid-distillates, are a general classification of refined petroleum products that includes heating oil, distillate fuel oil, jet aviation fuel, and kerosene. Generally speaking, the typical boiling range of mid-distillates is 300°F–750°F, and that for residuum is 600°F–1000°F.

These oils, except for residuum, were catalytically cracked to motor fuels and light hydrocarbons in a vapor-phase hydrogenation stage, which serves as the second stage of the process. Some argue that the severe conditions used in the original process might have been due to the fact that German coals are much more difficult to liquefy than U.S. coals. It is truly remarkable, from a technological standpoint, that the process, under the severe process conditions of a hydrogen atmosphere, was very successfully operated on a large scale in the 1940s.

The residence time for catalytic conversion was about 80–85 min, which was quite long, and hydrogen consumption was also quite significant—approximately 11% by mass of the dry ash-free (daf) coal.

### 3.3.2 H-Coal Process

The H-Coal process is a direct catalytic coal liquefaction process developed in 1963 by Hydrocarbon Research, Inc. (HRI), currently Hydrocarbon Technologies Inc. (HTI). The process development proceeded through several stages from conceptual, to bench scale (25 lb/day), to process development unit (PDU) (3 tons/day), and to a pilot plant in Catlettsburg, KY (200–600 tons/day) [7]. This pilot plant project received $300 million in funds from the U.S. Department of Energy (DOE), the Commonwealth of Kentucky, EPRI, Mobil, AMOCO, CONOCO, Ruhrkohle, Ashland Oil, SUN Oil, Shell, and ARCO.

A schematic of the H-Coal process is shown in Figure 3.7 [1,8]. Pulverized coal, recycle liquids, hydrogen, and a catalyst are brought together in the ebullated-bed reactor to convert coal into hydrocarbon liquids and gaseous products. The catalyst pellets are 0.8 to 1.5 mm diameter extrudates, and pulverized coal is of −60 mesh. The term −60 mesh denotes the particle fraction that passes through the 60-mesh screen, that is, the particle size in this fraction is smaller than the hole opening of the 60-mesh screen. Coal slurried with recycle oil is pumped to a pressure of up to 200 bars and introduced into the bottom of the ebullated-bed reactor. The H-Coal process development has contributed very significantly to the field of chemical reaction engineering in the areas of multiple-phase reactions as well as design of ebullated and liquid entrained reactors. The ebullated bed reactor is similar to a liquid entrained reactor, but much larger gas bubbles help fluidize the solid particles in a

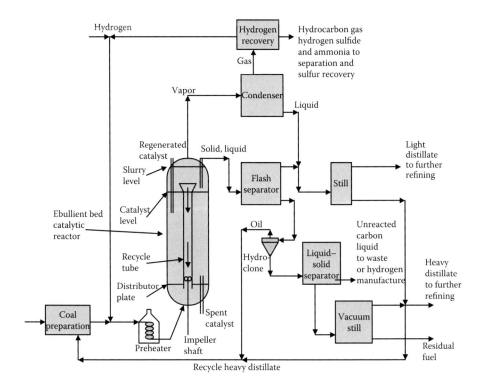

**FIGURE 3.7** A schematic of the H-Coal process.

gas–liquid–solid fluidized bed reactor. Therefore, relatively large coal particles can be used in the ebullated bed reactor, unlike a liquid entrained reactor, which operates on very fine particles.

The process temperature conditions, 345°C–370°C, may be altered appropriately to produce different product slates [1]. Table 3.4 shows typical product compositions from the H-Coal process [1]. A higher reaction temperature of 445°C–455°C has also been successfully demonstrated for the process [8].

Advantages and disadvantages of the H-Coal process are summarized in Table 3.5. Like most other single-stage processes, the H-Coal process is best suited for highly volatile bituminous coal.

### 3.3.3  SOLVENT REFINED COAL

In 1962, the Spencer Chemical Co. began to develop a process that was later taken up by Gulf Oil Co., which, in 1967, designed a 50-ton/day SRC pilot plant at Fort Lewis, WA [8]. The plant was operated in the solvent refined coal (SRC-I) mode from 1974 until late 1976. In 1972, Southern Services Co. (SSC) and Edison Electric Institute (EEI) designed and constructed a 6-ton/day SRC-I pilot plant at Wilsonville, AL.

The principal objective of the original SRC-I process was to produce a solid boiler fuel with a melting point of about 150°C and a heating value of 16,000 Btu/lb. In the interest of enhancing commercial viability, the product slate was expanded to include

## TABLE 3.4
## Product Compositions from the H-Coal Process

|                              | Illinois | | Wyodak Synthetic Crude |
|------------------------------|:-:|:-:|:-:|
|                              | Synthetic Crude | Low-Sulfur Fuel Oil | |
| Product (wt%)                |      |      |      |
| $C_1$–$C_3$ hydrocarbons     | 10.7 | 5.4  | 10.2 |
| $C_4$–200°C distillate       | 17.2 | 12.1 | 26.1 |
| 200°C–340°C distillate       | 28.2 | 19.3 | 19.8 |
| 340°C–525°C distillate       | 18.6 | 17.3 | 6.5  |
| 525°C + residual oil         | 10.2 | 29.5 | 11.1 |
| Unreacted ash-free coal      | 5.2  | 6.8  | 9.8  |
| Gases                        | 15.0 | 12.8 | 22.7 |
| Total (100+$H_2$ reacted)    | 104.9| 103.2| 106.2|
| Conversion (%)               | 94.8 | 93.2 | 90.2 |
| $H_2$ consumption (scf/ton)  | 18,600 | 12,200 | 23,600 |

*Source:*   Speight, J.G., *The Chemistry and Technology of Coal* (Rev. Ed.), Marcel Dekker, New York, 1994.

## TABLE 3.5
## Advantages and Shortcomings of the H-Coal Process

| Advantages | Disadvantages |
|---|---|
| 1. Coal dissolution and upgrading to distillates are accomplished in one reactor. | 1. High reaction temperature (445°C–455°C) results in high gas yields (12%–15%) due to excessive thermal cracking. |
| 2. Products have a high H/C ratio and low heteroatom content. | 2. Hydrogen consumption is relatively high. Some distillate product is gasified to supplement hydrogen need. |
| 3. High throughput of coal occurs due to fast reaction rates of catalytic hydrogenation. | 3. Product contains considerable vacuum gas oil (345°C–525°C, bp), which is difficult to upgrade by standard refinery process. |
| 4. Ash is removed by vacuum distillation, followed by gasification of vacuum tower bottoms to generate the hydrogen required for the process. | 4. Due to the considerable amount of vacuum gas oil, it has utility solely as a boiler fuel. |

liquids that were products of a coker/calciner, an expanded-bed hydrocracker, and a naphtha hydrotreater (HTR) [8].

SRC-I is a thermal liquefaction process in which solvent, coal, and hydrogen are reacted in a "dissolver" reactor to produce a nondistillable resid, which, upon deashing, can be used as a clean boiler fuel. Reaction conditions are slightly less severe than H-Coal process. The absence of a catalyst diminishes the hydrogenation rates, and the resid has an H/C ratio about the same as the coal feed. Again, this process is

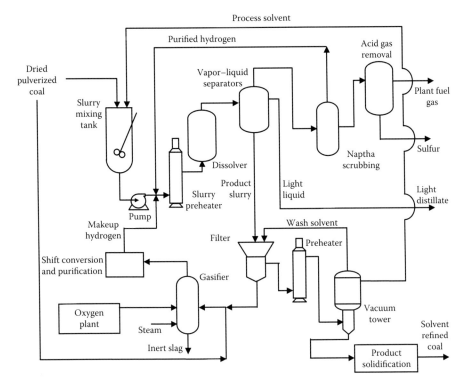

**FIGURE 3.8**  A schematic of the SRC-I process. (From Speight, J.G., *The Chemistry and Technology of Coal* (Rev. Ed.), Marcel Dekker, New York, 1994. With permission.)

also ideally suited for bituminous coals, especially those containing high concentrations of pyrite. The pyrite in this process is considered to be the liquefaction catalyst. A schematic of the SRC-I process is shown in Figure 3.8 [1].

Advantages and disadvantages of SRC-I are given in Table 3.6.

Nondistillable SRC-I resid products cannot be deashed by vacuum distillation. Extraction-type separation processes were developed specifically for this process

**TABLE 3.6**

**Advantages and Disadvantages of SRC-I**

| Advantages | Disadvantages |
|---|---|
| 1. A good boiler fuel with high heating value is obtained. | 1. Distillate solvent is of poor quality. |
| | 2. Solvent is frequently incorporated into the resid product. |
| 2. Reaction conditions are less severe. | 3. Due to no. 2, solvent balance cannot be achieved. |
| 3. The process is noncatalytic and easy to operate. | 4. Nondistillable SRC-I resid cannot be recovered by vacuum distillation. |

[8]. Typical of these is Kerr-McGee's critical solvent deashing (CSD). This deashing process uses a light aromatic solvent to precipitate the heaviest (toluene-insoluble) fraction of the resid, all of the ash, and unconverted coal. This process recovers a heavy but solid-free recycle solvent. The CSD process and its operational principle are very similar to that of the supercritical fluid extraction in which strong solvent power is achieved near or beyond the critical points of the solvents. CSD was also used for the two-stage liquefaction (TSL) processing that is discussed later in this chapter.

### 3.3.4 EXXON DONOR SOLVENT PROCESS

A schematic of the Exxon Donor Solvent (EDS) process is shown in Figure 3.9 [1]. The EDS process utilizes a noncatalytic hydroprocessing step for the liquefaction of coal to produce liquid hydrocarbons. Its salient feature is the hydrogenation of the recycle solvent, which is used as a hydrogen donor to the slurried coal in a high-pressure reactor. This process is also considered to be a single-stage process, as both coal dissolution and resid upgrading take place in one thermal reactor. The liquefaction reaction is carried out noncatalytically. The recycle solvent, however, is catalytically hydrogenated in a separate fixed-bed reactor [8]. This solvent is responsible for transferring hydrogen to the slurried coal in the high-pressure liquefaction reactor. In this regard, the solvent may be considered a hydrogen donor and the coal a hydrogen acceptor. Reaction conditions are similar to those of SRC-I and H-Coal.

EDS solvent must be well and easily hydrogenated to be an effective hydrogen donor. Some examples of the hydrogen donor solvents include tetralin (1,2,3,4-tetrahydronaphthalene) and decalin (decahydronaphthalene). The recycle solvent "donates"

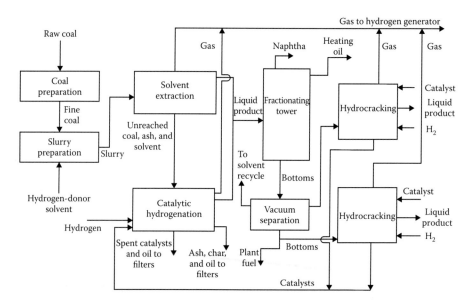

**FIGURE 3.9**   A schematic of the Exxon Donor Solvent (EDS) process.

hydrogen to effect rapid hydrogenation of primary liquefaction products. Thermal hydrogenation and cracking follow this step to produce distillates [8]. The product quality is slightly inferior to that of H-Coal, due to the absence of a hydrotreating catalyst. Distillate yields are also lower than the H-Coal process. Overall, its process economics are still about equal to the H-Coal process because of the less expensive thermal reactor and the simple solids removal process.

The EDS process development started with bench-scale research in the mid-1960s and then progressed to a pilot plant study in the 1970s and 1980s with a 1-ton/day scale. The initial program of process development was completely under Exxon's own responsibility, whereas the latter part of development was cosponsored as a joint venture between Exxon and the US DOE. In 1980, a large-scale (250 tons/day) demonstration type of installation, which was named the Exxon Coal Liquefaction Plant (ECLP), was constructed and put into operation. The plant was shut down and dismantled in 1982.

### 3.3.5 SRC-II Process

The SRC-II process uses direct hydrogenation of coal in a reactor at high pressure and temperature to produce liquid hydrocarbon products instead of the solid products in SRC-I. The 50-ton/day pilot plant at Fort Lewis, WA, which operated in the SRC-I mode from 1974 to 1976, was modified to run in the SRC-II mode, producing liquid products for testing [8]. The pilot plant was successfully operated from 1978 until 1981.

The SRC-II process is a thermal process and uses the mineral matter in the coal as the only catalyst. The mineral matter concentration in the reactor is kept high by recycling of the heavy oil slurry. The recycled use of mineral matter and the more severe reaction conditions distinguish the SRC-II operation from the SRC-I process and also account for the lighter products. The net product is −540°C distillate, which is recovered by vacuum distillation. The term *−540°C distillate* denotes the fraction that comes out below 540°C of distillation temperature. The vacuum bottoms including ash are sent to gasification to generate process hydrogen. The SRC-II process is limited to coals that contain catalytic mineral matter and therefore excludes all lower-rank coals and some bituminous coals. Pulverized coal of particle size smaller than 0.125 in. and a solvent-to-coal ratio of 2.0 are used for SRC-II, whereas the solvent-to-coal ratio is 1.5 for SRC-I. The liquid product quality is inferior to that of the H-Coal process. A schematic of the SRC-II process is shown in Figure 3.10.

### 3.3.6 Nonintegrated Two-Stage Liquefaction

Even though single-stage processes like EDS, SRC-I, SRC-II, and H-Coal are technologically sound, their process economics suffers for the following reasons:

1. The reaction severity is high, with temperatures of 430°C–460°C and liquid residence times of 20–60 min. These severe operating conditions were considered necessary to achieve coal conversions of over 90% (to tetrahydrofuran [THF]- or quinoline-solubles).

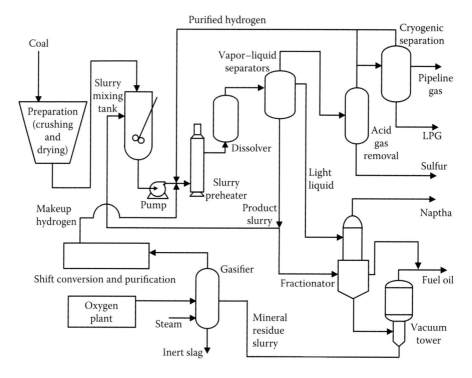

**FIGURE 3.10**    A schematic of the SRC-II process. (From Speight, J.G., *The Chemistry and Technology of Coal* (Rev. Ed.), Marcel Dekker, New York, 1994.)

2. Distillate yields are low, only about 50% for mineral matter and ash free (mmaf) bituminous coals and even lower for subbituminous coals.
3. Hydrogen efficiency is low due to high yields of hydrocarbon gases.
4. The costs associated with the SRC-I process or the like may be too high to produce a boiler fuel.

Based on these reasons, a coal liquefaction process is best applied to make higher value-added products, such as liquid transportation fuels [8]. To produce higher value-added products from the SRC-I process, the resid must first be hydrocracked to distillate liquids. Efforts made by Mobil and Chevron on fixed-bed hydrocracking were not entirely successful, due to the plugging of the fixed-bed ashes and rapid deactivation of the catalyst by coking.

The SRC-I resid was successfully hydrotreated by Lummus–Cities–Fining Chevron Lummas Global (CLG)'s LC-Fining technology, a variation of ebullated-bed technology developed by Cities Services research and development (R&D) [9]. As a result, a hydrocracking part was added to the SRC-I process to form nonintegrated two-stage liquefaction (NTSL). This rather unique name was given because the hydrocracking part did not contribute solvent to the SRC-I part. In other words, the NTSL process was a combination of two separate processes, namely, coal liquefaction and resid upgrading. A schematic of the NTSL process is shown in

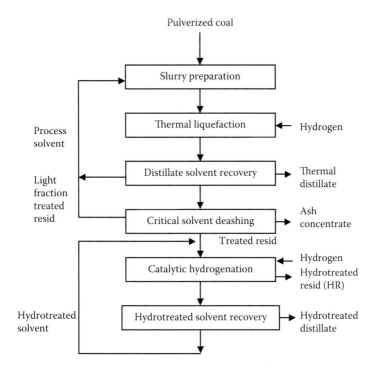

**FIGURE 3.11** A block diagram of the NTSL process. (From Schindler, H.D., Coal Liquefaction—A Research and Development Needs Assessment, COLIRN Panel Assessment, DOE/ER-0400,UC-108, Final Report, Vol. II, March 1989.)

Figure 3.11. Even with the addition of the hydrocracking section, NTSL was a somewhat inefficient process due to the shortcomings listed earlier. SRC-I product is a less reactive feed to hydrocracking, thus requiring high-temperature (over 430°C) and low space velocity (i.e., low productivity) for complete conversion to distillates. In order to keep the temperature and reactor size at reasonable levels, resid conversion was held below 80%. NTSL operation data at the Wilsonville facility are presented in Table 3.7. Yields were higher than those for H-Coal, but hydrogen consumption was still high due to the extensive thermal hydrogenation step in the SRC-I dissolver, which was renamed the thermal liquefaction unit (TLU) [8]. NTSL was short-lived, and a newer integrated approach was later developed.

### 3.3.7 THERMAL INTEGRATED TWO-STAGE LIQUEFACTION

Thermal coal dissolution studies by Consol, Mobil, and Wilsonville in the late 1970s had shown that coal conversion to THF-solubles is essentially complete in an extremely short time, 1–5 min. Within this short dissolution period, hydrogenation from the gas phase is negligible, and almost all hydrogen comes from the solvent in the liquid phase [8]. If hydrogen transfer from the solvent is insufficient to satisfy the liquefaction needs, the product will have a high concentration of toluene-insolubles, causing precipitation and plugging in the reactor or in downstream equipment.

## TABLE 3.7
## NTSL at Wilsonville Facility (Illinois No. 6 Coal)

| Operating Conditions | |
|---|---|
| Run ID | 241CD |
| Configuration | NTSL |
| Catalyst | Armak |
| Thermal stage | |
|     Average reactor temperature (°F) | 805 |
|     Coal space velocity (lb/h/ft$^3$ @ >700°C) | 20 |
|     Pressure (psig) | 2170 |
| Catalytic stage | |
|     Average reactor temperature (°F) | 780 |
|     Space velocity (lb feed/h/lb catalyst) | 1.7 |
|     Catalyst age (lb resid/lb catalyst) | 260–387 |
| **Yields (wt% mmaf coal)** | |
| $C_1$–$C_3$ gas | 7 |
| $C_4$+ distillate | 40 |
| Resid | 23 |
| Hydrogen consumption | 4.2 |
| Hydrogen efficiency | |
|     lb $C_4$+ distillate/lb $H_2$ consumed | 9.5 |
| Distillate selectivity | |
|     lb $C_1$–$C_3$/lb $C_4$+ distillate | 0.18 |
| Energy content of feed coal rejected to ash concentrate (%) | 20 |

*Source:* Schindler, H.D., Coal Liquefaction—A Research and Development Needs Assessment, COLIRN Panel Assessment, DOE/ER-0400,UC-108, Final Report, Vol. II, March 1989.

With a well-hydrogenated solvent, however, short-contact-time (SCT) liquefaction is the preferred thermal dissolution procedure because it eliminates the inefficient thermal hydrogenation inherent in the SRC-I. Cities Services R&D successfully hydrocracked the SRC-I resids by LC-Fining at relatively low temperatures of 400°C–420°C. Gas yield was low, and hydrogen efficiency was high. A combination of this process with SCT is certainly a good idea and provides a successful example of process integration. The low-temperature LC-Fining provides the liquefaction solvent to the first-stage SCT; thus, the two stages become integrated. This combination has the potential to liquefy coal to distillate products in a more efficient process than any of the single-stage processes [8].

### 3.3.7.1 Lummus ITSL (1980–1984)

A combination of SCT liquefaction and LC-Fining was made by Lummus in the ITSL process [10]. A process flow diagram of the Lummus ITSL process is given in Figure 3.12. Coal is slurried with recycled solvent from LC-Fining and is converted

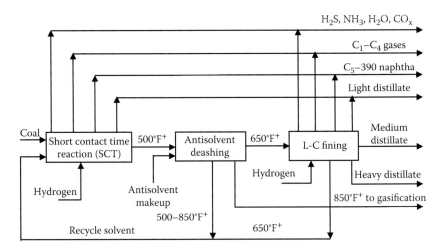

**FIGURE 3.12** A schematic of integrated two-stage liquefaction (ITSL). (From Speight, J.G., *The Chemistry and Technology of Coal*, Marcel Dekker, New York, 1983.)

to quinoline-solubles (or THF-solubles) in the SCT reactor. The resid is hydrocracked to distillates in the LC-Fining stage, where recycle solvent is also generated. The ash is removed by the Lummus antisolvent deashing (ASDA) process, which is similar to deasphalting operations with petroleum. In the ASDA, insoluble material such as ash is separated from a coal liquefaction product by gravity settling in the presence of promoting liquid. The settled underflow having a specified ash content is removed from the valuable hydrocarbon liquid products [11]. The net liquid product is either −340°C or −450°C distillate. The recycle solvent is hydrogenated 340°C+ atmospheric bottoms. It is the recycle of these full-range bottoms, including resid, which couples the two reaction stages and results in high yields of all distillate products [8].

Some of the features of the Lummus ITSL are summarized as follows:

1. The SCT reactor is actually the preheater for the dissolver in the SRC-I process, thus eliminating a long-residence-time high-pressure thermal dissolution reactor.
2. Coal conversion in the SCT reactor was 92% of mmaf coal for bituminous coals and 90% for subbituminous coals.
3. Molecular hydrogen gas consumption was essentially zero, and the hydrogen transferred from the solvent was equivalent to 1.2%–2.0% of the coal weight. Gaseous hydrocarbon yield was reduced to 1% for bituminous coal and to 5%–6% for subbituminous coal.
4. The SCT resid was more reactive to hydrocracking than SRC-I resid.
5. The LC-Fining second reactor as an HTR accomplishes two principal tasks: (1) to make essentially all distillate product and (2) to generate recycle solvent capable of supplying the hydrogen required by the SCT reactor.

6. All distillate products were produced as a result of full recycle of unconverted resid to the first stage.
7. A second-stage HTR temperature of 400°C provides sufficient hydrogenation and cracking activity to accomplish both tasks.
8. Catalyst deactivation was much slower than other processes operated at higher temperatures.
9. The SCT resid was more reactive, not only for conversion to distillate but also for heteroatom removal. Product quality surpassed that achieved by the preceding processes. Chevron successfully refined the ITSL products for specification transportation fuels.
10. The ash was removed by ASDA, which used process-derived naphtha as antisolvent to precipitate the heaviest components of the resid and the solids.
11. ASDA had the advantage of low-pressure (100–1000 psi) and low-temperature (260°C–282°C) operation.

Data for typical product yields by the Lummus ITSL process are given in Table 3.8 [8], and the product quality of the Lummus ITSL distillates is shown in Table 3.9 [8].

### 3.3.7.2 Wilsonville ITSL (1982–1985)

The Advanced Coal Liquefaction R&D Facility at Wilsonville, AL, sponsored by the DOE, the Electric Power Research Institute (EPRI), and AMOCO, was operated by Catalytica, Inc., under the management of Southern Company Services, Inc. The hydrotreater design was supplied by HRI, and the deashing technology was provided by Kerr-McGee Corporation.

**TABLE 3.8**
**Lummus ITSL Product Yields**

| Product | lb/100 lb mmaf Coal | |
|---|---|---|
| | Illinois No. 6 | Wyodak |
| $H_2S$, $H_2O$, $NH_3$, $CO_x$ | 15.08 | 23.08 |
| $C_1$–$C_4$ | 4.16 | 7.30 |
| Total gas | 19.24 | 30.38 |
| $C_5$–390°F | 6.92 | 1.25 |
| 390°F–500°F | 11.46 | 8.49 |
| 500°F–650°F | 17.26 | 22.46 |
| 650°F–850°F | 23.87 | 21.36 |
| Total distillate product | 59.51 | 53.56 |
| Organics rejected with ash | 26.09 | 20.22 |
| Grand total | 104.84 | 104.16 |
| Molecular hydrogen consumption | 4.84 | 4.16 |
| Hydrogen efficiency, lb distillates/lb $H_2$ | 12.28 | 12.86 |
| Distillate yield, bbl/ton mmaf coal | 3.52 | 3.08 |

**TABLE 3.9**

**Lummus ITSL Distillate Product Quality (Illinois No. 6 Coal)**

| °API | C | H | O | N | S | HHV, Btu/lb |
|------|------|------|------|------|------|------|
| | | | **Naphtha** | | | |
| 36.8 | 86.79 | 11.15 | 1.72 | 0.18 | 0.16 | 19,411 |
| 45.4 | 86.01 | 13.16 | 0.62 | 0.12 | 0.09 | 20,628 |
| | | **Light Distillates (390°F–500°F)** | | | | |
| 15.5 | 88.62 | 9.51 | 1.50 | 0.28 | 0.09 | 18,673 |
| 22.9 | 87.75 | 11.31 | 0.73 | 0.13 | 0.08 | 19,724 |
| | | **Medium Distillates (500°F–650°F)** | | | | |
| 7.5 | 90.69 | 8.76 | 0.27 | 0.25 | 0.03 | 18,604 |
| 12.9 | 89.29 | 10.26 | 0.28 | 0.12 | 0.05 | 19,331 |
| | | **Heavy Distillates (650°F–850°F)** | | | | |
| −1.5 | 91.47 | 7.72 | 0.26 | 0.50 | 0.05 | 18,074 |
| 1.8 | 90.77 | 8.47 | 0.45 | 0.23 | 0.08 | 18,424 |

*Source:*  Schindler, H.D., Coal Liquefaction—A Research and Development Needs Assessment, COLIRN Panel Assessment, DOE/ER-0400, UC-108, Final Report, Vol. II, March 1989.

The Wilsonville facility began operations as a 6-ton/day single-stage plant for SRC-I in 1974. In 1978, a Kerr-McGee CSD unit replaced the filtration equipment that had been used for solids removal from the SRC product. In 1981, an H-Oil ebullated-bed hydrotreater was installed for upgrading the recycle solvent and product. In 1985, a second ebullated-bed reactor was added in the hydrotreater area to allow operation with close-coupled reactors. A schematic for the integrated two-stage liquefaction (ITSL) configuration used at the Wilsonville facility for bituminous coal runs is shown in Figure 3.13. A distillate yield of 54%–59% of mmaf coal was confirmed, as shown in Table 3.10. It is noted that the hydrogen efficiency for the ITSL based on the distillate productivity per hydrogen consumption is substantially increased from that for the NTSL result.

Lummus enhanced the ITSL process by increasing the distillate yield by placing the deasher after the second stage, with no detrimental effect of ashy feed on catalyst activity. This enhanced process is called *reconfigured two-stage liquefaction* (RITSL), as illustrated in Figure 3.14. The process improvements were experimentally confirmed at the Wilsonville facility. The enhancements included higher distillate yield, lower resids, and less energy rejects.

With the deasher placed after the second-stage reactor and the two stages operating at about the same pressure, the two reactors were close-coupled to minimize holding time between the reactors and to eliminate pressure letdown and repressurizing between stages [8]. This enhancement was called *close-coupled ITSL*

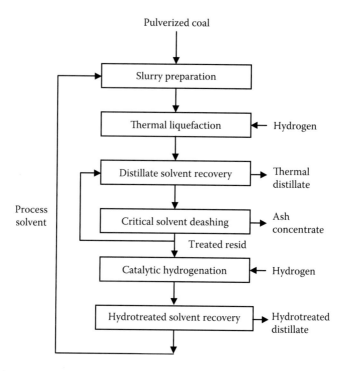

**FIGURE 3.13** A block diagram of the ITSL process. (From Schindler, H.D., Coal Liquefaction—A Research and Development Needs Assessment, COLIRN Panel Assessment, DOE/ER-0400, UC-108, Final Report, Vol. II, March 1989. With permission.)

(CC-ITSL). The improved results were evidenced by higher distillate yield, lower resids, and lower energy reject.

### 3.3.8 Catalytic Two-Stage Liquefaction

Beginning in 1985, all process demonstration unit (PDU) programs in the United States have used two catalyst stages. The two-stage liquefaction was found to be much more effective than the single-stage counterparts. As mentioned earlier, all single-stage liquefaction processes have faced difficulties in converting subbituminous coal into soluble liquids, though they can handle bituminous coals satisfactorily.

#### 3.3.8.1 HRI's Catalytic Two-Stage Liquefaction Process

In 1982, HRI (currently, HTI, a subsidiary of Headwaters Technology Innovation Group, Inc.) initiated the development of a catalytic two-stage concept, overcoming the drawbacks of H-Coal, which is inherently a high-temperature catalytic process [7]. The first-stage temperature was lowered to 400°C to more closely balance hydrogenation and cracking rates, and to allow the recycle solvent to be hydrogenated *in situ* to facilitate hydrogen transfer to coal dissolution. The second stage was operated

# TABLE 3.10
## ITSL and NTSL Operation Data at Wilsonville Facility (Illinois No. 6 Coal)

| Run ID | 241CD | 7242BC | 243JK/244B | 247D | 250D | 250G(a) |
|---|---|---|---|---|---|---|
| Configuration | NTSL | ITSL | ITSL | RITSL | CC-ITSL | CC-ITSL |
| Catalyst | Armak | Shell324M | Shell324M | Shell324M | Amocat IC | Amocat IC |
| **Operating Conditions** | | | | | | |
| Thermal stage | | | | | | |
| Average reactor temperature (°F) | 805 | 860 | 810 | 810 | 824 | 829 |
| Coal space velocity, lb/h/ft$^3$ @ >700°C | 20 | 43 | 28 | 27 | 20 | 20 |
| Pressure, psig | 2170 | 2400 | 1500–2400 | 2400 | 2500 | 2500 |
| Catalytic stage | | | | | | |
| Average reactor temperature (°F) | 780 | 720 | 720 | 711 | 750 | 750 |
| Space velocity, lb feed/h/lb catalyst | 1.7 | 1.0 | 1.0 | 0.9 | 2.08 | 2.23 |
| Catalyst usage, lb resid/lb catalyst | 260–387 | 278–441 | 380–850 | 446–671 | 697–786 | 346–439 |
| **Yields (wt% mmaf coal)** | | | | | | |
| C$_1$–C$_3$ gas | 7 | 4 | 6 | 6 | 7 | 8 |
| C$_4$+ distillate | 40 | 54 | 59 | 62 | 64 | 63 |
| Resid | 23 | 8 | 6 | 3 | 2 | 5 |
| Hydrogen consumption | 4.2 | 4.9 | 5.1 | 6.1 | 6.1 | 6.4 |
| Hydrogen efficiency | | | | | | |
| lb C$_4$+ distillate/lb H$_2$ consumed | 9.5 | 11 | 11.5 | 10.2 | 10.5 | 9.8 |
| Distillate selectivity | | | | | | |
| lb C$_1$–C$_3$/lb C$_4$+ distillate | 0.18 | 0.07 | 0.10 | 0.10 | 0.11 | 0.12 |
| Energy content of feed coal rejected to ash concentrate (%) | 20 | 24 | 20–23 | 22 | 23 | 16 |

*Source:* Schindler, H.D., Coal Liquefaction—A Research and Development Needs Assessment, COLIRN Panel Assessment, DOE/ER-0400, UC-108, Final Report, Vol. II, March 1989.

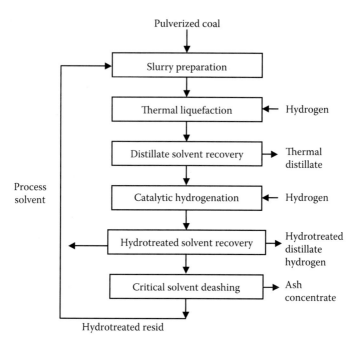

Pulverized coal

Slurry preparation

Thermal liquefaction ← Hydrogen

Distillate solvent recovery → Thermal distillate

Process solvent

Catalytic hydrogenation ← Hydrogen

Hydrotreated solvent recovery → Hydrotreated distillate hydrogen

Critical solvent deashing → Ash concentrate

Hydrotreated resid

**FIGURE 3.14** A block diagram of the RITSL process. (From Schindler, H.D., Coal Liquefaction—A Research and Development Needs Assessment, COLIRN Panel Assessment, DOE/ER-0400, UC-108, Final Report, Vol. II, March 1989.)

at higher temperatures (435°C–440°C) to promote resid hydrocracking and generate an aromatic solvent, which is then hydrogenated in the first stage [8]. The lower first-stage temperature provides better overall management of hydrogen consumption and reduced hydrocarbon gas yields [8,12]. A schematic of this process is shown in Figure 3.15 [8].

The HRI's catalytic two-stage liquefaction (CTSL) had three major changes in comparison to the H-Coal process. The first was the two-stage processing; the second was incorporation of a pressure filter to reduce resid concentration in the reject stream (filter cake) below the 45%–50% in the vacuum tower bottoms of the H-Coal process; and the third change was in the catalyst itself. The H-Coal process used a cobalt–molybdenum (CoMo)-on-alumina catalyst, American Cyanamid 1442 B, which had been effective in hydrocracking petroleum resids. In coal liquefaction, hydrogenation must occur first, followed by thermal cracking of hydroaromatics, whereas in petroleum applications, the contrary is true. Therefore, the H-Coal catalyst was found unsuitable due to its porosity distribution, which was designed for smaller molecules. For CSTL, the H-Coal catalyst was replaced by a nickel–molybdenum (NiMo) catalyst of a bimodal pore size distribution with larger micropores (115–125 Angstroms) as opposed to 60–70 Angstroms for the H-Coal catalyst. The nickel promoter is also more active for hydrogenation than cobalt. Table 3.11 shows a comparison between H-Coal and HRI CTSL [8].

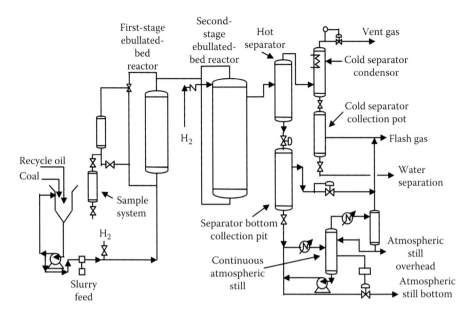

**FIGURE 3.15**   A schematic of HRI's CTSL.

---

**TABLE 3.11**
**CTSL versus H-Coal Demonstration Runs on Illinois No. 6 Coal**

|  | H-Coal | CSTL | |
|---|---|---|---|
|  | PDU-5 | (227-20) | (227-47) |
| Yields (wt% mmaf coal) | | | |
| C$_1$–C$_3$ | 11.3 | 6.6 | 8.6 |
| C$_4$–390°F | 22.3 | 18.2 | 19.7 |
| 390°F–650°F | 20.5 | 32.6 | 36.0 |
| 650°F–975°F | 8.2 | 16.4 | 22.2 |
| 975°F+ oil | 20.8 | 12.6 | 2.7 |
| Hydrogen consumption (wt% mmaf coal) | 6.1 | 6.3 | 7.3 |
| Coal conversion (wt% mmaf coal) | 93.7 | 94.8 | 96.8 |
| 975°F+ conversion (wt% mmaf coal) | 72.9 | 82.2 | 94.1 |
| C$_4$–975°F (wt% mmaf coal) | 51.0 | 67.2 | 77.9 |
| Hydrogen efficiency | 8.4 | 10.7 | 10.7 |
| C$_4$+ distillate product quality | | | |
| EP (°F) | 975 | 975 | 750 |
| °API | 26.4 | 23.5 | 27.6 |
| % hydrogen | 10.63 | 11.19 | 11.73 |
| % nitrogen | 0.49 | 0.33 | 0.25 |
| % sulfur | 0.02 | 0.05 | 0.01 |
| bbl/ton | 3.3 | 4.1 | 5.0 |

*Source:*  Speight, J.G., *The Chemistry and Technology of Coal*, Marcel Dekker, New York, 1983.

As shown in the table, the two-stage catalytic reaction produces a liquid with low heteroatom concentrations and a high H/C ratio, thus making the product closer to petroleum than other coal liquids made by earlier processes. Their later-version enhanced process is named the *HTI coal process* [7]. The modern version of this process uses HTI's proprietary GelCat catalyst, which is a dispersed, nanoscale, iron-based catalyst [7].

### 3.3.8.2 Wilsonville CTSL

A second ebullated-bed reactor was added at the Wilsonville Advanced Coal Liquefaction Facility in 1985. Since then, the plant has been operated in the CTSL mode. As in ITSL, Wilsonville preferred to have most of the thermal cracking take place in the first reactor and solvent hydrogenation in the second reactor [8]. Therefore, the first reactor was at a higher temperature (426°C–438°C), whereas the second reactor was kept lower at 404°C–424°C. A flow diagram of the Wilsonville CTSL is shown in Figure 3.16 [8]. Run data of the Wilsonville CTSL are summarized in Table 3.12. Distillate yields of up to 78% and reduced organic rejection to 8%–15% were achieved at Wilsonville operating over 4 tons of coal per day.

### 3.3.9  EVOLUTION OF LIQUEFACTION TECHNOLOGY

An extensive review by the Coal Liquefaction Research Needs (COLIRN) panel assessment [8] was published by the US DOE. Substantial technological innovations

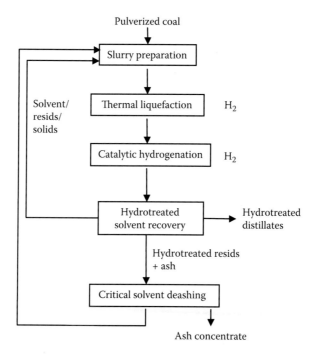

**FIGURE 3.16**  A flow diagram of CSTL with a solids recycle at Wilsonville.

**TABLE 3.12**
**CTSL Operation Data at Wilsonville Facility**

| Operating Conditions | | | |
|---|---|---|---|
| Run ID | 253A | 254G | 251-IIIB |
| Configuration | CTSL | CTSL | CTSL |
| Coal | Illinois No. 6 | Ohio No. 6 | Wyodak |
| Catalyst | Shell 317 | Shell 317 | Shell 324 |
| First stage | | | |
| Average reactor temperature (°F) | 810 | 811 | 826 |
| Inlet hydrogen partial pressure (psi) | 2040 | 2170 | 2510 |
| Feed space velocity (lb/h/lb catalyst) | 4.8 | 4.3 | 3.5 |
| Pressure (psig) | 2600 | 2730 | 2600 |
| Catalyst age (lb resid/lb catalyst) | 150–350 | 1003–1124 | 760–1040 |
| Catalytic stage | | | |
| Average reactor temperature (°F) | 760 | 790 | 719 |
| Space velocity (lb feed/h/lb catalyst) | 4.3 | 4.2 | 2.3 |
| Catalyst age (lb resid/lb catalyst) | 100–250 | 1166–1334 | 371–510 |
| Yield (wt.% of mmaf coal) | | | |
| $C_1$–$C_3$ gas | 6 | 8 | 11 |
| $C_4$+ distillate | 70 | 78 | 60 |
| Resid | ~1 | ~1 | 2 |
| Hydrogen consumption | 6.8 | 6.9 | 7.7 |
| Hydrogen efficiency | | | |
| lb $C_4$+ distillate/lb $H_2$ consumed | 10.3 | 11.3 | 7.8 |
| Distillate selectivity | | | |
| lb $C_1$–$C_3$/lb $C_4$+ distillate | 0.08 | 0.11 | 0.18 |
| Energy content of feed coal rejected to ash concentrate (%) | 20 | 10 | 15 |

and enhancements have been realized for the last several decades of the twentieth century, especially in the areas of process configurations and catalysts. Table 3.13 summarizes the history of process development improvements in the form of yields and distillate quality [8,13].

Distillate yields have increased roughly from 41% to 78%, resulting in equivalent liquid yields of about five barrels per ton of mmaf bituminous coal. The distillate quality was comparable to or better than no. 2 fuel oil with good hydrogen content and low heteroatom content.

Due to the severity of process conditions, significant energy input, use of costly catalyst, and high capital investment, commercial scale of direct liquefaction of coal has not been practiced in the United States. However, the R&D in this field has tremendously contributed to the advances in chemical and petrochemical process industries, namely, design and operation of multiphase reactor systems, separation processes, denitrification and desulfurization, energy integration and waste heat

**TABLE 3.13**
**History of Liquefaction Process Development for Bituminous Coal**

| Process | Configuration | Distillate (wt% mmaf Coal) | Yield (bbl/ton mmaf Coal) | °API Gravity | Nonhydrocarbon (wt%) | | |
|---|---|---|---|---|---|---|---|
| | | | | | S | O | N |
| SRC II (1982) | One-stage noncatalytic | 41 | 2.4 | 12.3 | 0.33 | 2.33 | 1.0 |
| H-Coal (1982) | One-stage catalytic | 52 | 3.3 | 20.2[a] | 0.20 | 1.0 | 0.50 |
| RITSL, Wilsonville (1985) | Integrated two-stage thermal–catalytic | 62 | 3.8 | 20.2[b] | 0.23 | 1.9 | 0.25 |
| CTSL, Wilsonville (1986) | Integrated close-coupled two-stage catalytic–catalytic | 70 | 4.5 | 26.8[b] | 0.11 | <1 | 0.16 |
| CTSL, Wilsonville (1987) | Integrated close-coupled two-stage low–ash coal | 78 | 5.0 | —[c] | —[c] | —[c] | —[c] |
| CTSL, HRI (1987) | Catalytic–catalytic | 78 | 5.0 | 27.6 | 0.01 | — | 0.25 |

*Source:* Schindler, H.D., Coal Liquefaction—A Research and Development Needs Assessment, COLIRN Panel Assessment, DOE/ER-0400, UC-108, Final Report, Vol. II, March 1989.

[a] Light product distribution, with over 30% of product in gasoline boiling range; less than heavy turbine fuel.

[b] Higher boiling point distribution, with 30% of product in gasoline fraction and over 40% in turbine fuel range.

[c] Data unavailable.

recovery, minimization of wastes and pollutants, catalyst development, energy materials, control and implementation, and much more.

## 3.4 INDIRECT LIQUEFACTION OF COAL

The indirect liquefaction of coal involves the production of synthesis gas mixture from coal as a first stage and the subsequent catalytic production of hydrocarbon fuels and oxygenates from the synthesis gas as a second stage. Indirect liquefaction can be classified into two principal areas [14]:

1. Conversion of syngas to light hydrocarbon fuels via Fischer–Tropsch synthesis (FTS).
2. Conversion of syngas to oxygenates such as methanol, higher alcohols, dimethyl ether (DME), and other ethers.

### 3.4.1 FTS FOR LIQUID HYDROCARBON FUELS

The synthesis process of converting a syngas mixture of $H_2$ + CO into liquid hydrocarbons was first developed in Germany by Franz Fischer and Hans Tropsch in 1925. The FTS process is currently being operated commercially. The largest-scale plants of FTS technology are operated by the South Africa Synthetic Oil Liquid (SASOL) Limited in South Africa. The SASOL plant in South Africa has been in operation since 1956. A generalized flow sheet for the SASOL plant is shown in Figure 3.17 [15].

#### 3.4.1.1 Reaction Mechanism and Chemistry

FTS follows a simple polymerization reaction mechanism, the monomer being a $C_1$ species derived from CO. This polymerization reaction follows a molecular-weight distribution (MWD) described mathematically by Anderson [16], Schulz [17], and Flory [18]. Recognizing these three independent groups' work, the description of the FTS product distribution is usually referred to as the Anderson–Schulz–Flory (ASF) distribution, which is generally accepted and frequently used. The ASF distribution equation is written as

$$\log(w_n/n) = n \log x + \log[(1-x)^2/x]$$

where $w_n$, $n$, and $x$ are the mass fraction, the carbon number, and the probability of chain growth, respectively. This equation can predict the maximum selectivity attainable by an FTS with an optimized process and catalyst, as shown in Table 3.14.

From a linear plot of $\log(w_n/n)$ versus $n$, the chain growth probability can be computed either from the slope ($\log x$) or from the intercept, $\log[(1-x)^2/x]$. These predicted values are valid for the products that are hydrocarbons only (i.e., paraffins and olefins) or hydrocarbons plus alcohols (i.e., a mixture of paraffins, olefins, and alcohols).

#### 3.4.1.2 Fischer–Tropsch Catalysis

According to the ASF equation, catalysts with a small value of $x$ (i.e., a lower chain growth probability) produce a high fraction of methane. On the other hand, a large $x$

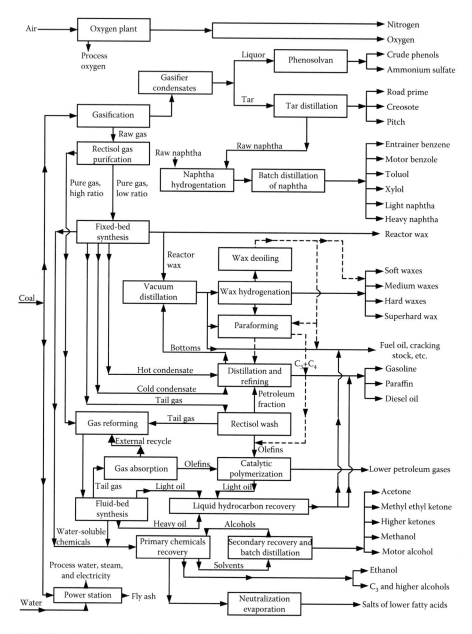

**FIGURE 3.17** A generalized flow sheet for the SASOL plant. (From Hoogendorn, J.C. and Salomon, J.M., *Br. Chem. Eng.*, 2, 1957. With permission.)

value indicates the production of heavier hydrocarbons. The latest FTS processes aim at producing high-molecular-weight products and very little methane and then cracking these high-molecular-weight substances to yield lower hydrocarbons. There have been numerous attempts to surpass or exceed the ASF distribution so that one could produce liquid fuels in yields that exceed those predicted by the ASF equation.

**TABLE 3.14**

**Maximum Selectivities Attainable by FTS**

| Product | Maximum Selectivity (wt%) |
|---|---|
| Methane | 100 |
| Ethylene | 30 |
| Light olefins ($C_2$–$C_4$) | 50 |
| Gasoline ($C_5$–$C_{11}$) | 48 |

*Source:* Schindler, H.D., Coal Liquefaction—A Research and Development Needs Assessment, COLIRN Panel Assessment, DOE/ER-0400, UC-108, Final Report, Vol. II, March 1989.

A wide variety of catalysts can be used for the FTS process, among which transition metals such as iron, cobalt, and ruthenium are most common. Cobalt-based catalysts are highly active, whereas iron-based catalysts are more suitable for conversion of CO-rich syngas due to their catalyzing effects on the water gas shift (WGS) reaction that supplies additional amounts of hydrogen to the synthesis process. Accordingly, inexpensive iron catalysts are popularly used for the FTS. These catalysts are prepared by fusing iron oxides such as mill scale oxides. In practice, either an alkali salt or one or more nonreducible oxides are added to the catalyst [8]. Group 1 alkali metals are promoters for iron catalysts, whereas they are poisons to cobalt catalysts. A great deal of literature data are available; however, very few share the common grounds in their catalyst pretreatment, catalyst ingredients, catalyst preparation, and reactor design and configuration, making direct comparison of the results very difficult, if not impossible. Since the hydrocarbon synthesis reaction from syngas is exothermic, the removal of reaction-generated heat is an essential task. There have been a number of different reactor configurations that have been designed and employed for the FTS process and they include multiple-tube fixed bed reactor, slurry reactor, entrained flow reactor, and circulating fluidized bed reactor.

In FTS, removal of wax, high-molecular-weight hydrocarbons, formed by the synthesis reaction is crucial as it can disable the catalytic activity. High molecular wax can condense on the catalyst and form agglomerates.

Large-scale FTS plants require high capital costs, while the production cost of liquid fuel is substantially lower for a large-scale plant than a smaller scale plant. The most notable of large-scale FTS plants include SASOL's plants and Shell's Pearl gas-to-liquid (GTL) facility in Ras Laffan, Qatar. The latter is primarily based on the utilization of natural gas-derived syngas as a feedstock. The SASOL plants furnish a major portion of South Africa's requirements for fuels and chemicals. South Africa is a nation with large coal reserves but with little oil. The first commercial FTS operation in South Africa started in 1952. Data on the existing SASOL plants are given in Table 3.15.

An approximate distribution of products from SASOL-2 operation is given in Table 3.16 [19].

**TABLE 3.15**
**SASOL Plants**

| Plant | Location | Start Date | Coal ton/day | Liquids bbl/day | Cost $ billion |
|-------|----------|-----------|--------------|-----------------|----------------|
| SASOL-1 | Sasolburg, S. Africa | 1935 | 6600 | 6000 | — |
| SASOL-2 | Secunda, S. Africa | 1981 | 30,000 | 40,000 | 2.9 |
| SASOL-3 | Secunda, S. Africa | 1982 | 30,000 | 40,000 | 3.8 |

*Source:* Schindler, H.D., Coal Liquefaction—A Research and Development Needs Assessment, COLIRN Panel Assessment, DOE/ER-0400, UC-108, Final Report, Vol. II, March 1989.

**TABLE 3.16**
**Product Distribution of SASOL-2**

| Product | Tons/Year |
|---------|-----------|
| Motor fuels | 1,650,000 |
| Ethylene | 204,000 |
| Chemicals | 94,000 |
| Tar products | 204,000 |
| Ammonia (as N) | 110,000 |
| Sulfur | 99,000 |
| Total saleable products | 2,361,000 |

*Source:* Wender, I., Review of Indirect Liquefaction, in Coal Liquefaction, USDOE Contract DE-AC01-87-ER 30110, Ed., Schindler, H.D. (Chairman of COLIRN), Final Report, March 1989.

### 3.4.1.3 Fischer–Tropsch Processes Other than SASOL

There have been a great number of publications and patents on other FTS catalysts, mainly cobalt, ruthenium, nickel, rhodium, and molybdenum. However, none of these has been commercially verified either by SASOL or by other efforts.

Modern gasifiers, as discussed in Chapter 2, produce syngas with low (0.6–0.7) $H_2/CO$ ratios, which may be labeled as CO-rich syngas. Iron is known as a good WGS catalyst, whereas neither cobalt nor ruthenium is active for the WGS reaction. In the absence of a WGS reaction, a syngas with an $H_2/CO$ ratio of 2 would be preferably needed to produce olefins or alcohols [19]. For synthesis of paraffins, an $H_2/CO$ ratio of larger than 2 would be required. When water is formed in the FTS reaction, it can react with CO to form more $H_2$ by the WGS reaction, so that syngas with a low $H_2/CO$ ratio can still be used with iron-based catalysts. This is the reason why the WGS reaction is very important in the FTS.

Recent efforts involve slurry Fischer–Tropsch reactors. Mobil (currently ExxonMobil), in the 1980s, studied upgrading a total vaporous Fischer–Tropsch reactor effluent over zeolite Socony Mobil-5 (ZSM-5) catalyst. Shell, in 1985, announced its Shell Middle-Distillate Synthesis (SMDS) process for the production of kerosene and gas oil from natural gas [20]. This two-stage process involves the production of long-chain hydrocarbon waxes and subsequent hydroconversion and fractionation into naphtha, kerosene, and gas oil. A commercial-scale Shell facility in Bintulu, Malaysia, converts natural gas into low-sulfur diesel fuels and food-grade wax, and its capacity is 12,000 barrels per day. UOP (formerly, Universal Oil Products) characterized Fischer–Tropsch wax and its potential for upgrading [20]. Dow has developed molybdenum catalysts with a sulfur tolerance up to about 20 ppm. The catalyst system is selective for the synthesis of $C_2$–$C_4$ hydrocarbons, especially when promoted with 0.5–4.0 wt% potassium [19].

Using a precipitated iron catalyst, the slurry Fischer–Tropsch reactor operating with a finely divided catalyst suspended in an oil reactor medium has been shown to yield high single-pass syngas conversion with low $H_2$/CO ratios [19–22]. A great number of studies involving three-phase slurry reactors have been published. Development of the FTS process has scientifically contributed to the design and analysis of multiphase reactor systems.

Most recently, China and South Africa announced a major collaborative project in indirect coal liquefaction [23]. Shenhua group, China's largest coal producer, and SASOL of South Africa are the main entities involved in this venture. The new Chinese plants by Shenhua CTL, once completed, would have a daily capacity of 80,000 barrels per day (or 13,000 $m^3$/day) [23,24].

### 3.4.2 CONVERSION OF SYNGAS TO METHANOL

The synthesis of methanol from syngas is a well-established technology. Because liquid oxygenated hydrocarbon of methanol is synthesized from coal via syngas as an intermediate, the coal-to-methanol synthesis process is classified as an indirect coal liquefaction process. Synthesis gas is typically produced via gasification of coal, by biomass gasification, or via steam reforming of natural gas. Therefore, the profitability of a methanol plant is made on a case-by-case basis to account for location-specific and infrastructural factors such as available energy resources, regional consumption infrastructure, environmental impact, and capital cost. Methanol plants exist where there are large reserves of competitively priced natural gas or coal, or where there are large captive uses for product methanol by neighboring chemical and petrochemical plants.

The advent of methanol synthesis has given a boost to the value of natural gas. Conventional methane steam reforming (MSR) produces hydrogen-rich syngas at low pressure. This process is well suited to the addition of carbon dioxide, which utilizes the excess hydrogen and hence increases the methanol productivity [25].

The global methanol demand increased about 8%/year from 1991 to 1995, then slowed down to 3%–4%/year over the next 10-year period following 1995. In 2005, the global demand for methanol amounted to about 32 million tons per year, whereas in 2010, the global demand for methanol totaled about 45.6 million metric tons and was expected to exceed 60 million metric tons in 2014. The growth rate of methanol demand has been far exceeding that of the gross domestic product (GDP). The worldwide combined production capacity as of 2013 is about 75 million metric tons of over 90 methanol plants.

Due to the decline and phaseout of methyl t-butyl ether (MTBE) as an oxygenated gasoline additive in recent years, the regional demand for methanol has suffered in some countries, especially in North America and Europe. Nonetheless, the global demand for methanol is still expected to grow steadily and strongly as a basic building-block chemical, transportation fuel, solvent, and so forth. Asia, especially China, will be the main driver for growth in the demand of methanol and its derivatives. Average growth rates for Asia are expected to be 3.8% for methanol, 4.8% for acetic acid, and 4.4% for formaldehyde [26]. The breakdown of the methanol demand is given in Table 3.17 [26]. A detailed analysis of the methanol market is published annually by Chemical Market Associates, Inc. (CMAI) [27]. This report provides information on supply, demand, production, history, and forecasts for methanol capacity, trade, and pricing.

All industrially produced methanol is made by the catalytic conversion of synthesis gas containing carbon monoxide, carbon dioxide, and hydrogen as the main components. Modern commercial methanol processes can be classified into vapor-phase low-pressure synthesis and liquid-phase low-pressure synthesis [25,28,29]. The former is more conventional and dominates in the current marketplace. This low-pressure vapor-phase synthesis process replaced its earlier version of high-pressure technology and is more suited for $H_2$-rich synthesis gas of typical $H_2/CO$ ratio ranging between 2 and 3 [25,30]. The liquid-phase methanol synthesis technology is more recent in its development and more suitable for CO-rich synthesis gas of the typical $H_2/CO$ ratio ranging from 0.6 to 0.9. As discussed in Chapter 2, CO-rich syngas is produced typically by advanced coal gasifiers.

Methanol productivity can be enhanced by synthesis gas enrichment with additional carbon dioxide to a certain limit [25]. There is an optimal concentration of carbon dioxide, which is dependent upon the process type (vapor phase vs. liquid

---

**TABLE 3.17**
**Methanol Demand by Chemicals and End Uses**

| Chemicals and End Uses | % |
|---|---|
| Formaldehyde | 36 |
| Methyl t-butyl ether (MTBE) | 25 |
| Acetic acid | 11 |
| Solvents | 4 |
| Methyl methacrylate (MMA) | 3 |
| Gas and fuels | 3 |
| Dimethyl terephthalate (DMT) | 2 |
| Others (manufacturing other chemicals) | 16 |
| Total | 100 |

*Source:* Web site by Lurgi on Methanol market and Technology, 2006; accessible through http://www.lurgi.de/lurgi_headoffice_kopie/english/nbsp/menu/products/gas_to_petrochemicals_and_fuels/methanol/markets/index.html. With permission.

phase), synthesis gas feed compositions, and operating temperature and pressure conditions. However, if too much $CO_2$ is present in the syngas, it accelerates catalyst deactivation, shortens its lifetime, adversely affects the syngas conversion, and produces an excessive amount of water, which adversely affects the catalyst matrix stability, resulting in crystallite growth via hydrothermal synthesis phenomena [25]. Although this statement is generally true for low-pressure synthesis of methanol, it was also found that a high concentration of $CO_2$ enhances the catalyst structural stability by formation of $ZnCO_3$ on the original $Cu/ZnO/Al_2O_3$ catalyst. As a different approach, a special catalyst has also been designed to operate under high $CO_2$ conditions. The catalyst's crystallites are located on energetic stable sites that lower the tendency to migrate. This stability also minimizes the influence of water *in situ* formed on the catalyst matrix. This catalyst preserves its higher activity due to a lower deactivation rate over long-term operations [25].

The basic stoichiometric reactions involved in methanol synthesis are

$$CO_2 + 3H_2 = CH_3OH + H_2O \quad \Delta H^\circ_{298} = -52.81 \text{ kJ/mol} \tag{3.1}$$

$$CO + 2H_2 = CH_3OH \quad \Delta H^\circ_{298} = -94.08 \text{ kJ/mol} \tag{3.2}$$

$$CO + H_2O = CO_2 + H_2 \quad \Delta H^\circ_{298} = -41.27 \text{ kJ/mol} \tag{3.3}$$

Of the three reactions, only two are stoichiometrically independent. In other words, material balance of the above reaction system would only require any two of the three stoichiometric equations. It should be clearly noted that this does not mean that the actual chemical reactions are proceeding via the two chosen reactions. Rather, it is a mere mathematical consequence of the linear independence of stoichiometric equations.

The chemical mechanism of the methanol synthesis over the $Cu/ZnO/Al_2O_3$ catalyst has been quite controversial [19,22,25]. The controversy involved is whether the synthesis of methanol over the $Cu/ZnO/Al_2O_3$ goes predominantly via $CO_2$ hydrogenation (Equation 3.1) or via CO hydrogenation (Equation 3.2). Along with the synthesis reaction, the second companion reaction was automatically in the middle of the controversy. In this case, the controversy was whether the WGS reaction proceeds forward or backward under normal synthesis conditions over the very same catalyst. However, more experimental evidence points toward the theory that methanol synthesis over the $Cu/ZnO/Al_2O_3$ proceeds predominantly via the $CO_2$ hydrogenation and the forward WGS reaction [20,25,28,31,32] as

$$CO_2 + 3H_2 = CH_3OH + H_2O$$

$$CO + H_2O = CO_2 + H_2$$

More detailed discussions on the methanol synthesis technology are available in Chapter 10.

### 3.4.3 CONVERSION OF METHANOL TO GASOLINE OR TARGET HYDROCARBONS

Methanol itself can be used as a transportation fuel just as liquefied petroleum gas (LPG) and ethanol. However, direct use of methanol as a motor fuel in passenger vehicles would require nontrivial engine modifications and substantial changes in the lubrication system. Even though methanol has a high octane rating and is, by molecular formula, an excellent candidate for oxygenated hydrocarbon, its use as a gasoline blending chemical is also limited due to its high Reid vapor pressure (RVP), which is a measure of affected volatility of blended gasoline (ASTM-D-323). This is one of the reasons why the conversion of methanol to gasoline is quite appealing [33,34].

The Mobil Research and Development Corporation developed the methanol-to-gasoline (MTG) process in the early 1970s. The process technology is based on the catalytic reactions using the zeolites of the ZSM-5 class [35,36]. MTG reactions can be written as

$$n\ CH_3OH \rightarrow (-CH_2-)_n + nH_2O$$

The detailed reaction path is described in the work of Chang [35]. The following simplified steps describe the overall reaction path:

$$2CH_3OH = (CH_3)_2O + H_2O$$

$$(CH_3)_2O \rightarrow \text{light olefins} + H_2O$$

$$\text{Light olefins} \rightarrow \text{heavy olefins}$$

$$\text{Heavy olefins} \rightarrow \text{paraffins}$$

$$\text{aromatics}$$

$$\text{naphthenes}$$

The MTG reactions are exothermic and go through the DME intermediate route. The conversion of methanol to DME is via a catalytic dehydration process.

Based on the shape-selective pore structure of the ZSM-5-class catalysts, the product hydrocarbons can be tailor-made to fall predominantly in the gasoline boiling range. The product distributions are generally influenced by the temperature, pressure, space velocity, reactor type, and Si/Al ratio of the catalyst [37]. Paraffins are dominated by isoparaffins, whereas aromatics are dominated by highly methyl-substituted aromatics. $C_9^+$ aromatics are dominated by symmetrically methylated isomers, reflecting the shape-selective nature of the catalyst. The $C_{10}$ aromatics are mostly durene (1,2,4,5-tetramethylbenzene), which has an excellent octane number, but the freezing point is very high at 79°C. Too high a durene content in the gasoline may impair automobile driving characteristics, especially in cold weather, due to its tendency to crystallize at a low temperature [14]. Mobil's test found no drivability loss at −18°C using a synthetic gasoline containing 4 wt% durene [14]. Mobil also developed a heavy gasoline treating (HGT) process to convert durene into other high-quality gasoline components by isomerization and alkylation [37].

Basically, three types of chemical reactors were developed for the MTG process: (1) adiabatic fixed bed, (2) fluidized bed, and (3) direct heat exchange. The first two were developed by Mobil and the last by Lurgi.

The adiabatic fixed-bed concept uses a two-stage concept, namely, (1) the first-stage DME reactor and (2) the second-stage DME conversion to hydrocarbons. The first commercial plant of 14,500 barrels per day gasoline capacity was constructed in New Zealand. The plant had been running successfully from its 1985 startup until 1995, when the facility was converted into chemical-grade methanol production only. The synthesis gas is generated via steam reforming of natural gas obtained from the offshore Maui fields. The successful MTG operation in New Zealand also achieved the reduction of durene content to 2 wt%. Therefore, the successful operation of MTG in New Zealand was a very important milestone in the history of synthetic fuels, as it made possible the chemical synthesis of gasoline from unlikely fossil fuel sources, like natural gas and coal. By this technology demonstration on a commercial scale, it has been firmly established that petroleum crude is no longer the sole source for gasoline or diesel. Although the plant for MTG in New Zealand ceased its operation in the late 1990s, a major plant complex for methanol-to-olefins (MTO) is underway in Nigeria by Singapore-based Eurochem with UOP as the technology provider [27,38].

A fluidized-bed MTG concept was concurrently developed by Mobil. The exothermic heat of reaction can be removed from the reactor either directly using a cooling coil or indirectly using an external catalyst cooler. The process research went through several stages involving a bench-scale, fixed bed, and fluidized bed, or in terms of production capacities, 4 bpd, 100 bpd cold-flow models, and a 100 bpd semiwork plant. Table 3.18 shows typical MTG process conditions and product yields [14].

During the MTG development, Mobil researchers found that the hydrocarbon product distribution can be shifted to light olefins by increasing the space velocity, decreasing the methanol partial pressure, and increasing the reaction temperature [39]. Typical yields [14] from the 4 bpd operation were $C_1$–$C_3$ paraffins, 4 wt%; $C_4$ paraffins, 4 wt%; $C_2$–$C_4$ olefins, 56 wt%; and $C_5^+$ gasoline, 35 wt%. Using olefins from the MTO or FTS processes, diesel and gasoline can be made via a process converting olefins to diesel and gasoline. Using acid catalysts, catalytic polymerization is a standard process and is being used at SASOL to convert $C_3$–$C_4$ olefins into gasoline and diesel. Recently, Mobil developed an olefins-to-gasoline-and-diesel (MOGD) process using their commercial zeolite catalyst [40,41]. Lurgi also developed its own version of the methanol-to-propylene (MTP) process [8].

Recently, an innovative process enhancement has been made by Lee and coworkers [42] under the sponsorship of the EPRI. Their process, called the DME-to-Gasoline (DTG) process (DME-to-Hydrocarbons [DTH] or DME-to-Olefins [DTO] process, depending upon the final product), is based on the direct conversion of DME to hydrocarbon over ZSM-5 type catalyst [42,43]. This process is built upon the novel, economical, single-stage synthesis process of DME from syngas, which produces methanol only as an intermediate for DME. By producing DME in a single stage, the intermediate methanol formation is no longer limited by chemical equilibrium,

**TABLE 3.18**

**Typical Process Conditions and Product Yields for the MTG Process**

| Conditions | Fixed-Bed Reactor | Fluid-Bed Reactor |
|---|---|---|
| Methanol/water charge (w/w) | 83/17 | 83/17 |
| Dehydration reactor inlet T (°C) | 316 | — |
| Dehydration reactor outlet T (°C) | 404 | — |
| Conversion reactor inlet T (°C) | 360 | 413 |
| Conversion reactor outlet T (°C) | 415 | 413 |
| P (kPa) | 2170 | 275 |
| Recycle ratio (mol/mol charge) | 9.0 | — |
| Weight hourly space velocity (WHSV) | 2.0 | 1.0 |
| Yields (wt% of MeOH charged) | | |
|    MeOH + dimethyl ether | 0.0 | 0.2 |
|    Hydrocarbons | 43.4 | 43.5 |
|    Water | 56.0 | 56.0 |
|    $CO$, $CO_2$ | 0.4 | 0.1 |
|    Coke, other | 0.2 | 0.2 |
| Total | 100.0 | 100.0 |
| Hydrocarbon product (wt%) | | |
|    Light gas | 1.4 | 5.6 |
|    Propane | 5.5 | 5.9 |
|    Propylene | 0.2 | 5.0 |
|    Isobutane | 8.6 | 14.5 |
|    $n$-Butane | 3.3 | 1.7 |
|    Butenes | 1.1 | 7.3 |
|    $C_5$+ gasoline | 79.9 | 60.0 |
| Total | 100.0 | 100.0 |
| Gasoline (including alkylate), RVP—62 kPa (9 psi) | 85.0 | 88.0 |
| LPG | 13.6 | 6.4 |
| Fuel gas | 1.4 | 5.6 |
| Total | 100.0 | 100.0 |
| Research octane number (RON) of gasoline | 93 | 97 |

*Source:* US DOE Working Group on Research Needs for Advanced Coal Gasification Techniques (COGARN) (S.S. Penner, Chairman), Coal Gasification: Direct Application and Synthesis of Chemicals and Fuels, DOE Contract No. DE-AC01-85 ER30076, DOE Report DE/ER-0326, June 1987.

thus substantially increasing the reactor productivity, in terms of total hydrogenation extent. This is especially true for the synthesis of methanol in the liquid phase. Furthermore, by feeding DME directly to the ZSM-5 reactor instead of methanol, the stoichiometric conversion and hydrocarbon selectivity increase substantially due to less water formation and its detrimental involvement. The difference between MTG and DTG, therefore, is in the placement of methanol dehydration reaction step (i.e., DME formation reaction). In the MTG, methanol-to-DME conversion takes place in the gasoline reactor, whereas methanol-to-DME conversion, in the DTG, takes place in the syngas conversion reactor. Therefore, methanol is an intermediate of the syngas conversion reactor for DTG, whereas DME is an intermediate for the gasoline synthesis reactor for MTG. The DTG process has not yet been tested on a large scale.

A different approach, an improved version of MTG process, the Topsoe Integrated Gasoline Synthesis Process (TIGAS), uses combined steam reforming and autothermal reforming for syngas production with a multifunctional catalyst system to produce an oxygenated hydrocarbon mixture rather than methanol [19]. The TIGAS process eliminates the requirement of upstream methanol production and intermediate storage [44].

### 3.4.4 HIGHER ALCOHOL SYNTHESIS

Mixtures of $C_1$–$C_6$ alcohols can be used as transportation fuels either as is or as an additive to gasoline. In the United States, however, selling new unleaded fuels or fuel additives in unleaded fuels has been prohibited by the Clean Air Act. Exceptions have been granted in the form of EPA waivers, and good examples have been the waivers granted to requests by DuPont, ARCO, SUN, and American Methyl in the late 1970s through the mid-1980s. Some of the more significant ones that are still impacting the current fuel market in the United States are related to the phased-out MTBE blending and the use of 10% ethanol in gasoline (E10). Due to the public health and environmental problems cited in a number of states in the United States and Europe, MTBE has been phased out completely in these countries.

Technical advantages of using $C_1$–$C_6$ alcohol blends with gasoline can be summarized as follows:

1. Enhancement of octane number
2. Enhancement in hydrocarbon solubility in comparison to methanol–gasoline blends
3. Enhanced water tolerance compared to unblended gasoline
4. Enhanced control of fuel volatility

Despite some obvious technological benefits, certain EPA restrictions, especially volatility specifications (evaporative index [EI] or RVP), have imposed serious economic penalties on alcohol blends (except ethanol), thus making them difficult to be accepted by refiners and blenders. Fuels containing higher alcohol blends have been in use in Germany at approximately 3–5 mol% for automobile transportation.

Higher alcohol synthesis (HAS) has been practiced in Germany since 1913 after BASF successfully developed cobalt- or osmium-catalyzed synthesis of a mixture of alcohols and other oxygenates at 10–20 MPa and 300°C–400°C [8]. This was followed by the Fischer–Tropsch Synthol process for alcohol mixtures in 1923–1924. It was also found that higher alcohols were coproducts of methanol synthesis over $ZnO/Cr_2O_3$ catalysts, alkalized $ZnO/Cr_2O_3$ catalysts, and alkalized Cu-based catalysts. Later, the Synthol process was further developed and enhanced to a process at a lower temperature of <200°C, medium pressure of 20 MPa, and inexpensive but potent iron catalysts. Later, the process incorporated several additional reactor stages with intermediate $CO_2$ removal and gas recycle [45]. In 1984, Dow Chemical Co.

announced a new process for HAS based on molybdenum sulfide ($MoS_2$) catalysis, and Union Carbide Corporation also revealed a new process [8]. (The two companies merged in 2000 for unrelated business reasons.) The Dow Chemical process is also known as Dow HAS. On the other hand, the technology for higher alcohol based on alkali-promoted $ZnO/Cr_2O_3$ methanol synthesis catalysts for the high-pressure methanol synthesis was further developed by Snamprogetti, Enichem, and Haldor Topsoe A/S (SEHT). This is often referred to as the SEHT HAS process.

## 3.5 COAL AND OIL COPROCESSING

Coprocessing is defined as the simultaneous reaction treatment of coal and petroleum resid, or crude oil, with hydrogen to produce distillable liquids. More strictly speaking, this technology may be classified under direct liquefaction as a variation. Petroleum liquids have often been used as a liquefaction solvent, mainly for startup or whenever coal-derived liquids were unavailable. However, some serious considerations have been recently given to the processing possibilities of hydrocracking petroleum resid while liquefying coal in the same reactor. In this sense, coprocessing has an ultimate objective of cobeneficiation.

An early coprocessing patent was granted to UOP, Inc., in 1972 for a process whereby coal is solvent-extracted with petroleum [46]. Another early patent on coprocessing was issued to HRI in 1977 for the single-stage ebullated-bed COIL process based on the HRI's H-Oil and H-Coal technology [47]. Consol R&D tested the use of a South Texas heavy oil for coal hydroextraction but found that, even after hydrogenation, the petroleum made a very poor liquefaction solvent [14]. The Canada Centre for Mineral and Energy Technology (CANMET) developed the CANMET hydrocracking process for petroleum resids. They found that small additions of coal (<5 wt%) to the petroleum feedstock significantly improved distillate product yields. A 5000 bpd plant using this process was started up in 1985 by Petro-Canada near Montreal, Quebec [48,49].

In summary, coprocessing has several potential economic and technological advantages relative to coal liquefaction or hydroprocessing of heavy petroleum residua. Synergisms and cobeneficiating effects can be achieved, especially in the area of (1) replacement of recycle oil, (2) sharing hydrogen between hydrogen-rich and hydrogen-deficient materials, (3) aromaticity of the product, (4) demetalation and catalyst life extension, and (5) overall energy efficiency [49]. For the current technology, temperatures of 400°C–440°C, 2000 psig hydrogen pressure, and alumina-supported cobalt, molybdenum, nickel, or disposable iron catalysts are frequently used. Various efforts in developing more selective and resilient catalysts are being executed.

## REFERENCES

1. J. G. Speight, *The Chemistry and Technology of Coal*. New York, NY: Marcel Dekker, 1994.
2. R. F. Probstein and R. E. Hicks, *Synthetic Fuels*. New York, NY: McGraw-Hill, 1982.
3. F. H. Schoemann, H. D. Terzian, J. F. Jones, L. J. Scott and N. J. Brunsvold, "Char oil energy development. vol. I. Final report," FMC Corp, Princeton, NJ, Tech. Rep. FE-1212-T-9, 1975.
4. S. Lee, *Oil Shale Technology*. Boca Raton, FL: CRC Press, 1991.

5. T. F. Johnson, K. C. Krupinski and R. J. Osterholm, "Clean coke process: Fluid-bed carbonization of Illinois coal," in *Symposium on New Coke Processes Minimizing Pollution, Division of Fuel Chemistry, ACS*, Chicago, IL, 1975, pp. 33–45.

6. Ralph M. Parsons Company, "Coalcon process review. R and D interim report no. 3." Pasadena, CA, Tech. Rep. ORNL/SUB-7186/16, 1978.

7. HTI direct coal liquefaction technology, a webpage. Available at http://www.htigrp.com/data/upfiles/pdf/DCL\%20.

8. H. D. Schindler, "Coal liquefaction—A research and development needs assessment, COLIRN panel assessment, final report, vol. II," Tech. Rep. DOE/ER-0400, UC-108, 1989.

9. J. D. Potts, K. E. Hastings and R. S. Chillingworth, "Expanded bed hydroprocessing of solvent refined coal (SRC) extract. Total reactor pressure and space velocity parameters and short contact time coal extract feedstocks (deashed and non-deashed). Interim technical progress report." Tech. Rep. FE-2038-42, 1980.

10. H. D. Schindler, J. M. Chen and J. D. Potts, Integrated Two-Stage Liquefaction, Final Tech. Report, DOE Contract DE-AC22-79 ET14804, June 1993.

11. G. J. Snell and M. C. Sze, "Antisolvent deashing process for coal liquefaction product solutions," Canadian Patent No. 1060828, August 21, 1979.

12. A. G. Comolli and J. B. McLean, "The low-severity catalytic liquefaction of Illinois no. 6 and wyodak coals," in *Proceedings of the 2nd Annual Pittsburgh Coal Conference*, Pittsburgh, PA, 1985.

13. W. Weber and N. Stewart, "EPRI Monthly Review," January 1987.

14. US DOE Working Group on Research Needs for Advanced Coal Gasification Techniques (COGARN) (S.S. Penner, Chair), Coal Gasification: Direct Application and Synthesis of Chemicals and Fuels, DOE Contract no. DE-AC01-85 ER30076, DOE Report DE/ER-0326, June 1987.

15. J. C. Hoogendoorn and J. M. Salomon, "Sasol: World's largest oil-from-coal plant II," *Br. Chem. Eng.*, vol. 2, pp. 238–244, 1957.

16. R. B. Anderson, H. Kölbel and M. Rálek, *The Fischer–Tropsch Synthesis*. Orlando, FL: Academic Press, 1984.

17. G. V. Schulz, *Z. Phys. Chem.*, vol. 32, p. 27, 1936.

18. P. J. Flory, "Molecular size distribution in linear condensation polymers," *J. Am. Chem. Soc.*, vol. 58, pp. 1877–1885, 1936.

19. I. Wender, "Review of indirect liquefaction, in coal liquefaction, final report," Schindler, H. D. (chairman of COLIRN), Tech. Rep. USDOE Contract DE-AC01-87-ER 30110, 1989.

20. M. J. Van der Burgt, J. V. Klinken and S. T. Sie, "The shell middle distillate synthesis process," in *Synfuels' 5th Worldwide Symposium: Shoreham Hotel, Washington, D.C., November 11–13, 1985*, K. J. Hamilton, Ed. New York, NY: McGraw-Hill, 1985.

21. H. Kölbel and M. Rálek, "The Fischer–Tropsch synthesis in the liquid phase," *Catal. Rev. Sci. Eng.*, vol. 21, pp. 225–274, 1980.

22. J. Humbach and N. W. Schoonover, *Indirect Liquefaction Contractors Meeting*, Pittsburgh Energy Technology Center, Pittsburgh, PA, 1985, pp. 29–38.

23. K. Silverstein, Coal liquefaction plants spark hope, daily issue alert. Available at http://www.utilipoint.com/issuealert/article.asp?id=2314, November 1, 2004.

24. A. Begum, "Sasol and Shenhua prime CTL pumps," *Upstream Online (NHST Media Group)*, June 22, 2009.

25. S. Lee, *Methanol Synthesis Technology*. Boca Raton, FL: CRC Press, 1990.

26. Methanol market and technology, a website by Lurgi GmbH. Available at http://www.lurgi.de/lurgi\_headoffice\_kopie/english/nbsp/menu/products/gas\_to\_.

27. Chemical Market Associates Inc, "2006 world methanol analysis," CMAI, Houston, TX, 2006.

28. A. Cybulski, "Liquid-phase methanol synthesis: Catalysts, mechanism, kinetics, chemical equilibria, vapor–liquid equilibria, and modeling—A review," *Catal. Rev. Sci. Eng.*, vol. 36, pp. 557–615, 1994.

29. S. Lee, "Research to support liquid phase methanol synthesis process development," Electric Power Research Institute, Palo Alto, CA, Tech. Rep. EPRI AP-4429, 1986.

30. S. S. Öztürk, Y. T. Shah and W. Deckwer, "Comparison of gas and liquid phase methanol synthesis processes," *Chem. Eng. J.*, vol. 37, pp. 177–192, 3, 1988.

31. K. G. Chanchlani, R. R. Hudgins and P. L. Silveston, "Methanol synthesis from H2, CO and CO2 over Cu/ZnO catalysts," *J. Catal.*, vol. 136, pp. 59–75, 1992.

32. G. D. Sizgek, H. E. Curry-Hyde and M. S. Wainwright, "Methanol synthesis over copper and ZnO promoted copper surfaces," *Appl. Catal. Gen.*, vol. 115, pp. 15–28, 1994.

33. G. A. Mills, "Status and future opportunities for conversion of synthesis gas to liquid fuels," *Fuel*, vol. 73, pp. 1243–1279, 1994.

34. J. M. Fox, "The different catalytic routes for methane valorization: An assessment of processes for liquid fuels," *Catal. Rev. Sci. Eng.*, vol. 35, pp. 169–212, 1993.

35. C. D. Chang, "Hydrocarbons from methanol," *Catal. Rev. Sci. Eng.*, vol. 25, pp. 1–118, 1983.

36. C. D. Chang and A. J. Silvestri, "The conversion of methanol and other O-compounds to hydrocarbons over zeolite catalysts," *J. Catal.*, vol. 47, pp. 249–259, 1977.

37. C. D. Chang and A. J. Silvestri, "The MTG process: Origin and evolution," in *21st State-of-the-Art ACS Symposium on Methanol as a Raw Material for Fuels and Chemicals*, Marco Island, FL, 1986, pp. 115–118.

38. A. Jagger, "Singapore's Eurochem delays Nigeria olefins project—UOP," *ICIS News*, January 2011.

39. R. F. Socha, C. T. W. Chu and A. A. Avidan, "An overview of methanol-to-olefins research at Mobil: From conception to demonstration plant," in *21st State-of-the-Art ACS Symposium on Methanol as a Raw Material for Fuels and Chemicals*, Marco Island, FL, 1986.

40. S. A. Tabak and F. J. Krambeck, "Shaping process makes fuel," *Hydrocarbon Process.*, vol. 64, pp. 72–74, 1985.

41. S. A. Tabak, A. A. Avidan and F. J. Krambeck, "MTO-MOGD process," in *21st State-of-the-Art ACS Symposium on Methanol as a Raw Material for Fuels and Chemicals*, Marco Island, FL, 1986.

42. S. Lee, M. R. Gogate, K. L. Fullerton and C. J. Kulik, "Catalytic process for production of gasoline from synthesis gas," U.S. Patent No. 5,459,166, October 17, 1995.

43. S. Lee, M. Gogate and C. J. Kulik, "Methanol-to-gasoline vs. DME-to-gasoline, II. Process comparison and analysis," *Fuel Sci. Technol. Int.*, vol. 13, pp. 1039–1057, 1995.

44. Gasoline—TIGAS. A website by Haldor-Topsoe, available at http://www.topsoe.com/ business_areas/gasification_based/Processes/Gasoline_TIGAS.aspx.

45. X. Xiaoding, E. B. M. Doesburg and J. J. F. Scholten, "Synthesis of higher alcohols from syngas—Recently patented catalysts and tentative ideas on the mechanism," *Catal. Today*, vol. 2, pp. 125–170, 1987.

46. J. G. Gatsis, "Solvent extraction of coal by a heavy oil," U.S. Patent No. 3,705,092, December 5, 1972.

47. M. C. Chervenak and E. S. Johanson, "Catalytic hydrogenation of blended coal and residual oil feeds," U.S. Patent No. 4,054,504, October 18, 1977.

48. J. F. Kelly and S. A. Fouda, "CANMET coprocessing: An extension of coal liquefaction and heavy oil hydrocracking technology," in *DOE Direct Liquefaction Contractors' Review Meeting*, Albuquerque, NM, 1984.

49. S. Lee, *Alternative Fuels*. Washington, DC: Taylor & Francis, 1996.

# 4 Coal Slurry Fuel

*Sunggyu Lee*

## CONTENTS

## 4.1 INTRODUCTION

Coal slurry fuels consist of finely ground coal particles dispersed into one or more liquids such as water, oil, or methanol. Slurry fuels have the advantages of being convenient to handle as liquid fuel (similar to heavy fuel oil) and possessing high energy density, as illustrated in Table 4.1.[1,2] Coal slurries have been investigated as a potentially efficient replacement for oil in boilers and furnaces, fuel in internal combustion engines, and recently, energy feedstock for cofiring of coal fines in utility boilers. Coal slurry is used around the world in countries such as the United States, Russia, Japan, China, and Italy.

Coal slurry fuels have been investigated since the nineteenth century, but economic constraints have kept it from becoming a major energy source. Typically, interest in coal slurry develops whenever regional or short-term oil availability is in doubt, such as periods during both world wars and again in the energy crises of 1973 and 1979.[3] Much of the work during these time periods was focused on coal–oil fuels, which could quickly and readily replace oil or liquid fuel in furnaces and boilers. However, recent research, since 1980, has concentrated more on coal–water slurry fuels (CWSFs) for the complete replacement of oil in industrial steam boilers, utility boilers, blast furnaces, process kilns, and diesel engines.[4,5]

The initial development of coal slurry was noted over a hundred years ago by Smith and Munsell.[4,6] By World War I, a full-scale slurry test was successfully made

**TABLE 4.1**

**Fuel Energy Densities**

| Fuel | Density (lb/gal) | Btu/lb | Btu/gal | Btu/ft³ |
|---|---|---|---|---|
| Coal in bulk (7% moisture) | 6.2–9.4 | 12,500 | 76,000–116,500 | 573,000–872,000 |
| Residual oil | 8.2 | 18,263 | 150,000 | 1,122,000 |
| 60% coal/40% water blend | 9.8 | 8000 | 78,700 | 589,000 |
| 70% coal/30% water blend | 10.2 | 9373 | 95,600 | 715,000 |

*Source:* Kesavan, S., Stabilization of Coal Particle Suspensions using Coal Liquids, M.S. thesis, University of Akron, 1985. With permission.

on a U.S. Navy scout ship. The test revealed some technical problems such as high ash content (of raw material coal), visible fluid track left in the ship's wake, and stability problems with the slurry (such as settling and sedimentation).[7]

In the 1930s, the Cunard ship company used coal–oil slurries on both land and sea trials in an attempt to reduce oil imports and further develop new markets for coal.[8–10] At about the same time in Japan, tests were conducted on coal-in-oil fuels at their National Fuel Laboratory.[11] Similar tests were also performed in Germany on a mixture of powdered coal (55 wt%) and tar oil (45 wt%) called "Fliesskhloe." The German tests showed that the coal–oil mixture (COM) burned well and had thermal efficiencies of 70%–75%.[12] Although the systems worked well from technological standpoints, economic limitations hindered further development.

Development during the Second World War consisted of two comprehensive programs at the Bureau of Mines and Kansas State College. The programs explored methods of preparation, flow, stability, and burning processes of coal–oil slurry.[7] After the war, development on coal–oil slurries ceased until the 1970s. On the other hand, work on coal–water fuels started in the Union of Soviet Socialist Republics in the 1950s,[13,14] and similar work was conducted in the United States and Germany on storage, pumping, and combustion properties.[15]

The energy crisis of 1973 again propelled research and development (R&D) of coal–oil slurry. A consortium of companies led by General Motors (GM) was formed in 1973 to develop the technology.[16] In 1975, the Department of Energy (DOE) joined support of the project, and by 1976, the program had expanded into switching utility gas and oil boilers to COMs. The initial comprehensive investigations were completed in 1977, and the DOE transferred GM projects to places such as New England Power and Service Co. (NEPSCO) and Pittsburgh Energy Technology Center (PETC).[7]

However, R&D efforts since 1980 have centered on CWSFs for replacement of oil.[4,17] The economic incentive for replacing oil with coal–oil slurries disappeared as oil prices stabilized in the 1980s and 1990s. This spurred development of coal–water slurries (CWSs) for complete replacement of oil or coburning of coal fines in a slurry. Record-high oil prices in the twenty-first century will rekindle interest in coal slurry fuel as a potential transportation fuel and as an alternative fuel for diverse applications. A significant amount of development has been accomplished on coal slurry rheology, characterization, atomization, combustion mechanisms, and transport techniques.[5]

## 4.2   COAL SLURRY CHARACTERIZATION

Coal slurry mixtures can be made from a combination of various liquids, the most common liquid ingredients being oil, water, and methanol. Detailed descriptions for various types of coal slurries are as follows:

1. *COMs*: a suspension of coal in fuel oil, also referred to as coal–oil dispersions (COD).
2. *Coal–oil–water (COW)*: a suspension of coal in fuel oil with less than 10 wt% water in which oil is the main ingredient.
3. *Coal–water–oil (CWO)*: a suspension of coal in fuel oil with more than 10 wt% water in which water is the main ingredient.
4. *Coal–water fuels (CWFs), coal–water mixtures (CWMs), CWSs, or CWSF*: a suspension of coal in water.
5. *Coal–methanol fuel (CMF)*: a suspension of coal in methanol.
6. *Coal–methanol–water (CMW)*: a suspension of coal in methanol and water.

The CMF and CMW slurries possess favorable properties; however, the cost of methanol has all but eliminated them from further development. COM, once actively investigated, has now been shelved for economic reasons. CWF has been investigated for complete oil replacement in boilers and furnaces and internal combustion engines, but fluctuating oil prices in the past decades as well as competition from biofuels have reduced the economic advantage and somewhat cooled the interest. However, CWF developed from waste streams and tailings has been investigated for cofiring in boilers and furnaces.[18] The comparative economics of CWS against the conventional fuel in the twenty-first century is undoubtedly far more favorable.[5] Furthermore, CWF provides a valuable mode of coal transportation through a long pipeline.

Important slurry characteristics are stability, pumping, atomizability, and combustion characteristics. These properties control the hydrodynamics and rheology of the coal slurry system. A coal slurry must have low viscosity at pumping shear rates ($10–200 \text{ s}^{-1}$) and at atomization shear rates ($5,000–30,000 \text{ s}^{-1}$). This allows for low pumping power requirements and increased boiler and furnace efficiencies through smaller droplets sizes.[19] There are several types of pumps that are developed for coal slurry pumping, e.g., a progressive cavity pump that is designed to pump the highly abrasive, viscous, and corrosive coal slurry.

In order to understand coal slurry hydrodynamics and rheology, an understanding of dispersed systems is required. Solid–liquid dispersed systems are classified into two types, based on their particle sizes, namely, *colloidal* and *coarse-particle* systems. Colloidal dispersed systems consist of particles smaller than 1 μm, whereas coarse-particle dispersion systems (suspension) consist of particles larger than 1 μm. In colloidal dispersion systems, sedimentation is prevented by Brownian motion (thermal activity). However, suspensions are thermodynamically unstable and will tend to precipitate owing to the overwhelming gravitational force on large-size particles.

In certain regions and industries, where abundant supply of petroleum coke (or, petcoke) is available, coal can be replaced by petroleum coke as the slurrying solid fuel. Petroleum coke–oil slurry (PCOS), which is a slurry mixture between petroleum coke and oil, has received growing interest, due to its similarity with coal–oil slurry as well as its acceptability in certain internal combustion engines and fuel boilers.[20]

### 4.2.1 Particle Size Distribution

Typical coal slurry fuels have a particle size distribution (PSD) with 10%–80% of the particles smaller than 74 μm (–200 mesh). Micronized CWF has a PSD with a mean particle diameter of less than 15 μm, and 98% of the particles are smaller than 44 μm (–325 mesh). This type of slurry is typically produced by coal beneficiation systems in the removal of mineral matter, mainly pyrites ($FeS_2$) and ash.

The sizing of coal is a multistep process consisting of coal crushing, pulverization, and finishing steps. Finishing steps encompass coarse, fine, and ultrafine crushing of coal. Coal crushing reduces coal size to 20–7.6 and 5–3.2 cm, depending on the application, and coarse pulverization further reduces the coal size to <3.2 mm. The finishing processes can be carried out by wet or dry grinding. They reduce the particle size to <1 mm for coarse, <250 μm for fine, and <44 μm for ultrafine grinding. In slurry preparation, wet grinding is often used to minimize oxidation of coal, which is normally detrimental to many beneficiation or treatment processes. Weathered coal is a kind of oxidized coal, which typically exhibits a higher content of sulfatic sulfur in addition to increased presence of oxygen in the coal's ultimate analysis.

Coal slurries are most economical when they have the maximum amount of coal (i.e., highest solid loading) at the lowest possible viscosity. To obtain the highest possible loading, a bimodal or multimodal PSD is utilized, as shown in Figure 4.1.[2] The finer coal particles fit into the interstices of the larger coal particles, forming a higher concentrated network of particles. These particles may also act as a lubricant, leading to a lower viscosity.[21] A unimodal slurry has a peak solid loading of ~65%, at which time the viscosity becomes infinite, whereas idealized multimodal systems offer a theoretically possible loading in excess of 80%, as shown in Figure 4.2.[22] The lowest viscosity of a CWM occurs at a fine-to-coarse ratio of 35 ± 5:65 ± 5, regardless of the ratio of the mean diameters.[21] Multimodal systems are commonly used because they can easily be generated by a common grinding scheme. A typical multimodal distribution formulated for minimum viscosity is shown in Figure 4.3.[22]

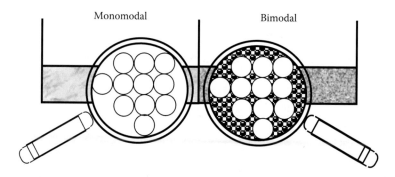

**FIGURE 4.1** PSD for unimodal and bimodal distributions. (From Kesavan, S., Stabilization of Coal Particle Suspensions using Coal Liquids, M.S. thesis, University of Akron, 1985.)

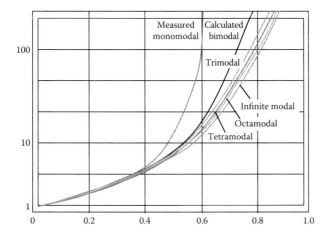

**FIGURE 4.2** PSD effect on viscosity. (From Hunter, R.J., *Foundations of Colloidal Science*, Vol. 1, Oxford University Press, Oxford, 1987.)

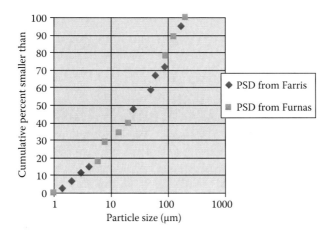

**FIGURE 4.3** PSD formulated for minimal viscosity. (From Hunter, R.J., *Foundations of Colloidal Science*, Vol. 1, Oxford University Press, Oxford, 1987.)

The settling of coal particles is a complex phenomenon from a theoretical point of view, involving hydrodynamic and physiochemical forces. The falling movement of fine coal particles in slurries has a Reynolds number (Re) << 1, which leads to the use of *Stokes' equation* for hindered settling rate, as shown in Equations 4.1 and 4.2.[3]

$$v = \frac{d^2 g (\rho_1 - \rho_2)}{18\mu} f(\phi) \tag{4.1}$$

$$f(\phi) = \frac{v}{v_t} \leq 1 \tag{4.2}$$

where
  $v$ = hindered settling rate
  $v_t$ = single-particle settling velocity
  $\mu$ = viscosity of dispersing medium
  $\rho_1$ = density of dispersed medium
  $\rho_2$ = density of dispersing medium
  $d$ = diameter of dispersed particles
  $g$ = gravitational acceleration
  $f(\phi)$ = function of volume fraction of suspended solids

Stokes' law suggests that to reduce the sedimentation velocity, the particle diameter should be reduced, the viscosity of the dispersing medium should be increased, and the difference in density between the solid and the liquid phase should be decreased. However, optimal slurry processing demands high loading at low viscosity for transportation and atomization requirements. Therefore, the low-viscosity slurry inherently promotes the sedimentation of the fines. To alleviate this obvious difference in the property requirements of the resultant slurry, various additives and surfactants have been developed.

### 4.2.2 RHEOLOGY

Rheology is the study of a system's response to a mechanical perturbation in terms of elastic deformations and viscous flow.[23,24] In most rheological systems, elastic response is associated with solids, whereas viscous response is associated with liquids. Therefore, a suspension system such as coal slurry exhibits behaviors of both elastic and viscous responses. These responses in coal slurry are a function of the type of coal, coal concentration, PSD, properties of dispersing phase, and additive package.[4,25]

Coal slurries, in general, exhibit non-Newtonian behavior; however, they do exhibit a wide range of responses including Newtonian, dilatant, pseudoplastic (shear-thinning), and plastic flow characteristics. Each of these responses is shown graphically in Figure 4.4, as a plot of the shear stress versus the rate of shear.[26] Newtonian is the simplest response and exhibits a linear functionality between shear stress and shear rate. Typically, slurries exhibit pseudoplastic or shear-thinning behaviors, meaning that as the shear stress is increased, the shear rate increases at a slower rate. This behavior is typical of a material that has a fragile internal structure that degrades with shearing stresses. The material does not have a yield point, but the apparent viscosity continually decreases with applied stress. An extension of this behavior is thixotropy, which exhibits shear thinning that requires significant periods of time to reform the internal structure. The period of time can range anywhere from minutes to several days. Dilatant behavior is the opposite of shear thinning, as the resistance to flow increases with the shear stress. In plastic behavior, a sufficient stress, higher than the yield stress, must be applied for flow to begin. Once the suspension yields, the shear stress is linear with shearing rate. The important rheological characteristics are yield stress, viscosity, and plasticity (thixotropy).[27] These values are determined experimentally.

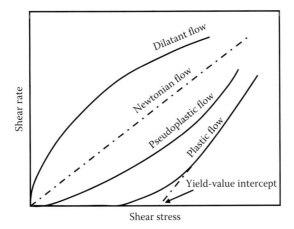

**FIGURE 4.4** Rheograms of various flow behaviors. (From Evans, D.F., and Wennerstrom, H., *The Colloidal Domain Where Physics, Chemistry, Biology, and Technology Meet*, VCH, New York, 1994.)

Viscosity has proved to be very difficult to model because it exhibits both elastic and viscous responses. The *Einstein equation* for viscosity in dilute suspensions is given by Equation 4.3.

$$\mu_r = \frac{\mu}{\mu_0} = 1 + 2.5\phi \tag{4.3}$$

where

$\mu_0$ = viscosity of dispersing phase
$\mu_r$ = relative viscosity
$\mu$ = absolute viscosity
$\phi$ = volume fraction of solids

This equation works well for dilute systems. To describe higher concentrations, a number of investigators have developed relations in the form of $\mu_r = f(\phi)$ that asymptotically reduce to Einstein's equation at low concentrations.[4] Modeling of these systems, however, utilizes only one parameter, the volume fraction of solids. This type of model is called a *one-parameter model*. This assumes that the particles are inert and, as such, interactions between particles are negligible.

The viscosity of the suspending medium not only affects the sedimentation velocity but also the rate of agglomeration. An empirical expression for $f(\phi)$ is given by

$$f(\phi) = e^{-5.9\phi} \tag{4.4}$$

This equation illustrates how rapidly an increase in solids' volume fraction, $\phi$, can reduce the sedimentation rate. However, the increased stability compromises slurry properties such as viscosity, combustion characteristics, and overall handling.[28]

Physicochemical forces are important in coal slurry because of the small size of the particles (most coal particles in slurry are smaller than 50 μm). Although bulk properties such as density and viscosity of coal and water are important, surface properties have a large effect on the slurry properties.

### 4.2.3 STABILITY

Slurry stability is classified into three broad categories, namely, *sedimentative* (static), *mechanical* (dynamic), and *aggregative*. The stability of a slurry is a crucial factor in its processability and applicability, which ultimately determine the value of the slurry. The factors that affect slurry stability are density, particle size, solid concentration, surface properties (relative hydrophilic nature), surface charge (zeta potential), morphology of coal, and type of slurrying liquid.[29]

The stability of a slurry against gravity is called "sedimentative stability." A statically unstable slurry will settle, but as the system becomes more stable, the degree of settling decreases. Static stability in a fluid requires a *yield stress* in the fluid sufficient to support the largest particle. The stability in a dynamic system is called *dynamic stability*. Dynamic stability involves the superposition of mechanical stresses; some examples are pumping and mixing.[30] The third stability type, *aggregative stability*, is a function of interparticle forces.

### 4.2.4 SUSPENSION TYPES

A suspension can be classified into three broad categories, namely, aggregatively stable, flocculated, and coagulated, as shown in Figure 4.5.[3] In an *aggregatively stable* suspension, repulsion forces do not allow particles to adhere to each other. They tend to settle owing to gravity, leading to a highly classified and compact sediment with coarse particles at the bottom and finest particles on the top.

In the second suspension type, *flocculated*, the particles weakly interact to form porous clusters called *flocs*. They tend to settle slowly because of increased drag forces from the floc structure. The formed sediments are very loose and occupy a large fraction of the original slurry volume. The slurry is easily brought back to original uniform concentrations with mild agitation.

In the last suspension type, *coagulated*, the particles interact strongly. The strong attractive interparticle forces promote the formation of compact and tightly bound clusters, which are difficult to break loose without significant agitation. These unstable slurries have fast settling rates and often display non-Newtonian behavior like thixotropic (time-dependent), pseudoplasticity (shear-thinning), or plastic behavior.

### 4.2.5 INTERPARTICLE INTERACTIONS

Recent studies have shown that stability in slurries is achieved by promoting networks through weak interparticle interactions.[29] The properties of coal slurries are governed by the nature of the forces between particles. Six important particle–particle interactions may exist in aqueous dispersions.[29] A more comprehensive analysis of these phenomena can be found in the literature.[31,32]

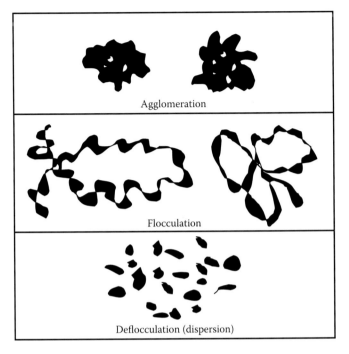

**FIGURE 4.5**  Illustrations of suspension types. (From Papachristodoulou, G., and Trass, O., *Can. J. Chem. Eng.*, 65, 1987.)

1. Interaction between *electrical double layers* (EDLs)
2. Van der Waals (VDW) attraction
3. Steric interactions
4. Polymer flocculation
5. Hydration- and solvation-induced interactions
6. Hydrophobic interactions

When a substance is brought into contact with an aqueous polar medium, it acquires a surface electrical charge through mechanisms such as ionization, ion adsorption, or ion dissolution. The surface charge influences the distribution of nearby ions in solution, that is, ions of opposite charge are attractive, whereas ions of like charges are repulsive. This, coupled with the mixing effects of thermal motion, leads to the formation of the EDL. The EDL consists of a surface charge with a neutralizing excess of counterions, and, further from the surface, co-ions distributed in a diffuse manner, as shown in Figure 4.6.[29] The EDL is important because the interaction between charged particles is governed by the overlap of their diffuse double layers. This creates a potential (Stern potential) at the interface of the Stern plane and the diffuse layer. Unfortunately, direct measurement of the Stern potential is impossible; however, it is possible to measure the zeta potential ($\zeta$), which corresponds to the shear plane adjacent to the Stern plane, as shown in Figure 4.7.[29] Although the

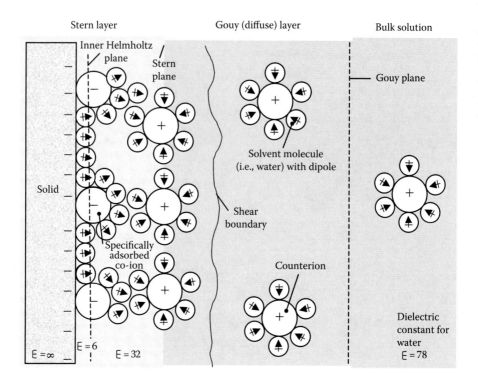

**FIGURE 4.6**  A schematic representation of EDL in the vicinity of a liquid–solid interface. (From Rowell, R.L., The Cinderella Synfuel, *CHEMTECH*, April 1989.)

zeta potential may not necessarily give a good indication of Stern potential, in certain cases, an expression such as Equation 4.5 can be formulated for the repulsive energy ($V_R$) of interaction between particles based on surface roughness, shape, and other factors.[33]

$$V_R = 2\pi\varepsilon a\zeta^2 e^{(-\kappa h)} \tag{4.5}$$

where
    $a$ = radius of two particles
    $\zeta$ = zeta potential
    $\kappa$ = inverse Debye length
    $\varepsilon$ = permissivity of the medium
    $h$ = distance between particles

The attractive force, VDW force, encourages aggregation between particles when the distance between particles is very small. These forces are due to spontaneous electric and magnetic polarizations giving a fluctuating electromagnetic field within the dispersed solids and aqueous medium separating the particles.[29] Two common methods are used

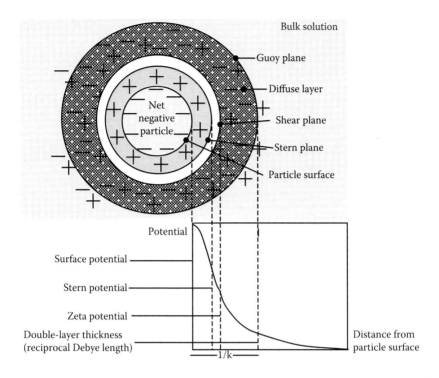

**FIGURE 4.7** Distribution of electric potential in the double-layer region surrounding a charged particle. (From Rowell, R.L., The Cinderella Synfuel, *CHEMTECH*, April 1989.)

for predicting these forces, namely, the Hamaker approach and Lifshitz approach. The Hamaker approach adds up the forces pairwise between the two bodies, whereas the Lifshitz method directly computes the attractive forces based on the electromagnetic properties of the media. Fundamental concepts of the Hamaker approach are explained in this chapter, though the more rigorous Lifshitz method is not covered.

The simpler method, Hamaker, for identical spheres is represented by

$$V_A = -\frac{A_{12}a}{12h} \tag{4.6}$$

where $A_{12}$ is the Hamaker constant.

Now, it is possible to predict the interactions of the EDL and VDW forces by

$$V_T = V_A + V_R \tag{4.7}$$

This forms the basis of the Derjaguin–Landau–Verwey–Overbeek (DLVO) theory of colloid stability.[27] The energy interactions for EDL and VDW forces are shown in Figure 4.8.[29] The subsequent total energy curve allows for prediction of aggregation

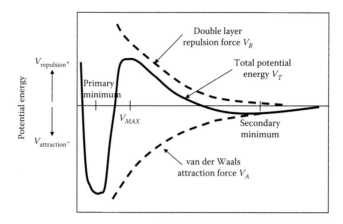

**FIGURE 4.8** Particle–particle interaction potential energy. (From Rowell, R.L., The Cinderella Synfuel, *CHEMTECH*, April 1989.)

at close distances (primary minimum) and the possibility of a weak and reversible aggregation in the secondary minimum.[29] The form of the curve depends on the size of the particles and surface charge.[34] For example, coarse particles are more likely to be vulnerable to aggregation at the secondary minimum.

Kinetic effects are important because thermodynamic prediction may not yield sufficient information. Coagulation rates are described in terms of the stability ratio. The stability ratio, $W$, can be thought of as the efficiency of interparticle collisions resulting in coagulations, as shown in Equation 4.8, for two identical particles brought together by diffusion.[29]

$$W = 2 \int_0^\infty \frac{\exp\left(\dfrac{V_T}{kT}\right)}{(2+s)^2} \, ds \tag{4.8}$$

where $s = h/a$.

The other types of interparticle interactions (steric interactions, polymer flocculation, hydration- and solvent-induced interactions, and hydrophobic interactions) represent special cases. The most significant of these are steric interactions and polymer flocculations. Steric interactions develop when molecules (usually polymers or macromolecues) are adsorbed onto the particle surface at high coverages. The polymer molecules protrude from the surface of the particle into the solution. When particles approach one another, the polymer chains overlap and often dehydrate, increasing the stability of the slurry.[35] Polymer flocculation occurs when the particle surface has low coverage of a high-molecular-weight polymer. The polymers bridge the particles and form flocs.[5,35]

Hydration- and solvation-induced interactions become important when interparticle distance is on the order of a solvent molecule, that is, very short. For aqueous systems, these solvation effects are clearly visible in structuring of the water near the interfacial surface (with their bound water), which interacts with hydrated ions from the solution. The net effect is an increased stability or net repulsion between particles as it becomes necessary for the ions to lose their bound water to allow the approach to continue.[29]

Hydrophobic interactions are analogous to hydration and solvation effects because stability can be enhanced by the attraction between two hydrophobic particles. Therefore, hydrophobic particles tend to associate with each other. The hydrophobic forces are greater than the VDW force and have a longer range.[29]

## 4.3   COAL–WATER SLURRY

CWSs attracted technologists' attention initially as a replacement fuel for oil in furnaces and boilers and recently for cofiring of boilers and furnaces in the use of coal fines.[5] CWM and CWSF have received a great deal of attention for use as fuel because of the relative ease of handling (similar to fuel oil and not explosive, unlike coal dust), storage in tanks, and injection into furnaces and boilers. CWMs typically have extremely high loadings in the range of 60–75 wt% coal, which leads to high energy densities per unit mass. Possible applications include gas turbines, diesel engines, fluid bed combustors, blast furnaces, and gasification systems.[2] CWM with a lower coal loading of 50 wt%, may be used for the internal combustion engine, especially when nonexplosive fuel is needed, for example, in military vehicles and helicopters. However, coal slurries for cofiring purposes are limited by economics to the use of minimal additives as well as lower coal concentrations (50 wt%).[36]

The physical properties of coal slurry are extremely important in the processing of the fuel. A slurry must be stable and exhibit low viscosity in the shear rates of pumping and atomization. The flow characteristics of CWS depend on (1) physicochemical properties of the coal; (2) the volume fraction, $\phi$, of the suspended solids; (3) the particle size range and distribution (PSD); (4) interparticle interactions (affected by the nature of surface groups, pH, electrolytes, and chemical additives); and (5) temperature.[17,37] Rheological and hydrodynamic behavior of CWSs varies from coal to coal. Each coal has a unique package of PSD, concentration, and additives to reach the desired processability.

A parameter that measures how well a coal will slurry, that is, *slurry capability* (or slurriability), is the equilibrium moisture content of coal. The equilibrium moisture content is a measure (index) of the hydrophilic nature of the coal. The equilibrium moisture content of a coal sample can be readily determined by proximate analysis. The more hydrophilic a coal is, the more water it will hold and the less likely it is to produce a highly concentrated slurry.[17,38] The CWS viscosity increases with the hydrophilicity of the coal. Therefore, a hydrophobic coal can more easily form a slurry of low viscosity at high solid loadings.

Conventional high-rank coals (black coal), except anthracite, have a hydrophobic nature from the lack of acid groups and will form a slurry of ~80 wt% (dry coal weight basis). Anthracite coals have low reactivity and are not very volatile, leading to poor ignition stability.[2] However, low-rank coals (brown coal) are hydrophilic from an abundance of oxygen functional groups and will form slurries of only 20–25 wt% (dry coal weight basis).[39] These slurries have low concentrations but form slurries that are nonagglomerating and have high reactivities. On the other hand, petroleum coke is hydrophobic by nature. Differences in coal slurry stability between types of coals are a function of the relative balance between acid and base groups on the coal surface.[40]

CWMs are loaded to the highest possible concentration at acceptable viscosities. However, viscosity increases with coal concentration (loading), though reducing the viscosity compromises the stability of the slurry.[3] The viscosity of the slurry increases gradually with increasing solid loading until a critical point is reached at which interparticle friction becomes important. Beyond this point, the viscosity increases very sharply until the slurry ceases to flow.[5,17] To properly stabilize CWSs, that is, to enhance the stability of the coal dispersion, additives such as surfactants and electrolytes are added.

Surfactants are used as dispersants to wet and separate coal particles by reducing the interfacial tension of the coal–water system. Surfactants are short-chain molecules containing both a hydrophobic group and a hydrophilic oxide (nonionic) or a charged ionic group (ionic). These molecules attach themselves to the coal particles through adsorption or ionic interaction. Generally speaking, most dispersants used in coal-water slurry are ionic. Some examples of such dispersants are sodium, calcium, and ammonium lignosulfates and the sodium and ammonium salts of naphthalene formaldehyde sulfonates.[4]

Ionic surfactants adsorb onto the alkyl groups at hydrophobic sites on the coal particle. This gives the coal particles a negative charge, which affects the EDL, enhancing the repulsive forces and thus preventing agglomeration.[19] Anionic surfactants decrease the viscosity of the slurry up to a critical loading. At this point, the coal adsorption sites are saturated, and the remaining surfactant forms micelles in the slurry, leading to an enhanced structure and increased viscosity.

Nonionic surfactants function by two different methods depending on the nature of the coal. On a hydrophilic coal surface, the hydrophobic end of the surfactant is toward the aqueous phase. The water then acts as a lubricating material between coal particles. The second method is via attachment of surfactant on a hydrophilic coal. The hydrophilic end of the surfactant attaches to the coal molecule, leaving the hydrophobic end in the aqueous medium. This increases the amount of water near the surface of the coal particle, producing a hydration layer or solvation shell. This prevents agglomeration by cushioning coal particles and lowers the viscosity.[19]

The ionic strength of water in the CWM is an important parameter in the rheological and hydrodynamic characteristics of a slurry. Because coal is a mixture of macromolecular carbonaceous materials and mineral matter rather than a uniformly homogeneous substance, the ionic strength of the water will affect the interaction with coal. In a hydrophobic colloidal system dispersed by electrically repulsive forces, the electrolyte concentration (and its ionic strength) has a considerable effect on the stability against flocculation of particles.[19] The cation concentration causes an increase in the viscosity of the slurry with decreasing pH.[41] Electrolytes strongly affect the degree of particle dispersion and, thus, rheology in CWM that uses anionic dispersant.[41] The addition of electrolyte to a slurry using nonionic dispersants has no appreciable effect on the viscosity.[41]

In highly concentrated slurries,[42] minimal settling is expected, but viscosity-reducing additives increase the settling rate. To stabilize the dispersion, flocculating agents are added, which produce a gel. Some examples of this are nonionic amphoteric polymers of polyoxyethylene, starches, natural gums, salts, clays, and water-soluble polymeric resins.[2,42]

Polymers have been used for drag reduction, that is, viscosity reduction.[42,43] Both ionic and nonionic polymer solutions show reduction in viscosity, although the reduction is more pronounced for anionic polymers.

## 4.4 COAL–OIL SLURRY

Coal–oil slurries have been investigated for over 100 years. Typically, interest peaks in times of high prices and shortages of oil. The most recent interest was fueled by the energy crises of 1973 and 1979, when tremendous effort was expended in finding a quick viable alternative to oil in boilers and furnaces. Since the mid-1980s, however, most slurry investigation has been directed toward CWM.

In general, coal–oil slurries exhibit non-Newtonian behavior, mostly pseudoplastic except at low coal loadings where the slurry is Newtonian (provided the oil is also Newtonian). The viscoelastic properties of the dispersion depend on coal concentration, PSD, coal type, oil type, and chemical additives. Rheological properties of COM are highly sensitive to coal concentration. At this critical concentration, a dramatic increase in viscosity occurs with incremental changes in concentration.

The standard PSD in COMs is listed in Table 4.2.[3] COMs are classified as lyophobic because the dispersed particles are not compatible with the dispersion medium. A lyophobic system is typically characterized by a lack of attraction between the colloid medium (i.e., dispersed coal particles) and the dispersion medium (i.e., oil) of a colloid system (i.e., coal–oil slurry). These systems are thermodynamically unstable and will separate into two continuous phases.

Ultrafine COMs with 95% of particles smaller than 325 mesh (44 μm) and slurry concentrations of 50 wt% have been investigated. These slurries reduce abrasiveness,

**TABLE 4.2**
**COM Particle Size Distribution**

| Percentage[a] | Particle Size (μm) |
|---|---|
| 100 | <200 |
| 80 | <74 |
| 65 | <44 |
| 15 | <10–20 |
| 1 | <1 (colloidal) |

*Source:* Papachristodoulou, G., and Trass, O., *Can. J. Chem. Eng.*, 65, 1987. With permission.

[a] Passing through screen opening size.

exhibit improved combustion characteristics, and do not contain additives.[4,44] However, the grinding cost becomes higher, and coal concentration is typically limited to 50 wt%.

Common chemical additives for COMs are surfactants and polymers. The surfactants add stability to the mixture by preventing agglomeration and enhancing flocculation. Cationic polymers are the most effective surfactants for stabilization,[45] though anionic polymers are used more frequently to reduce the drag.[42,46] Finding effective additives for various COMs often becomes a key factor in the R&D investigation.

In many instances, water is added to coal–oil slurries, forming COW or CWO, depending on the water concentration. Water, a flocculating agent, is added to increase the stability of the slurry as well as for cost savings. Water increases the viscosity of the resultant slurry through the formation of aggregates and particle bridging,[46] although the combustion properties are more or less retained.

## 4.5  ADVANCED TRANSPORTATION OF COAL SLURRY

The transportation of coal slurries can be accomplished by truck and railroad tanks, slurry tankers, and slurry pipelines. Historically, transportation schemes of CWSs have received the most attention and development. Although all transportation schemes and options have been investigated, few new processing developments have occurred in truck, railroad, or slurry tankers. In these cases, the slurry stability has been enhanced to minimize settling during transportation, or the slurry has been dewatered to maximize energy density and cost-effectiveness of transportation. On the other hand, coal–water pipeline systems across the world have undergone almost constant development since the 1950s. Coal pipelines can be broken into four different systems: (1) conventional fine coal, (2) conventional coarse coal, (3) stabilized-flow coal, and (4) CWM. These systems differ by the particle size of coal in the slurry, as shown in Table 4.3.[47]

Only two systems in the United States are conventional fine coal slurry pipelines. The first was the Ohio pipeline by the Consolidation Coal Company, which was built in 1957 and operated for several years until another competitive transportation mode, the unit train, became available. The slurry traveled at moderate velocities and had a coal content of 50 wt%. The second pipeline system built, the Black Mesa Pipeline, followed the same basic design of the earlier Consolidation pipeline. The pipeline began operation in 1970 and continued its commercial operation till December 2005. The pipeline operates at 4–6 mph and is optimized for minimal erosion and settling

---

**TABLE 4.3**
**Summary of Particles Sizes in Coal–Water Slurry Pipeline Systems**

| Coal–Water Slurry Systems | Particle Sizes |
|---|---|
| Conventional fine coal | Less than 1 mm |
| Conventional coarse coal | 50–150 mm |
| Coal–water mixtures | −30 and −150 μm (less than 200 mesh) |
| Stabilized-flow coal | Less than 0.2 mm (fine) and less than 50 mm (coarse) |

*Source:*  Hsu, B.D. et al., *J. Eng. Gas Turbines Power*, 110, 1988. With permission.

at a 50% coal loading.[47,48] The operational experience of this pipeline has significantly impacted other solid-transporting slurry pipelines in the world.

In conventional coarse coal slurry transportation, run-of-mine (ROM) coal is transported at high velocities in coal loadings ranging from 35% to 60%, depending on the coal particle size (50–150 mm).[47] In this pipeline, coal slurry preparation costs are minimal; however, energy requirements for the pipeline transportation are high. This type of system is limited to short distances where other transportation modes such as rail, barge, truck, or conveyor can be used.

Stabilized-flow coal slurry systems use smaller particles to support the coarse coal particles. The PSD for fine coal is <0.2 mm and for coarse coal, <50 mm.[47] This PSD was utilized to enhance the stability of the slurry for long transportation distances. Coal concentrations of up to 70 wt% can be used. Coarse coal is easier to dewater than fine coals because of the particle size.

CWM pipeline systems are designed for direct use at the final destination site, most likely in utility boilers and furnaces. The coal concentration is between 70 and 75 wt% with bimodal and polymodal PSDs of coal particles smaller than 200 mesh.[47] CWMs use approximately 1% surfactants and dispersants for slurry stability. These pipelines have yet to be investigated over long distances.

Many different coal processing configurations exist for coal use at the end of a coal slurry pipeline. The options include direct use of CWM, dewatering and conveying to a plant (Figure 4.9), piping to an offshore platform and dewatering into ship (Figure 4.10), dewatering into a railcar or truck, and piping into an offshore buoy system into a ship.[49–51]

Table 4.4 is a comparison of coal content in coal slurries for the pipeline systems discussed.[47] Each coal transportation mode has its own merits, and therefore, decisions have to be made on an individual basis. The transport systems are evaluated in Table 4.5.[47]

As mentioned earlier, one of the most significant and successful coal slurry pipelines is the Black Mesa Pipeline, which is a CWS pipeline 273 mi. (440 km) long and 18 in. (457 mm) in diameter, originating from Black Mesa in northeastern Arizona. The system was used to transport coal from Peabody Western Coal Company's open pit mine to the Mohave Generating Station, which is a 1580 MW steam-powered

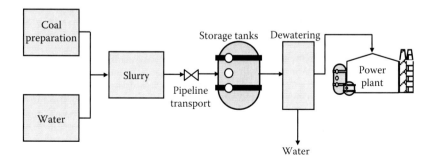

**FIGURE 4.9** Domestic combustion pipeline scheme. (From Hsu, B.D. et al., *J. Eng. Gas Turbines Power*, 110, 1988.)

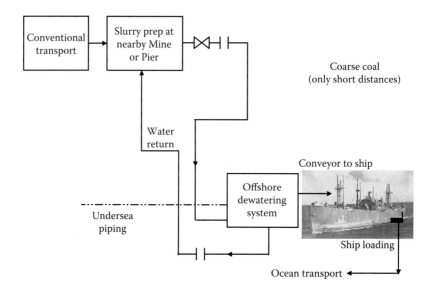

**FIGURE 4.10** Offshore coarse coal pipeline scheme with dewatering. (From Hsu, B.D. et al., *J. Eng. Gas Turbines Power*, 110, 1988.)

electric generating plant located in Laughlin, NV. The Black Mesa Pipeline began its commercial operation in November 1970 and continued till December 2005. During this period, the system transported over 120 million tons (as mined) with an availability factor of 98%. The pipeline delivered coal at a rate of 570 to 600 M/T (or 630–660 short tons) per hour, and the nominal capacity is over 4.5 million metric tons (5 million short tons) per year. The Black Mesa Pipeline has helped development of other solid-transportation pipelines in the world, which include slurry pipeline transportation of limestone, copper concentrates, iron concentrates, Gilsonite (a registered trade name for uintaite), and phosphate. Some of the major long-distance slurry pipeline projects for coal slurry and limestone slurry are shown in Table 4.6.

**TABLE 4.4**

**Comparison of Coal Content in Coal Slurries**

| Coal–Water Slurry Systems | Percentage of Coal in Mixture |
|---|---|
| Conventional fine coal | 50 (overland and ship loading) |
|  | 75 (aboard ship) |
| Conventional coarse coal | 30–60 (overland—short distances only and ship loading) |
|  | 90–92 (aboard ship) |
| Coal–water mixtures | 70 (overland) |
|  | 85–90 (ship loading and aboard ship) |
| Stabilized-flow coal | 70–75 (overland, ship loading, and aboard ship) |

*Source:* Hsu, B.D. et al., *J. Eng. Gas Turbines Power*, 110, 1988. With permission.

**TABLE 4.5**
**Coal–Water Slurry Pipeline System Selection**

| Objective | System Characteristics | | |
|---|---|---|---|
| | Length (mi.) | Type | Best Pipeline Selection |
| Rapid implementation | <5 | Domestic | Coarse coal |
| | | Export | Coarse coal |
| | 50–100 | Domestic | Conventional fine coal |
| | | Export | Conventional fine coal |
| | >100 | Domestic | Conventional fine coal |
| | | Export | Conventional fine coal |
| Lowest cost and water use | <5 | Domestic | Coarse coal |
| | | Export | Coarse coal |
| | 50–100 | Domestic | Stabilized flow |
| | | Export | Stabilized flow |
| | >100 | Domestic | Stabilized flow |
| | | Export | Stabilized flow |
| Oil displacement | All | Domestic | CWM |
| | | Export | CWM |

*Source:* Hsu, B.D. et al., *J. Eng. Gas Turbines Power*, 110, 1988. With permission.

**TABLE 4.6**
**Major Long-Distance Slurry Pipeline Projects**

| Material Transported | System Name Location | Length (mi.) | Annual Tonnage (millions) | Initial Operation |
|---|---|---|---|---|
| Coal | Consolidation, Ohio | 108 | 1.3 | 1957 |
| | Black Mesa, Arizona | 273 | 4.8 | 1970 |
| Limestone | Calaveras, California | 17 | 1.5 | 1971 |
| | Rugby, England | 57 | 1.7 | 1964 |
| | Gladstone, Australia | 15 | 2.0 | 1981 |

## 4.6  ENVIRONMENTAL ISSUES

Environmental issues play an important role in the implementation of coal slurry fuels. The transportation systems must take into account possible leaks and spills encountered with slurry handling. Combustion processes must be thoroughly investigated to find the differing combustion mechanisms in coal slurry, specifically, the mineral matter and sulfur content of the coal.

In the transportation of coal slurries, spills, leaks, and catastrophic disasters are important factors in safe handling. Transportation of CWMs has been widely investigated for both short and long distances. Slurries of water are considered nontoxic and nonhazardous. Historically, in the United States, when there have been handling accidents in pipeline systems, cleanup of the coal has not been necessary. The coal has been allowed to reenter the ground naturally. CWMs have a distinct advantage in that they are not readily combustible or pyroforic in accidents.

In the combustion of slurries, many processes will not be able to handle the high sulfur and mineral matter content of coal. For example, combustion processes in utility boilers and furnaces may exhibit slagging, fouling, and erosion. On the other hand, other applications such as cement and asphalt kilns and fluid bed combustors are able to easily handle the increases in inorganic material such as coal ash. The amount of ash that induces fouling in boilers varies from boiler to boiler. Table 4.7 shows the predicted acceptable ash levels for differing types of boilers.[2]

An important constituent in coal is sulfur, both in organic and inorganic forms. Sulfur is a precursor for ash formation and can form $SO_x$, a precursor to acid rain, which has regulated emissions. In many instances, coal beneficiation techniques are used to clean the coal after grinding. These processes are broken into physical and chemical cleaning methods. Physical cleaning methods are based on differences in density and surface properties between mineral matter and coal organics. Examples of physical cleaning processes are gravitational separation,[52] froth flotation,[53] selective agglomeration,[54] heavy medium separation,[55] high-gradient magnetic separation,[56] microbubble flotation,[57] and biological method.[58] Chemical cleaning methods have been widely investigated at the laboratory scale but are not commonly used on an industrial scale.[59]

Other environmentally regulated emissions from combustion of coal slurries are $NO_x$, hydrocarbons (volatile organic compounds [VOCs]), and particulate matter (PM). The burning of coal slurries reduces the amount of hydrocarbon emissions significantly in oil-designed boilers and furnaces. $NO_x$ emissions are a function of fuel's nitrogen content, fuel composition, combustion temperature, and amount of excess air.[4] $NO_x$ emissions are controllable by conventional methods such as low

**TABLE 4.7**
**Permissible Coal Ash Content for Utility Boilers**

| Ash Content (%) | Boiler Type |
|---|---|
| 5–7 | Coal designed/coal capable |
| 2–3 | Liberal oil design |
| <0.5 | Compact oil designed |

*Source:* Kesavan, S., Stabilization of Coal Particle Suspensions Using Coal Liquids, M.S. thesis, University of Akron, 1985. With permission.

of coal from the evaporative effects of water and the slower burning of coal.[53,70] Erosion of parts in the injection system as well as in the cylinders is a major concern for the life of the engine. Engine wear is six times higher in a CWM-fired engine than a diesel-fired one but can be reduced significantly to twice as much with special alloys and redesigned parts.[68] These findings hold great promise for the success of efforts to operate an internal combustion engine on CWSF.

## 4.8  RECENT ADVANCES AND THE FUTURE

CWFs, among all slurry fuels, have received the most process development recently. Coal slurry facilities have been built in Australia, Canada, China, Italy, Japan, Sweden, and the United States. The most active countries have been Japan, China, and Russia. China has built several slurry production facilities and boiler units. Japan has done considerable research and technological development on slurry processes and has converted several boilers. Russia has built several pipelines, production facilities, and boilers.

In Australia, a group of companies comprising Ube Industries, Nissho Iwai Corp., and Coal and Allied Industries have been working since 1987 on developing the production, transportation, and marketing of CWM from Australia to Japan.[71] In 1991, the first bilateral trading of coal slurry began between China and Japan.[72] A CWS plant was built in Rizhao located in Shandong province by Yanzhou Coal Mining Bureau, Nisshon Iwai Corp., and JWG Corporation. The plant has a capacity of 250,000 tons per year. The coal is mined and then transported by train to Rizhao. The coal is processed into CWM and shipped to an overseas terminal. Once in the relay terminal, the CWM is transferred to coastal shipping and then, finally, to end users. The manufacture, transportation, and combustion of coal slurry have involved only minimal technical problems.[72]

In the United States, the greatest potential use for CWM may be in the utilization of coal fines for cofiring boilers. Over 40 million tons of coal fines are discarded annually in the United States into slurry ponds, and some 2.3 billion tons of coal fines were estimated to reside in ponds in 1994.[18] Coal fine (−100 mesh) production has increased over the years with the increase in demand of cleaner coal. Increased demand for high-quality coal (often beneficiated) has led to rejecting 20%–50% of coal mined as coal fines.[73] The environmental impact of coal fines includes nonproductive use of land, loss of aesthetics, danger of slides, dam failure, significant permitting costs, and possible water pollution.[73]

Recently, GPU Energy (PENELEC), New York State Electric and Gas Corporation (NYSEG), Pennsylvania Electric Energy Development Authority (PEEDA), Pennsylvania Electric Energy Research Council (PEERC), and the Electric Power Research Institute (EPRI) commissioned a project to investigate the utilization of CWM developed for coal fines. The project investigated laboratory-scale as well as full-scale cofiring of a 32 MW boiler.[74]

CWM will reduce the coal-handling problems and eliminate the need for costly dryers and their associated environmental hazards. CWM will also eliminate the need for addition of oil at startup, which is used to stabilize the combustion. Moreover, the equipment life of the pulverizer will also be extended by reducing the equipment load. Using CWM from coal fines has the ability to stabilize the cost of fuel to the boiler because it is a lower-cost fuel, and fuel ratio can be controlled up to 50%.[75,76] Environmentally, the addition of coal slurry reduces the $NO_x$ emissions.[76]

To be economically more competitive, the lowest-cost slurry must be developed, which may lead to a "low-tech slurry," that is, one without stabilizers, dispersants, or the need for further grinding. This slurry can be developed from coal pond fines, from the fine coal fraction of existing coal supplies, or using advanced coal technologies in the future to deep-clean fine coal.[68] The slurry has been tested with solid loadings ranging between 54 and 67 wt% coal, and the test has demonstrated excellent handling and storage properties.[77]

However, slurry developed from coal fines is highly variable. The slurriability is a function of PSD, ash level, and extent of oxidation. Therefore, slurry preparation can vary from a minimal processing to a significant processing. Minimal processing is from coals derived from wet, fine coal that has been cleaned. Significant processing can come from coal fines that have high ash content and oxidation levels. High oxidation makes the coals more wettable through oxygen bonds, which, in turn, increases the viscosity and stability of the slurry. Typically, weathered coals are more oxidized.

Currently, slurry produced from coal fines at Homer City, PA is being cofired along with powdered coal in the 32 MW boiler #14 of the Seward Station. The plant has been operating without disruption. The $NO_x$ production has decreased by 10%–20%, while $CO_2$, $SO_2$, and particulates have remained essentially the same as for powder coal, but the CO level is highly variable.[77]

Coal slurry development finds itself adjusting once again to process economics. Since the mid-1980s, COM, once the most investigated slurry, has become uneconomical, and development has focused on CWM. In regions where coal is plentiful but transportation is lacking, such as China and Russia, slurry development continues on production, pipelines, and combustion. In other places such as Japan where natural resources and storage spaces are limited, the ease of transportation that CWM makes possible is now being exploited. In the United States, coal slurry development is now centered on coal utilization. CWM enables the use of higher amounts of coal after beneficiation and greatly reduces the environmental liability in discarded coal fines. Coal slurry development continues in internal combustion engines in injection systems, atomizer design, and materials for construction.[76] Coal slurry process development will continue and may be poised to represent a major form of coal energy.

## REFERENCES

1. Choudhury, R., Slurry fuels, *Prog. Energy Combust. Sci.*, 18, 409–427, 1992.
2. Kesavan, S., Stabilization of coal particle suspensions using coal liquids, M.S. thesis, University of Akron, Akron, OH, 1985.
3. Papachristodoulou, G. and Trass, O., Coal slurry fuel technology, *Can. J. Chem. Eng.*, 65, 177–201, 1987.
4. Smith, H.R. and Munsell, H.M., Liquid fuels, U.S. Patent No. 219181, 1879.
5. Kawatra, S.K., Coal–water slurries, in *Encyclopedia of Chemical Processing*, Vol. 1, Lee, S., Ed., Taylor & Francis, New York, 2005, pp. 495–503.

6. Lord, N.W., Ouellette, R.P. and Farah, O.G., *Coal–Oil Mixture Technology*, Ann Arbor Science, Ann Arbor, MI, 1982.

7. Manning, A.B. and Taylor, R.A.A., Colloidal fuel, *J. Inst. Fuel*, 9, 303, 1936.

8. Manning, A.B. and Taylor, R.A.A., Colloidal fuel, *Trans. Inst. Chem. Eng.*, 14, 45, 1936.

9. Adams, R.A., Holmes, F.C.V. and Perrin, A.W., Colloidal fuel using cracked oil and high carbon residue, U.K. Patent No. 396,432 August 2, 1933.

10. Barkley, J.F., Hersberger, A.B. and Burdick, L.R., *Trans. Am. Soc. Mech. Eng.*, 66, 185, 1944.

11. Nakabayashi, Y., Outline of COM R&D in Japan, *Proc. 1st Int. Symp. on Coal–Oil Mixture Combustion, M78-97*, St. Petersburg, FL, May 1978.

12. Basta, N. and D'Anastaio, M., The pulse quickens for coal-slurry projects, *Chem. Eng.*, 22–25, 1984.

13. Bergman, P.D., Kirkland, L. and George, T.J., Why coal slurries stir worldwide interest, *Coal Mining Process.*, 19(10), 24–42, 1982.

14. Marnell, P., Direct firing of coal water suspensions, state of the art review, *Proc. Coal Technol. '80*, Industrial Presentation Inc., Houston, TX, 1980.

15. Brown, A., Jr., Powdered COM Program, Final Report of the General Motors Corporation, October 1977.

16. Miller, B.G., Coal–water slurry fuel utilization in utility and industrial boilers, *Chem. Eng. Prog.*, 85(3), 29–38, 1989.

17. Kawatra, S.K. and Bakshi, A.K., The on-line pressure vessel rheometer for concentrated coal slurries, *Coal Prep.*, 22(1), 41–56, 2002.

18. Meyer, C.W., Stabilization of coal/fuel oil slurries, *Proc. 2nd Int. Symp. on Coal–Oil Mixture Combustion*, Vol. 2, CONF-791160, Danvers, MA, November 1979.

19. Williams, R.A., Characterization of process dispersions, in *Colloid and Surface Engineering: Applications in the Process Industries*, Williams, R.A., Ed., Butterworth Heinemann, Boston, MA, 1992.

20. He, Q., Wang, R., Wang, W., Xu, R. and Hu, B., Effect of particle size distribution of petroleum coke on the properties of petroleum coke–oil slurry, *Fuel*, 90(9), 2896–2901, 2011.

21. Barnes, H.A., Hutton, J.F. and Walters, K., *An Introduction to Rheology*, Elsevier, New York, 1989.

22. Hunter, R.J., *Foundations of Colloidal Science*, Vol. 1, Oxford University Press, Oxford, U.K., 1987.

23. Shaw, D.J., *Introduction to Colloid and Surface Chemistry*, 4th ed., Butterworth Heinemann, Boston, MA, 1991.

24. Napper, D.H., *Polymeric Stabilization of Colloidal Dispersions*, Academic Press, Waltham, MA, 1983.

25. Gregory, J., Flocculation of polymers and polyelectrolytes, in *Solid Liquid Dispersions*, Tadros, T.F., Ed., Academic Press, Waltham, MA, 1987, pp. 163–181.

26. Evans, D.F. and Wennerstrom, H., *The Colloidal Domain Where Physics, Chemistry, Biology, and Technology Meet*, VCH, New York, 1994.

27. Vossoughi, S. and Al-Husaini, O.S., Rheological characterization of the coal/oil/water slurries and the effect of polymer, *Proc. 19th Int. Tech. Conf. on Coal Utilization and Fuel Systems*, Clearwater, FL, March 21–24, 1994, pp. 115–122.

28. Ross, S. and Morrison, I.D., *Colloidal Systems and Interfaces*, John Wiley & Sons, New York, 1988.

29. Rowell, R.L., The Cinderella Synfuel, *CHEMTECH*, 244–248, April 1, 1989.

30. Turian, R.M., Fakhreddine, M.K., Avramidis, K.S. and Sung, D.-J., Yield stress of coal–water mixtures, *Fuel*, 72(9), 1305–1315, 1993.

31. Roh, N.-S., Shin, D.-H., Kim, D.-C. and Kim, J.-D., Rheological behavior of coal–water mixtures 1. Effects of coal type, loading and particle size, *Fuel*, 74(8), 1220–1225, 1995.

32. Woskoboenko, F., Siemon, S.R. and Creasy, D.E., Rheology of Victorian brown coal slurries, 1. Raw coal–water, *Fuel*, 66, 1299–1304, 1987.

33. Roh, N.-S., Shin, D.-H., Kim, D.-C. and Kim, J.-D., Rheological behavior of coal–water mixtures 2. Effect of surfactants and temperature, *Fuel*, 74(9), 1313–1318, 1995.

34. Kaji, R., Muranaka, Y., Miyadera, H. and Hishinuma, Y., Effect of electrolyte on the rheological properties of coal–water mixtures, *AIChE J.*, 33(1), 11–18, 1987.

35. Rowell, R.L., Vasconcellos, S.R., Sala, R.J. and Farinato, R.S., Coal–oil mixtures 2. Surfactant effectiveness on coal oil mixture stability with a sedimentation column, *Ind. Eng. Chem., Proc. Des. Dev.*, 20, 283–288, 1981.

36. Bertram, K.M. and Kaszynski, G.M., A comparison of coal–water slurry pipeline systems, *Energy*, 11(11/12), 1167–1180, 1986.

37. Brolick, H.J. and Tennant, J.D., Innovative transport modes: coal slurry pipelines, ASME Fuels and Combustion Division Pub. *FACT*, 8, 85–91, 1990.

38. Manford, R.K., Coal–water slurry: a status report, *Energy*, 11(11/12), 1157–1162, 1986.

39. Ng, K.L., Coal unloading system using a slurry system, ASME Fuels and Combustion Division Pub. *FACT*, 8, 281–287, 1990.

40. Hapeman, M.J., Review and update of the coal fired diesel engine, ASME Power Division Pub. *PWR*, 9, 47–50, 1990.

41. Likos, W.E. and Ryan, T.W., III, Experiments with coal fuels in a high-temperature diesel engine, *J. Eng. Gas Turbines Power*, 110, 444–452, 1988.

42. Zakin, J., Zhang, Y. and Yunying, Q., Drag reducing agents, in *Encyclopedia of Chemical Processing*, Vol. 2, Lee, S., Ed., Taylor & Francis, New York, 2005, pp. 767–785.

43. Rao, A.K., Melcher, C.H., Wilson, R.P., Jr., Balles, E.N., Schaub, F.S. and Kimberly, J.A., Operating results of the cooper-bessemer JS-1 engine on coal–water slurry, *J. Eng. Gas Turbines Power*, 110, 431–436, 1988.

44. Urban, C.M., Mecredy, H.E., Ryan, T.W., III, Ingalls, M.N. and Jetss, B.T., Coal–water slurry operation in an EMD diesel engine, *J. Eng. Gas Turbines Power*, 110, 437–443, 1988.

45. Hsu, B.D., Progress on the investigation of coal–water slurry fuel combustion in a medium speed diesel engine: Part 1—ignition studies, *J. Eng. Gas Turbines Power*, 110, 415–422, 1988.

46. Hsu, B.D., Progress on the investigation of coal–water slurry fuel combustion in a medium speed diesel engine: Part 2—preliminary full load test, *J. Eng. Gas Turbines Power*, 110, 423–430, 1988.

47. Hsu, B.D., Leonard, G.L. and Johnson, R.N., Progress on the investigation of coal–water slurry fuel combustion in a medium speed diesel engine: Part 3—accumulator injector performance, *J. Eng. Gas Turbines Power*, 110, 516–520, 1988.

48. Wenglarz, R.A. and Fox, R.G., Jr., Physical aspects of deposition from coal–water fuels under gas turbine conditions, *J. Eng. Gas Turbines Power*, 112, 9–14, 1990.

49. Veal, C.J. and Wall, D.R., Coal–oil dispersions—overview, *Fuel*, 60, 873–882, 1982.

50. Wenglarz, R.A. and Fox, R.G., Jr., Chemical aspects of deposition/corrosion from coal–water fuels under gas turbine conditions, *J. Eng. Gas Turbines Power*, 112, 1–8, 1990.

51. Dwyer, J.G., Australian coal water mixtures (CWM) plant development at Newcastle, NSW, *Proc. 19th Int. Tech. Conf. on Coal Utilization and Fuel Systems*, 1994, pp. 35–38.

52. Hamieh, T., Optimization of the interaction energy between particles of coal water suspensions, *Proc. 19th Int. Tech. Conf. on Coal Utilization and Fuel Systems*, 1994, pp. 103–114.

53. Vossoughi, S. and Al-Husaini, O.S., Rheological characterization of the coal/oil/water slurries and the effect of polymer, *Proc. 19th Int. Tech. Conf. on Coal Utilization and Fuel Systems*, 1994, pp. 123–131.

54. Tobori, N., Ukigia, T., Sugawara, H. and Arai, H., Optimization of additives for CWM commercial plant with a production rate of 500 thousand tons per year, *Proc. 19th Int. Tech. Conf. on Coal Utilization and Fuel Systems*, 1994, pp. 123–133.

55. Addy, S.N. and Considine, T.J., Retrofitting oil-fired boilers to fire coal water slurry: an economic evaluation, *Proc. 19th Int. Tech. Conf. on Coal Utilization and Fuel Systems*, 1994, pp. 341–352.

56. Kaneko, S., Suganuma, H. and Kabayashi, Y., Fundamental study on the combustion process of CWM, *Proc. 19th Int. Tech. Conf. on Coal Utilization and Fuel Systems*, 1994, pp. 403–414.

57. Takahashi, Y. and Shoji, K., Development and scale-up of CWM preparation process, *Proc. 19th Int. Tech. Conf. on Coal Utilization and Fuel Systems*, 1994, pp. 485–495.

58. Tu, J., Cefa, K., Zhou, J., Yao, Q., Fan, H., Cao, X., Qiu, Y., Huang, Z., Wu, X. and Liu, J., The comparing research on the ignition of the pulverized coal and the coal water slurry, *Proc. 19th Int. Tech. Conf. on Coal Utilization and Fuel Systems*, 1994, pp. 517–528.

59. Battista, J.J., Bradish, T. and Zawadzki, E.A., Test results from the co-firing of coal water slurry fuel in a 32 MW pulverized coal boiler, *Proc. 19th Int. Tech. Conf. on Coal Utilization and Fuel Systems*, 1994, pp. 619–630.

60. Miller, S.F., Morrison, J.L. and Scaroni, A.W., The formulation and combustion of coal water slurry fuels from impounded coal fines, *Proc. 19th Int. Tech. Conf. on Coal Utilization and Fuel Systems*, 1994, pp. 643–650.

61. Schimmeller, B.K., Jacobsen, P.S. and Hocko, R.E., Industrial use of technologies potentially applicable to the cleaning of slurry pond fines, *Proc. 19th Int. Tech. Conf. on Coal Utilization and Fuel Systems*, 1994, pp. 805–817.

62. Crippa, E.R., 50,000 HP coal slurry diesel engine, *Proc. 19th Int. Tech. Conf. on Coal Utilization and Fuel Systems*, 1994, pp. 821–828.

63. Bradish, T.J., Battista, J.J. and Zawadzki, E.A., Co-firing of water slurry in a 32 MW pulverized coal boiler, *Proc. 18th Int. Tech. Conf. on Coal Utilization and Fuel Systems*, 1993, pp. 303–313.

64. Yanagimacho, H., Matsumoto, O. and Tsuri, M., CWM production in China and CWM properties in all stages from production to combustion in the world's first bilateral CWM trade, *Proc. 18th Int. Tech. Conf. on Coal Utilization and Fuel Systems*, 1993, pp. 327–337.

65. Morrison, D.K., Melick, T.A. and Sommer, T.M., Utilization of coal water fuels in fire tube boilers, *Proc. 18th Int. Tech. Conf. on Coal Utilization and Fuel Systems*, 1993, pp. 339–347.

66. Pisupati, S.V., Britton, S.A., Miller, B.G. and Scarani, A.W., Combustion performance of coal water slurry fuel in an off-the-shelf 15,000 lb steam/h fuel oil designed industrial boiler, *Proc. 18th Int. Tech. Conf. on Coal Utilization and Fuel Systems*, 1993, pp. 349–360.

67. Morrison, J.L., Miller, B.G. and Scaroni, A.W., Preparing and handling coal water slurry fuels: potential problems and solutions, *Proc. 18th Int. Tech. Conf. on Coal Utilization and Fuel Systems*, 1993, pp. 361–368.

68. Battista, J.J. and Zawadzki, E.A., Economics of coal water slurry, *Proc. 18th Int. Tech. Conf. on Coal Utilization and Fuel Systems*, 1993, pp. 455–466.

69. Ohene, F., Luther, D. and Simon, U., Fundamental investigation of non-Newtonian behavior of coal water slurry on atomization, *Proc. 18th Int. Tech. Conf. on Coal Utilization and Fuel Systems*, 1993, pp. 607–617.

70. Kihm, K.D. and Kim, S.S., Investigation of dynamic surface tension of coal water slurry (CWS) fuels for application to atomization characteristics, *Proc. 18th Int. Tech. Conf. on Coal Utilization and Fuel Systems*, 1993, pp. 637–648.

71. Tu, J., Yao, Q., Cao, X., Cen, K., Ren, J., Huang, Z., Liu, J., Wu, X. and Zhao, X., Studies on thermal radiation ignition of coal water slurry, *Proc. 18th Int. Tech. Conf. on Coal Utilization and Fuel Systems*, 1993, pp. 659–668.

72. Hamich, T. and Siffert, B., Physical-chemical properties of coals in aqueous medium, *Proc. 18th Int. Tech. Conf. on Coal Utilization and Fuel Systems*, 1993, pp. 771–782.

73. Fullerton, K.L., Lee, S. and Kesavan, S., Laboratory scale dynamic stability testing for coal slurry fuel development, *Proc. 18th Int. Tech. Conf. on Coal Utilization and Fuel Systems*, 1993, pp. 799–808.

74. Zang, Z.-X., Zhang, L., Fu, X. and Jiang, L., Additive for coal water slurry made from weak slurrability coal, *Proc. 18th Int. Tech. Conf. on Coal Utilization and Fuel Systems*, 1993, pp. 821–833.

75. Hamich, T. and Siffert, B., Rheological properties of coal water highly concentrated suspensions, *Proc. 18th Int. Tech. Conf. on Coal Utilization and Fuel Systems*, 1993, pp. 809–820.

76. Sadler, L.Y. and Sim, K.G., Minimize solid–liquid mixture viscosity by optimizing particle size distribution, *Chem. Eng. Prog.*, 87(3), 68–71, 1991.

77. Everett, D.H., *Basic Principles of Colloid Science*, Royal Society of Chemistry, London, 1988.

# 5 Liquid Fuels from Natural Gas

*James G. Speight*

## CONTENTS

## 5.1 INTRODUCTION

*Natural gas* is the gaseous mixture associated with petroleum reservoirs and is predominantly methane. Like petroleum and coal, natural gas is derived from the remains of plants and animals and microorganisms that lived millions and millions of years ago.

Natural gas is used primarily as a fuel and as a raw material in manufacturing. It is used in home furnaces, water heaters, and cooking stoves. As an industrial fuel, it is used in brick, cement, and ceramic-tile kilns; in glass making; for generating steam in water boilers; and as a clean heat source for sterilizing instruments and processing foods. As a raw material in petrochemical manufacturing, natural gas is used to produce hydrogen, sulfur, carbon black, and ammonia. The ammonia is used in a range of fertilizers and as a secondary feedstock for manufacturing other chemicals, including nitric acid and urea. Ethylene, an important petrochemical, is also produced from natural gas.

Natural gas is considered an environmentally clean fuel, offering important environmental benefits when compared to other fossil fuels. Its superior environmental qualities over coal or crude oil are that emissions of sulfur dioxide are negligible and that the levels of nitrogen oxide and carbon dioxide emissions are lower. This helps to reduce problems of acid rain, ozone layer damage, or greenhouse gas emission. Natural gas is also a very safe source of energy when transported, stored, and used.

Natural gas, in itself, might be considered a very uninteresting gas; it is colorless and odorless in its pure form. Natural gas is a combustible mixture of hydrocarbon gases, and while the major constituent is methane, ethane, propane, butane, and pentane are also present, and the composition of natural gas varies widely. It is combustible and, when burned, produces energy. Unlike other fossil fuels, however, natural gas is clean burning and emits lower levels of potentially harmful by-products into the air.

Liquid fuels, on the other hand, are usually more complex mixtures of hydrocarbons than natural gas. Thus, for the purposes of this chapter, the term *liquid fuel* includes all liquids ordinarily and practically usable in internal combustion engines. Diesel and all aviation fuels are also included within this definition.

Natural gas can also be used to produce alternative fuels. The term *alternative fuel* includes methanol, ethanol, and other alcohols; mixtures containing methanol; and other alcohols with gasoline or other fuels, biodiesel, fuels (other than alcohol) derived from biological materials, and any other fuel that is substantially not a petroleum product.

The production of liquid fuels from sources other than petroleum broadly covers liquid fuels that are produced from tar sand (oil sand) bitumen, coal, oil shale, and natural gas. Synthetic liquid fuels have characteristics similar to those of liquid fuels generated from petroleum but differ because the constituents of synthetic liquid fuels do not occur naturally in the source material used for the production (Han and Chang 1994). Thus, the creation of liquids to be used as fuels from sources other than to crude petroleum broadly defines synthetic liquid fuels. For much of the twentieth century, for synthetic fuels, emphasis was on liquid products derived from coal upgrading or by extraction or hydrogenation of organic matter in coke liquids, coal tars, tar sands, or bitumen deposits. More recently, the potential for natural gas as a source of liquid fuels has been recognized, and attention is now on the development of natural gas as a source of liquid fuels.

Projected shortages of petroleum make it clear that, for the present century, alternative sources of liquid fuels are necessary. Such sources (for example, natural gas) are available, but the exploitation technologies are, in general, not as mature as for petroleum. The feasibility of the upgrading of natural gas to valuable chemicals, especially liquid fuels, has been known for years. However, the high cost of the steam-reforming and the partial oxidation processes, used for the conversion of natural gas to synthesis gas (syngas), has hampered the widespread exploitation of natural gas. Other sources include tar sand (also called oil sand or bituminous sand) (Speight 1990, 2008) and coal (Speight 2008, 2013a), which are also viable sources of liquid fuels.

Natural gas, which typically has 85%–95% methane, has been recognized as a plentiful and clean alternative feedstock to crude oil. Currently, the rate of discovery of proven natural gas reserves is increasing faster than the rate of natural gas production. Many of the large natural gas deposits are located in areas where abundant crude oil resources lie, such as in the Middle East. However, huge reserves of natural gas are also found in many other regions of the world, providing oil-deficient countries access to a plentiful energy source. The gas is frequently located in remote areas far from centers of consumption, and pipeline costs can account for as much as one-third of the total natural gas cost. Thus, tremendous strategic and economic incentives exist for gas conversion to liquids, especially if this can be accomplished on site or at a point close to the wellhead where shipping costs becomes a minor issue.

However, despite reduced prominence, coal technology continues to be a viable option for the production of liquid fuels in the future. World petroleum production is expected ultimately to level off and then decline, and despite apparent surpluses of natural gas, production is expected to suffer a similar decline. Coal gasification to syngas is utilized to synthesize liquid fuels in much the same manner as natural gas steam-reforming technology. But the important aspect is to use the natural gas reserves when they are available and to maximize the use of these reserves by conversion of natural gas to liquid fuels.

Liquid fuels possess inherent advantageous characteristics in terms of being more readily stored, transported, and metered than natural gas. Liquid fuels are also generally easy to process or clean by chemical and catalytic means. Also, liquid fuels are more compatible with the twentieth century and early twenty-first century world fuel infrastructure because most fuel-powered conveyances are designed to function only with relatively clean, low-viscosity liquids. Therefore, production of synthetic fuels from alternative feedstocks is based on adjusting the hydrogen-to-carbon ratio to the desired intermediate level.

This chapter covers the essential technical points of natural gas. The objectives of the chapter are to give the reader an introduction to natural gas by describing the origin and composition of natural gas, gas sources, phase behavior and properties, and transportation methods.

## 5.2 OCCURRENCE AND RESOURCES

Natural gas occurs in the porous rock of the Earth's crust either alone or with accumulations of petroleum (Mokhatab et al. 2006; Speight 2007). The gas is under pressure in the rock reservoirs in the Earth's crust, either in conjunction with and dissolved in higher-molecular-weight hydrocarbons and water or by itself. It is produced from the reservoir similarly to or in conjunction with crude oil. Like petroleum, natural gas has been formed by the degradation of organic matter accumulated in the past millions of years (Speight 2014).

Natural gas resources are typically divided into two categories: conventional and unconventional (Mokhatab et al. 2006; Speight 2007, 2014). Conventional gas typically is found in reservoirs with a permeability greater than 1 millidarcy (1 mD) and can be extracted via traditional techniques. A large proportion of the gas produced globally to date is conventional and is relatively easy and inexpensive to extract. In contrast, unconventional gas is found in reservoirs with relatively low permeability (<1 mD) and hence cannot be extracted by conventional methods.

When natural gas is associated with petroleum, the reservoir usually contains three main *fluids*: (1) natural gas, (2) oil, and (3) water, with minor constituents being acid gases (carbon dioxide and hydrogen sulfide). These components will vary greatly in combination and proportion within each reservoir, and in the case of heavy oil, the amount of gas will be substantially less than would be found in a conventional oil reservoir. The gas forms the *gas cap*, which is the mass of gas trapped between the liquid petroleum and the impervious cap rock of the petroleum reservoir. When the pressure in the reservoir is sufficiently high, the natural gas may be dissolved in the petroleum and is released upon penetration of the reservoir as a result of drilling operations.

There is an abundance of natural gas in North America, but it is a nonrenewable resource and essentially irreplaceable. Therefore, understanding the availability of the supply of natural gas is important as the use of natural gas is increased. Current estimates of natural gas availability throughout the world are substantial (BP 2012). Thus, there is a vast amount of natural gas estimated to still be in the ground. However, it is important to compare the different methodologies and the different systems of classification used in various estimates that are completed. Therefore, it is important to delve into the assumptions and methodology behind each study to gain a complete understanding of the estimate itself with a particular focus on proven and potential resources.

Constant revisions are being made to the estimates. New technology, combined with increased knowledge of particular areas and reservoirs, means that these estimates are in a constant state of flux. Further complicating the scenario is the fact that there are no universally accepted definitions for the terms that are used differently by geologists, engineers, and resource accountants.

Most of the natural gas that is found in North America is concentrated in relatively distinct geographical areas, or basins. The states that are located on top of a major basin have the highest level of natural gas reserves; U.S. natural gas reserves are very concentrated around Texas and the Gulf of Mexico.

Natural gas produced from geological formations varies widely in composition and can be broadly categorized into four distinct groups: (1) nonassociated gas, which occurs in conventional gas fields; (2) associated gas, which occurs in conventional petroleum reservoirs; (3) continuous gas or unconventional gas, which includes *tight gas* or *tight sands gas*, found in low-permeability rock (shale) (Speight 2013b); and (4) coal-bed methane (CBM), which is natural gas that has been formed along with the geological processes that formed coal (Speight 2013a), natural gas from geopressurized aquifers (which refers to gas dissolved under high pressure and at high temperatures in brine located deep beneath the surface of the Earth), gas hydrates (which are icelike structures of water and gas located under the permafrost), and deep gas (which is found at levels much deeper than conventional gas) (Speight 2014).

Briefly, shale formations and silt formations are the most abundant sedimentary rocks in the Earth's crust. In petroleum geology, organic shale formations are source rocks as well as seal rocks that trap oil and gas (Speight 2014). In reservoir engineering, shale formations are flow barriers. In drilling, the bit often encounters greater shale volumes than reservoir sands. In seismic exploration, shale formations interfacing with other rocks often form good seismic reflectors. As a result, seismic and petrophysical properties of shale formations and the relationships among these properties are important for both exploration and reservoir management.

Shale gas (*tight gas* or *tight sands gas*) is natural gas produced from shale formations that typically function as both the reservoir and source rocks for the natural gas. In terms of chemical makeup, shale gas is typically a dry gas composed primarily of methane (60%–95% v/v), but some formations do produce wet gas. The Antrim and New Albany shale formations have typically produced water and gas. Gas shale formations are organic-rich shale formations that were previously regarded only as source rocks and seals for gas accumulating in the strata near sandstone and carbonate reservoirs of traditional onshore gas development.

## 5.2.1 NONASSOCIATED GAS

Nonassociated gas (sometimes called *gas-well gas*) is produced from geological formations that typically do not contain much, if any, higher-boiling hydrocarbons (gas liquids) than methane (Speight 2007, 2014). Nonassociated gas can contain non-hydrocarbon gases such as carbon dioxide and hydrogen sulfide. Nonassociated gas is directly controllable by the producer; one just turns the valves. The gas flows up the well under the influence of the reservoir's own energy, through the wellhead control valves, and along the flow line to the treatment plant. Treatment requires the temperature of the gas to be reduced to a point dependent upon the pressure in the pipeline so that all liquids that would exist at pipeline temperature and pressure condense and are removed (Speight 2007, 2014).

## 5.2.2 ASSOCIATED GAS

Associated gas is produced during crude oil production and is the gas that is associated with crude oil (Speight 2007, 2014). Crude oil cannot be produced without producing some of its associated gas, which comes out of solution as the pressure is reduced on the way to and on the surface. Crude oil in the reservoir with minimal or no dissolved associated gas is rare and, as dead crude oil, is often difficult to produce as there is little energy to drive it.

After the production fluids are brought to the surface, they are separated at a tank battery at or near the production lease into a hydrocarbon liquid stream (crude oil or *gas condensate*), a produced water stream (brine or salty water), and a gaseous stream. The gaseous stream is traditionally very rich (*rich gas*) in *natural gas liquids* (NGLs); the latter include ethane, propane, butanes, and pentanes and higher-molecular-weight hydrocarbons ($C_{6+}$). The higher-molecular-weight hydrocarbon product is commonly referred to as *natural gasoline*.

There are several general definitions that have been applied to natural gas. Thus, *lean* gas is gas in which methane is the major constituent. *Wet gas* contains considerable amounts of the higher-molecular-weight hydrocarbons. *Sour gas* contains hydrogen sulfide, whereas *sweet gas* contains very little, if any, hydrogen sulfide. *Residue gas* is natural gas from which the higher-molecular-weight hydrocarbons have been extracted, and *casing head gas* is derived from petroleum but is separated at the separation facility at the wellhead. The terms *rich gas* and *lean gas*, as used in the gas-processing industry, are not precise indicators of gas quality, but only indicate the relative amount of NGLs in the gas stream (Speight 2007, 2014).

On the other hand, the terms *dry gas* and *wet gas* do have quantitative meaning—the term *dry gas* indicates that there is less than 0.1 gal. (1 gal., US, = 264.2 m$^3$) of gasoline vapor (higher-molecular-weight paraffins) per 1000 ft.$^3$ (1 ft.$^3$ = 0.028 m$^3$) while the term *wet gas* indicates that there are such paraffins present in the gas, in fact, more than 0.1 gal./1000 ft.$^3$.

## 5.3 COMPOSITION

Natural gas is colorless and odorless in its pure form. It is a combustible mixture of hydrocarbon gases. While natural gas is formed primarily of methane ($CH_4$), it can also include ethane ($C_2H_6$), propane ($C_3H_8$), butane ($C_4H_{10}$), and pentane ($C_5H_{12}$). The composition of natural gas can vary widely (Table 5.1) before it is refined. In its purest form, natural gas is almost pure methane ($CH_4$). Ethane, propane, and the other hydrocarbons are also commonly associated with natural gas (Table 5.1).

Natural gas is considered *dry* when it is almost pure methane, having had most of the other commonly associated hydrocarbons removed. When other hydrocarbons are present, the natural gas is *wet*.

Natural gas is commonly associated with petroleum reservoirs. Once brought from underground, the natural gas is refined to remove impurities such as water, other gases, sand, and various other compounds. Some of the constituent hydrocarbons (such as propane and butane) are removed and sold separately. After refining, the clean natural gas is transmitted through a network of pipelines to its point of use.

The amount of energy (energy content, heat content, calorific value) that is obtained from the burning of a volume of natural gas is measured in British thermal units (Btu) and is proportional to the amount of hydrocarbon constituents in the gas. Thus, the value of natural gas is calculated by its Btu content. The energy content of natural gas is variable and depends on its accumulations, which are influenced by the

---

**TABLE 5.1**
**Typical Composition of Natural Gas**

| Constituent | Formula | % v/v |
|---|---|---|
| Methane | $CH_4$ | 70–90 |
| Ethane | $C_2H_6$ | 0–5 |
| Propane | $C_3H_8$ | 0–5 |
| Butane | $C_4H_{10}$ | 0–5 |
| Pentane | $C_5H_{12}$ | 0–5 |
| Hexane (and higher) | $\geq C_6H_{14}$ | Trace–5 |
| Benzene (and higher) | $\geq C_6H_6$ | Trace |
| Carbon dioxide | $CO_2$ | 0–8 |
| Oxygen | $O_2$ | 0–0.2 |
| Nitrogen | $N_2$ | 0–5 |
| Hydrogen sulphide | $H_2S$ | 0–5 |
| Rare gases | He, Ne, A, Kr, Xe | Trace |
| Water | $H_2O$ | Trace–5 |

---

amount and types of energy gases they contain: the more noncombustible gases in a natural gas, the lower the Btu value. In addition, the volume mass of energy gases that are present in a natural gas accumulation also influences the Btu value of natural gas. The more carbon atoms in a hydrocarbon gas, the higher its Btu value. Btu analyses of natural gas are done at each stage of the supply chain. Gas chromatographic process analyzers are used in order to conduct fractional analysis of the natural gas streams, separating natural gas into identifiable components. The components and their concentrations are converted into a gross heating value in Btu–cubic foot.

The Wobbe index (calorific value divided by the specific gravity) gives a measure of the heat input to an appliance through a given aperture at a given gas pressure.

## 5.4  NATURAL GAS LIQUIDS

*NGLs*, often referred to as *natural gasoline*, are often used to designate the hydrocarbons having higher molecular weight than methane that also occur in natural gas (Speight 2007). Mixtures of liquefied petroleum gas, pentanes, and higher-molecular-weight hydrocarbons fall into this category. Caution should be taken not to confuse *natural gasoline* with the term *straight-run gasoline* (often also incorrectly referred to as natural gasoline), which is the gasoline distilled unchanged from petroleum.

## 5.5  CONVERSION OF NATURAL GAS TO LIQUIDS

Two routes can be used achieve the conversion of natural gas to liquid fuels via indirect technology, once the syngas has been produced (Figure 5.1) and both routes have been commercialized. One route involves use of the Fischer–Tropsch technology to produce liquid fuels directly or upon further processing (Chadeesingh 2011). The other route involves the production of methanol and, thereafter, the conversion of methanol to liquid fuels.

In general, the proven technology to upgrade methane is via steam reforming to produce syngas (carbon monoxide plus hydrogen). Such a gas mixture is clean and,

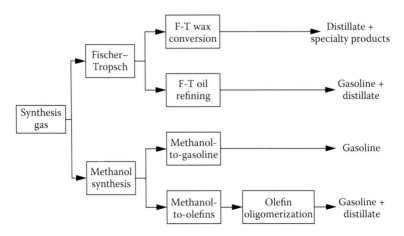

**FIGURE 5.1**   Natural gas to liquid fuels.

when converted to liquids, produces liquid fuels that are, with the exception of a trace amount, free of heteroatom compounds that contain sulfur and nitrogen.

The direct methane conversion technology, which has received considerable attention, involves the oxidative coupling of methane to produce higher hydrocarbons, such as ethylene. These olefin products (i.e., hydrocarbons containing the $-C = C-$function) may be upgraded to liquid fuels via catalytic oligomerization processes, as currently practiced in the petroleum and petrochemical industries (Speight 2014 and references cited therein).

A second trend in synthetic fuels is increased attention to oxygenate compounds as alternative fuels as a result of the growing environmental concern about burning fossil-based fuels. The environmental impact of the oxygenates, such as methanol, ethanol, and methyl *tert-butyl* ether (MTBE), requires very serious consideration since the environmental issues regarding their use are not fully understood or resolved.

Thus, the use of natural gas for production of synthetic fuels (synfuels) and chemicals offers and a clean and economic alternative to conventional fuels and chemicals. For chemicals production, the most promising scenarios involve manufacturing olefins and also the ensuing polymer products at a remote location and shipping them to developed markets.

Several technologies have been, and continue to be, evaluated by various companies, including gas-to-liquids (GTL), methanol-to-gasoline (MTG), methanol-to-olefins (MTO), methanol-to-propylene (MTP), olefins-to-gasoline and distillates (MOGD), dimethyl ether (DME), and large-scale methanol processes, and power generation from methanol.

The commercially proven technologies by Shell (middle-distillate synthesis [MDS] process) and Sasol for the production of middle distillates via GTL processes show a great potential for fuel alternatives and higher-value products.

Fischer–Tropsch naphtha and gas oils, produced by the various GTL processes, are attractive for steam cracking applications because of their high concentration of normal paraffin components. The high paraffinic content of the Fischer–Tropsch liquids allows them to be cracked at very high severities that are not usual for conventional feedstocks. When compared to conventional naphtha cracking yields, the yields for the Fischer–Tropsch naphtha show a higher selectivity to ethylene and less to heavier products, such as butanes.

### 5.5.1 Syngas Production

*Syngas* is a mixture of carbon monoxide (CO) and hydrogen ($H_2$) that is manufactured by steam-reforming natural gas. Sometimes, carbon dioxide is almost always inevitably produced.

$$CH_4 + H_2O \rightarrow CO + 3H_2$$

The name *synthesis gas* originates from its use as an intermediate in creating synthetic natural gas (SNG) and for producing ammonia and/or methanol. Syngas is also used as an intermediate in producing synthetic fuels via the Fischer–Tropsch reaction. In the strictest sense, syngas consists primarily of carbon monoxide and hydrogen, although carbon dioxide and nitrogen may also be present. The chemistry of syngas production is relatively simple, but the reactions are often much more complex than indicated by simple chemical equations (Speight 2008, 2013a, 2014; Chadeesingh 2011).

This gas is an extremely important precursor for the synthesis of methanol, hydrogen, ammonia, and other products. In particular, methane can be converted to syngas and gasoline can then be produced from this mixture by the Fischer–Tropsch process. Gasoline is currently produced in Sasol, South Africa, and in Malaysia using this technology. Methanol can be produced from syngas using the MTG process:

$$CO + 2H_2 \rightarrow CH_3OH$$

The syngas produced using existing methods accounts for about 60% of the total conversion cost of natural gas to gasoline. Thus, natural gas is an excellent candidate as an abundant alternative energy source.

The trend to increase the number of hydrogenation (*hydrocracking* and/or *hydrotreating*) processes in refineries coupled with the need to process the heavier oils, which require substantial quantities of hydrogen for upgrading, has resulted in vastly increased demands for this gas. Part of the hydrogen needs can be satisfied by hydrogen recovery from catalytic reformer product gases, but other external sources are required (Hsu and Robinson 2006; Gary et al. 2007; Speight 2014). Most of the external hydrogen is manufactured either by steam-methane reforming or by oxidation processes. However, other processes, such as steam–methanol interaction or ammonia dissociation, may also be used as sources of hydrogen. Electrolysis of water produces high-purity hydrogen, but the power costs may be prohibitive.

The steam-methane reforming is a continuous catalytic process that has been employed for syngas production and, in some refineries, hydrogen production over a period of several decades. The major reaction is the formation of carbon monoxide and hydrogen from methane and steam,

$$CH_4 + H_2O \rightarrow CO + 3H_2,$$

but higher-molecular-weight feedstocks (such as propane and other hydrocarbon constituents of natural gas) may also yield carbon monoxide and hydrogen that can be adjusted to fit the required ratio for syngas:

$$C_3H_8 + 3H_2O \rightarrow 3CO + 7H_2$$

That is,

$$C_nH_m + nH_2O \rightarrow nCO + (0.5m + n)H_2$$

In the actual process, the feedstock is first desulfurized by passage through activated carbon, which may be preceded by caustic and water washes. The desulfurized material is then mixed with steam and passed over a nickel-based catalyst ($730°C–845°C$ [$1350–1550°F$] and 400 psi). Effluent gases are cooled by the addition of steam or condensate to about 370°C (700°F), at which point (if hydrogen is the desired product) carbon monoxide reacts with steam in the presence of iron oxide in a shift converter to produce carbon dioxide and hydrogen. The carbon dioxide is removed by amine washing; the hydrogen is usually a high-purity (>99%) material.

Another syngas generation process is a continuous, noncatalytic process that produces carbon monoxide and hydrogen by partial oxidation of gaseous or liquid hydrocarbons. A controlled mixture of preheated feedstock and oxygen is fed to the top of the reactor, where carbon dioxide and steam are the primary products. A secondary reaction between the feedstock and the gases forms carbon monoxide and hydrogen.

If necessary for hydrogen production only, the effluent is then led to a shift converter with high-pressure steam, where carbon monoxide is converted to carbon dioxide with the concurrent production of hydrogen at the rate of 1 mole of hydrogen for every mole of carbon dioxide. Reactor temperatures vary from 1095°C to 1480°C (2000°F–2700°F), and pressures vary from atmospheric to more than 1500 psi. Gas purification depends on the use of the gas.

### 5.5.2 Fischer–Tropsch Process

The hydrocarbon resources of the world are not evenly distributed, and in particular, a substantial proportion of known reserves are situated in locations remote from areas of high consumption. Transportation of liquid hydrocarbons from source to consumer is a task for which a large and flexible infrastructure exists. However, where natural gas deposits in remote locations are to be exploited, the transportation task becomes a major challenge—particularly if geography, economics, or a combination of both precludes the possibility of a pipeline.

There are two routes for the production of synthetic fuels from natural gas (via syngas): (1) the hydrocarbons route and (2) the methanol route. Both routes often fall under the general umbrella description of the Fischer–Tropsch synthesis (Chadeesingh 2011).

The first step for both routes is the conversion of natural gas into syngas—a mixture of hydrogen and carbon monoxide with some carbon dioxide. The proportions of these components in the mixture vary according to the individual synthesis process selected and also according to the product slate desired. Typical values of the principal characteristic, the $H_2/CO$ ratio, for different processes cover a wide range from below 1 to nearly 3. The range of $H_2/CO$ ratios required for the different synthesis processes means that considerable effort is required to match the syngas generation and synthesis process so as to ensure the optimum overall conversion rate. In addition, varying amounts of pure hydrogen may be required for hydrogenation of the crude product from the synthesis.

#### 5.5.2.1 General Process Description

Fischer–Tropsch synthesis is a well-known process for conversion of syngas to synthetic fuels and raw materials for the chemical industry. The process is versatile in its raw materials consumption, that is, it can use any type of coal, natural gas, or similar carbon-containing feedstock as its material and similarly, the product distribution can also be changed. The increase of mineral oil prices has caused intense efforts to develop the Fischer–Tropsch process on a commercial scale. The details of the Fischer–Tropsch reactions and process are still not entirely understood, even though the reaction was discovered 90 years ago.

Synthetic fuels produced by the Fischer–Tropsch process are nowadays more expensive than natural oil-based hydrocarbon fuels. However, under certain conditions and in the next century, the process economics may become favorable. The process deserves special attention (1) where coal reserves are significant and available at a cheaper rate and (2) where natural gases are abundant. In the long run, production of hydrocarbon synthesis based on coal will exceed that of oil.

Syngas (a mixture of CO and $H_2$) can be obtained from coal/coke, natural gas, or similar carbon-containing feed stock by steam-reforming or gasification processes (above). Catalysts are used for the production of hydrocarbons. Metals such as ruthenium (Ru), iron (Fe), and cobalt (Co) are usually deposited on an inert support, such as silica, alumina, or aluminosilicates, to increase the surface area of the catalyst (Espinoza et al. 1999; Steynberg et al. 1999). Addition of promoters to improve properties of the catalyst or selectivity are mostly applied. Typical reaction conditions of the Fischer–Tropsch process are P = 10 – 40 bar and T = 200°C–300°C.

### 5.5.2.2 Chemistry

In the Fischer–Tropsch synthesis, hydrocarbons ($C_xH_y$) are synthesized from carbon monoxide (CO) and hydrogen ($H_2$) (syngas). The Fischer–Tropsch process is an established technology and already applied on a large scale.

The number of chemical reactions involved in the manufacture of syngas is very large. The most important of these (limited to methane since it is the major constituent of natural gas), given the objective of producing carbon monoxide and hydrogen from the methane, and the most desirable reactions are those of reforming (Equation 5.1) and partial oxidation (Equation 5.3) producing hydrogen/carbon monoxide ratios of 3 and 2, respectively. If a source of carbon dioxide is available (or natural gas rich in carbon dioxide), reforming with carbon dioxide (Equation 5.2) provides a hydrogen/carbon monoxide ratio of 1. The figures for higher hydrocarbons in natural gas are correspondingly lower. The final hydrogen/carbon monoxide ratio is influenced further by the carbon monoxide shift reaction (Equation 5.5).

1. Reforming (strongly endothermic)

$$CH_4 + H_2O \rightarrow CO + 3H_2 \tag{5.1}$$

$$CH_4 + CO_2 \rightarrow 2CO + 2H_2 \tag{5.2}$$

2. Combustion (strongly exothermic)

$$2CH_4 + O_2 \rightarrow 2CO + 4H_2 \tag{5.3}$$

$$CH_4 + 2O_2 \rightarrow CO_2 + 2H_2O \qquad\qquad (5.4)$$

3. Shift conversion (mildly exothermic)

$$CO + H_2O \rightarrow CO_2 + H_2 \qquad\qquad (5.5)$$

Carbon

$$CH_4 \rightarrow 2H_2 + C \qquad\qquad (5.6)$$

$$2CO \rightarrow CO_2 + C \qquad\qquad (5.7)$$

The reforming reactions (Equations 5.1 and 5.2) are strongly endothermic and must be supported by the strongly exothermic reactions of partial oxidation (Equation 5.3) and/or complete combustion (Equation 5.4). The latter reaction is, however, in principle, less desirable since neither $H_2$ nor CO is produced.

Thus, in the (exothermic) Fischer–Tropsch reaction, 1 mole of carbon monoxide reacts with 2 moles of hydrogen to afford a hydrocarbon chain extension ($-CH_2-$). The oxygen from the CO is released as product water:

$$CO + 2H_2 \rightarrow -CH_2 + H_2O \quad \Delta H = -165 \text{ kJ/mol}$$

The reaction implies a hydrogen/carbon monoxide ratio of at least 2 for the synthesis of the hydrocarbons. When the ratio is lower, it can be adjusted in the reactor with the catalytic water gas shift reaction:

$$CO + H_2O \rightarrow CO_2 + H_2 \quad \Delta H = -42 \text{ kJ/mol}$$

When catalysts with water gas shift activity are used, the water produced in the reaction can react with carbon monoxide to form additional hydrogen. In this case, a minimal $H_2$/CO ratio of 0.7 is required, and the oxygen from the CO is released as $CO_2$:

$$2CO + H_2 \rightarrow -CH_2 + CO_2 \quad \Delta H = -204 \text{ kJ/mol}$$

The reaction affords mainly aliphatic straight-chain hydrocarbons ($C_xH_y$). Besides these straight-chain hydrocarbons, also, branched hydrocarbons, unsaturated hydrocarbons (olefins), and primary alcohols are formed in minor quantities. The kind of liquid obtained is determined by the process parameters (temperature, pressure), the kind of reactor, and the catalyst used. Typical operation conditions for the Fischer–Tropsch synthesis are a temperature range of 200°C–350°C and pressures of 15–40 bar, depending on the process.

The Fischer–Tropsch process is a very complicated process that requires a well-defined choice of reactors, catalysts, and operating conditions to synthesize the desired products. Even then, a mixture of compounds is obtained. The Fischer–Tropsch synthesis was originally operated in packed bed reactors. These reactors have several drawbacks that can be overcome by a slurry reactor. In these slurry reactors, the syngas is bubbled through a suspension of catalyst particles (typically 30–50 μm) in an inert liquid. The heat of reaction is removed by circulating the slurry through external or internal heat exchangers. The slurry reactor can be operated at higher temperatures and at low $H_2$/CO ratios without problems due to efficient heat transfer and uniform temperatures. The application of very small catalyst particles caused no occurrence of intraparticle heat and mass transfer resistances.

Economic studies have shown that the Fischer–Tropsch synthesis in a slurry bubble column has several advantages over the fixed-bed reactors. Because the Fischer–Tropsch process offers a number of advantages in the slurry phase over the two-phase processes, special attention is paid to the effects of an inert liquid phase on the reaction rate, mass transfer, and product distribution.

### 5.5.2.3 Products

The subsequent chain growth in the Fischer–Tropsch process is comparable with a polymerization process, resulting in a distribution of chain lengths of the products. In general, the product range includes the light hydrocarbons methane ($CH_4$), ethane ($C_2H_6$), propane ($C_3H_8$), and butane ($C_4H_{10}$); naphtha ($C_5H_{12}$ to $C_{12}H_{26}$); kerosene–diesel fuel ($C_{13}H_{28}$ to $C_{22}H_{46}$); low-molecular-weight wax ($C_{23}H_{48}$ to $C_{32}H_{66}$); and high-molecular-weight wax (>$C_{33}H_{68}$). Linear alpha-olefins are also produced, but the distribution of the products depends on the catalyst and the process operation conditions (temperature, pressure, and residence time).

The (theoretical) chain length distribution can be described by means of the following equation (Chadeesingh 2011):

$$\log W_n/n = 2\log (\ln \alpha) + n\log \alpha$$

$W_n$ is the weight fraction of chains with $n$ carbon atoms and the chain growth probability factor ($\alpha$), which is defined by

$$\alpha = k_p/k_p + k_t$$

where $k_p$ is the propagation rate and $k_t$ is the termination rate.

### 5.5.2.4 Catalysts

Several types of catalysts can be used for the Fischer–Tropsch synthesis (Ponec 1982; Snel 1987; Van der Laan and Beenackers 1999). The most important catalysts are based on iron (Fe) or cobalt (Co). Cobalt catalysts have the advantage of a higher conversion rate and a longer life (over 5 years). The cobalt catalysts are, in general, more reactive for hydrogenation and, therefore, produce less unsaturated hydrocarbons and alcohols compared to iron catalysts. Iron catalysts have a higher tolerance for sulfur, are cheaper, and produce more olefin products and alcohols. The lifetime of the iron catalysts is short and, in some commercial installations, may be measured in weeks.

### 5.5.2.5 Commercial Processes

The three main industrially proven processes involve various versions of tubular steam reforming, catalytic autothermal reforming, and noncatalytic partial oxidation.

In tubular steam reforming, the methane–steam reaction (Equation 5.1) takes place over a catalyst in a tube that is externally heated. A large steam surplus is required to suppress carbon formation in the catalyst. This tends to drive the carbon monoxide–steam shift reaction (Equation 5.5) to the right, resulting in a hydrogen-rich syngas. The heat is supplied largely by the undesirable complete combustion reaction of methane to carbon dioxide and water (Equation 5.4) outside of the tubes.

In catalytic autothermal reforming, oxygen is added to the feed. The heat requirement for the methane–steam reaction (Equation 5.1) is largely met by the methane–oxygen partial oxidation reaction (Equation 5.1), thus producing a lower hydrogen/carbon monoxide ratio in the syngas. As in tubular reforming, considerable amounts of steam are required to suppress carbon formation. The absence of the metallurgical limitations of the catalyst tubes of a steam reformer allows higher operating temperatures, thus reducing methane slip. At these higher temperatures, the carbon monoxide shift equilibrium is also more favorable to carbon monoxide than in the case of the tubular steam reformer.

In noncatalytic partial oxidation, the reaction of methane with oxygen (Equation 5.3) is dominant. The absence of any catalyst means that the process is tolerant of a small degree of carbon formation and allows even higher operating temperatures. It is thus possible to operate partial oxidation without any steam addition, and the resulting syngas is rich in carbon monoxide.

The art of selecting the right syngas generation process (or combination of processes) consists of ensuring the correct gas specification as required by the selected synthesis while simultaneously minimizing certain inherent inefficiencies of the individual processes. In the case of tubular reforming, this inherent inefficiency lies in the use of external complete combustion, requiring an expensive heat recovery train and still involving substantial losses in the stack gas. In the case of autothermal reforming and partial oxidation, the inefficiency lies in the energy requirement and investment for the oxygen plant.

Lurgi GMBH is currently involved in the design and supply of syngas production units for two major synfuel projects—one based on Sasol's Synthol process and the other using the Shell middle-distillate synthesis (SMDS) (Sie et al. 1991). There is a substantial difference in the hydrogen/carbon monoxide ratios required by the two processes, and this has led to the selection of different syngas production routes.

The *Synthol* process has been operated at commercial scale since the 1950s by Sasol and has undergone some evolution. The initial process, the Arge process, involved low temperatures (200°C–250°C), medium pressures, and a fixed catalyst bed. The process primarily produced a linear paraffin wax, which had use as petrochemical feedstock and also for transport fuels after further processing. This was the only process available until the 1950s and 1960s, when the Sasol Synthol process was developed, and the process involved higher temperatures (300°C–360°C) and medium pressures, but used a circulating fluidized bed to produce light olefins for chemicals production and gasoline components. This process has recently been updated to the Advanced Synthol process.

The latest development of Fischer–Tropsch technology is the Sasol slurry phase reactor, which is an integral part of the Sasol slurry phase distillate (SSPD) process, and carries out the synthesis reaction at low temperatures (200°C–250°C) and low pressures.

The process involves bubbling hot syngas through a liquid slurry of catalyst particles and liquid reaction products. Heat is removed from the reactor via coils within the bed, producing steam. Liquid products are removed from the reactor, and the liquid hydrocarbon wax is separated from the catalyst. The gas stream from the top of the reactor is cooled to recover light hydrocarbons and reaction water. The Sasol slurry phase technology has undergone several developments primarily concerned with catalyst formulations. Initial development used an iron-based catalyst, but recent developments have used a cobalt-based catalyst, giving greater conversion.

The *SMDS* process is a two-step process (Figure 5.2) that involves Fischer–Tropsch synthesis of paraffinic wax called heavy paraffin synthesis (HPS). The wax is subsequently hydrocracked and isomerized (in the presence of hydrogen) to yield a middle-distillate-boiling-range product in the heavy paraffin conversion (HPC) (Sie et al. 1991; Eisenberg et al. 1998). In the HPS stage, wax is maximized by using a proprietary catalyst having high selectivity toward heavier products and by the use of a tubular, fixed-bed reactor. The HPC stage employs a commercial hydrocracking catalyst in a trickle-flow reactor. The HPC step allows for production of narrow-range hydrocarbons not possible with conventional Fischer–Tropsch technology.

The products manufactured are predominantly paraffinic, free from sulfur, nitrogen, and other impurities, and have excellent combustion properties. The very high cetane number and smoke point indicate clean-burning hydrocarbon liquids having reduced harmful exhaust emissions. The process has also been proposed to produce chemical intermediates, paraffinic solvents, and extra-high-viscosity-index lube oils.

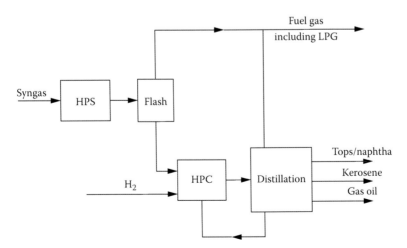

**FIGURE 5.2**  Shell middle-distillate synthesis (SMDS) process.

The *Lurgi combined reforming* process was originally developed for large-scale methanol production, and it is with an example from this application that it is described here. In the context of the production of liquid fuels, it is suitable as a building block for the MTG process or the MOGD process. For the high-pressure Synthol process, carbon dioxide needs to be purged from the system. The conventional tubular steam-reforming process as used for methanol syngas production produces a hydrogen/carbon monoxide ratio of over 4 and a stoichiometric ratio of 2.6 to 2.9 (i.e., a hydrogen-rich gas) depending on the quality of the natural gas feedstock.

Autothermal reforming or partial oxidation produces carbon monoxide-rich gases with a hydrogen/carbon monoxide ratio in the range of 1.5 to 3.5 and a stoichiometric number of approximately 1.8.

Approximately half the feed is processed in the tubular primary reformer. The other half, together with the primary reformer effluent, is autothermally reformed with pure oxygen in the secondary reformer. Besides matching hydrogen-rich and carbon monoxide-rich process steps to produce an optimum stoichiometric ratio, the combined reforming process has beneficial effects, such as the following:

- The methane slip of the overall reforming process is governed by the temperature of the secondary reformer that is not subject to the same limitations of tube metallurgy as the tubular reformer. The combined process can thus provide a lower methane slip.
- Less syngas of optimized quality is required per ton of methanol, reducing both the syngas compressor load and the capital cost of the synthesis unit.
- The operating temperature of the primary reformer need no longer be chosen to minimize methane slip. The reformer can be operated under mild conditions. The higher operating pressure thus enables the syngas compressor load to be further reduced.
- The reduced throughput through the primary reformer and the lower operating temperature combine to reduce the tubular reformer to about 25% of the size of that required for the single-stage process. This reduces the stack gas losses referred to earlier by the same amount.

The Shell SGP process is a much older process, the basic development having been made in the 1950s, and more than 150 units have been built in the last 50 years. With natural gas feedstock, the unit produces syngas with a hydrogen/carbon monoxide ratio of 1.7 to 1.8 and a carbon dioxide content of 1.7 to 3 volume percent, depending on the steam addition rate.

In the unit, the gas feed is preheated with the raw gas to a temperature of about 380°C for desulfurization prior to being fed to the SGP reactor with oxygen. The partial oxidation reaction takes place at about 1300°C to 1400°C in the refractory lined reactor. The sensible heat of the hot gas is used to generate high-pressure steam, with or without superheat as required. The noncatalytic partial oxidation reactor produces small amounts of soot that are washed out in a scrubber. The carbon is concentrated in the reaction water that is discharged to the wastewater treatment.

Comparing the syngas quality with that produced by combined reforming shows that there is a considerably lower hydrogen/carbon monoxide ratio of 1.86, compared with 3.14 for combined reforming, making the SGP process a better match for, for instance, the SMDS process. The amount of natural gas required to produce the syngas is some 3.5% lower than for the case of combined reforming. These advantages may be offset by higher oxygen requirements.

One feature common to all Fischer–Tropsch processes is the inherent lack of selectivity. The actual selectivity will depend on the desired product slate as well as on the catalyst and the operating conditions. In all cases, however, substantial quantities of gaseous hydrocarbons including methane are produced.

In principle, it is desirable to recycle these gaseous hydrocarbons to produce more syngas. On the other hand, it is necessary to provide a purge of inert gases, principally argon and nitrogen. The other aspect of selectivity that requires recognition is that a proportion of higher-molecular-weight products, including waxes, is produced, and this may require hydrotreating for conversion to a saleable product.

The Syntroleum process is a cost-effective refinement of GTL technology that has been in use for several decades. A major advantage is that the process uses compressed air instead of pure oxygen to facilitate the conversion reaction, substantially reducing the capital costs and vastly improving the safety of the process plants.

The Syntroleum process consists of three major reaction steps: (1) natural gas is first partially oxidized with air to produce syngas; (2) the syngas is then reacted in a Fischer–Tropsch reactor to produce liquid hydrocarbons of various chain lengths; and (3) the higher-boiling fraction of the products is separated and hydrocracked to transportation fuels. The liquid synthetic fuels (synfuels) produced are middle distillates consisting of naphtha, kerosene, diesel, and other hydrocarbon-based products.

### 5.5.3 OTHER PROCESSES

In the ExxonMobil AGC-21 process, there are three key steps (Lopez et al. 2003). In the first step, syngas is generated by contacting methane with steam and a limited amount of oxygen in a high-capacity catalytic reactor. Hydrocarbons are synthesized in the second step at high alpha as described by a distribution in a novel slurry reactor using new, high-productivity catalysts operating at high levels of syngas conversion (Chadeesingh 2011). The full-range, primarily normal paraffin product contains significant 650°F+ waxy material that is a solid at room temperature and melts above 250°F, unsuitable for pipelining or transporting in conventional crude carriers. The final step, accomplished with proprietary catalysts in a packed bed reactor, converts wax to high-quality liquids that make excellent feeds for refineries and chemical plants, and directly marketable products in some instances, such as lube base stocks or specialty solvents.

The chemistry of each step is straightforward yet becomes more complex as processes go to high yield and selectivity. Oxygen, methane, and steam ratios are carefully controlled to produce syngas (carbon monoxide and hydrogen) at stoichiometric proportions of about 2.1 to 1 hydrogen to carbon monoxide.

Methanol is produced catalytically from syngas, and by-products such as ethers, formic acid esters, and higher hydrocarbons are formed in side reactions and are found in the crude methanol product. Whereas for many years, methanol was produced from coal, after World War II, low-cost natural gas and light petroleum fractions replaced coal as the feedstock. Following from this, one of the most significant developments in synthetic fuel technology since the discovery of the Fischer–Tropsch process is the Mobil MTG process (Chang 1983). Methanol is efficiently transformed into hydrocarbons ranging from ethane to decane by a reaction that is catalyzed by synthetic zeolites.

$$n\mathrm{CH_3OH} \rightarrow \mathrm{C}_n\mathrm{H}_{2n+2} + n\mathrm{H_2O}$$

Olefins and aromatic compounds are also produced, and the chemistry of the reaction is more complex than this simple equation illustrates. For accuracy and economics, chemical equations should be derived for each particular aspect of the process.

In the process (which can use a fixed-bed reactor or a fluid bed reactor), the methanol feed, vaporized by heat exchange with reactor effluent gases, is converted in a first-stage reactor containing an alumina catalyst to an equilibrium mixture of methanol, DME, and water. This is combined with recycled light gas, which serves to remove reaction heat from the highly exothermic MTG reaction, and enters the reactors containing zeolite catalyst, and reaction conditions are 360°C–415°C at a pressure of approximately 300–350 psi.

Because the MTG process produces primarily gasoline, a variation of that process has been developed that allows for production of gasoline and distillate fuel. The combined process (MTO and MOGD) produces gasoline and distillate in various proportions and, if needed, can be terminated at a point to produce olefin by-products.

In the MTO process, methanol is converted over a zeolite catalyst to give high yields of olefins with some ethylene and low-boiling light saturated hydrocarbons. Generally, catalyst and process variables that increase methanol conversion decrease olefin yield. In the MTO process, typical conversions exceed 99.9%. The coked catalyst is continuously withdrawn from the reactor and burned in a regenerator.

In the MOGD process, low-molecular-weight olefins produce oligomers in the gasoline boiling range using a zeolite catalyst. Other distillate products are also produced. Gasoline/distillate product ratios can vary, depending on process conditions, from 0.2 to >100.

The direct conversion of natural gas (methane) to liquid fuels involves conversion of methane to the desired liquid fuels while bypassing the syngas step. Direct upgrading routes that have been extensively studied include direct partial oxidation to oxygenates, oxidative coupling to higher hydrocarbons, and pyrolysis to higher hydrocarbons.

In a series of patents, the Rentech process takes advantage of the synergism between iron-based Fischer–Tropsch synthesis and plasma-based syngas production. For example, in one process (Yakobson et al. 2002), the plasma-based syngas production is a front-end conversion process whereby a hydrocarbon feedstock, such as natural gas, is fed to a high- or low-temperature plasma torch or electrical arc reactor. The arc converts the feedstock into a hydrogen and carbon monoxide syngas. This syngas can then be converted by the Fischer–Tropsch process into liquid hydrocarbons. During the Fischer–Tropsch synthesis using iron-based catalysts, a portion of the syngas is converted into carbon dioxide. By recycling the carbon dioxide extracted from the process tail gas back to the plasma torch, it can be efficiently converted into carbon monoxide, which is then fed to the Fischer–Tropsch reactor.

The plasma can reform natural gas with only the addition of carbon dioxide from the Fischer–Tropsch reactor, producing a syngas consisting almost exclusively of carbon monoxide and hydrogen. In addition to the carbon dioxide produced in the Fischer–Tropsch reactor, the reactor tail gas components can be converted into additional carbon monoxide and hydrogen for the Fischer–Tropsch reactor feedstock. This synergy between iron-based Fischer–Tropsch synthesis and plasma-based syngas production allows for significantly improved carbon conversion efficiency and is an excellent technique for reducing carbon dioxide emissions while producing cleaner fuels that are sulfur and aromatic free. Additionally, plasma-based syngas production offers distinct advantages over other methods for producing syngas as it does not need an air separation plant and has essentially no moving parts.

Oxygen permeable membranes can potentially substitute for the expensive cryogenic oxygen plants for oxygen production. When combined with an appropriate catalyst on the oxygen-lean side, the membrane can be used to convert natural gas to syngas via a partial oxidation reaction:

$$2CH_4 + O_2 \rightarrow CO + 2H_2$$

## 5.6 THE FUTURE

Fischer–Tropsch diesel has a high cetane number and can be manufactured free of both aromatics and sulfur. Fischer–Tropsch diesel can be used in existing distribution infrastructure and diesel engines, and compared to petroleum diesel, it has lower emissions of oxides of nitrogen and particulate matter. The near-zero sulfur content of Fischer–Tropsch diesel may also enable exhaust aftertreatment, resulting in further emission reductions. Thus, Fischer–Tropsch diesel, could help displace the diesel fuel that is typically used to power medium- and heavy-duty vehicles and meet specifications for extremely low-sulfur diesel fuel.

The syngas production routes described above are based on proven technologies and provide us with reliable starting points for the development of processes that offer the potential for further reduction of both capital and operating costs.

For some syntheses, the use of straight catalytic autothermal reforming can be of advantage. Lurgi has used this process for both methanol production and treating Fischer–Tropsch tail gases. Whereas in a secondary reformer configuration, the hot hydrogen-rich primary reformer effluent is self-igniting, ignition of a straight autothermal reformer requires the use of a noble metal-promoted ignition catalyst. Velocities in the ignition-catalyst area must be kept high to eliminate any possibility of back burning. These high velocities lead, over a period of time, to mechanical attrition of the expensive ignition catalyst. As part of the ongoing development of its reforming processes, the process arrangement has been modified to bring the operating conditions closer to that of a secondary reformer, thus dispensing with the need for the ignition catalyst.

The HOT reforming technology is based on the use of the so-called HCT reformer tube. In principle, this is a normal centrifugally cast reformer tube, catalyst filled, heated from outside and normally designed for downflow of the process gas through the catalyst bed. But on the inside, it encloses a double helix made of tubes of suitable material embedded in the catalyst. The reformed process gas passes this double helix in counterflow to the process gas flow through the catalyst bed, thus transferring a part of its sensible heat to the reforming process. Calculations and practical experience have shown that based on an inlet temperature of 450°C and reaction end temperature of 860°C, this internal heat transfer covers up to 20% of the sensible and reaction heat of the process gas. In addition to the resultant saving of fuel, an investment savings of approximately 15% can be expected, the bulk of which is attributable to the smaller convection bank required.

Based on current conditions, the outlook for future energy supplies and conversion technologies indicates a growing reliance on natural gas as the economic fuel of choice for the generation of electric power. Although relatively low-cost, abundant supplies of natural gas are projected to be available for the foreseeable future, the committee also recognizes that the future is inherently uncertain and long-term projections of price trends and fuel mixtures could easily be wrong. The role of coal, while not necessarily declining rapidly, could suffer a temporary slowdown in use if future environmental policies (in response to concerns about global climate change) require sizeable reductions in emissions of carbon dioxide from coal-fired power plants (Speight 2013c). Therefore, maintaining a diverse, low-cost energy supply will require controlling these emissions through carbon management (NRC 2000).

## REFERENCES

BP. 2012. *BP Statistical Review of World Energy*. British Petroleum, 1 St. James's Square, London, England. See also: http://www.bp.com/centres/energy/index.asp.

Chadeesingh, R. 2011. The Fischer–Tropsch process, Part 3, Chapter 5. In *The Biofuels Handbook*, J.G. Speight (Editor). The Royal Society of Chemistry, London, United Kingdom, pp. 476–517.

Chang, C.D. 1983. Hydrocarbons from Methanol. *Catalysis Reviews—Science and Engineering*, 25: 1.

Eisenberg, B., Fiato, R.A., Mauldin, C.H., Ray, G.R., and Soled, R.L. 1998. Advanced Gas Conversion Technology. *Studies in Surface Science and Catalysis*, 119: 943.

Espinoza, R.L., Steynberg, A.P., Jager, B., and Vosloo, A.C. 1999. Low temperature Fischer-Tropsch synthesis from a Sasol perspective. *Applied Catalysis A: General*, 186: 13–29.

Gary, J.G., Handwerk, G.E., and Kaiser, M.J. 2007. *Petroleum Refining: Technology and Economics*, 5th Edition. CRC Press, Taylor & Francis Group, Boca Raton, Florida.

Han, S., and Chang, C.D. 1994. Synthetic (liquid) fuels. In *Kirk-Othmer Encyclopaedia of Chemical Technology*, 4th Edition, J.I. Kroschwitz and M. Howe-Grant (Editors). John Wiley & Sons Inc., New York, 12: 155.

Hsu, C.S., and Robinson, P.R. (Editors). 2006. *Practical Advances in Petroleum Processing*, Volume 1 and Volume 2. Springer Science, New York.

Lopez, A.M., Fiato, R.A., Ansell, I.L., Quinlan, C.W., and Ramage, M.P. 2003. *Hydrocarbon Asia*, July/August, p. 56.

Mokhatab, S., Poe, W.A., and Speight, J.G. 2006. *Handbook of Natural Gas Transmission and Processing*. Elsevier, Amsterdam, Netherlands.

NRC. 2000. *Vision 21: Fossil Fuel Options for the Future Committee on R&D Opportunities for Advanced Fossil-Fueled Energy Complexes, Board on Energy and Environmental Systems, National Research Council*. National Academy Press, Washington, DC.

Ponec, V. 1982. Fischer-Tropsch Synthesis and Some Related Heterogeneous Reactions. In *Metal Support and Metal Additive Effects in Catalysis*, B. Imelik (Editor). Elsevier, Amsterdam, The Netherlands, p. 63 et seq.

Sie, S.T., Senden, M.M.G., and van Wechum, H.M.H. 1991. Conversion of Natural Gas to Transportation Fuels. *Catalysis Today*, 8: 371.

Snel, R. 1987. Fischer-Tropsch Products. *Catalysis Reviews—Science and Engineering*, 29: 361.

Speight, J.G. 1990. Natural Gas. In *Fuel Science and Technology Handbook*, Part V, J.G. Speight (Editor). Marcel Dekker Inc., New York.

Speight, J.G. 2007. *Natural Gas: A Basic Handbook*. GPC Books, Gulf Publishing Company, Houston, Texas.

Speight, J.G. 2008. *Synthetic Fuels Handbook: Properties, Processes, and Performance*. McGraw-Hill, New York.

Speight, J.G. 2013a. *The Chemistry and Technology of Coal*, 3rd Edition. Marcel Dekker Inc., New York.

Speight, J.G. 2013b. *Shale Gas Production Processes*. Gulf Professional Publishing, Elsevier, Oxford, United Kingdom.

Speight, J.G. 2013c. *Coal-Fired Power Generation Handbook*. Scrivener Publishing, Salem, Massachusetts.

Speight, J.G. 2014. *The Chemistry and Technology of Petroleum*, 5th Edition. CRC Press, Taylor & Francis Group, Boca Raton, Florida.

Steynberg, A.P., Espinoza, R.L., Jager, B., and Vosloo, A.C. 1999. High temperature Fischer–Tropsch synthesis in commercial practice. *Applied Catalysis A: General*, 186: 41–54.

Van der Laan, G.P., and Beenackers, A.A.C.M. 1999. Fischer-Tropsch Synthesis. *Catalysis Reviews—Science and Engineering*, 41: 255.

Yakobson, D.L., Vavruska, J.S., Bohn, E., and Blutke, A. 2002. United States Patent 6,380,268, April 30.

# 6 Resids

*James G. Speight*

## CONTENTS

## 6.1  INTRODUCTION

Approximately 2000 years ago, Arabian scientists developed methods for the distillation of petroleum that led to an interest in the thermal product of petroleum (nafta, naphtha) when it was discovered that this material could be used as an illuminant and as a supplement to asphalt incendiaries in warfare. The discovery also led to the production of resids that may have been used when the supply of bitumen or natural asphalt from natural seepages became limited.

A *residuum* (plural *residua*, also shortened to *resid*, plural *resids*) is the residue obtained from petroleum after nondestructive distillation has removed all the volatile materials. The temperature of the distillation is usually maintained below the temperature at which the rate of thermal decomposition of petroleum constituents is minimal.

*Residua* are black, viscous materials and are obtained by distillation of a crude oil under atmospheric pressure (atmospheric residuum) in an atmospheric distillation unit (atmospheric tower, atmospheric pipe still) or under reduced pressure (vacuum residuum) in a vacuum distillation unite (vacuum tower, vacuum pipe still) (Figure 6.1). They may be liquid at room temperature (generally atmospheric residua) or almost solid (generally vacuum residua) depending on the nature of the petroleum from which the resid was obtained (Table 6.1) or depending upon the cut point of the distillation (Table 6.2).

Resids contain very-high-molecular-weight molecular polar species called asphaltenes that are soluble in carbon disulfide, pyridine, aromatic hydrocarbons, and chlorinated hydrocarbons (Van Gooswilligen 2000; Speight and Ozum 2002; Hsu and Robinson 2006; Gary et al. 2007; Speight 2014). They are the nonvolatile fractions of petroleum that are isolated from the atmospheric distillation unit and form the vacuum distillation unit. Resids derive their characteristics from the nature of their crude oil precursor, with some variation possible by choice of the end point of the distillation.

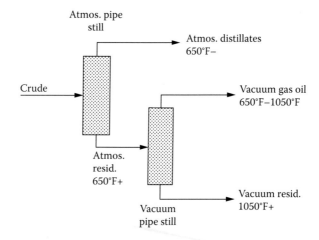

**FIGURE 6.1**  Schematic of resid production.

# TABLE 6.1
## Properties of Atmospheric and Vacuum Resids from Different Crude Oils

| Crude Oil Origin | Kuwait | | Khafji | | Bachaquero | | West Texas | | Safaniya | | Alaska (North Slope) | | Tia Juana (Light) | |
|---|---|---|---|---|---|---|---|---|---|---|---|---|---|---|
| Residuum Type | Atmospheric | Vacuum | Atmospheric | Vacuum | Atmospheric | Vacuum | Atmospheric | Vacuum | Atmospheric | Vacuum | Atmospheric | Vacuum | Atmospheric | Vacuum |
| Fraction of crude, vol% | 42 | 21 | – | – | 34 | – | – | – | 40 | 22 | 58 | 22 | 49 | 18 |
| Gravity, °API | 13.9 | 5.5 | 14.4 | 6.5 | 17 | 2.8 | 9.4 | 18.4 | 11.1 | 2.6 | 15.2 | 8.2 | 17.3 | 7.1 |
| Viscosity SUS, 210°F | – | – | – | – | – | – | 313 | 86 | – | – | 1281 | – | 165 | – |
| SPS, 122°F | 553 | 500,000 | 429 | – | – | – | – | – | – | – | – | – | 172 | – |
| SPS, 210°F | – | – | – | – | – | – | – | – | – | – | – | – | – | – |
| cSt, 100°F | – | – | – | – | – | – | – | – | – | – | 42 | 1950 | 890 | – |
| cSt, 210°F | 55 | 1900 | – | – | – | – | – | – | – | – | 75 | – | 35 | 7959 |
| Pour point, °F | 65 | – | – | – | – | – | – | – | – | – | 1.6 | 2.2 | 1.8 | 2.6 |
| Sulfur, wt% | 4.4 | 5.45 | 4.1 | 5.3 | 2.4 | 3.7 | 3.3 | 2.5 | 4.3 | 5.3 | 1.6 | 2.2 | 1.8 | 2.6 |
| Nitrogen, wt% | 0.26 | 0.39 | – | – | 0.3 | 0.6 | 0.5 | 0.6 | 0.4 | 0.4 | 0.36 | 0.63 | 0.3 | 0.6 |
| Metals, ppm | | | | | | | | | | | | | | |
| Nickel | 14 | 32 | 37 | 53 | 450 | 100 | 27 | 11 | 26 | 46 | 18 | 47 | 25 | 64 |
| Vanadium | 50 | 102 | 89 | 178 | – | 900 | 57 | 20 | 109 | 177 | 30 | 82 | 185 | 450 |
| Asphaltenes, wt% | | | | | | | | | | | | | | |
| Pentane insolubles | – | 11.1 | – | 12.0 | 10 | – | – | – | 17.0 | 30.9 | 4.3 | 8.0 | – | – |
| Hexane insolubles | – | – | – | – | – | – | – | – | – | – | – | – | – | – |
| Heptane insolubles | 2.4 | 7.1 | – | – | – | – | – | – | – | – | 31.5 | – | – | – |
| Resin, wt% | – | 39.4 | – | – | – | – | – | – | – | – | – | – | – | – |
| Carbon residue, wt% | | | | | | | | | | | | | | |
| Ramsbottom | 9.8 | – | – | – | – | – | 16.9 | 6.6 | 14.0 | 25.9 | 8.4 | 17.3 | – | – |
| Conradson | 12.2 | 23.1 | – | 21.4 | 12 | 27.5 | – | – | – | – | – | – | 9.3 | 21.6 |

*Source:* From Speight, J.G., *The Desulfurization of Heavy Oils and Residua,* 2nd ed., Marcel Dekker, New York, 2000. With permission.

**TABLE 6.2**

**Properties of Tia Juana Crude Oil and Resids Produced Using Deferent Cut Points**

| Boiling Range | | | | | | | | | |
|---|---|---|---|---|---|---|---|---|---|
| °F | Whole crude | >430 | >565 | >650 | >700 | >750 | >850 | >950 | >1050 |
| °C | Whole crude | >220 | >295 | >345 | >370 | >400 | >455 | >510 | >565 |
| Yield on crude, vol% | 100.0 | 70.2 | 57.4 | 48.9 | 44.4 | 39.7 | 31.2 | 23.8 | 17.9 |
| Gravity, °API | 31.6 | 22.5 | 19.4 | 17.3 | 16.3 | 15.1 | 12.6 | 9.9 | 7.1 |
| Specific gravity | 0.8676 | 0.9188 | 0.9377 | 0.9509 | 0.9574 | 0.9652 | 0.9820 | 1.007 | 1.0209 |
| Sulfur, wt% | 1.08 | 1.42 | 1.64 | 1.78 | 1.84 | 1.93 | 2.12 | 2.35 | 2.59 |
| Carbon residue (Conradson), wt% | – | 6.8 | 8.1 | 9.3 | 10.2 | 11.2 | 13.8 | 17.2 | 21.6 |
| Nitrogen, wt% | – | – | – | 0.33 | 0.36 | 0.39 | 0.45 | 0.52 | 0.60 |
| Pour point, °F | –5 | 15 | 30 | 45 | 50 | 60 | 75 | 95 | 120 |
| Viscosity: | | | | | | | | | |
| Kinematic, cSt @ 100°F | 10.2 | 83.0 | 315 | 890 | 1590 | 3100 | – | – | – |
| @ 210°F | 10.2 | 83.0 | 315 | 890 | 1590 | 3100 | – | – | – |
| Furol (SFS) sec | – | 9.6 | 19.6 | 35.0 | 50.0 | 77.0 | 220 | 1010 | 7959 |
| @ 122°F | – | – | 70.6 | 172 | 292 | 528 | – | – | – |
| @ 210°F | – | – | – | – | 25.2 | 37.6 | 106 | 484 | 3760 |
| Universal (SUS) sec @ 210°F | – | 57.8 | 96.8 | 165 | 234 | 359 | 1025 | – | – |
| Metals: | | | | | | | | | |
| Vanadium, ppm | – | – | – | 185 | – | – | – | – | 450 |
| | – | – | – | 25 | – | – | – | – | 64 |
| | – | – | – | 28 | – | – | – | – | 48 |

*Source:* From Speight, J.G., *The Desulfurization of Heavy Oils and Residua*, 2nd ed., Marcel Dekker, New York, 2000. With permission.

When a residuum is obtained from a crude oil and thermal decomposition has commenced, it is more usual to refer to this product as *pitch*. The differences between parent petroleum and the residua are due to the relative amounts of various constituents present, which are removed or remain by virtue of their relative volatility.

The chemical composition of a residuum from an asphaltic crude oil is complex and subject to the method of production (i.e., the temperature at which the distillation is carried out). Physical methods of fractionation usually indicate high proportions of asphaltenes and resins, even in amounts up to 50% (or higher) of the residuum. In addition, the presence of ash-forming metallic constituents, including such organo-metallic compounds as those of vanadium and nickel, is also a distinguishing feature of residua and heavy oils. Furthermore, the deeper the *cut* into the crude oil, the greater is the concentration of sulfur and metals in the residuum and the greater the deterioration in physical properties (Speight 2000 and references cited therein).

Even though a resid is a manufactured product, the constituents do occur naturally as part of the native petroleum, assuming that thermal decomposition has not taken place during distillation. Resids are specifically produced during petroleum refining, and the properties of the various residua depend upon the cut point or boiling point at which the distillation is terminated.

## 6.2  RESID PRODUCTION

Resid production involves distilling everything possible from crude petroleum until a nonvolatile residue remains. This is usually done by stages (Speight and Ozum 2002; Hsu and Robinson 2006; Gary et al. 2007; Speight 2014) in which distillation at atmospheric pressure removes the lower-boiling fractions and yields an atmospheric residuum (*reduced crude*) that may contain higher-boiling (lubricating) oils, wax, and asphalt.

The *atmospheric distillation* tower is divided into a number of horizontal sections by metal trays or plates, and each is the equivalent of a still. The more trays, the more redistillation, and hence, the better the fractionation or separation of the mixture fed into the tower. A tower for fractionating crude petroleum may be 13 feet in diameter and 85 feet high with 16 to 28 trays. The feed to a typical tower enters the vaporizing or flash zone, an area without trays. The majority of the trays are usually located above this area. The feed to a bubble tower, however, may be at any point from top to bottom with trays above and below the entry point, depending on the kind of feedstock and the characteristics desired in the products.

However, the usual permissible temperature in the vaporizing zone to which the feedstock can be subjected is usually considered to be 350°C (660°F). The rate of thermal decomposition increases markedly above this temperature: if decomposition occurs within a distillation unit, it can lead to coke deposition in the heater pipes or in the tower itself, with the resulting failure of the unit. Some distillation units use temperatures in the vaporizing zone up to 393°C (740°F) and reduce the residence time in the hot zone. Caution is advised when using a higher temperature because units that are unable to continually attain the petroleum flow-through as specified in the design can suffer from resid decomposition (due to longer residence times in the vaporizing zone) and poorer quality asphalt.

Distillation of the reduced crude under vacuum removes the oils (and wax) as overhead products, and the asphalt remains as a bottom (or residual) product. The majority of the polar functionalities and high-molecular-weight species in the original crude oil, which tend to be nonvolatile, concentrate in the vacuum residuum (Speight 2000), thereby conferring desirable or undesirable properties on the asphalt.

At this stage, the residuum is frequently and incorrectly referred to as pitch and has a softening point (ASTM D-36, ASTM D-61, ASTM D-2319, ASTM D-3104, ASTM D-3461) related to the amount of oil removed and increases with increasing overhead removal. In character with the elevation of the softening point, the pour point is also elevated: the more oil distilled from the residue, the higher the softening point.

*Vacuum distillation* has seen wide use in petroleum refining since the boiling point of the heaviest cut obtainable by distillation at atmospheric pressure is limited by the temperature in the vaporizing zone (ca. 350°C–390°C, 660°F–740°F) at which the residue starts to decompose or *crack*, unless *cracking distillation* is preferred. When the feedstock is required for the manufacture of lubricating oils, further fractionation without cracking is desirable, and this can be achieved by distillation under vacuum conditions.

The fractions obtained by vacuum distillation of reduced crude include the following:

1. *Heavy gas oil*, an overhead product and used as catalytic cracking stock or, after suitable treatment, a light lubricating oil
2. *Lubricating oil* (usually three fractions: light, intermediate, and heavy), obtained as a sidestream product
3. *Vacuum residuum*, the nonvolatile product that may be used directly as asphalt or to asphalt

The residuum may also be used as a feedstock for a coking operation or blended with gas oils to produce a heavy fuel oil.

## 6.3 PROPERTIES

Resids exhibit a wide range of physical properties, and several relationships can be made between various physical properties (Speight 2014). Whereas the properties such as viscosity, density, boiling point, and color of petroleum may vary widely, the ultimate or elemental analysis varies, as already noted, over a narrow range for a large number of samples. The carbon content is relatively constant, while the hydrogen and heteroatom contents are responsible for the major differences between petroleum. The nitrogen, oxygen, and sulfur can be present in only trace amounts in some petroleum, which, as a result, consists primarily of hydrocarbons.

The properties of resids are defined by a variety of standard tests that can be used to define quality (Table 6.3). And remembering that the properties of residua vary with the cut point (Table 6.2), that is, the vol% of the crude oil, helps the refiner produce asphalt of a specific type or property.

Properties such as the API gravity and viscosity also help the refinery operator to gain an understanding of the nature of the material that is to be processed. The

---

**TABLE 6.3**
**Analytical Inspections for Petroleum and Resids**

| Petroleum | Heavy feedstocks |
|---|---|
| Density (specific gravity) | Density (specific gravity) |
| API gravity | API gravity |
| Carbon (wt%) | Carbon (wt%) |
| Hydrogen (wt%) | Hydrogen (wt%) |
| Nitrogen (wt%) | Nitrogen (wt%) |
| Sulfur (wt%) | Sulfur (wt%) |
| | Nickel (ppm) |
| | Vanadium (ppm) |
| | Iron (ppm) |
| Pour point | Pour point |
| Wax content | |
| Wax appearance temperature | |
| Viscosity (various temperatures) | Viscosity (various temperatures) |
| Carbon residue of residuum | Carbon residue[a] |
| | Ash (wt%) |
| Distillation profile | Fractional composition |
| All fractions plus vacuum residue | Asphaltenes, wt% |
| | Resins (wt%) |
| | Aromatics (wt%) |
| | Saturates (wt%) |

[a] Conradson carbon residue or microcarbon residue.

---

products from high-sulfur feedstocks often require extensive treatment to remove (or change) the corrosive sulfur compounds. Nitrogen compounds and the various metals that occur in crude oils will cause serious loss of catalyst life. The carbon residue presents an indication of the amount of thermal coke that may be formed to the detriment of the liquid products.

## 6.3.1 ELEMENTAL (ULTIMATE) ANALYSIS

The analysis of resids for the percentages of carbon, hydrogen, nitrogen, oxygen, and sulfur is perhaps the first method used to examine the general nature and perform an evaluation. The atomic ratios of the various elements to carbon (i.e., H/C, N/C, O/C, and S/C) are frequently used for indications of the overall character of the resid. It is also of value to determine the amounts of trace elements, such as vanadium and nickel, in a resid since these materials can have serious deleterious effects on process and product performance.

Resids are not composed of a single chemical species but, rather, are a complex mixture of organic molecules that vary widely in composition and are composed of carbon, hydrogen, nitrogen, oxygen, and sulfur as well as trace amounts of metals, principally vanadium and nickel. The heteroatoms, although a minor component compared to the hydrocarbon moiety, can vary in concentration over a wide range depending on the source of the asphalt and can be a major influence on asphalt properties.

Generally, most resids contain 79%–88% w/w carbon, 7%–13% w/w hydrogen, trace–8% w/w sulfur, 2%–8% w/w oxygen, and trace–3% w/w nitrogen. Trace metals such as iron, nickel, vanadium, calcium, titanium, magnesium, sodium, cobalt, copper, tin, and zinc occur in crude oils. Vanadium and nickel are bound in organic complexes and, by virtue of the concentration (distillation) process by which asphalt is manufactured, are also found in resids.

Thus, elemental analysis is still of considerable value to determine the amounts of elements in resids, and the method chosen for the analysis may be subject to the peculiarities or character of the resid under investigation and should be assessed in terms of accuracy and reproducibility. The methods that are designated for elemental analysis are as follows:

1. *Carbon* and *hydrogen content* (ASTM D-1018, ASTM D-3178, ASTM D-3343, ASTM D-3701, ASTM D-5291, ASTM E-777, IP 338)
2. *Nitrogen content* (ASTM D-3179, ASTM D-3228, ASTM D-3431, ASTM E-148, ASTM E-258, ASTM D-5291, and ASTM E-778)
3. *Oxygen content* (ASTM E-385)
4. *Sulfur content* (ASTM D-124, ASTM D-129, ASTM D-139, ASTM D-1266, ASTM D-1552, ASTM D-1757, ASTM D-2622, ASTM D-2785, ASTM D-3120, ASTM D-3177, ASTM D-4045 and ASTM D-4294, ASTM E-443, IP 30, IP 61, IP 103, IP 104, IP 107, IP 154, IP 243)

The most pertinent property in many contexts is the *sulfur content*. This and the API gravity represent the two properties that have the greatest influence on the value of a resid.

The sulfur content varies from about 2% to about 6% for residua (Speight 2000 and references cited therein). In fact, the nature of the distillation process by which residua are produced (removal of distillate without thermal decomposition) dictates that the majority of the sulfur, which is predominantly in the higher-molecular-weight fractions, be concentrated in the resid (Speight and Ozum 2002; Hsu and Robinson 2006; Gary et al. 2007; Speight 2014).

### 6.3.2 METAL CONTENT

Resids contain relatively high proportions of metals either in the form of salts or as organometallic constituents (such as the metalloporphyrins), which are extremely difficult to remove from the feedstock. Indeed, the nature of the process by which residua are produced virtually dictates that all the metals in the original crude oil are concentrated in the residuum (Speight 1999). Those metallic constituents that may actually *volatilize* under the distillation conditions and appear in the higher-boiling distillates are the exceptions here.

Determination of metals can be determined by direct methods (ASTM 2003) and also by indirect methods in which the sample is combusted so that only inorganic ash remains. The ash can then be digested with an acid and the solution examined for metal species by atomic absorption (AA) spectroscopy or by inductively coupled argon plasma (ICP) spectrometry.

### 6.3.3 DENSITY AND SPECIFIC GRAVITY

*Density* is defined as the mass of a unit volume of material at a specified temperature and has the dimensions of grams per cubic centimeter (a close approximation to grams per milliliter). *Specific gravity* is the ratio of the mass of a volume of the substance to the mass of the same volume of water and is dependent on two temperatures, those at which the masses of the sample and the water are measured. The *API gravity* is also used.

For clarification, it is necessary to understand the basic definitions that are used: (1) *density* is the mass of liquid per unit volume at 15.6°C (60°F); (2) *relative density* is the ratio of the mass of a given volume of liquid at 15.6°C (60°F) to the mass of an equal volume of pure water at the same temperature; and (3) *specific gravity* is the same as the relative density, and the terms are used interchangeably.

Although there are many methods for the determination of density due to the different nature of petroleum itself, the accurate determination of the API gravity of petroleum and its products (ASTM D-287) is necessary for the conversion of measured volumes to volumes at the standard temperature of 60°F (15.56°C). Gravity is a factor governing the quality of crude oils. However, the gravity of a petroleum product is an uncertain indication of its quality. Correlated with other properties, gravity can be used to give approximate hydrocarbon composition and heat of combustion. This is usually accomplished though use of the API gravity that is derived from the specific gravity,

$$\text{API gravity, deg} = (141.5/\text{sp gr } 60/60°F) - 131.5,$$

and is also a critical measure for reflecting the quality of petroleum.

API gravity or density or relative density, can be determined using one of two hydrometer methods (ASTM D-287, ASTM D-1298). The use of a digital analyzer (ASTM D-5002) is finding increasing popularity for the measurement of density and specific gravity.

Most resids have a specific gravity of petroleum higher than 1.0 (10° API), with most resids having an API gravity on the order of 5° to 10° API.

### 6.3.4 VISCOSITY

*Viscosity* is the most important single fluid characteristic governing the motion of resids and is actually a measure of the internal resistance to motion of a fluid by reason of the forces of cohesion between molecules or molecular groupings. It is generally the most important property for monitoring resid behavior in pipes when it is moved from one unit to another.

A number of instruments are in common use with resids for this purpose. The vacuum capillary (ASTM D-2171) is commonly used to classify paving asphalt at

60°C (140°F). Kinematic capillary instruments (ASTM D-2170, ASTM D-4402) are commonly used in the 60°C to 135°C (140°F to 275°F) temperature range for both liquid and semisolid asphalts in the range of 30 to 100,000 cSt. Saybolt tests (ASTM D-88) are also used in this temperature range and at higher temperatures (ASTM E-102). At lower temperatures, the cone-and-plate instrument (ASTM D-3205) has been used extensively in the viscosity range of 1000 to 1,000,000 poises.

The viscosity of resids varies markedly over a very wide range and is dependent upon the cut point at which the distillation is terminated. Values vary from less than several hundred centipoises at room temperature to many thousands of centipoises at the same temperature. In the present context, the viscosity of vacuum resids is at the higher end of this scale, where a relationship between viscosity and density has been noted.

## 6.3.5 CARBON RESIDUE

The carbon residue presents indications of the *coke-forming propensity* of a resid. There are two older well-used methods for determining the carbon residue: the Conradson method (ASTM D-189) and the Ramsbottom method (ASTM D-524). A third, more modern thermogravimetric method (ASTM D-4530) is also in use.

Resids having high carbon residues and resids that contain metallic constituents will have erroneously high carbon residues. The metallic constituents must first be removed from the resid, or they can be estimated as ash by complete burning of the coke after carbon residue determination.

The *carbon residue* of a resid serves as an indication of the propensity of the sample to form carbonaceous deposits (thermal coke) under the influence of heat. What is produced is also often used to provide thermal data that give an indication of the composition of the asphalt (Speight 1999, 2001).

Tests for Conradson carbon residue (ASTM D-189, IP 13), the Ramsbottom carbon residue (ASTM Test Method D524, IP 14), the microcarbon carbon residue (ASTM D4530, IP 398), and asphaltene content (ASTM D-2006, ASTM D-2007, ASTM D-3279, ASTM D-4124, ASTM D-6560, IP 143) are often included in inspection data for resids. All three methods are applicable to resids. The data give an indication of the amount of coke that will be formed during thermal processes as well as an indication of the amount of asphaltenes in the resid.

The data produced by the microcarbon test (ASTM D4530, IP 398) are equivalent to those produced by the Conradson carbon method (ASTM D-189 IP 13). However, this microcarbon test method offers better control of test conditions and requires a smaller sample. Up to 12 samples can be run simultaneously.

Other test methods (ASTM D-2416, ASTM D-4715) that indicate the relative coke-forming properties of tars and pitches might also be applied to resids. Both test methods are applicable to resids having an ash content ≤0.5% (ASTM D-2415). The former test method (ASTM D-2416) gives results close to those obtained by the Conradson carbon residue test (ASTM D-189 IP 13). However, in the latter test method (ASTM D-4715), a sample is heated for a specified time at 550°C ± 10°C (1022°F ± 18°F) in an electric furnace. The percentage of residue is reported as the coking value.

Resids that contain ash-forming constituents will have an erroneously high carbon residue, depending upon the amount of ash formed.

### 6.3.6 HEAT OF COMBUSTION

The gross heat of combustion of crude oil and its products are given with fair accuracy by the equation

$$Q = 12,400 - 2100d^2,$$

where $d$ is the 60/60°F specific gravity; deviation is generally less than 1%.

For thermodynamic calculation of equilibria useful in petroleum science, combustion data of extreme accuracy are required because the heats of formation of water and carbon dioxide are large in comparison with those in the hydrocarbons. Great accuracy is also required of the specific heat data for the calculation of free energy or entropy. Much care must be exercised in selecting values from the literature for these purposes, since many of those available were determined before the development of modern calorimetric techniques.

### 6.3.7 MOLECULAR WEIGHT

For resids that have little or no volatility, *vapor pressure osmometry* (*VPO*) has been proven to be of considerable value.

The molecular weights of resids are not always (in fact, rarely) used in the determination of process behavior. Nevertheless, there may be occasions when the molecular weight of asphalt is desired.

Currently, of the methods available, several standard methods are recognized as being useful for determining of the molecular weight of petroleum fractions, and these methods are as follows:

- ASTM D-2224: Test Method for Mean Molecular Weight of Mineral Insulating Oils by the Cryoscopic Method (discontinued in 1989 but still used by some laboratories for determining the molecular weight of petroleum fractions up to and including gas oil)
- ASTM D-2502: Test Method for Estimation of Molecular Weight (Relative Molecular Mass) of Petroleum Oils from Viscosity Measurements
- ASTM D-2503: Test Method for Estimation of Molecular Weight (Relative Molecular Mass) of Hydrocarbons by Thermoelectric Measurement of Vapor Pressure
- ASTM D-2878: Method for Estimating Apparent Vapor Pressures and Molecular Weights of Lubricating Oils
- ASTM D-3593: Test Method for Molecular Weight Averages/Distribution of Certain Polymers by Liquid Size Exclusion (Gel Permeation) Chromatography—GPC—Using Universal Calibration (has also been adapted to the investigation of molecular weight distribution in petroleum fractions)

Each method has proponents and opponents because of assumptions made in the use of the method or because of the mere complexity of the sample and the nature of the intermolecular and intramolecular interactions. Before application of any one or more of these methods, consideration must be given to the mechanics of the method and the desired end result.

Methods for molecular weight measurement are also included in other more comprehensive standards (ASTM D-128, ASTM D-3712), and several indirect methods have been proposed for the estimation of molecular weight by correlation with other, more readily measured physical properties (Speight 1999, 2001).

The molecular weights of the individual fractions of resids have received more attention, and have been considered to be of greater importance, than the molecular weight of the resid itself (Speight 1999, 2001). The components that make up the resid influence the properties of the material to an extent that is dependent upon the relative amount of the component, the molecular structure of the component, and the physical structure of the component, which includes the molecular weight.

Asphaltenes have a wide range of molecular weights, from 500 to at least 2500, depending upon the method (Speight 1994). Asphaltenes associate in dilute solution in nonpolar solvents, giving higher molecular weights than is actually the case on an individual molecule basis. The molecular weights of the resins are somewhat lower than those of the asphaltenes and usually fall within the range of 500 to 1000. This is due not only to the absence of association but also to a lower absolute molecular size. The molecular weights of the oil fractions (i.e., the resid minus the asphaltene fraction and minus the resin fraction) are usually less than 500, often 300 to 400.

A correlation relating molecular weight, asphaltene content, and heteroatom content with the carbon residues of whole residua has been developed and has been extended to molecular weight and carbon residue (Schabron and Speight 1997a,b).

## 6.3.8 OTHER PROPERTIES

Resids may be liquid or solid at ambient temperature, and in the latter case, the *melting point* is a test (ASTM D-87 and D-127) that is widely used to know the melting point to prevent solidification in pipes.

The *softening point* of a resid is the temperature at which the resid attains a particular degree of softness under specified conditions of test.

There are several tests available to determine the softening point of resids (ASTM D-36, ASTM D-61, ASTM D-2319, ASTM D-3104, ASTM D-3461, IP 58). In the test method (ASTM D-36, IP 58), a steel ball of specified weight is laid on a layer of sample contained in a ring of specified dimensions. The softening point is the temperature, during heating under specified conditions, at which the asphalt surrounding the ball deforms and contacts a base plate.

The *pour point* is the lowest temperature at which the resid will pour or flow under prescribed conditions. For residua, the pour points are usually high (above 0°C, 32°F) and are more an indication of the temperatures (or conditions) required to move the material from one point in the refinery to another.

The thermal cracking of resid is a first-order reaction, but there is an induction period before the coke begins to form.

The focus in such studies has been on the asphaltene constituents and the resin constituents (Speight 2014 and references cited therein). Several chemical models (Speight 1994, 2014) describe the thermal decomposition of asphaltene constituents and, by inference, the constituents of the resin fraction. The prevalent thinking is that the polynuclear aromatic fragments become progressively more polar as the paraffinic fragments are stripped from the ring systems by scission of the bonds (preferentially) between the carbon atoms alpha and beta to the aromatic rings.

The environmental impact and toxicological profile of petroleum residua is high because of the high content of polynuclear aromatic hydrocarbon derivatives (typically in the asphaltene fraction), some of which contain heteroaromatic ring systems. Soil contamination has been a growing concern, because it can be a source of groundwater (drinking water) contamination; contaminated soils can reduce the usability of land for development; and weathered petroleum residua that are particularly recalcitrant may stay bound to soil for years, if not decades (Speight and Arjoon 2012).

The biodegradation of residua requires a complex metabolic pathway, which usually can be observed in a microbial community. Many studies have been carried out on the biodegradation of petroleum hydrocarbons using a consortium of microorganisms (Morais and Tauk-Tornisielo 2009). In the literature, it is revealed that in soils that are permeated with the heavy hydrocarbons, bacteria including indigenous ones would survive and function after contaminations (Whyte et al. 1997; Foght et al. 1990) seeped through the soil. Selection of bacterial communities for petroleum substances occurs rapidly even after short-term exposures of soil to petroleum hydrocarbons (Van der Meer et al. 1992; Van der Meer 1994). During adaptation of microbial communities to hydrocarbon components, particularly complex ones, genes for hydrocarbon-degrading enzymes that are carried on plasmids or transposons may be exchanged between species, and new catabolic pathways eventually may be assembled and modified for efficient regulation (Rabus et al. 2005). Other cell adaptations leading to new ecotypes may include modifications of the cell envelope to tolerate solvents (Ramos et al. 2002) and development of community-level interactions that facilitate cooperation within consortia (Kim and Crowley 2007).

One of the factors that limit the biodegradation of residua is the bioavailability of the constituents, which is directly related to the complex chemical. One of the options to increase bioavailability of polynuclear aromatic systems (as are found in residue) is the use of surfactants to enhance the oil mobility, improving the biodegradation.

## 6.4   COMPOSITION

Determination of the composition of resids has always presented a challenge because of the complexity and high molecular weights of the molecular constituents. The principle behind composition studies is to evaluate resids in terms of composition and process behavior.

Physical methods of fractionation usually indicate a high proportions of asphaltenes and resins, even in amounts up to 50% (or higher) of the residuum. In addition, the presence of ash-forming metallic constituents, including such organometallic compounds as those of vanadium and nickel, is also a distinguishing feature of resids. Furthermore, the deeper the *cut* into the crude oil, the greater the concentration of sulfur and metals in the residuum and the greater the deterioration in physical properties (Speight 2014 and references cited therein).

## 6.4.1 Chemical Composition

The chemical composition of resids is, in spite of the large volume of work performed in this area, largely speculative (Speight 2014). Acceptance that petroleum is a continuum of molecular types that continues from the low-boiling fractions to the nonvolatile fractions (Speight 1999 and references cited therein) is an aid to understanding the chemical nature of the heavy feedstocks.

### 6.4.1.1 Hydrocarbon Compounds

On a molecular basis, resids contain hydrocarbons as well as the organic compounds of nitrogen, oxygen, and sulfur; metallic constituents are also present. Even though free hydrocarbons may be present and the hydrocarbon skeleton of the various constituents may appear to be the dominating molecular feature, it is, nevertheless, the nonhydrocarbon constituents (i.e., nitrogen, oxygen, and sulfur) that play a large part in determining the nature and, hence, the processability of resids.

In the *asphaltene fraction* (a predominant fraction of resids), free condensed naphthenic ring systems may occur, but general observations favor the occurrence of combined aromatic–naphthenic systems that are variously substituted by alkyl systems. There is also general evidence that the aromatic systems are responsible for the polarity of the asphaltene constituents. Components with two aromatic rings are presumed to be naphthalene derivatives, and those with three aromatic rings may be phenanthrene derivatives. Currently, and because of the consideration of the natural product origins of petroleum, phenanthrene derivatives are favored over anthracene derivatives. In addition, trace amounts of peri-condensed polycyclic aromatic hydrocarbons such as methylchrysene, methylperylenes, and dimethylperylenes, and benzofluorenes have been identified in crude oil. Chrysene and benzofluorene homologues seem to predominate over those of pyrene.

The polycyclic aromatic systems in the *asphaltene fraction* are complex molecules that fall into a molecular weight and boiling range where very little is known about model compounds (Speight 1994, 1999). There has not been much success in determining the nature of such systems in resids. In fact, it has been generally assumed that as the boiling point of a petroleum fraction increases, so does the number of condensed rings in a polycyclic aromatic system. To an extent, this is true, but the simplicities of such assumptions cause an omission of other important structural constituents of the petroleum matrix, the alkyl substituents, the heteroatoms, and any polycyclic systems that are linked by alkyl chains or by heteroatoms.

### 6.4.1.2 Sulfur Compounds

Sulfur compounds are perhaps the most important nonhydrocarbon constituents of resids and occur as a variety of structures (Speight 1999 and references cited therein). During the refining sequences involved to convert crude oils to salable products, a great number of the sulfur compounds that occur in petroleum are concentrated in the resids.

The major sulfur species are alkyl benzothiophene derivatives, dibenzothiophene derivatives, benzonaphthothiophene derivatives, and phenanthrothiophene derivatives.

### 6.4.1.3  Nitrogen Compounds

The presence of nitrogen in resids is of much greater significance in refinery operations than might be expected from the relatively small amounts present.

Nitrogen in resids may be classed arbitrarily as *basic* and *nonbasic*. The basic nitrogen compounds (Speight 1999 and references cited therein), which are composed mainly of pyridine homologues and occur throughout the boiling ranges, have a decided tendency to exist in the higher-boiling fractions and residua (Speight and Ozum 2002; Hsu and Robinson 2006; Gary et al. 2007; Speight 2014). The nonbasic nitrogen compounds, which are usually of the pyrrole, indole, and carbazole types, also occur in the higher-boiling fractions and residua.

Typically, about one-third of the compounds are basic, that is, pyridine and its benzologs, while the remainder are present as neutral species (amides and carbazole derivatives). Although benzoquinoline and dibenzoquinoline derivatives found in petroleum are rich in sterically hindered structures, hindered and unhindered structures have also been found.

Porphyrins (nitrogen–metal complexes) are also constituents of petroleum and usually occur in the nonbasic portion of the nitrogen-containing concentrate (Reynolds 1998). Pyrrole, the chief constituent of the porphyrin molecule, is marked by high stability due to its aromatic character.

The presence of vanadium and nickel in resids, especially as metal porphyrin complexes, has focused much attention in the petroleum refining industry on the occurrence of these metals in feedstocks (Reynolds 1998). Only a part of the total nickel and vanadium in crude oil is recognized to occur in porphyrin structures.

### 6.4.1.4  Oxygen Compounds

The total oxygen content of petroleum is usually less than 2% w/w (although larger amounts have been reported) but does increase with the boiling point of the fractions. In fact, resids may have oxygen content up to 8% w/w. Though these high-molecular-weight compounds contain most of the oxygen, little is known concerning their structure, but those of lower molecular weight have been investigated with considerably more success and have been shown to contain carboxylic acids (R-CO$_2$H) and phenols (Ar-OH, where Ar is an aromatic moiety).

### 6.4.1.5  Metallic Compounds

The occurrence of metallic constituents in crude oil is of considerably greater interest to the petroleum industry than might be expected from the very small amounts present.

Distillation concentrates the metallic constituents in the resids (Table 6.2); some can appear in the higher-boiling distillates, but the latter may, in part, be due to entrainment. The majority of the vanadium, nickel, iron, and copper in resids may be precipitated along with the asphaltenes by low-boiling alkane hydrocarbon solvents. Thus, removal of the asphaltenes with *n*-pentane reduces the vanadium content of the oil by up to 95%, with substantial reductions in the amounts of iron and nickel.

### 6.4.2 FRACTIONATION

Understanding the chemical transformations of the macromolecular constituents during conversion is limited by the diversity (over a million chemical structures) of the complex macromolecules in a resid.

One way to sample this molecular diversity is to separate the resid and its conversion products into fractions using solubility/insolubility in low-boiling liquid hydrocarbons as well as adsorption followed by desorption on solids such as silica gel and/or alumina and/or clay (Figure 6.2). Following the application of such a procedure, it can be shown that, in the simplest sense, resids from the same crude differ in the relative amounts of these fractions that are present (Figure 6.3) as they often do when

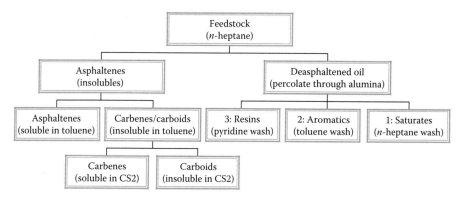

**FIGURE 6.2**   Resid fractionation.

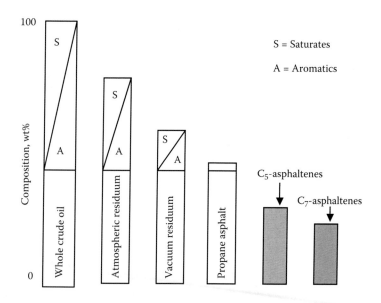

**FIGURE 6.3**   Comparison of petroleum and resid composition showing the asphaltene content.

produced from different crude oil. Thus, the *physical composition* (or *bulk composition*) that refers to the composition of a resid is determined by these various physical techniques (Speight 1999 and references cited therein).

There are also two other operational definitions that should be noted at this point, and these are the terms *carbenes* and *carboids*. Both such fractions are, by definition, insoluble in benzene (or toluene), but the *carbenes* are soluble in carbon disulfide, whereas the *carboids* are insoluble in carbon disulfide. Only traces, if any, of these materials occur in natural resids. On the other hand, resids that have received some thermal treatment (such as visbroken resids) may have considerable quantities of these materials present as they are also considered to be precursors to coke.

Resids, especially vacuum resids (*vacuum bottoms*), are the most complex fractions of petroleum. Few molecules are free of heteroatoms; and the molecular weight of the constituents extends from 400 to >2000, and at the upper end of this molecular weight range, characterization of individual species is virtually impossible. Separations by group type become blurred by the shear frequency of substitution and by the presence of multiple functionalities in single molecules.

Resids can be separated into a variety of fractions using a myriad of different techniques that have been used since the beginning of petroleum science (Speight 1999). In general, the fractions produced by these different techniques are called *saturates, aromatics, resins,* and *asphaltenes* (Figure 6.2). And much of the focus has been on the asphaltenes because of their high sulfur content and high coke-forming propensity (Speight 1994, 1999).

The methods employed can be conveniently arranged into a number of categories: (1) fractionation by precipitation; (2) fractionation by distillation; (3) separation by chromatographic techniques; (4) chemical analysis using spectrophotometric techniques (infrared, ultraviolet, nuclear magnetic resource, x-ray fluorescence, emission, neutron activation), titrimetric and gravimetric techniques, elemental analysis; and (5) molecular weight analysis by mass spectrometry, VPO, and size-exclusion chromatography.

### 6.4.2.1 Asphaltene Separation

By definition, the *asphaltene fraction* is that portion of the feedstock that is precipitated when a large excess (40 volumes) of a low-boiling liquid hydrocarbon (e.g., *n*-pentane or *n*-heptane) is added to (1 volume of) the crude oil (Speight 1994, 1999). *n*-Heptane is the preferred hydrocarbon, with *n*-pentane still being used (Speight et al. 1984; Speight 1994; ASTM 2003).

The asphaltene fraction (ASTM D-2006, ASTM D-2007, ASTM D-3279, ASTM D-4124, ASTM D-6560, IP 143) is the highest-molecular-weight and most complex fraction in petroleum. The asphaltene content gives an indication of the amount of coke that can be expected during processing (Speight 1999, 2001; Speight and Ozum 2002).

In any of the methods for the determination of the asphaltene content, the resids are mixed with a large excess (usually >30 volumes hydrocarbon per volume of sample) of low-boiling hydrocarbon such as *n*-pentane or *n*-heptane. For an extremely viscous residuum, a solvent such as toluene may be used prior to the addition of the low-boiling hydrocarbon, but an additional amount of the hydrocarbon (usually >30 volumes hydrocarbon per volume of solvent) must be added to compensate for the presence of the solvent. After a specified time, the insoluble material (the asphaltene fraction) is separated (by filtration) and dried. The yield is reported as percentage (% w/w) of the original sample.

It must be recognized that, in any of these tests, different hydrocarbons (such as *n*-pentane or *n*-heptane) will give different yields of the asphaltene fraction, and if the presence of the solvent is not compensated by use of additional hydrocarbon, the yield will be erroneous. In addition, if the hydrocarbon is not present in large excess, the yields of the asphaltene fraction will vary and will be erroneous (Speight 1999).

The *precipitation number* is often equated to the asphaltene content, but there are several issues that remain obvious in its rejection for this purpose. For example, the method to determine the precipitation number (ASTM D-91) advocates the use of naphtha for use with black oil or lubricating oil, and the amount of insoluble material (as a % v/v of the sample) is the precipitating number. In the test, 10 mL of sample is mixed with 90 mL of ASTM precipitation naphtha (that may or may not have a constant chemical composition) in a graduated centrifuge cone and centrifuged for 10 min at 600 to 700 rpm. The volume of material on the bottom of the centrifuge cone is noted until repeat centrifugation gives a value within 0.1 mL (the precipitation number). Obviously, this can be substantially different from the asphaltene content.

In another test method (ASTM D-4055), pentane-insoluble materials above 0.8 μm in size can be determined. In the test method, a sample of oil is mixed with pentane in a volumetric flask, and the oil solution is filtered through a 0.8 μm membrane filter. The flask, funnel, and filter are washed with pentane to completely transfer the particulates onto the filter, which is then dried and weighed to give the yield of pentane-insoluble materials.

Another test method (ASTM D-893) that was originally designed for the determination of pentane- and toluene-insoluble materials in used lubricating oils can also be applied to resids. However, the method may need modification by first adding a solvent (such as toluene) to the resid before adding pentane.

### 6.4.2.1.1 Carbon Disulfide–Insoluble Constituents

The component of highest carbon content is the fraction termed *carboids* and consists of species that are insoluble in carbon disulfide or in pyridine. The fraction that has been called *carbenes* contains molecular species that are soluble in carbon disulfide and soluble in pyridine but that are insoluble in toluene (Figure 6.2).

Resids are hydrocarbonaceous materials that are composed of constituents (containing carbon, hydrogen, nitrogen, oxygen, and sulfur) that are completely soluble in carbon disulfide (ASTM D-4). Trichloroethylene or 1,1,1-trichloroethane has been used in recent years as a solvent for the determination of asphalt solubility (ASTM D-2042).

The carbene and carboid fractions are generated by thermal degradation or by oxidative degradation and are not considered to be naturally occurring constituents of asphalt. The test method for determining the toluene-insoluble constituents of tar and pitch (ASTM D-4072, ASTM D-4312) can be used to determine the amount of carbenes and carboids in resids.

## 6.4.2.2   Fractionation of Deasphaltened Oil

After removal of the asphaltene fraction, further fractionation of resids is also possible by variation of the hydrocarbon solvent.

However, fractional separation has been the basis for most resid composition analysis (Figure 6.2). The separation methods that have been used divide a resid into operationally defined fractions. Three types of resid separation procedures are now in use: (1) chemical precipitation, in which *n*-pentane separation of asphaltenes is followed by chemical precipitation of other fractions with sulfuric acid of increasing concentration (ASTM D-2006); (2) adsorption chromatography using a clay-gel procedure where, after removal of the asphaltenes, the remaining constituents are separated by selective adsorption/desorption on an adsorbent (ASTM D-2007 and ASTM D-4124); and (3) size-exclusion chromatography, in which gel permeation chromatographic (GPC) separation of asphalt constituents occurs based on their associated sizes in dilute solutions (ASTM D-3593).

The fractions obtained in these schemes are defined operationally or procedurally. The solvent used for precipitating them, for instance, defines the amount and type of asphaltenes in a resid. Fractional separation of a resid does not provide well-defined chemical components. The materials separated should only be defined in terms of the particular test procedure (Figure 6.2). However, these fractions generated by thermal degradation are not considered to be naturally occurring constituents of resid. The test method for determining the toluene-insoluble constituents of tar and pitch (ASTM D-4072, ASTM D-4312) can be used to determine the amount of carbenes and carboids in resid.

Many investigations of relationships between composition and properties take into account only the concentration of the asphaltenes, independently of any quality criterion. However, a distinction should be made between the asphaltenes that occur in straight-run residua and those that occur in cracked residua. Remembering that the asphaltene fraction is based on solubility and the constituents collectively are a solubility class rather than a distinct chemical class means that vast differences occur in the makeup of this fraction when it is produced by different processes.

For example, liquefied gases, such as propane and butane, precipitate as much as 50% by weight of the resid. The precipitate is a black, tacky, semisolid material, in contrast to the pentane-precipitated asphaltenes, which are usually brown, amorphous solids. Treatment of the propane precipitate with pentane then yields the insoluble brown, amorphous asphaltenes and soluble, near-black, semisolid resins, which are, as far as can be determined, equivalent to the resins isolated by adsorption techniques.

Separation by adsorption chromatography essentially commences with the preparation of a porous bed of finely divided solid, the adsorbent. The adsorbent is usually contained in an open tube (column chromatography); the sample is introduced at one end of the adsorbent bed and induced to flow through the bed by means of a suitable solvent. As the sample moves through the bed, the various components are held (adsorbed) to a greater or lesser extent depending on the chemical nature of the component. Thus, those molecules that are strongly adsorbed spend considerable time on the adsorbent surface rather than in the moving (solvent) phase, but components that are slightly adsorbed move through the bed comparatively rapidly.

There are three ASTM methods that provide for the separation of a feedstock into four or five constituent fractions (Speight 2001 and references cited therein). It is interesting to note that as the methods have evolved, there has been a change from the use of pentane (ASTM D-2006 and D-2007) to heptane (ASTM D-4124) to separate asphaltenes. This is, in fact, in keeping with the production of a more consistent fraction that represents the higher-molecular-weight complex constituents of petroleum (Speight et al. 1984; Speight 1994, 2001, 2014).

Two of the methods (ASTM D-2007 and D-4124) use adsorbents to fractionate the deasphaltened oil, but the third method (ASTM D-2006) advocates the use of various grades of sulfuric acid to separate the material into compound types. Caution is advised in the application of this method since the method does not work well with all feedstocks. For example, when the *sulfuric acid* method (ASTM D-2006) is applied to the separation of heavy feedstocks, complex emulsions can be produced.

## 6.5 USE OF DATA

It has been asserted that more needs to be done in correlating analytical data obtained for resids with processability (Speight 2001; Speight and Ozum 2002; Hsu and Robinson 2006; Gary et al. 2007; Speight 2014). Currently, coke yield predictors are simplistic and specific to the feedstock (Speight and Ozum 2002; Hsu and Robinson 2006; Gary et al. 2007; Speight 2014).

Standard analyses on resids, such as determinations of elemental compositions and molecular weight, have not served to be reliable predictors of processability, and determining average structural features also does not appear to be greatly helpful.

The data derived from any one or more of the evaluation techniques described here give an indication of resid behavior. The data can also be employed to give the refiner a view of the differences between different residua, thereby presenting an indication of the means by which the resids should be processed as well as for the prediction of product properties (Dolbear et al. 1987; Speight and Ozum 2002; Hsu and Robinson 2006; Gary et al. 2007; Speight 2014).

Because of resid complexity, there are disadvantages of relying upon the use of bulk properties as the sole means of predicting behavior. Therefore, fractionation of resids into components of interest and study of the components appears to be a better approach than obtaining data on whole residua. By careful selection of a characterization scheme, it may be possible to obtain a detailed overview of feedstock composition that can be used for process predictions.

The use of composition data to model resid behavior during refining is becoming increasingly important in refinery operations (Speight 1999).

In the simplest sense, a resid can be considered a composite of four major operational fractions, and this allows different resids to be compared on a relative basis to provide a very simple but convenient feedstock *map* (Figure 6.4). However, such a map does not give any indication of the complex interrelationships of the various fractions, although predictions of feedstock behavior are possible using such data. It is necessary to take the composition studies one step further using subfractionation of the major fractions to obtain a more representative indication of petroleum composition.

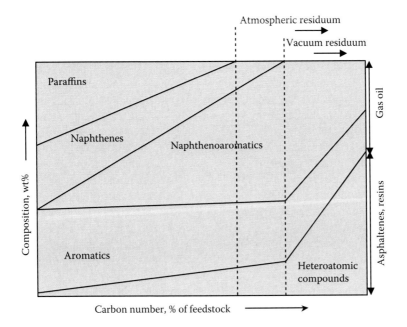

**FIGURE 6.4** Simplified representation and resid composition.

Further development of this concept (Long and Speight 1989, 1990, 1998; Speight and Long 1996) involved the construction of a different type of compositional map using the molecular weight distribution and the molecular type distribution as coordinates. The separation involved the use of an adsorbent such as clay, and the fractions were characterized by *solubility parameter* as a measure of the polarity of the molecular types. The molecular weight distribution can be determined by GPC.

Using these two distributions, a map of composition can be prepared using molecular weight and solubility parameter as the coordinates for plotting the two distributions. Such a composition map can provide insights into many separation and conversion processes used for resid processing.

## 6.6 RESID CONVERSION

Resid quality is a relevant issue with regard to the selection and efficiency of the conversion technology. High levels of metals (vanadium and nickel) are a well-known characteristic of many resids. The Conradson carbon and asphaltene content of resids are also high and so represents challenge for the upgrading technology.

### 6.6.1 VISBREAKING

Visbreaking is a low-conversion thermal process used originally to reduce the resid viscosity to meet the specification for heavy fuel oil applications. Currently, the visbreaking process converts resids (15%–20% v/v conversion) to produce some liquid fuel boiling-range liquids, and visbroken resid is used to meet heavy fuel oil specifications.

The process is not designed to produce coke and, therefore, operates with the induction period prior to coke formation.

A visbreaker reactor may be similar to a delayed coker, with a furnace tube followed by a soaker drum. However, the drum is much smaller in volume to limit the residence time with the entire liquid product flowing overhead. Alternatively, the entire visbreaker may be a long tube coiled within a furnace. Differences in resid properties can cause coke to form in the vessel, and some coke removal protocols may be necessary.

Visbreaking may be applied to atmospheric resids, vacuum resids, and also solvent deasphalter bottoms (asphalt). A common operation involves visbreaking the atmospheric residue in combination with a thermal cracker to minimize fuel oil while producing additional light distillates.

However, visbreaking is typically applied to vacuum resids and is frequently used as a mild vacuum residue conversion process when feedstock for fluid catalytic cracking is sought. The lower-boiling distillates are recovered for processing to transportation fuels, and the higher-boiling distillates are recovered as feedstock for the fluid catalytic cracking unit. A high-temperature coil visbreaker allows a high boiling distillate to be recovered from the fractionator. For lower-boiling distillates and with soaking drum technology, a separate vacuum flasher is usually required.

Conversion in a visbreaker is limited by the requirement to produce a stable fuel oil. As the resid is thermally cracked, reactions occur that increase the asphaltene content that, coupled with the reduction (by thermal cracking) of the resins that hold the asphaltene constituents in solution, leads to precipitation of the asphaltenes.

### 6.6.2 COKING

Coking is a high-temperature (450°C–500°C, 842°F–932°F) process and is a most popular conversion choice for resids that are usually have a high content of polynuclear aromatic systems (low-hydrogen, high-heteroatom, high-Conradson carbon). Coking converts the polynuclear aromatic systems to coke (a relatively low-value product) and overhead (relatively high-value distillates) that can be upgraded further to liquid fuels and other products.

*Delayed coking* is the oldest and most popular choice for resid conversion. The resid is heated by flowing through a long tube in a furnace and then reacted by flowing into the bottom of a high, cylindrical, insulated drum. The drums are used in pairs, with one on-stream and the other off-stream. Volatile (overhead) products pass to a fractionator, and coke accumulates in the drum. High-boiling liquid products may be recycled to the furnace and pass through the coke drum again.

When the drum fills up with coke, the hot feed is switched off-stream, and the second drum is switched on-stream. The coke is removed from the off-stream drum using high-pressure water, after which the on-stream/off-stream cycle (usually about 16 h) will be reversed.

The *fluid coking and flexicoking* processes are also employed for resid conversion. In the *fluid coking* process, the hot resid is sprayed on a hot, fluidized bed of coke particles in a reactor. The volatile products (overhead) pass to a fractionator, while the coke particles are removed from the bottom of the reactor and transferred to another vessel, the burner or regenerator. Here, the excess coke is partially burned with air to provide the heat for the process, and the coke then is recirculated back to the reactor.

In the flexicoking process, which is very similar to the fluid coking process, a third vessel (the gasifier) is added to the fluid coking flow. The gasifier coke is used to gasify excess coke with steam and air to produce a low-BTU gas containing hydrogen, carbon monoxide, nitrogen, and hydrogen sulfide. After removal of the hydrogen sulfide, the low-BTU gas is burned as a clean fuel within the refinery and/or in a nearby power plant.

### 6.6.3   RESID CATALYTIC CRACKING

Resid catalytic cracking has much better selectivity to desired products (high gasoline and low gas yields) than for coking or hydroconversion.

In fluid catalytic cracking, feed is sprayed on zeolite catalyst in a short-contact-time, riser reactor. The vaporized product flows to a fractionator while the catalyst with coke and adsorbed hydrocarbons flow to a fluidized bed regenerator where the coke and hydrocarbons are burned off the catalyst. Fluid catalytic cracking requires much higher-quality feeds than coking or hydroconversion. This is because of expensive zeolite catalysts; intolerance to sodium, nickel, vanadium, and basic nitrogen; as well as limitations on the amount of coke that can be burned in the regeneration step by cooling capacity. As a result, the feed in resid catalytic cracking is, at worst, an excellent-quality atmospheric resid, but mixtures of vacuum gas oil and atmospheric resids are more common. The total feed is limited to Conradson carbon residues of 3–8 wt%, depending on the cooling capacity.

### 6.6.4   HYDROCONVERSION

Hydroconversion combines thermal cracking with hydrogenation. The addition of hydrogen increases the coke induction period by lowering the solubility parameter by hydrogenating polynuclear aromatic systems but primarily by terminating free radicals and reducing the frequency of aromatics from combining to form larger polynuclear aromatic systems. Thus, hydroconversion of vacuum resids to volatile liquids can be over 85% as opposed to 50%–60% for coking. However, one has to deal with the cost of hydrogen and catalyst, high-pressure vessels, poisoning of catalysts, the difficulty of asphaltenes to diffuse through small pores, and the intolerance to coke and sediment formation. The active catalyst needs to be a transition metal sulfide because the high level of sulfur in resid feeds will poison other hydrogenation catalysts.

Atmospheric and vacuum residue desulfurization units are commonly operated to desulfurize the residue as a preparatory measure for feeding low-sulfur vacuum gas-oil feed to cracking units (fluid catalytic cracking units and hydrocracking units), low-sulfur residue feed to delayed coker units, and low-sulfur fuel oil to power stations. Two different types of processing units are used for the direct catalytic hydroprocessing of residue (Speight 2000; Ancheyta and Speight 2007).

These units are either a (1) downflow, trickle phase reactor system (fixed catalyst bed) or (2) liquid recycle and back-mixing system (ebullating bed).

### 6.6.4.1   Fixed-Bed Units

Since metals removal is one of the fastest reactions and since the metals accumulate in the pores of supported catalysts, it is common to have a guard bed in front of the fixed bed. When insufficient metals removal occurs in the guard bed, the feed is switched to a second guard bed with fresh catalyst, and the catalyst is replaced in the first guard bed. Thus, the fixed bed is protected from metals deposition. In order to hydrogenate the largest macromolecules in the resid, the asphaltenes, some or all of the catalysts need to have pores 50 to 100 μm in diameter. Even with these precautions, it is difficult to get longer than 1-year run lengths on fixed-bed hydro-conversion units with vacuum resid feeds and conversions to volatile liquids of 50% or more. This is because of catalyst deactivation with coke or by coke and sediment formation downstream of the reactor.

### 6.6.4.2   Ebullating-Bed Units

The LC Finer and H-Oil units use a mechanically ebullated catalyst so that it can be mixed and replaced on-stream. The conversion is greatly dependent on the feed, but conversions of vacuum resids to volatile liquids of the order of 70% are possible. Often, these units are limited by the deposition of coke and sediment downstream of the reactor in hot and cold separators.

### 6.6.4.3   Dispersed Catalyst Processes

If one cannot diffuse the asphaltenes to the catalyst, why not diffuse the catalyst to asphaltenes? Dispersed catalysts also can be continuously added in low-enough amounts (i.e., 100 ppm) to consider them throwaway catalysts with the carbonaceous by-product. However, economics usually dictates some form of catalyst recycle to minimize catalyst cost. Nevertheless, by designing the reactor to maximize the solubility of the converted asphaltenes, the conversion of vacuum resids to gas and volatile liquids can be above 95%, with greater than 85% volatile liquids. However, the last 5%–10% conversion may not be worth the cost of hydrogen and reactor volume to produce hydrocarbon gases and very aromatic liquids from this incremental conversion.

### 6.6.5   Solvent Deasphalting

Solvent deasphalting, while not strictly a conversion process, is a separation process that represents a further step in the minimization of resid production.

The process takes advantage of the fact that maltenes are more soluble in light paraffinic solvents than asphaltenes. This solubility increases with solvent molecular weight and decreases with temperature. As with vacuum distillation, there are constraints with respect to how deep a solvent deasphalting unit can cut into the residue or how much deasphalted oil can be produced.

In the case of solvent deasphalting, these constraints are typically (1) the quality of the deasphalted oil required by conversion units and (2) the residual fuel oil stability and quality.

Solvent deasphalting has the advantage of having the flexibility to meet a wide range of deasphalted oil quality. The process has very good selectivity for the rejection of asphaltene constituents and metal constituents along with varied selectivity (depending on the feedstock) for other coke precursors but less selectivity for sulfur constituents and for nitrogen constituents. The disadvantages of the process are that it performs no conversion, it produces a very-high-viscosity by-product pitch, and where high-quality deasphalted oil is required, the solvent deasphalting process is limited in the quality of feedstock that can be economically processed.

The viability of the solvent deasphalting process is dependent on the potential for upgrading the deasphalted oil and the differential between the value of the cutter stocks and the price of high-sulfur residual fuel oil. If there is an outlet for asphalt (from the unit) and the conversion capacity exists to upgrade the deasphalted oil, solvent deasphalting can be a highly attractive option for resid conversion.

### 6.6.6 FUTURE PROCESSES

We are still on the steep part of the learning curve as to the characterization, phase behavior, and conversion chemistry of petroleum resids. As a result, much room remains to improve resid conversion processes. However, the rate of construction of new resid conversion units in refineries has been decreasing. The future growth appears to be at or near heavy crude production sites to decrease heavy crude viscosity and improve the quality to ease transportation and open markets for crude oils or resids that are of marginal value. There remains room for improving coking and hydroconversion processes by reducing hydrocarbon gas formation, by inhibiting the formation of polynuclear aromatic systems not originally present in the resid, and by separating an intermediate-quality (low in cores) fraction before or during conversion. Both of these processes would benefit if a higher-valued by-product, such as carbon fibers or needle coke, could be formed from the polynuclear aromatic systems. In addition, the challenge for hydroconversion is to take advantage of the nickel and vanadium in the resid to generate an *in situ* dispersed catalyst and to eliminate catalyst cost. Finally, resid catalytic cracking needs to move to poorer-quality and lower-cost feeds by making more tolerant catalysts, by processing only the saturates and small ring aromatics portion, and/or by improved methods to remove heat from the regenerator.

In recent years, the *average quality* of crude oil has deteriorated and continues to do so as more heavy oil and tar sand bitumen are being sent to refineries (Speight 2008, 2011, 2014). This has caused the nature of crude oil refining to change considerably. Indeed, the declining reserves of lighter crude oil have resulted in an increasing need to develop options to desulfurize and upgrade the heavy feedstocks, specifically high-resid feedstocks such as heavy oil and bitumen. This has resulted in a variety of process options that specialize in sulfur removal during refining.

In addition, with the general trend throughout refining to produce more distillates from each barrel of feedstock, there has been a need to process the distillate products in different ways to meet the product specifications for use in modern engines. Overall, the demand for gasoline has rapidly expanded, and demand has also developed for gas oils and fuels for domestic central heating and fuel oil for power generation, as well as for light distillates and other inputs, derived from crude oil, for the petrochemical industries (Speight 2011).

As the need for the lower-boiling products developed, petroleum yielding the desired quantities of the lower-boiling products became less available, and refineries had to introduce conversion processes to produce greater quantities of lighter products from the higher-boiling fractions. The means by which a refinery operates in terms of producing the relevant products depends not only on the nature of the petroleum feedstock but also on its configuration (i.e., the number of types of the processes that are employed to produce the desired product slate), and the refinery configuration is, therefore, influenced by the specific demands of a market.

Therefore, refineries need to be constantly adapted and upgraded to efficiently process residua and remain viable as well as responsive to ever-changing patterns of crude supply and product market demands. As a result, refineries have been introducing increasingly complex and expensive processes to gain higher yields of lower-boiling products from the higher-boiling fractions and residua.

Finally, the yields and quality of refined petroleum products produced by any given oil refinery depend on the mixture of crude oil used as feedstock and the configuration of the refinery facilities. Light/sweet crude oil is generally more expensive and has inherent great yields of higher-value low-boiling products such naphtha, gasoline, jet fuel, kerosene, and diesel fuel. Heavy sour crude oil is generally less expensive and produces greater yields of lower-value higher-boiling products that must be converted into lower-boiling products.

The configuration of refineries may vary from refinery to refinery. Some refineries may be more oriented toward the production of gasoline (large reforming and/or catalytic cracking), whereas the configuration of other refineries may be more oriented toward the production of middle distillates such as jet fuel and gas oil. Thus, changes in the characteristics of the feedstock will trigger changes in refinery configurations and corresponding investments. The future crude slate will consist of larger fractions of both heavier, sourer crudes and extra-light inputs, such as natural gas liquids. There will also be a shift toward heavier feedstocks, such tar sand bitumen, with the recognition that tar sand bitumen is *not* the environmental disaster that various organizations would have us believe (Levant 2010). These changes will require investment in upgrading, either at the field level to process tar sand bitumen and oil shale into synthetic crude oil shale, either at a field site or at the refinery level (Speight 2014).

A change in feedstock quality requires a very substantial increase in upgrading capacity. Where this upgrading capacity will be built is likely to be strongly influenced by greenhouse gas policy. In fact, the petroleum and petrochemical industries are coming under increasing pressure not only to compete effectively with global competitors utilizing more advantaged hydrocarbon feedstocks, but also to ensure that its processes and products comply with increasingly stringent environmental legislation (Speight 2005).

There is also the need for a refinery to be able to accommodate *opportunity crude oils* and/or *high acid crude oils* (Speight 2011, 2014).

Opportunity crude oils are often dirty and need cleaning before refining by removal of undesirable constituents such as high-sulfur, high-nitrogen, and high-aromatics (such as polynuclear aromatic) components. A controlled visbreaking treatment would *clean up* such crude oils by removing these undesirable constituents (which, if not removed, would cause problems further down the refinery sequence) as coke or sediment. On the other hand, high acid crude oils cause corrosion in the atmospheric and vacuum distillation units. In addition, overhead corrosion is caused by the mineral salts, magnesium, calcium, and sodium chloride, which are hydrolyzed to produce volatile hydrochloric acid, causing a highly corrosive condition in the overhead exchangers. Therefore, these salts present a significant contamination in opportunity crude oils. Other contaminants in opportunity crude oils that are shown to accelerate the hydrolysis reactions are inorganic clays and organic acids.

An option for resids that is often ignored and not given the recognition publication it deserves is the concept of a gasification refinery, one that would have, as the centerpiece, gasification technology and would produce synthesis gas (syngas) from the resid from which liquid fuels would be manufactured using the Fischer–Tropsch synthesis technology. Furthermore, such a refinery that could produce synthesis gas can proceed from other carbonaceous feedstocks, including biomass (Speight 2008, 2011, 2013a,b). Inorganic components of the feedstock, such as metals and minerals, are trapped in an inert and environmentally safe form such as char, which may have use as a fertilizer.

In a gasifier, the carbonaceous material undergoes several different processes: (1) pyrolysis of carbonaceous fuels, (2) combustion, and (3) gasification of the remaining char. The process is very dependent on the properties of the carbonaceous material and determines the structure and composition of the char, which will then undergo gasification reactions (Speight 2013a,b, 2014).

As high-resid feedstocks increase, the desirability of coproducing gas from these feedstocks and other carbonaceous feedstocks will increase, allowing gasification to compete as an economically viable process. The conversion of the gaseous products of gasification processes to synthesis gas, a mixture of hydrogen ($H_2$) and carbon monoxide (CO), in a ratio appropriate to the application needs additional steps after purification. The product gases—carbon monoxide, carbon dioxide, hydrogen, methane, and nitrogen—can be used as fuels or as raw materials for chemical or fertilizer manufacture.

Finally, in addition to taking preventative measures for the refinery to process high-resid feedstocks without serious deleterious effects on the equipment, refiners will need to develop programs for detailed and immediate evaluation of high-resid feedstocks so that the qualities of a crude oil can be identified quickly and it can be valued appropriately. There is also the need to assess the potential impact of contaminants, like metals or acidity, in crudes so that the feedstock can be correctly valued and management of the crude processing can be planned.

## REFERENCES

Ancheyta, J., and Speight, J.G. 2007. *Hydroprocessing of Heavy Oils and Residua*. CRC-Taylor & Francis Group, Boca Raton, Florida.

Dolbear, G.E., Tang, A., and Moorehead, E.L. 1987. Metal Complexes in Fossil Fuels. In: *Metal Complexes in Fossil Fuels*, R.H. Filby and J.F. Branthaver (Editors). Symposium Series No. 344. American Chemical Society, Washington, DC, p. 220.

Foght, J.M., Fedorak, P.M., and Westlake, D.W.S. 1990. Mineralization of [$^{14}$C] hexadecane and [$^{14}$C] phenanthrene in crude oil: Specificity among bacterial isolates. *Can. J. Microbiol.*, 36: 169–175.

Gary, J.G., Handwerk, G.E., and Kaiser, M.J. 2007. *Petroleum Refining: Technology and Economics*, 5th Edition. CRC Press, Taylor & Francis Group, Boca Raton, Florida.

Hsu, C.S., and Robinson, P.R. (Editors). 2006. *Practical Advances in Petroleum Processing*, Volume 1 and Volume 2. Springer Science, New York.

Kim, J.-S., and Crowley, D.E. 2007. Microbial diversity in natural asphalts of the Rancho La Brea Tar Pits. *Appl. Environ. Microbiol.*, 73: 4579–4591.

Levant, E. 2010. *The Case for Canada's Oil Sands*. McClelland and Stewart, Toronto, Ontario.

Long, R.B., and Speight, J.G. 1989. Studies in petroleum composition. II: Scale-up studies for separating heavy feedstocks by adsorption. *Ind. Eng. Chem. Res.*, 28: 1503.

Long, R.B., and Speight, J.G. 1990. Studies in petroleum composition. III: The distribution of nitrogen species, metals, and coke precursors during high vacuum distillation. *Rev. Inst. Fr. Pét.*, 45: 553.

Long, R.B., and Speight, J.G. 1998. The composition of petroleum. In: *Petroleum Chemistry and Refining*. Taylor & Francis Publishers, Washington, DC, Chapter 1.

Morais, E.B., and de Tauk-Tornisielo, S.M. 2009. Biodegradation of oil refinery residues using mixed-culture of microorganisms isolated from land farming. *Brazilian Arch. Biol. Technol.*, 52(6): 1571–1578.

Rabus, R., Kube, M., Heider, J., Beck, A., Heitmann, K., Widdel, F., and Reinhardt, R. 2005. The genome sequence of an anaerobic aromatic-degrading denitrifying bacterium, strain Ebn1. *Arch. Microbiol.*, 183: 27–36.

Ramos, J.L., Duque, E., Gallegos, M.T., Godyoy, P., Ramos-Gonzalez, M.I., Rojas, A., Teran, W., and Segura, A. 2002. Mechanisms of solvent tolerance in gram-negative bacteria. *Ann. Rev. Microbiol.*, 56: 743–768.

Reynolds, J.G. 1998. Metals and heteroatoms in heavy crude oils. In: *Petroleum Chemistry and Refining*, J.G. Speight (Editor). Taylor & Francis Publishers, Washington, DC, Chapter 3.

Schabron, J.F., and Speight, J.G. 1997a. An evaluation of the delayed coking product yield of heavy feedstocks using asphaltene content and carbon residue. *Rev. Inst. Fr. Pét.*, 52: 73.

Schabron, J.F., and Speight, J.G. 1997b. Correlation between carbon residue and molecular weight. *Prepr. Div. Fuel Chem. Am. Chem. Soc.*, 42(2): 386.

Speight, J.G., Long, R.B., and Trowbridge, T.D. 1984. Factors influencing the separation of asphaltenes from heavy petroleum feedstocks. *Fuel*, 63: 616.

Speight, J.G. 1994. Chemical and physical studies of petroleum asphaltenes. In: *Asphaltenes and Asphalts. I*, Developments in Petroleum Science, 40, T.F. Yen and G.V. Chilingarian (Editors). Elsevier, Amsterdam, Netherlands, Chapter 2.

Speight, J.G., and Long, R.B. 1996. The concept of asphaltenes revisited. *Fuel Sci. Technol. Int.*, 14: 1.

Speight, J.G. 2000. *The Desulfurization of Heavy Oils and Residua*, 2nd Edition. Marcel Dekker Inc., New York.

Speight, J.G. 2001. *Handbook of Petroleum Analysis*. John Wiley & Sons Inc., New York.

Speight, J.G., and Ozum, B. 2002. *Petroleum Refining Processes*. Marcel Dekker Inc., New York.

Speight, J.G. 2005. *Environmental Analysis and Technology for the Refining Industry*. John Wiley & Sons Inc., Hoboken, New Jersey.

Speight, J.G. 2008. *Synthetic Fuels Handbook: Properties, Processes, and Performance*. McGraw-Hill, New York.

Speight, J.G. 2011. *The Refinery of the Future*. Gulf Professional Publishing, Elsevier, Oxford, United Kingdom.

Speight, J.G., and Arjoon, K.K. 2012. *Bioremediation of Petroleum and Petroleum Products*. Scrivener Publishing, Salem, Massachusetts.

Speight, J.G. 2013a. *The Chemistry and Technology of Coal*, 3rd Edition. CRC Press, Taylor & Francis Group, Boca Raton, Florida.

Speight, J.G. 2013b. *Coal-Fired Power Generation Handbook*. Scrivener Publishing, Salem, Massachusetts.

Speight, J.G. 2014. *The Chemistry and Technology of Petroleum*, 5th Edition. CRC Press, Taylor & Francis Group, Boca Raton, Florida.

Van Der Meer, J.R., Devos, W.M., Harayama, S., and Zehnder, A.J.B. 1992. Molecular mechanisms of genetic adaptation to xenobiotic compounds. *Microbiol. Rev.*, 56: 677–694.

Van Der Meer, J.R. 1994. Genetic adaptation of bacteria to chlorinated aromatic compounds. *FEMS Microbiol. Rev.*, 15: 239–249.

Van Gooswilligen, G. 2000. Bitumen Manufacture. In: *Modern Petroleum Technology, Volume 2: Downstream*, A.G. Lucas (Editor). John Wiley & Sons Inc., New York.

Whyte, L.G., Bourbonnière, L., and Greer, C.W. 1997. Biodegradation of petroleum hydrocarbons by psychrotrophic *pseudomonas* strains possessing both alkane (*alk*) and naphthalene (*nah*) catabolic pathways. *Appl. Environ. Microbiol.*, 63(9): 3719–3723.

# 7 Liquid Fuels from Oil Sand

*James G. Speight*

## CONTENTS

## 7.1 INTRODUCTION

Liquid fuels are produced from several sources—the most common source is petroleum (Speight and Ozum 2002; Hsu and Robinson 2006; Gary et al. 2007; Speight 2014). However, other sources such as oil sand and coal are also viable sources of liquid fuels (Speight 1990, 2008, 2013a). When liquid fuels are produced from a source such as oil sand, the initial product is often referred to as *synthetic crude oil* or *syncrude*. Thus, *synthetic crude oil* (also referred to as *syncrude*) is, in the present context, a liquid fuel that does not occur naturally.

*Oil sand*, also called *tar sand* in the United States and many parts of the world, or the more geologically correct term *bituminous sand*, is commonly used to describe a sandstone reservoir that is impregnated with a heavy, viscous bituminous material. Oil sand is actually a mixture of sand, water, and bitumen, but many of the oil sand deposits in countries other than Canada lack the water layer that is believed to facilitate the hot-water recovery process. The heavy bituminous material has a high viscosity under reservoir conditions and cannot be retrieved through a well by conventional production techniques.

Geologically, the term *tar sand* is commonly used to describe a sandstone reservoir that is impregnated with bitumen, a naturally occurring material that is solid or near solid and is substantially immobile under reservoir conditions. The bitumen cannot be retrieved through a well by conventional production techniques, including currently used enhanced recovery techniques (Speight 2009). In fact, oil sand is defined (FE-76-4) in the United States as follows:

> The several rock types that contain an extremely viscous hydrocarbon which is not recoverable in its natural state by conventional oil well production methods including currently used enhanced recovery techniques. The hydrocarbon-bearing rocks are variously known as bitumen-rocks oil, impregnated rocks, tar sands, and rock asphalt.

By inference, conventional petroleum and heavy oil are recoverable by well production methods (i.e., primary and secondary recovery methods) (Speight 2013c) and by currently used enhanced oil recovery (EOR) methods (methods) (Speight 2013c, 2014).

However, the term *tar sand* is actually a misnomer; more correctly, the name *tar* is usually applied to the heavy product remaining after the destructive distillation of coal or other organic matter (Speight 2005, 2008, 2011a, 2013a, 2014). Current recovery operations of bitumen in oil sand formations have been focused predominantly on mining technique, but thermal *in situ* processes are now showing success.

In addition to the above definitions, there are several tests that must be carried out to determine whether or not, in the first instance, a resource is an oil sand deposit. Most of all, a core taken from an oil sand deposit and the bitumen isolated therefrom are certainly not identifiable by the preliminary inspections (sight and touch) alone. In the United States, the final determinant is whether or not the material contained therein can be recovered by primary, secondary, or tertiary (enhanced) recovery methods (US Congress 1976).

In addition, the names *oil sand* and *tar sand* are scientifically incorrect since oil sand does not contain oil and tar is most commonly produced from bituminous coal and is generally understood to refer to the product from coal, although it is advisable to specify coal tar if there is the possibility of ambiguity (Speight 2013a). Thus, technically, *oil sand* should be called *bituminous sand* since the hydrocarbonaceous material is bitumen (soluble in carbon disulfide) and is not oil. The term *oil sand* is used in reference to the synthetic crude oil that can be manufactured from the bitumen.

Oil sand is a mixture of sand, water, and bitumen, and the sand component is predominantly quartz in the form of rounded or subangular particles, each of which (as far as is known for the Athabasca deposit) is wet with a film of water. Surrounding the wetted sand grains and somewhat filling the void among them is a film of bitumen. The balance of the void volume is filled with connate water plus, sometimes, a small volume of gas. High-grade oil sand contains about 18% w/w of bitumen that may be equivalent in consistency (viscosity) to an atmospheric or vacuum petroleum residuum.

The definition of bitumen has been very loosely and arbitrarily based on API gravity or viscosity and is quite arbitrary and too general to be technologically accurate. There have been attempts to rationalize the definition based upon viscosity,

API gravity, and density, but they also suffer from a lack of technical accuracy. For example, 10° API is the generally used line of demarcation between oil sand bitumen and heavy oil. But one must ask if the difference between 9.9° API gravity oil and 10.1° API oil is really significant. Both measurements are within the limits of difference for standard laboratory test methods. Similarly, the use of viscosity data is also subject to question, since the difference between oil having a viscosity of 9,950 centipoises and oil having a viscosity of 10,050 centipoises is minimal and, again, both measurements are within the limits of difference for a standard laboratory test method.

More appropriately, oil sand *bitumen* in oil sand deposits is a highly viscous hydrocarbonaceous material that *is not recoverable in its natural state through a well by conventional oil well production methods, including currently used enhanced recovery techniques*, as specified in the US government regulations. Thus, it is not surprising that the properties of bitumen from oil sand deposits are significantly different from the properties of conventional crude oil (recoverable by primary and secondary techniques) and heavy oil (recoverable by EOR techniques).

Chemically, the material should perhaps be called bituminous sand rather than oil sand since the organic matrix is bitumen, a hydrocarbonaceous material that consists of carbon and hydrogen with smaller amounts of nitrogen, oxygen, sulfur, and metals (especially nickel and vanadium).

The bitumen in various oil sand deposits represents a potentially large supply of energy. However, many of the reserves are available only with some difficulty, and optional refinery scenarios will be necessary for conversion of these materials to liquid products because of the substantial differences in character between conventional petroleum, heavy oil, and oil sand bitumen (Table 7.1). However, because of the diversity of available information and the continuing attempts to delineate the various world oil sand deposits, it is virtually impossible to present accurate numbers that reflect the extent of the reserves in terms of the barrel unit. Indeed, investigations into the extent of many of the world's deposits are continuing at such a rate that the numbers vary from one year to the next. Accordingly, the data quoted here must be recognized as approximate with the potential of being quite different.

Current commercial recovery operations of bitumen in oil sand formations involve use of a mining technique. This is followed by bitumen upgrading and refining to produce a synthetic crude oil. Other methods for the recovery of bitumen from oil sand are based either on mining combined with some further processing or on operation on the oil sands *in situ*.

The API gravity of oil sand bitumen varies from 5° API to approximately 10° API depending upon the deposit, viscosity is very high, and volatility is low. The viscosity of bitumen is high, being on the order of several thousand to 1 million centipoises, with higher viscosities being recorded. Bitumen volatility is low, and there are very little of the naphtha and kerosene constituents present.

In addition, bitumen is relatively hydrogen deficient and therefore requires that there be substantial hydrogen addition during refining. Bitumen is currently commercially upgraded by a combination of carbon rejection (coking) and product hydrotreating. Coking, the process of choice for residua, is also the process of choice for bitumen conversion. Bitumen is currently converted commercially by delayed

**TABLE 7.1**

**Properties of Bitumen and Conventional Crude Oil**

| Property | Bitumen | Conventional Crude Oil |
|---|---|---|
| Gravity, °API | 8 | 35 |
| Viscosity | | |
| Centipoise at 100°F (38°C) | 500,000 | 10 |
| Centipoise at 210°F (99°C) | 1700 | |
| SUS at 100°F (38°C) | 35,000 | 30 |
| SUS at 210°F (99°C) | 500 | |
| Pour point, °F | 50 | 0 |
| Elemental analysis, % w/w | | |
| Carbon | 83 | 86 |
| Hydrogen | 10.6 | 13.5 |
| Sulfur | 4.8 | 0.1 |
| Nitrogen | 0.4 | 0.2 |
| Oxygen | 1 | 0.2 |
| Fractional composition, % w/w | | |
| Asphaltenes | 19 | 5 |
| Resins | 32 | 10 |
| Aromatics | 30 | 25 |
| Saturates | 19 | 60 |
| Metals, ppm | | |
| Vanadium | 250 | 10 |
| Nickel | 100 | 5 |
| Carbon residue, % w/w | 14 | 5 |
| Heating value, btu/lb | 17,500 | 19,500 |

coking and by fluid coking. In each case, the bitumen is converted to distillate oils, coke, and light gases. The coker distillate is a partially upgraded material and is a suitable feed for hydrodesulfurization to produce a low-sulfur synthetic crude oil.

The only commercial operations for the recovery and upgrading of bitumen occur in northeast Alberta, Canada, near the town of Ft. McMurray, where bitumen from the Athabasca deposit is converted to a synthetic crude oil. Therefore, most of the data available for inspection of bitumen and determination of behavior originate from studies of these Canadian deposits. The work on bitumen from other sources is fragmented and spasmodic. The exception is the bitumen from deposits in Utah, United States, where ongoing programs have been in place at the University of Utah for more than four decades.

Currently, the potential for the production of liquid fuels from the Canadian oil sand deposits is being realized, and the liquid fuels produced from these reserves offers a means of alleviating shortfalls in the supply of liquid fuels (Speight 2011a,b). Thus, the purpose of this chapter is to describe the occurrence, production, and properties of oil sand bitumen and the methods used to convert the bitumen to synthetic crude oil. Properties of the synthetic crude oil are also given.

## 7.2 OCCURRENCE AND RESERVES

The occurrence and reserves of oil sand bitumen that are available for production of liquid fuels are known to an approximation, but the definitions by which these reserves are estimates need careful consideration. Best estimates are all that are available.

Thus, the world reserves of conventional petroleum (arbitrarily defined as having a gravity equal to or greater than 20° API) are reported to be composed of approximately 1195 billion barrels (1195 × 10$^9$ bbl) or 30% by volume of the total reserves (of petroleum plus heavy oil plus bitumen), heavy oil (arbitrarily defined as having a gravity grater than 10° API but less than 20° API) is reported to be at 690 billion barrels (690 × 10$^9$ bbl) or 15% by volume of the total reserves, and oil sand bitumen (arbitrarily defined as having a gravity equal to or less than 10° API) is reported to be at 1920 billion barrels (1920 × 10$^9$ bbl) or 55% by volume of the total reserves. However, the bitumen reserves contain what is termed *extra heavy oil* that is sometimes used to describe bitumen. The API gravity of this material is less than 10° API, but the viscosity may fall into a different range when compared to bitumen viscosity. And in such reserves estimations, there is often no mention of the method of recovery on which the definition of oil sand bitumen hinges.

Therefore, estimations of bitumen availability must be placed in the correct definitional context and also, more particularly, in the context of the available recovery method.

Oil sand deposits are widely distributed throughout the world in a variety of countries, and the various deposits have been described as belonging to one of two types: (1) stratigraphic traps and (2) structural traps, although gradations between the types of deposit invariably occur (Meyer and Dietzman 1981). In terms of specific geological and geochemical aspects of the formation, the majority of the work has, again, been carried out on the Athabasca deposit. For this reason, the focus of this chapter must be on the work carried out on the Canadian oil sand deposits.

Nationally, the largest oil sand deposits are in Alberta and Venezuela, with smaller oil sand deposits occurring in the United States (mainly in Utah), Peru, Trinidad, Madagascar, the Soviet Union, the Balkan states, and the Philippines. Oil sand deposits in northwestern China (Xinjiang Autonomous Region) also are larger; at some locations, the bitumen appears on the land surface around the town of Karamay.

In Canada, the Athabasca deposit along with the neighboring Wabasca, Peace River, and Cold Lake deposits have been estimated to contain approximately 2 trillion barrels (2 × 10$^{12}$ bbl) of bitumen. The Venezuelan deposits may at least contain 1 trillion barrels (1.0 × 10$^{12}$ bbl) of bitumen. Deposits of oil sand, each containing approximately 20 million barrels (20 × 10$^6$ bbl) of bitumen, have also been located in the United States, Albania, Italy, Madagascar, Peru, Romania, Trinidad, Zaire, and the Union of Soviet Socialist Republics (USSR). The oil sand deposits in the United States are contained in a variety of separate deposits in various states (Speight 1990), but since many of these deposits are small, information on most is limited.

The Californian deposits are concentrated in the coastal region west of the San Andreas Fault. The largest deposit is the Edna deposit, which is located midway between Los Angeles and San Francisco. The deposit occurs as a stratigraphic trap, extends over an area of about 7000 acres, and occurs from outcrop to a depth of 100 feet (30 m). The Sisquoc deposit (Upper Pliocene) is the second-largest in California, and the total thickness of the deposit is about 185 feet (56 m), occurring over an area of about 175 acres, with an overburden thickness between 15 and 70 feet (4.6 and 21 m). The third California deposit at Santa Cruz is located approximately 56 miles (90 km) from San Francisco. The Kentucky oil sand deposits are located at Asphalt, Davis-Dismal Creek, and Kyrock. Oil sand deposits in New Mexico occur in the Triassic Santa Rosa sandstone. Finally, the oil sand deposits in Missouri occur over an area estimated at 2000 square miles, and the individual bitumen-bearing sands are approximately 50 feet (15 m) in thickness except where they occur in channels, which may actually be as thick as 250 feet (76 m).

Oil sand deposits in Venezuela occur in the Officina/Tremblador tar belt, which is believed to contain bitumen-impregnated sands of a similar extent to those of Alberta, Canada. The organic material is bitumen having API gravity less than 10°.

The Bemolanga (Madagascar) deposit is the third-largest oil sand deposit presently known and extends over some 150 square miles in western Madagascar, with a recorded overburden from 0 to 100 feet (0 to 30 m). The average pay zone thickness is 100 feet (30 m), with a total bitumen in place quoted at approximately 2 billion barrels (approximately $2 \times 10^9$ bbl).

The largest oil sand deposit in Europe is that at Selenizza Albania. This region also contains the Patos oil field, throughout which there occurs extensive bitumen impregnation.

The Trinidad Asphalt Lake (situated on the Gulf of Paria, 12 miles west-southwest of San Fernando and 138 feet [43 m] above sea level) occupies a depression in the Miocene sheet sandstone.

The Rumanian deposits are located at Derna and occur (along with Tataros and other deposits) in a triangular section east and northeast of Oradia between the Sebos Koros and Berrettyo rivers.

Oil sands occur at Cheildag, Kobystan, and outcrop in the south flank of the Cheildag anticline; there are approximately 24 million barrels ($24 \times 10^6$ bbl) of bitumen in place. Other deposits in the former USSR occur in the Olenek anticline (northeast of Siberia), and it has been claimed that the extent of bitumen impregnation in the Permian sandstone is of the same order of magnitude (in area and volume) as that of the Athabasca deposits. Oil sands have also been reported from sands at Subovka, and the Notanebi deposit (Miocene sandstone) is reputed to contain 20% w/w bitumen. On the other hand, the Kazakhstan occurrence, near the Shubar-Kuduk oil field, is a bituminous lake with a bitumen content that has been estimated to be in the order of 95% w/w of the deposit.

Oil sand occurrences also occur in the Southern Llanos of Colombia, Burgan in Kuwait, and the Inciarte and Bolivar coastal fields of the Maracaibo Basin, but very little in known about the deposits. There are also small deposits in the Leyte Islands (Philippines), in the Mefang Basin in Thailand, in Chumpi, and near Lima (Peru). Oil sand deposits have also been recorded in Spain, Portugal, Cuba, Argentina, Thailand, and Senegal, but most are poorly defined and are considered to contain (in place) less than 1 million barrels ($1 \times 10^6$ bbl) of bitumen.

The fact that commercialization has taken place in Canada does not mean that commercialization is imminent for other oil sand deposits. There are considerable differences between the Canadian deposits and the deposits in the United States and the rest of the world that could preclude across-the-board application of the principles applied to the Canadian oil sand deposits to the other oil sand deposits.

## 7.3  BITUMEN PROPERTIES

Bitumen can be assessed in terms of sulfur content, carbon residue, nitrogen content, and metal content. Properties such as the API gravity and viscosity also help the refinery operator to gain an understanding of the nature of the material that is to be processed. The products from high-sulfur feedstocks often require extensive treatment to remove (or change) the corrosive sulfur compounds. Nitrogen compounds and the various metals that occur in crude oils will cause serious loss of catalyst life. The carbon residue presents an indication of the amount of thermal coke that may be formed to the detriment of the liquid products.

### 7.3.1  ELEMENTAL (ULTIMATE) COMPOSITION

The elemental analysis of oil sand bitumen has been widely reported, and of the data that are available, the proportions of the elements vary over fairly narrow limits (Speight 1990 and references cited therein):

Carbon, 83.4% ± 0.5%
Hydrogen, 10.4% ± 0.2%
Nitrogen, 0.4% ± 0.2%
Oxygen, 1.0% ± 0.2%
Sulfur, 5.0% ± 0.5%
Metals (Ni and V), >1000 ppm

Bitumen from oil sand deposits in the United States has a similar ultimate composition to the Athabasca bitumen. However, to note anything other than the hydrogen/carbon atomic ratio (which is an indicator of the relative amount of hydrogen needed for upgrading) or the amount of nitrogen is beyond the scope of general studies.

### 7.3.2  CHEMICAL COMPOSITION

The precise chemical composition of bitumen, despite the large volume of work performed in this area, is largely speculative. In very general terms (and as observed from elemental analyses), heavy oil and bitumen are complex mixtures of (1) hydrocarbons, (2) nitrogen compounds, (3) oxygen compounds, (4) sulfur compounds, and (5) metallic constituents. However, this general definition is not adequate to describe the composition of bitumen as it relates to conversion to liquid products.

It is therefore convenient to divide the hydrocarbon components of bitumen into the following four classes:

1. Paraffins, which are saturated hydrocarbons with straight or branched chains but without any ring structure. The occurrence of such chemical species in heavy oil and bitumen is rare.
2. Naphthenes, which are saturated hydrocarbons containing one or more rings, each of which may have one or more paraffinic side chains (more correctly known as alicyclic hydrocarbons).

3. Aromatics, which are hydrocarbons containing one or more aromatic nuclei, such as benzene, naphthalene, and phenanthrene ring systems, which may be linked up with (substituted) naphthene rings and/or paraffinic side chains.
4. Heteroatom compounds, which include organic compounds of nitrogen, oxygen, sulfur, and porphyrins (metallo-organic compounds). These are, by far, the major class of compounds contained in bitumen and play a major role in conversion processes.

On a molecular basis, bitumen is a complex mixture of hydrocarbons with varying amounts of organic compounds containing sulfur, oxygen, and nitrogen, as well as compounds containing metallic constituents, particularly vanadium nickel, iron, and copper (Reynolds 2000). Compared to the more conventional crude oils where the hydrocarbon content may be as high as 97% w/w, bitumen (depending upon the source) may contain as little as 50% w/w hydrocarbon constituents, with the remainder being compounds that contain nitrogen, oxygen, sulfur, and metals.

### 7.3.3 FRACTIONAL COMPOSITION

Fractional composition is an important property of bitumen and bitumen, can be separated into a variety of fractions called saturates, aromatics, resin constituents, and asphaltene constituents (Speight 2000; Ancheyta and Speight 2007; Speight 2014). Much of the focus has been on the constituents of the asphaltene fraction because of its high sulfur content, high coke-forming propensity, and the complexity of the cracking reactions (Chakma 2000; Yen 2000).

Data that define the composition are extremely important to refining processes. The data give the refiner an indication of the potential behavior of the bitumen in refinery processes and the potential yields of products that might be expected. The data also provide guidelines for the mining operation that is deigned to produce an average feedstock for further processing and so maintains a product balance.

### 7.3.4 THERMAL REACTIONS

The thermal reactions of bitumen have received considerable attention and provide valuable information about the potential chemical conversion that can be performed. Thus, bitumen constituents can be thermally decomposed under conditions similar to those employed for visbreaking (viscosity breaking, ca. 470°C, 880°F) to afford, on the one hand, light oils that contain higher paraffins and, on the other hand, coke:

$$\text{Bitumen} \rightarrow H_2, + CO, + CO_2, + H_2S, + SO_2, + H_2O, + CH_2 = CH_2, + CH_4,$$

$$+ CH_3CH_3, + (CH_3)_3CH, + CH_3(CH_2)_nCH_3, \text{ and so forth.}$$

The reaction paths are extremely complex; spectroscopic investigations indicate an overall dealkylation of the aromatics to methyl (predominantly) or ethyl (minority) groups. In fact, the thermal decomposition of heavy oil and bitumen constituents affords light oil and a hydrocarbon gas composed of the lower paraffins. Coke is also produced. The constituents of bitumen may also be hydrogenated to produce resins and oils at elevated temperatures (>250°C).

### 7.3.5 Physical Properties

The specific gravity of bitumen shows a fairly wide range of variation. The largest degree of variation is usually due to local conditions that affect material lying close to the faces, or exposures, occurring in surface oil sand deposits. There are also variations in the specific gravity of the bitumen found in beds that have not been exposed to weathering or other external factors.

Bitumen gravity primarily affects the upgrading requirements needed because of the low hydrogen content of the produced bitumen. The API gravity of known United States oil sand bitumen ranges downward from about 14° API (0.973 specific gravity) to approximately 2° API (1.093 specific gravity). Although only a vague relationship exists between density (gravity) and viscosity, very-low-gravity bitumen generally has very high viscosity.

Bitumen is relatively nonvolatile (Speight 2000; Ancheyta and Speight 2007; Speight 2014), and nondestructive distillation data (Table 7.2) show that oil sand bitumen is a high-boiling material. There is usually little or no gasoline (naphtha) fraction in bitumen, and the majority of the distillate falls in the gas oil-lubrication

**TABLE 7.2**
**Distillation Data (Cumulative% w/w Distilled) for Bitumen and Crude Oil**

| Cut Point | | Cumulative% w/w Distilled | | |
|---|---|---|---|---|
| °C | °F | Athabasca | PR Spring | Leduc (Canada) |
| 200 | 390 | 3 | 1 | 35 |
| 225 | 435 | 5 | 2 | 40 |
| 250 | 480 | 7 | 3 | 45 |
| 275 | 525 | 9 | 4 | 51 |
| 300 | 570 | 14 | 5 | |
| 325 | 615 | 26 | 7 | |
| 350 | 660 | 18 | 8 | |
| 375 | 705 | 22 | 10 | |
| 400 | 750 | 26 | 13 | |
| 425 | 795 | 29 | 16 | |
| 450 | 840 | 33 | 20 | |
| 475 | 885 | 37 | 23 | |
| 500 | 930 | 40 | 25 | |
| 525 | 975 | 43 | 29 | |
| 538 | 1000 | 45 | 35 | |
| 538+ | 1000+ | 55 | 65 | |

distillate range (greater than 260°C; greater than 500°F). Typically, 50% w/w of oil sand bitumen is nondistillable under the conditions of the test; this amount of nonvolatile material responds very closely to the amount of asphaltenes plus resins of the feedstock.

The pour point is the lowest temperature at which the bitumen will flow. The pour point for oil sand bitumen can exceed 300°F—far greater than the natural temperature of oil sand reservoirs. The pour point is important to consider because for efficient production, a thermal extraction process to increase the reservoir temperature to beyond the pour point temperature must supply supplementary heat energy.

## 7.4  BITUMEN RECOVERY

Current commercial operations involve open-pit mining of oil sand, after which the sand is transported to a processing plant, the bitumen is extracted, and disposal of the waste sand (Speight 2013b).

There are two approaches to open-pit mining of oil sand. The first uses a few mining units of custom design, for example, bucket-wheel excavators and large draglines in conjunction with belt conveyors. In the second approach, a multiplicity of smaller mining units of conventional design is employed, for example, scrapers and truck-and-shovel operations have been considered. Each method has advantages and risks.

Underground mining options have also been proposed but, for the moment, have been largely discarded because of the fear of collapse of the formation onto any operators or equipment. This particular option should not, however, be rejected out of hand, because a novel aspect or the requirements of the developer (which remove the accompanying dangers) may make such an option acceptable.

Once the oil sand is mined, there remains the issue of recovering the bitumen. This is accomplished by application of the hot-water process. To date, the hot-water process is the only successful commercial process to be applied to bitumen recovery from mined oil sand in North America. Many process options have been tested with varying degrees of success, and one of these options may even supersede the hot-water process. The hot-water process utilizes the linear and nonlinear variation of bitumen density and water density, respectively, with temperature so that the bitumen, which is heavier than water at room temperature, becomes lighter than water at approximately 80°C (180°F). Surface-active materials in the oil sand also contribute to the process.

The oil sand deposits in the United States and the rest of the world have received considerably less attention than the Canadian deposits. Nevertheless, approaches to recover the bitumen from the US oil sands have been made. In the present context, an attempt has been made to develop the hot-water process for the Utah sands (Miller and Misra 1982). The process differs significantly from that used for the Canadian sands because of the oil-wet Utah sands in contrast to the water-wet Canadian sands. This necessitates disengagement by hot-water digestion in a high-shear force field under appropriate conditions of pulp density and alkalinity. The dispersed bitumen droplets can also be recovered by aeration and froth flotation.

The other aboveground method of separating bitumen from oil sands after the mining operation involves direct heating of the oil sand without previous separation of the bitumen. Thus, the bitumen is not recovered as such but is an upgraded overhead product. Although several processes have been proposed to accomplish this, the common theme is to heat the oil sand to separate the bitumen as a volatile product. At this time, however, it must be recognized that the volatility of the bitumen is extremely low and that what actually separates from the sand is a cracked product with the coke remaining on the sand.

The coke that is formed as a result of the thermal decomposition of the bitumen remains on the sand, which is then transferred to a vessel for coke removal by burning in air. The hot flue gases can be used either to heat incoming oil sand or as refinery fuel.

A later proposal suggested that the Lurgi process might have applicability to bitumen conversion (Rammler 1970). A more modern approach has also been developed that also cracks the bitumen constituents on the sand (Taciuk 1981a,b). The processor consists of a large, horizontal, rotating vessel that is arranged in a series of compartments. The two major compartments are a preheating zone and a reaction zone. Product yields and quality are reported to be high.

## 7.5   LIQUID FUELS FROM OIL SAND

Liquid fuels are produced from oil sand bitumen in the form of synthetic crude oil (syncrude) that undergoes further refining at a conventional refinery to produce the liquid fuels.

Synthetic crude oil is a complex mixture of hydrocarbons, somewhat similar to petroleum but differs in composition from petroleum insofar as the constituents of synthetic crude oil are not found in nature.

As a feedstock, the quality of oil sand bitumen is low compared to that of conventional crude oil and heavy oil. The high carbon residue of bitumen dictates that considerable amounts of coke will be produced during thermal refining (Table 7.3; Figure 7.1). Thus, production of liquid fuels from oil sand bitumen has included options for coke use.

Technologies for the production of liquid fuels from bitumen can be broadly divided into *carbon rejection* processes and *hydrogen addition* processes (Figure 7.2).

Carbon rejection processes redistribute hydrogen among the various components, resulting in fractions with increased hydrogen/carbon atomic ratios and fractions with lower hydrogen/carbon atomic ratios. On the other hand, hydrogen addition

## TABLE 7.3
### Predicted Coke Yields from Various Feedstocks

| API Gravity of Feedstock | Carbon Residue, % w/w | Coke Yield | |
|---|---|---|---|
| | | Delayed Coking | Fluid Coking |
| 2 | 30 | 45 | 35 |
| 6 | 20 | 36 | 23 |
| 10 | 15 | 28 | 17 |
| 16 | 10 | 18 | 12 |
| 26 | 5 | 9 | 3 |

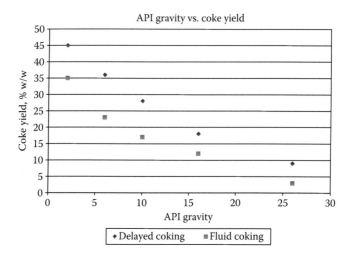

**FIGURE 7.1**    Comparison of coke produced by delayed coking and fluid coking processes.

processes involve reaction heavy crude oils with an external source of hydrogen and result in an overall increase in hydrogen/carbon ratio. Within these broad ranges, all upgrading technologies can be subdivided as follows:

1. Carbon rejection: for example, visbreaking, steam cracking, fluid catalytic cracking, and coking.

    Carbon rejection processes offer attractive methods of conversion of bitumen because they enable low operating pressure, while involving high operating temperature, without requiring expensive catalysts.

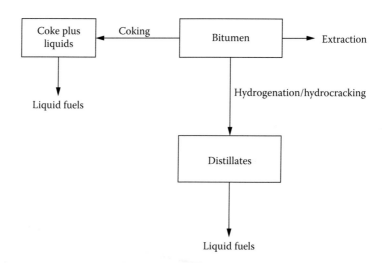

**FIGURE 7.2**    Production of liquid fuels from oil bitumen by carbon rejection processes and hydrogen addition processes.

2. Hydrogen addition: catalytic hydroconversion (hydrocracking), hydrovisbreaking, and donor solvent processes.

   Bitumen hydrotreating processes (with the attendant process parameters, Table 7.4) offer desulfurization to low-sulfur feedstocks for other processes or hydrocracking to kerosene and gas oil.

3. Separation processes: distillation and deasphalting.

   Solvent deasphalting allows removal of sulfur and nitrogen compounds as well as metallic constituents in the high-carbon asphalt and would be more appropriate for use in combination with other processes.

Currently, the overall upgrading process by which bitumen is converted to liquid fuels is accomplished in two steps. The first step is the *primary upgrading* or *primary conversion* process (Figure 7.3), which improves the hydrogen/carbon ratio by either carbon removal or hydrogen addition, cracking bitumen to produce distillable products that are more easily processed downstream to liquid fuels.

### TABLE 7.4
### Hydrotreating Processing Parameters

| Parameter | Naphtha | Bitumen |
|---|---|---|
| Temperature, °C | 300–400 | 340–450 |
| Pressure, atm | 35–70 | 50–200 |
| LHSV | 4.0–10.0 | 0.2–1.0 |
| $H_2$ recycle rate, scf/bbl | 400–1000 | 3000–5000 |
| Catalysts life, years | 3.0–10.0 | 0.5–1.0 |
| Sulfur removal, % | >95 | <80 |
| Nitrogen removal, % | >95 | <40 |

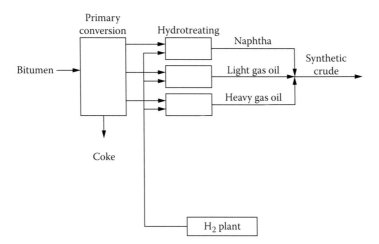

**FIGURE 7.3**  Bitumen conversion to liquid fuels by primary upgrading and secondary processes.

The *secondary upgrading* process involves hydrogenation of the primary products and is the means by which sulfur and nitrogen are removed from the primary products. The upgraded or synthetic crude can then be refined to a variety of liquid fuels such as gasoline, diesel fuel, and jet fuel.

Bitumen is hydrogen deficient and is upgraded by carbon removal (coking) or hydrogen addition (hydrocracking). There are two methods by which bitumen conversion can be achieved: (1) by direct heating of mined oil sand and (2) by thermal decomposition of separated bitumen. The latter is the method used commercially, but the former deserves mention here since the potential for commercialization remains open.

Although this improvement in properties may not appear to be too drastic, it usually is sufficient to have major advantages for refinery operators. Any incremental increase in the units of hydrogen/carbon ratio can save amounts of costly hydrogen during upgrading. The same principles are also operative for reductions in the nitrogen, sulfur, and oxygen content. This latter occurrence also improves catalyst life and activity as well as reduces the metal content. In short, *in situ* recovery processes (although less efficient in terms of bitumen recovery relative to mining operations) may have the added benefit of leaving some of the more obnoxious constituents (from the processing objective) in the ground.

The low proportion of volatile constituents (i.e., those constituents boiling below 200°C, 390°F) in bitumen precludes refining by distillation, and it is recognized that refining by thermal means is necessary to produce liquid fuel streams. A number of factors have influenced the development of facilities that are capable of converting bitumen to a synthetic crude oil.

Visbreaking has been considered as one process for the primary upgrading step (Table 7.5). However, a visbreaking product is still high in sulfur and nitrogen, with some degree of unsaturation. This latter property enhances gum formation with the accompanying risk of pipeline fouling and similar disposition problems in storage facilities and fuel oil burners. A high sulfur content of the finished products is environmentally unacceptable. In addition, high levels of nitrogen cause problems in the downstream processes, such as in catalytic cracking, where nitrogen levels in excess of 3000 parts per million (ppm) will cause rapid catalyst deactivation; metals (nickel and vanadium) cause similar problems.

The higher-boiling constituents (i.e., those boiling in the range of 200°C–400°C, 390°F–750°F) can be isolated by distillation, but in general terms, more than 40% w/w of oil sand bitumen boils above 540°C (1000°F). Nevertheless, the low proportion of volatile constituents (i.e., those constituents boiling below 200°C, 390°F) in bitumen precludes complete refining by distillation, although a vacuum distillation unit has been included in the latest plant (Syncrude) operation. However, it is recognized that refining by thermal means (i.e., thermal cracking) is necessary to produce liquid fuel streams. A number of factors have influenced the development of facilities that are capable of converting bitumen to a synthetic crude oil. A visbreaking product would be a hydrocarbon liquid that is still high in sulfur and nitrogen with some degree of unsaturation. This latter property enhances gum formation with the accompanying risk of pipeline fouling and similar disposition problems in storage facilities and fuel oil burners—finished products with high sulfur content are environmentally unacceptable.

**TABLE 7.5**

**Examples of Product Yields and Properties for Visbreaking Athabasca Bitumen and Similar-API Feedstocks**

| Feedstock | Arabian Light Vacuum Residuum | Arabian Light Vacuum Residuum | Iranian Light Vacuum Residue | Athabasca Bitumen |
|---|---|---|---|---|
| API gravity | 7.1 | 6.9 | 8.2 | 8.6 |
| Carbon residue[a] | 20.3 | | 22.0 | 13.5 |
| Sulfur, wt% | 4.0 | 4.0 | 3.5 | 4.8 |
| Product yields,[b] vol% | | | | |
| Naphtha (<425°F, <220°C) | 6.0 | 8.1 | 4.8 | 7.0 |
| Light gas oil (220°C–340°C, 425°F–645°F) | 16.0 | 10.5 | 13.1 | 21.0 |
| Heavy gas oil (340°C–540°C, 645°F–1000°F) | | 20.8 | —[b] | 35.0 |
| Residuum | 76.0 | 60.5 | 79.9 | 34.0 |
| API gravity | 3.5 | 0.8 | 5.5 | |
| Carbon residue[a] | | | | |
| Sulfur, wt% | 4.7 | 4.6 | 3.8 | |

[a] Conradson.

[b] A blank product yield line indicates that the yield of the lower-boiling product has been included in the yield of the higher-boiling product.

Thus, a product of acceptable quality could be obtained by distillation to an appropriate cut point, but the majority of the bitumen would remain behind to be refined by whichever means would be appropriate, remembering, of course, the need to balance fuel requirements and coke production. It is therefore essential that any bitumen-upgrading program convert the nonvolatile residuum to a lower-boiling, low-viscosity, low-molecular-weight product that also has a high hydrogen/carbon ratio.

## 7.5.1 COKING PROCESSES

Coking processes are the primary upgrading processes by which bitumen is converted to distillable products.

In the early stages of oil sand development, coking became the process of choice for bitumen conversion, and bitumen is currently converted commercially by delayed coking (Suncor) and by fluid coking (Syncrude). In each case, the charge is converted to distillate oils, coke, and light gases. The coke fraction and product gases can be used for plant fuel. The coker distillate is a partially upgraded material in itself and is a suitable feed for hydrodesulfurization to produce a low-sulfur synthetic crude oil.

Delayed coking (Table 7.6) is a semibatch process in which feed bitumen is heated before being fed to coking drums that provide sufficient residence time for the cracking reactions to occur.

## TABLE 7.6
## Examples of Product Yields and Product Properties for Delayed Coking of Athabasca Bitumen and Similar-API Feedstocks

| Feedstock | Kuwait Residuum | West Texas Residuum | Tia Juana Residuum | Alaska North Slope Residuum | Arabian Light Residuum | Athabasca Bitumen |
|---|---|---|---|---|---|---|
| API gravity | 6.7 | 8.9 | 8.5 | 7.4 | 6.9 | 7.3 |
| Carbon residue[a] | 19.8 | 17.8 | 22.0 | 18.1 | | 17.9 |
| Sulfur, wt% | 5.2 | 3.0 | 2.9 | 2.0 | 4.0 | 5.3 |
| Product yields, vol% | | | | | | |
| Naphtha (35°C–220°C, 95°F–425°F) | 26.7 | 28.9 | 25.6 | 12.5 | 19.1 | 20.3 |
| Light gas oil (220°C–340°C, 425°F–645°F) | 28.0 | 16.5 | 26.4 | —[b] | —[b] | —[b] |
| Heavy gas oil (340°C–540°C, 645°F–1000°F) | 18.4 | 26.4 | 13.8 | 51.2 | 48.4 | 58.8 |
| Coke | 30.2 | 28.4 | 33.0 | 27.2 | 32.8 | 21.0 |
| Sulfur, wt% | 7.5 | 4.5 | | 2.6 | 5.6 | 8.0 |

[a] Conradson.
[b] A blank product line indicates that the yield of the lower-boiling product has been included in the yield of the higher-boiling product.

The Suncor plant (in operation since 1967) involves a delayed coking technique followed by hydrotreating of the distillates to produce synthetic crude oil that has properties that are substantially different from the original bitumen and that are close to the properties of conventional petroleum (Table 7.7). The selection of delayed coking over less severe thermal processes, such as visbreaking, was based (at the time of planning, from 1960 to 1964) on the high yields of residuum produced in these alternate processes. The yields of coke from the residuum would have exceeded the plant fuel requirements, especially if the distillate had to be shipped elsewhere for hydrogen treating as well as more favorable product distribution and properties. Alternate routes for the disposal of the excess coke would be needed.

In the Suncor operation, bitumen conversion to liquids is on the order of 75% v/v, with fluid coking giving a generally higher yield of liquids compared to delayed coking (Table 7.3; Figure 7.1). The remainder appears as coke (approximately 15% w/w) and gases.

Fluid coking is a continuous process employing two vessels with fluid coke. It provides a better yield of overhead products than delayed coking. Feed oil flows to the reactor vessel where cracking and formation of coke occur; coke is combusted in the burner. Fluid transfer lines between these vessels provide the coke circulation necessary for heat balance. The proportion of coke burned is just sufficient to satisfy heat losses and provide the heat for the cracking reactions.

In the fluid coking process, whole bitumen (or topped bitumen) is preheated and sprayed into the reactor, where it is thermally cracked in the fluidized coke bed at temperatures typically between 510°C and 540°C (950°F and 1000°F) to produce light products and coke. The coke is deposited on the fluidized coke particles while the light products pass overhead to a scrubbing section in which any high-boiling products are condensed and recombined with the reactor fresh feed. The uncondensed scrubber overhead passes into a fractionator in which liquid products of suitable boiling ranges for downstream hydrotreating are withdrawn. Cracked reactor

## TABLE 7.7
### Properties of Synthetic Crude Oil from Athabasca Bitumen

| Property | | Bitumen | Synthetic Crude Oil | Crude Oil |
|---|---|---|---|---|
| Gravity, °API | | 8 | 32 | 35 |
| Sulfur, % w/w | | 4.8 | 0.2 | 0.1 |
| Nitrogen, % w/w | | 0.4 | 0.1 | 0.2 |
| Viscosity, centipoise at 100°F | | 500,000 | 10 | 10 |
| Distillation profile, % w/w (cumulative) | | | | |
| °C | °F | | | |
| 0 | 30 | 0 | 5 | 5 |
| 30 | 85 | 0 | 30 | 30 |
| 220 | 430 | 1 | 60 | 40 |
| 345 | 650 | 17 | 90 | 70 |
| 550 | 1020 | 45 | 100 | 90 |
| Residuum | | 100 | | 100 |

gases containing butanes and lower-molecular-weight hydrocarbon gases pass overhead to a gas recovery section. The propane material ultimately flows to the refinery
gas system, and the condensed butane and butenes may (subject to vapor pressure
limitations) be combined with the synthetic crude. The heat necessary to vaporize
the feed and to supply the heat of reaction is supplied by hot coke that is circulated
back to the reactor from the coke heater. Excess coke that has formed from the fresh
feed and deposited on hot circulating coke in the fluidized reactor bed is withdrawn
(after steam stripping) from the bottom of the reactor.

Sulfur is distributed throughout the boiling range of the delayed coker distillate,
as with distillates from direct coking. Nitrogen is more heavily concentrated in the
higher-boiling fractions but is present in most of the distillate fractions. Raw coker
naphtha contains significant quantities of olefins and diolefins that must be saturated
by downstream hydrotreating. The gas oil has a high aromatic content typical of
coker gas oils.

The limitations of processing oil sand bitumen depend, to a large extent, on the
tendency for coke formation and the deposition of metals and coke on the catalyst
due to the higher molecular weight (low volatility) and heteroatom content. In fact,
the increasing supply of heavy crude oil and oil sand bitumen is a matter of serious
concern for the petroleum industry. In order to satisfy the changing pattern of product demand, significant investments in refining conversion processes will be necessary to profitably utilize these heavy crude oils. The most efficient and economical
solution to this problem will depend, to a large extent, on individual country and
company situations.

Depending on the properties, the option for oil sand bitumen is to subject the
feedstock to delayed coking, fluid coking, LC-Fining, or H-Oil hydrocracking as the
*primary upgrading* step (Figure 7.1), with some prior distillation or topping (Speight
2005, 2009, 2011a, 2013c, 2014). In ebullated-bed processes, such as the LC-Fining
and H-Oil hydrocracking process and variations thereof (Speight 2013c, 2014), the
catalyst within the reactor bed is not fixed. In such a process, the feedstock enters the
bottom of the reactor and flows upward through the catalyst—the catalyst is kept in
suspension by the pressure of the fluid feed. Ebullating-bed reactors are capable of
converting the most problematic feeds, such as atmospheric residua, vacuum residua,
heavy oil, and tar sand bitumen (all of which have a high content of asphaltene constituents as well as metal constituents, sulfur constituents, and constituents ready
to form sediment) to lower-boiling, more valuable products while simultaneously
removing the contaminants.

In any process where the feedstock has high metal content or has a propensity for
early coke formation, a demetallization reactor (*guard reactor, guard-bed reactor*)
may be required. This is a reactor that is placed on-stream in front of the hydrocracking reactor to remove contaminants and typically employs an inexpensive catalyst
to remove metals from the hydrocracker feed. A catalyst support having large pores
preferentially demetallized with a low degree of desulfurization and high tolerance
for metals is preferred.

Feedstocks that have relatively high metal contents (>300 ppm) substantially
increase catalyst consumption because the metals poison the catalyst, thereby requiring frequent catalyst replacement. The usual desulfurization catalysts are relatively

expensive for these consumption rates, but there are catalysts that are relatively inexpensive and can be used in the first reactor to remove a large percentage of the metals. Subsequent reactors downstream of the first reactor would use normal catalysts. Since the catalyst materials are proprietary, it is not possible to identify them here, but it is understood that such catalysts contain little or no metal promoters, that is, nickel, cobalt, and molybdenum. Metals removal on the order of 90% has been observed with the guard-bed catalysts.

Thus, one method of controlling demetallization is to employ separate smaller *guard reactors* just ahead of the hydrocracking reactor. The preheated feed and hydrogen pass through the guard reactors that are filled with an appropriate catalyst for demetallization. The advantage of this system is that it enables replacement of the most contaminated catalyst (*guard bed*), where pressure drop is highest, without having to replace the entire inventory or shut down the unit. The feedstock is alternated between guard reactors while catalyst in the idle guard reactor is being replaced.

Finally, there is not one single upgrading technology that will fit all refineries. Bitumen properties, existing refinery configuration, and desired product slate all can have a significant effect on the final configuration. Furthermore, a proper evaluation, however, is not a simple undertaking for an existing refinery. The evaluation starts with an accurate understanding of the nature of the bitumen along with an assessment of the conversion chemistry. Once the options have been defined, development of the optimal configuration for refining the incoming feedstocks can be designed.

### 7.5.2 Product Upgrading

After primary upgrading, the product streams are hydrotreated and combined to form a *synthetic crude oil* that is shipped to a conventional refinery for further processing to liquid fuels.

The primary liquid product is then hydrotreated (secondary conversion or refining) to remove sulfur and nitrogen (as hydrogen sulfide and ammonia, respectively) and to hydrogenate the unsaturated sites exposed by the conversion process. It may be necessary to employ separate hydrotreaters for light distillates and medium-to-heavy fractions; for example, the heavier fractions require higher hydrogen partial pressures and higher operating temperatures to achieve the desired degree of sulfur and nitrogen removal. Commercial applications have, therefore, been based on the separate treatment of two or three distillate fractions at the appropriate severity to achieve the required product quality and process efficiency.

Hydrotreating is generally carried out in downflow reactors containing a fixed bed of cobalt–molybdate catalysts. The reactor effluents are stripped of the produced hydrogen sulfide and ammonia. Any light ends are sent to the fuel gas system, and the liquid products are recombined to form synthetic crude oil.

Finishing and stabilization (hydrodesulfurization and saturation) of the liquid products is achieved by hydrotreating the liquid streams, as two or three separate streams. This is necessary because of the variation in conditions and catalysts necessary for treatment of a naphtha fraction relative to the conditions necessary for treatment of gas oil. It is more efficient to treat the liquid product streams separately and then to blend the finished liquids to a synthetic crude oil. In order to take advantage of optimum operating conditions for various distillate fractions, the Suncor coker distillate is treated as three separate fractions: naphtha, kerosene, and gas oil. In the operation used by Syncrude, the bitumen products are separated into two distinct fractions: naphtha and mixed gas oils. Each plant combines the hydrotreated fractions to form synthetic crude oil that is then shipped by pipeline to a refinery. The upgraded or synthetic crude oil has properties that are quite different from the original feedstock and are closer to the properties of a conventional high-API-gravity crude oil (Table 7.7), and the product can be sent by pipeline to a refinery for further upgrading.

## 7.5.3 OTHER PROCESSES

There are several other processes that have received some attention for bitumen upgrading. These processes include partial upgrading (a form of thermal deasphalting), flexicoking, the Eureka process, and various hydrocracking processes.

Direct coking of tar sand with a fluid-bed technique has also been tested. In this process, tar sand is fed to a coker or still, where the tar sand is heated to approximately 480°C (approximately 895°F) by contact with a fluid bed of clean sand from which the coke has been removed by burning. Volatile portions of the bitumen are distilled. Residual portions are thermally cracked, resulting in the deposition of a layer of coke around each sand grain. Coked solids are withdrawn down a standpipe, fluid with air, and transferred to a burner or regenerator (operating at approximately 800°C, 1470°F) where most of the coke is burned off the sand grains. The clean, hot sand is withdrawn through a standpipe. Part (20% to 40%) is rejected, and the remainder is recirculated to the coker to provide the heat for the coking reaction. The products leave the coker as a vapor, which is condensed in a receiver. Reaction off-gases from the receiver are recirculated to fluidize the clean, hot sand that is returned to the coker.

An early process involved a coker for bitumen conversion and a burner to remove carbon from the sand. A later proposal suggested that the Lurgi process might have applicability to bitumen conversion (Rammler 1970). Another approach has also been developed that also cracks the bitumen constituents on the sand. The processor consists of a large, horizontal, rotating vessel that is arranged in a series of compartments. The two major compartments are a preheating zone and a reaction zone (Taciuk 1981a,b).

A partial coking or thermal deasphalting process provides a minimal upgrading of bitumen. In partial coking, the hot-water process froth is distilled at atmospheric pressure, and mineral matter and water are removed. A dehydrated mineral-free bitumen product is obtained that contains most of the asphaltenes and coke precursors. The process is carried out in batch equipment in laboratory tests over periods ranging from 30 min to 4 h. Thermal cracking begins as the liquid temperature passes 340°C (645°F). The distillation is continued into the range of 370°C to 450°C (700°F to 840°F). With slow heating (10°C, 18°F, temperature rise per hour) the coke production rate is approximately 1% w/w of feed per hour. As the coke forms about the entrained mineral particles, 1%–4% w/w coke, up to 50% v/v of the feedstock is recovered as distillate. After this treatment, the residue may be filtered to yield an essentially ash-free production suitable for applications such as metallurgical coke or production of bituminous paints, for which the original mineral content would have disqualified it.

In flexicoking, a gasifier vessel is added to the system in order to gasify excess coke with a gas–air mixture to a low-heating-value gas that can be desulfurized and used as a plant fuel. The Eureka process is a variant of delayed coking and uses steam stripping to enhance yield and produce a heavy pitch rather than a coke by-product.

Another option includes the presence of steam as an agent to reduce coke formation. For example, thermal cracking of Athabasca bitumen at various reaction conditions with and without the presence of steam showed that the presence of steam decreased coke yield, decreased sulfur removal, and reduced the H/C ratio of the liquid products.

Hydrocracking has also been proposed as a means of bitumen upgrading, that is, asphaltene conversion to liquid fuels (Solari 2000). The overall liquid yield of direct hydrogenation or hydrocracking of bitumen is substantially higher than that of coking, and significant amounts of sulfur and nitrogen are removed. Currently, however, large quantities of external fuel or hydrogen-plant feedstock are required.

Most hydrocracking processes start with an upflow reactor system in which the 524°C (975°F) material is cracked or converted. To prevent coking, the processes operate at high pressure with direct contact between bitumen feed and circulating hydrogen. Hydrocracking processes include, predominantly, the H-Oil process (Table 7.8) and the LC-Fining process (Table 7.9). In fact, LC-Fining is now an on-stream process for bitumen conversion to liquid fuels.

The hydrocracker products, as expected, have higher hydrogen and lower sulfur and nitrogen contents than those from the coking route and require less secondary upgrading. However, disadvantages of the hydrogen route include relatively high hydrogen consumption and high-pressure operation. Processes that use conventional (e.g., Co-Mo or Ni-Mo) catalysts are susceptible to metals poisoning, which may limit applicability to, or economics of, operation on feeds high in metals such as bitumen.

### 7.5.4 THE FUTURE

Of the Canadian oil sand deposits, the Athabasca deposit is the only oil sand deposit with reserves shallow enough to be mined. Since 90% of the Canadian oil sand deposits lie deep below the surface and cannot be recovered by open-pit (surface) mining techniques, *in situ* processes are being developed to access the deeper deposits. One of the most promising *in situ* techniques is referred to as steam-assisted gravity drainage (SAGD). This involves injecting steam though a series of wells into the deposit, after which the hot bitumen migrates by draining to the production wells, but this is not the answer for all oil sand deposits (Speight 2013c). The properties of the bitumen and associated sand typically dictate that each deposit is a site specific in terms of recovery processes and upgrading processes.

The future of upgrading bitumen lies in the development of new processes and the evolution of refinery operations to meet the challenge of these heavy feedstocks. In fact, the essential step required of refineries is the upgrading of heavy oil and bitumen, particularly residua. In fact, the increasing supply of heavy crude oil is a matter of serious concern for the petroleum industry. In order to satisfy the changing

**TABLE 7.8**

**H-Oil Process Feedstock and Product Data for Athabasca Bitumen and Low-API Feedstocks**

| Feedstock | Arabian Medium Vacuum Residuum | Athabasca Bitumen |
|---|---|---|
| API gravity | 4.9 | 8.3 |
| Sulfur, wt% | 5.4 | 4.9 |
| Nitrogen, wt% | | 0.5 |
| Carbon residue, wt% | | |
| Metals, ppm | 128.0 | |
| Ni | | |
| V | | |
| Residuum (>525°C, >975°F), wt% | | 50.3 |
| % Conversion | 0.9 | |
| Products, wt% | | |
| Naphtha (C5-204°C, C5-400°F) | 23.8 | 16.0 |
| Sulfur, wt% | | 1.0 |
| Distillate (204°C–343°C, 400°F–650°F) | 36.5 | 43.0 |
| Sulfur, wt% | | 2.0 |
| Vacuum gas oil (343°C–534°C, 650°F–975°F) | 37.1 | 26.4 |
| Sulfur, wt% | | 3.5 |
| Residuum (>534°C, >975°F) | 9.5 | 16.0 |
| Sulfur, wt% | | 5.7 |

pattern of product demand, significant investments in refining conversion processes will be necessary to profitably utilize these heavy crude oils. The most efficient and economical solution to this problem will depend, to a large extent, on individual country and company situations.

New processes for bitumen conversion may involve pretreatment using visbreaking (or hydrovisbreaking) with hydrocracking as a follow-up step. Other processes may replace or augment the deasphalting processes in many refineries. Conceivably, other heavy oils and bitumen might be upgraded in the same manner and, depending upon the upgrading facility, upgraded further for sales.

The limitations of processing oil sand bitumen depend, to a large extent, on the amount of higher-molecular-weight constituents (i.e., asphaltene constituents) that contain the majority of the heteroatom constituents. These constituents are responsible for high yields of thermal and catalytic coke. The majority of the metal constituents in crude oils are associated with asphaltenes. Part of these metals forms organometallic complexes. The rest are found in organic or inorganic salts that are soluble in water or in crude. In recent years, attempts have been made to isolate and to study the vanadium present in petroleum porphyrins, mainly in asphaltene fractions.

## TABLE 7.9
## LC-Fining Process Feedstock and Product Data for Athabasca and Low-API Feedstocks

| Feedstock | Gach Saran Vacuum Residuum | Arabian Heavy Vacuum Residuum | AL/AH Vacuum Residuum | Athabasca Bitumen |
|---|---|---|---|---|
| API gravity | 6.1 | 7.5 | 4.7 | 9.1 |
| Sulfur, wt% | 3.5 | 4.9 | 5.0 | 5.5 |
| Nitrogen, wt% | | | | 0.4 |
| Carbon residue, wt% | | | | |
| Metals | | | | |
| Ni | | | 39.0 | |
| V | | | 142.0 | |
| Products, wt%[a] | | | | |
| Naphtha (C5–205°C, C5–400°F)[a] | 9.7 | 14.3 | 23.9 | 11.9 |
| Sulfur, wt% | | | | 1.1 |
| Nitrogen, wt% | | | | |
| Distillate (205°C–345°C, 400°F–650°F)[a] | 14.1 | 26.5 | 64.8 | 37.7 |
| Sulfur, wt% | | | | 0.7 |
| Nitrogen, wt% | | | | |
| Heavy distillate (345°C–525°C, 650°F–975°F)[a] | 24.1 | 31.1 | 11.9 | 30 |
| Sulfur, wt% | | | | 1.1 |
| Nitrogen, wt% | | | | |
| Residuum (>525°C, >975°F)[a] | 47.5 | 21.3 | 5.0 | 12.9 |
| Sulfur, wt% | | | | 3.4 |
| Nitrogen, wt% | | | | |
| Carbon residue, wt% | | | | |

*Note:* AL/AH = Arabian light crude oil blended with Arabian heavy crude oil.
[a] Distillation ranges may vary by several degrees because of different distillation protocols.

In addition, when catalytic processes are employed, complex molecules, such as those that occur in the original asphaltene fraction or those formed during the process, are not sufficiently mobile (or are too strongly adsorbed by the catalyst) to be saturated by hydrogenation. The chemistry of the thermal reactions of some of these constituents dictates that certain reactions, once initiated, cannot be reversed and proceed to completion, and coke is the eventual product. These deposits deactivate the catalyst sites and eventually interfere with the hydroprocess.

Although it has been deemed necessary to continue the construction of extraction and upgrading plants in the immediate vicinity of the mining operation, bitumen is now shipped a *dilbit*—a blend of naphtha (produced from the bitumen) and bitumen, is acceptable to many refineries. Transportation of the undiluted bitumen is still a matter of economics and bitumen properties.

To develop the present concept of liquid fuels from oil sands, it is necessary to combine three operations, each of which contributes significantly to the cost of the venture: (1) a mining operation capable of handling 2 million tons, or more, of oil sand per day; (2) an extraction process to release the bitumen from the sand; and (3) an upgrading plant to convert the bitumen oil to a synthetic crude oil.

Obviously, there are many features to consider when development of oil sand resources for the production of liquid fuels is given consideration. It is more important to recognize that what are important features for one resource might be less important in the development of a second resource. Recognition of this facet of oil sand development is a major benefit that will aid in the production of liquid fuels in an economic and effective manner.

In the United States, oil sand economics is still very much a matter for conjecture. The estimates that have been published for current and proposed Canadian operations are, in a sense, not applicable to operations in the United States because of differences in the production techniques that may be required.

During the next 20 to 30 years, the evolution and future of heavy feedstock refining and the current refinery layout will be primarily on process modification with some new innovations coming on-stream (Speight 2007, 2011a). The industry will move predictably on to (1) deep conversion of heavy feedstocks, (2) higher hydrocracking and hydrotreating capacity, and (3) more efficient catalysts.

High-conversion refineries will also move to gasification of feedstocks for the development of alternative fuels and to enhance equipment usage. A major trend in the refining industry market demand for refined products will be in synthesizing fuels from simple basic reactants (e.g., synthesis gas) when it becomes uneconomical to produce super clean transportation fuels through conventional refining processes. Fischer–Tropsch plants together with IGCC systems will be integrated with or even into refineries, which will offer the advantage of high-quality products.

## REFERENCES

Ancheyta, J., and Speight, J.G. 2007. *Hydroprocessing of Heavy Oils and Residua*. CRC Press, Taylor & Francis Group, Boca Raton, Florida.

Chakma, A. 2000. Kinetics and mechanisms of asphaltene cracking during petroleum recovery and processing operations. In: *Asphaltenes and Asphalts 2*, Developments in Petroleum Science, 40B, T.F. Yen and G.V. Chilingarian (Editors). Elsevier, Amsterdam, Netherlands, Chapter 6.

Gary, J.G., Handwerk, G.E., and Kaiser, M.J. 2007. *Petroleum Refining: Technology and Economics*, 5th Edition. CRC Press, Taylor & Francis Group, Boca Raton, Florida.

Hsu, C.S., and Robinson, P.R. (Editors). 2006. *Practical Advances in Petroleum Processing*, Volume 1 and Volume 2. Springer Science, New York.

Meyer, R.F. (Editor). 1991. *Heavy Crude and Oil sands—Hydrocarbons for the 21st Century*. Petróleos de Venezuela S.A., Caracas, Venezuela.

Meyer, R.F., and Dietzman, W.D. 1981. Tar sand deposits. In: *The Future of Heavy Crude Oil and Tar Sands*, R.F. Meyer and C.T. Steele (Editors). McGraw-Hill, New York, p. 16.

Miller, J.C., and Misra, M. 1982. Hot-water process development for Utah tar sand. *Fuel Process. Technol.* 6: 27–59.

Rammler, R.W. 1970. The production of synthetic crude oil from oil sand by application of the lurgi-ruhrgas-process. *Can. J. Chem. Eng.* 48: 552–560.

Reynolds, J.G. 2000. Understanding metals in fossil fuels: A perspective of contributions by T.F. Yen. In: *Asphaltenes and Asphalts 2*, Developments in Petroleum Science, 40B, T.F. Yen and G.V. Chilingarian (Editors). Elsevier, Amsterdam, Netherlands, Chapter 3.

Solari, R.B. 2000. Asphaltene hydroconversion. In: *Asphaltenes and Asphalts 2*, Developments in Petroleum Science, 40B, T.F. Yen and G.V. Chilingarian (Editors). Elsevier, Amsterdam, Netherlands, Chapter 7.

Speight, J.G. 1990. Tar sand. In: *Fuel Science and Technology Handbook*, J.G. Speight (Editor). Marcel Dekker Inc., New York, Part II, Chapters 12–16.

Speight, J.G. 2000. *The Desulfurization of Heavy Oils and Residua*, 2nd Edition. Marcel Dekker Inc., New York.

Speight, J.G. 2005. Upgrading and refining of natural bitumen and heavy oil, in coal, oil shale, natural bitumen, heavy oil and peat, from *Encyclopedia of Life Support Systems (EOLSS)*. Developed under the Auspices of the UNESCO, EOLSS Publishers, Oxford, UK. Available at http://www.eolss.net.

Speight, J.G. 2008. *Synthetic Fuels Handbook: Properties, Processes, and Performance*. McGraw-Hill, New York.

Speight, J.G. 2009. *Enhanced Recovery Methods of Heavy Oil and Tar Sands*. Gulf Publishing Company, Houston, Texas.

Speight, J.G. 2011a. *The Refinery of the Future*. Gulf Professional Publishing, Elsevier, Oxford, United Kingdom.

Speight, J.G. 2011b. *An Introduction to Petroleum Technology, Economics, and Politics*. Scrivener Publishing, Salem, Massachusetts.

Speight, J.G. 2013a. *The Chemistry and Technology of Coal*, 3rd Edition. CRC Press, Taylor & Francis Group, Boca Raton, Florida.

Speight, J.G. 2013b. *Oil Sand Production Processes*. Gulf Professional Publishing, Elsevier, Oxford, United Kingdom.

Speight, J.G. 2013c. *Heavy and Extra Heavy Oil Upgrading Technologies*. Gulf Professional Publishing, Elsevier, Oxford, United Kingdom.

Speight, J.G. 2014. *The Chemistry and Technology of Petroleum*, 5th Edition. CRC Press, Taylor & Francis Group, Boca Raton, Florida.

Speight, J.G., and Ozum, B. 2002. *Petroleum Refining Processes*. Marcel Dekker Inc., New York.

Taciuk, W. 1981a. Apparatus and process for recovery of hydrocarbons from inorganic host materials. United States Patent 4,285,773, August 25.

Taciuk, W. 1981b. Process for recovery of hydrocarbons from inorganic host materials. United States Patent 4,406,961, November 22.

US Congress. 1976. Public Law FEA-76-4. United States Congress, Library of Congress, Washington, DC.

Yen, T.F. 2000. The realms and definitions of asphaltenes. In: *Asphaltenes and Asphalts 2*, Developments in Petroleum Science, 40B, T.F. Yen and G.V. Chilingarian (Editors). Elsevier, Amsterdam, Netherlands, Chapter 3.

# 8 Shale Oil from Oil Shale

*Sunggyu Lee*

## CONTENTS

## 8.1 OIL SHALE AS A SYNTHETIC FUEL (SYNFUEL) SOURCE

Interest in retorting oil from oil shale to produce a competitively priced synfuel has intensified since the oil embargo of the 1970s. Commercial interest, once very high in the 1970s and 1980s, substantially declined in the 1990s owing to the stable and low oil price. However, interest in oil shale as a clean liquid fuel source has been renewed in the twenty-first century, mainly triggered by the skyrocketing petroleum prices as well as the shortage of oil in the global market. Very serious commercial exploitation of oil shale as a major fuel resource has begun in the 21st century, as technological breakthroughs have been achieved in forms of horizontal drilling and hydraulic fracturing technologies, which has enabled deep-deposited oil shale resources to become more readily available. The rapid growth in oil shale fuel production has brought a so-called "shale gas boom" to the world, which is largely considered as a major game changer in the conventional energy and fuel market. However, it should be noted that oil shales have been used as liquid and solid fuels in certain areas for a long time, and its research also has quite a long history.

Mixed with a variety of sediments over a lengthy geological time period, shale forms a tough, dense rock ranging in color from light tan to black. Shales are often called *black shale* or *brown shale*, depending on the color. Oil shales have also been given various names in different regions. For example, the Ute Indians, on observing that some outcroppings burst into flames upon being hit by lightning, referred to oil shale as *the rock that burns*.

Oil shales are widely distributed throughout the world, with known major deposits in every continent. In this regard, oil shale is quite different from petroleum, which is more concentrated in certain regions of the world. Table 8.1 shows some published information regarding worldwide oil shale reserves.[1] In general, oil shale reserves refer to oil shale resources that are recoverable under given technoeconomic constraints, whereas oil shale deposits include all ranges (from uneconomical to very exploitable) of oil shale resources. Furthermore, the reported values are inevitably subjected to the geological and subsurface exploration and resource estimation technology as well as the accessibility and levels of potential commercial interests of

**TABLE 8.1**

**Oil Shale Reserves of the World (Reported in the Pre-Fracking Era)**

| IFP (1973) | | BP (1978) | | WEC (2002) | | USGS (2003) | | USDOE (2005) | |
|---|---|---|---|---|---|---|---|---|---|
| USA | 66 | USA | 63 | USA | 78 | USA | 70 | USA | 72 |
| Brazil | 24 | Brazil | 23 | Russia | 7.4 | Russia | 15 | Brazil | 5.4 |
| USSR | 3.4 | USSR | 3.3 | Brazil | 2.5 | Zaire | 3.3 | Jordan | 4.2 |
| Congo | 3.0 | Zaire | 2.9 | Jordan | 1.0 | Brazil | 2.7 | Morocco | 3.5 |
| Canada | 1.3 | — | — | Australia | 1.0 | Italy | 2.4 | Australia | 2.1 |
| Italy | 1.1 | — | — | Estonia | 0.5 | Morocco | 1.8 | China | 1.5 |
| China | 0.8 | — | — | China | 0.5 | Jordan | 1.1 | Estonia | 1.1 |
| Sweden | 0.1 | — | — | France | 0.2 | Australia | 1.0 | Israel | 0.3 |
| Germany | 0.1 | — | — | — | — | Estonia | 0.5 | — | — |
| Burma | 0.1 | — | — | — | — | China | 0.5 | — | — |
| — | — | — | — | — | — | Canada | 0.5 | — | — |
| — | — | — | — | — | — | France | 0.2 | — | — |

*Source:* Laherrere, J., Review on oil shale data, September 2005. Available at http://www.oilcrisis.com/laherrere/OilShaleReview200509.pdf.

*Note:* Figures are percentages; IFP = Institut Français du PÈtrole; BP = British Petroleum; WEC = World Energy Council; USGS = United States Geological Survey.

surveyed regions. Therefore, depending on the data source and the year of reporting, the statistical values vary significantly. Shales have been used in the past as a source of liquid fuel throughout the world, including Scotland, Sweden, France, South Africa, Australia, the USSR, China, Brazil, and the United States. However, the conventional oil shale industry has experienced several fluctuations on account of political, socioeconomic, market, and environmental reasons. One may find that the data shown in Table 8.1 are drastically different from the more recent data reported in the literature, in both the reporting format and the focused fuel sources/types desired from oil shale. One may also find that the modern listing is mostly for technically recoverable reserves in forms of shale gas and shale oil crude from oil shale deposits. For example, Advanced Resources International, Inc. (ARI) provides one of the most comprehensive data sets for technically recoverable shale gas and shale oil resources based on 137 shale formations in 41 countries outside the United States, in "EIA/ARI World Shale Gas and Shale Oil Resource Assessment" (June 2013). A principal difference in the oil shale data between the pre-2005 and post-2005 eras comes from a critical shift in economically feasible and favorable technology for oil shale exploitation, which has changed from largely "retorting-based technology" to "fracking-based technology." Also, the primary fuel type from oil shale for the pre-2005 era was shale oil crude from pyrolysis of oil shale kerogen, whereas that for the post-2005 era is shale gas and/or shale oil directly recoverable. This chapter deals with the oil shale technology that is geared toward the conventional technology mostly based on *ex situ* and *in situ* retorting technologies, whereas Chapter 9 is solely devoted to the modern oil shale fracking technology. Even though fracking is currently booming while retorting remains subdued in the current world, it should be clearly noted that both technologies provide excellent means for utilization of oil shale resources. Furthermore, the invaluable scientific information and database of the chemistry, petrochemistry, geology, and physicochemical properties of oil shale have been extensively studied and developed on worldwide oil shales in conjunction with the retorting type of process technologies investigated and developed.

It is believed (though documented historical evidence is lacking) that oil shales have been used directly as solid fuels in various regions, especially in areas with rich shales readily available near the earth's surface. For instance, an oil shale deposit at Autun, France, was commercially exploited as early as 1839.[2] As early as the 1850s, shale oil was promoted as being a replacement for wood, which America depended on for its energy. Logically, the oil shale industry in the United States was an important part of the U.S. economy prior to the discovery of crude oil in 1859. As Colonel Drake drilled his first oil well in Titusville, Pennsylvania, shale oil and its commercial production were gradually forgotten about and virtually disappeared with the availability of vast supplies of inexpensive liquid fuel, that is, petroleum. Similarly, Scotland had a viable shale industry from 1850 to 1864, when the low price of imported crude oil forced it to cease operation. It is interesting to note that British Petroleum (BP) was originally formed as a shale oil company. Likewise in Russia, oil shale from Estonia once supplied fuel gas for Leningrad.

In 1912, the president of the United States, by executive order, established the Naval Petroleum and Oil Shale Reserves. The Office of Fossil Energy of the U.S.

Department of Energy has been overseeing the U.S. strategic interests in oil shale since that time. The U.S. interest in oil shale revived briefly in the 1920s as domestic reserves of crude oil declined. But subsequent discoveries of large quantities of oil deposits in Texas again killed the hopes of an embryonic oil shale industry. Serious interest in oil shale commercialization and development revived once again in the 1970s and the 1980s, as the Arab oil embargo affected world energy supply and, consequently, the world economy.

In 1974, Unocal developed their *Union B* retort process and in 1976 planned for a commercial-scale plant at Parachute Creek to be built when investment would be economical. Many other companies, like Exxon, Shell, Dow Chemical, SOHIO, TOSCO, ARCO, AMOCO, Paraho, and others, initiated their own versions of oil shale development. In 1981, Unocal began construction of their Long Ridge 50,000 bbl/day plant based on their *Union B* retorting technology. AMOCO completed their *in situ* retorting demonstration of 1900 and 24,400 bbl of shale oil in 1980 and 1981, respectively. In 1980, Exxon purchased ARCO's Colony interest and in 1981 began Colony II construction, aiming at a production level of 47,000 bbl/day based on the TOSCO II process. In 1982, Exxon announced the closure of their Colony II project due to low demand and high cost. This event was known as *Exxon Black Sunday*. Meanwhile, Shell continued with their *in situ* experiments at Red Pinnacle until 1983. To make matters worse, Congress abolished the Synthetic Liquid Fuels Program after 40 years of operation and an investment of $8 billion. Paraho reorganized itself as New Paraho and began shale oil-modified asphalt (SOMAT) production. In 1991, Occidental closed their C-b (Rio Blanco County, Colorado) tract project without actual operation. *Unocal* operated their last large-scale experimental mining and retorting facility in western United States from 1980 until the shutdown of its Long Ridge (San Miguel County, Colorado) project in 1991. Unocal produced a total of 4.5 million barrels of shale oil from oil shale with an average of about 34 gal of shale oil per ton of rock over the life of the project.[2] After Unocal's shutdown in 1992, there has been no oil shale production in the United States. In the 1980s and 1990s, the stable crude oil price once again served as the principal reason for the diminishing interest in oil shale. Shell continued with some efforts in oil shale, particularly in the area of *in situ* heating technology at their property in Mahogany, Colorado. A notable experiment on *in situ* heating was done in 1997. Although oil shale activities based on the retorting technology in the United States have all but halted, some significant efforts continued in other countries such as Estonia, Australia, and Brazil.

There is again a strongly renewed interest in oil shale in the twenty-first century, as unstable and high energy prices, including those for natural gas and petroleum products, were experienced in most developed regions of the world at the onset of the new century. This strong interest has been successfully satisfied with the technological breakthroughs achieved in horizontal drilling technology and hydraulic fracture technology. Examples of energy-related crises are (1) California blackouts in 2001, (2) gasoline price surges in various regions of the United States in 2000 and 2001, (3) very high gasoline price owing to short supply of crude oil in 2004, (4) very high crude oil price in 2005 and 2006, and (5) sharp increases in residential energy costs in 2000, 2001, 2005, and 2006. This strong interest in oil shale exploitation has been successfully satisfied with the technological breakthroughs achieved in horizontal drilling technology and hydraulic fracture technology. It is truly remarkable to note that this technological breakthrough has created, in a very short period of time, a shale gas boom in the world, especially in the U.S., and many states in the U.S. view the shale energy industry as a major driver and game changer for the state's economy. This chapter deals with the conventional oil shale utilization technology and the fundamentals of oil shale, while Chapter 9 deals with the currently booming fracking technology.

At present, the conventional oil shale retorting is commercially exploited in several countries, such as Brazil, China, Estonia, and Australia. Brazil has a long history of oil shale development and exploitation since the late nineteenth century. In 1935, shale oil was produced at a small plant in São Mateus do Sul in the state of Paraná.[3] A more serious developmental effort was imitated by Petrobras, which developed the *Petrosix process* for oil shale retorting. The Semiworks retort was developed in 1972 and operated on a limited commercial scale, and then a larger Industrial Module Retort was brought into service for commercial production in December 1991. The total annual production of shale oil in Brazil was 195,200 tons in 1999.[3]

In China, the total annual production of shale oil in Fushun, Liaoning province, amounted to 80,000 tons in 2001.[4] They used 80 new retorts, which are known as *Fushun retorts*. Fushun used to produce as much as 780,000 tons of shale oil a year using the earlier retorts, and the production peaked in 1959.[4] Another major developmental effort is also being planned at Jilin province by China Power Investment Corp. (CPIC), one of the country's major power producers. The estimated oil shale deposit in Jilin province is 17 billion tons, which is about 56% of the total deposits in China. Some reports[4] claim that China has the fourth largest oil shale deposits in the world after the United States, Brazil, and Russia. However, more recent data as of May 2013 according to the U.S. Energy Information Administration (EIA) list that Russia, U.S., China, and Argentina are the top 4 countries in technically recoverable shale oil reserves with 75, 58, 32, and 27 billion barrels, respectively, and also show that China, Argentina, Algeria, and U.S. are the top 4 countries in technically recoverable shale gas reserves with 1115, 802, 707, and 665 trillion cubic feet, respectively. Almost certainly, both the ranking and estimated reserves will change in later years, as more exploration and discoveries will be made.

Estonia also has a long history of oil shale development and commercial exploitation. Its deposits are situated in the west of the Baltic Oil Shale Basin, and their oil shale is of high quality. Permanent mining of oil shale began in 1918 and continues to date.[3] The oil shale output in Estonia peaked at 31.35 million tons in 1980. In 1999, output was 10.7 million metric tons, out of which 1.3 million tons was retorted to produce 151,000 tons of shale oil.[3]

In the United States, the Energy Security Act, S.932, was legislated under the Carter administration on June 30, 1980. This legislation was intended to help create 70,000 jobs a year to design, build, operate, and supply resources for synfuel plants and for production of biomass fuels. The act established the Synthetic Fuels Corporation. New directions under President Reagan along with relatively stable oil prices made the synfuel industry less attractive to the public. Under President Bush's administration, production and development of synfuels became strategically less important than clean coal technology (CCT) and acid rain control. Under the Clinton administration, when energy prices were very stable and low, this deemphasizing trend further intensified in favor of national budget deficit reduction, which received public support and was based on projected long-term stability in energy supply and cost. Environmental protection received strong governmental and public support. In the twenty-first century, under President Bush, because of record-high energy prices and frequently experienced shortages, a renewed interest in energy self-sufficiency and development of commercial oil fields has been revived, but at the expense of some potential environmental disturbances. In 2004, the Office of Naval Petroleum and Oil Shale Reserves of the U.S. Department of Energy initiated a study on the significance of America's oil shale resources. The U.S. government also launched a new oil shale program with the *Oil Shale Development Act of 2005* to establish a leasing program in 2006. Since then, hydraulic fracture technology along with horizontal drilling methodology has matured and made deep-deposited shale fuel resources affordably available for commercial production and a "shale gas boom" has been experienced during the Obama administration period. Likewise, interest in alternative energy also intensified worldwide.

Oil and gas, that is, fluids from fossil fuels, accounted for one-third of the total energy consumed in the United States by the late 1920s. By the mid-1940s, oil and gas began to provide half of U.S. energy needs. They account for approximately three-fourths of U.S. energy needs today. The strong demand for oil and gas is likely to persist for a while, even though *bioenergy* and *hydrogen* fuels are rapidly gaining popularity, and are generally perceived as the principal energy sources for the future. Consequently, modern society's unprecedented appetite for fluid-type energy sources—without any new discoveries of major petroleum deposits in sight—will make it necessary to supplement supplies of domestic energy with synfuels such as those derived from oil shale or coal, as well as alternative fuel sources such as biomass, crops (e.g., soy and corn), and recycled materials. Hydrogen and ethanol will undoubtedly play very important roles as new gaseous and liquid fuels in the future energy market.

Market forces based on supply and demand will greatly affect the commercial development of oil shale. Besides competing with conventional crude oil and natural gas, conventional shale oil will have to compete with coal-derived fuels for a similar market.[5] Liquid fuels derived from coal are methanol, additional products of indirect liquefaction, Fischer–Tropsch hydrocarbons, or oxygenates. Table 8.2 shows synfuel products and their corresponding market characteristics.

Table 8.3 summarizes various countries involved in major types of synfuel development.

Depending on the relative level of success in synfuel or alternative energy development, the energy consumption patterns in the twenty-first century may be significantly affected. More emphasis will be undoubtedly placed on clean and renewable energy development, as well as environmentally clean utilization of conventional fuel.

**TABLE 8.2**

**Synfuel Products and Markets**

| Product | Technology Status | Market | Commercialization |
|---|---|---|---|
| Retorted shale oil[a] | Pilot plants up to 2000 tons/day | Mid-distillates (jet fuel, diesel fuel) | Regionally operated; fracking-produced shale oil, not a synfuel, being produced in large quantities |
| Oil sand | Small-, medium-scale, large-scale plants | Synthetic crudes (transportation fuels including gasoline and diesel) | In active commercial production; production increased sharply, making Canada as a major crude oil producing nation |
| Coal liquids | Direct liquefaction in pilot plants (250 tons/day) | Light and mid-distillates; petrochemical feedstocks | Small scale; specific application oriented; large-scale plants planned |
| Coal hydrocarbons | Indirect liquefaction of coal via Fischer–Tropsch synthesis; technology proved on large scales by SASOL | Petrochemical feedstocks | In commercial production; large-scale plants are being constructed |
| Methanol from natural gas | Low-pressure methanol synthesis technology actively used worldwide using syngas derived from natural gas reformation. Syngas derived from shale gas reformation is also likely to be used as feedstock for methanol synthesis | Chemical and petrochemical feedstocks; gas turbine; MTBE/ETBE/TAME production; dimethylether (DME) production; fuel gasoline market; off-peak energy generation and storage; fuel cell application | Very active and in large capacities; commercially used since the 1920s; in commercial production and worldwide product demand is steadily growing |
| Methanol from coal | Coal gasification proved in large-scale plants; liquid-phase methanol (LPMeOH) process proved for coal-derived syngas; single-stage synthesis process of dimethylether (DME) from coal-derived syngas being developed | Chemical and petrochemical feedstock; DME synthesis; once-through methanol (OTM); integrated gasification combined cycle (IGCC) application | Large-scale plant commercialized in the mid-1990s; large-scale plants are also being planned |

[a] Shale oil crude produced by modern hydraulic fracture technology is not included under synfuel products.

**TABLE 8.3**
**Synfuel Developmental Efforts in Various Countries**

| Nation | Coal | Oil Shale | Oil Sand | Biomass |
|--------|------|-----------|----------|---------|
| Australia | ✓ | ✓ | | ✓ |
| Brazil | ✓ | ✓ | | ✓ |
| Canada | ✓ | ✓ | ✓ | ✓ |
| China | ✓ | ✓ | | ✓ |
| Europe | ✓ | ✓ | | ✓ |
| Israel | | ✓ | | |
| Japan | ✓ | | | ✓ |
| Korea | ✓ | | | ✓ |
| Russia | ✓ | ✓ | | ✓ |
| South Africa | ✓ | | | |
| USA | ✓ | ✓ | | ✓ |

*Note:* A prediction of production and development cannot be made, because of the high uncertainty in this field. More effort in synfuel production from agricultural sources (e.g., crops) are expected in various regions of the world.

## 8.2   CONSTRAINTS IN COMMERCIAL PRODUCTION OF RETORTING-BASED SHALE OIL

During commercial exploitation of retorting-based shale oils, one can be faced with various constraints that represent possible deterring factors. These constraints originate from a variety of sources: technological, economical (or financial), institutional, environmental, socioeconomical, political, and water availability. The Office of Technology Assessment (OTA) analyzed the requirements for achieving each of the production goals by 1990, given the state of knowledge and the regulatory structure of the early 1980s.[6] Table 8.4 shows the factors that could hinder attainment of the goals, as assessed by OTA. In this table, their original target year of 1990 has been used without any alteration or modification. Taking into consideration the fact that no serious commercialization activity was realized in the 1990s for a variety of reasons, the readers should use their own discretion in interpreting the given information. The constraints judged to be "moderate" will hamper, but not necessarily preclude, development; those judged to be "critical" could become more serious barriers; and when it was unclear whether or to what extent certain factors would impede development, they were called "possible" constraints. Even though the information contained in this table may be currently outdated, it is still very informative and somewhat applicable to the present situation, considering the relative inactivity in this field during the last decade of the twentieth century. Due to the high and fluctuating prices of energy, especially those of liquid fuels, in the twenty-first century, there is rekindled interest in our commercialization of oil shale processes throughout the world, both retorting-based and fracking-based.

## TABLE 8.4
## Constraints to Implementing Retorting-Based Shale Oil Production Targets (as of 1981)

| Potential Deterring Factors | Severity of Impediment to 1990 Production Target (bbl/day) | | | |
|---|---|---|---|---|
| | 100,000 | 200,000 | 400,000 | 1,000,000 |
| **Technological** | | | | |
| Readiness | None | None | None | Critical |
| **Economic** | | | | |
| Availability of private capital | None | None | None | Moderate |
| Marketability of shale oil | Possible | Possible | Possible | Possible |
| Investor participation | None | Possible | Possible | Possible |
| **Institutional** | | | | |
| Availability of land | None | None | Possible | Critical |
| Permitting procedures | None | None | Possible | Critical |
| Major pipeline capacity | None | None | None | Critical |
| Design and construction services | None | None | Moderate | Critical |
| Equipment availability | None | None | Moderate | Critical |
| **Environmental** | | | | |
| Compliance with regulations | None | None | Possible | Critical |
| **Water Availability** | | | | |
| Availability of surplus surface water | None | None | None | Possible |
| Adequacy of existing supply systems | None | None | Critical | Critical |
| **Socioeconomic** | | | | |
| Adequacy of community facilities and services | None | Moderate | Moderate | Critical |

*Source:* Office of Technology Assessment.

## 8.2.1 TECHNOLOGICAL CONSTRAINTS

In the sense of the conventional shale energy utilization, oil shale can be retorted by either aboveground (*ex situ*) or underground (*in situ*) processing. In aboveground processing, shale is mined, transported to a processing facility, and then heated in retorting vessels. Underground retorting processes can be classified into two large categories: (1) In *true in situ (TIS) processing*, an oil shale deposit is first fractured by explosives and then retorted underground, and (2) a *modified in situ (MIS) processing* is a more advanced *in situ* technology in which a portion of the deposit is mined and the rest rubblized by explosives and retorted underground. The crude shale oil can be burned as a boiler fuel, or it can be further converted into syncrude by adding hydrogen.

Critical issues that need to be answered include the following:

1. What are the advantages and disadvantages of different mining and processing methods?
2. Are the technologies ready for large-scale commercial applications?
3. What are the major areas of uncertainty or unforeseen challenges in these technologies?
4. Are the technologies for process optimization available?
5. Are there sufficient scale-up data obtained from pilot and demonstration-scale plant operation?
6. What is the possibility of further technological breakthroughs in the overall process scheme?
7. Are there sufficient data regarding the physical, chemical, and geological properties of oil shale and shale oil?

## 8.2.2 ECONOMIC AND FINANCIAL CONSTRAINTS

Even though an oil shale retorting plant having a significant capacity is quite costly to build, the product oil will still have to be competitive for the current and future energy price structure. World petroleum prices have been fluctuating for the past four decades, and the crude oil price has been sharply rising in the early part of the twenty-first century. However, long-term profitability of the industry could be impacted by future pricing strategies of competing fuels. This concern elevates the risk level of an oil shale retorting and processing industry. Considering the shortage of clean liquid fuel sources in the world energy market and the general trend of increasing prices, marketability of oil shale is improving, with good future prospects. This may be especially true in countries that do not produce sufficient petroleum but possess vast deposits of oil shale. In this regard, there are three issues possible: (1) the involvement of the government can improve the economic scenario by providing the industry with incentives and credits in a variety of forms and tying the industry to the economic development of a region; (2) specialization of products and diversification of by-products can contribute to the profitability of the industry; and (3) securing captive use of shale oil in strategically developed energy-intensive industries also can contribute to the stability of the industry. Even though these statements were made for retorting-based shale oil production, they are somewhat applicable to the fracking-based shale oil since the gross margin of the fracking based crude oil is relatively low in comparison to the prevailing petroleum crude price on the global market.

Even though a generalized cost breakdown of conventional shale oil production is very difficult to make, a typical cost distribution for an oil shale project may be estimated as shown in Table 8.5. It can be seen from the table that the mining cost takes up a good share of the total operating cost. The cost to obtain shale oil crude, which includes the mining and retorting cost, is approximately 70% of the total operating cost. *Energy efficiencies* of most oil shale processes range from 58% to 63%, which

**TABLE 8.5**
**Typical Cost Distribution for an Oil Shale Retorting Project**

| Cost Factor | Construction (%) | Operation (%) |
|---|---|---|
| Mining (shale crushing and spent shale disposal) | 16 | 43 |
| Retorting | 37 | 28 |
| Upgrading | 22 | 29 |
| Utilities and off-sites | 25 | — |
| Total | 100 | 100 |

*Source:* Taylor, R.B., *Chemical Engineering*, 59, 1981.

can be further improved by utilizing efficient motors, adopting creative energy integration schemes, and exploiting waste energy. Table 8.6 shows some comparative information regarding the energy efficiencies of various synfuel processes. It should be noted that the efficiencies are very difficult to compare on a fair basis because efficiencies reported for the same process can be quite different from one another, depending on who reports it, how it is measured, on what basis it is calculated, etc. Improving energy efficiency without increasing the capital and operational cost is, therefore, a very important task that has to be undertaken by the process development team.

Issues related to the economics of oil shale processing are as follows:

1. What are the economic and energy-supply benefits of oil shale development?
2. What are the environmental and ecological impacts of oil shale development?
3. What are the economic impacts of establishing an oil shale industry?
4. How many different shale oil products can be used in the local region?
5. Is there enough capacity for pipeline transportation of retorted shale oil?
6. Is there an upgrading facility operating in the vicinity, or is a separate upgrading facility going to be built as part of the project?

**TABLE 8.6**
**Energy Efficiencies of Various Synfuel Processes**

| Process | Efficiency (%) |
|---|---|
| Lurgi high-pressure gasification | 70 |
| Lurgi high-pressure gasification followed by shift methanation | 63 |
| Shale oil retorting processes | 58–63 |
| IGCC | 57 |
| Fischer–Tropsch synthesis | 40 |
| Low-pressure methanol synthesis | 49 |
| Methanol synthesis followed by methanol-to-gasoline (MTG) | 45 |

7. How much will an oil shale retorting and upgrading facility cost?
8. For how many years can the facility be operated in the original location?
9. What is the rate of return on investment (ROI), especially for the long term?
10. At what level of petroleum crude price is shale oil competitive?
11. When the petroleum crude price goes up, what is the impact on the production cost of shale oil crude? Is it going to be more competitive? Is there any threshold value for the petroleum crude price for shale oil to be a strong competitor?
12. Overall, is oil shale retorting economically competitive without any tax credits and incentives and at what level?

## 8.2.3 Environmental and Ecological Constraints

The oil shale deposits found in the *Green River Formation* in the states of Colorado, Wyoming, and Utah are the largest in terms of size of deposit and most studied in the United States. The oil contained in these deposits is estimated at about 1800 billion barrels of recoverable shale oil. Owing to the vast resources and high oil content of shales, this region has long been the most attractive to oil shale industries. However, the technology used in mining and processing oil shale has aroused environmental and ecological concerns. The *Devonian–Mississippian eastern black shale* deposits are widely distributed between the Appalachian and Rocky Mountains. Even though these oil shales also represent a vast resource of fossil fuel, they are generally lower in grade (oil content per unit mass of shale rock) than Green River Formation oil shales.[7]

Several factors affecting the environmental constraints in commercially exploiting Green River Formation oil shale are discussed hereafter. This analysis is provided as an example and may also serve as a guideline for other similar projects. Therefore, similar analyses can be conducted for other oil shale deposits worldwide.

### 8.2.3.1 Region of Oil Shale Field and Population

The Upper Colorado region, which is the upper half of the Colorado River Basin, is traditionally "western rural" and consists of sparsely vegetated plains. The population density is also low, approximately three persons per square mile. The region is therefore less sensitive to disturbances on land, changes in traffic patterns, and construction and operation noises.

### 8.2.3.2 Water Availability

A rate-limiting factor in further development of the area is the availability of water, which may not be a problem in other regions. Water of the Colorado River could be made available for depletion by oil shale. An important factor that must be taken into consideration in any water use plan is the potential salt loading of the Colorado River. With oil shale development near the river, the average annual salinity is anticipated to increase, unless some preventive measures or treatment methods are implemented. The ecological damages associated with these higher salinity levels could be significant and have been the subject of extensive ecological studies. While these statement are made for the case of retorting-based shale oil production technology, it is also true for modern fracking technology (please refer to Chapter 10).

### 8.2.3.3 Other Fossil Energy and Mineral Resources

The Green River Formation oil shale area has extensive fossil fuel resources other than oil shale. Natural gas recoverable from this area is estimated at 85 trillion ft³, crude oil reserves are estimated at 600 million barrels, and coal deposits are estimated at 6–8 billion tons. These nonshale energy resources are not trivial, and they can also be developed together with oil shale. Furthermore, 27 billion tons of alumina and 30 billion tons of nahcolite are present in the central Piceance Creek Basin. These minerals may be mined in conjunction with oil shale. Such an effort can potentially enhance the profitability of the combined venture. This combined mineral processing is largely unrealistic with the adoption of modern shale gas fracking technology in other areas in the U.S.

### 8.2.3.4 Regional Ecology

Ecologically, the tristate region is very valuable. Owing to the sparse population density, the region has retained its natural character, of which the community is proud. Fauna include antelopes, bighorn sheep, mule deer, elks, black bears, moose, and mountain lions. However, there is little fishery habitat in the oil shale areas, even though the Upper Colorado region includes 36,000 acres of natural lakes.

### 8.2.3.5 Fugitive Dust Emission and Particulate Matter Control

Operations such as crushing, sizing, transfer conveying, vehicular traffic, and wind erosion are typical sources of fugitive dust. Control of airborne particulate matters (PMs) could pose a challenge. Compliance with regulations regarding PM control must be factored in.

### 8.2.3.6 Hazardous Air Pollutants

Gaseous emissions such as $H_2S$, $NH_3$, CO, $SO_2$, $NO_x$, and trace metals are sources of air pollution. Such emissions are at least conceivable in oil shale-processing operations. However, the level of severity is far less than that of other types of fossil fuel processing. The same argument can be made for the emission of carbon dioxide, which is a major greenhouse gas.

### 8.2.3.7 Outdoor Recreation and Scenery

Outdoor recreation in the tristate oil shale region has always been considered of high quality, because of the vastness of the essentially pristine natural environment and the scenic and ecological richness of the area. Maintaining the beauty of the area and preserving the high-quality natural resources must be taken into serious consideration, when oil shale in the region is commercially exploited. Such considerations generally hold true for other oil shale regions in the world.

### 8.2.3.8 Groundwater Contamination

Control of groundwater contamination is an important and nontrivial task. It is generally true for all types of oil shale operations including *ex situ* retorting, *in situ* pyrolysis, spent shale disposal and reburial, and upgrading. Without appropriate preventive measures, the groundwater could be contaminated by heavy metals, inorganic salts, organics such as polycyclic aromatic hydrocarbons (PAHs), etc. Both prevention and treatment must be fully investigated. Since the retorting operation generally requires a substantially lower amount of water consumption, the severity of ground water contamination is conceivably much lower than that for the fracking operation.

## 8.3 RESEARCH AND DEVELOPMENT NEEDS IN OIL SHALE

The synthetic crude reserves in oil shale, in terms of their crude oil production potential, are sufficient to meet U.S. consumption for several centuries at the current rate of liquid fuel utilization. Raw shale oil in the conventional sense is the crude oil product of the retorting process and is highly *paraffinic*, that is, containing mostly straight-chained hydrocarbons. However, it also contains fairly high levels of sulfur, nitrogen, and oxygen, as well as olefins; and it requires substantial upgrading before it can be substituted for refinery feed. Sulfur removal down to a few parts per million (ppm) is necessary to protect multimetallic reforming catalysts. Removing nitrogen from condensed heterocyclics, which also poisons cracking catalysts, requires an efficient technology that uses less hydrogen. The technologies developed for petroleum crude upgrading can be adopted for oil shale upgrading with relatively minor or no modifications.

This problem stresses the need for understanding the properties of oil shale and shale oil on a molecular level. Research needs can be broadly classified into six general categories: (1) chemical characterization of the organic and inorganic constituents in oil shale, (2) correlations of physical properties, (3) enhanced recovery processes of synfuels, (4) refining of crude retorted shale oil, (5) process design with efficient energy integration schemes, and (6) environmental and toxicological problems.

### 8.3.1 CHEMICAL CHARACTERIZATION

Improved analytical techniques must be developed to obtain the information needed to better understand oil shale chemistry, as well as to develop new technologies for utilizing shale. Most analytical methods developed and used for petroleum analytical chemistry have a long history of successful application, but their validity to the shale application is often questionable. Basic questions that need to be answered include the following:

1. In what forms do the organic heteroatoms exist?
2. How are they bonded into the basic carbon structure?
3. What can serve as the model compounds for sulfur and nitrogen sources in oil shale?
4. What is the aromaticity level of shale oil?
5. What is the ratio of alkanes to alkenes?
6. What are the effects of different retorting processes on the boiling range distribution of the shale oil crude?
7. What are the inorganic ingredients in oil shale and shale oil?

One way of characterizing oil shale is via separation of organics by extraction. The most commonly used techniques include gel permeation chromatography (GPC) for molecular weight distribution, gas chromatography (GC) using both packed columns and glass capillary columns for product oil distribution, simulated distillation using GC for boiling range determination, and molecular identification by liquid chromatography (LC).

Mass spectrometry (MS) can be extremely useful. In particular, GC–MS is good for product identification and model compound studies. Pyrolysis GC–MS can also be used for examining shale decomposition reactions.

Elemental analysis of oil shale is quite similar to that of petroleum and coal. More stringent requirements for C, H, O, N, and S analysis are needed for oil shale, as organic carbon in the shale should be distinguished from inorganic carbon in carbonate materials. Analysis for C, H, N, and S, which are very frequently used for coal analysis, can also be used for oil shale.

### 8.3.2 CORRELATION OF PHYSICAL PROPERTIES

Physical properties of oil shale should be characterized via electrical and conductive measurements, scanning microscopy, spectroscopic probes, and all other conventional methods. Especially, the correlation between physical properties and pyrolysis conversion is useful for designing a pilot-scale or commercial retort. The $^{13}$C-NMR work should be expanded to provide a detailed picture of the various chemical forms encountered. Electron spin resonance (ESR) studies of carbon radicals in oil shale could also provide invaluable clues about the conversion process.

Correlations of physical properties that can be used for a variety of oil shales are especially useful. Predictive forms of correlation are a powerful tool in engineering design calculations, as well.

### 8.3.3 MECHANISMS OF RETORTING REACTIONS

Kinetics of oil shale retorting has been studied by various investigators. However, the details of reaction mechanism have not been generally agreed upon. This may be one of the reasons why problems associated with *in situ* pyrolysis require primarily field experiments rather than small laboratory-scale experiments. The retorting process itself can be improved and optimized when its chemical reaction mechanisms are fully elucidated. Similarly, a more efficient retort design can be accomplished.

### 8.3.4 HEAT AND MASS TRANSFER PROBLEMS

Oil shale rocks are normally low in both porosity and permeability. Therefore, it is very important to know the combined heat and mass transfer processes of a retort system. The processes of heat and mass transfer in a retort operation affect the process operating cost significantly, because the thermal efficiency is directly related to the total energy requirement, and mass transfer conditions directly affect the recovery of oil and gas from a retort. The various process technologies may differ from one another only in their modes of heat and mass transfer, and there is always room for further improvement. Therefore, understanding the transport processes in oil shale retorting is essential in mathematical modeling of a retort system, as well as in design of an efficient retort system. The analysis of heat and mass transfer requires a vast amount of information, such as physical properties of inorganic and organic ingredients of oil shale; bed and rock porosity and its distribution; permeability; dolomite and other carbonate compositions in the shale, etc.

### 8.3.5 Catalytic Upgrading of Retorted Shale Oil Crudes

As mentioned earlier, prerefining of crude shale oil is necessary to reduce sulfur and nitrogen levels, and contamination by mineral particulates. Because large portions of nitrogen and sulfur species in shale oil are present as heteroaromatics, research opportunities of great significance exist in the selective removal of heteroaromatics, final product quality control, and molecular weight reduction.

Raw retorted shale oil has a relatively high pour point of 75°F to 80°F, compared to 30°F for Arabian Light. Olefins and diolefins may account for as much as one-half of the low-boiling fraction of 600°F or lower and lead to the formation of gums by polymerization.

Raw shale oil typically contains 0.5% to 1.0% oxygen, 1.5% to 2.0% nitrogen, and 0.15% to 1.0% sulfur. As can be seen, the nitrogen level in shale oil is very high, whereas the sulfur level is in a similar range to other fossil fuel liquids. Sulfur and nitrogen removal must be very complete as their compounds poison most of the catalysts used in refining; their oxides ($SO_x$ and $NO_x$) are also well-known air pollutants.

As raw shale oil is a condensed overhead product of pyrolysis, it does not contain the same kinds of macromolecules found in petroleum and coal residuum. Conventional catalytic cracking, however, is an efficient technique for molecular weight reduction. It is crucially important to develop a new cracking catalyst that is more resistant to basic poisons (nitrogen and sulfur compounds).[8] At the same time, research should also focus on reduction of molecular weight of retorted shale oil crude with low consumption of hydrogen.

### 8.3.6 By-Product Minerals from U.S. Oil Shale

Many different kinds of carbonate and silicate minerals occur in oil shale formations. Trona beds [$Na_3(CO_3)(HCO_3)_3$] in Wyoming are a major source of soda ash (sodium carbonate [$Na_2CO_3$]), whereas nahcolite ($NaHCO_3$) is a potential by-product of oil shale mining from Utah and Colorado.

Precious metals and uranium are contained in good amounts in eastern U.S. shales. For Marcellus shale and Utica shale of the eastern U.S., radioactive signature has been effectively traced and its detection has also been incorporated in the shale gas exploration technique. It is unlikely that in the near future, recovery of these mineral resources will be possible, as a commercially favorable recovery process has not yet been developed. However, it should be noted that there are many patents on recovery of alumina from Dawsonite-bearing beds by leaching, precipitation, calcination, etc. The chemical formula for dawsonite is $NaAl(CO_3)(OH)_2$. It should be noted that commercial co-production of mineral matters from oil shale is more or less limited to the conventional retorting-based oil shale industries which actually mine oil shale rocks for *ex situ* processing.

### 8.3.7 CHARACTERIZATION OF INORGANIC MATTERS IN OIL SHALE

The analytical techniques applied to the characterization of inorganic constituents in oil shale include x-ray powder diffraction (XRD), thermogravimetric analysis (TGA), scanning electron microscopy (SEM), and transmission electron microscopy (TEM). Further, electron microprobe analysis (EMPA) may provide a useful tool for studying crystals at the molecular level and generate valuable information on the phases encountered. Most oil shale is also rich in inorganic matter, and these inorganic ingredients also go through the same process treatment as the organic constituents of the shale. The most significant of this inorganic matter include dolomite ($CaCO_3·MgCO_3$) and calcite ($CaCO_3$), both of which decompose upon heating and liberate gaseous carbon dioxide. This carbonate decomposition reaction is endothermic in nature, that is, absorbing heat thereby reducing the thermal efficiency of kerogen (oil shale hydrocarbon) pyrolysis, that is, oil shale retorting.

## 8.4  PROPERTIES OF OIL SHALE AND RETORTED SHALE OIL

To develop efficient retorting processes as well as to design a cost-effective commercial-scale retort, physical, chemical, and physicochemical properties of oil shales (raw material) and shale oils (crude liquid products) must be fully known. However, difficulties exist in the measurement of various physical and chemical properties of a variety of oil shales on a consistent basis. In this section, the properties that are essential in designing an efficient retort, as well as in understanding the oil shale retorting process, will be discussed. Even though a good deal of literature and data have been presented in this section, it is not solely intended to build a data bank of oil shale properties; rather, appropriate utilization (and interpretation) of data and their measurements are stressed. More extensive treatment of property data can be found in references 7, 15, 16, 18, and 35.

### 8.4.1  PHYSICAL AND TRANSPORT PROPERTIES OF OIL SHALE

#### 8.4.1.1  Fischer Assay

The nominal amount of condensable oil that can be extracted from oil shale by retorting is commonly denoted by the term *Fischer assay* of the oil shale. The Fischer assay is a simple and representative quantity that can be obtained quite easily for all kinds of oil shale by following *the standardized retorting procedure* under nitrogen atmosphere. The actual oil content in the oil shale, both theoretically and nominally, exceeds the Fischer assay. Depending on the treatment processes as well as the type of oil shale, oil yield from oil shale often exceeds the Fischer assay value by as much as 80%. Examples of such extraction processes include retorting in a hydrogen-rich environment, retorting in a $CO_2$ sweep gas environment, supercritical fluid extraction of oil shale, etc. The procedure for Fischer assay of oil shale is modified from a Fischer assay procedure for carbonization of coal at a low temperature. A brief description of the *Fischer assay procedure* is as follows: take a crushed sample of oil shale of 100 g and subject it to a preprogrammed (such as linear ramping) heating schedule in an inert (such as nitrogen) environment. The oil shale is heated from 298 to 773 K very linearly over a 50 min period while being purged with nitrogen. The linear heating rate is 9.5°C/min. Following the heat-up period, the sample is held at 773 K for an additional 20 to 40 min, and the oil collected, typically in a condenser tube, is measured. This recovered oil amount is then recorded in a unit of liters per ton (l/ton) or gallons per ton (gal/ton).

This procedure cannot recover all the organic matter originally contained in shale and leaves char associated with ash in the rock matrix, as well as larger-molecular-weight hydrocarbons blocking the pores. Nevertheless, the Fischer assay is used as a very quick and convenient measure of recoverable organic hydrocarbon content and provides a common basis for comparison among various oil shales. If this value is higher than 100 l/ton, it is typically considered a *rich shale*; if the value is less than 30 l/ton, it is a *lean shale*. Typically, Colorado oil shale has a Fischer assay exceeding well over 100 l/ton and is classified as rich shale.

### 8.4.1.2 Porosity

Porosity of porous materials can be defined in a number of different ways, depending on what specific pores are looked at and how the void volumes are measured. They include interparticle porosity, intraparticle porosity, internal porosity, porosity by liquid penetration, porosity by saturable volume, porosity by liquid absorption, superficial porosity, total open porosity, bed porosity (bed void fraction), packing porosity, etc.

Porosity of the mineral matrix of oil shale cannot be determined by the methods used for determining porosity of petroleum reservoir rocks, because the organic matter in the shale exists in solid form and is essentially insoluble. However, the results of a laboratory study at the Laramie Center[9] (currently, Western Research Institute, Laramie, Wyoming) have shown that inorganic particles contain a micropore structure, about 2.36 to 2.66 vol%. Although the particles have an appreciable surface area, about 4.24 to 4.73 m²/g for shale assaying of about 29 to 75 gal/ton, it seems to be limited mainly to the exterior surface rather than the pore structure. Measured porosities of the raw oil shales are shown in Table 8.7.[10]

As noted, except for the two low-yield oil shales, naturally occurring porosities in the raw oil shales are almost negligible, and they do not afford access to gases. Porosity may exist to some degree in the oil shale formation where fractures, faults, or other structural defects occurred. It is also believed that a good portion of the pores are either blind or very inaccessible. Crackling and fractures or other structural defects often create new pores and also break up some of the blind pores. It should be noted here that *closed* or *blind* pores are normally not accessible by mercury porosimetry even at high pressures or by other fluid penetration measurement techniques. Owing to the severity of mercury poisoning, the instrument based on pressurized mercury penetration through pores is no longer used.

### 8.4.1.3 Permeability

Permeability is the ability, or measurement of a rock's ability, to transmit fluids and is typically measured in darcies or millidarcies. Permeability is part of the proportionality constant in *Darcy's law*, which relates the flow rate of the fluid and the fluid viscosity to a pressure gradient applied to the porous media. Darcy's law is a phenomologically derived constitutive equation based on the conservation of

**TABLE 8.7**

**Porosities and Permeabilities of Raw and Treated Oil Shale**

| Fischer Assay | Porosity | | Plane | Permeability | |
| | Raw | Heated to 815°C | | Raw | Heated to 815°C |
|---|---|---|---|---|---|
| 1.0[a] | 9.0[b] | 11.9 | A | — | 0.36[c] |
| | | | B | — | 0.56 |
| 6.5 | 5.5 | 12.5 | A | — | 0.21 |
| | | | B | — | 0.65 |
| 13.5 | 0.5 | 16.4 | A | — | 4.53 |
| | | | B | — | 8.02 |
| 20.0 | <0.03 | 25.0 | A | — | — |
| | | | B | — | — |
| 40.0 | <0.03 | 50.0 | A | — | — |
| | | | B | — | — |

*Source:* Reprinted from *Bitumens, Asphalts, and Tar Sands*, Chilingarian, G.V. and Yen, T.F., Chapter 1, Copyright 1978, with permission from Elsevier.

*Note:* A = perpendicular to the bedding plane; B = parallel to the bedding plane.

[a] Fischer assay in gal/ton.

[b] Numbers in percentages of the initial bulk volume. Porosity was taken as an isotropic property, that is, property that is independent of measurement direction.

[c] Units in millidarcy.

momentum that describes the flow of a fluid through a porous medium. A simple relationship relates the *instantaneous discharge rate* (local volumetric flow rate) through a porous medium to the *local hydraulic gradient* (change in hydraulic head over a distance, that is, $\Delta h/L$, $dh/dL$, or $\nabla h$) and the *hydraulic conductivity* ($k$) at that point.

$$Q = -kA \frac{h_a - h_b}{L}$$

Or dividing both sides by the area ($A$) yields

$$q = -k\nabla h$$

where $q$ is the *Darcy flux*, that is, the discharge rate per unit area, expressed in terms of [length/time]. Even though the final unit of the Darcy flux is the same as that of velocity, a clear conceptual difference between the two must be understood. Based on the analogy between Darcy's and Poiseuille's law, the hydraulic conductivity term can be factored out in terms of intrinsic permeability and the fluid properties as

$$k = (k')(\rho g/\mu)$$

where $k'$ is the *intrinsic permeability*, which has the dimension of [length$^2$]. Whereas the term [$\rho g/\mu$] describes the penetrating fluid properties, the intrinsic permeability ($k'$) summarizes the properties of the porous medium. The usual unit for permeability is the *darcy* (D), or more commonly the *millidarcy* or *mD* (1 darcy $\approx$ 10$^{-12}$ m$^2$ and 1 millidarcy = 0.001 darcy). Converted to SI units, precisely speaking, 1 D is equivalent to 0.986923 $\mu$m$^2$. The discussion presented here is a simplified one, in which unidirectional homogeneous fluid permeation without explicit inclusion of any external force terms is considered. However, it still may be evident that permeability should be expressed in a multidirectional manner; in other words, it is most adequately expressed as a *permeability tensor*. Exhaustive discussions about Darcy's law can be found in most textbooks on transport phenomena,[11] fluid mechanics, and hydraulogy.

The permeability of raw oil shale is essentially zero because the pores are filled with a nondisplaceable organic material. Tisot[10] showed that gas permeability, either perpendicular or parallel to the bedding plane, was not detected in most oil shale samples at a pressure differential across the cores of 3 atm of helium for 1 min. In general, oil shale constitutes a highly impervious system. Thus, one of the major challenges of any *in situ* retorting project is the creation of a suitable degree of permeability in the formation. This is why an appropriate rubblization technique is essential to the success of an *in situ* pyrolysis project.

Of practical interest is the dependency of porosity or permeability on temperature and organic content. Upon heating to 510°C, an obvious increase in oil shale porosity is noticed. These porosities, which vary from 3 to 6 vol%[10] of the initial bulk oil shale volume, represented essentially the volumes occupied by the organic matter before the retorting treatment. Therefore, oil shale porosity increases as the pyrolysis reaction proceeds or retorting progresses. In the low-Fischer-assay oil shales, that is, lean oil shales, structural breakdown of the cores is insignificant, and the porosities are those of intact porous structures. However, in the high-Fischer-assay oil shales, that is, rich oil shales, this is not the case because structural breakdown and mechanical disintegration due to retorting treatment become extensive and the mineral matrices no longer remain intact. Thermal decomposition of the mineral carbonates, such as magnesium and calcium carbonates ($MgCO_3$ and $CaCO_3$), actively occurring around 380°C–900°C also results in an increase in porosity. The increase in porosity from low- to high-Fischer-assay oil shales varies from 2.82% to 50%, as shown in Table 8.7.[12] These increased porosities constitute essentially the combined spaces represented by the loss of the organic matter and the decomposition of the mineral carbonates. Crackling of particles is also due to the devolatilization of organic matter, which increases the internal vapor pressure of large nonpermeable pores to such an extent that the mechanical strength of the particle can no longer retain the gas inside. Liberation of carbon dioxide from mineral carbonate decomposition also contributes to the pressure buildup in the oil shale pores.

Gas permeability[10] is low in both planes of the mineral matrices from the three low-Fischer-assay oil shales heated to 815°C. As noted in Table 8.7, the mineral matrix from the 13.5 gal/ton oil shale has the highest permeability of 8.02 mD. This value may be somewhat higher than the primary permeability of the mineral matrix as the permeabilities created by removing the organic matter via devolatilization and pyrolysis, as well as by thermally decomposing the mineral carbonates, are also included. Even though the oil shale cores used for these measurements have no visible fractures, minute fractures may have formed during heating up to 815°C, which probably contributed to some secondary permeability. Permeabilities after heating up to 815°C, for the oil shales whose Fischer assay exceeded 13.5 gal/ton, are not given. In these oil shales, structural breakdown of the mineral matrices under a stress-free environment was so extensive as to preclude measurements of the permeabilities in high-Fischer-assay oil shales.[10] Dolomite ($CaCO_3 \cdot MgCO_3$) decomposition via half-calcination and full-calcination reactions becomes very active at temperatures higher than 380°C, when magnesium carbonate ($MgCO_3$) starts to decompose readily, releasing carbon dioxide. Once the temperature is raised beyond 890°C, decomposition of calcite ($CaCO_3$) via calcination reaction becomes quite active and thermodynamically favored. At these two temperatures (380°C and 890°C), the equilibrium constant for decomposition of magnesium carbonate and calcium carbonate, respectively, becomes unity, or $K_p = 1$.

In the case of eastern U.S. shales, especially Devonian oil shales, decomposition of kerogen produces lighter hydrocarbons than those from other shales. This often results in a substantial increase in volatile pressure in the solid matrix, which leads to cracking and mechanical disintegration of solid structure. This is also the reason why the oil yield from eastern oil shale pyrolysis via a procedure similar to Fischer assay is not necessarily an accurate measure of the organic content of the shale.[7]

### 8.4.1.4  Compressive Strength

The raw oil shales have high compressive strength, both perpendicular and parallel to the bedding plane.[6] After heating, the inorganic matrices of low-Fischer-assay oil shales retain high compressive strength in both perpendicular and parallel planes. This indicates that a high degree of inorganic cementation exists between the mineral particles comprising each lamina and between adjacent laminae. With an increase in organic matter of oil shale, the compressive strength of the respective organic-free mineral matrices decreases and it becomes very low in those rich oil shales, as shown in Figure 8.1.[13]

It is also noteworthy that a structural transition point exists. Gradual expansion (volume swelling) of oil shale under a stress-free environment was noted immediately upon application of heat. Around 380°C, the samples are seen to undergo drastic changes in compressive strength. The greater loss of compressive strength at the yield point and the low recovery on reheating for the richer oil shales are both attributed to extensive *plastic deformation* effects. The degree of plastic deformation thus seems to be directly proportional to the amount of organic matter in oil shale. The discontinuities in the pressure plot at temperatures below the yield point presumably arise from the evolution of pore vapor from the oil shale matrix. The well-defined transition point at 380°C, therefore, represents a pronounced change

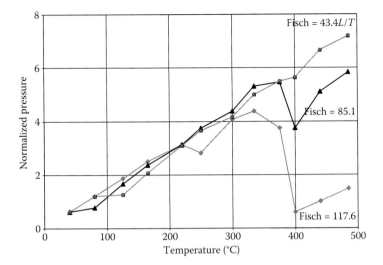

**FIGURE 8.1**    Variation of compressive strength of oil shale as a function of Fischer assay of oil shale. (From Wang, Y., M.S. thesis, University of Akron, Akron, OH, 1982.)

in the compressive strength of richer oil shale. It is interesting to note that near this temperature, most coals also exhibit similar plastic behaviors. Similarity in plastic properties between oil shale and coal may be attributed to the macromolecular structure of their organic matter.

### 8.4.1.5    Thermal Properties

The term *thermal* is used here to represent those parameters that are directly or indirectly related to the transport, absorption, or release of heat, that is, thermal energy. Properties such as thermal conductivity, thermal diffusivity, enthalpy, density, and heat capacity fall into this category. For materials that undergo thermal decomposition or phase transformation (this is the case with oil shales in general), it is necessary to characterize their thermal behavior by thermoanalytical techniques such as *TGA*[14] and *differential thermal analysis* (*DTA*).

*8.4.1.5.1    Thermal Conductivity*

Measurements of thermal conductivity of oil shale show that blocks of oil shale are anisotropic about the bedding plane. The measurements were made by techniques such as the transient probe method,[10] the thermal comparator technique,[15] and the line source method.[16] The range of temperature and shale grades investigated in some instances is, however, quite limited.[7,17] Some earlier studies did not focus on the anisotropic nature of the heat conduction. However, later studies have shown that the thermal conductivity as a function of temperature, oil shale assay, and direction of heat flow, parallel to the bedding plane (parallel to the earth's surface for a flat oil shale bed), was slightly higher than with thermal conductivity perpendicular to the bedding plane. As layers of material were laid to form the oil shale bed over a very long period of geological years, the resulting continuous strata have slightly higher resistance to heat flow perpendicular to the strata than parallel to the strata. A summary of the literature data on the thermal conductivity of Green River oil shale is given in Table 8.8.

The results shown in Table 8.8 indicate that the thermal conductivities of retorted and burnt shales are lower than those of the raw shales from which they are obtained. This is attributable to the fact that mineral matter is a better conductor of heat than the organic matter; on the other hand, organic matter is still a far better conductor than the voids created by its removal.[16] Whereas the first of these hypotheses is well justified when one takes into account the contribution of the lattice conductivity to the overall value, the effect of the amorphous carbon formed from the decomposition of the organic matter could also be important in explaining the differences in thermal conductivity values for retorted shales and the corresponding burnt samples. The role of voids in determining the magnitude of the effective thermal conductivity is likely to be significant only for samples with high organic content. The data for thermal conductivity measured by Tihen et al.[16] are presented in Figure 8.2. On the figure, the ordinate is the thermal conductivity in J/m·s·°C and the abscissa is in Fischer assay, gal/ton. Other measurements[18] were made of the effect of oil shale assay on thermal conductivity. These data also show that thermal conductivity decreases with an increase in oil shale assay.[20]

**TABLE 8.8**

**Comparison between Thermal Conductivity Values for Green River Oil Shales**

| Temperature Range (°C) | Fischer Assay (gal/ton) | Plane | Thermal Conductivity (J/m·s·°C) | Reference |
|---|---|---|---|---|
| 38–593 | 7.2–47.9 | — | 0.69–1.56 (raw shales) | [16] |
| | | | 0.26–1.38 (retorted shales) | |
| | | | 0.16–1.21 (burnt shales) | |
| 25–420 | 7.7–57.5 | A | 0.92–1.92 | [22] |
| | | Average | 1.00–1.82 (burnt shales) | |
| 38–205 | 10.3–45.3 | A | 0.30–0.47 | [19] |
| | | B | 0.22–0.28 | |
| 20–380 | 5.5–62.3 | A | 1.00–1.42 (raw shales) | [15] |
| | | B | 0.25–1.75 (raw shales) | |

*Note:* A = parallel to the bedding plane; B = perpendicular to the bedding plane; Average = average of both directions.

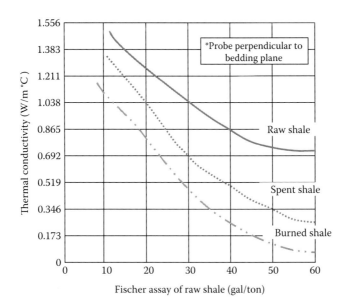

**FIGURE 8.2**  Thermal conductivity of raw, spent, and burnt oil shale as a function of Fischer assay (1 gal/ton = 4.18 cm³/kg). (From Smith, J.W., *U.S. Bur. Mines Rep. Invest.*, 7248, 1969.)

The thermal conductivities of oil shales are, in general, only weakly dependent on temperature;[20] most studies report a gradual decrease with increasing temperature.[17,21] However, extreme caution needs to be exercised in the interpretation of results at temperatures close to the decomposition temperature of the shale organic matter. This is because the kerogen decomposition reaction (or pyrolysis reaction) is endothermic in nature, and, as such, the temperature transients can be confounded between the true rate of heat conduction and the rate of heat of reaction.

For example, thermal conductivity values reported at temperatures around 400°C normally include the thermal effects of decomposition of the shale organic matter, and, therefore, they may not be the intrinsic values.

The thermal conductivity values of oil shales show an inverse dependence on organic matter content.[16,22] Equations that have been proposed by various authors, relating thermal conductivity to the three parameters, namely, temperature, organic content, and extent of kerogen conversion, are shown as follows:

$$K = c_1 + c_2 x_1 + c_3 x_2 + c_4 x_3 + c_5 x_1 x_2 + c_6 x_2 x_3 + c_7 x_3 x_1 + c_8 x_1 x_2 x_3 \tag{8.1}$$

where $x_1$, $x_2$, and $x_3$ denote the organic content, kerogen conversion, and temperature, respectively. The equation by Tihen et al.,[16] which was taken as the average for the perpendicular and parallel thermal conductivities of raw and spent shales, is given by

$$K = (1 - x_2)\{1.9376 - 4.739 \times 10^{-2} x_1$$

$$+ 1.776 \times 10^{-3}(x_3 - 273) + 4.371 \times 10^{-4} x_1^2$$

$$- 4.885 \times 10^{-6}(x_3 - 273)^2$$

$$- 1.671 \times 10^{-5} x_1(x_3 - 273)\} + x_2\{1.680 - 5.204 \times 10^{-2} x_1 \qquad (8.2)$$

$$- 1.003 \times 10^{-4}(x_3 - 273) + 4.951 \times 10^{-4} x_1^2$$

$$- 1.468 \times 10^{-9}(x_3 - 273)^2$$

$$+ 0.0667 \times 10^{-5} x_1(x_3 - 273)\} \quad \text{J/m·s·°C}$$

where $x_1$ is the shale's organic content in mass fractions, $x_2$ is fractional conversion of kerogen, and $x_3$ is the temperature in degrees Kelvin.

Prats and O'Brien[22] proposed a second-order polynomial in $(x_3 - 25)$ of the form

$$K = c_1[1 - D_1(x_3 - 25) + D_2(x_3 - 25)^2] \exp(c_2 F) \qquad (8.3)$$

where $F$ is the Fischer assay (in L/ton), $K$ is the thermal conductivity (in W/m °C), and $x_3$ is the shale temperature (in °C). In Equation 8.3, $c_1$, $c_2$, $D_1$, and $D_2$ are empirically determined. By Equation 8.3, the thermal conductivity is a relatively simple function of the temperature and the organic content of shale. An even simpler equation was proposed for Baltic shales[23] as

$$K = 1.30/F + 0.06 + 0.003\ T \qquad (8.4)$$

where $F$ is the Fischer assay (in L/ton), $T$ is temperature (in °C), and $K$ is the thermal conductivity (in W/m °C).

When using simple correlations, one has to realize their limitations, especially in the case of extrapolation or interpolation of experimental data. Most simple expressions normally have narrower ranges of validity.

The problem of thermal conduction through a bed of oil shale rubble is quite complex. It is similar to that of packed beds of randomly sized and oriented particles. It is also difficult to generalize the size distribution of fractured underground beds of oil shale as well as to accurately control the size and shape of particles when *in situ*, underground rubblization takes place. To improve process efficiency, various process ideas of rubblization and heating shales have been generated.

*Thermal diffusivity*, $\alpha$, is defined as

$$\alpha = \frac{k}{\rho C_p} \qquad (8.5)$$

where $k$, $\rho$, and $C_p$ denote the thermal conductivity, density, and heat capacity, respectively. Therefore, the thermal diffusivity has a dimension of $L^2 t^{-1}$ (as in the unit of $cm^2/s$), similar to mass and momentum diffusivities. For an oil shale particle to reach a predetermined temperature throughout the particle dimension, the required time may be estimated by[7]

$$t = \frac{\rho C_p L_{ch}^2}{k} \times 0.9 \tag{8.6}$$

where $L_{ch}$ is the characteristic length of the oil shale particle. Equation 8.6 is based on the isothermality criterion of $t_{ch}^* = 0.9$, in which the characteristic time becomes 0.9. The characteristic time $t^*$ is defined as $t^* = \alpha \cdot t/L_{ch}^2$. The underlying idea for the characteristic length may be explained by the *penetration depth* or *a representative linear dimension* for conduction. For a sphere, the characteristic length may be calculated by

$$L_{ch} = \frac{\text{volume}}{\text{surface area}} = \frac{\frac{4}{3}\pi R^3}{4\pi R^2} = \frac{R}{3} = \frac{D}{6} \tag{8.7}$$

The same calculation can be carried out for a regular cylinder whose diameter is the same as its length:

$$L_{ch} = \frac{\text{volume}}{\text{surface area}} = \frac{2\pi R^3}{[2\pi R(2R) + 2\pi R^2]} = \frac{R}{3} = \frac{D}{6} \tag{8.8}$$

An analogous calculation can be made for determination of a characteristic dimension for other geometries.[7]

As can be readily seen from Equation 8.6, the heat-up time required is proportional to the square of the characteristic dimension. In other words, a successful operation of *in situ* retort using the combustion retorting process depends quite strongly on how finely the rubblization of oil shale bed can be achieved. If the particle size of rubblized oil shale is large, the heat-up period would be quite long, thus making the retort inefficient.

The dependence of the thermal diffusivity of Green River oil shale perpendicular to the bedding plane on temperature and Fischer assay was measured by Dubow et al.[18] and is shown in Figure 8.3. The thermal diffusivity values show the same broad trends as the thermal conductivities, with variations in temperature and shale grade.[20] As expected, the thermal diffusivity decreases with increasing temperature and organic content in the oil shale. Thus, the retorted and burnt shales show reduced thermal diffusivities relative to those for the raw shales,[20] as shown in Table 8.9. Oil shale samples containing large amounts of pyrites ($FeS_2$) are likely to show high thermal diffusivities, as the thermal diffusivity of pyrite itself is quite high.

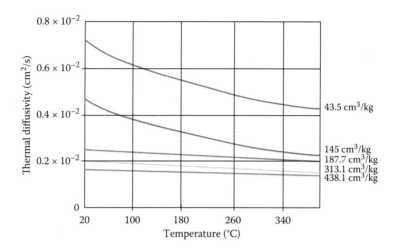

**FIGURE 8.3** Dependence of oil shale thermal diffusivity perpendicular to the bedding plane on temperature and Fischer assay. (From Dubow, J. et al., *11th Oil Shale Symposium Proc.*, Colorado School of Mines, Golden, CO, 1978, p. 350.)

**TABLE 8.9**
**Thermal Diffusivity of Green River Oil Shale**

| Temperature Range (°C) | Fischer Assay (gal/ton) | Shale Type | Thermal Diffusivity (cm²/s) | Measurement Technique | Reference |
|---|---|---|---|---|---|
| 38–260 | 6.7–48.4 | Raw shale | 0.26–0.98 | Transient line probe | [16] |
| 38–482 | Low | Retorted shale | 0.13–0.88 | Transient line probe | [16] |
| 38–593 | ~0 | Burnt shale | 0.10–0.72 | Transient line probe | [16] |
| 25–350 | 5.0–82.2 | Raw shale | 0.10–0.90 | Laser flash | [24] |

### 8.4.1.5.2 Heat Capacity of Oil Shale

Earlier work by McKee and Lyder[25] on specific heat of U.S. oil shales is restricted to limited ranges of temperatures and shale grades. Later studies by Wang et al.[26] reported heat capacity dependencies on temperatures and shale grade, characterized by the following type of equation:

$$c = c_1 + c_2 x_1 + c_3 x_2 + c_4 x_1 x_2 \qquad (8.9)$$

Again, it is very difficult to generalize the heat capacity of oil shale in any simple functional form, because of the vast heterogeneity of oil shales even within the same formation, as well as among different formations.

Considerable increases in the values of heat capacity with increasing organic content have been observed,[25,26] although relative contributions of various oil shale constituents to the overall values are somewhat uncertain. Values of heat capacity for most oil shales are not readily available. Although actual measurement of heat capacity of a solid sample is not a complicated task, it may not be a bad idea to measure it when the information is needed.

The heat capacity correlation given by Shih and Sohn[27] is

$$C_{px} = \{(907.09 + 505.85x_1)(1 - x_2) + 827.06x_2\}$$
$$+ \{(0.6184 + 5.561x_1)(1 - x_2) + 0.92x_2\}(x_3 - 298) \quad \text{J/kg°C} \tag{8.10}$$

where $x_1$ is the organic content of shale in mass fraction, $x_2$ is the fractional conversion of kerogen, and $x_3$ is the temperature in degrees Kelvin.

### 8.4.1.5.3 Enthalpy and Heat of Retorting

Wise et al.[28] measured the enthalpy of raw, spent, and burnt oil shale from the Green River formation. Oil was removed from the spent shale by conventional retorting, but it still retained small amounts of char residue. However, virtually all organic matter was removed from the burnt shale, as shown in Figure 8.4.[29] The *enthalpy* data may be represented by a function of temperature and Fischer assay of oil shale as

$$\Delta H = a + bT + cT^2 + dT^3 + eT^4 + fT^2F \tag{8.11}$$

where $T$ is the temperature and $F$ is the oil shale Fischer assay. All the coefficients from $a$ to $f$ are also given in reference 28. The *specific heat* can be determined by differentiation of the enthalpy with respect to the temperature:

$$C_{p,s} = \left(\frac{\partial H_s}{\partial T}\right)_p \tag{8.12}$$

These data are in good agreement with earlier data by Shaw[30] and Sohns et al.[31]

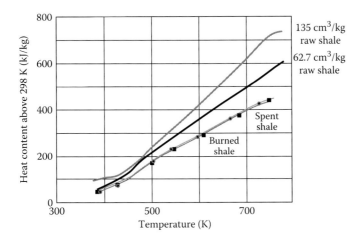

FIGURE 8.4   Heat capacity of raw, spent, and burnt oil shale. (From Johnson, W.F. et al., *Q. Colo. Sch. Mines*, 70(3), 237, 1975.)

**TABLE 8.10**

**Values Reported for Heat of Retorting of Green River Oil Shales**

| Heat of Retorting (kJ/kg) | Fischer Assay (gal/ton) | Reference |
|---|---|---|
| 238–878 | 23.5–46.7 | [31] |
| 581–699 | 8.0–32.8 | [32] |
| 335 | 25.6 | [63] |

*Source:* From Wang, Y., M.S. thesis, University of Akron, Akron, OH, 1982.

The available data for *heat of retorting* of Green River oil shales are listed in Table 8.10. The reported values for the heat of retorting (endothermic) show the expected increase with increasing shale assay and temperature. The disparity in the range of values observed by different investigators possibly reflects differences in the composition of shale samples. It should be borne in mind that the presence of minerals, which decompose at temperatures below the range at which the organic matter is thermally extracted, would increase the energy requirements for processing shales. Thus, it has been estimated that shales containing nahcolite and dawsonite would require an additional 117 cal/g (490 J/g) and 215 cal/g (890 J/g), respectively.[32] Heat requirements for retorting oil shales containing 17% analcite would be increased by about 6%.[33]

### 8.4.1.5.4 Density or Specific Gravity

The density of Green River oil shale was measured by Tisot[10] and found to be in the range of 1.8 to 2.0 g/cm³. Later, some efforts were made to correlate the oil yield from oil shale (such as the Fischer assay) with the specific gravity.[21] This idea may have some practical significance, as the oil yield is usually a constant fraction of the organic content and the oil shale density is dependent upon the organic content.

### 8.4.1.5.5 Self-Ignition Temperature

The *self-ignition temperature* (SIT) is the temperature at which an oil shale sample spontaneously ignites in the presence of atmospheric oxygen or under any other prescribed oxidative conditions. There is no standardized procedure generally adopted for this measurement. However, if a consistent measurement of this temperature is made for oil shale, it can provide very valuable information regarding the fuel characteristics of the shale.

The spontaneous ignition temperature of oil shale has been measured and characterized under a variety of conditions by Allred.[34] The information regarding SIT is very important, as it governs not only the initiation of the combustion retorting process but also the dynamics of oil shale retorting by the advancing oxidation zone. Branch[35] also explains its importance in his article. In the countercurrent combustion retorting process, the combustion front moves toward the injected oxidizer, whereas in the cocurrent process, the front moves in the same direction as the oxidizer.

Therefore, the spontaneous ignition temperature of the raw oil shale should be below the oil shale retorting temperature in the countercurrent process. In the cocurrent combustion process, char remaining in the shale after retorting is burned to sustain the retort by providing the necessary thermal energy.

The SITs of Colorado oil shale have been measured over a wide range of total pressure and oxygen partial pressure.[34] With nitrogen as a diluent, the ignition temperature was found to depend on the oxygen partial pressure, but not significantly on the total pressure. The temperature at which ignition could occur was also found to correlate to the temperature at which methane and other light hydrocarbons are devolatilized from the oil shale. As shown in Figure 8.5, in the case for raw Colorado oil shale, higher ignition temperatures were required for lower oxygen partial pressures, and the lowest ignition temperature of 450 K was considerably lower than the oil production temperature of about 640 K. More detailed information is available from the original text[21] or from other sources.[7,34,35] This experimental evidence strongly suggests that ignition of oil shale may be associated with oxidation of gaseous hydrocarbons that evolved from oil shale.

Joshi[36] took a different approach to studying the SIT. He defined the SIT as the temperature at which the shale bursts into flames in air within 360 s of introduction into a preheated isothermal retorter. His SIT measurements were made at an oxygen partial pressure of 0.21 atm, that is, under atmospheric conditions. Joshi's approach was different from that of Allred's in the sense that different oil shales can be characterized by the SIT. In other words, the SIT defined by Joshi can be used more like a physical property that is easy to measure. It was found that the SIT depends very strongly on the Fischer assay of oil shale. Generally speaking, the higher the Fischer assay, the lower the SIT. Figure 8.6 graphically illustrates the relation between SIT and Fischer assay for several different shales.

Joshi's results also suggest that the ignition of oil shale is associated with the oxidation of gaseous hydrocarbons evolved from oil shale. The ignition of the

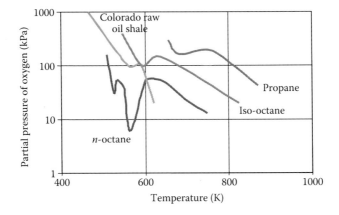

**FIGURE 8.5** SIT of raw oil shale, *i*-octane, *n*-octane, and propane. (From Smith, J.W., *U.S. Bur. Mines Rep. Invest.*, 7248, 1969.)

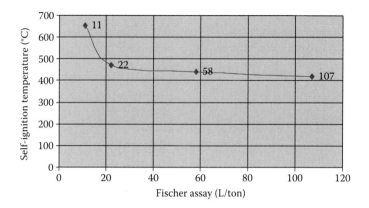

**FIGURE 8.6** SIT of oil shale as a function of Fischer assay. (From Joshi, R., M.S. thesis, University of Akron, Akron, OH, 1983.)

shale samples were always preceded by a slight exploding (crackling) sound, which strongly suggests that ignition occurred only when the vapor pressure of the gaseous hydrocarbons evolved reached a certain level. Further tests by Joshi[36] revealed that the SIT is also a function of particle size. Tests were carried out on particles of size −4+8 mesh to −60 mesh. It was also found that Colorado shale particles of size −40+60 mesh and smaller did not burst into flames under the defined conditions. The same phenomenon was observed for Cleveland shale no. 2 with the limiting particle size being −20+40 mesh, which did not burst into flames. It has been stated that as the particle size becomes smaller, the diffusional limitations of the product vapor decrease significantly. As a result, there is a continuous diffusion of product vapor (gaseous hydrocarbons) from the rock matrix to the surface and from the surface to the boundary layer.

Consequently, heating induces a buildup of gaseous hydrocarbons inside the rock matrix, which ultimately causes cracking of the shale particles. The concentration of the gaseous hydrocarbons near the surface immediately after internal cracking is high enough to stimulate ignition. As mentioned earlier, the SIT data can be very useful, as ignition temperature governs the initiation of the combustion process and also the dynamics of *in situ* oil shale retorting. Further, these data are also valuable because they provide an indication of the explosivity of oil shale dust during oil shale-mining operations. However, this hazard is considerably less likely to take place in oil shale mining than during coal mining.[10,22]

### 8.4.2 Thermal Characteristics of Oil Shale and Its Minerals

Thermal or thermoanalytical methods, such as TGA and DTA, are particularly useful for characterization of thermal behavior of oil shales and oil shale minerals. The use of both TGA and DTA has been well established in the areas of coal and polymer research, as a relatively simple procedure generates valuable information about the sample's thermal behavior.

### 8.4.2.1 Thermoanalytical Properties of Oil Shale

Figures 8.7 and 8.8 show the effects of the surrounding atmosphere on the thermal behavior of Green River oil shale.[20] Figure 8.7 shows the DTA curve in an inert atmosphere of flowing $N_2$, whereas Figure 8.8 shows the DTA curve in the presence of air, that is, in an oxidative atmosphere.

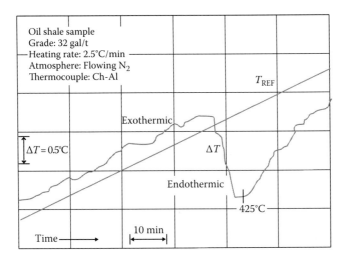

**FIGURE 8.7** Effect of the surrounding atmosphere on the thermal behavior of Green River oil shale—DTA in an inert atmosphere of flowing nitrogen. (From Rajeshwar, K. et al., *J. Mater. Sci.*, 14, 2025–2052, 1979.)

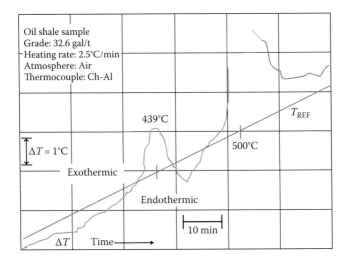

**FIGURE 8.8** Effect of the surrounding atmosphere on the thermal behavior of Green River oil shale—DTA in the presence of air. (From Rajeshwar, K. et al., *J. Mater. Sci.*, 14, 2025–2052, 1979.)

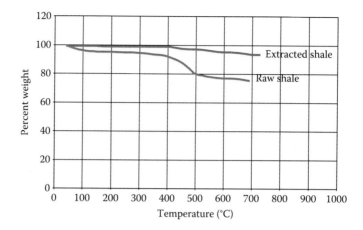

**FIGURE 8.9** TGA of stuart shale before and after oil extraction. (From Kesavan, S.K. and Lee, S., *Fuel Sci. Technol. Int.*, 6(5), 505, 1988.)

The peak corresponding to kerogen decomposition is seen to be endothermic in nature, that is, absorbing heat. This endothermic nature is very much expected, as all pyrolysis reactions require input of thermal energy. In the presence of air, however, two exotherms are apparent as shown in Figure 8.8, the first peak at 439°C and the second at approximately 500°C. While the first exothermic peak may be attributed to the combustion of light hydrocarbon fractions from the shale organic matter, the second exotherm appears to be from the burn-off of carbonaceous char.[20]

Figure 8.9 shows the TGA curve of Stuart oil shale of Queensland, Australia.[37] The sample used in this experiment was obtained from Kerosene Creek member and supplied by Southern Pacific Petroleum NL. The total mass loss of the raw shale by the TGA from 25°C to 600°C was 12.2%, which includes (1) evaporative loss of moisture, (2) kerogen decomposition and devolatilization, and (3) thermal decomposition of mineral carbonates and resultant $CO_2$ liberation. It is not surprising to observe a further mass loss from preextracted shale, which is attributable principally to decomposition of mineral carbonates.

### 8.4.2.2 Thermochemical Properties of Oil Shale Minerals

The identification and quantification of various carbonates existing in the Green River oil shales, such as ferroan (or ferroan dolomite), ankerite [calcium iron magnesium manganese carbonate or $Ca(Fe, Mg, Mn)(CO_3)_2$], and dawsonite, were accomplished using thermal analysis techniques.[38] A quantitative determination method for nahcolite [$NaHCO_3$] and trona [$Na_3(CO_3)((HCO_3)\cdot 2H_2O)$] in Colorado oil shales was proposed by Dyni et al.[39]

Rajeshwar et al.[20] summarized the thermochemical properties of minerals commonly found in oil shale deposits, and their results are shown in Table 8.11.

DTA involves heating or cooling a test sample and an inert reference under identical conditions and recording any temperature difference between the sample and the reference. This differential temperature ($\Delta T$) is then plotted against time or against temperature. Changes in the sample that lead to the absorption or release of heat can be detected relative to the inert reference. The DTA peak temperature is the temperature indicated at

## TABLE 8.11
## Thermochemical Properties of Common Minerals in Oil Shale Deposits

| Minerals | Chemical Formula | Type of Chemical Reaction | DTA Peak Temperature (°C) |
|---|---|---|---|
| Calcite | $CaCO_3$ | Dissociation | 860–1010 |
| Dolomite | $CaCO_3 \cdot MgCO_3$ | Dissociation | 790, 940 |
| Analcite | $NaAlSi_2O_6 \cdot H_2O$ | Dehydration; dissociation | 150–400 |
| Shortite | $Na_2Ca_2(CO_3)_3$ | Dissociation | 470 |
| Trona | $Na_2CO_3 \cdot NaHCO_3 \cdot 2H_2O$ | Dissociation; dehydration | 170 |
| Pyrite | $FeS_2$ | Oxidation; dissociation | 550 |
| Potassium feldspar | $KAlSi_3O_8$ | Dissociation | — |
| Gaylussite | $CaNa_2(CO_3)_2 \cdot 5H_2O$ | Dehydration; crystallographic transformation; melting | 145, 175, 325, 445, 720–982 |
| Illite | $K_{0.6}(H_3O)_{0.4}Al_{1.3}Mg_{0.3}Fe^{2+}_{0.1}Si_{3.5}O_{10}(OH)_2 \cdot$ $(H_2O)$ (empirical formula) | Dehydroxylation | 100–150, 550, 900 |
| Plagioclase | $NaAlSi_3O_8–CaAl_2Si_2O_8$ | Dissociation | — |
| Nahcolite | $NaHCO_3$ | Dissociation | 170 |
| Dawsonite | $NaAl(OH)_2CO_3$ | Dehydroxylation; dissociation | 300, 440 |
| Gibbsite | $\gamma\text{-}Al(OH)_3$ | Dehydroxylation | 310, 550 |
| Ankerite | $Ca(Mg, Mn, Fe)(CO_3)_2$ | Dissociation | 700, 820, 900 |
| Siderite | $FeCO_3$ | Oxidation; dissociation | 500–600, 830 |
| Albite | $NaAlSi_3O_8$ | Dissociation | — |
| Quartz | $SiO_2$ | Crystallographic transformation | ~575 |

*Source:* Branch, M.C., *Prog. Energy Combust. Sci.,* 5, 193, 1979.

the time of the maximum peak value of the DTA curve, which is a characteristic property of the material[40] and is not dependent on the size of the material sample.

Regarding the decomposition of *dawsonite*, there are some conflicting theories.[20] One belief is that dawsonite decomposes at 370°C according to the following reaction:

$$NaAl(OH)_2CO_3 \rightarrow H_2O + CO_2 + NaAlO_2 \qquad (8.13)$$

Another is based on the investigation of thermal behavior of dawsonite at temperatures between 290°C and 330°C, and the chemical reaction that is taking place is believed to be[41]

$$2NaAl(OH)_2CO_3 \rightarrow Na_2CO_3 + Al_2O_3 + 2H_2O + CO_2 \qquad (8.14)$$

Yet another belief is that dawsonite decomposes in two steps.[42] In the first step, it has been found that between 300°C and 375°C, crystalline dawsonite decomposes with the evolution of all the hydroxyl groups and two-thirds of the carbon dioxide, leaving an amorphous residue. In the second step, the balance $CO_2$ is released between 360°C and 650°C, producing crystalline $NaAlO_2$.

*Plagioclase* is a form of feldspar that has a chemical composition of $NaAlSi_3O_8$. Plagioclase is usually white in color but can also be gray and greenish white. This mineral was found to be abundant in the moon rock samples.

The dominant mineral constituent of oil shale is dolomite. Dolomite is approximately a one-to-one mixture of magnecite ($MgCO_3$) and calcite ($CaCO_3$). Therefore, the chemical formula of dolomite is often expressed by either $CaCO_3 \cdot MgCO_3$ or $CaMg(CO_3)_2$. Upon heating, dolomite undergoes a two-stage thermal decomposition reaction generally known as *calcination*.

$$CaCO_3 \cdot MgCO_3 \overset{\Delta}{\leftrightarrow} CaCO_3 \cdot MgO + CO_2 \qquad (8.15)$$

$$CaCO_3 \cdot MgO \overset{\Delta}{\leftrightarrow} CaO \cdot MgO + CO_2 \qquad (8.16)$$

The first reaction is called *half-calcination of dolomite*, whereas the entire reaction combining both reactions 8.15 and 8.16 is called *full calcination of dolomite*. Equilibrium decomposition temperature, that is, where $K_a = 1$, is 380°C for magnecite decomposition and 890°C for calcite decomposition, respectively.[7] It can be readily seen that dolomite decomposes quite actively in typical retorting conditions.

### 8.4.3  Electric Properties of Oil Shale

Electric properties also change as functions of temperature and other variables. Both alternating current (AC) and direct current (DC) methods can be employed in measuring electric properties of oil shale. In general, AC techniques are preferable in view of their capability to detect and resolve various polarization mechanisms in the material.

### 8.4.3.1  Electric Resistivity

Measurements on various types of oil shales in DC electric fields have shown an exponential decrease in resistivity values as a function of temperature.[13,20,43] The trend is typically characteristic of ionic solids, which conduct currents by a thermally activated transport mechanism. The presence of various minerals in the oil shale rock matrix makes it difficult to conclusively identify the current-carrying ions in the material. However, the close correspondence of activation energies at high temperatures (>380°C) with those typically observed for carbonate minerals seems to indicate that *carbonate ions* could be a major current-carrying species.[13,20] However, estimates made from such data are at best speculative and must be used with due caution.[20] The chemical change in oil shale material due to heating could also influence its conduction property. Thus, changes in the resistivity (from $10^{10}$ $\Omega$-cm at room temperature to 10 $\Omega$-cm at 900°C) of Russian shales were attributed to the thermal decomposition of oil shale kerogen.[23,43] Figure 8.10 shows the frequency-dependent behavior of

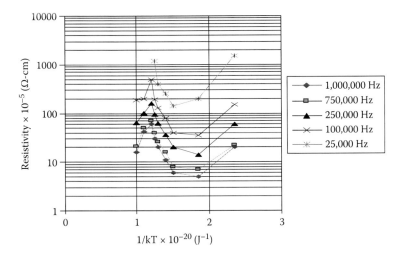

**FIGURE 8.10** Frequency-dependent behavior of electrical resistivity as a function of reciprocal temperature for a 117 L/ton shale sample. (From Rajeshwar, K. et al., *J. Mater. Sci.*, 14, 2025–2052, 1979.)

electric resistivity as a function of reciprocal temperature ($1/T$) for a sample (117 L/ton Fischer assay) of raw Green River oil shale. The minima in the resistivity curves are observed at temperatures ranging from 40°C to 210°C and are due to the gradual loss of free moisture and bonded water molecules to the clay particles in the shale matrix. Figure 8.11 shows the same behavior for the reheated materials; however, the trends for these curves are quite different from those for raw shales. In the figure, k is the Boltzmann constant which is equal to $1.3806488 \times 10^{-23}$ J·K$^{-1}$.

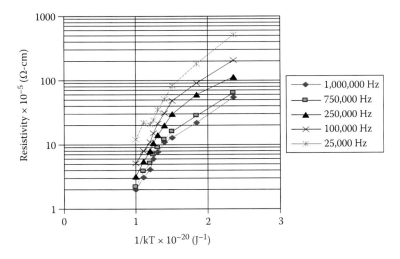

**FIGURE 8.11** Frequency-dependent behavior of electrical resistivity as a function of reciprocal temperature for a 117 L/ton shale sample reheated in a second cycle. (From Rajeshwar, K. et al., *J. Mater. Sci.*, 14, 2025–2052, 1979.)

For these reheating experiments, the shales were cooled back to room temperature and reheated back to approximately 500°C. The curves of Figure 8.11 exhibit the usual Arrhenius behavior typical of ionic solids, as mentioned before. It can be seen that there is no minimum or peak in the resistivity data and that the results are attributable to thermally activated conduction.

### 8.4.3.2 Dielectric Constants

The dielectric constant ($k$) is a number that is a characteristic property relating the ability of a material to carry AC to the ability of vacuum to carry AC. The capacitance created by the presence of the material, therefore, is directly related to the dielectric constant of the material.

An extensive review of the dielectric constant of oil shales is presented in the work of Rajeshwar et al.[20] The dielectric constant of oil shales also exhibits a functional dependency on temperature and frequency. Anomalously high dielectric constants are observed for oil shales at low temperatures, and these high values are attributed to electrode polarization effects, according to Scott et al.[44] A more likely explanation is the occurrence of interfacial polarization (e.g., Maxwell–Wagner type) in these materials arising from the presence of moisture and as a result of accumulation of charges at the sedimentary varves in the shale.

Figure 8.12 shows the variation of dielectric constant with the number of heating cycles for several grades of Green River oil shales.[45] Each heating cycle consisted of heating the sample at 110°C for 24 h and cooling back to room temperature prior to testing. A noticeable decrease in dielectric constant with each subsequent drying cycle is very evident.

Figure 8.13 shows the variation of dielectric constant with frequency and thermal treatment for Green River oil shales.[45] The degree of frequency dispersion at each heating cycle attests to an appreciable effect of moisture on the interfacial polarization mechanisms in oil shale.

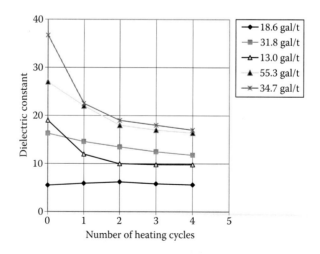

**FIGURE 8.12**  Variations of dielectric constant ($\varepsilon'$) with number of heating cycles for several grades of Green River shales. (From Rajeshwar, K. et al., *J. Mater. Sci.*, 14, 2025–2052, 1979.)

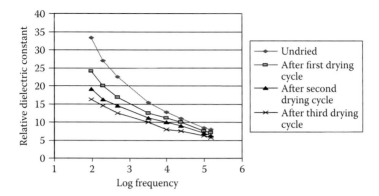

**FIGURE 8.13** Variation of dielectric constant ($\varepsilon'$) with frequency and thermal treatment for Green River oil shales. (From Rajeshwar, K. et al., *J. Mater. Sci.*, 14, 2025–2052, 1979.)

Figures 8.14 and 8.15 show the variation of dielectric constant of Green River oil shales as a function of frequency and temperature at low temperatures (<250°C) as well as at high temperatures (>250°C).[46] A general explanation may be that the dielectric constant decreases with increasing temperatures up to 250°C and thereafter increases again, attaining values comparable to those observed initially for the raw shales. The initial decrease may be due to the gradual release of absorbed moisture and chemically bonded water from the shale matrix. However, the subsequent increase may be due to more complex factors including (1) increased orientational freedom of the kerogen molecules, (2) buildup of carbon in the shale, and (3) presence of a space charge layer in the material at high temperatures. Further details can be found from the work by Rajeshwar et al.[20]

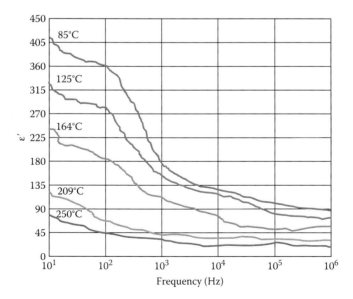

**FIGURE 8.14** Frequency and temperature dependence of dielectric constant ($\varepsilon'$) at low temperatures (<250°C). (From Rajeshwar, K. et al., *J. Mater. Sci.*, 14, 2025–2052, 1979.)

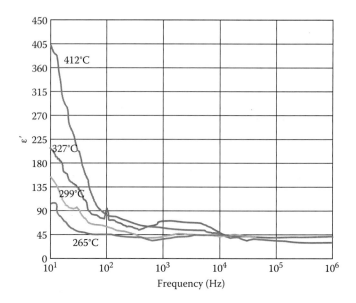

**FIGURE 8.15** Frequency and temperature dependence of dielectric constant ($\varepsilon'$) at high temperatures (>250°C). (From Rajeshwar, K. et al., *J. Mater. Sci.*, 14, 2025–2052, 1979.)

### 8.4.4 MOLECULAR CHARACTERIZATION OF KEROGEN

In applying chemical and engineering principles to the decomposition of oil shale, difficulties are encountered owing to the lack of structural and molecular understanding of kerogen. A *chemical model of kerogen structure* has been proposed by Yen[47] as $C_{220}H_{330}O_{18}N_2S_4$. If we use this formula as a representative chemical formula for kerogen molecule, its molecular weight becomes 3414. Kerogen has a macromolecular structure that gives a fairly large molecular (formula) weight.

#### 8.4.4.1 Derivation of Stoichiometric Coefficient

As the structure of kerogen cannot be represented as a uniquely defined chemical species, a stoichiometric equation for kerogen decomposition is often impractical. The kerogen decomposition reaction is frequently expressed by the following descriptive equation:

$$\text{Kerogen} \rightarrow \text{condensable oil vapors} + \text{gaseous products} + \text{residual char} \quad (8.17)$$

This equation cannot be taken as a stoichiometric equation, as the atomic balance on constituent atoms is not established. The following analysis is therefore intended to provide a theoretical bridge between a qualitative expression and a stoichiometric equation.[48]

The initial mass of organic carbon per cubic meter of particle volume is

$$m_i = \rho_s \omega, \text{ kg/m}^3$$

$$= \rho_s \omega 1000/M_k, \text{ mol/m}^3 \qquad (8.18)$$

where $\rho_s$ is the density of oil shale particle (in kg/m³), $\omega$ is the mass fraction of kerogen in the particle, and $M_k$ is the molecular weight (or formula weight) of kerogen (in g/mol). The total mass of oil recoverable (conventionally) from oil shale per cubic meter of particle volume is

$$m_2 = F\left[\frac{cm^3}{kg}\right]\rho_s\left[\frac{kg}{m^3}\right]\rho_o\left[\frac{g}{cm^3}\right] = F\rho_s\rho_o\left[\frac{g}{m^3}\right]$$

$$= F\rho_s\rho_o/M_p \ mol/m^3$$

(8.19)

where $F$ is the Fischer assay of oil shale (in cm³/kg), $\rho_o$ is the density of shale oil (not oil shale; in g/cm³), and $M_p$ is the average molecular weight of condensable product (in g/mol).

Assume that the kerogen decomposes into oil vapors and gaseous products completely at moderate decomposition temperatures. Then, the stoichiometric coefficient equivalent may be expressed by

$$\alpha = \frac{F\rho_s\rho_o/M_p}{\rho_s\omega 1000/M_k}$$

$$= \frac{F\rho_o M_k}{\omega M_p 1000}$$

(8.20)

where $\alpha$ is a theoretically obtained stoichiometric coefficient (or its equivalent) for reaction 8.17.

### 8.4.4.2 Relation between Fischer Assay and Mass Fraction of Kerogen

An empirical correlation between the Fischer assay of oil shale and the weight percentage of kerogen is proposed by Cook:[49]

$$F = 2.216 \ wp - 0.7714, \ gal/ton$$

(8.21)

where $F$ is the Fischer assay estimated in gallons of oil recoverable per ton of shale, and $wp$ is the weight percentage of kerogen in the shale. However, caution must be exercised in using this equation, as the mass fraction of kerogen in oil shale is strongly dependent on the measurement technique. For certain oil shales, the maximum recoverable oil amount via supercritical extraction or by $CO_2$ retorting process is significantly higher than the Fischer assay value.[7]

It is also conceivable that such a correlation can be sensitive to the types of oil shale retorted as well. Nonetheless, there is little doubt that the Fischer assay of any oil shale is strongly correlated against the kerogen content of the shale.

### 8.4.4.3 Nitrogen Compounds in Shale Oil

Nitrogen compounds in shale oil cause technological difficulties in the downstream processing of shale oil, in particular, poisoning the refining and upgrading

**TABLE 8.12**

**Predominant Classes of Nitrogen Compounds in a Shale Oil Light Distillate**

| Nitrogen Type | Total Nitrogen |
|---|---|
| Alkylpyridines | 42 |
| Alkylquinolines | 21 |
| Alkylpyrroles (N-H) | 8 |
| Alkylindoles (N-H) | 7 |
| Cyclic amides (pyridones, quinolines) | 3 |
| Anilides | 2 |
| Unclassified very weak bases (*N*-alkylpyrroles and *N*-alkylindoles) | 4 |
| Other unclassified very weak bases (reduced to nontitratable types and not sulfoxides) | 3 |
| Nonbasic (nontitratable) nitrogen in original light distillate | 8 |
| Analytical loss | 2 |
| Total | 100 |

*Source:* Poulson, R.E. et al., *ACS Div. Pet. Chem. Prepr.*, 15(1), A49–A55, 1971.

catalysts. Needless to say, these nitrogen compounds originate from oil shale, and the amount and specific type depend heavily on the petrochemistry of the oil shale deposits. Although direct analysis and determination of molecular forms of nitrogen-containing compounds in oil shale rock are very difficult, analysis of the shale oil extracted by retorting processes provides valuable information regarding the organonitrogen species in the oil shale. Poulson[50] reported the breakdown of organonitrogen compounds in shale oil based on a preliminary study of shale oil light distillates, and the results are shown in Table 8.12. The major compound classes Poulson identified were pyridines, quinolines, pyrroles, and indoles.[50]

## 8.4.5 Boiling Range Distributions of Various Shale Oils

Shale oils or oil shale crudes are obtained by various oil shale retorting and extraction processes. The characteristics of shale oils are very important in devising a process for upgrading shale oils, as well as in identifying the market for them. In particular, distillation properties of shale oil are crucially important for its refining and upgrading.

### 8.4.5.1 Analytical Methods

In characterizing hydrocarbon mixtures for specification or for other purposes, a precise analytical distillation may be needed. Actual distillation may require 25 to 100 plates. It is extremely costly and tedious to carry out such a distillation on a reasonable scale. Instead, GC with a separating column at a constant temperature, can be effectively used to obtain a *boiling point analysis*. However, this technique is somewhat restricted to a rather narrow boiling range, as lighter components elute too soon and tend to overlap, and heavy components emerge very late, producing relatively wide bands or still remaining in the column.

A newer technique of temperature programming of the separating column makes a wide-range, single-stage analysis possible.[51] By using a column packing that separates according to the boiling point and by precise programming of the column temperature, the boiling range for various peaks can be determined from elution times or temperatures of emergence.

A standard method for *boiling range distribution* of petroleum fractions by GC has been adopted by the American Society for Testing and Materials, West Conshohocken, Pennsylvania (ASTM), and is given in the ASTM D2887[52] Procedure, which is briefly described in the following section.

### 8.4.5.1.1 ASTM D2887 Procedure[52]

1. *Scope*: This standard method determines the boiling range distribution of petroleum products. It is generally applicable to petroleum fractions with a final boiling point of 1000°F (538°C) or lower at atmospheric pressure.

2. *Summary of the method*: The sample is introduced into a gas chromatographic column, which separates hydrocarbons in the order of their boiling points. The column temperature is raised at a reproducible (preprogrammed) rate, and the area under the chromatogram is recorded throughout the run. Boiling temperatures are assigned to the elution time axis from a calibration curve obtained under the same conditions by running a known mixture of hydrocarbons (ranging from $C_m$ to $C_n$) covering the boiling range expected from the sample. From these data, the boiling range distribution of the sample may be obtained.

3. *Initial and final boiling points*: The *initial boiling point* (*IBP*) is the point at which a cumulative area count equals 0.5% of the total area under the chromatogram. On the other hand, the *final boiling point* (*FBP*) is the point at which a cumulative area count equals 99.5% of the total area under the chromatogram. The *normal boiling point* (*NBP*) is the point at which the vapor pressure reaches 760 mmHg or 1 atm.

4. *Apparatus for boiling range distribution*: A gas chromatograph equipped with a thermal conductivity detector (TCD) is typically used for the experiment of gasoline fractions. For all other types of samples, either a TCD or a flame ionization detector (FID) may be used. The detector must have sufficient sensitivity to detect 1% dodecane ($C_{12}H_{26}$) with a peak height of at least 10% of full scale on the recorder under the conditions prescribed in this method and without loss of resolution. Practically any column can be used, provided that under the conditions of the test, separations are in order of boiling points and the column resolution (*R*) is at least 3 and not more than 8. As a stable baseline is essential for the accuracy of this method, matching dual columns are required to compensate for column bleed, which cannot be eliminated completely by conditioning alone.

The temperature programming must be done over a range that is sufficient to establish a retention time of at least 1 min for the IBP to elute the entire sample. A microsyringe is needed for sample injection, and a flow controller is also required for holding carrier gas flow constant to ±1% over the full operating temperature range. The carrier gas used is either helium or hydrogen for a TCD, whereas nitrogen, helium, or

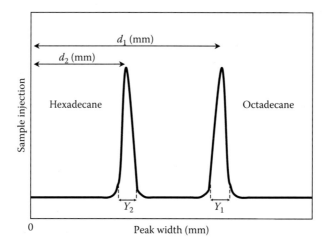

**FIGURE 8.16** Column resolution ($R$). (From ASTM Standards D2887.)

## TABLE 8.13
## Typical Operating Conditions

|                                      | 1      | 2      | 3       | 4      |
| ------------------------------------ | ------ | ------ | ------- | ------ |
| Column length (ft)                   | 4      | 5      | 2       | 2      |
| Column ID (in)                       | 0.188  | 0.090  | 0.188   | 0.188  |
| Liquid phase                         | OV-1   | SE-30  | UC-W98  | SE-30  |
| Liquid phase (%)                     | 3      | 5      | 10      | 10     |
| Support material                     | S[a]   | G[b]   | G[b]    | P[c]   |
| Support mesh size                    | 60/80  | 60/80  | 60/80   | 60/80  |
| Initial column temperature (°C)      | 20     | 40     | 50      | 50     |
| Final column temperature (°C)        | 360    | 350    | 350     | 390    |
| Programming rate (°C/min)            | 10     | 6.5    | 8       | 7.5    |
| Carrier gas                          | He     | He     | $N_2$   | He     |
| Carrier gas flow rate (ml/min)       | 40     | 30     | 60      | 60     |
| Detector                             | TCD    | FID    | FID     | TCD    |
| Detector temperature (°C)            | 360    | 360    | 350     | 390    |
| Injection port temperature (°C)      | 360    | 370    | 200     | 390    |
| Sample size (µL)                     | 4      | 0.3    | 1       | 5      |
| Column resolution ($R$)              | 5.3    | 6.4    | 6.5     | 3      |

*Source:* ASTM Standards D2887.

*Note:* OV-1 = methyl silicone polymer liquid phase; SE-30 = dimethyl silicone elasto-
mer liquid phase; UC-W98 = silicone liquid phase; TCD = thermal conductivity
detector; FID = flame ionization detector.

[a] Diatoport S, silane-treated.

[b] Chromosorb G (AW-DMS); AW = acid washed; DMS = treated with
dimethylchlorosilane.

[c] Chromosorb P, acid-washed.

argon may be used with an FID. The calibration mixture to be used is a mixture of hydrocarbons of known boiling point covering the boiling range of the sample. At least one compound in the mixture must have a boiling point lower than the IBP of the test sample in order to obtain an accurate distribution of the boiling range. Both integral and differential plots can be constructed from the chromatographic data. Most retorted shale oils show density functions that are close to a normal distribution.

To test the column resolution, a mixture of 1% each of $C_{16}$ and $C_{18}$ n-paraffin in a suitable solvent such as octane ($C_8$) needs to be prepared. Inject the same volume of this mixture as used in analyses of samples and obtain the chromatogram. As shown in Figure 8.16, calculate the resolution ($R$) from the distance between the $C_{16}$ and $C_{18}$ n-paraffin peaks at the peak maxima ($d$) and the width of the peaks at the baseline ($Y_1$ and $Y_2$):

$$R = [2(d_1 - d_2)]/(Y_1 + Y_2) \tag{8.22}$$

Resolution ($R$), based on Equation 8.22, must be at least 3 but not more than 8.

Tables 8.13 and 8.14 show the typical operating conditions and the boiling points of n-paraffins, respectively. These tables provide very valuable information regarding the boiling range properties of paraffinic hydrocarbons.

## TABLE 8.14
### Boiling Points of n-Paraffins

| Carbon Number | Boiling Point (°C) | Carbon Number | Boiling Point (°C) | Carbon Number | Boiling Point (°C) |
|---|---|---|---|---|---|
| 2 | 89 | 17 | 302 | 32 | 468 |
| 3 | 42 | 18 | 317 | 33 | 476 |
| 4 | 0 | 19 | 331 | 34 | 483 |
| 5 | 36 | 20 | 344 | 35 | 491 |
| 6 | 69 | 21 | 356 | 36 | 498 |
| 7 | 98 | 22 | 369 | 37 | 505 |
| 8 | 126 | 23 | 380 | 38 | 512 |
| 9 | 151 | 24 | 391 | 39 | 518 |
| 10 | 174 | 25 | 402 | 40 | 525 |
| 11 | 196 | 26 | 412 | 41 | 531 |
| 12 | 216 | 27 | 422 | 42 | 537 |
| 13 | 235 | 28 | 432 | 43 | 543 |
| 14 | 253 | 29 | 441 | 44 | 548 |
| 15 | 271 | 30 | 450 | | |
| 16 | 287 | 31 | 459 | | |

Source: ASTM Standards D2887.

Note: $C_1$ to $C_{20}$ values taken from Selected Values of Hydrocarbons and Related Compounds, API Project 44, Loose-Leaf data sheet: Table 20a–e (Part 1), April 30, 1956. $C_{21}$ to $C_{44}$ values taken from Vapor Pressures and Boiling Points of High Molecular Weight Hydrocarbons, $C_{21}$ to $C_{100}$, report of Investigation of API Project 44, August 15, 1965.

## 8.5   OIL SHALE EXTRACTION AND RETORTING PROCESSES

The organic matter in oil shale typically contains both bitumen and kerogen. The bitumen fraction is soluble in most organic solvents, and it is not difficult to extract directly from oil shale. The readily soluble bitumen content in oil shale occupies only a minor portion, whereas insoluble kerogen accounts for the major portion of oil shale organic matter. Furthermore, kerogen is nearly inert to most chemicals owing to its macromolecular and complex structure, therefore making most reactive processes less effective and making extraction more difficult.

Several different approaches are possible for the extraction of oil (organic matter) from the mineral matrix (inorganic rock matrix): (1) to drastically break the chemical bonds of the organics, (2) to mildly degrade or depolymerize the organics, and (3) to use solvents that have extraordinarily strong solvating power. The first approach is widely used in industrial applications, as high-temperature pyrolysis decisively cleaves the chemical bonds of the organics. *Retorting* processes belong to this category and have a long history. The second approach may be achieved by a biochemical process or a controlled oxidative process. The third approach can be accomplished by potent extraction methods, such as a supercritical fluid extraction process that is based on the strong solvating power of a fluid in its supercritical region.

During the process of extraction of shale oil from oil shale, both chemical and physical properties of oil shale play important roles. The low porosity, low permeability, and tough mechanical strength of the oil shale rock matrix make the extraction process less efficient by making the mass transfer of reactants and products much harder. Both heat and mass transfer conditions of a process also crucially affect the process economics as well as process efficiency.

The oil shale retorting processes can be classified as *ex situ* (or, above ground, off the sites) and *in situ* (or subsurface, within the existing formation) processes. As the names imply, *ex situ* processes are carried out above the ground after the shale is mined and crushed, whereas *in situ* processes are carried out under the ground, thus not requiring mining of shale entirely.

In this section, various processes developed and demonstrated for oil shale extraction are reviewed in terms of engineering and technological aspects.

### 8.5.1   *Ex Situ* Retorting Processes

In an *ex situ* process, oil shale rock is mined, either surface or underground, crushed, and then conveyed to a retorter that is subjected to temperatures around 500°C to 550°C. At this temperature, chemical bonds of the organic compounds are broken, and the kerogen molecules are pyrolyzed, yielding simpler and lighter hydrocarbon molecules.

The advantages of *ex situ* processes are as follows:

1. Efficiency of organic matter recovery has been demonstrated to be high, about 70% to 90% of the total organic content of the shale. Therefore, the amount of wasted organic matter, that is, the unextracted portion or the organic residue, can be minimized by keeping the efficiency high.

2. Control of process operating variables is relatively straightforward. Therefore, the effects of undesirable process conditions can be minimized.
3. Once oil is formed, product recovery becomes relatively simple.
4. Process units can be used repeatedly for a large number of retorting operations.

However, the disadvantages of *ex situ* processes are as follows:

1. Operating cost is usually high, as oil shale has to be first mined, crushed, transported, and then heated. Mining and transportation costs may become quite significant.
2. Spent shale disposal, underground water contamination, and revegetation problems are yet to be solved convincingly and effectively.
3. The process is somewhat limited to rich shale resources accessible to surface or shallow-subsurface mining. In this regard, process economics plays an important role.
4. The capital investment for large-scale units may become very high as reusability is typically limited. Once the mine is depleted, a part of the investment may be lost forever.

The liberated compounds from oil shale retorting include gas and oil, which is collected, condensed, and upgraded into a liquid product that is roughly equivalent to crude oil. This oil can be transported by a pipeline or by a tanker to a refinery, where it is refined into the final product.

Various *ex situ* retorting processes are discussed in the following sections.

### 8.5.1.1   U.S. Bureau of Mines' Gas Combustion Retort

The first major experimental retort was built and operated by the U.S. Bureau of Mines (USBM). From 1944 to 1956, pilot plant investigations of oil shale retorting were carried out by USBM at facilities several miles west of Rifle, Colorado. Among the numerous types of retorts, the *gas combustion retort* gave the most promising results and was studied extensively.[53] This retort is made of a vertical, refractory-lined vessel through which crushed shale moves downward by gravity, countercurrent to the retorting gases. Recycled gases enter the bottom of the retort and are heated by the hot retorted shale as they pass upward through the vessel. Air and additional recycle gas (labeled as dilution gas) are injected into the retort through a distributor system at a location approximately one-third of the way up from the bottom and are mixed with rising hot recycled gas. Figure 8.17 shows a schematic of a gas combustion retort. Combustion of the gases and of some residual carbon provides thermal energy to heat the shale immediately above the combustion zone to the retorting temperature. The incoming shale cools oil vapors and gases, and the oil leaves the top of the retort as a mist. The oil vapors and mists are subsequently chilled to produce liquid oil products that have to be upgraded.

This retort is similar to a *moving bed reactor* popularly used for coal gasification in its operating concepts.

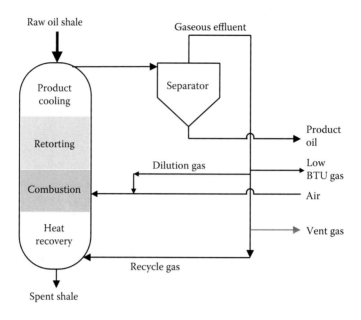

Raw oil shale

Gaseous effluent

**FIGURE 8.17**   Gas combustion retort.

## 8.5.1.2  TOSCO II Oil Shale Process

The TOSCO II oil shale retorting process was developed by The Oil Shale Corp (TOSCO). An article by Whitcombe and Vawter[54] describes the process in detail and also presents economic projections for production of crude shale oil and hydrotreated shale oil. As an oil shale process, the TOSCO process is one of the few complete processes for production of retorted shale oil.

### 8.5.1.2.1  Process Description

Oil shale is crushed and heated to approximately 480°C by direct contact with *heated ceramic balls*. At this temperature, the organic material (kerogen and bitumen) in oil shale rapidly decomposes to produce hydrocarbon vapor. Subsequent cooling of this vapor yields crude shale oil and light hydrocarbon gases. Figure 8.18 represents a schematic diagram of the process.

The pyrolysis reaction takes place in a retorting kiln (also referred to as the pyrolysis reactor or retort) shown in the central portion of the schematic. The feed streams to the retort are 1/2 in. diameter ceramic balls heated to about 600°C and preheated (substantially lower than 450°C) shales crushed to a size of 1/2 in. or smaller. The rotation of the retort mixes the feed materials and facilitates a high rate of heat transfer from the ceramic balls to the shale. At the discharge end of the retort, the ceramic balls and shale are at an isothermal (or near-isothermal) temperature by the time the shale is fully retorted.

The hydrocarbon vapor formed by the pyrolysis reaction flows through a cyclone separator to remove entrained solids, and then into a fractionation system, which

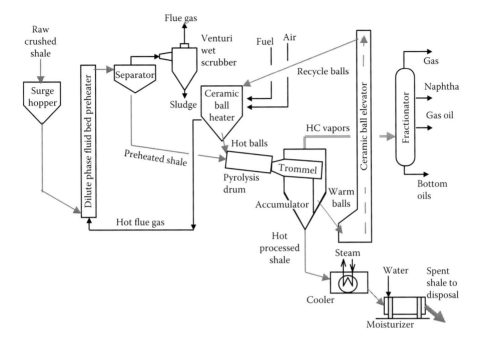

**FIGURE 8.18**   TOSCO II process.

is similar to the primary fractionator of a catalytic cracking unit. From this stage onward, oil vapor produces heavy oil, distillate oils, naphtha, and light hydrocarbon gases.

The ceramic balls and spent shale are moved from the retort into a cylindrical trammel screen. Spent shale passes through the screen openings into a surge hopper. The ceramic balls move across the screen and into a bucket elevator for transport to the ball heater, where they are reheated by direct contact with flue gas. The ceramic balls are then recycled back to the retort.

Spent shale, discharged from the retort at 480°C, is first cooled in a rotating vessel containing tubes in which water is vaporized to produce high-pressure steam. The shale then flows into another rotating vessel in which it is further cooled by direct contact with water. The water flow is controlled so that the spent shale from the vessel contains 12% moisture by mass. The moisture is added to control dust emissions to make the spent shale suitable for compaction before disposal.

The preheating of oil shale is achieved by direct contact between the crushed shale and the flue gas effluent from the ball heater. The gaseous effluent from the process is the flue gas used to heat the ceramic balls and to preheat the shale. The process includes a wet scrubber system to control the particulate content of the gas and an incinerator to control its hydrocarbon content in the flue gas. Emissions of $SO_x$ and $NO_x$ are controlled by the choice of fuels used in the process, as well as the firing temperatures of process heaters. The process effectively uses the concept of energy integration for minimization of energy cost (in particular, heating shales by contacting hot ceramic balls, producing steam using the residual heat of the spent shale, and managing and recovering flue gas energy).

### 8.5.1.2.2 Process Yield of TOSCO

Tests in a pilot plant and semiworks have shown that the TOSCO II process recovers nearly 100% of the recoverable hydrocarbon in oil shale, as determined by the Fischer assay procedure. It is remarkable that the TOSCO process may be regarded as an effective scale-up of the Fischer assay procedure. Table 8.15 shows results from a 7-day, continuous operation of the semiworks plant.[54]

The average plant yield during this period was 161.1 kg of hydrocarbons per metric ton of oil shale processed, approximately 1.7% higher than by Fischer assay of the average shale sample used for the period.[54]

### 8.5.1.2.3 Gaseous and Crude Shale Oil Product from TOSCO Process

Table 8.16 shows a typical analysis of the $C_4$ and higher hydrocarbons produced by the TOSCO II retort. The effluent gas is practically free of nitrogen and contains a good amount of carbon dioxide produced by pyrolysis. However, a relatively high amount of hydrogen sulfide is also present, which has to be removed in the gas cleanup stage.

Table 8.17 shows the properties of shale oil ($C_5$ and heavier fractions) produced by the TOSCO II retorting process. The average sulfur level in the liquid oil product is 0.7%, whereas the average nitrogen content is 1.9%, which is high compared to that of conventional crude oils. The nitrogen content of conventional crude oil very seldom exceeds 1.0% by mass. The high level of nitrogen in shale oil may be attributed to the geological reason based on the original formation of this fossil fuel, that is, oil shale deposits having come from protein-containing sources. The principal objective of the hydrotreating process is removal of nitrogen compounds that are poisonous to catalysts of many upgrading processes, including reforming, cracking, and hydrocracking.

---

**TABLE 8.15**

**TOSCO II Semiworks Plant Yield Data**

| Hydrocarbons | Plant Yield (kg/t) | Fischer Assay Yield (kg/t) |
|---|---|---|
| Total hydrocarbons | 161.1 | 158.3 |
| $C_1$–$C_4$ | 24.8 | 12.1 |
| $C_5$ and heavier fractions | 136.3 | 146.2 |
| Other gaseous products | | |
|   $H_2$ + CO | 2.25 | 1.85 |
|   $CO_2$ + $H_2O$ | 16.35 | 15.65 |

*Source:* Whitcombe, J.A. and Vawter, R.G., in *Science and Technology of Oil Shale*, Yen, Y.F., Ed., Ann Arbor Science MI, 1976, P.51, Table 4.1.

---

**TABLE 8.16**

**Typical Analysis of $C_4$ and Lighter Gases from the TOSCO Semiworks Plant**

| Component/Fraction | Mass% |
|---|---|
| $H_2$ | 1.50 |
| CO | 3.51 |
| $CO_2$ | 33.08 |
| $H_2S$ | 5.16 |
| $CH_4$ | 11.93 |
| $C_2H_4$ | 8.67 |
| $C_2H_6$ | 8.43 |
| $C_3H_6$ | 11.08 |
| $C_3H_8$ | 5.45 |
| $C_4 \cdot s$ | 11.19 |
| Total | 100.00 |

*Source:* Whitcombe, J.A. and Vawter, R.G., The TOSCO-II oil shale process, in *Science and Technology of Oil Shale*, Yen, Y.F., Ed., Ann Arbor Science, Ann Arbor, MI, 1976, Chapter 4.

**TABLE 8.17**

**Properties of Crude Shale Oil from TOSCO II Retorting Process**

| Boiling Ranges and Components | Vol% | °API | Mass% S | N |
|---|---|---|---|---|
| $C_5$–204 | 17 | 51 | 0.7 | 0.4 |
| 204–510 | 60 | 20 | 0.8 | 2.0 |
| 510+ | 23 | 6.5 | 0.7 | 2.9 |
| Total | 100 | 21 | 0.7 | 1.9 |

*Source:* Whitcombe, J.A. and Vawter, R.G., The TOSCO-II oil shale process, in *Science and Technology of Oil Shale*, Yen, Y.F., Ed., Ann Arbor Science, Ann Arbor, MI, 1976, Chapter 4.

*Note:* Boiling ranges are given in °C.

### 8.5.1.2.4 TOSCO Process Units

Figure 8.19 represents a schematic of the TOSCO II process for commercial operation.[54] The commercial plant involves two hydrotreating units. The first one is the distillate hydrotreater that processes the 200°C–500°C oil formed in the retorter plus similar boiling range components formed in the coker. The second hydrotreater

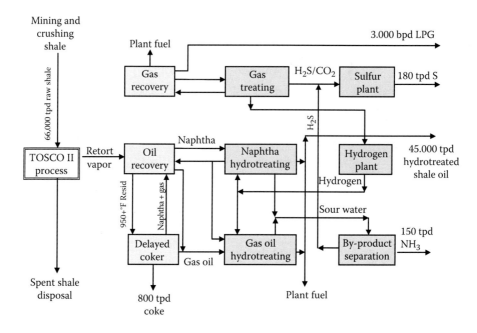

**FIGURE 8.19**   Block flow diagram—shale oil hydrotreating plant. (From Thumann, A., *The Emerging Synthetic Fuel Industry*, Fairmont Press, 1981.)

processes $C_5$ to 200°C naphtha formed in the retort, the coker, and the distillate hydrotreater. The distillate hydrotreater is designed to reduce the nitrogen content of the 200°C-plus product from the unit to a level lower than 1000 ppm. The naphtha hydrotreater is designed to reduce the nitrogen content about 1 ppm or below. Sulfur removal is nearly complete in each of the hydrotreating units.[54] The product compositions shown in Table 8.14 can be altered by changing the fuels chosen for burning in the process facilities. The production of $C_5^+$ fractions can be increased by burning the $C_3^-$ products instead of hydrotreated oil. Table 8.18 shows the properties of the $C_5$–510°C fractions of hydrotreated shale oil, which is a blend of sulfur-free distillate products.

Refining such hydrotreated oils is relatively straightforward and requires an atmospheric distillation unit and a reformer. It would produce gasoline, sulfur-free light distillate fuels (Nos. 1 and 2 heating oils, as well as diesel fuel), and a sulfur-free heavier distillate fuel oil that is suitable for use as industrial fuel oil.[54]

### 8.5.1.2.5 Spent Shale Disposal

The typical spent shale produced by the TOSCO II process is a fine-grained dark material comprising approximately 80 mass% of the raw oil shale fed. It contains an average 4.5% of organic carbon via char formation in a hydrogen-deficient environment. The mineral constituents of spent shale consisting of principally dolomite, calcite, silica, and silicates are mostly unchanged by the retorting process treatment, except that some carbonate minerals such as dolomite decompose to oxides,

**TABLE 8.18**

**Properties of Typical Hydrotreated Shale Oil**

| Boiling Ranges and Components | Vol% | °API | Nitrogen (ppm) |
|---|---|---|---|
| $C_5$–204 | 43 | 50 | 1 |
| 204–361 | 34 | 35 | 800 |
| 361–EP | 23 | 30 | 1200 |
| Total | 100 | 40[a] | |

*Source:* Whitcombe, J.A. and Vawter, R.G., The TOSCO-II oil shale process, in *Science and Technology of Oil Shale*, Yen, Y.F., Ed., Ann Arbor Science, Ann Arbor, MI, 1976, Chapter 4.

*Note:* Boiling ranges are given in °C.

[a] API for the total amount of the product.

liberating carbon dioxide. During the retorting process, significant size reduction also takes place, yielding the particle (grain) size of most spent shale finer than 8 mesh.

The technology for spent shale disposal was seriously tested in 1965 after completion of the 1000 ton/day semiworks plant at Parachute Creek. Small revegetation test plots were constructed in 1966 to evaluate both plant growth factors and plant species. In 1967, the first field demonstration revegetation plot was constructed and seeded. Extensive off-site investigations have been carried out, including the spent shale permeability, quality analysis of water runoff from spent shale embankments, etc.

### 8.5.1.3  Union Oil Retorting Process

In the Union Oil process, the heat needed for retorting is provided by *combustion of coke* inside the retort. The shale is fed from the bottom of the retort and conveyed (pumped) upward by means of a specially developed *rock pump*. The product oil is siphoned out from the bottom of the retort and fully recovered. The process is quite unique and innovative, utilizing well-designed rock pumps and adopting a number of designs for heating shales in the retort. According to a process described by Deering,[55] coke-containing spent shale derived from a gas-heated reduction zone is passed through a *combustion–gasification zone* countercurrently to an upflowing mixture of steam and oxygen-containing gas to effect partial combustion of the coke on the spent shale, that is, $C + 1/2\ O_2 = CO$. The resulting heat (exothermic heat) of combustion is used to support concurrent endothermic gasification reactions between steam and unburnt coke, that is, $C + H_2O = CO + H_2$. Figure 8.20 shows a schematic of the Union Oil process. The oil shale feed rate can be varied considerably depending on the size of the retort and the desired retention time.

The recycled water gas ($CO + H_2O + H_2 + CO_2$) contains hydrogen, and it must pass through the combustion–gasification zone in which hydrogen-burning temperatures prevail. A significant aspect of the process is that the overall yield of hydrogen,

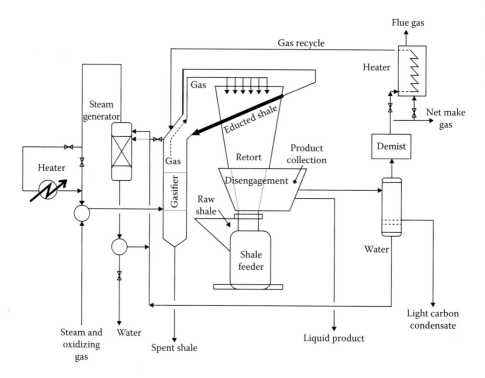

**FIGURE 8.20**    Union Oil's retorting-gasification process. (Based on U.S. patent 4,010,092.)

even with such passage, is still not significantly affected. This retort uses lump shales of about the same size range that the Gas Combustion Retort uses. Another important feature of the process is that it does not require cooling water.

The Union Oil Co. developed three different process designs as follows:

1. *The A retort* is the one in which internal combustion of gas and residual char from shale provides the energy required for the process. This design is based on direct heating.
2. *The B retort* is the one in which the oil shale is heated indirectly by a recycled stream of externally heated gas.
3. *The Steam–Gas Recirculation (SGR) retort* is the one in which the heat carrier for the process is generated in a separate vessel by gasifying the residual char with air and steam.

Union Oil had accumulated pilot plant experiences with the preceding three designs, which included a 2 ton/day prototype and a 50 ton/day pilot plant using the A retort design at Wilmington, California, and a nominal 6 ton/day pilot retort using the B mode and a pilot using the SGR mode at the Union Research Center, Brea, California. A larger-scale pilot plant was built near Parachute Creek in Western Colorado, where the oil shale could be mined readily from an outcrop of the Mahogany zone. The pilot plant had a capacity of 1000 tons/day, and its operation was completely successful. Unocal ceased operation in 1991. This operational experience is undoubtedly very valuable for future commercialization of the process.

### 8.5.1.4 Lurgi–Ruhrgas Process

The *Lurgi–Ruhrgas (LR)* process distills hydrocarbons from oil shale by bringing raw shale in contact with hot fine-grained solid heat carrier. The ideal heat carrier for this process is the spent shale. However, if rich shales are used in the process as raw shales, they typically deteriorate into a fine powder during process treatment and must be supplemented by more durable materials like sand for use as heat carriers. A schematic of the L-R process is presented in Figure 8.21.

The pulverized oil shale and heat carriers are brought into contact in a mechanical mixer such as a screw conveyer. In pilot plant tests, the shale was first crushed to a maximum size of one-fourth to one-third of an inch, but larger commercial units might process particles as large as half an inch.[55] The oil vapor and gaseous products are cleaned of dust in a hot cyclone, and the liquid oil is separated by condensation.

Retorted shale from the mixer passes through a hopper to the bottom of a lift pipe with the dust from the cyclone. Preheated air introduced at the bottom of the pipe carries the solids up to the surge bin. Solids are heated by the combustion of the residual char in the shale to approximately 550°C. In case residual char is not sufficient for this process, fuel gas is also added. In the surge bin, the hot solids separate from the combustion gases and return to the mixer, where they are brought in contact with fresh oil shale, completing the cycle. As an improvement, a new design of

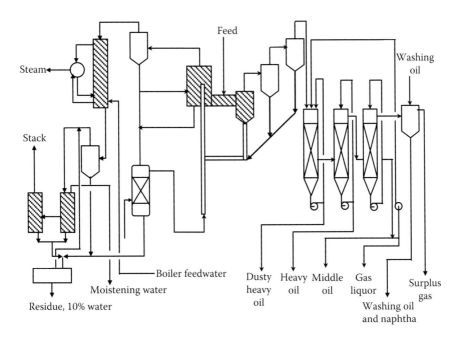

**FIGURE 8.21** Lurgi–Ruhrgas retort system. (From Matar, S., *Synfuels: Hydrocarbons of the Future*, Pennwell Publishing Co., Tulsa, OK, 1982.)

surge bin was introduced by Kennedy and Krambeck.[56] This improved surge bin has baffles that facilitate uniform flow of feed material through the surge bin.

Pilot plant tests have produced high yields, exceeding the Fischer assay value of the raw Colorado shale, at approximately 30 gal/ton of shale. As no combustion occurs in the mixer retort during this process, the product gas from the mixer has a relatively high calorific value (CV). The L-R process can operate with a wide range of particle sizes (very fine to medium), and, therefore, it can be modified for a variety of shale feedstocks. The process hardware is mechanically simple except for the mixer, which may be difficult to design because it must operate reliably in a harsh environment. However, the movement of dust through the system potentially causes two major problems of concern. One is the accumulation of combustible dust in the transfer lines, increasing the likelihood of fires and plugging. The other is entrainment of dust in the oil produced. (Even though most of the dust is removed in the hot cyclone, some is inevitably carried over to the retort product.) When the crude oil from the process is fractionated, the dust concentrates in the heaviest fraction, requiring an additional processing step. This heavy fraction can be diluted and filtered or recycled to the mixer.

The L-R process, originally developed in the 1950s for the low-temperature flash-carbonization of coal, was tested on European and Colorado oil shale in a 20 ton/day pilot plant at Herten, Germany. Two 850 ton/day pilot plants for carbonizing brown coal were built in Yugoslavia in 1963, and a large plant that uses the L-R process to produce olefins by cracking light oils was built in Japan.[55,57]

As an improvement over this process, *time domain reflectometry* (TDR) was evaluated and developed by Reeves and Elgezawi[58] to monitor volumetric water content ($\theta_v$) in oil shale solid waste retorted and combusted by the Lurgi–Ruhrgas process. A TDR probe was designed and tested that could be buried and compacted in waste embankments and provide *in situ* measurements for $\theta_v$ in the high-saline and high-alkaline conditions exhibited by the spent oil shale solid waste.[58]

### 8.5.1.5 Superior's Multimineral Process

The multimineral process was developed by the Superior Oil Co. In addition to synthetic gas (syngas) and oil, it also produces minerals such as nahcolite ($NaHCO_3$), alumina ($Al_2O_3$), and soda ash.[59] A schematic of Superior Oil's multimineral retort process is shown in Figure 8.22. The process is basically a four-step operation for oil shale that contains recoverable concentrations of oil, nahcolite, and dawsonite (a sodium–aluminum salt [$Na_3Al(CO_3)_3 \cdot Al(OH)_3$]). Superior Oil operated a pilot plant of this process in Cleveland, Ohio.

Nahcolite is in the form of discrete nodules that are more brittle than shale. It is recovered by secondary crushing and screening, followed by a specialized process called *photosorting* that recovers nahcolite product of >80% purity. After removal of nahcolite, shale is retorted using the *McDowell–Wellman* process.[55] The process was originally developed as a stirred-bed, low-Btu coal gasifier. The unique, continuously fed, circular-moving grate retort used in this process is a proven, reliable piece of hardware that provides accurate temperature control, separate process zones, and a water seal that eliminates environmental contamination.

Nahcolite has been tested as a dry scrubbing agent to absorb sulfurous and nitrous oxides. Dawsonite in shale is decomposed in the retort to alumina and soda

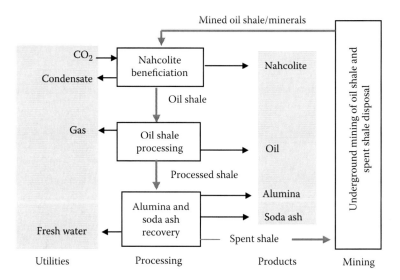

**FIGURE 8.22** Superior oil's multimineral retort system.

ash. After shale is leached with recycled liquor and makeup water from the saline subsurface aquifer, the liquid is seeded and pH is lowered to recover the alumina. This alumina can be extracted and recovered at a competitive price with alumina from bauxite. Soda ash is recovered by evaporation and can be used for a variety of industrial applications, such as neutralizing agents. The leached spent shale is then returned to the process. The by-products that this process generates may make the process economically even more attractive.

### 8.5.1.6  Paraho Gas Combustion Process

The Paraho retort is a stationary, vertical, cylindrical, and refractory-lined kiln of mild steel developed by Paraho Development Corp. Raw shale enters at the top and is brought to the retorting temperature by a countercurrent flow of hot combustion gases. A schematic of the Paraho gas combustion retort system is shown in Figure 8.23. Shale is fed at the top along a rotating "pantsleg" distributor and moves downward through the retort. The rising stream of hot gas breaks down the kerogen to oil, gas, and residual char. The oil and gas are drawn off, and the residual char burns in the mixture of air and recycled gas. By injecting a part of the gas–air mixture through the bottom of the kiln, much of the sensible heat in the spent shale is recovered. The retort temperature is controlled by adjusting the compositions of the gas–air mixtures to the preheat and combustion zones.[55] Shale oil vapors flow upward and pass at a moderate temperature to an oil recovery unit. The end products are shale oil and low-Btu gas. A typical analysis of crude shale oil from this process is shown in Table 8.19. The shale oil produced can be upgraded to a crude feedstock.

The heavy naphtha cut (88°C–178°C) from the treated oil has a higher octane rating and lower sulfur than a comparable Arabian crude fraction. The diesel fraction (178°C–341°C) is identical to comparable fractions from other sources, so the heavy cut can be used as a feedstock for cracking units.[59]

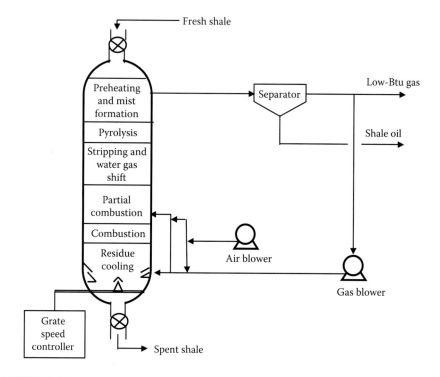

**FIGURE 8.23**    Paraho process for shale oil extraction.

**TABLE 8.19**
**Typical Analysis of Paraho Retort Shale Oil**

| | |
|---|---|
| API gravity (°API) | 19.70 |
| Nitrogen (wt%) | 2.18 |
| Conradson carbon (wt%) | 4.50 |
| Ash (wt%) | 0.06 |
| Sulfur (wt%) | 0.74 |
| Pour point (°C) | 26.0 |
| Viscosity, cp, 38°C | 256.0 |

Research and development on the Paraho retort, initiated by the company in August 1973, continued until April 1976, under the sponsorship of 17 energy and engineering companies. In 1978, Paraho delivered 100,000 bbl of raw shale oil to the U.S. Navy for defense testing purposes.[55]

The Paraho process can handle shale particle sizes of at least 3 in., keeping crushing and screening costs to a minimum, yet achieving a high conversion of better than 90% of Fischer assay. By burning the residual char and also recovering sensible heat from the spent shale, a high thermal efficiency can be achieved. The process is mechanically simple, requiring little auxiliary equipment. Also, no water is required for product cooling.

In 1987, Paraho reorganized as New Paraho Corp. at Boulder, Colorado, and began production of SOMAT[11] additives, which is used in test strips in five states in the United States. In 1991, New Paraho reported successful tests of SOMAT shale oil asphalt additives.

### 8.5.1.7 Petrosix Retorting Process

The Petrosix retorting process was developed in Brazil by Petrobras to use oil shale deposits in the Irati belt, which extends up to 1200 km.[59] The estimated deposit for this area is 630 billion barrels of oil, 10 million tons of sulfur, 45 million tons of liquefied gas, and 22 billion m³ of fuel gas.

This process has an external heater that raises the recycle gas temperature to approximately 700°C. The fuel for the heater can be gas, liquid, or solid. Shale is crushed in a two-stage stem that incorporates a fine rejection system. This retort has also three zones, that is, high, middle, and low.

Crushed shale is fed by desegregation feeders to the retort top and then it is forced downward by gravity, countercurrent to the hot gas flow. In the middle zone, hot recycle gas is fed at 700°C. Shale oil in mist form is discharged from the upper zone and is passed to a battery of cyclones and onto an electric precipitator to coalesce. The shale is then recovered, and some gas is recycled to the lower zone to adjust the retort temperature.[59] The remaining gas is treated in a light-ends recovery section sweetened (desulfurized) and discharged as liquefied petroleum gas (LPG). Figure 8.24 shows a schematic of the Petrosix retort process, and Table 8.20 shows some properties of shale oil produced by this process.[59]

The most up-to-date version of the Petrosix process has the primary characteristic of being easy to operate.[60] After the oil shale is taken out by open-cut mining, the shale goes to the grinder to reduce the particle size of the rocks (shales) that vary between 6 and 70 mm. These ground shale rocks (stones) are then transferred to a retort, where they are heated at a temperature of approximately 500°C, thereby releasing organic matter from the shale in the form of oil and gas.[60]

### 8.5.1.8 Chevron Retort System

A small pilot unit with a shale-feed capacity of 1 ton/day was developed by Chevron Research. This process used a catalyst and a fractionation system. The pilot operated on a staged, turbulent-flow bed process that reportedly used the shale completely. Figure 8.25 shows a schematic of this process, which is also called *shale oil hydrofining* process. The heart of this process is in the shale oil upgrading part rather than in the retorting part.

### 8.5.1.9 Moving Bed Retorting Process

A U.S. patent by Barcellos[61] describes a moving bed retorting process for obtaining oil, gas, sulfur, and other products from oil shale. The process comprises drying, pyrolysis, gasification, combustion, and cooling of pyrobituminous shale or similar rocks in a single passage of the shale continuously in a moving bed. The charge and discharge of oil shale are intermittent, and the maximum temperature of the bed is

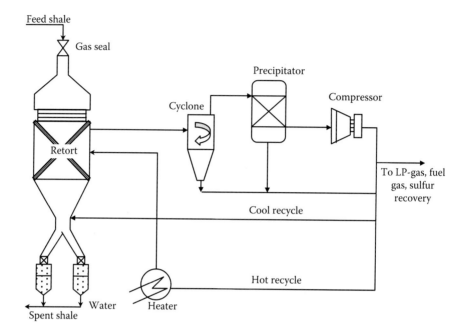

**FIGURE 8.24**   Petrosix retort system.

---

**TABLE 8.20**
**Typical Shale Oil Properties by Petrosix**

| | |
|---|---|
| Gravity (°API) | 19.60 |
| Sulfur (wt%) | 1.06 |
| Nitrogen (wt%) | 0.86 |
| Pour point (°F) | 25.0 |

*Source:*   Matar, S., *Synfuels: Hydrocarbons of the Future*,
Pennwell Publishing Co., Tulsa, OK, 1982.

---

maintained within a range of about 1050°C–1200°C or even higher. The temperature employed in this process is much higher than any other retorting processes discussed in this chapter. At this high temperature, the shale is essentially completely freed from the organic matter, fixed carbon, and sulfur resulting in a clean solid residue, which can be disposed of without harming the ecology, according to the inventor's claim. The advantage of this process is claimed to be in its retorting efficiency, as the process operates at high temperatures. However, a main concern may be in its energy efficiency and minimization of waste energy. No pilot plant research data have been published for this process. No large-scale demonstration has been done on this process.

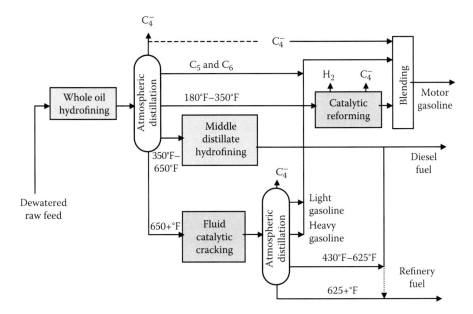

**FIGURE 8.25** Chevron shale oil hydrotreating system.

## 8.5.1.10 Carbon Dioxide Retorting Process

Lee and Joshi[62] developed a unique retorting process that differs from other retorting processes in its chemistry by use of a different sweep gas, that is, carbon dioxide. They claim that pyrolysis products of kerogen can be swept out far better in the carbon dioxide medium, and, therefore, the richest oil shales can be retorted with significantly higher yields than their Fischer assay values when carbon dioxide is used as a process gas. Oil shale kerogen exhibits softening and swelling in a carbon dioxide medium, which facilitates an enhanced recovery of shale oil by this process. The phenomenon of swelling and softening was observed to be quite significant even at low pressures and relatively low temperatures. If oil shale kerogen is subjected to supercritical carbon dioxide at typical retorting temperatures, oil shale kerogen macromolecules exhibit enhanced solubility and fluidity, thus making the subsequent pyrolytic decomposition reaction proceed in a fluid phase. In both cases, the end result is an enhanced recovery of shale oil. In particular, if supercritical carbon dioxide is used, the retorting process can be operated at a substantially lower temperature, thereby increasing the overall energy efficiency of the process. Other advantages of the processes include the suppression of dolomite and calcite decomposition reaction because of the higher partial pressure of $CO_2$ in the system. Suppression of dolomite decomposition (both half-calcination and calcination) represents an increased energy efficiency, by alleviating or avoiding wasteful endothermic reactions of mineral decomposition.

The *carbon dioxide retorting* process can be adopted in various retort designs with little or no system configuration change. It is also applicable to both *ex situ* and *in situ* processes. The preliminary experimental data show that the $CO_2$ retorting process substantially enhances the oil yields from Colorado and Australian shales over the conventional process (Fischer assay) up to 75%. However, if the $CO_2$ retorting process is applied to a lean, low-permeability shale, like Ohio Devonian shale, the swollen kerogen blocks the narrow porepaths, resulting in poorer oil yields than the case with nitrogen as sweep gas as in Fischer assay. This process is scientifically interesting in the sense that kerogen swells and softens in both subcritical and supercritical $CO_2$ medium. Since this patented process was developed in early 1980s when the oil shale interest had cooled off, no large-scale demonstration has been done on this process concept.

### 8.5.2 IN SITU RETORTING PROCESSES

Oil shale retorting can also be achieved underground, that is, without mining the shale. Such a process is called an *in situ oil shale retorting process* or *subsurface retorting*. In a typical *in situ* process, the shale is fractured by either explosives or hydrostatic pressure. A portion of the oil shale organic matter is then burned to obtain heat necessary for retorting. The retorted shale oil is pumped out of the production zone in a manner similar to the extraction of crude petroleum.

The advantages of *in situ* processing include the following:

1. Oil can be recovered from deep deposits of oil shale formation.
2. Mining costs can be eliminated or minimized.
3. There is no solid waste disposal problem, as all operations are conducted through well bores. Therefore, the process may be environmentally more desirable, as long as mineral leaching or harmful side effects of the processed shales are absent or controlled.
4. Shale oil can be extracted from leaner shale, for example, deposits containing <15 gal/ton of oil.
5. The process is ultimately more economical owing to elimination or reduction of mining, transportation, and crushing costs.

However, the disadvantages of *in situ* processing are as follows:

1. It is difficult to control subsurface combustion because of insufficient permeability within the shale formation.
2. Drilling cost is still high.
3. Recovery efficiencies are generally low.
4. It is difficult to establish the required permeability and porosity in the shale formation.
5. There is a concern for possible contamination of aquifers. If not controlled or treated, effects may linger for an extended period of time even after the project completion. Tests and control may require extensive efforts.

The *in situ* technology for production of shale oil from shales, in general, optimizes recovery process economics while minimizing environmental impact. This is why considerable emphasis has also been placed on these processes.

*In situ* retorting processes can be roughly classified into two types, that is, MIS and TIS. MIS retorting, the brainchild of Occidental Petroleum, involves partial mining of the oil shale deposit to create a void space and rubblizing the rest into this space so as to increase the overall permeability of the shale. The underground rubblized shale is then ignited using an external or internal fuel source.[63] TIS retorting is similar to MIS, but no mining is done in this process. The shale deposits are rubblized to increase the permeability, and then the underground burning is begun.

A review of the oil shale literature indicates that all *in situ* oil shale processes can be classified into the following categories:[64]

1. Subsurface chimney
   A. Hot gases
   B. Hot fluids
   C. Chemical extraction
2. Natural fractures
   A. Unmodified
   B. Enlargements by leaching
3. Physical induction
   A. No subsurface voids

Other ways of classifying the *in situ* oil shale retorting processes are as follows:

1. Formation of retort cavities
   A. Horizontal sill pillar
   B. Columnar voids
   C. Slot-shaped columnar voids
   D. Multiple zone design
   E. Multiple horizontal units
   F. Multiple adjacent production zones
   G. Multiple gallery-type retort zones
   H. Spaced-apart upright retort chambers
   I. Permeability control of rubble pile
   J. Formation of rich and lean zones
   K. Successive rubblization and combustion
   L. Thermomechanical fracturing
   M. Water leaching and explosive fracturing
   N. Inlet gas means
   O. Fluid communication
   P. Cementation to minimize plastic flow
   Q. Near-surface cavity preparation
   R. Dielectric heating
2. Retorting techniques
   A. Ignition techniques
   B. Multistage operation
   C. Steam leaching and combustion
   D. Pressure swing recovery
   E. Multistratum reservoir
   F. Production well throttling

    G. Combined combustion techniques
    H. Laser retorting
    I. Low-heat fans for frontal advance units
    J. Gas introduction and blockage
    K. Water injection
    L. Oil collection system
    M. Handling system for feed and products
    N. Uniform gas flow
    O. Postretorting flow
    P. Sound monitoring
    Q. Underground weir separator
    R. Emulsion breaking technique
    S. Offgas recycling
    T. Prevention of offgas leakage
  3. Others
    A. Molecular sulfur and benzene recovery
    B. Hydrogen sulfide and carbon dioxide capture
    C. Hot-fluid injection into solvent-leached shale
    D. Steam treatment and extended soak period
    E. Steam-driven excavating unit
    F. Anaerobic microorganisms
    G. Hot aqueous alkaline liquids and fluid circulation
    H. Plasma arc

In the second half of the twentieth century, extensive research and development efforts were devoted to the commercialization of the *in situ* pyrolysis of oil shale. However, most of these efforts were either shelved, halted, or scaled down in the late twentieth century because of the unfavorable process economics in the short term. As mentioned earlier in the chapter, the comparative economics of shale oil has become substantially more favorable in the twenty-first century. Several noteworthy *in situ* retorting processes are described in the following sections.

### 8.5.2.1 Sinclair Oil and Gas Company Process

In 1953, Sinclair Oil and Gas Co. performed one of the earliest experiments on *in situ* oil shale retorting. Their process concept was similar to that shown in Figure 8.26. Their study found that (1) communication between wells could be established through induced or natural fracture systems, (2) wells could be ignited successfully, and (3) combustion could be established and maintained in the oil shale bed. They also realized that high pressures were required to maintain injection rates during the heating period. These tests were conducted near the outcrop in the southern part of the Piceance Creek Basin.[65] Additional tests were done several years later at a depth of about 365 m in the north-central part of the Piceance Creek Basin with some limited success, which was believed to be due to the inability to obtain the required surface area for the heat transfer. However, their experiments established the basic technology required for *in situ* retorting of oil shale and suggested further study areas.

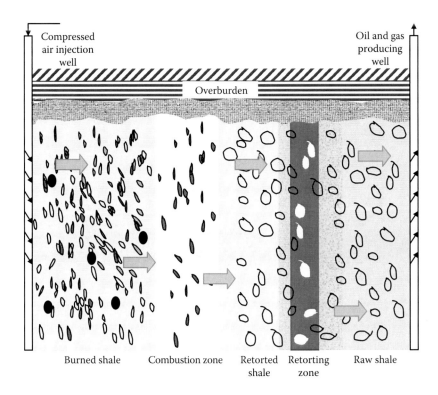

**FIGURE 8.26** *In situ* oil shale retorting process.

### 8.5.2.2 Equity Oil Co. Process

The Equity Oil Co. of Salt Lake City[66] studied an *in situ* process that is somewhat different from the Sinclair process. This process involves injecting hot natural gas into the shale bed to retort the shale. One injecting well and four producing wells were drilled into the oil shale formation in an area of the Piceance Creek Basin. The natural gas was compressed to about 85 atm, heated to approximately 480°C, and delivered through insulated tubing to the retorting zone. Based on the experimental results and a mathematical model developed from them, it was concluded that this technique was feasible and potentially an economically viable method for extracting shale oil. However, the process economics is undoubtedly strongly dependent on the cost of natural gas and the amount required for makeup of natural gas.

### 8.5.2.3 Occidental Petroleum Process

Occidental Petroleum developed a MIS process in which conventional explosives are used to expand solid blocks of shale into a vertical mined-out cavity, creating underground chimneys of fractured shale. Figure 8.27 shows a schematic of this retort design. To improve the fluid communication, about 10% to 25% of the shale in the chimney is removed. Air is then blown down through the remaining crushed shale, and the top ignited with a burner that can be fueled with shale oil or offgas from other retorts. On ignition, the burner is withdrawn and the steam is mixed with

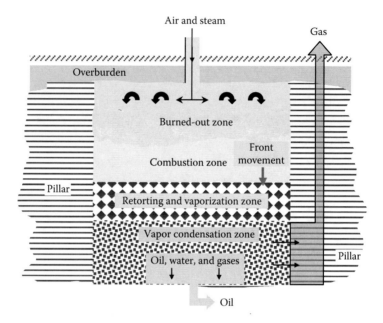

**FIGURE 8.27**    MIS retort of Occidental Petroleum.

the inlet air to control the process.[55] The liquid and gaseous products flow to the bottom of the chimney, leaving the char in the shale behind as the main source of fuel for the slowly advancing flame front.

Occidental Petroleum did a series of field tests on this process at Logan's Wash in Debeque, Colorado, with three retorts—30 ft. across and 72 ft. deep—each containing 6000 to 10,000 tons of oil shale. Based on their experimental success, full-scale production of 57,000 bbl/day was very seriously considered.

### 8.5.2.4  LETC Process (LERC Process)

Laramie Energy Technology Center (LETC), currently Western Research Institute, has been sponsoring several field projects to demonstrate the technical and economic feasibility of shale oil recovery by *in situ* technology.

LETC initiated their study on *in situ* retorting in the early 1960s with laboratory tests, simulated pilot plant tests on 10 and 150 ton retorts, and field tests at Rock Spring, Wyoming. The test results demonstrated that it was possible to move a self-sustaining combustion zone through an oil shale formation and to produce shale oil.

The underground shale bed is prepared for the LETC process by first boring injection and production wells into the shale, and then increasing the permeability of formation by conventional fracturing techniques. Based on the LETC tests, the sequential use of hydraulic fracturing and explosives worked best. Once the formation is fractured, hot gases are forced into it to heat the area surrounding the injection point. As the desired temperature is reached and air is substituted for the hot gas, combustion begins and becomes self-sustaining across a front that gradually moves through the bed. As retorting progresses, oil and gas products are pumped out through the pre-drilled production wells. A schematic[67] for the LETC process is shown in Figure 8.28.

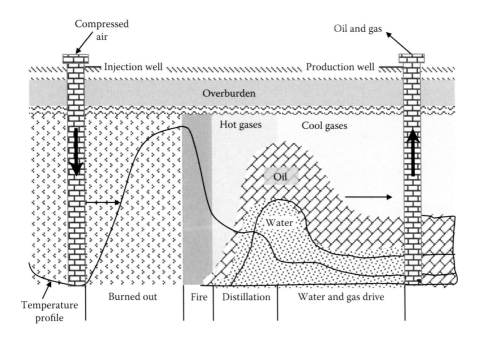

**FIGURE 8.28** Laramie Energy Technology Center (LETC) *in situ* retorting process. (From Cook, W.E., *Q. Colo. Sch. Mines*, 65(4), 133, 1970.)

### 8.5.2.5 Dow Chemical Co.'s Process

The Dow Chemical Co., under contract with the U.S. DOE, conducted a 4-year research program to test the feasibility of deep *in situ* recovery of low-heat content gas from Michigan Antrim Shale.[68]

The Antrim shale is part of the eastern and midwestern oil shale deposits, formed some 260 million years ago during the Devonian and Mississippian ages. These oil shales underlie an area of 400,000 mi.² ($1.07 \times 10^{12}$). In Michigan, the oil shale is approximately 61 m thick and is in a basin at depths ranging from about 0.8 km to outcroppings in three northern counties. The Michigan Antrim shale is believed to contain an equivalent hydrocarbon volume of 2500 million bbl. Even applying a 10% recovery factor, this resource is about nine times the amount of the U.S.-proven oil reserves.

Extensive fracturing (rubblizing) of the oil shale is considered essential for adequate *in situ* retorting and recovery of energy from the Antrim shale. Two wells were explosively fractured using 19,000 kg of metalized ammonium nitrate slurry. Their test facility was located 75 mi. northeast of Detroit, Michigan, over 1 acre of field. The process used was TIS retorting.

Combustion of the shale was started using a 440 V electric heater (52 kW) and a propane burner (250,000 Btu/h). The special features of this process include shale gasification and tolerance to severe operating conditions. Their tests also showed that explosive fracturing in mechanically underreamed wells did not produce extensive rubblization. They also tested hydrofracturing, chemical underreaming, and explosive underreaming.

### 8.5.2.6 Talley Energy Systems Process

Talley Energy Systems Inc. carried out a U.S. DOE–Industry Cooperative oil shale project at 11 mi. west of Rock Springs, Wyoming. The shale in this area is part of the Green River shale formation, which is about 50 million years old. This process is also based on TIS processing and uses explosive fracturing, additional hydraulic fracturing, and no mining.

### 8.5.2.7 Geokinetics Process

Geokinetics Inc. developed an *in situ* process that may be best described by modified horizontal technology. Explosive fracturing is used, and the process can be used even for the shallow thin-seam recovery. This was one of the U.S. DOE–Industry Cooperative oil shale projects and was located 61 mi. northwest of Grand Junction, Colorado. The geological formation used for this process study was Green River formation, Parachute Creek member, and Mahogany zone, which was the same as for the Occidental Oil process. The only difference between the two is that the Geokinetics process is *modified horizontal*, whereas the Occident oil process is *modified vertical*. Therefore, the Geokinetics process is good for shallow thin-seam recovery, whereas the Occidental Oil process is better suited for deep thick-seam recovery.

### 8.5.2.8 Osborne's *In Situ* Process

This process was developed by Osborne in 1983; a U.S. patent[69] describing the process has been assigned to Synfuel (an Indiana limited partnership). The process is unique, and enhanced oil recovery is achieved by forming generally horizontal electrodes from the injection of molten metal into preheated or unheated fractures of formation. A nonconductive spacing material is positioned in the casing of the borehole between the electrodes. A fracture horizontally intermediate between the metallic electrodes is propped up with a nonconductive granular material. Unterminated standing waves from a radio frequency (RF) generator are passed between the electrodes to heat the oil shale formation. The hydrocarbons in the formation are vaporized and recovered at the surface by their transport through the intermediate fracture and tubing. By this method, radial metallic electrodes can be formed at various depths throughout a subterranean oil shale formation to devolatilize the hydrocarbons contained within the oil shale formation.[69]

One advantage of this process is in the uniform heating of the rock formation that can be achieved by using RF electrical energy that corresponds to the dielectric absorption characteristics of the rock formation. An example of such techniques is described in U.S. Patent numbers 4,140,180 and 4,144,935, in which many vertical conductors are inserted into the rock formation and bound to a particular volume of the formation. A frequency of electrical excitation is selected to attain a relatively uniform heating of the rock formation. The energy efficiency of the process is very good; however, the economics of the process strongly depends on the cost of the electrodes and RF generation. The other merits of the process include the relative ease of controlling the retort size.

The difficulty, however, with this process is in the necessity of implanting an electrode within the subterranean rock formation at a precise distance. A schematic of this process is shown in Figure 8.29.

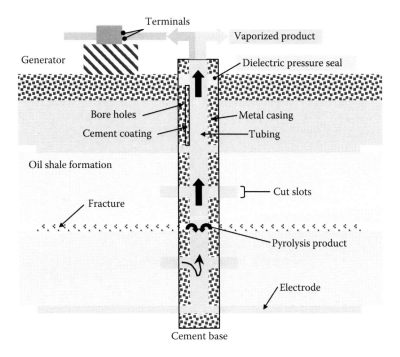

**FIGURE 8.29** Vertical sectional view of a borehole penetrating a subterranean oil shale formation in completed condition for the recovery of hydrocarbons from oil shale—The Osborne process. (From Osborne, J., U.S. Patent No. 4,401,162, August 30, 1983.)

### 8.5.2.9 Shell Oil's Thermally Conductive *In Situ* Conversion Process

Shell Oil is currently developing an *in situ* retorting process known as *thermally conductive in situ conversion*.[70] This process involves *in situ* heating of underground oil shale using electric heaters placed in deep vertical holes drilled through a section of oil shale. The entire volume of oil shale is heated over a period of 2 to 3 years until it reaches 650°F to 700°F, at which point oil is released from the shale. The released product is gathered in collection wells positioned within the heated zone. Shell's current plan also involves use of ground-freezing technology to establish an underground barrier called a *freeze wall* around the perimeter of the extraction zone. The freeze wall is created by pumping refrigerated fluid through a series of wells drilled around the extraction zone. The freeze wall prevents groundwater from entering the extraction zone and keeps hydrocarbons and other products generated by the *in situ* retorting from leaving the project perimeter and contaminating the surrounding soil.

In 1997, Shell Oil successfully conducted small-scale field tests of this novel *in situ* process based on slow underground heating via thermal conduction, on Mahogany property. After deferring further tests because of economic reasons, Shell returned to Mahogany for further tests in 2000, and the R&D program is currently in operation. Larger-scale operations need to be conducted to establish technical viability, especially with regard to eliminating or alleviating any adverse impacts on groundwater quality.[70] The process has a number of merits that can contribute to lowering the processing cost of oil shale, as well as toward environmentally benign processing of this vast energy resource. The Shell Oil process is technologically classified as one of the TIS retorting processes, as there is no mining of shale involved.

### 8.5.2.10   TIS and MIS Retorting

*In situ* retorting of oil shale is often classified into TIS and MIS cases, as briefly mentioned earlier. In this section, these two terms are further clarified.

TIS retorting involves drilling wells and fracturing oil shale rock to increase its permeability. It, however, does not involve any mining of oil shale. Typically, a hot gas mixture is used to heat the oil shale rubble. Forced air then helps burn the oil shale. A flame front is formed and gradually moves through the bed, and the produced oil and gas are drawn through the production wells to the surface. As mentioned earlier, the Shell *in situ* process does not use a hot-gas heating technique to initiate retorting but is classified as TIS process.

In TIS modified underground retorting, a blocked-out area is mined to remove approximately 10% to 25% of the oil shale. Vertical or horizontal wells are drilled through the remaining portion and are detonated. The produced voids help fracture and rubblize the oil shale. This is a modification of the TIS conversion process and was first developed by Occidental Oil.

### 8.5.3   SHALE OIL REFINING AND UPGRADING

As the demand for light hydrocarbon fractions constantly increases, there is much interest in developing economical methods for recovering liquid hydrocarbons from oil shale on a commercial scale. However, the recovered hydrocarbons from oil shale are not yet economically competitive against the petroleum crude produced. Furthermore, the value of hydrocarbons recovered from oil shale is diminished because of the presence of undesirable contaminants. The major contaminants are sulfurous, nitrogenous, and metallic (and organometallic) compounds, which cause detrimental effects to various catalysts used in the subsequent refining processes. These contaminants are also undesirable because of their disagreeable odor, corrosive characteristics, and combustion products that further cause environmental problems.

Accordingly, there is great interest in developing more efficient methods for converting the heavier hydrocarbon fractions obtained in a form of shale oil into lighter-molecular-weight hydrocarbons. The conventional processes include catalytic cracking,[8,71] thermal cracking,[72] coking, etc.

It is known that heavier hydrocarbon fractions and refractory materials can be converted to lighter materials by hydrocracking. These processes are most commonly used on liquefied coals or heavy residual or distillate oils for the production of substantial yields of low-boiling saturated products, and to some extent on intermediates that are used as domestic fuels, and still heavier cuts that are used as lubricants. These destructive hydrogenation or hydrocracking processes may be operated on a strictly thermal basis or in the presence of a catalyst. Thermodynamically speaking, larger hydrocarbon molecules are broken into lighter species when subjected to heat. The H-to-C ratio of such molecules is lower than that of saturated hydrocarbons, and abundantly supplied hydrogen improves this ratio by saturating reactions, thus producing liquid species. These two steps may occur simultaneously.

However, the application of the hydrocracking process has been hampered by the presence of certain contaminants in such hydrocarbons. The presence of sulfur- and nitrogen-containing compounds along with organometallics in crude shale oils and various refined petroleum products has long been considered undesirable. Desulfurization and denitrification processes have been developed for this purpose.

### 8.5.3.1   Thermal Cracking Process

Gulf Research and Development[73,74] developed a process for the noncatalytic thermal cracking of shale oil in the presence of a gaseous diluent and an entrained stream of inert heat carrier solids. The cracking process is directed toward the recovery of gaseous olefins as the primarily desired cracked product, in preference to gasoline-range liquids. By this process, it is claimed that at least 15% to 20% of the feed shale oil is converted to ethylene, which is the most common gaseous product. Most of the feed shale oil is converted to other gaseous and liquid products. Other important gaseous products are propylene, 1,3-butadiene, ethane, and other $C_4$'s. Hydrogen is also recovered as a valuable nonhydrocarbon gaseous product. Liquid products can comprise 40 to 50 wt% or more of the total product. Recovered liquid products include benzene, toluene, xylene, gasoline-boiling-range liquids, and light and heavy oils.

Coke is a solid product of the process and is produced by polymerization of unsaturated materials. Coke is typically formed in an oxygen-deficient environment via dehydrogenation and aromatization.[72] Most of the formed coke is removed from the process as a deposit on the entrained inert heat carrier solids.

The thermal cracking reactor does not require a gaseous hydrogen feed. In the reactor, entrained solids flow concurrently through the thermal riser at an average riser temperature of 700°C to 1400°C. The preferred high L-to-D ratio is in the range of a high 4:1 to 40:1, or 5:1 to 20:1, preferably.

### 8.5.3.2   Moving Bed Hydroprocessing Reactor

This process was developed by Universal Oil Products Co.[75] for deriving crude oil from oil shale or tar sands containing large amounts of highly abrasive PM, such as rock dust and ash. The hydroprocessing takes place in a dual-function moving bed reactor, which simultaneously removes PM by the filter action of the catalyst bed. The effluent from the moving bed reactor is then separated and further hydroprocessed in fixed bed reactors with fresh hydrogen added to the heavier hydrocarbon fraction to promote desulfurization.

A preferred way of treating the shale oil involves using a moving bed reactor followed by a fractionation step to divide the wide-boiling-range crude oil produced from the shale oil into two separate fractions. The lighter fraction is hydrotreated for the removal of residual metals, sulfur, and nitrogen, whereas the heavier fraction is cracked in a second fixed bed reactor normally operated under high-severity conditions.

Hydrotreating (HDT) reactions are generally carried out at high pressures (100–3000 psi) and high temperatures (270°C–350°C). During hydrotreating, the following reactions take place: hydrodesulfurization,[76] hydrodenitrogenation, hydrodemetallization, hydrodeoxygenation, and hydrogenation.[8]

### 8.5.3.3 Fluidized Bed Hydroretort Process

This process was developed by Cities Service Co.[77] in 1978. The process eliminates the retorting stage of conventional shale upgrading by directly subjecting crushed oil shale to a hydroretorting treatment in an upflow, fluidized bed reactor such as that used for the hydrocracking of heavy petroleum residues. This process is a *single-stage retorting and upgrading* process. Therefore, the process involves (1) crushing oil shale, (2) mixing the crushed oil shale with a hydrocarbon liquid to provide a pumpable slurry, (3) introducing the slurry along with a hydrogen-containing gas into an upflow, fluidized bed reactor at a superficial fluid velocity sufficient to move the mixture upwardly through the reactor, (4) hydroretorting the oil shale, (5) removing the reaction mixture from the reactor, and (6) separating the reactor effluent into several components.[78]

The mineral carbonate decomposition is minimized, as the process operating temperature is lower than that used in retorting. Therefore, the gaseous product of this process has a greater heating value than that of other conventional methods. In addition, owing to the exothermic nature of the hydroretorting reactions, less energy input is required per barrel of product obtained. Furthermore, there is practically no upper or lower limit on the grade of oil shale that can be treated.

### 8.5.3.4 Hydrocracking Process

Hydrocracking is essentially a cracking process in which higher-molecular-weight hydrocarbons pyrolyze to lower-molecular-weight paraffins and olefins in the presence of hydrogen.[71] The hydrogen saturates the olefins formed during the cracking process. Hydrocracking is used to process low-value stocks with a high heavy metal content. It is also suitable for highly aromatic feeds that cannot be processed easily by conventional catalytic cracking. Shale oils are not highly aromatic, whereas coal liquids are very highly aromatic.

Middle-distillate (often called *mid-distillate*) hydrocracking is carried out with a noble metal catalyst. The average reactor temperature is 480°C, and the average pressure is around 130 to 140 atm. The most common form of hydrocracking is carried out as a two-stage operation.[71] The first stage is to remove nitrogen compounds and heavy aromatics from the raw crude, whereas the second stage is to carry out selective hydrocracking reactions on the cleaner oil from the first stage. Both stages are processed catalytically. Once the hydrocracking stages are over, the products go to a distillation section that consists of a hydrogen sulfide stripper and a recycle splitter. Commercial hydrocracking processes include Gulf HDS, H-Oil, IFP Hydrocracking, Isocracking, LC-Fining, Microcat-RC (also known as M-Coke), Mild Hydrocracking, Mild Resid Hydrocracking (MRH), Residfining, Unicracking, and Veba Combi-Cracking (VCC).[71]

## REFERENCES

1. Laherrere, J., Review on oil shale data, September 2005, http://www.oilcrisis.com/laherrere/OilShaleReview200509.pdf.
2. Energy Minerals Division, American Association of Petroleum Geologists, http://emd.aapg.org/technical_areas/oil_shale.cfm, accessed 2009.

3. World Energy Council, Survey of Energy Resources, http://www.worldenergy.org/wec-geis/publications/reports/ser/shale/shale.asp, accessed 2007.
4. Qian, J., Wang, J. and Li, S., Oil shale development in China, *Oil Shale*, 20(e Special), 356–359, 2003.
5. Taylor, R.B., Oil shale commercialization: The risks and the potential, *Chemical Engineering*, 59, 1981.
6. Thumann, A., *The Emerging Synthetic Fuel Industry*, Fairmont Press, Lilburn, GA, 1981.
7. Lee, S., *Oil Shale Technology*, CRC Press, Boca Raton, FL, 1991.
8. Kundu, A., Dwivedi, N., Singh, A. and Nigam, K.D.P., Hydrotreating catalysts and processes—current status and path forward, in *Encyclopedia of Chemical Processing (EChP)*, Vol. 2, Lee, S., Ed., Taylor & Francis, New York, 2006, pp. 1357–1366.
9. Tisot, P.R. and Murphy, W.I.R., *Chem. Eng. Prog. Symp. Ser.*, 61(54), 25, 1965.
10. Tisot, P.R., *J. Chem. Eng. Data*, 12(3), 405, 1967.
11. Lukens, L.A., Asphalt rejuvenater and recycled asphalt composition, U.S. Patent No. 5,755,865, May 1998.
12. Chilingarian, G.V. and Yen, T.F., *Bitumens, Asphalts, and Tar Sands*, Elsevier, Amsterdam, 1978, Chapter 1.
13. Nottenburg, R., Rajeshwar, K., Rosenvold, R. and Dubow, J., *Fuel*, 58, 144, 1979.
14. Hill, J.O., Thermogravimetric analysis, in *Encyclopedia of Chemical Processing (EChP)*, Vol. 5, Lee, S., Ed., Taylor & Francis, New York, 2005, pp. 3017–3029.
15. Nottenburg, R., Rajeshwar, K., Rosenvold, R. and Dubow, J., *Fuel*, 57, 789, 1978.
16. Tihen, S.S., Carpenter, H.C. and Sohns, H.W., Thermal conductivity and thermal diffusivity of Green River oil shale, Conf. Thermal Conductivity Proc. 7th, NBS Special Publ. 302, p. 529, September 1968.
17. Barnes, A.L. and Ellington, R.T., *Q. Colo. Sch. Mines*, 63(4), 827, 1968.
18. Dubow, J., Nottenburg, R., Rajeshwar, K. and Wang, Y., The effects of moisture and organic content on the thermophysical properties of Green River oil shale, *11th Oil Shale Symposium Proc.*, Colorado School of Mines, Golden, CO, 1978, p. 350.
19. Sladek, T., Ph.D. thesis, A determination of the composition and temperature dependencies of thermal conductivity factors for Green River oil shale, Colorado School of Mines, Golden, CO, 1970.
20. Rajeshwar, K., Nottenburg, R. and Dubow, J., *J. Mater. Sci.*, 14, 2025–2052, 1979.
21. Smith, J.W., *U.S. Bur. Mines Rep. Invest.*, 7248, 1969.
22. Prats, M. and O'Brien, S.M., *J. Pet. Technol.*, 97, 1975.
23. Skrynnikova, G.N., Avdonina, E.S., Golyand, M.M. and Akhmedova, L. Ya., Trudy Vsesoyuz Nauch-Issledovatel Inst. *Pererab. Slants.*, 7, 80, 1959.
24. Wang, Y., Dubow, J., Rajeshwar, K. and Nottenburg, R., *Thermochim. Acta*, 28, 23, 1979.
25. McKee, R.H. and Lyder, E.E., *J. Ind. Eng. Chem.*, 13, 613, 1921.
26. Wang, Y., Rajeshwar, K., Rosenvold, R. and Dubow, J., *Thermochim. Acta*, 30, 141, 1979.
27. Shih, S.-M. and Sohn, H.Y., *Fuel*, 57, 662, 1978.
28. Wise, R.L., Miller, R.C. and Sohns, H.W., *U.S. Bur. Mines Rep. Invest.*, 7482, 1971.
29. Johnson, W.F., Walton, D.K., Keller, H.H. and Couch, E.J., *Q. Colo. Sch. Mines*, 70(3), 237, 1975.
30. Shaw, R.J., *U.S. Bur. Mines Rep. Invest.*, 4151, 1947.
31. Sohns, H.W., Mitchell, L.E., Cox, R.J., Burnet, W.I. and Murphy, W.I.R., *Ind. Eng. Chem.*, 43, 33, 1951.
32. Cook, W.E., *Q. Colo. Sch. Mines*, 65(4), 133, 1970.
33. Johnson, P.R., Young, N.B. and Robb, W.A., *Fuel*, 54, 249, 1975.
34. Allred, V.D., *Q. Colo. Sch. Mines*, 59(3), 47, 1964.

35. Branch, M.C., *Prog. Energy Combust. Sci.*, 5, 193, 1979.
36. Joshi, R., M.S. thesis, A comparative study between the kinetics of retorting of Ohio and Colorado shale, University of Akron, Akron, OH, 1983.
37. Kesavan, S.K. and Lee, S., *Fuel Sci. Technol. Int.*, 6(5), 505, 1988.
38. Johnson, D.R., Young, N.B. and Smith, J.W., LERC/RI, Laramie Energy Research Center, Laramie, Wyoming, June 1977.
39. Dyni, J.R., Mountjoy, W., Hauff, P.L. and Blackman, P.D., U.S. Geological Survey, Professional Paper No. 750B, 1971.
40. Hill, J.O., Thermal analysis techniques, in *Encyclopedia of Chemical Processing (EChP)*, Vol. 5, Lee, S., Ed., Taylor & Francis, New York, 2005, pp. 2965–2974.
41. Loughman, F.C. and See, G.T., *Am. Miner.*, 52, 1216, 1967.
42. Huggins, C.W. and Green, T.E., *Am. Miner.*, 58, 548, 1973.
43. Agroskin, A.A. and Petrenko, I.G., *Zavodskaya Lab.*, 14, 807, 1948.
44. Scott, J.H., Carroll, R.D. and Cunningham, D.R., *J. Geophys. Res.*, 72, 5101, 1967.
45. Nottenburg, R., Rajeshwar, K., Freeman, M. and Dubow, J., *Thermochim. Acta*, 31, 39, 1979.
46. Rajeshwar, K., Nottenburg, R., Dubow, J. and Rosenvold, R., *Thermochim. Acta*, 27, 357, 1978.
47. Yen, Y.F., Structural investigations on Green River oil shale, in *Science and Technology of Oil Shale*, Yen, Y.F., Ed., Ann Arbor Publishers, Ann Arbor, MI, 1976.
48. Wang, Y., M.S. thesis, A single particle model for pyrolysis of oil shale, University of Akron, Akron, OH, 1982.
49. Cook, E.W., *Fuel*, 53, 16, 1976.
50. Poulson, R.E., Jensen, H.B. and Cook, G.L., *ACS Div. Pet. Chem. Prepr.*, 15(1), A49–A55, 1971.
51. Eggertsen, F.T., Groennings, S. and Holst, J.J., *Anal. Chem.*, 32(8), 904, 1960.
52. ASTM Standards D2887-06, Standard Test Method for Boiling Range Distribution of Petroleum Fractions of Gas Chromatography, ASTM International, 2006.
53. Matzick, A., Dannenburg, R.O., Ruark, J.R., Phillips, J.E., Lankford, J.D. and Guthrie, B., *U.S. Bur. Mines Bull.*, 635, 99, 1966.
54. Whitcombe, J.A. and Vawter, R.G., The TOSCO-II oil shale process, in *Science and Technology of Oil Shale*, Yen, Y.F., Ed., Ann Arbor Science, Ann Arbor, MI, 1976, Chapter 4.
55. National Research Council (U.S.), Panel on R&D Needs in Refining of Coal and Shale Liquids, *Refining Synthetic Liquids from Coal and Shale*, National Academy Press, Washington, D.C., 1980, Chapter 5, pp. 78–135.
56. Kennedy, C.R. and Krambeck, F.J., Surge bin retorting solid feed material, U.S. Patent No. 4,481,100, 1984.
57. Rammler, R.W., *Q. Colo. Sch. Mines*, 65(4), 141–168, 1970.
58. Reeves, T.L. and Elgezawi, S.M., Time domain reflectometry for measuring volumetric water content in processed oil shale waste, *Water Resour. Res.*, 28(3), 769–776, 1992.
59. Matar, S., *Synfuels: Hydrocarbons of the Future*, Pennwell Publishing Co., Tulsa, OK, 1982.
60. Petrobras Web site about Petrosix process, 2006, http://www2.petrobras.com.br/minisite/refinarias/ingles/six/conheca/ProcPetrosix.html.
61. Barcellos, E.D., U.S. Patent No. 4,060,479, November 29, 1977.
62. Lee, S. and Joshi, R., U.S. Patent No. 4,502,942, March 5, 1985.
63. Gregg, M.L., Campbell, J.H. and Taylor, J.R., *Fuel*, 60, 179, 1981.
64. Yen, T.F., Oil shales of United States: A review, in *Science and Technology of Oil Shale*, Yen, T.F., Ed., Ann Arbor Science, Ann Arbor, MI, 1976, pp. 1–17.
65. Dinneen, G.U., Retorting technology of oil shale, in *Oil Shale*, Yen, T.F. and Chilingarran, G.V., Eds., Elsevier, Amsterdam, Netherlands, 1976, Chapter 9, pp. 181–197.

66. Dougan, P.M., Reynolds, F.S. and Root, P.J., The potential for in situ retorting of oil shale in the Piceance Creek Basin of northwestern Colorado, *Q. Colo. Sch. Mines*, 65(4), 57–72, 1970.
67. Sladek, T.A., Recent trends in oil shale—Part 2: Mining and shale oil extraction processes, *Colo. Sch. Mines, Miner. Ind. Bull.*, 18(1), 1–20, 1975.
68. McNamara, P.H. and Humphrey, J.P., Hydrocarbons from eastern oil shale, *Chemical Engineering Progress*, 75, 88, 1979.
69. Osborne, J., U.S. Patent No. 4,401,162, August 30, 1983.
70. Bartis, J.T., LaTourrette, T., Dixon, L., Peterson, D.J. and Cecchine, G., *Oil Shale Development in the United States: Prospects and Policy Issues*, MG-414-NETL, report prepared for National Energy Technology Laboratory, U.S. Department of Energy, part of RAND Corporation, *Monograph Series*, 2005.
71. Speight, J.G., Hydrocracking, in *Encyclopedia of Chemical Processing (EChP)*, Vol. 2, Lee, S., Ed., Taylor & Francis, New York, 2005, pp. 1281–1288.
72. Tsai, T.C. and Albright, L.F., Thermal cracking of hydrocarbons, in *Encyclopedia of Chemical Processing (EChP)*, Vol. 5, Lee, S., Ed., Taylor & Francis, New York, 2006, pp. 2975–2986.
73. Wynne, F.E., Jr., U.S. Patent No. 4,057,490, November 8, 1977.
74. McKinney, J.D., Sebulsky, R.T. and Wynne, F.E., Jr., U.S. Patent No. 4,080,285, March 21, 1978.
75. Anderson, R.F., U.S. Patent No. 3,910,834, October 7, 1975.
76. Song, C. and Turaga, U.T., Desulfurization, in *Encyclopedia of Chemical Processing (EChP)*, Vol. 1, Lee, S., Ed., 2005, pp. 651–661.
77. Gregoli, A.A., U.S. Patent No. 4,075,081, February 21, 1978.
78. Ranney, M.W., *Oil Shale and Tar Sands Technology-Recent Developments*, Noyes Data Corporation, NJ, 1979, p. 238.

# 9 Shale Gas and Shale Fuel

*Sunggyu Lee, Amber Tupper, Barbara Wheelden,*
*Ryan Tschannen, Aaron Gonzales, and*
*Maxwell Tobias Tupper*

## CONTENTS

## 9.1  WHAT IS SHALE GAS?

Shale gas is natural gas extracted from oil shale rock formations, or plays, located deep below the surface of the earth. The amount or magnitude of shale gas reserves available under the earth's surface is very vast and extremely difficult to estimate. It has long been known that natural shale gas and shale oil reserves were deposited and trapped in hard dense shale rocks formed from ancient ocean basins several hundred million years ago. Until recently, the oil shale resources that were economically useful, technologically feasible, and environmentally manageable had been limited to the oil shale deposits that are shallow and more easily approachable from the earth's surface. In other words, we did not have all the enabling technologies and machineries to be able to tap the resources that are trapped very deep under the terrain surface. Initially, most efforts in commercial oil shale utilization were focused on shale natural gas production. The popularly used terminology "shale gas" is reflecting this trend and historic event. However, increasingly more efforts have been spent in oil-focused drilling for production of shale crude oil, as well evidenced in North Dakota, United States. To distinguish the crude oil product extracted from oil shale deposits, people use the terminology "shale fuel," which includes both gas and liquid products, or "shale oil crude," which is only for the liquid product.

Even though major commercial exploitations of shale gas and shale fuel have been publicized as a "shale gas boom" for the United States, the oil shale deposits are distributed all over the world. Therefore, global interest in oil shale commercial development is currently very active. It is believed to be one of the most viable game changers in the energy independence and sustainability, not only for the United States but also for many nations in the world.

## 9.2 BACKGROUND INFORMATION

### 9.2.1 DISTRIBUTION OF SHALE GAS AND SHALE OIL DEPOSITS

Exploration of shale gas deposits deep under the earth's surface is not a simple task. Techniques and devices for shale gas exploration have been constantly developed and improved. The estimates and resultant statistics for the worldwide reserves for shale fuel are far from complete or reliable. Therefore, the deposit data information has been evolving in nature, i.e., being updated constantly. The U.S. Energy Information Administration (EIA) published a report on analysis and projections of shale oil resources in June 2013 as an update of its April 2011 report. The updated 2013 report covers 137 shale formations in 41 countries (including the United States), while the 2011 report covered 69 formations in 33 countries. The 2013 report estimates the technically recoverable shale gas resources at 7299 trillion cubic feet and the technically recoverable shale oil (and tight oil) resources at 345 billion barrels, which are updated numbers from 6622 and 32, respectively, from the 2011 report. While the available resource values are quite significant, these values will undoubtedly change over time as additional and/or new information becomes available.

Table 9.1 shows the data information for technically recoverable shale gas and shale oil unproven resources for the world and the United States. The data may be interpreted as currently estimated values for the total recoverable world resources. The data will be constantly and inevitably updated and revised based on new discoveries, more rigorous research drillings, new geologic and reservoir data, revised assessment, and technological advances.

As shown in Table 9.1, the currently estimated oil shale resources make up good percentages of the total oil and gas reserves of the world. Considering that the estimates are likely to be updated with significantly larger figures over time and that much of the numbers in the current estimates are available in near terms, the fuels recoverable from oil shale are truly remarkable and may be considered "a game changer in the future energy."

The oil shale resources are widely distributed in all continents of the world. The top 10 countries with technically recoverable shale oil resources and their estimated recoverable resources, according to the U.S. EIA, are listed in Table 9.2.

The top 10 countries with technically recoverable shale gas resources and their estimated recoverable resources, according to the U.S. EIA, are listed in Table 9.3. While the U.S. estimate for recoverable shale gas reserves by the EIA is 665 trillion

**TABLE 9.1**

**Technically Recoverable Shale Gas and Shale Oil Unproved Resources**

| | Crude Oil, Billion Barrels | Wet Gas, Trillion Standard Cubic Feet (scf) | Further References |
|---|---|---|---|
| **Outside of the United States** | | | [1] |
| Shale oil and shale gas unproven resources | 287 | 6634 | |
| Other proven reserves | 1617 | 6521 | [2] |
| Other unproven resources | 1230 | 7296 | [3,4] |
| Total for U.S. oil shale resources | 3134 | 20451 | |
| Increase in total oil and gas reserves due to oil shale | 10% | 48% | |
| Percentage of oil shale resources of total | 9%[a] | 32%[b] | |
| **United States** | | | [1] |
| EIA shale/tight oil and shale gas proven resources | n/a | 97 | [5] |
| EIA shale/tight oil and shale gas unproven resources | 58 | 567 | [6] |
| EIA proven reserves | 25 | 220 | |
| EIA unproven resources | 139 | 1546 | [6] |
| Total for U.S. oil shale resources | 223 | 2431 | |
| Increase in total oil and gas reserves due to oil shale | 35% | 38% | |
| Percentage of oil shale resources of total | 26%[a] | 27%[b] | |
| **Total World** | | | [1] |
| Shale/tight oil and shale gas proven resources | n/a | 97 | |
| Shale/tight oil and shale gas unproven resources | 345 | 7201 | |
| Other proven reserves | 1642 | 6741 | |
| Other unproven resources | 1370 | 8842 | |
| Total for world oil shale resources | 3357 | 22882 | |
| Increase in total oil and gas reserves due to oil shale | 11% | 47% | |
| Percentage of oil shale resources of total | 10%[a] | 32%[b] | |

*Source:* U.S. Energy Information Administration, Analysis and Projections—Technically Recoverable Shale Oil and Shale Gas Resources: An Assessment of 137 Shale Formations in 41 Countries outside the United States, U.S. Department of Energy, Washington, DC, 2013.

[a] The numbers represent the percentages of the shale gas reserve portion out of the total conventional natural and shale gas resources.

[b] The numbers denote the percentages of the shale oil crude reserve portion out of the total conventional crude and shale crude oil resources.

scf, Advanced Resources International (ARI) estimates it as 1161 trillion scf. Even though these data provide some rough ideas about country-specific reserve data, the information listed here is far short of being comprehensive. The resource reserve data for the Middle East, Central Africa, and other regions are currently unavailable.

It may be quite easily noticeable from Table 9.3 that the worldwide shale gas distribution is quite widely scattered in all continents, and the amount that is technically recoverable is also significant.

**TABLE 9.2**

**Top 10 Countries with Technically Recoverable Oil Resources**

| Rank | Country | Recoverable Shale Oil, Billion Barrels |
|------|---------|----------------------------------------|
| 1 | Russia | 75 |
| 2 | United States | 58 |
| 3 | China | 32 |
| 4 | Argentina | 27 |
| 5 | Libya | 26 |
| 6 | Australia | 18 |
| 7 | Venezuela | 13 |
| 8 | Mexico | 13 |
| 9 | Pakistan | 9 |
| 10 | Canada | 9 |
| | Subtotal for top 10 nations | 280 |
| | World total | 345 |

*Source:* U.S. Energy Information Administration, Analysis and Projections—Technically Recoverable Shale Oil and Shale Gas Resources: An Assessment of 137 Shale Formations in 41 Countries outside the United States, U.S. Department of Energy, Washington, DC, 2013.

**TABLE 9.3**

**Top 10 Countries with Technically Recoverable Shale Gas Resources**

| Rank | Country | Recoverable Shale Gas, Trillion scf |
|------|---------|-------------------------------------|
| 1 | China | 1115 |
| 2 | Argentina | 802 |
| 3 | Algeria | 707 |
| 4 | United States | 665 |
| 5 | Canada | 573 |
| 6 | Mexico | 545 |
| 7 | Australia | 437 |
| 8 | South Africa | 390 |
| 9 | Russia | 285 |
| 10 | Brazil | 245 |
| | Subtotal for top 10 nations | 5764 |
| | World total | 7298 |

*Source:* U.S. Energy Information Administration, Analysis and Projections—Technically Recoverable Shale Oil and Shale Gas Resources: An Assessment of 137 Shale Formations in 41 Countries outside the United States, U.S. Department of Energy, Washington, DC, 2013.

## 9.2.2 Commercially Noteworthy Shale Deposits in the United States

The United States shale fuel production has rapidly grown in recent years, and its shale gas reserve estimates have also increased significantly based on new drilling and exploration data in the shale fields. The U.S. EIA provides a detailed analysis and statistics as well as geological maps [1,7]. Therefore, discussions on the U.S. shale fuel resources as well as the specific shale plays and deposits are not repeated in this chapter. Further, the information available is being constantly revised based on the new information.

The following shale formations in the United States are being seriously developed and have received significant investments in recent years:

*Bakken shale*, North Dakota and Montana: Bakken shale is a formation from the late Devonian (382.7–372.2 million years ago) to early Mississippian age (358.9–323.2 million years ago). Bakken shale has a relatively long history of commercial development, which started with vertical drilling and later switched to horizontal drilling. Its application of horizontal drilling of an earlier version even started as early as the 1980s. The modern hydraulic fracking technology was applied in 2000 and 2001, and its initial production was mostly in the State of Montana and then expanded and shifted to the State of North Dakota. The principal fossil fuel product from Bakken shale is crude shale oil, while shale gas is a coproduct. As of July 2013, daily oil production from North Dakota exceeded 800,000 barrels a day from over 6000 wells [8]. The crude oil production in the Bakken shale region of North Dakota and Montana is estimated to have topped one million barrels per day in December 2013. As evidenced, this is a remarkable speed of production growth.

*Barnett shale*, Northern Texas: This is one of the most active shale formations in the United States and is estimated to hold 43.4 trillion standard cubic feet of natural gas. Most of this shale formation is about a mile and half deep and is below the most populated regions of Northern Texas, which includes the Dallas/Fort Worth area. Barnett shale has greatly contributed to the state economy in both job creation and revenue generation. It consists of sedimentary rocks of Mississippian age (354–323 million years ago) and is known as a "tight gas" reservoir. Active production of natural gas has been ongoing, and in 2013, Barnett shale produced more than 6% of the U.S. total natural gas production of about 24 trillion cubic feet of dry gas (2013).

*Fayetteville shale*, Arkansas: This shale formation is of Mississippian age and the depth of the formation is about 1500 to 6500 feet from the ground surface. Due to the relatively shallow depth, the natural gas production was originally through vertical drilled wells and has now switched to the modern fracking technology coupled with horizontal drilling.

*Haynesville shale*, Louisiana: Haynesville shale is of the Jurassic period (201.3–145 million years ago) and mostly located in northwest Louisiana. Haynesville shale in northeast Texas is also known as Bossier shale, even though most geologists believe the two are distinct and not the same. The productive interval of Haynesville shale is greater than 10,000 feet deep below the ground surface. Commercial development started recently after successful operation in other areas.

*Marcellus shale*, Pennsylvania, West Virginia, Ohio, and New York: The Marcellus shale gas formation is very rich in natural gas and is one of the largest shale deposits in the United States. It is also estimated to be the second-largest natural gas find in the world, and the U.S. Geological Survey estimates the formation's total area to be around 95,000 miles, with the formation depth ranging from about 4000 to 8000 feet from the ground surface. The current estimate (as of 2013) of the deposit size is more than 410 trillion cubic feet [9]. A good portion of the geographical region for Marcellus shale (of Middle Devonian, 393.3–382.7 million years ago) overlaps that of Utica shale (of Lower Devonian; 419.2–393.3 million years ago), which is a few thousand feet below the Marcellus.

*Utica shale*, New York, Ohio, Kentucky, Maryland, West Virginia, Virginia, Tennessee, and Ontario (Canada): Utica shale is situated even deeper by several thousand feet than Marcellus shale, that is, it is an older deposit than Marcellus shale. Utica shale is a stratigraphical unit of Middle Ordovician age (470–458.4 million years ago) in the Appalachian Basin. It has come to attention only in recent years. It has great potential to become an enormous natural gas resource. The Utica shale formation is thicker than the Marcellus and is geographically more extensive. Based on early testing, it has already proven the ability to support profitable commercial production. Early commercial development of Utica shale for shale gas production started in Quebec in 2006 and it is rapidly becoming a major player in the State of Ohio. If the Utica is proven commercially viable throughout the extent of formation, it would be the single largest natural gas field known today.

*Eagle Ford shale*, Southwest Texas: Eagle Ford shale formation was discovered in 2008 and it is a sedimentary rock formation from Late Cretaceous age (100.5–66 million years ago). Unlike most other shale formations, Eagle Ford is rich in both natural gas and oil. Eagle Ford is estimated to have 20.8 trillion cubic feet of gas and 3.35 billion barrels of oil, based on 2013 estimates. The formation is about 5700 to 10,200 feet below the terrain surface.

*Other shale formations in the United States*: Other shale basins in the United States that have active and potential development interest include Antrim, Chattanooga, Woodford, Carney, Pearsal, Bend, Pierre, Lewis, Hermosa, Monterey, Mancos, Baxter, Hillard, Niobrara, Cody, Mowry, Gammon, Excello-Mulky, New Albany, and more [10]. The list is likely to grow as new discoveries are made.

### 9.2.3 DIFFERENCE BETWEEN TECHNICALLY RECOVERABLE AND ECONOMICALLY AVAILABLE RESOURCES

As explained earlier, the abundant presence of oil shale resources throughout the world has long been known. However, only small fractions of these resources, most in reasonably shallow depth from the earth's surface, have been considered for potential commercial exploitation. The oil shale resources situated deep under the earth's surface had been considered "technically unrecoverable" rather strictly based on the levels of available technologies, irrespective of the prevailing oil and gas prices on the market. Due to the advances in enabling technologies and availability of highly capable machinery, significantly greater fractions of these oil shale resources are now technically recoverable. These resources are labeled as "technically recoverable resources," which are simply recoverable using currently available technologies without considering economic profitability of the production operation or other socioeconomic implications of such activity. For example, the data reported in Tables 9.1 through 9.3 refer to technically recoverable resources. Therefore, the designation of technical recoverability is subject to change with time, as more technological advances are achieved in related fields. Also, all technically recoverable resources are not necessarily economically prudent and profitable on the current market and/or in specific regions with other constraints.

The economic recoverability of oil and gas from oil shale needs to be clearly distinguished from its technical recoverability. In general, the economic recoverability depends upon three principal factors: (1) the cost of drilling and completing wells; (2) the amount of oil and gas produced from an average well over its lifetime; and (3) the prices received for oil and gas produced [1,4]. Based on recent experience with shale fuel production in the United States, it has been learned that the above-the-ground factors as well as geology significantly influence the economic recoverability of shale gas and shale oil. One such above-the-ground advantage in the United States and Canada is the private ownership of subsurface rights that provide a very strong incentive for commercial exploitation. The subsurface mineral rights may not be applicable to other countries. The other above-the-ground advantages for the United States include (1) availability of many highly capable independent operators and supporting contractors with relevant expertise and special machinery, (2) preexisting infrastructure including pipeline networks, (3) the availability of water and sand resources for use in hydraulic fracturing, and (4) state-level support for regional economic boost.

### 9.2.4 REMARKABLE SPEED OF COMMERCIAL DEVELOPMENT

Technological breakthroughs in horizontal drilling and hydraulic fracturing in recent years as well as the market conditions with persistently high energy prices have made much of the previously unrecoverable oil shale resources not only technically recoverable but also economically recoverable and profitable. Natural gas and oil have proven to be very quickly producible from these resources in large volumes at a relatively low cost. It is truly remarkable that tight oil and shale gas resources have revolutionized the U.S. oil and natural gas production, and all this game-changing turnaround has happened in a matter of a few years in the twenty-first century. In 2012, 29% of the total U.S. crude oil production was from shale/tight oil, while 40% of the total U.S. natural gas production was recorded from shale gas. Analysts predict that shale gas production would account for more than 50% of North American gas production by 2016 and about two-thirds by 2035. In history, no other energy source has made such a dramatic production increase at a revolutionary speed. Even though the amount of fossil energy production is already very significant, it is still too premature to say that the shale fuel industry is mature and well stabilized like the conventional petroleum and natural gas industry. The supporting and enabling technologies are still being actively developed and constantly enhanced.

## 9.3 HYDRAULIC FRACTURE TECHNOLOGY

### 9.3.1 WHAT IS HYDRAULIC FRACTURE?

Hydraulic fracture, hydraulic fracturing, "fracking" or "fracing" has been used in the development of U.S. oil and gas resources since the late 1940s, even though the word *fracking* has been popularly mentioned in the field of energy and fuels only recently. It is estimated that over 1 million wells have been hydraulically fractured over the past 60 years, and the fracking technology has gradually matured as lessons have been learned from ample practical experience. The oil and gas industry have used these lessons to minimize the environmental and societal impacts with the resource development while maximizing the efficiency of the technology [11]. The key component of hydraulic fracture is *horizontal drilling*, which has improved drastically over the years in terms of required machinery, operating depth, achievable length, drilling speed, bore sizes, well casing, and more. The advances in horizontal drilling and hydraulic fracture technology make it possible to drill vertically several miles deep and reaching out horizontally to additional several miles.

The current technology of hydraulic fracture uses highly viscous fluid to fracture the soil and rock matrix via all three forms of fracture mechanisms, namely, tensile, blasting, and impact fractures. Due to the extreme depth of shale fuel formation and its weight of soil overburden, a very high pressure has to be applied to the hydraulic fluid, in this case, fracking fluid mixture or frac fluid.

### 9.3.2 HOW IS HYDRAULIC FRACTURE DONE?

Figure 9.1 shows an illustrative summary of a shale fuel fracking operation.

A typical hydraulic fracture operation for shale fuel production is accomplished in the following sequence:

1. First, a drilling rig has to be installed at a pre-determined site. Secure the supply of water, sands, and additive chemicals. Perform a pressure test on the high-pressure fracturing pump and flow line for operational and environmental safety.
2. Prepare the fracking fluid mixture, which is usually a water-based slurry of fine sands and other chemical additives.
3. Then, a steel pipe known as surface casing is cemented into place at the uppermost portion of a well. This surface casing is very important to protect the groundwater. Therefore, the depth of the surface casing has to be determined based on location-specific factors, among others.
4. As a well is drilled deeper, additional casing is needed and installed to isolate the formation from which oil and gas is to be produced. This isolating casing further protects groundwater from the developing formation in the well.

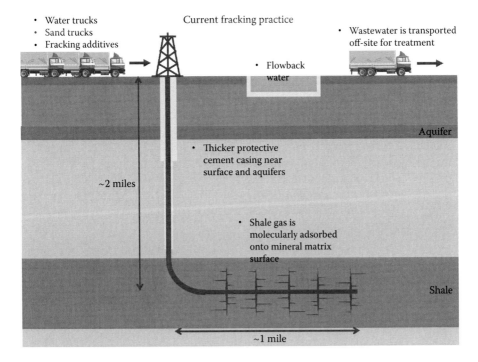

**FIGURE 9.1**    An illustration of the current shale gas fracking technology.

5. The initial injection step is to inject a large volume of specially formulated fluid(s) into the producing formation at an injection rate that will place sufficient stress on the shale rock to cause it to fracture and split into smaller pieces while creating fissures, channels, and pathways in its structure. This initial volume of special fluid is called the "pad" and typically accounts for about 20% of the total fluid volume. The pad fluid is pumped under very high pressure, typically 8000 pound-force per square inch (psi) or higher, to create enough fracture width to accommodate proppant particles. The required pressure depends upon the drilling depth of the formation; that is, the deeper the formation, the higher the applied pressure. Proppant is typically comprised of round white sand, which is size-graded. Other solid particles, including man-made particles like polymer resins, can also be used.

6. Proppant particles are mixed into additional fracking fluids, and this special fracking slurry is pumped into the well, thereby propping open the created fissures and channels so that they will remain open and permeable long after the pump pressure is relieved. The created fissures allow oil and/ or gas to escape freely from the pores of rock that it is trapped inside.

7. After injecting the slurry, a tubular volume of clean flush fluid is pumped to clear and replace tubular of proppant slurry out of the way.

8. The pump gets shut down.

9. The well pressure is bled off so that the fractures are allowed to close on the inserted or interjected proppant in its freshly fissured structure.

10. The final step is to recover the injected fluid by flowing or lifting the well. This step is called load recovery or frac load recovery.
11. Repeat the high-pressure injection of fracking fluids and take out flowback water.

Generally speaking, the fracking fluid mixture consists of water, sand, and chemical additives; water accounts for approximately 90%, sand 9.5%, and chemicals about 0.5%. There are many different methods and formulations the fracking companies employ and use to manage their frac fluid mixtures. The information is often proprietary to the specific companies and depends upon a large number of factors, including geological data on the deposits; aboveground locations; availability of water resources; availability of sands and sand types; ecological considerations specific to the geographical locations; availability of chemicals; local, state and federal regulations; size of the well drilled; and more.

### 9.3.3 Chemicals and Additives for Hydraulic Fracture Fluids

Fracking fluid compositions are widely varying and depend upon many factors that are specific to the producing formation(s) and also to the recipe of the operating industry. These chemicals perform a number of essential functions and deliver desired physical, chemical, and transport properties to the resultant fluid mixtures. It has been claimed that chemicals account for less than 1%, approximately 0.5% by volume, of the total fracking fluid mixtures that are currently used [12]. Considering a very large amount of the fracking fluid mixture needed per gas well, the total amount of chemicals used in the fracking operation is quite significant. The number would be even more significant when the amounts for the entire nation or a shale-active state are summed up. Some of the chemicals that have been used in the fracking fluid formulations are either known as potential carcinogens or posing other risks to human health [13]. This is why the chemicals used in fracking operation as well as the potential dangers these chemicals could render are the hottest debate topics in the development and utilization of shale gas and shale oil. Many state and local governments have started to require that the list of chemicals used in fracking operation be disclosed. As an example, a list of toxic chemicals used by Ohio shale drillers must be made available locally to governments, first responders, and residents under a new state directive of 2013.

Chemicals used in typical fracking fluid formulations can be classified based on their desired or intended functions as follows [14,15]:

- Friction reducers
- Mineral acids
- Corrosion inhibitors
- Biocides
- Iron control agents
- Cross-linking agents
- Oxidizers or peroxides
- pH adjusters or buffer solutions

- Scaling inhibitors and antiscaling agents
- Gelling agents
- Clay stabilizers
- Surfactants or nonemulsifiers
- Others

It should be noted that this is neither an exhaustive list of all chemical types, nor all of these listed additives are used in the current frac formulations, and also that the specific formulations are widely varying depending upon the fracking industry, shale formation, geographical region, and more. Some details on these additives used in the current technology are explained below.

1. *Friction reducers.* Addition of friction reducers helps minimize the friction and frictional energy loss during the injected flow of fracking fluid mixture through the tube, thereby minimizing the pressure drop as well as avoiding clogging. High-boiling mineral oils based on $C_{15}$–$C_{40}$ hydrocarbons whose average carbon numbers range between $C_{18}$ and $C_{20}$ are frequently used. Both paraffinic (straight hydrocarbon) and naphthenic (aromatic) kinds as well as their mixtures can be used. Ethylene glycol is also used for this purpose. However, vegetable oils are deemed unsuitable for this purpose for a number of valid reasons, including (1) active and/or potential growth of microorganisms, (2) valuable food ingredients, and (3) prohibitive cost.

2. *Mineral acids.* Hydrochloric acid (HCl aqueous solution) and phosphoric acid ($H_3PO_4$) are frequently used. These acids seep into the microfissures that are created due to hydraulic fracture operations and expand them. It is interesting to note that these two acids have been frequently found in the chemical formulations used for concrete and driveway cleaning solutions.

3. *Corrosion inhibitors.* The fracking fluids and operations create corrosive environments for the machineries employed as well as the well installations. Due to the large number of functionally diverse additives that are used in the fracking fluid formulations, it is very difficult to find one or two corrosion inhibitors that will do the magic of corrosion protection. Dimethylformamide (DMF), which has been widely used for a long time in corrosion prevention industry, is frequently used in the fracking formulations. DMF has a chemical structure of

Methanol, formaldehyde, formic acid, and isopropanol also have corrosion inhibition functions and are used in some formulations.

4. *Biocide.* Biocide is used to control biofouling of industrial water systems such as power plant cooling water, upstream gas production, and more. Biocides are added in order to prevent microorganism or algae growth in water. Various aldehydes known as biocidal aldehydes, such as

glutaraldehyde, are frequently used. Quaternary ammonium chloride and tetrakis hydroxymethyl phosphonium sulfate are also used as biocides that eliminate bacteria in water, which creates corrosive by-products. However, chlorine dioxide ($ClO_2$) is believed to be more suitable for gas production and fracking operations due to its high effectiveness, corrosion integrity, and sustainable microbial resistance. Chlorine dioxide is currently an on-site generation chemical; however, it can also be cost-effectively produced on a small scale. In particular, pure chlorine dioxide [16–18] that is manufactured using environmentally friendly acids and without using direct chlorine gas or strong mineral acids is ideal for shale gas operation. Recently, developed chlorine dioxide gel [19,20] could be of great significance, since the fracking fluid involves a similar kind of gelling material.

5. *Iron control agents.* Iron control agents prevent the formation of iron oxides such as $Fe_2O_3$ and $Fe_3O_4$ as well as iron hydroxide, $Fe(OH)_3$. These iron compounds, if present, in the fluid mixture would cause a problem of gel hardening, which results in a very hard gel that is difficult to pump. As an iron control agent, a mild acid such as citric acid or lactic acid is frequently used. Acetic acid, thioglycolic acid, and sodium hydroborate are also used to prevent the formation and precipitation of iron oxides [15].

6. *Cross-linking agent.* Cross-linkers are needed to maintain the cross-links and sustain the high viscosity of the fluid mixture, especially in a high-temperature environment, as expected deep underground. The geothermal temperature gradient, that is, a temperature increase with respect to increasing depth in the earth's interior, is roughly 25°C per kilometer. Therefore, it is not difficult to imagine that a shale gas well that is 2 miles deep would be quite easily in the temperature range of 90°C to 100°C. There are a wide variety of chemicals and polymeric materials that can be used for this desired function, and they include the following:

   • Borate salts, Me ($BO_3$), or boric acid: Boron would be at the center of cross-linking. This chemical helps maintain the fluid viscosity as temperature increases. Borate salts are usually carried by petroleum distillate.
   • Polyacrylamide, $[-CH_2-C(NH_2)H-]_n$, which is a water-soluble polymer and easily forms a polymeric gel. This chemical also helps minimize the friction of the fluid more as a friction reducer and is often called a "slick" agent.
   • Other water-soluble polymeric resins can also be used for this purpose, as long as the cost is not prohibitive.

7. *Peroxides, strong oxidizers, or breakers.* These are added to prevent the breakdown of the cross-linked polymers, gels, and/or foams in a high-temperature environment. This type of chemical is similar to free radical initiators (FRIs) in polymer synthesis and has highly oxidizing properties. Ammonium persulfate, $(NH_4)_2S_2O_8$, is frequently used due to its handling convenience (compared to peroxides) and reasonable cost. Ammonium persulfate allows a delayed breakdown of the gel. Various peroxides such as magnesium peroxide can also be used.

8. *pH adjuster.* There is a need for adjustment of the pH of the fracking fluid mixture for a number of practical reasons, which include maintaining the effectiveness of other essential ingredients, such as cross-linkers. For pH control, buffer solutions are used and/or created. Frequently used chemicals include the following:
   - Soda ash, $Na_2CO_3$, and potassium ash, $K_2CO_3$
   - Sodium bicarbonate, $NaHCO_3$
   - Sodium bisulfate, $NaHSO_4$
   - Potassium hydroxide, KOH
   - Acetic acid, $CH_3COOH$

9. *Antiscaling agent or scaling inhibitor.* To inhibit scaling of the tube wall, ethylene glycol (EG) or propylene glycol (PG) is frequently used. These chemicals, commonly known as antifreeze chemicals, have been used in the automotive industry for similar purposes. Acrylamide copolymers, sodium acrylate, sodium polycarboxylate, and phosphonic acid salt can also be used to prevent scale deposit formation in the pipes [15].

10. *Gelling agent.* The fracking fluid mixture needs to have very high viscosity to be able to deliver or transport sand particles through the tube. Since the sand density is about 2.5 to 2.7 $g/cm^3$, it would easily settle in aqueous solution and start to clog the channel, if not properly gelled or immobilized. In general, a true gelling agent can be cost prohibitive. Superabsorbent polymers (SAPs), such as polyacrylic acid (PAA), mildly cross-linked polyacrylic acid sodium, or potassium salts, have excellent properties that are very effective in forming gels in very low polymer concentrations. Natural guar gum and hydroxyl ethyl cellulose are popularly used. Both can be easily derived from naturally available resources. Polysaccharide blends can also be used to thicken the fluid to help suspend the proppant sand.

11. *Clay stabilizer.* Clay stabilizer, which is also known as a clay antiswelling agent, is very important in ensuring that the fractured fissures stay open and permeable for natural gas flow. If clay swells, then the clay nanostructure would be involved in complex mechanistic and morphological interactions of intercalation and exfoliation that result in closure of microfissures in the shale formation. The chemical also helps prevent clay shifting. Potassium chloride (KCl), sodium chloride (NaCl), and tetramethyl ammonium chloride (TMAC) are very popularly used as clay stabilizers. The molecular structure of TMAC is shown below.

12. *Surfactant and nonemulsifier.* Surfactant is typically used in a situation where the interfacial tension of the system needs to be lowered. This chemical helps prevent the formation of emulsion in the frac fluid, and the use

of lauryl sulfate is a good example. However, this is not the only reason why surfactant is being used in the fracking fluid formulation. The main function of this additive is to keep the viscosity of the fluid high at low temperatures. Isopropyl alcohol (IPA) is frequently used for this purpose. The viscosity of water at room temperature is about 1 centipoise, whereas that of IPA is around 3 centipoise. The difference becomes even higher at freezing temperatures. As a benefit of using this additive, the gel holds up nicely at low temperatures. Therefore, these chemicals are also known as winterizing agents or product stabilizers. The name *surfactant* may not be the most adequate for this additive. Methanol ($CH_3OH$), ethanol ($C_2H_5OH$), 2-butoxyethanol, naphthalene, and ethylene glycol can also be used for the same purpose.

The above list is far from being a complete list, since most drillers have their own proprietary formulations as well as adjusting their frac fluid compositions based on many factors that are specific to the shale formation and geology and the geographic location of the drilling operation. The Ground Water Protection Council (GWPC) [21] and the Interstate Oil and Gas Compact Commission (IOGCC) [22] have created a publicly accessible website, FracFocus [15], which is a chemical disclosure registry website for hydraulic fracturing chemicals. Drilling companies have been supportive of this registry website where people can freely search for information about the chemicals used in hydraulic fracturing of shale gas and oil wells.

### 9.3.4 PROPPANT SAND

In the current technology of hydraulic fracture, a proppant is an important ingredient for the process. A proppant is a solid material whose role is to keep induced hydraulically fractured fissures open during and following a hydraulic fracturing treatment. Since this solid material has to be inserted between the fractured creases, round or spherical particles are intuitively preferable to random or planar-shaped particles. These proppant solid particles have to be delivered to the fissures by viscous fracking fluid. Round-shaped particles would induce the smallest pressure drop in the pipe, thus requiring the lowest pumping energy as well as minimizing the pipe clogging possibilities. Furthermore, these proppant particles should be structurally strong and tough to withstand the compression stress that would be exerted on them by the weight of miles-high overburden. Typically, treated sand or man-made ceramic materials are used in the modern hydraulic fracture technology.

As explained earlier in this chapter, a typical frac fluid mixture includes about 9%–10% sands by mass. Considering that a typical shale gas well requires a total of 200–800 water trucks at 20,000 liters of water per truck, the amount of proppant sand required by this well is quite significant. In recent years, the silica sand industry has seen a dramatic increase in the silica sand demand thanks to the fracking boom. This sand is often called frac sand or proppant sand. Frac sand must be of greater than 99% quartz or silica. However, suitable U.S. deposits of silica sand are limited, and geology plays an important role in what makes the grade as proppant sand. Proppant sand specifications are established by the American Petroleum Institute (API).

The API recommends specifications on particle size, sphericity, roundness, crush resistance, and mineralogy [23].

- *Grain or particle size*: A size fraction of −20+40 mesh or 0.84–0.42 mm is most widely used.
- *Sphericity and roundness*: The API recommends a comparison chart that relates to sphericity and roundness, devised by Krumbein and Sloss in 1955 [24]. The chart provides a guide for visual estimation of sphericity and roundness that are assigned numbers from 0 to 1 on an *x–y* chart. The scale for sphericity is on the *y*-axis, while the roundness scale is given on the *x*-axis. For example, (0.1, 0.1) would be far from being spherical or round, whereas (0.7, 0.7) is quite close to a spherical shape. The API further recommends sphericity and roundness of 0.6 or higher for proppant particles [23].
- *Crush resistance*: Crush resistance is defined as the resistance of a quartz grain under compressive loading. The API specifications recommend the silica sand to withstand compressive stresses of 4000 to 6000 psi before it breaks apart or ruptures. The API also specifies the test method in a uni-axial compression cylinder by specifying a maximum loss percentage for each size fraction specified.
- *Solubility*: The solubility test measures the amount of nonquartz miner-als present in the proppant sand. The test measures the mass loss of a 5 g sample that has been added to a 100 mL solution of 12 ppm hydrochloric acid (HCl) and 3 ppm hydrofluoric acid (HF) and heated at 150°F in a water bath for 30 min. The API requires that the mass loss be less than 2% for the −6 + 12 mesh fraction to the −30 + 50 mesh fraction, and less than 3% for the −40 + 70 to the −70 + 140 fraction, and so forth [23].
- *Turbidity*: Turbidity is the amount of silt/clay-sized minerals in the sand. However, turbidity is generally not a problem, since the material is washed during the processing of proppant sand.

Silica sand deposits in the United States are well identified, and high-purity quartz sands are common throughout the country. However, some of these deposits have envi-ronmental factors that limit or restrict mining, and others are located too far from the fracking operations to be economically feasible. Furthermore, due to the tight and narrow specifications, in particular, roundness and sphericity, the fracking industry requires, many deposits are not deemed suitable for the current hydraulic fracture technology. Due to the relative scarcity of high-quality proppant sand as well as the technological evolution and demands, man-made materials such as man-made ceramic materials and polymeric proppants are being developed and/or explored.

### 9.3.5 FRAC WATER

Hydraulic fracture requires a large amount of water for the frac fluid mixture that needs to be injected into the deep underground formation. The exact amount of water depends upon the size of the well bore, depth of the formation, lateral dimension of the fractured formation, specific fracking method employed, and type and specific properties of the formation. In general, about 200–800 water tankers at 20,000 liters per tanker, or roughly 1 to 4 million gallons of water, are needed per shale gas well using the current fracking technology. Well sourcing is not always a simple matter. The current practice is to obtain this water from freshwater resources such as rivers, lakes, and aquifers. This water is hauled in by trucks to the well site, and so is the frac sand supply. Needless to say, water supply and management is a significant part of the operation and also of the operating cost.

As easily imagined, some of the injected water will flow back out of the well, and this flowback water that contains frac chemicals and rock contaminants needs to be properly managed and/or controlled.

### 9.3.6 FLOWBACK WATER CONTROL AND MANAGEMENT

Flowback water is also referred to as backflow water. It consists of two principal portions, which are (1) frac fluid returning to the ground surface and (2) produced water. The former is also known as frac load recovery, whereas the latter is naturally occurring water produced by the formation. The frac load recovery is about 15%–40% of the initial volume of frac fluid injected, which means that more than half of the injected fluid remains in the formation [25]. A typical flowback frac fluid is about 25,000–40,000 barrels.

Returning frac fluid flows back over a period of 3–4 weeks after fracking, most of it in a matter of 7–14 days. Figure 9.2 shows an illustrative sketch of typical flowback water rate as a function of time. The area under the curve would represent the total amount of flowback frac water returned to the ground surface.

After a certain period of flowback of frac fluid, there is a transition point between the frac load recovery (frac water flowback) and the flow of water produced naturally by the formation. It is generally hard to clearly determine this demarcation point of transition. Some chemical signatures that are unique to the specific formation may be possibly detected from the recovered fluid, thus indicating the presence of the formation water.

Flowback frac water may be classified as having high salinity and total dissolved solids (TDS). It is also loaded with the same fracking chemicals used in the injected fluid mixture and some unique contaminants that are present in the rock formation water very deep below the ground surface. The level of salinity of flowback fluid is

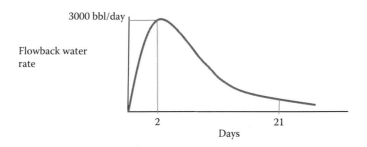

**FIGURE 9.2**    A sketch of typical flowback frac water rate trend.

in general very high due to the additional dissolution of salts in the formation. Since many chemical ingredients that are used in frac fluid formulations, as well as ones dissolved from the formation, are environmentally hazardous and potentially polluting, the returning fluid and its postrecovery fate need to be strictly managed, treated, controlled, and/or reused.

The flowback fluid is generally recovered and collected in metal tanks, or otherwise in open pools, lagoons, or pits lined with a multiple layers of plastic materials [25]. This recovered fluid is then subjected to removal of solids and dissolved matters as well as appropriate treatment, and water is trucked off-site to a wastewater disposal facility or recycled for reuse in fracking operations. Even though recycled use of frac water is logically more sound and environmentally acceptable, most drillers currently do not recycle the frac water. The principal reasons that drillers hesitate to recycle frac water are (1) high cost due to the lack of suitable technology that is economically affordable; (2) technological inflexibility due to the lack of versatile and robust technology applicable to all types of wells and capable of handling the variance in flowback water compositions; and (3) too many roadblocks such as difficulties to obtain permits required for wastewater treatment; and so forth. The Pennsylvania Department of Environmental Protection (PA DEP) has formed a partnership with the oil and gas industry to research and develop new and innovative methods of treating and recycling flowback water. With recycling of flowback water, the overall process strives toward a closed-loop system with near-zero discharge in terms of water utilization and management.

A wide variety of water treatment methods are being studied, applied, and developed for flowback water treatment, some aiming at recycled use on-site and others targeting safe disposal, decontamination, and disinfection. Some of the noteworthy water treatment techniques that are potentially applicable include (1) membrane filtration, reverse osmosis, microfiltration, and ultrafiltration [26]; (2) ultraviolet (UV) and ozone ($O_3$); (3) supercritical water oxidation (SCWO) and supercritical water partial oxidation (SCWPO) [27]; (4) chlorine dioxide treatment [17,28]; and (5) agglomeration and agglo-flotation; and more.

## 9.4 ENVIRONMENTAL CONSIDERATIONS

Potential environmental and ecological concerns and issues with respect to shale fuel development and production that must be properly and decisively addressed include the following:

- Potential groundwater contamination
- Potential underground water contamination
- Potential impact on the surrounding air quality
- Problems associated with or triggered by potential well leakage
- Erosion of lined pits and soil contamination due to the leached chemicals
- Problems caused or induced by excessive depletion of sand mine
- Potential aggravation of acid mine drainage (AMD) problem
- Greenhouse gas (GHG) emission
- Noise and vibration to the neighboring community

- Ecological impacts on animal and wildlife in the region
- Heavy traffic, in particular, heavy trucks for transportation of equipment, water, and sand

State and local governments are working very closely with the shale fuel and drilling industries to address all the issues listed above as well as to eliminate or mitigate the identified problems and concerns. Concerted efforts are being made in the areas of (1) setting regulatory standards, (2) mandating compliance requirements, (3) developing and providing specifications, (4) developing advanced technologies and machinery, (5) making publicly available the registry of frac chemicals, (6) enhancing public awareness and gaining public acceptance, (7) staffing and training first responders, (8) improving infrastructure, and (9) strengthening workforce education, and more.

## 9.5 FUTURE RESEARCH AND DEVELOPMENT NEEDS

Even though the hydraulic fracture technology has long been used in the oil and gas industry, the modern version of hydraulic fracture technology applied to shale gas and oil production has only a short history of commercial exploitation in terms of drilling depth, scope of venture, capacity and magnitude, level of sophistication, environmental impacts, and technological challenges. Therefore, the grand technology of shale fuel fracking is an integrated technology involving deep vertical drilling, extensive horizontal drilling, risk-free groundwater protection, efficient frac fluid formulation, hydraulic fracture, load recovery, recovered frac water treatment, and associated environmental technology. Due to its rapid commercial success represented by the shale gas boom on the market, commercial production, technological innovations and advances, infrastructure development, and establishment of regulatory standards are concurrently taking place. This is an integrated strategy based on a combination of (1) adoption of the best possible practices; (2) feedback assessment; (3) on-site application of new ideas, machinery, and method; (4) implementation of new advances in the subsequent preparation of wells; and (5) providing additional measures of safety and environmental protection.

The following are the areas for current and future research and development in the field of shale fuel technology:

- Efficient and cost-effective exploration techniques for shale fuel deposits
- Added mobility in rig design and machinery development
- Development of energy-efficient and noise-reduced machinery
- Postfracture pretreatment of well for maximum recovery of hydrocarbons over the well life
- New frac fluid formulations using environmentally friendly chemicals
- Development and implementation of new biocide
- Reduction in freshwater consumption
- Development of waterless fracking technology and associated additives
- Development and implementation of new materials replacing sands, gellants, and more
- Development and implementation of new quick-drying cement formulation

- Development and adoption of new plastic liner materials
- Development and implementation of corrosion protection processes and products
- Elimination or mitigation of causes and circumstances that could trigger, or be linked with, earthquakes and mini-earthquakes near the shale fuel production region
- Development and implementation of water treatment and recycling technology
- Elimination of source chemicals and/or causes for trihalomethane (THM) precursor generation in wastewater
- Development and implementation of downstream water treatment to achieve the highest-quality drinking water
- Development and adoption of efficient remediation of contaminated soil
- Development and implementation of odor control technology
- Development and implementation of enhanced fuel recovery technology
- Development of more flexible and robust technologies that are applicable to worldwide shale deposits
- Technology modification toward commercial development of natural gas hydrates

## REFERENCES

1. U.S. Energy Information Administration (EIA), "Analysis and projections—Technically recoverable shale oil and shale gas resources: An assessment of 137 shale formation in 41 countries outside the United States," U.S. Department of Energy, Washington, DC, 2013.
2. Oil and Gas Journal, "Worldwide report," December 3, 2012.
3. U.S. Geological Survey, "An estimate of undiscovered conventional oil and gas resources of the world, 2012," Tech. Rep. Fact Sheet 2012-3028, 2012.
4. U.S. Geological Survey, "Assessment of potential additions to conventional oil and gas resources of the world (outside the United States) from reserve growth, 2012," Tech. Rep. Fact Sheet 2012-3052, 2012.
5. U.S. Energy Information Administration (EIA), "U.S. crude oil, natural gas, and NG liquids proved reserves with data for 2010," U.S. Department of Energy, Washington, DC, 2012.
6. U.S. Energy Information Administration (EIA), "Annual energy outlook 2013," U.S. Department of Energy, Washington, DC, Tech. Rep. DOE/EIA-0383, 2013.
7. U.S. Energy Information Administration (EIA), "Review of emerging resources: U.S. shale gas and shale oil plays," U.S. Department of Energy, Washington, DC, 2011.
8. North Dakota Department of Mineral Resources. North Dakota monthly bakken oil production statistics. North Dakota Industrial Commission, 2013. Available at https://www.dmr.nd.gov/oilgas/stats/historicalbakkenoilstats.pdf.
9. Energy from Shale. Marcellus shale, 2013. Available at http://www.energyfromshale.org/hydraulic-fracturing/marcellus-shale-gas.
10. U.S. Energy Information Administration (EIA). Lower 48 states shale plays, May 9, 2011. Available: http://www.eia.gov/oil_gas/rpd/shale_gas.pdf.
11. Energy from Shale. What is fracking? October 4, 2013. Available at http://www.energyfromshale.org/hydraulic-fracturing/what-is-fracking.
12. R. D. Vidic, "Novel techniques for managing frack water," in *Marcellus Shale Natural Gas Stewardship: Understanding the Environmental Impact. A Temple University Summit*, Philadelphia, PA, 2010, pp. 1–26.

13. N. Kusnetz, "Fracking chemicals cited in congressional report stay underground," ProPublica, April 18, 2011. Available at http://www.propublica.org/article/fracking-chemicals-cited-in-congressional-report-stay-underground.

14. R. LaFollette, "Key considerations for hydraulic fracturing of gas shales," in *American Association of Petroleum Geologists Webinar Presentation*, AAPG, Tulsa, OK, 2010.

15. FracFocus, Chemical disclosure registry managed by the Ground Water Protection Council and Interstate Oil and Gas Compact Commission, 2013. Available at http://frac focus.org/.

16. S. Lee, "Manufacturing aqueous chlorine dioxide," U.S. Patent No. 5,855,861, January 5, 1999.

17. S. Lee and H. B. Lanterman, "Novel chlorine dioxide manufacturing process—Product delivery and treatment system," U.S. Patent 6,051,135, April 18, 2000.

18. S. Lee, "Method and apparatus for making aqueous chlorine dioxide," U.S. Patent No. 5,855,861, January 5, 1999.

19. S. Lee and P. Roberts, "Chlorine dioxide gel and associated methods," U.S. Patent No. 7,229,647, June 12, 2007.

20. S. Lee and P. Roberts, "Chlorine dioxide gel and associated methods," U.S. Patent No. 8,202,548, June 19, 2012.

21. Ground Water Protection Council. Homepage for Ground Water Protection Council (GWPC), October 2013. Available at http://www.gwpc.org/.

22. IOGCC. Homepage for Interstate Oil and Gas Compact Commission, 2013. Available at http://www.iogcc.state.ok.us/.

23. M. Zdunczyk, "The facts of frac," *Drilling Minerals*, pp. 58–61, January 2007.

24. W. C. Krumbein and L. L. Sloss, *Startigraphy and Sedimentation*. San Francisco: W.H. Freeman and Co., 1963.

25. WikiMarcellus. Flowback water, December 15, 2011. Available at http://waytogoto.com/wiki/index.php/Flowback_water.

26. M. A. Kuehne, N. N. Li, R. Q. Song, M. Tsai and J. C. Li, "Water reclamation," in *Encyclopedia of Chemical Processing*, S. Lee, Ed. New York: Taylor & Francis, 2007, pp. 3217–3226.

27. S. Lee, A. Garcia-Ortiz and J. Wootton, "Systems for water purification through super-critical oxidation," U.S. Patent No. 7,186,345, March 6, 2007.

28. DuPont. Chlorine dioxide solutions, 2013. Available at http://www2.dupont.com/Chlorine_Dioxide/en_US/index.html?src=gg_clo2_na_chlorine-dioxide.

# 10 Methanol Synthesis from Syngas

*Sunggyu Lee*

## CONTENTS

## 10.1 INTRODUCTION

The commercial synthesis of methanol has been widely practiced since the 1920s, when BASF announced a catalytic synthesis process that later has been called "high-pressure synthesis of methanol." Methanol has long been used as a building-block chemical, solvent, processing fluid, combustion fuel, and more. Over the decades, scientists and engineers have tried to develop better catalysts and more cost-effective processes that would enable the synthesis reaction to be carried out at less severe conditions with higher efficiencies. Efforts have also been made to diversify the raw material sources including conventional natural gas, shale gas, coal, biomass, wastes, and more. Owing to the enormous volume of methanol demands in a variety of industrial sectors, the size of commercial production units has been ever increasing, thus making process efficiency of utmost importance for process economics and viability. According to the Methanol Institute,[1] over 90 methanol plants worldwide have a combined production capacity of about 100 million metric tons a year, which is roughly 33 billion gallons a year. The environmental constraints on the process have also played a major role in the production and utilization of methanol.[2]

Even though methanol is very widely used in chemical, petrochemical, pharmaceutical, and polymer industries as starting materials for synthesis reactions as well as solvents for other chemicals, market demand and interest level have historically been tied to global energy outlooks and contemporary issues. Whenever the conventional transportation fuel price in the marketplace is high, the interest in methanol also becomes intense as an alternative fuel or its precursor. It may be recalled that the movement toward oxygenated fuel for cleaner air pushed methanol demand very high, for the manufacture of *methyl tert-butyl ether (MTBE)*[3]; however, this boom did not last too long after the harmful health effects of MTBE were discovered and challenged. This chemical, MTBE, as a gasoline-blending oxygenate, was once the fastest-growing chemical commodity of the world market, in the 1990s.

Methanol is very toxic and fatal if taken internally by humans or animals. Methanol is far more toxic than ethanol, which is a $C_2$ homologue of the former. Even though the methanol molecule contains only one carbon and has a low molecular weight (of only 32), which is about the same as an oxygen molecule, its synthesis chemistry is quite complex and controversial. Methanol itself has a high octane value, ~105, and burns cleanly. Similar to ethanol, methanol raises the octane rating of gasoline and reduces engine "knock" or "ping" without affecting the efficiency of the conventional catalytic converter. A 5% blend of methanol in unleaded gasoline may raise the octane rating at the pump by 1–1.5. However, its use as a gasoline-blending fuel to enhance the oxygenate content of gasoline has not been popular, owing to the relatively high volatility of the methanol-blended gasoline, as characterized by the Reid vapor pressure (RVP). Methanol's RVP is 32 kPa,[4] whereas that of ethanol is 13.8 kPa. Methanol has outstanding chemical properties as an excellent solvent as well as chemical reactant in a number of important chemical syntheses. Methanol is an essential chemical in manufacturing biodiesel using plant oil via transesterification reaction. In recent years, it has become a popular choice for the development of fuel cell technologies, in particular, direct methanol fuel cells (DMFCs).[5] Methanol has a very low freezing point of −97.6°C,[4] which makes it an ideal chemical ingredient for a windshield washer fluid formulation for cold climates.

The *methanol economy*[6] is a hypothetical economy in which methanol fuel would replace fossil fuels as a means of transportation of energy. It offers an alternative to the hydrogen economy and the ethanol economy. Many arguments are offered for preferring the methanol economy against the hydrogen economy, in terms of the cost of energy generation; cleanness of conversion processes; continued dependence on fossil fuel sources; volumetric power density; infrastructural and transformational cost; safety associated with the fuel in various aspects of synthesis, distribution, and storage, and so forth. Methanol has been playing a major role in the biodiesel industry by providing a significant amount of raw material as a reactant in transesterification reactions, in which 1 mol of triglyceride reacts with 3 mols of methanol in the presence of an alkali catalyst. *DMFCs* are being very actively developed to power portable electronics. They can be a very viable power source in many applications if their power density and energy conversion efficiency can be increased.[5] As such, methanol in DMFC can be a contributing player in consumer electronics and many other domestic applications. Methanol is widely recognized as a representative chemical of $C_1$ oxygenates that include methanol, formaldehyde, and formic acid. Essentially, all of the world's formaldehyde production is based on the direct conversion of methanol.

Methanol synthesis is also a good subject for academic research and teaching. A number of process design and stoichiometry problems have been developed for students in capstone design courses as well as for textbook examples in chemical reaction kinetics and material and energy balances. Methanol synthesis involves a great deal of process modeling and practical engineering problems in a variety of topics including classical thermodynamics, condensed phase thermodynamics, reaction kinetics, reactor design, reactor modeling, reactor thermal stability, reactor configuration, catalyst design, catalyst life management, pore diffusion and mass transfer, recycling of unreacted feed stream, separation, waste heat recovery, process and energy integration, unit operations, process economics, environmental engineering, and cost accounting.

In this chapter, a comprehensive overview of methanol chemistry and synthesis technology is presented with a particular emphasis placed on its value as an alternative fuel and petrochemical feedstock.

## 10.2 CHEMISTRY OF METHANOL SYNTHESIS

The catalytic synthesis of methanol has been commercially available since 1923, when the first commercial plant for the synthesis of methanol from syngas was built by BASF.[2] The technology of manufacturing methanol has gone through constant improvements and major modifications, among which the biggest change was undoubtedly a transition from high-pressure synthesis to low-pressure synthesis. Both process technologies adopted heterogeneous catalytic conversion to methanol from synthesis gas typically originated from natural gas or, alternatively, from coal. The quality and composition of synthesis gas differ very widely, depending on the process of conversion as well as the type and quality of the feedstock. Therefore, a variety of commercial process designs reflected and encompassed these differences. Accordingly, it is imperative that the chemistry of synthesis gas conversion be fully elucidated in the synthesis of methanol and further conversion of methanol into other petrochemicals, including alternative hydrocarbon fuels.

### 10.2.1 CONVERSION OF SYNGAS TO METHANOL

Synthesis gas is a mixture that contains hydrogen, carbon monoxide, and carbon dioxide as principal components, and methane and steam (moisture) as secondary components. Synthesis gas is also called *syngas*. Syngas is typically produced via steam reforming of natural gas (or methane steam reforming [MSR]), gasification or partial oxidation of coal, gasification of biomass, gasification of municipal solid wastes (MSWs) and coke oven gas, and so forth. The synthesis of methanol from syngas is typically conducted over a heterogeneous catalyst system, most popularly coprecipitated $Cu/ZnO/Al_2O_3$ catalyst, which is a reduced and active form of $CuO/ZnO/Al_2O_3$. In such a catalyst formulation, alumina ($Al_2O_3$) is a support that can be replaced by other similar supports such as $ThO_2$. The catalyst system used for this chemical system is commonly referred to as "Cu-based catalyst," while the catalyst system employed for its predecessor technology of high-pressure methanol synthesis is referred to as "Zn-based catalyst," which does not have copper in its formulation. The principal stoichiometric reactions considered in this chemical conversion are

$$CO_2 + 3H_2 = CH_3OH + H_2O$$

$$CO + 2H_2 = CH_3OH$$

$$CO + H_2O = CO_2 + H_2$$

It should be clearly noted that there are a great number of other chemical reactions also taking place in the reaction system, which are less significant in terms of their selectivity and extent of reaction. Among the preceding stoichiometric representations, only two of these three reaction equations are stoichiometrically independent. Stoichiometric independence can be very easily verified either by the Gauss elimination type of mathematical procedure as well as by showing the derivability of the third equation from a linear combination of the other two. In this specific case, a linear combination of any two stoichiometric equations would result in the third equation, thus leaving the system with only two linearly independent stoichiometric reactions. If the stoichiometry and material balances are the only matters of interest, there is very little difference with regard to which two reaction equations are to be chosen as principal reactions. However, if the mechanistic view of the process synthesis is involved, as in the cases with catalyst design and process development, then the scientifically accurate choice of the two principal reactions must be in line with real-world situations. This is where a controversy exists regarding the synthesis of methanol over a $Cu/ZnO/Al_2O_3$ catalyst system.[2,7–10] There are two major mechanistic views that are conflicting with each other as to the principal reactions in methanol synthesis from syngas over a $Cu/ZnO/Al_2O_3$ catalyst system.

### 10.2.1.1 CO Hydrogenation as Principal Reaction for Synthesis of Methanol

In this mechanistic view, the principal reactions have been taken as

$$CO + 2H_2 = CH_3OH$$

$$CO_2 + H_2 = CO + H_2O$$

According to this view, methanol is predominantly synthesized via direct hydrogenation of carbon monoxide. The second reaction is the reverse water–gas shift reaction (RWGS), which proceeds in the reverse direction. Thus, the direction of the water–gas shift (WGS) reaction is determined from material balance considerations, not necessarily from chemical thermodynamic considerations. Experimental reaction data involving typical syngas mixtures that contain 3%–9% $CO_2$ show a decrease in carbon dioxide concentration in the reactor effluent stream; thus, we intuitively infer that the WGS reaction proceeds in the direction of reducing carbon dioxide concentration, that is, in the reverse direction. However, it must be noted that this explanation involving the RWGS reaction is consistent only when the principal reaction for methanol synthesis is taken as the hydrogenation of carbon monoxide.

It should also be noted that the first reaction of the methanol synthesis is exothermic, whereas the second reaction of RWGS is endothermic. According to this mechanism, via depletion of carbon dioxide in the reverse WGS reaction, more reactant carbon monoxide is produced to boost the synthesis of methanol. The role of carbon dioxide in the overall synthesis was crucially important for reasons other than participation in the WGS reaction, as evidenced consistently by various investigators in the laboratory as well as researchers in the field. Deficiency of carbon dioxide in the feed composition can be extremely detrimental to the overall synthesis, very rapidly deactivating the catalysts and immediately lowering methanol productivity by the catalytic process. Typically, 2%–4% of carbon dioxide is mandatorily present in the syngas mixture for the vapor-phase synthesis of methanol, whereas this value is somewhat higher, 4%–8%, for the liquid-phase synthesis.[2,7,11]

### 10.2.1.2   $CO_2$ Hydrogenation as Principal Reaction for Methanol Synthesis

In this view, the principal chemical reactions that lead to the synthesis of methanol are

$$CO_2 + 3H_2 = CH_3OH + H_2O$$

$$CO + H_2O = CO_2 + H_2$$

It should be noted that according to this view, the synthesis of methanol proceeds predominantly via direct hydrogenation of carbon dioxide, not carbon monoxide. It should also be noted that the WGS reaction proceeds in the forward direction in this mechanism, consuming carbon monoxide to produce the principal reactants of carbon dioxide and hydrogen, thus boosting and sustaining the eventual methanol productivity. A number of different authors have conducted a variety of reaction experiments to elucidate the true reaction pathways or mechanistic pathways, including isotope labeling studies and kinetic studies involving complete absence of one of the syngas components.[2,7–10] More evidence, including isotope labeling studies,[13,14] extreme feed condition analysis,[10,11,15] and innovative catalyst characterization results[16] points toward the $CO_2$ hydrogenation as the principal reaction mechanistic step and $Cu^0$ as the active phase of the copper-based catalyst. In this view, the role of CO in the feed syngas is to help maintain the reduced state of the catalyst, and the reactant $CO_2$ is continuously supplied via concurrent WGS reaction involving abundant CO in the feed, which is effectively catalyzed by the methanol synthesis catalyst.[11]

### 10.2.1.3 Chemical Reactions under Extreme Syngas Conditions

Controlled experiments with extreme syngas compositions help elucidate the pathways of complex reaction systems that otherwise involve strongly intertwined reactions between primary reactants as well as between reactants and intermediate reaction products. As a case study, the following extreme conditions were examined using the aforementioned mechanistic postulates.[10] Two of the more obvious choices for extreme syngas conditions for methanol synthesis are CO-free feed syngas and $CO_2$-free feed syngas, thus effectively isolating one of the main chemical reactants from the beginning.

#### 10.2.1.3.1 CO-Free Syngas Feed

If the feed syngas is free of carbon monoxide, any involvement of carbon monoxide in the reaction system would have originated from carbon dioxide. Experimental observations from a lab-scale investigation show that the methanol productivity is very low from the beginning and slowly decreases even further. This kinetic reaction phenomenon can be explained, using the $CO_2$ hydrogenation reaction mechanism, as

$$CO_2 + 3H_2 = CH_3OH + H_2O$$

$$CO_2 + H_2 = CO + H_2O$$

Owing to the total absence of CO, the WGS reaction proceeds in the reverse direction, that is, in the direction that will generate some CO. Thus, the main reactant, $CO_2$, is wanted by both reactions. Considering that the RWGS reaction is a faster reaction than the methanol synthesis reaction over the catalyst used, the methanol production rate will have to suffer. Furthermore, both the RWGS reaction and the methanol synthesis reaction produce $H_2O$, whose concentration buildup in the system, particularly in the catalyst pores, adversely affects the conversion of $CO_2$ toward methanol by pushing the chemical system closer to the thermodynamic equilibrium condition. Moreover, too high a water concentration in the catalyst pore is detrimental to the longevity of the catalyst.[17] Therefore, methanol productivity further decreases. Other experimental evidence as for the essential need for carbon monoxide in the feed syngas is to keep the reductive environment for stabilization of the active phase of copper in the catalyst, that is, as $Cu^0$.[11]

The same experimental observation can also be explained differently by using the CO hydrogenation mechanism as

$$CO + 3H_2 = CH_3OH$$

$$CO_2 + H_2 = CO + H_2O$$

In this case, assuming that the CO hydrogenation mechanism is valid for methanol synthesis, the WGS reaction proceeds also in the reverse direction, because of the total lack of carbon monoxide. According to the CO hydrogenation mechanism, carbon monoxide is the essential reactant for methanol formation; however, the only source for this reactant would be coming secondarily from the RWGS reaction, because there is no CO in the feed syngas. Therefore, the reaction is very seriously limited by a lack of the essential reactant, that is, CO in this mechanistic view.

As shown, both the mechanisms could explain the situation more or less consistently with the experimental observations. Therefore, the experiments conducted under these conditions alone do not confirm which of the two mechanisms is the correct one for the synthesis of methanol over the $Cu/ZnO/Al_2O_3$ catalyst.

### 10.2.1.3.2  $CO_2$-Free Syngas Feed Conditions

If methanol synthesis is practiced over the $Cu/ZnO/Al_2O_3$ catalyst system using the $CO_2$-free syngas, methanol productivity is also significantly lower than under normal syngas feed conditions, and it rapidly decreases even further.[10]

According to the $CO_2$ hydrogenation mechanism, the following stoichiometric equations can be written as

$$CO_2 + 3H_2 = CH_3OH + H_2O \tag{10.1}$$

$$CO + H_2O = CO_2 + H_2 \tag{10.2}$$

Owing to the total absence of $CO_2$ in the feed syngas, the WGS reaction proceeds in the forward direction, resulting in carbon dioxide, which is the essential reactant for the methanol synthesis reaction, based on this mechanism. Because of the unavailability and limited supply of carbon dioxide, the principal reaction of methanol synthesis does not proceed sufficiently, resulting in poor methanol productivity. Further, the lack of carbon dioxide could make the Boudouard reaction also proceed in its reverse direction as

$$2CO(g) = CO_2(g) + C(s)$$

As can be expected, this reaction takes place on heterogeneous surfaces such as the catalyst surface and essentially results in carbon deposition. This reaction is responsible for catalyst deactivation via fouling. This may be one of the reasons for the rapid decrease of methanol productivity. However, the conditions promoting carbon deposition may be quite different between vapor-phase and liquid-phase synthesis processes. In this regard, carbon dioxide is a crucially important ingredient of the syngas mixture for the stability of catalytic activity. It is also found experimentally that the $CO_2$ deficiency in the feed syngas composition can be supplemented by $H_2O$ input to a certain degree.[18] Because $H_2O$ is directly involved in generation of $CO_2$ in $CO_2$-starved conditions via WGS reaction, this is also explainable. Nonetheless, complete absence of $CO_2$ in the feed syngas has been known to result in irreversible damage to the catalyst.

If CO hydrogenation is taken as the mechanism, the reaction of $CO_2$-free syngas would be represented by

$$CO + 2H_2 = CH_3OH \tag{10.3}$$

$$CO + H_2O = CO_2 + H_2 \tag{10.2}$$

As can be seen, carbon monoxide is required by both reactions. The WGS reaction is faster under these conditions and proceeds in the forward direction as long as there is some $H_2O$ in the system. Because carbon monoxide is abundantly available in the system, this alone would not explain the low methanol productivity. However, the explanation using the Boudouard reaction still holds for this mechanism. When there is no $CO_2$ in the feed gas, carbon deposition via Boudouard reaction can be promoted. Water promotes and participates in the WGS reaction, thereby producing $H_2$, which is a key reactant in methanol synthesis, and $CO_2$, which inhibits the carbon deposition reaction. The very low methanol productivity from the very beginning of the reaction is still not explained by this chemical mechanism alone. Rather, it strengthens the claim that the presence of carbon dioxide in the feed gas is essential for the methanol synthesis chemistry.

### 10.2.1.3.3  $H_2O$-Free Syngas Feed Conditions

In this case, let us assume that the syngas mixture still contains typical amounts of $H_2$, CO, and $CO_2$, but with total absence of $H_2O$.

First, if we adopt $CO_2$ hydrogenation as the principal reaction for methanol formation, the reaction system under the $H_2O$-free syngas condition may be described as follows:

$$CO_2 + 3H_2 = CH_3OH + H_2O$$

$$CO_2 + H_2 = CO + H_2O$$

As written, the WGS reaction proceeds in the reverse direction, at least in the beginning, when $H_2O$ is totally absent in the feed. Accordingly, $CO_2$ is the reactant for both the methanol synthesis reaction via $CO_2$ hydrogenation as well as the RWGS reaction. The two reactions occur in a competitive manner, thus resulting in a lower productivity of $CH_3OH$. Further, both reactions generate $H_2O$, whose concentration builds up in the reactor and in the catalyst pores and eventually approaches the reaction equilibrium of the two reactions. Once the $H_2O$ concentration reaches a certain level in the system, the WGS reaction is likely to go in the forward rather than in the reverse direction. The system will quickly restore order. As such, the role of water in the system can be rather easily compensated for by the adequate presence of carbon dioxide.

The same phenomenon can also be explained by the CO hydrogenation mechanism. The reactions under this mechanism would involve

$$CO + 2H_2 = CH_3OH$$

$$CO_2 + H_2 = CO + H_2O$$

In this case, it is rather obvious that the WGS reaction proceeds in the reverse direction, at least in the beginning, until it reaches WGS equilibrium and, eventually, the two-reaction equilibrium. Other than that, hydrogen is a reactant for both reactions, at least in the initial stage, and it is not very clear how the final methanol productivity will be impacted. The reaction is likely to proceed without much difficulty. The role of $H_2O$ in the syngas feed, if any, can be rather easily compensated for by the presence of carbon dioxide, as was the case with the other mechanism.

Water in the reformer effluent gas needs to be removed for initiation of and high conversion in the methanol synthesis reaction. As implied by the reaction chemistry, water in the reactant mixture is detrimental to conversion, regardless of whichever mechanism we may choose to explain the chemistry. However, if the $CO_2$ level in the syngas is excessively low, water does exhibit some compensating and complementing functions.[2,17]

Therefore, there are several factors of significance among most of the low-pressure methanol (LPM) synthesis technologies. They are as follows[2,8]:

1. The presence of carbon dioxide in the feed syngas mixture is essential. Different designs and processes may set this $CO_2$ concentration differently. However, there is a minimum threshold value of this concentration for the process to be functional. If $CO_2$ is absent or deficient in the system, the catalyst deactivation is greatly promoted. However, excess carbon dioxide, beyond the optimal range, under the current catalytic system is of no use at best or hurting the process efficiency, since most of it comes out of the reactor unconverted while affecting the catalyst system in a mildly negative manner.

2. The presence of carbon monoxide in the syngas feed composition is also very important. Lack of CO in the syngas feed results in not only low methanol productivity but also a continuous decrease in methanol productivity.

3. There is an optimal value for the temperature of the methanol synthesis reaction from the standpoint of optimal conversion of syngas as well as the kinetic reaction rate. The rate of reaction is increased with an increase in the temperature by following the Arrhenius type of temperature dependency, whereas the equilibrium conversion is thermodynamically unfavored with an increase in the reaction temperature. Furthermore, there is also a limit for the maximum temperature at which the process can be operated. This ceiling is mostly governed by the temperature tolerance of catalyst ingredients, in particular, the copper component of the catalyst. This temperature is about 280°C–300°C. Beyond this temperature, the catalyst would be subjected to sintering and fusing, which would result in permanent and irreversible damage to the catalyst. The temperature condition for a methanol synthesis reactor needs to be very carefully chosen and managed to allow for the digestion of the exothermic heat of reaction, which could cause local hot spots in the catalyst and/or create thermal stability issues for the reactor.

Even though the sensitivity of overall methanol productivity to $CO_2$ concentration variation is not as pronounced as that of a typical principal reactant of general chemical reaction systems, it has to be noted that the true picture regarding the role of carbon dioxide is inevitably masked by the presence of a very active WGS reaction. The methanol catalyst of $Cu/ZnO/Al_2O_3$ is also a very good catalyst for the WGS reaction, and many of the industrial low-temperature shift (LTS) catalysts have catalyst compositions that are very similar to that of a methanol synthesis catalyst. Conversely, a total lack of $CO_2$ in the feed syngas would cause irreversible damage to the catalyst. There are different views regarding the mechanism of catalyst deactivation in cases when carbon dioxide is lacking in the syngas feed.[7]

### 10.2.2 ACTIVE FORM OF METHANOL SYNTHESIS CATALYST

The commercial methanol synthesis catalyst is prepared by a coprecipitation technique in which both CuO and ZnO are precipitated onto a porous structure of support, typically alumina, $Al_2O_3$. The formula of this catalyst is most frequently expressed as $CuO/ZnO/Al_2O_3$. This is in an oxidized form that is stable upon exposure to air or other oxidizing environments, which is a reason why all catalysts of this type are shipped in an oxidized form for safety and storage. Therefore, this catalyst must be converted to a reduced form before use in a hydrogenation reaction such as methanol synthesis.[19] Since this reduced form is an active form of the catalyst, the procedure of reduction is also called an activation procedure of the catalyst. If the shipped form of the catalyst is not reduced before use in the reactor, hydrogen in the syngas as a reactant in the hydrogenation reactor would be first consumed in reduction of this oxidized form of catalyst. During this unintended reaction process, which is highly exothermic, sintering of the catalyst would occur and induce an irreversible damage to the catalyst. This is why an appropriate catalyst reduction procedure must be adopted and practiced before the reaction process is started.[19]

However, reduction of this catalyst also provides a couple of possibilities that are the source of another controversy regarding which is the active phase of copper in the catalyst, $Cu^0$ or $Cu^{+1}$, or more descriptively, $Cu/ZnO/Al_2O_3$ or $Cu_2O/ZnO/Al_2O_3$.[2]

An effective reduction procedure of the methanol synthesis catalyst in a laboratory reactor is described in the work of Lee[2] and Sawant et al.[19] The basic procedure follows a stepwise reduction strategy, thereby preventing the exothermic heat of reaction for reduction treatment from sintering or thermally annealing the catalyst. The procedure is basically the same for the vapor-phase or the liquid-phase process. Copper oxide in the oxidized (i.e., shipped) form of the catalyst is reduced to a lower oxidation status, such as $Cu^{+1}$ or $Cu^0$, from its original $Cu^{+2}$.

Many pieces of evidence and counterevidence have been presented in the literature as for the active form of copper in the reduced catalyst. Among them, the most striking evidence may be the one obtained by Lee and his coworkers.[16,20] They used a liquid-phase reduction treatment on their $CuO/ZnO/Al_2O_3$ catalyst in a mechanically agitated slurry reactor, and the active catalyst was analyzed by x-ray powder diffraction (XRD) technique. Because the analyzed catalyst, freshly recovered from the reactor, was coated by the protective film of high-boiling white mineral oil (Witco-70 or Freezene-100 oil), the catalyst analyzed using the XRD was not allowed to go through any atmospheric reoxidation to either CuO or $Cu_2O$ before the intended analysis. They found ample presence of $Cu^0$ and, at the same time, total absence of $Cu^{+1}$. By this physical evidence, the active phase of copper in the methanol synthesis catalyst is proven to be $Cu^0$. This finding also tends to support that the need for CO in the feed syngas is essential for keeping a reductive environment for the active phase of the catalyst.[11]

Deactivation of methanol synthesis catalysts may be attributed to the following four principal causes: (1) chemical poisoning by sulfur or carbonyls; (2) sintering, thermal deactivation, or annealing; (3) copper crystallite size growth[21]; and (4) catalyst fouling by carbon deposition and high hydrocarbons. Advances have been made, and all or most of these causes are avoidable in well-designed processes. They include sulfur- and carbonyl-resistant formulations, sintering-resistant formulations for higher temperature of operation, enhanced mass transfer conditions via optimized morphological design, enhanced selectivity via modified catalyst formulations, and more.

Another form of methanol catalyst deactivation is so-called thermal aging, which is a relatively slow process in a well-tuned reactor system. Phenomenologically, the process of thermal aging can be witnessed by the slow but persistent growth of copper crystallite size in the methanol catalyst during the reactive utilization. The phenomenon can be more effectively explained by the theory of crystal size, more specifically, hydrothermal crystallite size growth. It has also been found that the growth of copper crystallite size can be accelerated in syngas conditions that are far from the nominal compositions, such as CO-free or $CO_2$-free gas.[22] When a normally reduced catalyst system was used in a $CO_2$-free environment as an extreme condition for 100 h, the copper crystallite size was found to have grown to 109 Å from 37 Å of the freshly reduced catalyst. The change in the average crystallite size of copper was accompanied by a reduced methanol productivity of 20%, when the very same catalyst was reutilized in a normal syngas condition. While the copper crystallite size growth is relatively slow in a well-tuned reaction system, it is definitely present and contributes to a reduction in the methanol productivity. A regeneration process was developed for the deactivated catalyst whose crystallite size has grown. The regeneration process is based on repeated oxidation and reduction steps that constitute renucleation and redispersion of catalyst crystallites via successive phase changes, namely, $Cu^0$ to $Cu^{+2}$ to $Cu^0$ and so on. Lee and his coworkers[21] reported that their process is able to recover most of the lost activity due to crystallite size growth via the repeated oxidation–reduction cycles. In most cases, once or twice as a repeating cycle was sufficient for the recovery of lost activity back to more than 90% of the original.

### 10.2.3 CHEMICAL EQUILIBRIUM

Methanol synthesis reactions, both CO and $CO_2$ hydrogenation, are thermodynamically not favored at low pressure and high temperature, as shown by a plot of $K_p$ versus $T$ in Figure 10.1. Reducing the temperature of the synthesis reaction is kinetically undesirable because it significantly reduces the reaction rate. Therefore, the synthesis reaction must be carried out at a relatively high temperature, which further pushes the pressure requirement even higher. Higher-pressure operation, on the other hand, may represent higher capital investment, greater energy demands, and more severe operational conditions. Furthermore, a higher temperature increases the potential likelihood for thermal deactivation of the catalyst, and the risk is further complicated by the exothermic heat of the reaction, which could render a thermal stability problem for the reactor. Owing to the unfavorable equilibrium nature, the once-through conversion (or single-pass conversion) of the synthesis reaction is typically low, thus making the recycle duty of the reactor higher. All the commercial processes recycle

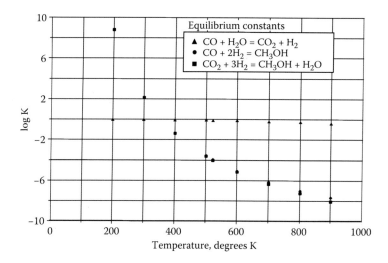

**FIGURE 10.1**   Temperature dependence of equilibrium constants for principal reactions.

the unconverted syngas back to the methanol converter for enhancement of overall conversion, thus improving the process economics. The typical operating conditions for the methanol synthesis reaction are 220°C to 270°C and 50 to 100 bars. It is quite interesting to note that the LPM synthesis process conditions are in the vicinity of the critical point of pure methanol, that is, about 240°C and 80 atm.

The WGS reaction is particularly worthy of note in all processes dealing with any syngas mixture in reactive environments. It has been understood that the reaction takes place predominantly on the heterogeneous surfaces of the catalyst rather than homogeneously, whenever a catalytic system is employed. This does not mean, by any means, that the WGS reaction does not take place homogeneously in all systems. It should be also noted that the WGS reaction has a relatively flat functional relation between the equilibrium constant and the temperature, that is, the temperature dependency of the equilibrium constant for the WGS reaction is weak compared to other syngas reactions, as shown in Figure 10.1. This means that the WGS reaction has a very wide temperature range of significance, namely, from room temperature to as high as 1000°C, thus affecting nearly all syngas-related processes. Another significance of its relatively low equilibrium constant over a wide range of temperatures is that the reaction equilibrium can be easily reversed in direction by changing the compositions (or partial pressures) of the involved species. More often than not, this fact obscures the true picture of the intrinsic mechanism of process chemistry, which is the case with methanol synthesis.

Furthermore, typical WGS reaction catalysts also have very similar compositions to those of the methanol synthesis reaction. This also means that the catalysts used for methanol synthesis will also catalyze the WGS reaction effectively. Owing to its reversibility over a wide range of process conditions as well as its high impact on the final product compositions, the WGS reaction equilibrium is an important issue in designing both methane steam reformers (MSRs) as well as methanol synthesis reactors.

## 10.2.4 Properties of Methanol

Methanol is also known as methyl alcohol, carbinol, methyl hydroxide, methylol, mono-hydroxymethane, wood alcohol, colonial spirit, Columbian spirit, hydroxymethane, or wood naphtha. The CAS number for methanol is 67-56-1. The Resource Conservation and Recovery Act (RCRA) waste number of methanol is U154 and UN 1230.

The density of methanol at room temperature, 25°C, is 0.7918 g/cm$^3$. The heat of formation for methanol as gas ($\Delta H^{\circ}_{f,gas}$) is −201.1 ± 0.2 kJ/mol, whereas that for liquid methanol ($\Delta H^{\circ}_{f,liquid}$) is −239.5 ± 0.2 kJ/mol. The constant-pressure heat capacity of methanol as gas ($C_{p,gas}$) at room temperature is 44 J/K-mol, whereas that for liquid methanol is 80 ± 1 J/K-mol. The Henry's law constant ($K^{\circ}_H$) for solubility of methanol in water at 298.15 K is 210 ± 10 mol/kg-bar.

The critical temperature and pressure of methanol are 513 ± 1.2 K and 81 ± 1.0 bars, respectively, whereas the normal boiling and melting temperatures are 337.8 ± 0.3 and 176.0 ± 1.0 K, respectively. The triple point of methanol is 175.5 ± 0.5 K. The flash point of methanol is 11°C. The enthalpy of vaporization[23] of methanol at room temperature ($\Delta H^{\circ}_{vap}$) is 37.83 kJ/mol. Methanol can be used as a supercritical solvent or cosolvent for a variety of modern separation processes. The hydroxyl group in its molecular structure makes possible unique properties that are normally not attainable from supercritical carbon dioxide (sc-CO$_2$) alone. Its critical point is harsher than CO$_2$'s but milder than H$_2$O's.

Methanol's acidity is pKa = ~15.5, and viscosity is 0.59 mPa-s at 20°C. Owing to its low freezing point, methanol has been popularly used as a cold-weather windshield washer fluid. A concentration of 30% by weight can provide antifreezing protection of −20°C. Because of the hydroxyl group in methanol, it has a strong tendency to hydrogen-bond. A hydrogen bond is not a true bond but a particularly strong form of dipole–dipole interaction. The O-H bonds are strongly polarized, leaving the hydrogen atom with a partially positive charge, which is electrophilic hydrogen. This hydrogen has a strong affinity for nonbonding electrons and, as such, it forms intermolecular attachments with the nonbonding electrons on the oxygen atom. Comparing the two isomers between ethanol (C$_2$H$_5$OH) and dimethylether (DME, CH$_3$OCH$_3$), both of which have a formulation of C$_2$H$_6$O, reveals a striking fact: ethanol has a much higher boiling point (78°C) than DME (−25°C). This significant difference of about 100°C in their boiling points is because ethanol has O-H hydrogen, which is extensively hydrogen-bonded, whereas DME has no O-H hydrogen.

Methanol also reacts easily with carboxylic acids to produce esters, from which water is a by-product of the condensation-type reaction. Methanol is an essential reactant in the manufacture of biodiesel via transesterification reaction, which reacts triglyceride with methanol in the presence of an alkali catalyst such as NaOH to produce a fatty acid methyl ester. With the rapid growth of biodiesel industry and wide utilization of biodiesel as transportation fuel, especially in Europe, the demand for methanol has grown substantially.

The presence of the methyl group in the molecular structure provides chemical affinity toward hydrocarbons, and as such, methanol shows excellent solubility toward a variety of organic materials, while the hydroxyl group in the structure promotes excellent water miscibility and necessary hydrophilicity. The dual nature of solubility makes methanol a good candidate for oxygenate fuel as well as a water remover from fuel systems. As a water remover in gasoline fuel systems, methanol creates a miscible ternary mixture (gasoline–methanol–water) instead of an immiscible binary mixture (gasoline–water).

Methanol has an octane rating of 105 (Research Octane Number [RON]) and can be used as an octane enhancer. Because high performance with a higher compression ratio requires a higher octane rating than its regular counterparts, methanol is used as a racing fuel. Because of the high oxygen content (50% by mass) in the methanol molecular structure, it does have a merit as an excellent oxygenate fuel. However, it should be noted that its blending vapor pressure increase is substantially higher than other competing oxygenate blends such as ethanol and MTBE, the latter of which has been phased out in the United States due to health and environmental risks, thus making it unsuitable as a gasoline-blending fuel. The vapor pressure increase due to gasoline blending is typically measured by *RVP*. High vapor pressure of blended gasoline can increase the chances for evaporative emission of the fuel as well as the risk of having "vapor lock" in the fuel line. Furthermore, methanol, if inhaled or consumed, is far more toxic than ethanol. Evaporative fuel emission is particularly of concern to environmental air quality during summer months, when the emitted hydrocarbon is directly linked with the environmental health problem of "high ozone level in the air."

The common method for measuring vapor pressure of petroleum products is the RVP test. There are basically two methods approved by ASTM, that is, the Reid method (ASTM D323-99a) and the Dry method (ASTM D4953-99a). The Reid method covers experimental measurements of vapor pressure of gasoline, volatile crude oil, and other volatile petroleum products, which include petroleum crude, gasoline, MTBE-blended gasoline, aviation fuel, and other petroleum products. There are four procedures provided by this method, depending on the types and vapor pressure ranges of the tested fuel. However, this method is not applicable to liquefied petroleum gas (LPG) and oxygenated gasoline except MTBE-blended gasoline. Determination of the vapor pressure of LPG is covered in ASTM D1267, whereas determination of the vapor pressure of gasoline–oxygenate blends is treated in ASTM D4953. The latter method is referred to as *Dry method*, which is applicable to most oxygenated gasoline, except MTBE-blended gasoline, with a vapor pressure range from 35 to 100 kPa (5 to 15 psi). There are thermodynamic algorithms developed for estimation of RVP without performing actual measurement. An algorithm developed by Vazquez-Esparragoza et al.[24] used the Gas Processors Association (GPA) Soave–Redlich–Kwong equation of state and assumed that liquid and gas volumes are additive. They found excellent agreement between model prediction and experimental data.

### 10.2.5 Reaction with Methanol

Methanol is used to produce formaldehyde via oxidation reaction:

$$CH_3OH + 1/2 \ O_2 = HCHO + H_2O$$

About 40% of methanol is converted to formaldehyde (HCHO) and, from there, into products as diverse as thermosetting polymers, plywood, paints, explosives, and permanent press textiles.

Methanol is also used for making a variety of ethers including MTBE, ethyl *t*-butyl ether (ETBE), *t*-amyl methyl ether (TAME), DME, and so forth. The first three are normally synthesized by catalytic distillation (CD), whereas the fourth is produced by catalytic dehydration. As the name implies, CD is a hybrid process between a well-established unit operation of distillation and catalytic chemical reaction. Methanol also readily reacts with carboxylic acids to produce methyl esters of these acids. These esters, in particular, fatty acid esters, are quite important as principal ingredients for biodiesel, fuel additives, fuel system cleaners, octane enhancers, and so forth. Triglyceride in plant oil can be reacted with methanol with the aid of alkali catalyst such as sodium hydroxide to produce biodiesel via transesterification reaction.

DME is gaining importance as an alternative fuel. DME is widely used in Europe as direct fuel for vehicles and farm equipment. Dimethyl ether has a high cetane rating (or cetane number [CN]) of 55–60, which makes it a good cetane enhancer as well. Typical diesel engines operate fuels having a CN of 40–50. DME has always been considered a major derivative of methanol, and its synthesis is based on the following stoichiometric equation:

$$2\,CH_3OH = CH_3OCH_3 + H_2O$$

Methanol serves as a direct reactant or as an intermediate in the synthesis of DME, depending on the reaction routes.[25] The first process route is based on conventional dehydration of methanol over a dehydration catalyst, where methanol is a direct reactant, and DME and water are recovered as products. This process is carried out as a stand-alone type of process.

In recent years, however, the second process option, in which methanol is nothing more than an intermediate in the conversion of syngas to DME, that is, single-stage synthesis of DME from synthesis gas, has attracted a fair amount of attention. A single-stage DME synthesis process[26] that bypasses the step of methanol synthesis is worthy of special note, with technological merits and economic advantages including its enhanced once-through productivity and resultant process economics. The reaction stoichiometry can be expressed as:

$$CO_2 + 3H_2 = CH_3OH + H_2O$$

$$CO + H_2 = CO_2 + H_2$$

$$2\,CH_3OH = CH_3OCH_3 + H_2O$$

The process benefits have been twofold: one, the direct synthesis of DME from syngas can overcome the thermodynamic equilibrium limitation imposed by methanol synthesis reaction, and two, the subsequent reaction converting DME into hydrocarbons beneficially realigns the grand scheme of syngas-to-hydrocarbon conversion processes, which is represented by DME-to-hydrocarbons (DTH), DME-to-gasoline (DTG),[30,32] and DME-to-olefins (DTO).[33] The conventional and large-scale proven processes for syngas-to-hydrocarbons[27,28] were represented by *methanol-to-olefin* (MTO), *methanol-to-gasoline* (MTG), and so forth. Even if DME itself is considered a final product, the direct route via single-stage synthesis is far more advantageous from the standpoints of both production cost and process efficiency. The single-stage DME synthesis exploits the equilibrium-unlimited dehydration of methanol to DME to its fullest extent by alleviating the equilibrium-limited nature of the metha-nol synthesis reaction.[10,25,26,29] This is accomplished via a dual-catalytic system in which methanol produced in situ is selectively removed from the reaction mixture by converting the product methanol to DME in the very same reactor. In this process, methanol concentration in the reactor is kept always low, thus keeping the methanol synthesis portion of the overall conversion process far from its chemical equilibrium. Catalyst 1, such as $Cu/ZnO/Al_2O_3$, catalyzes the first reaction of methanol synthesis, whereas catalyst 2, for example, $\gamma$-$Al_2O_3$, catalyzes dehydration of methanol. Both reactions are carried out in the very same reactor, and therefore, such a reaction system may be called a *dual-catalytic system*.[25] Obviously, as an advantage, a higher once-through conversion of syngas to DME can be attained. This enhancement can also be successfully exploited for realignment of the syngas-to-hydrocarbon synthe-sis process.[30]

Consider MTG's alignment, which is basically composed of syngas conversion reactor and gasoline synthesis reactor:

$$\underset{\text{Syngas conversion}}{\{\text{Syngas-to-methanol}\}} + \underset{\text{Gasoline synthesis}}{\{\text{Methanol-to-DME-to-gasoline}\}}$$

where methanol-to-DME conversion takes place in the second reactor, that is, the gasoline reactor. This methanol-to-DME reaction generates water, which is detrimental to hydrocarbon (such as gasoline) synthesis. However, the reaction step of methanol-to-DME conversion can be very advantageously moved from the gasoline reactor to the syngas reactor; thereby, the methanol conversion can be synergistically enhanced by alleviating the equilibrium limitation of metha-nol synthesis reaction. The resultant process alignment for syngas-to-gasoline[30] becomes as follows:

$$\underset{\text{Syngas conversion}}{\{\text{Syngas-to-methanol-to-DME}\}} + \underset{\text{Gasoline synthesis}}{\{\text{DME-to-gasoline}\}}$$

This process concept was introduced by Gogate et al.[29] and Lee et al.[30] Other benefits include catalyst life management, especially for zeolite catalysts such as Mobil's ZSM-5. The new process technology involving the DTH/DTG/DTO con-cept has not yet been demonstrated on a large scale. Even though the MTG plant in New Zealand terminated its successful commercial production in the late 1990s, the idea of production of clean liquid motor fuel via the synthesis route is remark-able. It not only diversifies the feedstock for hydrocarbon fuels such as gasoline, diesel, and jet fuel, but also provides additional and alternative options for energy policy and planning.

DMFCs are unique in their low-temperature, atmospheric-pressure operation, allowing them to be miniaturized to an unprecedented degree. DMFC, combined with the relatively easy and safe storage and handling of methanol, may contribute to the fuel cell-powered consumer electronics.

## 10.3   METHANOL SYNTHESIS TECHNOLOGY

Pure methanol was first isolated as a chemical form in 1661 by Robert Boyle, who called it *spirit of box*, because he produced it via the distillation of boxwood.

However, the systematic synthesis of methanol has a history of about 100 years dating back to the early 1900s, when methanol was almost exclusively produced by the destructive distillation of wood wastes. This is why methanol was called *wood alcohol*. On the other hand, ethanol and isopropyl alcohol are called *grain alcohol* and *rubbing alcohol*, respectively. In 1923, BASF developed a catalytic synthesis process based on a $ZnO/Cr_2O_3$ catalyst. This commercial synthesis technology, later termed *high-pressure methanol synthesis technology*, was popularly adopted by a number of industries for about 50 years. This process was quite successfully operated at a pressure of 250 to 350 atm and a temperature of 350°C to 450°C. Because the operating pressure required by this catalytic process was substantially higher than that for the later version of the synthesis process using a different catalytic system, the process was given the name *high-pressure methanol synthesis*.

In 1963, Imperial Chemical Industries (ICI, the company no longer in existence; their methanol technology now owned by Johnson Matthey) developed a new methanol synthesis technology termed low-pressure methanol *synthesis technology*, which has become an industrial successor to high-pressure synthesis.[2] This process is operated at a pressure of 50 to 100 atm and a temperature of 225°C to 275°C over a catalyst of $Cu/ZnO/Al_2O_3$. The catalyst system used for this process is often referred to as *Cu-based catalyst*, whereas that for high-pressure synthesis is called *Zn-based catalyst*.

A number of different versions of low-pressure methanol synthesis technology have also been developed and are being successfully operated; however, most of these are based on very similar process concepts in terms of catalysts, synthesis chemistry, incorporation of steam reforming, and so forth. It is interesting to note that conventional low-pressure synthesis of methanol is carried out under conditions close to the critical temperature and pressure of pure methanol.

The reactants for methanol synthesis are carbon dioxide, carbon monoxide, and hydrogen. These species constitute the main ingredients of synthesis gas. Synthesis gas may be obtained from diverse sources including natural gas, coal, municipal wastes, coke oven gas, biomass, other hydrogen sources, and so forth. Industries have been predominantly using the steam reforming of methane to generate hydrogen and carbon monoxide in a proportion of $H_2/CO = 2$–$3$, as shown in the following stoichiometry:

$$CH_4 + H_2O = 3H_2 + CO$$

It should be noted here that hydrogen produced by steam reforming comes from both water and hydrocarbon. To synthesize methanol over a Cu-based catalytic system, carbon dioxide ($CO_2$) may have to be added to the syngas originating from the steam reforming of methane. The presence of carbon dioxide in the syngas composition is crucial because it directly affects the catalytic activity and life. In the steam reformer, the WGS reaction also takes place:

$$CO + H_2O = CO_2 + H_2$$

As mentioned earlier, the WGS reaction is easily affected by chemical equilibrium over a wide range of temperatures, thus affecting the reformer product compositions significantly. The optimal percentage of $CO_2$ in the syngas for methanol synthesis varies from process to process. However, it is generally in the range of 2%–6%, with some exceptions that use substantially higher $CO_2$ contents in the feed syngas.

In methanol synthesis, scientists and engineers often use a term called *balanced gas* in referring to a 2:1 mixture of $H_2$ and CO. This 2:1 mixture is not an endorsement of a CO hydrogenation mechanistic path but, rather, a stoichiometric assessment of species balance that methanol as a product has a molecular formula of direct addition of 1 molecule of CO and 2 molecules of $H_2$. This is also consistent with the experimental facts that the number of moles of hydrogen consumed versus the number of moles of carbon monoxide consumed is slightly more than 2.[2,7] Although natural gas is an excellent source for such syngas, owing to its high hydrogen content, other carbonaceous and hydrocarbon-rich resources can also be viable sources, depending on the availability and local economy.

Some of the newer processes for methanol synthesis adopt CO-rich syngas as the feed syngas rather than $H_2$-rich syngas. The CO-rich syngas is typically generated from a resource that has a low H-to-C ratio, such as coal. Depending on the geological region, this option may prove to be an economically better option. This type of syngas is often termed *unbalanced gas*, which means that the $H_2$-to-CO ratio is much lower than 2:1. Second-generation coal gasifiers such as *Texaco* and *Shell* gasification processes, as well as Koppers-Totzek gasifiers, yield syngas with low $H_2$-to-CO ratios, typically, 0.75–1.0. The *liquid-phase methanol synthesis process*, also known as the LPMeOH process, originally developed by Chem Systems Inc. in 1975, is a good example of the processes that are targeting CO-rich syngas as the feed syngas. Owing to its substoichiometric $H_2$ content in the feed gas, its $H_2$ conversion per pass is relatively high, thus making such a process more suitable for a once-through synthesis technology such as the *once-through methanol (OTM)* process. As the name implies, this process is based on the synthesis of methanol without any recycling of unreacted syngas. The rest of the unreacted syngas can be used for power generation.

The pressure and temperature of the synthesis reaction are obviously two of the most important operating parameters from chemical kinetics and equilibrium conversion considerations. If the temperature is raised for the synthesis reaction, the kinetic rate of reaction increases while the conversion of carbon monoxide decreases. If the pressure of the reaction is increased, the conversion of CO also increases, but the increase is not very substantial above the pressure of 80 atm. The situation is also further complicated by the presence of the WGS reaction, which is also limited by chemical equilibrium at a typical methanol synthesis reaction condition. This WGS equilibrium can be very easily reversed in terms of its direction by the concentration of water in the reactor feed stream. Even 5 mol% of water in the feed syngas stream of $H_2/CO = 3:1$ is more than sufficient to affect the direction of the WGS reaction at 250°C and 75 atm. Under this specific circumstance, the WGS reaction proceeds in the forward direction. In the case of $H_2O$-free feed, otherwise under nominally the same condition, the WGS reaction proceeds in the reverse direction.

A typical methanol process technology in the modern era involves several common process steps, namely, (1) feed purification, (2) steam reforming, (3) syngas compression, (4) catalytic synthesis, (5) crude methanol distillation, and (6) recycle and recovery. If the syngas is prepared from coal, coal gasification becomes an important step for the overall process design and economics. Although the Lurgi gasifier yields hydrogen-rich syngas, Texaco, Shell, and Koppers-Totzek gasifiers yield carbon monoxide-rich syngas.

As for the steam reforming, there are typically two types of processes, that is, two-stage reforming and autothermal reforming (ATR). Although the two processes are very similar to each other, ATR has a thermal balance over the reactor by adopting a hybrid combination of sacrificial partial oxidation with the subsequent steam reformation reaction. The former reaction of partial combustion generates exothermic heat, which can be utilized by the endothermic reforming reaction; as such, an energy balance is achieved for the reactor. There are modified process technologies that can also be used for reformation of hydrocarbon feedstocks into syngas. They include supercritical water reformation, dry reformation, and tri-reformation.

The process design varies largely based on the availability of feedstocks of different types, process energy efficiency and local energy economics, and financial restrictions related to capital investment. In the ensuing subsections, a number of methanol synthesis process technologies are explained and compared.

### 10.3.1 Conventional ICI's 100 atm Methanol Synthesis Process

Even though it had long been desired in the mid-1900s to reduce the operating pressure of the methanol synthesis process, it was found that a process with a much lower pressure would not be ideal for large-capacity units. A simple reason for this finding may be that under low-pressure conditions, the equipment has to be very large, and the chemical reaction is not as fast as desired. With the advancement and development of better materials and equipment design, a newly focused objective was the search for a catalytic synthesis system that would be more active at about 100 atm. The result of this effort was highlighted by the ICI's announcement of a Cu/ZnO/Al$_2$O$_3$ catalyst system. This process was tried in August 1972 with great success. This milestone process has been enhanced and modified with subsequent designs and enhancements, especially in the areas of energy efficiency of the process as well as process optimization. The ICI's methanol synthesis technology is now owned by Johnson Matthey. A schematic of this process is given in Figure 10.2.

The original flow sheet includes two parts of the process, namely, natural gas reforming (for syngas generation) and synthesis (syngas conversion into methanol) sections. Even though unelaborated in Figure 10.2, process economics depends very heavily on the heat recovery and energy integration, recycle schemes, and refining and

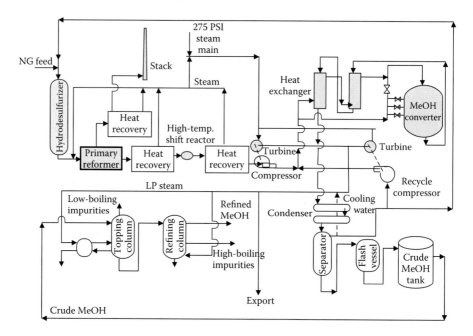

**FIGURE 10.2**   A schematic of ICI's low-pressure methanol synthesis process.

separation. Along with the efficient design, management of catalyst life has always been the principal issue of process maintenance and enhancement. Needless to say, catalyst life and efficiency are directly tied to the productivity of the plant. The ICI process has provided a basis and foundation for the later developments and designs. Most of commercial designs for methanol synthesis include both syngas manufacturing and methanol conversion sections, since capital investment of syngas manufacturing parts is typically higher than that of methanol synthesis, and energy saving for the plant can be achieved by efficient integration among systems involved in these two stages.

## 10.3.2   HALDOR TOPSOE A/S LOW-PRESSURE METHANOL SYNTHESIS PROCESS

This process is designed to produce methanol from natural or associated gas feedstocks, utilizing a two-step reforming process to generate a feed syngas mixture for the methanol synthesis.[31] Associated gas is natural gas produced with crude oil from the same reservoir. It is claimed that the total investment for this process is lower than with the conventional flow scheme based on straight steam reforming of natural gas by approximately 10%, even after considering an oxygen plant.

As shown in Figure 10.3, *two-stage reforming* is conducted by primary reforming, in which a preheated mixture of natural gas and steam is catalytically reacted, followed by secondary reforming, which further converts the exit gas from the primary reformer with the aid of oxygen that is fed separately. The amount of oxygen required as well as the balance of conversion between the primary and secondary reformers need to be properly adjusted so that a balanced syngas, that is, in a stoichiometric ratio of 2:1 of $H_2/CO$, is obtained with a low inert content.

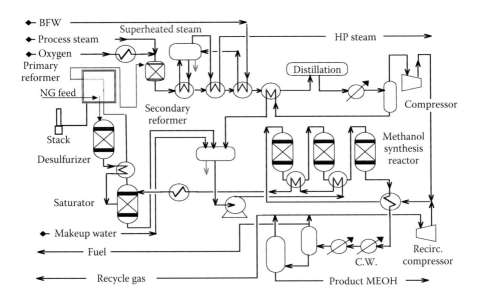

**FIGURE 10.3**    A schematic of the Haldor Topsoe A/S methanol synthesis process.

As an energy integration step, the heat content of flue gas is recovered for pre-heating reformer feed. Similarly, the heat content of the process gas is utilized for producing superheated high-pressure steam, preheating boiler feed water (BFW), and preheating process condensate before its entry into the saturator and reboiler of the distillation section. The synthesis section is composed of three adiabatic reactors with heat exchangers between the reactors; thus, exothermic heat of reaction is recovered and used for heating saturator water. Another energy integration is accomplished by cooling the effluent from the last reactor by preheating the feed to the first reactor. The total energy consumption for the process is claimed to be about 7.0 Gcal/ton of product methanol, including oxygen production. The process technology is suited for smaller as well as very large methanol plants up to 10,000 tpd. However, it has to be noted here that it is often very difficult to compare flowsheets for methanol synthesis among various commercial designs and processes because it is difficult to establish the common unbiased bases to compare reported design and operating data.

### 10.3.3   KVAERNER METHANOL SYNTHESIS PROCESS

This process developed by Kvaerner Process Technology/Synetix, United Kingdom, is based on an low-pressure methanol synthesis process and two-stage steam reforming, similar to the Haldor Topsoe process. Figure 10.4 shows a schematic of the Kvaerner methanol synthesis process. The gas feedstock may be natural or associated gas. In this process, however, carbon dioxide can be used as a supplementary feedstock to adjust the stoichiometric ratio of the syngas. Recent plants based on this process have an energy efficiency of 7.2–7.8 Gcal/ton of product methanol,[26] which is lower than Topsoe's published energy efficiency. However, this process is more suited for regions with high availability of low-cost gas such as $CO_2$-rich natural gas and financial restrictions of low

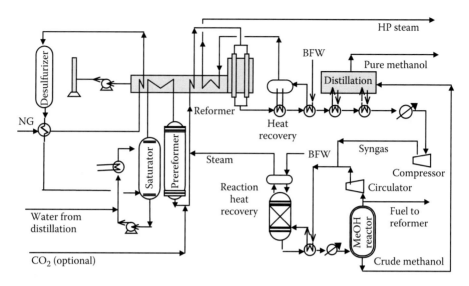

**FIGURE 10.4**   A schematic of the Kvaerner methanol synthesis process.

capital investment. There are a number of commercial plants currently in operation based on this design, and their typical sizes range from 2000 to 3000 mtpd.

### 10.3.4   KRUPP UHDE'S METHANOL SYNTHESIS TECHNOLOGY

The process, developed by Krupp Uhde GmbH, is also based on the low-pressure synthesis chemistry of methanol as well as steam reforming for synthesis gas generation. A unique feature of this process is its flexibility of feedstock choice, which includes natural gas, LPG, or heavy naphtha.[31]

The steam reformer is uniquely designed by Krupp Uhde and is a top-fired box-type furnace with a cold outlet header system. The steam reforming reaction takes place heterogeneously over a nickel catalyst system. The reformer effluent gas containing $H_2$, CO, $CO_2$, and $CH_4$ is cooled from 880°C to ambient temperature eventually, and most of the heat content is recovered by steam generation, BFW preheating, preheating of demineralized water, and heating of crude methanol for three-column distillation.

The typical energy consumption, including feed and fuel, ranges from 7 to 8 Gcal per metric ton of methanol and is very much dependent on individual plant concepts and designs. Eleven plants were built until 2005, using this technology. Figure 10.5 shows a schematic of Krupp Uhde's methanol synthesis process.

### 10.3.5   LURGI ÖL-GAS-CHEMIE GMBH PROCESS

This process is also designed to produce methanol in a single-train plant starting from natural gas or oil-associated gas with capacities up to 10,000 mtpd.[31] It can be used to increase the capacity of an existing methanol plant based on steam reforming. Figure 10.6 shows a schematic of this process. Steam reforming of natural gas is

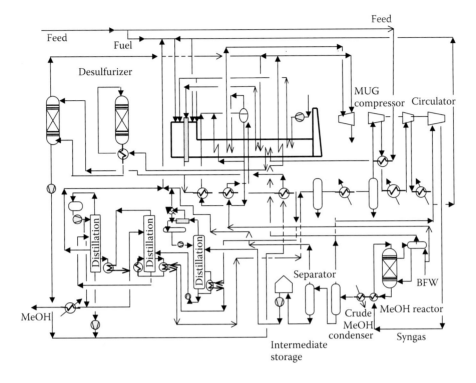

**FIGURE 10.5** A schematic of Krupp Uhde's methanol synthesis process. MUG, make-up gas.

accomplished in two stages, that is, prereforming and autothermal reforming (ATR). In the prereformer, the mixture gas of desulfurized natural gas and steam is converted to $H_2$, CO, $CO_2$, and $CH_4$, whereas in the autothermal reformer, the gas is reformed with oxygen and steam, producing product gas containing $H_2$, CO, $CO_2$, and a small amount of unconverted $CH_4$, in addition to low-pressure steam. Oxygen is involved in this reaction scheme to generate exothermic heat by partial oxidation of sacrificial natural gas to provide the necessary endothermic heat for the subsequent reforming reaction.

The reformed gas, that is, syngas, is mixed with hydrogen from the pressure swing adsorption (PSA) to increase the $H_2$-to-CO ratio. The produced synthesis gas is pressurized and mixed with recycled gas from the synthesis loop. The reaction takes place under near-isothermal conditions in the Lurgi water-cooled methanol reactor, which houses a fixed bed of catalyst in vertical tubes surrounded by boiling water. The reactor effluent gas is cooled to 40°C to separate methanol and water from the unreacted syngas. Methanol and water are separated in distillation units, whereas the major portion of the gas is recycled back to the methanol synthesis reactor for higher overall conversion. As mentioned in the earlier subsection, the single-pass conversion to methanol is typically low for all processes; therefore, recycling of the gas is imperative. Enhancements have been made, especially in the efficiency of the Lurgi combined converter (LCC), to reduce the recycle ratio down to about 2. The process

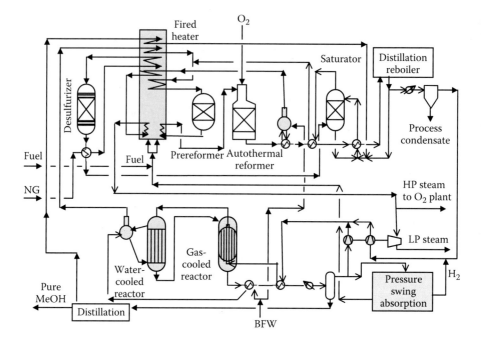

**FIGURE 10.6**  A schematic of the Lurgi Öl-Gas-Chemie GmbH process.

water is preheated in a fired heater and used as makeup water for the saturator, thus minimizing unnecessary water usage and treatment.

The reformed gas from the second-stage reformer contains a considerable amount of thermal energy that is recovered as high-pressure steam for energy required for the preheater and reboiler. The energy consumption for the process, including utilities and oxygen plant, is reported as about 7.1–7.2 Gcal/ton of product methanol.

## 10.3.6  SYNETIX LPM PROCESS

This is an improved version of the ICI's original low-pressure methanol (LPM) process.[31] This process is designed to produce a refined, high-purity methanol from natural gas, but it can also handle a variety of other hydrocarbon feedstocks, including naphtha, coal, and other petrochemical off-gas streams. This process is ideal for large capacities where conventional methanol processes may not be suitable or viable. Figure 10.7 shows a schematic of the Synetix LPM process.

The process consists of three principal sections, namely: (1) syngas preparation, (2) methanol synthesis, and (3) methanol purification. The process used for generation of syngas is the steam reforming process, whose product gas contains steam, hydrogen, carbon monoxide, and carbon dioxide. The reforming catalyst is nickel based, and therefore, the feed gas must be desulfurized before entering the reformer. The reforming catalyst is very sensitive and susceptible to sulfur poisoning. The syngas leaving the reformer is typically at 880°C and up to 20 atm, similar to the Lurgi process. The peak temperature of the gas mixture in the reformer is significantly higher than this temperature.

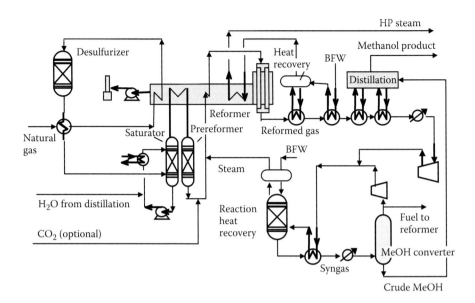

**FIGURE 10.7**    A schematic of the Synetix LPM process.

The methanol synthesis section involves a circulator, methanol reactor, heat recovery and cooling unit, and methanol separator. The synthesis catalyst is copper based, and typical operating conditions are 200°C to 290°C and 50 to 100 atm. From the reaction engineering standpoint, a temperature of close to 200°C is too low for a meaningful rate of chemical reaction, whereas a temperature of 290°C may be too high and too close to the potential catalyst-sintering range. As mentioned earlier, methanol formation is limited by chemical equilibrium, thus limiting the exit concentration only up to 7%. This value is higher at a lower reaction temperature, as can be predicted by the equilibrium constant for the conversion reaction. After condensing product methanol out by chilling, the unreacted syngas is recycled back to the methanol reactor for higher overall conversion.

The crude product methanol contains water and small amounts of undesired by-products, which are separated in a two-column distillation system. The two columns are a topping and a refining column. The former removes all light ends including dissolved gases, light hydrocarbons, and low-molecular-weight ethers, esters, and acetone, whereas the latter separates methanol from water and also eliminates higher hydrocarbons and alcohols by a side discharge from the column.

The total energy consumption for a self-contained plant is typically around 7.8 Gcal/ton. This figure also depends on the type of feedstock used.

### 10.3.7  LIQUID-PHASE METHANOL PROCESS

The liquid-phase methanol process was originally developed by Chem Systems Inc. in 1975. The research and development of this process was sponsored by the U.S. Department of Energy and Electric Power Research Institute (EPRI).

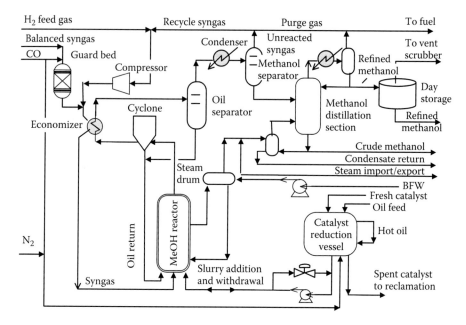

**FIGURE 10.8** A schematic of the liquid-phase methanol synthesis process.

Commercialized by Air Products and Chemicals Inc. and Eastman Chemical Co. in the 1990s, the process is based on the low-pressure methanol synthesis process concept. The chemical reaction is carried out in a slurry reactor using $Cu/ZnO/Al_2O_3$ catalyst at 230°C to 260°C and 50 to 100 atm. The commercial reactor used is a liquid entrained reactor (LER) in which fine powder catalyst is slurried in inert high-boiling oil, typically white mineral oil such as, but not limited to, Witco-70 or Freezene-100. Fed gaseous reactants are dissolved in the inert oil, and the dissolved molecular species, that is, reactants, are reacted on the catalytic surface in a slurry. The process has enhanced heat transfer characteristics owing to the higher thermal capacity of inert oil when compared to comparable vapor-phase processes. The heat transfer enhancements were achieved at some expense of mass transfer conditions due to the presence of an inert oil phase. To enhance the mass transfer properties of the process, specially developed, fine-powder catalysts are used, thereby effectively minimizing pore diffusional resistances and alleviating external mass transfer limitation. Special features of this process include its capability to handle unbalanced CO-rich syngas and high single-pass conversion of syngas. As a result, the process can be packaged with advanced gasification technology such as the Texaco gasifier and Shell gasifier. Figure 10.8 shows a schematic of the liquid-phase methanol synthesis process.

## 10.4   FUTURE OF METHANOL

Methanol can be produced from conventional natural gas, shale gas, biogas, coal, mixed wastes, and biomass. The process technology based on natural gas reforming is mature and practiced on a large scale. The starting material for syngas production is expected to change at least partially from conventional natural gas to shale gas, coal, biomass, wastes, and mixed cofeed. In addition, comprehensive multifuel generation as well as cogeneration process concepts involving methanol synthesis as one of the key processes are becoming more attractive. Methanol can also be used efficiently as an integrated gasification combined cycle (IGCC) companion fuel as well. Development of efficient engines utilizing methanol or methanol blends without rendering environmental concerns is also highly conceivable. With advances in internal combustion engines (ICEs) based on methanol fuel, it is also conceivable that methanol can be used more popularly as *transportation fuel* for passenger vehicles, not just for race cars. Following development of the efficient *DME* synthesis process as well as utilization of DME in a variety of internal combustion engines, methanol has good potential in the alternative fuel market. In the context of the ever-rising cost of liquid transportation fuel throughout the world, processes like *MTG*[27,28] and MTO as well as DTG[30,32] and DTO[33] make much more sense now and in the future than before, thus making methanol a more valuable chemical and fuel commodity. As a result of recent developments in DMFC and its use in consumer electronics, methanol has great potential in the future consumer electronics market. Further, methanol can play a more important role in the petrochemical industry as a *building-block chemical* of nonpetroleum origin.

Methanol also has a unique potential that could provide a very productive and decisive answer to the greenhouse gas (GHG) problem triggered by carbon dioxide in the atmosphere. The conversion of carbon dioxide and carbon dioxide-rich syngas into methanol on an industrial scale would be a remarkable breakthrough, which could essentially make carbon dioxide a valuable chemical feedstock instead of an environmental culprit.

## REFERENCES

1. Methanol Institute website, The methanol industry, http://www.methanol.org/Methanol-Basics/The-Methanol-Industry.aspx, last accessed July 31, 2013.
2. Lee, S., *Methanol Synthesis Technology*, CRC Press, Boca Raton, FL, 1990.
3. Lee, S., *Methane and Its Derivatives*, Marcel Dekker, New York, 1997.
4. Methanex Corporation website, www.methanex.com/methanol/techsafetydata.htm, last accessed October 13, 2013.
5. Larminie, J., and Dicks, A., *Fuel Cell Systems Explained*, 2nd ed., John Wiley & Sons, New York, 2003.
6. Olah, G.A., Beyond oil and gas: The methanol economy, *Angew. Chem. Int. Ed.*, 44(18), 2636–2639, 2005.
7. Cybulski, A., Liquid-phase methanol synthesis: Catalysts, mechanism, kinetics, chemical equilibria, vapor-liquid equilibria, and modeling—Review, *Catal. Rev.: Sci. Eng.*, 36(4), 557–615, 1994.
8. Lee, S., Parameswaran, V., Wender, I., and Kulik, C.J., The roles of carbon dioxide in methanol synthesis, *Fuel Sci. Technol. Int.*, 7(8), 1021–1057, 1989.
9. Chinchen, G.C., Mansfield, K., and Spencer, M.S., The methanol synthesis-how does it work?, *CHEMTECH*, 20(11), 692–699, 1990.
10. Lee, S., and Parameswaran, V., Reaction mechanism in liquid-phase methanol synthesis, EPRI-ER/GS-6715, 1–206, Electric Power Research Institute, Palo Alto, CA, 1990.
11. Wender, I., Reactions of synthesis gas, *Fuel Process. Technol.*, 48, 189–297, 1996.

12. Klier, K., Methanol synthesis, *Adv. Catal.*, 30, 243–313, 1982.

13. Chinchen, G.C., Denny, P.J., Parker, D.G., Spencer, M.S., and Whan, D.A., Mechanism of methanol synthesis from $CO_2/CO/H_2$ mixtures over copper/zinc oxide/alumina catalysts: Use of 14C-labelled reactants, *Appl. Catal.*, 30, 333–338, 1987.

14. Rozovskii, A., New data on the mechanism of catalytic reductions with the participation of carbon dioxide, *Kinet. Katal.*, 21, 97–107, 1980.

15. Lee, S., Lee, B.G., Gogate, M.R., and Parameswaran, V., Fundamentals of methanol synthesis, *Proceedings of DOE Annual Contractors' Conference on Indirect Coal Liquefaction*, U.S. DOE/PETC, Pittsburgh, PA, 1989.

16. Sawant, A.V., Lee, S., and Kulik, C.J., Phases in the active liquid phase methanol synthesis catalyst, *Fuel Sci. Technol. Int.*, 6, 151–164, 1988.

17. Sawant, A., Rodrigues, K., Kulik, C.J., and Lee, S., The effects of carbon dioxide and water on the methanol synthesis catalyst, *Energy Fuels*, 3(1), 2–7, 1989.

18. Parameswaran, V.R., Lee, S., and Wender, I., The role of water in methanol synthesis, *Fuel Sci. Technol. Int.*, 7, 899–918, 01/01; 1989.

19. Sawant, A., Parameswaran, V., Lee, S., and Kulik, C.J., In-situ reduction of a methanol synthesis catalyst in a three-phase slurry reactor, *Fuel Sci. Technol. Int.*, 5(1), 77–88, 1987.

20. Lee, S., Parameswaran, V., Sawant, A., and Ko, M.K., Potential improvements in methanol synthesis, *Proceedings: Twelfth Annual EPRI Contractors Conference on Fuel Science and Conversion*, AP-5460-SR, 6, 1-18, Electric Power Research Institute, Palo Alto, CA, 1988.

21. Lee, S., Sawant, A., and Kulik, C.J., Process for methanol catalyst regeneration using crystallite redispersion, U.S. Patent No. 5,004,717, April 2, 1991.

22. Sawant, A., Lee, S., and Foos, A., Crystal size growth in the liquid phase methanol synthesis catalyst, *Fuel Sci. Technol. Int.*, 6, 569–589, 1988.

23. Majer, V., and Svoboda, V., *Enthalpies of Vaporization of Organic Compounds: A Critical Review and Data Compilation*, Blackwell Scientific Publications, Oxford, 1985, p. 300.

24. Vazquez-Esparragoza, J.J., Iglesias-Silva, G.A., Hlavinka, M.W., and Bullin, J.A., How to estimate Reid vapor pressure (RVP) of blends, in *Encyclopedia of Chemical Processing and Design*, Vol. 47, McKetta, J.J., Ed., Marcel Dekker, New York, 1994, pp. 415–424.

25. Lee, S., and Gogate, M.R., Development of a single-stage liquid-phase synthesis process of dimethylether from syngas, EPRI-TR-100246, 1–179, Electric Power Research Institute, Palo Alto, CA, 1992.

26. Lee, S., and Sardesai, A., Liquid phase methanol and dimethyl ether synthesis from syngas, *Top. Catal.*, 32(3&4), 197–207, 2005.

27. Chang, C.D., Hydrocarbons from methanol, *Catal. Rev.: Sci. Eng.*, 25(1), 1–118, 1983.

28. Chang, C.D., and Silvestri, A.J., MTG origin, evolution, operation, *CHEMTECH*, 10, 624–631, 1987.

29. Gogate, M.R., Kulik, C.J., and Lee, S., A novel single-step dimethyl ether (DME) synthesis in a three-phase slurry reactor from CO-rich syngas, *Chem. Eng. Sci.*, 47(13–14), 3769–3776, 1992.

30. Lee, S., Gogate, M.R., and Fullerton, K.L., Catalytic process for production of gasoline from synthesis gas, U.S. Patent No. 5,459,166, 1995.

31. Petrochemical processes special report, *Hydrocarbon Process.*, 82(3), 105, 2003.

32. Lee, S., Gogate, M.R., and Kulik, C.J., Methanol-to-gasoline vs. DME-to-gasoline: Process comparison and analysis, *Fuel Sci. & Tech. Intl.*, 13(8), 1039–1058, 1995.

33. Sardesai, A. and Lee, S., Alternative source of propylene, *Energy Sources*, 27(6), 489–500, 2005.

# 11 Ethanol from Corn

*Sunggyu Lee*

## CONTENTS

## 11.1 INTRODUCTION

As a monohydric $C_2$ alcohol, ethanol or ethyl alcohol is one of the simplest alcohols and has long been consumed throughout the world in human history. The principal use of ethanol has always been in the preparation of alcoholic beverages till very recently. Commonly, ethanol has been called simply "alcohol" and "spirits." Ethanol is readily produced by fermentation of simple sugars that are obtained from sugar crops or can be easily converted from starch crops. Needless to say, fermentation is one of the earliest developed biochemical processes in human history. This method of alcohol preparation has long been practiced and refined throughout the world. Starting materials for such fermentation include a wide variety of biological products: starchy grains such as corn, barley, potato, rice, and wheat; sugar crops such as sugarcane and sugar beets; fruit crops such as grapes, apples and plums; and lignocellulosic materials such as wood chips, corn stover, and grasses. Ethanol produced by fermentation using starchy grains may be called *grain ethanol*, whereas ethanol produced from cellulosic biomass may be called *bioethanol* or *biomass ethanol*. Both grain ethanol and bioethanol are produced via biochemical processes, whereas *chemical ethanol* is synthesized by chemical synthesis routes that do not involve fermentation or biological product as starting material.

Ethanol, ethyl alcohol ($C_2H_5OH$), is a clear, colorless, volatile, and flammable liquid. As a fuel, ethanol burns cleanly because of the oxygen content in its molecular structure and because it possesses a high octane rating by itself, having a research octane number (RON) is 108–109. Therefore, ethanol is most commonly used to increase the octane rating of blend gasoline as well as to improve the emission quality of the gasoline engine. Owing to the presence of oxygen in it structure, ethanol is classified as an oxygenated fuel. In many regions of the United States, ethanol is blended up to 10% with conventional gasoline. The blend of 10% ethanol and 90% conventional gasoline is conveniently called *E10 blend* or simply *E10*. Ethanol is quite effective as an oxygenated blending fuel, because its Reid vapor pressure (RVP) is relatively low compared to methanol and other simple oxygenates, and its overall effect on the environment and public health is minimal. The U.S. Environmental Protection Agency (EPA) has promulgated regulations for fuel volatility limitations, under Section 211(h) of the Clean Air Act, which prohibits the sale of gasoline (and blended gasoline) with an RVP that exceeds 9.0 psi (or, 62 kPa) in "volatility attainment areas" and 7.8 psi (or 53.8 kPa) in "volatility nonattainment areas." The act includes an important exception to the volatility limitation that fuel containing "gasoline and 10 percent denatured anhydrous ethanol" can exceed the applicable RVP limitation by 1.0 psi (or 6.8 kPa). This exception has allowed E10 to be sold widely on the U.S. fuel market.

As mentioned earlier, ethanol can also be produced from any biological materials that contain appreciable amounts of sugar or feedstocks that can be readily converted into sugar. The former group includes sugar beets and sugarcanes, whereas the latter group includes starch and cellulose. For example, corn contains starch that can be easily converted into sugar and is, therefore, an excellent feedstock for ethanol fermentation. Because corn can be grown and harvested repeatedly, this feedstock eminently qualifies as a *renewable feedstock*; that is, this feedstock will not be easily depleted simply by consumption. The same argument of renewable feedstock can be made for cellulosic materials as well. Cellulosic materials are generally more difficult to convert into ethanol but possess unparalleled merits as their being nonfood crops and inexpensively available as raw materials.

Fermentation of sugars produces ethanol, and this process technology has been practiced for well over 2000 years in practically all regions of the world. Sugars can also be derived from a variety of sources. In Brazil, as an example, sugar from sugarcane is the primary feedstock for the country's ethanol industry, which has been very active and successful. In North America, the sugar for ethanol production is usually obtained via enzymatic hydrolysis of starch-containing crops such as corn or wheat. The enzymatic hydrolysis of starch is a simple, inexpensive, and effective process and is a commercially mature technology. Therefore, this process is used as a baseline or a benchmark that other hydrolysis processes can be compared against. Although the principal merit of ethanol production by fermentation of sugar and starch is in its technological simplicity, high efficiency, and renewability, its disadvantage is that the feedstock tends to be expensive on the marketplace and also competitively used for other principal applications such as food. Therefore, "food versus fuel" or "food versus oil" is an unavoidable critical issue addressing the risk of diverting farmland and/or food crops for production of biofuels such as corn ethanol to the potential detriment of the invaluable food supply on a global or regional scale. It has also contributed directly and indirectly to the escalation of food price, including corn price, which, in turn, raises the cost of feedstock and hurts the profit margin of the ethanol industry. Technoeconomically speaking, this high cost of feedstock can be offset to a certain extent by the sale of value-added by-products and coproducts such as dried distillers grains (DDGs). Many corn refineries produce both ethanol and other corn by-products such as cornstarches, sweeteners, and DDGs so that the capital and manufacturing costs can be kept as low as possible by maximizing the overall process revenue. While they are manufacturing ethanol, corn refiners also produce valuable coproducts such as corn oil, corn syrups, and corn gluten feed. The North American ethanol industry is, therefore, investing significant efforts in developing new value-added by-products (and coproducts) that are higher in value and minimizing the process wastes, thus constantly making the grain ethanol industry more cost-competitive. Efforts are also being made to utilize corn cob and stalks for profitable uses, thus minimizing the wastes from corn refineries.

Corn refining in the United States has a relatively long history going back to the time of the Civil War, with the development of cornstarch hydrolysis process. Before this milestone, the main sources for starch had been coming from wheat and potatoes. In 1844, the Wm. Colgate and Company's wheat starch plant in Jersey City, NJ, became, unofficially, the first dedicated cornstarch plant in the world. By 1857, the cornstarch industry accounted for a significant portion of the U.S. starch industry. However, for this early era of corn processing, cornstarch was the only principal product of the corn refining industry, and its largest customer was the laundry business. Cornstarch has also been used as a thickening agent in liquid-based foods such as soups, sauces, and gravies.

The industrial production of dextrose (or, D-glucose) from cornstarch started in 1866. This industrial application and subsequent scientific developments in the chemistry of sugars served as a major breakthrough in the starch technology and processing. Other product developments in corn sweeteners followed and took place with the first manufacture of refined corn sugar, or anhydrous sugar, in 1882. In the 1920s, corn syrup technology advanced significantly with the introduction of enzymatically hydrolyzed products. Corn syrups contain varying amounts of maltose (a disaccharide formed by a condensation reaction of two glucose molecules joined with an $\alpha$ $(1 \rightarrow 4)$ bond) and higher oligosaccharides (a polymeric saccharide with a small number, about 3 to 10, of monosaccharides). Even though the production of ethanol by corn refiners had begun as early as after World War II, major quantities of ethanol via this process route were not produced until the 1970s, when several corn refiners began fermenting dextrose to make beverage and industrial alcohol. The corn refiners' entry into the fermentation business has become a significant milestone for the major changes and transformation of the corn processing industry, especially in the fuel ethanol industry. The corn refining industry seriously began to develop an expertise in industrial microbiology, fermentation technology, separation process technology, energy integration technology, process equipment design, by-product and coproduct development, waste utilization and minimization, and environmental technology.

As of today, cornstarch and glucose are still important products of the corn wet milling industry. However, the products of microbiology and biochemical technology including ethanol, fructose, food additives, and target chemicals have gradually overtaken them. New research and developments have significantly expanded and diversified the industry's product/by-product/coproduct portfolio, thus making the industry more profitable, flexible to market demands, competitive and technologically advanced, and economically sustainable as a healthy business.

Lignocellulosic materials such as agricultural, hardwood, and softwood residues are also good sources of sugars for ethanol production. The cellulose and hemicellulose components of these materials are essentially long and high-molecular-weight chains of sugars. They are protected by components called lignin, which functions more like "glue" that holds all of these materials together in the structure by cross-linking different polysaccharides. Therefore, the liberation of simple sugars from lignocellulosic materials is not as simple and straightforward as that from sugar crops or starch crops and therefore involves well-conceived biological and/or chemical treatments. However, the biggest undeniable advantage of cellulosic ethanol is its use of nonfood feedstocks and no detrimental use of arable land for fuel production. Details of cellulosic ethanol technology are covered in Chapter 12 and, therefore, not repeated here.

Fuel ethanol, in particular, corn ethanol, plays three principal roles in today's economy and environment:

1. Ethanol in the United States replaces a significant amount of imported oil with a renewable domestic fuel. The 10% blend wall, as in the case with E10, represents the replacement of 10% of the total gasoline consumption in the United States by ethanol, which is roughly equivalent to over 13 billion gallons of ethanol in 2012.
2. Ethanol is an important oxygenated component of gasoline reformulation to reduce air pollution in many U.S. metropolitan areas, which are not achieving air quality standards mandated by the Clean Air Act Amendments (CAAA) of 1990. Ethanol is a cleaner-burning fuel due to its oxygen-containing molecular structure and also a superior gasoline blend fuel due to its renewability as a fuel and relatively low RVP of the blended fuel.
3. Ethanol provides a major income boost to farmers and agricultural communities where most ethanol feedstock is produced. Global corn prices have escalated more sharply than other crops due to the increased demand, and

higher corn prices have, in turn, motivated farmers to increase corn acreage at the expense of other crops, such as soybeans and cotton, raising or affecting their prices as well.

Ethanol, blended with gasoline at a 10% level (E10) or in the form of ethyl tertiary-butyl ether (ETBE) synthesized from ethanol, is effective in reducing carbon monoxide (CO) emission levels, ozone ($O_3$) pollution, and $NO_x$ emissions from automobile exhaust. However, two of the major barriers to the wide acceptance of ethanol as a gasoline blend fuel are: (1) its RVP being not low enough to allow for higher blend levels and (2) its high moisture-absorbing (hygroscopic) characteristics. As mentioned earlier, the RVP of ethanol is lower than that for methanol; however, it is still marginally high.

In the early years of the U.S. fuel ethanol industry, it expanded to meet the increased demand for oxygenated fuel that resulted from a withdrawal of methyl t-butyl ether (MTBE) from the domestic gasoline marketplace. In response to sharply rising national concern about the presence of MTBE in groundwater as well as potential risk to public health and the environment, the U.S. EPA convened a Blue Ribbon Panel to assess policy options regarding MTBE. The Blue Ribbon Panel recommended that the use of MTBE be dramatically reduced or eliminated. The EPA has subsequently stated that MTBE should be removed from all gasoline. Many U.S. states including California and New York mandated their own schedules of MTBE phaseouts and bans. As of September 2005, 25 states had signed legislation banning MTBE. According to a survey conducted in 2003, 42 states reported that they had action levels, cleanup levels, or drinking water standards for MTBE [1]. The EPA has issued a nonbinding advisory level of 20 to 40 parts per billion (ppb) in drinking water, to prevent bad odor and taste in the water, while many states set their own limits on levels of MTBE allowed in their drinking water. As part of the Energy Policy Act of 2005 (EPAct 2005), the U.S. Congress voted to remove the oxygen content requirement for reformulated gasoline (RFG), and this also contributed to the rapid phaseout of MTBE from the U.S. market.

It is a remarkable turnaround in the chemical and petrochemical marketplace considering that MTBE used to be the fastest-growing chemical in the United States in the 1990s. Even with a rapid decline and disappearance of MTBE in the U.S. market, the global demand for MTBE has remained relatively strong at 19.3 (2000), 18.0 (2005), and 12.1 (2011) million tons/year. The largest demand has been from the Asian markets, where the use of ethanol or other oxygenated replacements is not established and ethanol subsidies and/or tax incentives are not provided.

U.S. fuel ethanol production increased very rapidly during the first decade of the twenty-first century. According to the Renewable Fuels Association (RFA) [2], the U.S. ethanol production in 2002, 2003, and 2004 was 2.13, 2.80, and 3.40 billion U.S. gallons, respectively. Considering the production level of 2000 being 1.63 billion gallons, this is more than a twofold increase over 5 years. The U.S. production of ethanol in 2006, 2007, 2008, 2009, 2010, and 2012 was 4.9, 6.5, 8.9, 10.75, 13.2, and 13.2 billion U.S. gallons, respectively. The 2012 production data were lower than the initial projection of 13.8 billion gallons due to widespread drought in the corn-producing states. Comparing the U.S. production statistics between 2007 and 2010, it took only 4 years to double the production. However, the U.S. production of ethanol has remained virtually unchanged since 2010. The trend in ethanol production in the United States is shown in Figure 11.1, which shows an exponential growth in ethanol production in the United States for the first decade of the twenty-first century. Due to the high cost of petroleum crude in recent years, the role of ethanol has expanded far beyond the oxygenated fuel additive realistically into that of the true alternative renewable transportation fuel. The increased use of ethanol in the United States has significantly contributed to the alleviation of dependence on imported petroleum.

Corn refinery has also become America's premier by-products industry, and its success has set a desirable business model for future biofuels industries. Increased production of amino acids, proteins, antibiotics, and biodegradable plastics has added further value to the U.S. corn crop. In addition to cornstarches, sweeteners, and grain ethanol, corn refiners also produce corn oils as well as a variety of important feed products. Corn products in the modern world are found in a large variety of fields and applications: (1) livestock feed grains; (2) food ingredients including sweeteners, starches, and polyols; (3) oil products including corn oil, acid oil, middlings, and corn wax oil; (4) cornstarches for papermaking and corrugated products; (5) personal care products utilizing natural polymers; (6) health and nutrition including sugar-free and low-sugar foods; (7) animal feeds including corn gluten meal (CGM), corn germ meal, and steepwater grain solubles; (8) pharmaceutical products including anhydrous dextrose; and (9) manufacture of biodegradable polymer, poly(lactic acid) (PLA), using corn starch, and more. Corn is the most traded crop product in the world, with the United States being the leading exporter and Japan being the largest importer. The U.S. annual export of corn was about 50 million metric tons in the fiscal year 2010. Since 1980, the annual amount of U.S. export of corn has been fluctuating between 35 and 60 million metric tons. The maximum peak (~60 million M/T) was recorded in 2007/2008, while the lowest (~18 million M/T) was recorded in 2012/2013 due to severe drought endured in the corn-producing states. Even though the United States dominates the global trading market of corn, it accounts for about 15.2% (2010) of the total U.S. corn production. As such, corn prices are largely determined by supply-and-demand relationships in the U.S. market. The U.S.

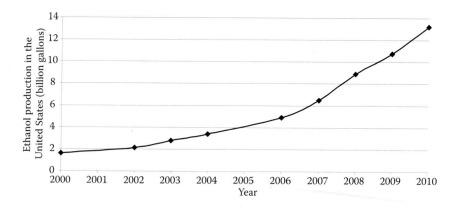

**FIGURE 11.1**  Ethanol production in the United States.

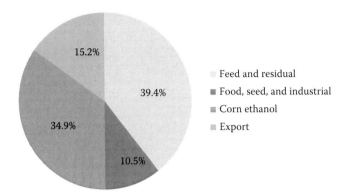

Feed and residual

Food, seed, and industrial

Corn ethanol

Export

**FIGURE 11.2** Breakdown by categories of 2010 end uses of corns in the United States. (From U.S. Grains Council, *World Corn Production and Trade*.)

corn crop was valued at $66.7 billion in the fiscal year 2010, and the production for the year was 331 million metric tons, which is equivalent to 12.1 billion bushels. The U.S. corn growth/production accounted for about 39% of the world growth/production [3]. About 80 million acres were planted to grow corn, and most of the corn production was in the heartland of the United States. Of the total corn produced in the United States in 2010, about 34.9% or 116 million metric tons was used for corn ethanol production. Figure 11.2 shows the breakdown of 2010 end uses of corns by end-use categories and sectors in the United States.

The production of ethanol from starch and sugar-based resources in the United States reached 13.2 billion U.S. gallons in 2010. The amount of gasoline used by the United States for transportation was approximately 134 million U.S. gallons in 2012, with ethanol used as a blend stock of up to 10% in marketed gasoline (E10) and also with a smaller E85 (85% ethanol and 15% gasoline) market. As ethanol production has increased, the demand for ethanol in the fuel supply chain has already reached the 10% "blend wall" of 13.4 billion gallons. It is not coincidental that the ethanol production in the United States has remained more or less flat at the level of 13–14 billion U.S. gallons a year since 2010. The Energy Independence and Security Act (EISA) of 2007 requires a mandatory Renewable Fuel Standard (RFS) requiring transportation fuels sold in the United States to contain a minimum of 36 billion gallons of renewable biofuels by 2022, including advanced and cellulosic biofuels and biomass-based diesel. The EISA further specifies that 21 billion U.S. gallons of the 2022 total biofuel blends in gasoline must be derived from noncornstarch products. Certainly, biomass-derived methanol, biomass-derived diesel, cellulosic ethanol, algae biodiesel, vegetable oil biodiesel, and others qualify as noncornstarch-based biofuels.

## 11.2 CORN ETHANOL AS OXYGENATED FUEL

### 11.2.1 Industrial Significances of Grain Ethanol

Ethanol production and utilization as automotive fuel received a major boost with the enforcement of the *CAAA of 1990*. Blending gasoline with ethanol has become a popular method for gasoline producers to meet the new oxygenate requirements mandated by the CAAA. Provisions of the CAAA established the *Oxygenated Fuels Program (OFP)* and the *Reformulated Gasoline (RFG) Program* in an attempt to control CO emission and ground-level $O_3$ problems. Both programs require certain oxygen levels in gasoline, namely, 2.7% by weight for oxygenated fuel and 2.0% by weight for RFG. In May 2006, the U.S. EPA officially lifted the oxygenate requirement for RFG based on the *EPAct 2005*, thus allowing refiners to produce gasoline without an oxygenate, as long as gasoline continues to meet Clean Air Act standards.

Public policies aimed at encouraging ethanol development/production were largely motivated by the nation's desire to improve air quality as well as to enhance future energy supply security. In addition, agricultural policymakers keenly see the expansion of the ethanol industry as a means of stabilizing farm income and reducing farm subsidies. Increasing ethanol production has induced a higher demand for corn crops and raised the average corn price. Higher corn prices and stronger demand for corns have contributed to reduction in Farm Commodity Program payments and the participation rate in the Acreage Reduction Program. From technical and scientific viewpoints as well as energy supply and environmental viewpoints, use of ethanol as transportation fuel or blend fuel makes sense, since ethanol can be produced in a renewable manner, that is, as a nondepleting energy source.

The rapidly growing corn ethanol industry in the first decade of the twenty-first century has increased the demand for corn very highly, which resulted in a significant price increase of corn and some other food crops. While the market's corn prices are determined largely by the supply-and-demand relationship, several important observations about the market responses involving the corn prices can be made, from both technoeconomic and socioeconomic viewpoints.

1. Tight supply and demand of corn pushed the corn price on the market to escalate faster than other crops, thus affecting the affordability of corn for food purposes and instigating a fuel-versus-food dilemma.
2. The escalation of corn price on the market has affected other food prices to varying extents. When corn price went up, poultry price also went up, due to the increase in the price of poultry feed, which was also corn based. When corn prices went up, prices of other grain products were also affected, because higher corn prices have motivated farmers to increase corn acreage at the expense of other food crops.
3. When the crude oil price on the market went up, so did corn ethanol price, although this trend is not unique for corn ethanol only. Furthermore, higher corn demand by the ethanol industry pushed the corn price higher, which, in turn, made ethanol production costlier due to higher feedstock prices [4].
4. Due to the large global market share of the U.S. corn trade, the worldwide corn prices were largely affected by the U.S. domestic corn prices.

### 11.2.2 Clean Air Act Amendments of 1990

The *CAAA* targets automobile emissions as a major source of air pollution. The act mandates the use of cleaner-burning fuels in the U.S. cities with smog and air pollution problems. The oxygen requirements of the CAAA spurred a market for oxygenates and created new market opportunities for ethanol. The OFP targets 39 cities that do not meet National Ambient Air Quality Standards (NAAQS) for CO. The CAAA mandates the addition of oxygen to gasoline to reduce CO emissions. It requires an oxygen level in gasoline of 2.7% by weight. Control periods vary by city because most CO violations occur during the winter season. The average control period is about 4 months. The most widely used oxygenate in the market has been a methanol-derived ether, MTBE, which was made mostly from natural gas feedstock and has been rapidly phased out from the U.S. market in the twenty-first century, as explained in the previous section.

Most major gasoline refiners are, more and more, using ethanol to meet gasoline oxygenate content requirements. In 1993, about 300–350 million gallons of ethanol were blended with gasoline and sold in markets covered by the OFP. In 2004, the fuel ethanol consumption reached 3.4 billion gallons in the United States. In the 11 years from 1993 to 2004, the U.S. production of grain ethanol has seen a tenfold increase. In 2010, the corn ethanol production in the United States reached 13.2 billion gallons, which is nearly a fourfold increase in the 6-year period from 2004 to 2010. The CAAA also requires the use of oxygenated fuels as part of the RFG program for controlling ground-level ozone formation. This program requires an oxygen level in gasoline of 2.0% by weight. Beginning in January 1995, RFG was required to be sold in nine ozone nonattainment areas year-round. Other provisions in the act allow as many as 90 other cities with less severe ozone pollution to "opt in" to the RFG program. Under a total opt-in scenario, as much as 70% of the nation's gasoline could be reformulated. The oxygenate requirement in RFG was officially lifted in 2006 based on the EPAct 2005. However, this change has not resulted in any reduction in the use of ethanol in RFG, since ethanol has already become an established and valuable component in gasoline.

An oxygen level of 2.0% by weight in gasoline means that at least 5.75% by weight of ethanol needs to be blended in gasoline, based on the stoichiometric calculation of $2.0 \times (46/16) = 5.75$. Therefore, the 2.7% oxygen requirement pushes the required level of ethanol in gasoline to 7.76 wt.% as a minimum. Thus, 10% ethanol blended gasoline, E10, sold in gas stations is consistent with the result of such calculation. Even though ethanol is clean burning and has a relatively low RVP of blending, it has a substantially lower heating value than the conventional gasoline. The higher heating value (HHV) and lower heating value (LHV) of gasoline are 47.3 and 44.4 kJ/g, respectively, whereas those for ethanol are 29.7 and 28.9 kJ/g, respectively. However, at a level of 10% ethanol blending, the reduced energy output is much less appreciable and could be compensated for by better performance of engines.

## 11.2.3 Energy Independence and Security Act of 2007

In 2007, the U.S. Congress enacted the *EISA* as a result of the President Bush's *Advanced Energy Initiative (AEI)* challenging the United States to change the way its citizens fuel their vehicles to improve the nation's energy security. The important message delivered by the AEI was "Keeping America competitive requires affordable energy." The EISA of 2007 is an energy policy act designed to improve the energy efficiency and to increase the supply of clean renewable fuels, thereby reducing the U.S. energy consumption by 7% and greenhouse gas (GHG) emissions by 9% by 2030. The initiative requires a mandatory *RFS* requiring transportation fuels sold in the United States to contain a minimum of 36 billion gallons of renewable biofuels by 2022, including advanced and cellulosic biofuels and biomass-based diesel. The EISA further specifies that 21 billion gallons of the 2022 total biofuel blends in gasoline must be derived from noncornstarch products. Additionally, the EISA requires that the Corporate Average Fuel Economy (CAFE) standard increase to 35 miles per gallon by the year 2020. While the EISA somewhat appears to discourage further growth of conventional corn-based ethanol as a blend fuel for the future and instead promotes rapid growth and market expansion of cellulosic ethanol, it still provides ample room for advancement for the U.S. ethanol industry.

### 11.2.4 NET ENERGY BALANCE OF CORN ETHANOL PRODUCTION

In analyzing and discussing the overall energy balance of alternative fuel production, a term called *net energy value (NEV)* is often used. The NEV is the difference between the energy content of product ethanol and the total energy used/consumed in producing and distributing ethanol. Higher corn yields of modern agricultural industry, lower energy consumption per unit of output in the fertilizer industry, and recent advances in fuel conversion technologies have significantly enhanced the economic and technical feasibility of producing ethanol from corn, when compared with the same just a decade ago. Therefore, studies based on the older data may tend to overestimate energy use (energy input), because the efficiency of growing corn as well as converting it to fuel ethanol has improved significantly over the past decade [5]. A large number of studies have been conducted to estimate the NEV of ethanol production. However, variations in input data and model assumptions resulted in a widely differing range of estimated values (conclusions), ranging from a very positive to a negative value. A negative NEV would mean that it takes more energy to produce the energy content of product ethanol.

According to a study by Shapouri et al. [5] of the U.S. Department of Agriculture (USDA), the NEV of corn ethanol was calculated as +16,193 Btu/gal, assuming that fertilizers were produced by modern (1995 or so) processing plants, corn was converted in modern (also of 1995) ethanol facilities, farmers achieved normal corn yields, and energy credits were allocated to coproducts. Updated values for the NEV of corn ethanol by Shapouri et al. show +21,205 Btu/gal (July 2002) [6] and +30,528 Btu/gal (October 2004) [7], respectively. The first value of 21,205 Btu/gal was based on the HHV of ethanol, whereas the second value of 30,528 Btu/gal was based on the LHV of ethanol. However, another study conducted by Pimentel and Patzek [8] resulted in a negative value of NEV and showed that the NEV of ethanol was −1,467 kcal/L (equivalent to −16,152 Btu/gal), which was based on the LHV. While this negative NEV instigated some controversial stir, a recent study thoroughly conducted by Argonne National Laboratory [9] showed that ethanol generates 35% more energy than it takes to generate.

As shown, sharp differences in the calculated NEV of ethanol production among various studies still existed, and they were largely stemming from several factors, which were comprehensively identified and directly compared in a report by MathPro

Inc. [10]. According to the MathPro's analysis, the differences in the NEV reflected sharp discrepancies in four energy usage categories, and they are as follows [10]:

1. Energy used in corn production: the USDA estimates (20.2 K Btu/gal in 2002 and 18.7 K Btu/gal in 2004) are about half that of Pimentel and Patzek (37.9 K Btu/gal).
2. Energy used in corn transport: the USDA estimates (2.1 K Btu/gal in 2002 and 2004) are less than half that of Pimentel and Patzek (4.8 K Btu/gal).
3. Energy used in ethanol production: the USDA estimates (46.7 K Btu/gal in 2002 and 49.7 K Btu/gal in 2004) are substantially lower than that of Pimentel and Patzek (56.4 K Btu/gal).
4. Coproduct energy credit: the 2002 USDA estimate (−13.5 K Btu/gal) is twice that of Pimentel and Patzek (−6.7 K Btu/gal). The 2004 USDA estimate (−26.3 K Btu/gal) is twice the 2002 USDA estimate and four times Pimentel and Patzek's estimate.

The "true" and "actual" value of ethanol's NEV would depend on various factors that involve the geographical region, agricultural productivity, collection and transportation, efficiency of the ethanol production process, energy efficiency of the fertilizer manufacture, and much more. It has been observed that most of the ethanol proponents have claimed positive NEVs, while many of the ethanol critics have referred to negative NEVs. As such, this subject has been controversial, from analytical, technoeconomical, and political standpoints. However, it is certain that modern corn ethanol plants use substantially less energy and produce more ethanol per bushel of corn than older plants, and it also appears certain that the claims of negative NEVs have been based on obsolete material and energy balance data of the corn ethanol industry [11]. In 2008, Mueller conducted a very extensive milestone survey of the nation's ethanol plants in terms of new energy use and coproduct data as well as land use, and the conclusions clearly showed significant improvements over the 2001 data [12].

A 2001 survey by BBI International found that dry mill ethanol plants use, on average, 36,000 Btu of thermal energy per gallon of ethanol produced and 1.09 kWh of electrical energy per gallon of ethanol produced, while producing an average of 2.64 gallons of ethanol per bushel of corn. However, ethanol plants in 2008 used an average of 25,859 Btu of thermal energy and 0.74 kWh of electricity per gallon of ethanol produced, which is 28% and 32.1% lower than the values of 2001, respectively. Ethanol produced per bushel of corn, meanwhile, increased by 5.3% to 2.78 gallons per bushel in 2008 [12]. This survey clearly also supports that the NEV of ethanol production based on modern technology data is clearly on the positive side. It is also foreseeable that the NEV value of the corn ethanol manufacture in the United States is going to further improve based on the innovations and enhancements made in agricultural, logistical, and ethanol production technologies.

### 11.2.5   FOOD VERSUS FUEL

While corn is an excellent source for starch and is heavily grown in the United States, its traditional use and value as a major food resource inevitably triggers a controversial debate of "food versus fuel." The supply-and-demand dynamics of corn on the U.S. marketplace for both fuel and food end uses has significantly contributed to the recent escalation of corn price, which, in turn, increased the production cost of corn ethanol as well as the price of corn-derived foods and goods. This has been one of the principal reasons that drive the commercialization efforts of cellulosic ethanol production, which is based on inedible renewable feedstocks such as trees, grasses, and wastes.

### 11.2.6 PROCESS TECHNOLOGIES OF CORN ETHANOL PRODUCTION

#### 11.2.6.1 Dry Mill Process versus Wet Mill Process

Ethanol production facilities can be classified into two broad types, that is, *wet milling* and *dry milling* operations. As the term "dry" implies, the dry milling process first grinds the entire corn kernel into flour, which is referred to as "meal" or "cornmeal." Dry mills are usually smaller in size (capacity) and are built primarily to manufacture ethanol only. The remaining stillage from ethanol purification undergoes a different process treatment to produce a highly nutritious livestock feed. On the other hand, wet milling facilities are called "corn refineries," also producing a list of high-valued coproducts such as high-fructose corn syrup (HFCS), dextrose, and cornstarch. Both wet and dry milling operations are currently used to convert corn to ethanol. Wet milling is usually a larger (scale) and more versatile process and could be valuable for coping with volatile energy markets. Wet milling can be used to produce a greater variety of products such as cornstarch, corn syrup, ethanol, DDGs, Splenda sweetener, and more. Although wet milling is a more versatile process and offers a more diverse product portfolio than dry milling, when producing fuel ethanol, dry milling has a higher efficiency and lower capital and operating costs than wet milling. Most of the recent ethanol plants built in the United States are based on dry milling operations [13]. As of the end of 2008, a total of 86% of corn ethanol in the United States was commercially produced using the dry mill process, and there were 150 dry milling plants in the United States [14].

#### 11.2.6.2 Energy Generation and Supply for Ethanol Plants

Thermal energy and electricity are the main types of energy used in both types of milling plants. Dry milling corn ethanol plants have traditionally used natural gas as their process fuel for production operation. Natural gas is used to generate steam for mash cooking, distillation, and evaporation; it is also used directly in DGS dryers and in thermal oxidizers that destroy the volatile organic compounds (VOCs) present in the dryer exhaust [15]. DGS stands for "distillers grain with solubles." Due to the increased production efficiencies and expanded fuel capabilities, *combined heat and power (CHP)* has become increasingly popular as an efficient energy option for many new ethanol plants [16]. CHP is an efficient, clean, and reliable energy services alternative, based on cogeneration of electricity and thermal energy on-site. Therefore, CHP achieves avoidance of line losses, increases reliability, and captures much of the thermal energy, otherwise normally wasted, in power generation to supply steam and other thermal energy needs at the plant site. A CHP system typically achieves a total system efficiency of 60%–80% compared to only about 50% for conventional separate generation of electricity and thermal energy [15]. By efficiently providing electricity and thermal energy from the same fuel source at the point of use, CHP significantly reduces the total fuel usage for a commercial ethanol plant, along with reductions in corresponding emissions of carbon dioxide ($CO_2$) and other pollutants. Generally speaking, electrical energy is used mostly for grinding and drying corn, while thermal energy is used for fermentation, ethanol recovery, and dehydration. On the other hand, flue gas is used for drying and stillage processing as part of waste heat recovery and energy integration efforts. The carbon dioxide generated from the fermentation process is also recovered and utilized to make carbonated beverages as well as to aid in the manufacturing of dry ice as a by-product of the ethanol process.

As mentioned earlier, based on the 2008 survey of 150 dry milling corn ethanol plants in the United States [14], ethanol plants in 2008 used an average of 25,859 Btu of thermal energy and 0.74 kW·h of electricity per gallon of ethanol produced, which was 28% and 32.1% lower than the 2001 values of 36,000 Btu and 1.09 kW·h, respectively. Ethanol productivity per bushel of corn also increased by 5.3% from 2.64 gallons in 2001 to 2.78 gallons per bushel in 2008 [12,14]. It was also found that on average, 5.3 pounds of DDGs and 2.15 pounds of wet distillers grains (WDGs) as well as 0.06 gallons of corn oil per every gallon of ethanol are also produced as process coproducts. One U.S. bushel as an agricultural volume unit is equivalent to 35.23907 L. Even though the U.S. corn ethanol industry has been considered a mature industry, the recent enhancements made on their process and energy efficiencies as well as the overall profitability are quite remarkable.

### 11.2.6.3 Feedstocks and Organisms for Ethanol Fermentation

The ethanol fuel manufacturing process is a combination of biochemical and physical processes based on traditional unit operations. Ethanol is produced by fermentation of sugars with yeast. The fermentation crude product is concentrated to fuel-grade ethanol via distillation. The organisms of primary interest to industrial fermentation of ethanol include *Saccharomyces cerevisiae, Saccharomyces uvarum, Schizosaccharomyces pombe,* and *Kluyueromyces* sp., among which *S. cerevisiae* is most commonly utilized.

Feedstocks for ethanol fermentation are either sugar- or starch-containing crops. These "biomass fuel crops" (tubers and grains) typically include sugar beets, sugarcanes, potatoes, corn, wheat, barley, Jerusalem artichokes, and sweet sorghum. Sugar crops such as sugarcane, sugar beets, or sweet sorghum are extracted to produce a sugar-containing solution or syrup that can be directly fermented by yeast. Starch feedstocks, however, must go through an additional step that involves *starch-to-sugar conversion*, as is the case for grain ethanol. Needless to say, sugar crops are simpler to convert to ethanol than starch crops. Therefore, the ethanol production cost, excluding the feedstock cost, is substantially lower for the sugar crops than starch crops.

### 11.2.6.4 Starch Hydrolysis

Starch may be regarded as a long-chain polymer of glucose (i.e., many glucose molecular units are bonded in a polymeric chain similar to a condensation polymerization product [17]). As such, macromolecular starches cannot be directly fermented to ethanol via the conventional fermentation technology. They must first be broken down into simpler and smaller glucose units through a chemical process called *hydrolysis*. Hydrolysis is generally defined as cleavage reaction of chemical bonds by the addition of water. In the hydrolysis step, starch feedstocks are ground and mixed with water to produce a mash typically containing 15%–20% starch. The mash is then cooked at or above its boiling point and treated subsequently with two enzyme preparations. The first enzyme hydrolyzes starch molecules to short-chain molecules, while the second enzyme hydrolyzes the short-chain molecules to glucose. The first enzyme is amylase. Amylase liberates "maltodextrin" by the liquefaction process. Such maltodextrins are not very sweet as they contain dextrins (a group of low-molecular-weight carbohydrates) and oligosaccharides (a saccharide polymer containing a small number of simple sugars, monosaccharides). The dextrins and oligosaccharides are further hydrolyzed by enzymes such as pullulanase and glucoamylase in a process known as *saccharification*. Complete saccharification converts all the limit dextrans (complex, branched polysaccharides of many glucose molecules) to glucose, maltose, and isomaltose. The mash is then cooled to about 30°C, and at this point, yeast is added for fermentation.

## 11.2.6.5 Yeast Fermentation

Yeasts are capable of converting sugar into alcohol by a biochemical process called fermentation. The yeasts of primary interest to industrial fermentation of ethanol include *S. cerevisiae, S. uvarum, S. pombe,* and *Kluyueromyces* sp. Under anaerobic conditions, yeasts metabolize glucose to ethanol primarily via the Embden–Meyerhof pathway. The *Embden–Meyerhof pathway* of glucose metabolism is the series of enzymatic reactions in the anaerobic conversion of glucose to ethanol, resulting in energy in the form of adenosine triphosphate (ATP) [18]. The overall net reaction represented by a stoichiometric equation involves the production of 2 moles of ethanol and 2 moles of carbon dioxide from each mole of glucose, as shown below. However, the yield attained in practical fermentation processes does not usually exceed 90%–95% of the theoretical value. In this case, the theoretical value, that is, 100% of yield, means that exactly 2 moles of ethanol is produced from each mole of glucose input to the fermenter. Therefore, the 100% yield would be equivalent to the mass conversion efficiency of 51%, which is defined later in this section. The following stoichiometric equation shows the basic biochemical reaction by fermentation of glucose to ethanol and carbon dioxide and also involves an endothermic heat of reaction.

$$C_6H_{12}O_6 = 2\,C_2H_5OH + 2\,CO_2 \quad \left(\Delta H_{298}^0\right) = 92.3 \text{ kJ/mol}$$

Theoretically, the maximum conversion efficiency of glucose to ethanol is 51% on a weight basis, which comes from a stoichiometric mass balance calculation of

2*(molecular wt. of ethanol)/(molecular wt. of glucose) = (2*46)/(180) = 0.51

However, some glucose has to be inevitably used by the yeast for production of cell mass and for metabolic products other than ethanol, thus reducing the conversion efficiency from its theoretical maximum of 51% to about 40%–48%. Assuming

46% fermentation efficiency, 1000 kg of fermentable sugar would produce about 583 L of pure ethanol, after taking into account of the density of ethanol (specific gravity at 20°C = 0.789). Or

(1000 kg sugar)*(0.46 kg ethanol/kg sugar)/(0.789 kg ethanol/L) = 583 L ethanol

Conversely, about 1716 kg fermentable sugar is required to produce 1000 L of ethanol, when 46% mass conversion efficiency is assumed. Mash typically contains between 50 and 100 g of ethanol per liter (about 5%–10% by weight) when the fermentation step is complete. This is called *distilled mash* or *stillage*, which still contains a large amount of nonfermentable portions of fibers or proteins.

### 11.2.6.6 Ethanol Purification and Product Separation

Ethanol is separated from mash by distillation, in which water and ethanol of the product solution are separated by differences in their boiling points (or individual vapor pressures). Separation is technically limited by the fact that ethanol and water form an *azeotrope*, or a constant boiling solution, of 95.63 wt.% alcohol and 4.37 wt.% water. This azeotrope is a minimum boiling mixture (or, a positive azeotrope), for which the boiling temperature of the azeotrope is lower than those of the individual pure components, that is, water and ethanol. The minimum boiling temperature at the azeotropic concentration is 78.2°C, while the normal boiling points of ethanol and water are 78.4°C and 100°C, respectively.

The 5%, more precisely, 4.37 wt.%, of water cannot be separated by conventional distillation, since the minimum boiling temperature is attainable at the azeotropic concentration, not at the pure ethanol concentration. Therefore, production of pure, water-free (anhydrous) ethanol requires an additional unit operation step following distillation. *Dehydration*, a relatively complex step in ethanol fuel production, is accomplished in one of the two typical processes. The first method uses a third liquid, most commonly benzene, which is added to the ethanol/water mixture. This third component changes the boiling characteristics of the solution (now, a ternary system instead of a binary system), allowing separation of anhydrous ethanol. In other words, this third component is used to break the azeotrope, thereby enabling conventional distillation to achieve the desired goal of separation. This type of distillation is also called *azeotropic distillation*, since the operation targets separating mixtures that form azeotropes. The second method employs *molecular sieves* that selectively absorb water based on the molecular size difference between water and ethanol. Molecular sieves are crystalline metal aluminosilicates having a three-dimensional interconnecting network of silica and alumina tetrahedra. Molecular sieves have long been known for their drying capacity (even to 90°C). There are different forms of molecular sieves that are based on the dimension of effective pore opening, and they include 3A, 4A, 5A, and 13X (pore diameter of 10A). Commercial molecular sieves are typically available in powder, bead, granule, or extrudate forms.

### 11.2.6.7 By-Products and Coproducts

The nonfermentable solids in distilled mash (stillage) still contain variable amounts of fiber and protein, depending on the feedstock. The liquid also contains soluble protein and other nutrients and, as such, is still valuable. The recovery of the protein and other nutrients in stillage for use as livestock feed can be essential to the overall profitability of ethanol fuel production. Protein content in stillage varies with feedstock. Some grains such as corn and barley yield solid by-products, *DDG*. Protein content in DDGs typically ranges from 25%–30% by mass and makes an excellent feed for livestock. Production and marketing of DDGs contributes significantly to the overall profitability of corn ethanol plant.

### 11.2.6.8 Potential Environmental Issues of Liquid Effluents

The production of ethanol also generates liquid effluent, which may render a potential pollution problem or concern of environmental stress on water systems. About 9 L of liquid effluents are generated for each liter of ethanol produced, which varies depending on the specific process adopted. Some of the liquid effluent may be recycled. Effluent can have a high level of biochemical oxygen demand (BOD), which is a measure of organic water pollution potential, and it is also acidic. Therefore, the liquid effluent must be treated before being discharged into the water stream. Specific treatment requirements depend on both feedstock quality (and type) as well as local pollution control regulations. Due to the acidity of the effluent, precautions and care must be taken if the effluent is directly spread over fields [19].

## 11.3 CHEMISTRY OF ETHANOL FERMENTATION

### 11.3.1 Sugar Content of Biological Materials

A much-simplified view of plant cell wall composition is shown in Figure 11.3. The base molecules that give plants their structure can be processed to produce *sugars*, which can be subsequently fermented to ethanol. Therefore, feedstocks that are able to generate sugars more readily and cost-effectively automatically become prime candidates for ethanol fermentation.

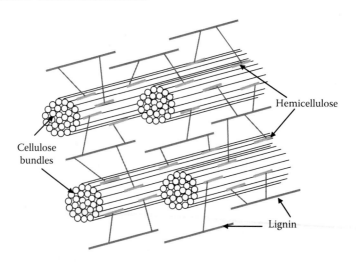

**FIGURE 11.3** A generalized description of a plant cell wall.

The principal components of most plant materials are commonly described as *lignocellulosic biomass*. This type of biomass is mainly composed of three types of compounds, called cellulose, hemicellulose, and lignin. *Cellulose* is a primary component of most plant cell walls and is made up of long chains of the six-carbon sugar *glucose* that are arranged in bundles (often described as crystalline bundles). The cellulose molecules in the plant cell wall are interconnected by another molecule called *hemicellulose*. The hemicellulose is primarily composed of the five-carbon sugar *xylose*. Besides cellulose and hemicellulose, another macromolecule called *lignin* is also present in significant amounts and provides the structural strength for the plant. Lignin is not easily converted into sugars or starches by the currently available technology and, therefore, has not been a target for alcohol fermentation. However, economically beneficial utilization of lignin is still quite important in enhancing the overall process economics of ethanol production as well as minimizing the process wastes. Technological developments have recently introduced a variety of processes of extracting and dissolving the cellulose and hemicellulose to produce sugars in such a form that can be readily fermented to ethanol. Efficient pretreatment can liberate the cellulose and hemicellulose from the plant material. Further treatment using chemicals, enzymes, or microorganisms can also be applied to liberate simple sugars from the cellulose and hemicellulose, thus making them available to microorganisms for fermentation to ethanol.

## 11.3.2 CONVERSION OF SUGARS TO ETHANOL

Figure 11.4 illustrates the stoichiometric conversion of cellulose to ethanol [20]. The first step involves cellulose hydrolysis that is essentially cleavaging the chemical bonds in the cellulose to produce glucose.

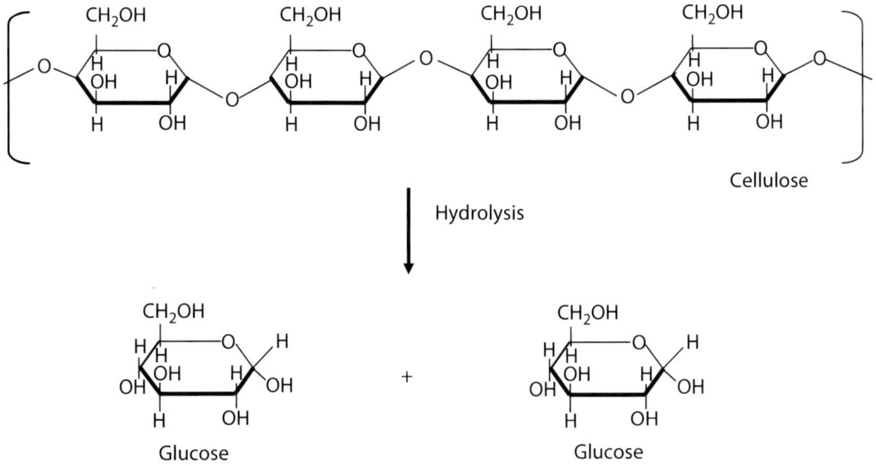

**FIGURE 11.4**   Hydrolysis of cellulose.

Once the large molecules are extracted from plant cells, they can be broken down into their component sugars using enzymes or acids. The sugars can be subsequently converted to ethanol using appropriately selected microorganisms via fermentation. The fermentation of ethanol from six-carbon sugars (such as D-glucose) follows the following stoichiometric equation:

$$\text{D-Glucose} \xrightarrow{\text{Fermentation}} 2\ CH_3CH_2OH + 2\ CO_2$$

According to the stoichiometric equation, 1 mole of D-glucose produces 2 moles of ethanol and 2 moles of carbon dioxide. Considering the molecular weights of glucose, ethanol, and carbon dioxide being 180, 46, and 44, respectively, the maximum theoretical yield of ethanol by wt% from the process would be $92/180 = 51\%$. Nearly half the weight of the glucose $88/180$ (49%) is converted to carbon dioxide at its theoretical maximum. Therefore, a significant amount of carbon dioxide is generated by the fermentation step, which needs to be captured and/or utilized for economically beneficial purposes. Otherwise, it would represent generation of GHG.

Hemicellulose is made up of the *five-carbon sugar xylose*, arranged in chains with other minor five-carbon sugars interspersed as side chains. Similarly to the cellulose case, the hemicellulose can also be extracted from the plant material and treated to liberate xylose, which, in turn, can be fermented to produce ethanol. However, xylose fermentation is not as straightforward or established as glucose fermentation, based on the currently available technology. Depending on the microorganism and conditions employed, a number of different fermentation paths are possible or conceivable. The array of products can include ethanol, carbon dioxide, and water as

$$\text{Xylose} \xrightarrow{\text{Fermentation}} 2\ CH_3CH_2OH + CO_2 + H_2O$$

Actually, three different reactions have been documented with yields of ethanol ranging from 30% to 50% of the weight of xylose as the starting material (that is, weight ethanol produced/weight xylose). They are

$$3\ \text{xylose} \rightarrow 5\ \text{ethanol} + 5\ \text{carbon dioxide}$$

$$3\ \text{xylose} \rightarrow 4\ \text{ethanol} + 7\ \text{carbon dioxide}$$

$$\text{Xylose} \rightarrow 2\ \text{ethanol} + \text{carbon dioxide} + \text{water}$$

The first reaction yields a maximum of 51% [= 5*46/(3*150)], the second 41% [= 4*46/(3*150)], and the third 61% (= 2*46/150), respectively. While the maximum theoretical ethanol yields from these fermentation reactions range between 41% and 61%, the practical yields of ethanol from xylose as starting material are in the range of 30%–50%.

In the discussion of potential yields of ethanol from various starting materials, two different ranges of efficiencies of hemicellulose-to-xylose conversion and xylose-to-ethanol conversion have been combined to provide an overall conversion efficiency of hemicellulose to ethanol of about 50%. Just as with glucose fermentation, the conversion of carbon dioxide to value-added products would vastly improve the overall process economics of ethanol production, since the yield of carbon dioxide is not only significant in amounts but also inevitable. It must be noted that even though xylose fermentation to ethanol is also mentioned in this chapter, the main focus of this chapter is on glucose fermentation, more particularly, corn sugars into ethanol. Ethanol-from-corn technology involves glucose fermentation, not xylose fermentation that is required in cellulosic ethanol technology.

## 11.4 CORN-TO-ETHANOL PROCESS TECHNOLOGY

Fermentation of sugars to ethanol, using commercially available fermentation technology, provides a fairly simple, straightforward means of producing ethanol with little technological risk. The system modeled assumes that the molasses are clarified and then fermented via cascade fermentation with yeast recycle. The stillage is concentrated by multiple-effect evaporation, and a molecular sieve is used to dehydrate the ethanol. Corn ethanol is commercially produced in one of two ways, using either the wet mill or dry mill process. The wet milling process involves separating the grain kernel into its component parts (germ, fiber, protein, and starch) prior to yeast fermentation. On the other hand, ICM-designed plants utilize the dry milling process, where the entire grain kernel is ground into flour form first. The starch in the flour is converted to ethanol during the fermentation process, also creating carbon dioxide and distillers grain as principal by-products.

### 11.4.1 WET MILLING CORN ETHANOL TECHNOLOGY

For the past two centuries in the United States, corn refiners have been developing, improving and perfecting the process of separating corn into its component parts to create a variety of value-added corn products and by-products. The *corn wet milling process* separates corn into its four basic components, namely, *starch, germ, fiber,* and *protein*. There are eight basic steps involved to accomplish this corn refining and alcohol fermentation process [21].

1. First, the incoming corn is *visually inspected and cleaned.* Corn refiners use #2 yellow dent corn, which is removed from the cob during harvesting. One bushel of yellow dent corn weighs about 56 pounds on average. Arriving corn shipments are visually inspected and cleaned two to three times to remove cob, dust, chaff, and any other foreign unwanted materials

before the next processing stage of *steeping*. An effective screening process can save a great deal of effort in the subsequent stages. The inspected and screened corn is then conveyed to storage silos holding up to 350,000 bushels.

2. Second, it is *steeped* to initiate bond cleavage of starch and protein molecules into simpler molecules. *Steeping* is typically carried out in a series of stainless steel tanks at warm temperatures. Each steep tank (or steeping tank) may hold about 2000–13,000 bushels of corn soaked in water at 50°C–52°C for 28–48 h. During steeping, the kernels (as shown in Figure 11.5) absorb water, thereby increasing their moisture levels from 15% to 45% by weight and also more than doubling in size by swelling [21]. The addition of 0.1% *sulfur dioxide (SO₂)* to the water suppresses excessive bacterial growth in the warm water environment. As the corn swells and softens, the mild acidity of the steeping water begins to loosen the gluten bonds within the corn and eventually release the starch [21]. A bushel is a unit of volume measure used as a dry measure of grains and produce. *A bushel of corn* or milo weighs about 56 pounds, a bushel of wheat or soybeans weighs about 60 pounds, and a bushel of sunflowers weighs about 25 pounds. Or, a U.S. bushel is equivalent to 35.23907 L as a volume unit.

3. The third step is the *germ separation*. It starts with coarse grinding of corn in the slurry to separate/break the germ from the rest of the kernel. Germ is the embryo of a kernel of grain, as shown in Figure 11.5. This germ separation is accomplished in *cyclone separators*, which spin the low-density corn germ out of the slurry. Therefore, this cyclone separator is called a *germ separator*. Or it is called a degerminating mill. The germs, which contain about 85% of corn's oil, are pumped onto screens and washed repeatedly to remove any starch left in the mixture [21]. A combination of mechanical and solvent processes extracts the oil from the germ. The oil is then refined

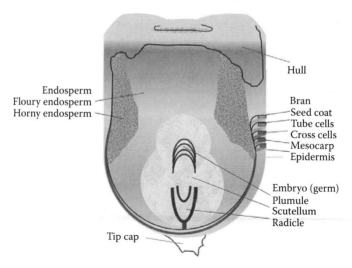

Endosperm
Floury endosperm
Horny endosperm

Hull

Bran
Seed coat
Tube cells
Cross cells
Mesocarp
Epidermis

Embryo (germ)
Plumule
Scutellum
Radicle

Tip cap

**FIGURE 11.5** Corn kernel.

and filtered into finished *corn oil*. The germ residue is saved as another useful component of animal feeds. Both corn oil and germ residues are important by-products of this process.

4. As the fourth step, the remaining slurry, consisting of fiber, starch, and protein, is *finely ground and screened* to separate the fiber from the starch and protein. After the germ separation described in step 3, corn and water slurry goes through a more thorough grinding in an impact or attrition-impact mill to release the starch and gluten from the fiber in the kernel. The suspension of starch, gluten, and fiber flows over fixed concave screens, which catch fiber but allow starch and gluten to pass through. The fiber is collected, slurried, and screened again to reclaim any residual starch or protein and then piped or sent to the feed house as a major ingredient of animal feeds. The starch–gluten suspension, called *mill starch*, is piped or sent to the starch separators [21].

5. Fifth, starch is separated from the remaining slurry in *hydrocyclones*. By centrifuging mill starch, the gluten is readily spun out due to the density difference between starch and gluten. Starch is denser than gluten. Separated gluten, a type of protein composite, can be used for animal feeds. *CGM* is a by-product of corn processing and is used as animal feed. CGM can also be used as an organic herbicide. The starch, now with just 1%–2% protein remaining, is diluted, washed 8 to 14 times, rediluted, and rewashed in hydrocyclones to remove the last trace of protein and produce high-quality starch, typically more than 99.5% pure. Some of the starch is dried and marketed as *unmodified cornstarch*, some other portion is modified into *specialty starches*, but most is converted into corn syrups and dextrose [21]. Cornstarch has a variety of industrial and domestic uses. All these are important by-products of the process that contribute to the corn distillers' profitability.

6. Sixth, the cornstarch then is converted to syrup (corn syrup), and this stage is called the *starch conversion or starch-to-sugar conversion* step. The starch–water suspension is liquefied in the presence of acid and/or enzymes. Enzymes help convert the starch to *dextrose*, which is soluble in water as an aqueous solution. Treatment with another enzyme is usually carried out, depending upon the desired process outcome. The process of acid and enzyme reactions can be stopped or terminated at key points throughout the process to produce a right mixture of sugars like dextrose (a monosaccharide, $C_6H_{12}O_6$) and maltose (a disaccharide, $C_{12}H_{22}O_{11}$) for syrups to meet desired specifications [21]. For example, in some cases, the conversion of starch to sugars can be halted at an early stage to produce low- to medium-sweetness syrups. In other situations, however, the starch conversion process is allowed to proceed until the syrup becomes nearly all dextrose. After this conversion process, the syrup is then refined in filters, centrifuges, or ion-exchange columns, and excess water is evaporated to result in *concentrated syrup*. Syrup can be sold directly as is, crystallized into pure dextrose, or processed further to produce *HFCS*. Across the corn wet milling industry, about 80% of starch slurry goes to corn syrup, sugar, and alcohol fermentation.

7. Seventh, the concentrated syrups can be made into several other products through a *fermentation* process. Dextrose is one of the most fermentable forms of all of the sugars. Dextrose is also called *corn sugar* and *grape sugar*, and dextrose is a naturally occurring form of glucose, that is, *D-glucose*. Dextrose is better known today as glucose. Following the conversion of starch to dextrose, dextrose is piped and sent to fermentation reactors/units/facilities where dextrose is converted to ethanol by traditional yeast fermentation. Using a *continuous* process, the fermenting mash is allowed to flow, or cascade, through several fermenters in series until the mash is fully fermented and then leaves the final tank. In a *batch* fermentation process, the mash stays in one fermenter for about 48 h before the distillation process for alcohol purification is initiated. Generally speaking, a continuous mode is more effective with a higher fermenter throughput, whereas a higher-quality product may be obtained from a batch mode.

8. As the eighth step, *ethanol separation or purification* follows the fermentation step. The resulting broth is distilled to recover ethanol or is concentrated through membrane separation to produce other by-products. Carbon dioxide generated from fermentation is recaptured for sale as dry ice, and nutrients still remaining in the broth after fermentation are used as components of animal feed ingredients. These by-products also contribute significantly to the overall economics of the corn refineries.

The *wet milling corn-to-alcohol process* can be summarized in a schematic process diagram, as shown in Figure 11.6.

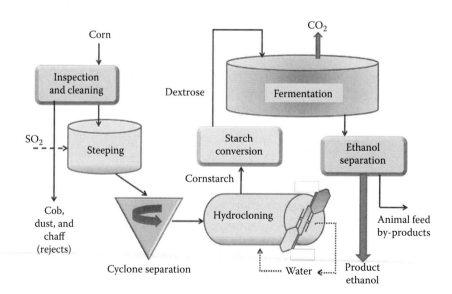

**FIGURE 11.6** A schematic of a typical wet milling corn-to-ethanol process.

## 11.4.2  DRY MILLING CORN ETHANOL PROCESS

In comparison to the wet milling ethanol process, where the corn kernel has to be separated into its components of germ, fiber, gluten, and starch prior to the fermentation step, the dry milling ethanol process first grinds the entire corn kernel into coarse flour form and then ferments the starch in the flour directly into ethanol. The dry milling corn ethanol process by ICM, Inc., is outlined as below [22].

1. Corn receiving and storage
   Corn grain is delivered by truck or rail to the corn ethanol fermentation plant. Grains are loaded in storage bins (silos) designed to hold sufficient amounts of grains to supply the plant operation continuously for 7–12 days.
2. Milling
   The grain is inspected and screened to remove debris (including corn cob, stalks, finer materials, stones, and foreign objects) and ground into coarse flour. The screening is usually done using a blower and screen. Coarse grinding is typically performed using a hammer mill. The feed rate from the milling step to the next stage of hot slurrying is typically controlled by the use of weighing tanks.
3. Cooking (hot slurry, primary liquefaction, and secondary liquefaction):
   During the so-called cooking process, starch in the flour is physically prepared and chemically modified for fermentation.
   3a. Hot slurry
       The coarsely ground grain is soaked in hot process water, the pH of the solution is adjusted to about 5.8, and an α-*amylase* enzyme is added. The slurry is heated to 180°F–190°F (82°C–88°C) for 30–60 min to reduce its viscosity. Agitation needs to be provided.
   3b. Primary liquefaction
       The slurry is then pumped through a pressurized jet cooker at 221°F (105°C) and held for 5 min. The jet cooker is also known as a steam injection heater. The mixture is then cooled by an atmospheric or vacuum flash condenser. The jet cooker is a critical component as steam helps to evenly hydrolyze and rapidly heat the slurry. The fluid dynamic relationship between the jet cooker's steam injector and condensing tube produces a pressure drop to help maximize shear action to improve starch conversion [23].
   3c. Secondary liquefaction
       After the flash condensation cooling, the mixture is held for 1–2 h at 180°F–190°F (82°C–88°C) to give the α-amylase enzyme a sufficient time to hydrolyze (or, depolymerize, decompose) the starch into short-chain, low-molecular-weight dextrins. This chemical conversion is called *gelatinization*. Generally speaking, during the gelatinization step, there is a sharp increase in the slurry viscosity that is rapidly decreased as the α-amylase hydrolyzes the starch into lower-molecular molecular-weight dextrins. Dextrins are a group of low-molecular-weight carbohydrates produced by the hydrolysis of starch

and are mixtures of polymers of D-glucose units linked by $\alpha$-(1 → 4) or $\alpha$-(1 → 6) glycosidic bonds.

After pH and temperature adjustment, a second enzyme, glucoamylase, is added as the mixture is pumped into the fermentation tanks. Glucoamylase is an amylase enzyme that cleaves the last $\alpha$-1,4-glycosidic linkages at the nonreducing end of amylase and amylopectin to yield glucose. In other words, glucoamylase is an enzyme that cleaves the chemical bonds near the ends of long-chain starches (carbohydrates) and releases maltose and free glucose. Maltose, or malt sugar, is a disaccharide that is formed from two units of glucose joined with an $\alpha$(1 → 4)bond.

4. Simultaneous saccharification fermentation

Once inside the fermentation tanks, the mixture is now referred to as mash, since it is an end product of mashing (which involves mixing of the milled kernel and water followed by mixture heating). The glucoamylase enzyme breaks down the dextrins, oligosaccharides, to form simple sugars, that is, monosaccharides. Yeast is added at this stage to convert the sugar to ethanol and carbon dioxide via alcohol fermentation reaction. The mash is then subjected to fermentation for 50–60 h, resulting in a mixture that contains about 15% ethanol as well as the solids from the grain and added yeast [22,24].

5. Distillation

The fermented mash is pumped into a multicolumn distillation system. The distillation columns utilize the boiling point difference between ethanol and water to distill and separate the ethanol from the solution. By the time the product stream is ready to leave the distillation columns, it contains about 95% ethanol by volume (which is 190-proof). This point is just immediately below the azeotropic concentration of the ethanol–water binary system, as explained earlier. To overcome this azeotropic limitation of maximum achievable ethanol concentration via straight distillation, several optional methods are being used, namely, jumping over the azeotropic point or bypassing the distillation. The residue from this process, called stillage, contains nonfermentable solids and water and is pumped out from the bottom of the distillation columns into the centrifuges.

6. Dehydration

The 190-proof ethanol still contains about 5 vol.% water. This near-azeotropic binary mixture is passed through a molecular sieve to physically separate the remaining water from the ethanol based on the size difference between the two molecules [22]. This step of dehydration produces 200-proof anhydrous (waterless) ethanol, that is, near 100% ethanol.

7. Product ethanol storage

Before the purified ethanol is sent to storage tanks, a small amount of denaturant chemical is added, making it unsuitable for human consumption. There are so many different kinds of denaturants available on the market for diverse purposes other than fuel ethanol. However, only certain gasoline-compatible blend stocks are suitable as denaturants for fuel ethanol. Some ethanol refineries also sell their denaturants for other ethanol industries. The ASTM D4806-11a specification covers nominally anhydrous denatured fuel ethanol intended for blending with unleaded or leaded gasoline for use as a spark-ignition automotive engine fuel. According to this specification, the only denaturants used for fuel ethanol would be natural gasoline, gasoline components, or unleaded gasoline at the minimum concentration prescribed. Methanol, pyrroles, turpentine, ketones, and tars are explicitly listed as prohibited denaturants for fuel ethanol meant to be used as gasoline blend stock [25]. Most ethanol plants' storage tanks are sized to allow storage of 7–12 days' production capacity.

8. Coproduct processing

   During the dry milling ethanol production process, two valuable coproducts are created: carbon dioxide and distillers grains. Their recoverable values are very important to the overall process economics, and this is why they are called coproducts rather than simply by-products.

During yeast fermentation, a large amount of carbon dioxide gas is generated. Since $CO_2$ is a major greenhouse chemical, its release into the atmosphere is not desirable. The carbon dioxide generated by fermentation is also of high concentration, and its purification is relatively straightforward. Therefore, carbon dioxide from ethanol fermentation is commonly captured and purified with a scrubber, so it can be marketed to the food processing industry for use in carbonated beverages and flash-freezing applications. Dry ice is a common coproduct of the ethanol refineries.

The stillage from the bottom of the distillation columns contains solids derived from the grain and added yeast as well as liquid from the water added during the process. The stillage is sent to centrifuges for separation into thin stillage (a liquid with 5%–10% solids) and WDG [22].

Some of the thin stillage is recycled back to the cook/slurry tanks as makeup water, reducing the amount of fresh water required by the cook (hot slurry) process. The rest is sent through a multiple-effect evaporation system, where it is concentrated into a syrup containing 25%–50% solids. This syrup, which is high in protein and fat contents, is then mixed back in with the WDG [22]. This is a step intended to recover most of the nutritive components from the stillage. With the added syrup, the WDG still contains most of the nutritive value of the original feedstock plus the added yeast, and as such, it makes excellent cattle feed. After the addition of the syrup, it is conveyed to a wet cake pad, where it is loaded for transport.

Many ethanol refinery facilities do not have enough nearby cattle farms or established markets to utilize all of their WDG products. However WDG must be used soon after it is produced, because it gets spoiled rather easily. Therefore, WDG is often sent through an energy-efficient drying system to remove moisture and extend its shelf life. This DDG is commonly used as a high-protein ingredient in cattle, swine, poultry, and fish diets. Modified forms of DDGs are also being researched for human consumption due to their outstanding nutritive values. In

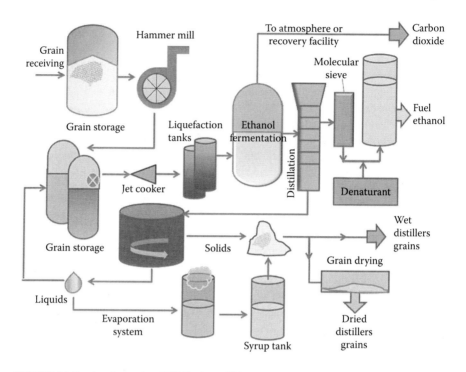

**FIGURE 11.7** A schematic of ICM's dry milling ethanol production system. (From *ICM's Dry Milling Ethanol Production*, http://www.icminc.com/ethanol/production_process/. Lee, S., and Shah, Y.T., *Biofuels and Bioenergy: Processes and Technologies,* CRC Press, Boca Raton, FL, 2012.)

more practical senses, DDG is better known as a corn ethanol coproduct than WDG.

A schematic of ICM's dry milling corn ethanol process [22] is shown in Figure 11.7.

### 11.4.3 INDUSTRIAL CLEANING OF ETHANOL PLANT

Periodic maintenance shutdowns and emergency stoppages for repairs are inevitable to all fuel and chemical production plants. Certain pieces of machineries involving ducts, pipes, headers, stacks, and valves tend to foul more quickly and frequently than other pieces, thereby plugging up the associated equipment, developing leaks, and potentially disabling the entire process. Therefore, highly efficient industrial cleaning of ethanol plants during periods of shutdown becomes an important issue. Several cleaning options are available and also conceivable, including chemical-based cleaning, hydro-cleaning, compressed air cleaning, and dry ice blasting. Dry ice blasting utilizes the $CO_2$ blasting process to clean the ethanol plant and its vital components during the shutdown periods [26]. One of its advantages over hydro-cleaning is that there is no water left behind where the system has to be dry, and other advantages include the nontoxicity of carbon dioxide as well as the availability of carbon dioxide on the plant site, if captured during the fermentation stage. The need for ethanol plant cleaning is universally applicable to both wet milling and dry milling plants.

## 11.5  BY-PRODUCTS AND COPRODUCTS OF CORN ETHANOL PROCESS

Ethanol by-products (coproducts) include *DDG, distillers dried grains with solubles (DDGS)*, WDG, corn bran, corn gluten feed, *CGM*, corn germ meal, and condensed fermented corn extractives. A bushel of corn produces about 2.78 gallons of ethanol and about 5.3 pounds of DDGs; 2.15 pounds of WDGs and 0.06 gallons of corn oil are also produced as coproducts per gallon of ethanol [14]. In other words, the amounts of DDGs and WDGs coproduced per bushel of corn are, on average, about 14.7 and 6.0 pounds, respectively. As of 2010, nearly 3.8 million tons of DDGs (including both DDG and DDGS) are produced in domestic dry grind ethanol production, that is, dry milling ethanol production. This accounts for more than 98% of the total U.S. DDG and DDGS production, and the remaining 1%–2% comes from the alcohol beverage industries. DDG is nearly identical to DDGS except that the former does not contain the distillers solubles, which is a "sticky" syrup.

Carbon dioxide is also becoming an important coproduct, as mentioned earlier. Carbon dioxide generated in the alcohol fermentation reaction can be relatively easily captured due to its high concentration at the source and subsequently purified for manufacture of dry ice or compressed carbon dioxide gas for food and beverage industries. Carbon dioxide capture and conversion to a value-added by-product is considered significant both economically and environmentally, since it is a GHG.

Corn oil is very valuable in both food and fuel applications. A corn kernel contains only about 3.6–4.0 wt.% of corn oil/fat. Due to this low level, any process targeting direct extraction of corn oil alone from corn kernels would not be a cost-effective solution. In this regard, recent dry milling ethanol plants potentially offer a good opportunity as potentially large sources of corn oil, as long as an economical separation process can be developed and implemented. Distillers corn oil (DCO) can also be used for biodiesel manufacture, and biodiesel made using DCO exhibits a lower cloud point( CP), a more desirable property, than that made using yellow grease (YG) or waste cooking oil. DCO-based biodiesel produced in the United States was about 76 million gallons in 2012, which is a sharp increase from 40.5 million gallons in 2011. Coproduction of corn oil is one of the promising options for the corn ethanol refinery to improve their gross margin of the industry.

Commercial processes for separation of corn oil are currently being developed [27,28]. POET, the largest ethanol producer in the world, has been producing corn oil since the beginning of 2011 for the biodiesel and feed markets. As of January 2013, twenty-five of twenty-seven POET's biorefineries have installed its patented corn oil technology, which has brought its total production capacity of corn oil to approximately 250,000 tons/year, enough to produce 68 million gallons of biodiesel a year [29]. The SunSource technology produces corn oil as an additional coproduct available to ethanol producers, most likely to be used in biodiesel production. The process uses centrifuge technology to extract the oil from the distillers grains in the evaporation step [28]. They also claim that removing the corn oil from the distillers grains does not lower the value of the grain feed coproduct and instead makes it easier to handle. They also claim that the process also reduces VOC emission from the dryers [28].

Research and development efforts in cost-effective corn oil extraction and purification are under way. Corn oil can be extracted from dry-milled germ or wet-milled germ by crushing for \$35–45/metric ton or by hexane extraction for \$20–40/metric ton, which are significant after considering all other associated costs as well as the market price for unrefined corn oil [30]. Quick germs [31] and enzymatically milled germs [32] have been successfully produced in laboratory quantities with 30% and 39% oil, respectively. Oil yields of 65 wt.% can be recovered from wet-milled or dry-milled germ by expeller pressing [33]. Oil separation from corn germ using aqueous extraction (AE) and aqueous enzymatic extraction (AEE) was studied, and the efficiency of the process was evaluated by Dickey et al. [34]. A recent AEE study [35] reports 90 wt.% oil recovery from wet-milled corn germ, at a 24 g scale.

## 11.6 ETHANOL AS OXYGENATED AND RENEWABLE FUEL

*Oxygenated fuel* is conventional gasoline that has been blended with an oxygenated hydrocarbon to achieve a certain desired concentration level of oxygen in the blended fuel. Oxygenated fuel is required by the CAAA of 1990 for areas that do not meet the federal air quality standards, especially that for carbon monoxide. The oxygen present in the blended fuel helps the engine to burn the fuel more completely, thus emitting less carbon monoxide. Extra oxygen already present *in situ* in the oxygenated fuel formulation helps efficient conversion into carbon dioxide rather than carbon monoxide. Gasoline blends of at least 85% ethanol are considered alternative fuels under the *Energy Policy Act of 1992 (EPAct 1992)*. E85 is used in flexible fuel vehicles (FFVs) that are currently offered by most major automobile manufacturers. FFVs can run on 100% gasoline, E85, or any combination of the two and qualify as alternative fuel vehicles (AFVs) under EPAct regulations.

*RFG* is a formulation of gasoline that has lower controlled amounts of certain chemical compounds that are known to contribute to the formation of $O_3$ and toxic air pollutants. It is less evaporative than conventional gasoline during the summer months, thus reducing evaporative fuel emission and leading to reduced VOC emission. It also contains oxygenates, which increase the combustion efficiency of the fuel and reduces carbon monoxide emission. The *CAAA of 1990* requires RFG to contain oxygenates and have a minimum oxygen content of 2.0% oxygen by weight. RFG is required in the most severe ozone nonattainment areas of the United States. Other areas with ozone problems have voluntarily opted into the program. The U.S. EPA has implemented the RFG program in two phases, that is, Phase I from 1995 to 1999 and Phase II beginning in 2000.

To be more specific, the CAAA mandated the sale of RFG in the nine worst ozone nonattainment areas beginning January 1, 1995. Initially, the U.S. EPA determined the nine regulated areas to be the metropolitan areas of Baltimore, Chicago, Hartford,

Houston, Los Angeles, Milwaukee, New York City, Philadelphia, and San Diego. The important parameters for RFG by the CAAA of 1990 were the following:

1. At least 2% oxygen by weight
2. A maximum benzene content of 1% by volume
3. A maximum of 25% by volume of aromatic hydrocarbons

In 2006, the U.S. EPA amended the RFG regulation to remove the oxygen content requirement and its associated compliance requirements, based on the *EPAct 2005* [36]. Even though the oxygen content requirement of the RFG was removed, the use of ethanol in gasoline was not adversely affected at all. By the time of the amended regulation, ethanol had already become an important component of gasoline and had proven its effectiveness. As of 2011, in the United States, RFG is required in cities with high smog levels and is optional elsewhere. RFG is currently used in 17 states and the District of Columbia. About 30% of gasoline sold in the United States as of 2011 is reformulated.

*MTBE* had been one of the most commonly used oxygenated blend fuels, till recent claims of health and environmental problems associated with MTBE use as a blending fuel. Tertiary-amyl methyl ether (TAME), ETBE, and ethanol have also been used in oxygenated and reformulated fuels. Responding to the rapid phaseout of MTBE in the United States, ethanol has gained the most popularity as a blending fuel, based on its clean burning, relatively low *RVP*, renewable nature of the fuel, minimal or no health concerns, and relatively low cost.

The RFG should have no adverse effects on vehicle performance or the durability of engine and fuel system components. However, there may be a slight decrease in fuel mileage (1%–3% or 0.2–0.5 mile/gal) in the case of well-tuned automobiles due to the higher concentrations of oxygenates that inherently have lower heating values. However, RFG burns more completely, thereby reducing formation of engine deposits and often boosting the actual gas mileage, particularly for older engines.

The *RVP* is crucially important information for blended gasoline from practical and regulatory standpoints. Evaporated gasoline compounds combine with other pollutants on hot summer days to form ground-level ozone, commonly referred to as *smog*. Ozone pollution is of particular concern because of its harmful effects on lung tissue and breathing passages. Therefore, the government, both federal and state, imposes an upper limit as a requirement, which limits the maximum level RFG can have as its RVP. By such regulations, the government not only controls the carbon monoxide emission level but also limits the evaporative emission of the fuel. Due to this limit, certain oxygenates may not qualify as a gasoline blending fuel even if they may possess excellent combustion efficiency and high octane rating. One such example is methanol, which has a high RVP. It must be clearly noted that the RVP of pure ethanol is lower than that of methanol but substantially higher than that of TAME, ETBE, or MTBE.

Further, the legal limits for the RVP depend upon many factors, including current environmental conditions, geographical regions, climates, time of the year (such as summer months vs. winter months), and so forth. It should also be noted that ground-level ozone is harmful to humans, whereas stratospheric ozone is so essential and beneficial for global environmental safety.

The *oxygenated fuel program (OFP)* is a wintertime program for areas with problems of carbon monoxide air pollution. The oxygenated winter fuel program uses normal gasoline with oxygenates added. On the other hand, the RFG program is for year-round use to help reduce ozone, CO, and air toxins. While both programs use oxygenates to reduce CO, RFG builds on the benefit of oxygenated fuel and uses improvements in the actual formulation of gasoline to reduce pollutants, including *VOCs* [37].

Although *MTBE* was once credited with significantly improving the nation's air quality, it has been found to be a major contributor to groundwater pollution. Publicity about the leaking of MTBE from gasoline storage tanks into aquifers as well as its adverse health effects has prompted legislators from the Midwestern United States to push for a federal endorsement of corn-derived ethanol as a substitute oxygenate. Many U.S. states including California and New York mandated their own schedules of MTBE phaseouts and bans. This MTBE phaseout has served as an incentive for corn ethanol industries for marketing their products as being environmentally more acceptable than other alternatives and, at the same time, renewable.

The *EPAct* 2005 (P.L. 110-58), established the first-ever *Renewable Fuels Standard (RFS)* in federal law, requiring increasing volumes of ethanol and biodiesel to be blended with the U.S. fuel supply between 2006 and 2012. The *EISA* of 2007 (P.L. 110-140, H.R. 6) amended and increased the RFS, requiring 9 billion gallons of renewable fuel use in 2008, stepping up to 36 billion gallons by 2022, as shown in Figure 11.8. A major portion of the increase is expected to be coming from cellulosic ethanol.

Considering the annual gasoline consumption in the United States is approximately 130–140 billion gallons, and also assuming that all gasoline sold in the United States is blended with ethanol up to 10%, that is, E10, the total annual demand for ethanol by E10 in the United States would be about 14.3–15.5 million gallons. One can readily notice that this estimated saturation point for ethanol demand in the United States for E10 blend is not far from the 2010 total United States ethanol production from corn, which was 13.2 million gallons. Thus, it is evident that the RFS numbers for the future years are based on (1) expanded use of nonethanol

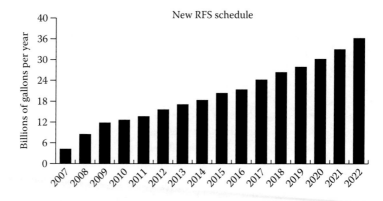

**FIGURE 11.8** New Renewable Fuels Standard indicating the total amount of renewable fuel use for 2008 through 2022. (From American Coalition for Ethanol, *All About Ethanol*, http://www.ethanol.org/.)

renewable fuels such as biodiesel, (2) increased availability of cellulosic ethanol, and (3) expanded adoption of AFVs and FFVs, and more.

## 11.7  ETHANOL VEHICLES

Fuel ethanol is most commonly used as a fuel for internal combustion, four-cycle, spark-ignition engines in transportation and agriculture. It can be used as a direct replacement fuel for gasoline or can be blended with gasoline as an extender and octane enhancer. Gasoline extender is a component in blended gasoline added exclusively for volume. The *RON* of ethanol is about 111–116, and its *motor octane number (MON)* is about 92–111. Therefore, ethanol blending enhances the octane rating of the blend gasoline [38]. The octane number is a quantitative measure of the maximum compression ratio at which a particular fuel can be utilized in an engine without some of the fuel/air mixture "knocking." By defining an octane number of 100 for iso-octane and 0 for *n*-heptane, linear combinations of these two components are used to measure the octane number of a particular fuel. Therefore, a fuel with an octane number of 90 would have the same ignition characteristics at the same compression ratio as a 90/10 mixture of iso-octane and *n*-heptane. It should be noted that there are several different rating schemes for octane numbers of fuels, namely, *RON, MON, and the average of the two, [(R + M)/2]*, which is often called antiknock index (*AKI*) or pump octane number (*PON*). The RON (or F1) simulates fuel performance under low-severity engine operation, whereas the MON (or F2) simulates more severe operation that might be incurred at high speed or high load. Therefore, nearly always, RON is higher in value than MON for the very same fuel. In the United States, the octane rating of a gasoline is usually reported as the average of RON and MON, that is, (R + M)/2, and therefore, this average value is often called the PON.

The use of ethanol to replace gasoline requires modifications to the carburetor, fuel injection system components, and often, the compression ratio. Therefore, efficient and safe conversion of existing gasoline engines is a complex matter. Therefore, engines specifically designed and manufactured to operate on ethanol fuel, or predominantly ethanol fuel, will generally be more efficient than modified gasoline engines. Ethanol concentrations of between 80% and 95% can be used as fuel, which eliminates the need for the cumbersome dehydration processing step, thus simplifying the distillation step. This complication comes from the fact that an ethanol–water solution makes an azeotropic mixture at 95.63% ethanol (by mass), that is, a minimum boiling mixture or a positive azeotrope. In many practical cases, the conversion of internal combustion engines to operate on azeotropic ethanol may be simpler and more cost-effective than ethanol dehydration as an effort to produce 99+% purity of ethanol.

In the United States, *E85* is a federally designated alternative fuel that contains 85% ethanol and 15% gasoline. In 2003, there were hundreds of thousands of E85 vehicles on the roads in the United States. In 2010, almost 8 million vehicles on the U.S. highways were FFVs [39], and by 2013, the number has grown to about 11 million. E85 vehicles are *FFVs*, which can run on a very wide range of fuels, ranging from 100% gasoline (with 0% ethanol) to 85% ethanol (with 15% gasoline); however, they run best on E85 [40]. Nearly all the major automobile makers offer many models of passenger cars and sports utility vehicles (SUVs) with E85 engines.

In the United States, the National Ethanol Vehicle Coalition (*NEVC*) is actively promoting expanded use of 85% ethanol (E85) motor fuel. NEVC is advocating E85, based on its clean burning as well as renewability of the fuel. E85 fuel can achieve a very high octane rating of 105. As an extra incentive plan for the E85 users and providers, the U.S. federal government provides several tax incentives for the use of E85 as well as for installation of E85 pumps at fueling stations. The E85 vehicles undoubtedly help alleviate the petroleum dependence of the world by using a renewable alternative fuel source.

In unmodified engines, it is generally believed that ethanol can replace up to 20% of the gasoline, that is, E20. However, E20 would substantially increase the blend fuel volatility measured by RVP, which significantly exceeds the permissible limit. In the United States, up to 10% blend of ethanol, that is, E10, is quite popularly used. Blending ethanol with gasoline extends the gasoline supply and improves the quality of gasoline by increasing its octane value as well as imparting clean-burning properties of oxygenates to the blend fuel. There are advantages to using gasoline/ethanol blends rather than pure (or very-high-concentration) ethanol. The most significant of these is that blends do not require engine modification. Therefore, ethanol can be integrated rapidly with the existing infrastructure, including gasoline supply and distribution systems.

Even though the use of ethanol in specially designed two-cycle engines has been demonstrated on a number of occasions, it is not yet commercialized. One of the major issues has been in the fact that ethanol does not mix well with lubricating oil typically used for such engines. Therefore, development of lubricating oils that are not affected by ethanol is an important step for this application.

Similarly, the use of ethanol in diesel-fueled engines is quite feasible but is not practiced much, due to a number of technical difficulties. These limitations are based on ethanol's inability to ignite in compression ignition engines as well as the poor miscibility with diesel. However, ethanol can be used in supercharged diesel engines for up to about 25% of the total fuel, the rest preferably being diesel. This can be achieved by delivering ethanol from a separate fuel tank and injecting it into the diesel engine through a supercharger airstream. This mode of fuel delivery system may be called "dual-fuel system," in comparison to blended fuel that is delivered as a preblended fuel from a single fuel tank. Ethanol can also replace aviation fuel in aircraft engines, even though this potential is not commercially exploited.

As a recent effort, a dual-fuel internal combustion engine technology has been developed and demonstrated, in which ethanol is used as a cofuel with acetylene ($C_2H_2$), which is the principal fuel in this specific application. The dual-fuel system has been favorably demonstrated on modified gasoline and diesel engines originally designed for cars, trucks, forklifts, tractors, and power generators. Up to 25% of ethanol in acetylene-based dual-fuel systems has been successfully tested. The role of ethanol was found very effective in eliminating knocking/pinging and lowering the combustion temperatures, thus reducing $NO_x$ emissions from combustion [41,42].

## 11.8 NONFUEL USES OF ETHANOL

Ethanol reacts with carboxylic acids to produce ethyl esters in the presence of an acid catalyst such as sulfuric acid. The two largest-volume ethyl esters are ethyl acrylate (from ethanol and acrylic acid) and ethyl acetate (from ethanol and acetic acid).

*Ethyl acetate* is as a common solvent used in paints, coatings, and the pharmaceutical industry. The most familiar application of ethyl acetate in the household is as a solvent for nail polish. A typical reaction that synthesizes ethyl acetate is based on esterification:

$$C_2H_5OH + CH_3COOH = C_2H_5O\cdot OCCH_3 + H_2O$$

This chemical reaction follows very closely a second-order reaction kinetics, as often used as an example problem for second-order elementary reactions in the chemical reaction engineering textbooks.

Recently, *Kvaerner Process Technology* developed a process that produces ethyl acetate directly from ethanol without acetic acid or other cofeeds. Considering that both acetic acid and formaldehyde can also be produced from ethanol, this ethanol-to-ethyl acetate process idea is quite significant. The Kvaerner process allows the use of fermentation ethanol, produced from bio-renewable feedstocks, as a sustainable single-source feed, which is remarkable. Further, the process elegantly combines both dehydrogenation and selective hydrogenation in its process scheme, thus producing hydrogen as a process by-product, which makes the process economics even better.

*Ethyl acrylate*, which is synthesized by reacting ethanol and acrylic acid, is a monomer used to prepare acrylate polymers for use in coatings and adhesives. It is also a reagent used to prepare various intermediates in the pharmaceutical industry.

Ethanol is a reactant for *ETBE*, as is the case for methanol to MTBE. ETBE is produced by reaction between isobutylene and ethanol as

$$C_2H_5OH + CH_3C(CH_3) = CH_2 \rightleftharpoons C(CH_3)_3\,OC_2H_5$$

Vinegar is a dilute aqueous solution of acetic acid prepared by the action of *Acetobacter* on ethanol solutions. Ethanol is also used to manufacture ethylamines by reacting ethanol and ammonia over a silica- or alumina-supported nickel catalyst at 150°C–220°C. First, ethylamine with a single amino group in the molecule is formed, and further reactions create diethylamine and triethyamine. The ethylamines are used in the synthesis of pharmaceuticals, agricultural chemicals, and surfactants. Diethylamine is totally different from diethanolamine (DEA or DEOA), which is typically produced by reacting ethylene oxide with aqueous ammonia. The latter is used to remove hydrogen sulfide and carbon dioxide from natural gas.

Ethanol can also be used for *transesterification of triglycerides* in biodiesel production using vegetable oils or algae oils. In the United States, methanol is currently more popularly used for this purpose, mainly due to its more favorable process economics. However, if corn oil is to be used for biodiesel production, corn ethanol would be a natural choice, since both are coproducts of the very same corn ethanol industry.

Ethanol can also be used as feedstock to synthesize petrochemicals that are usually derived from a petroleum source. Such chemicals include ethylene and butadiene but are not limited to these. This option may become viable for regions and countries where petrochemical infrastructure is weak but agricultural produce is vastly abundant. This is particularly true for the times when petroleum price is very high. Ethanol can also be converted into *hydrogen* via reforming reaction, that is, chemical reaction with water at an elevated temperature, typically with the aid of a noble metal catalyst. Even though this method of hydrogen generation may be economically less favorable than either steam reforming of methane or electrolysis of water, the process can be used for special applications, where specialty demands exist or other infrastructure is lacking. Hydrogen produced using bioethanol may also be considered biohydrogen.

More recently, supercritical water reformation of crude ethanol beer was developed for hydrogen production [43,44]. The process utilizes *supercritical water* (T > 374°C and P > 218 atm) functioning both as a highly energetic reforming agent and as a supercritical solvent medium, thus effectively eliminating the service of any noble metal catalyst or the need of pure ethanol. Further, its direct noncatalytic reformation of unpurified crude ethanol beer alleviates the need for any energy-intensive predistillation or distillation of water–ethanol solution, thereby achieving overall energy savings. The process technology has not yet been demonstrated on a large scale [44].

*Poly(lactic acid)* or polylactide (PLA) is a thermoplastic aliphatic polyester derived from cornstarch. PLA is one of the leading biodegradable polymers, which is derived from renewable bio-sources, more specifically corn in the United States. A variety of applications utilizing PLA are being developed, wherever biodegradability of plastic materials is desired. PLA can be used by itself, blended with other polymeric materials, or used as composites. As the biodegradable polymer technology further develops, the PLA market is also expected to grow, and so is the cornstarch market [44].

# REFERENCES

1. U.S. Environmental Protection Agency (EPA), "Regulatory determinations support document for CCL 2: Chapter 13. MTBE," Tech. Rep. EPA—OGWDW, June 2008.
2. F. O. Lichts, Industry statistics: 2010 world fuel ethanol production. Renewable Fuels Association, 2011. Available at http://www.ethanolrfa.org/pages/statistics#E.
3. U.S. Grains Council, World corn production and export. Available at http://www.grains.org/index.php/buying-selling/corn, 2014.
4. Energy Policy Research Inc. (EPRINC), Implementation issues for the renewable fuel standard—part I: Rising corn costs limit ethanol's growth in the gasoline pool. April 2011. Available at http://eprinc.org/pdf/EPRINC-CornLimitsEthanol.pdf (accessed March 26, 2014).
5. H. Shapouri, J. A. Duffield and M. S. Graboski, "Estimating the net energy balance of corn ethanol," United States Department of Agriculture, Tech. Rep. Agricultural Economic Report Number 721, July 1995.

6. H. Shapouri, J. A. Duffield, A. McAloon and M. Wang, "The 2001 net energy balance of corn ethanol," U.S. Department of Agriculture, Tech. Rep. AER-814, July 2002.
7. H. Shapouri, J. A. Duffield and M. Wang, "The energy balance of corn ethanol: An update," U.S. Department of Agriculture, October 2004.
8. D. Pimentel and T. W. Patzek, "Ethanol production using corn, switchgrass, and wood," *Nat. Resourc. Res.*, vol. 14, pp. 65–75, 2005.
9. H. Shapouri, J. A. Duffield and M. Wang, "The net energy value of corn ethanol," 2005. Available at http://www.ncga.com/public_policy/PDF/03_28_05ArgonneNatlLabEthanol Study.pdf.
10. I. MathPro, "The net energy value of corn ethanol: Is it positive or negative?" November 2005. Available at http://www.mathproinc.com/pdf/2.1.6_Ethanol_NEV_Comparison. pdf (accessed March 26, 2014).
11. Nebraska Corn Board, Corn ethanol plants using less energy but producing more ethanol per bushel. June 2010. Available at http://www.nebraskacorn.org/news-releases/corn-ethanol-plants-using-less-energy-but-producing-more-ethanol-per-bushel/ (accessed March 23, 2014).
12. S. Mueller and K. Copenhaver, "News from corn ethanol: Energy use, co-products, and land use," in *Near-Term Opportunities for Biorefineries Symposium*, Champaign, IL, October 11–12, 2010.
13. R. T. Dale and W. E. Tyner, "Economic and technical analysis of ethanol dry milling: Model description," Purdue University, College of Agriculture, Department of Agricultural Economics, West Lafayette, IN, 2006.
14. S. Mueller, Detailed report: 2008 national dry mill corn ethanol industry. University of Illinois-Chicago. May 4, 2010. Available at http://www.ethanolrfa.org/page/-/Ethanol SurveyReport.pdf?nocdn=1 (accessed March 26, 2014).
15. Energy and Environmental Analysis, Inc., "Impact of combined heat and power on the energy use and carbon emissions in the dry mill ethanol process," Report to U.S. Environmental Protection Agency, Combined Heat and Power Partnership, November 2007.
16. U.S. Environmental Protection Agency (EPA), 2008. Combined heat and power partnership: Dry mill ethanol. Available at http://www.epa.gov/chp/markets/ethanol.html (accessed March 26, 2004).
17. G. Odian, *Principles of Polymerization*. Hoboken, NJ: Wiley-Interscience, 2004.
18. W. A. N. Dorland, *Dorland's Illustrated Medical Dictionary*, 30th Edition. Philadelphia, PA: W. B. Saunders Company, Elsevier Health Sciences Division, 2003.
19. C. Bradley and K. Runnion, "Understanding ethanol fuel production and use," Volunteers in Technical Assistance (VITA), Arlington, VA, 1984.
20. P. Laenui, State of Hawaii, "Ethanol production in Hawaii report," 1994 and updated August 18, 2009. Available at http://www.docstoc.com/docs/10100171/The-Hawaii-Report (accessed December 21, 2013).
21. The corn refining process. Available at http://www.corn.org/.
22. ICM's dry milling ethanol production. 2013. Available at http://www.icminc.com/ innovation/ethanol/ethanol-production-process.html (accessed March 26, 2014).
23. Prosonix, AP-40 Starch processing for wet milling. December 2008. Available at http:// www.pro-sonix.com/files/pdf/AP-40_Starch_-_Wet_Milling_20101210.pdf (accessed March 26, 2014).
24. M. Knauf and K. Krau, "Specific yeasts developed for modern ethanol production," *Sugar Industry*, vol. 131, pp. 753–775, 2006.
25. ASTM International, "ASTM standards D4806-11a," in *ASTM Standards: Petroleum Standards*, ASTM Technical Committees, Ed. West Conshohocken, PA: ASTM International, 2011.

26. Midwest Dry Ice Blasting, Dry ice industrial cleaning for ethanol plants. April 2010. Available at http://www.prweb.com/releases/ethanolindustry/cleaningservices/prweb 3897754.htm (accessed March 26, 2014).

27. S. R. Schill. Plymouth oil to extract corn oil, germ at Iowa plant. *Ethanol Producer Mag.* August 4, 2008. Available at http://ethanolproducer.com/articles/4590/plymouth-oil-to-extract-corn-oil (accessed March 26, 2014).

28. C. M. Rendleman and H. Shapouri, New technologies in ethanol production. February 2007. Available at http://www.usda.gov/oce/reports/energy/aer842_ethanol.pdf (accessed March 26, 2014).

29. POET, POET producing corn oil at 25 biorefineries. February 2013. Available at http://www.poet.com/pr/poet-producing-corn-oil-at-25-biorefineries (accessed March 26, 2014).

30. G. Foster, "Corn fractionation for the ethanol industry," *Ethanol Producer Mag.*, vol. 11, pp. 76–78, 2005.

31. V. J. Singh and S. Eckhoff, "Effect of soak time, soak temperature and lactic acids on germ recovery parameters," *Cereal Chem.*, vol. 73, pp. 716–720, 1996.

32. D. Johnston, A. J. McAloon, R. A. Moreau, K. B. Hicks and V. J. Singh, "Composition and economic comparison of germ fractions from modified corn processing technologies," *J. Am. Oil Chem. Soc.*, vol. 82, pp. 603–608, 2005.

33. L. C. Dickey, P. H. Cooke, M. J. Kurantz, A. J. McAloon, N. Parris and R. A. Moreau, "Using microwave heating and microscopy to study optimal corn germ yield with a bench-scale press," *J. Am. Oil Chem. Soc.*, vol. 84, pp. 489–495, 2007.

34. L. C. Dickey, M. J. Kurantz and N. Parris, "Oil separation from wet-milled corn germ dispersions by aqueous oil extraction and aqueous enzymatic oil extraction," *Ind. Crop Prod.*, vol. 27, pp. 303–307, 2008.

35. R. A. Moreau, D. B. Johnston, M. J. Powell and K. B. Hicks, "A comparison of commercial enzymes for the aqueous enzymatic extraction of corn oil from corn germ," *J. Am. Oil Chem. Soc.*, vol. 81, pp. 77–84, 2004.

36. U.S. Environmental Protection Agency (EPA), Regulatory announcement EPA420-F-06-020: Removal of regulatory gasoline oxygen content requirement and revision of commingling prohibition to address non-oxygenated reformulated gasoline. February 2006. Available at http://www.epa.gov/otaq/regs/fuels/rfg/420f06020.pdf (accessed March 26, 2014).

37. J. G. Speight and S. Lee, *Handbook of Environmental Technologies*. New York: Taylor & Francis, 2000.

38. American Coalition for Ethanol, All about ethanol. 2013. Available at http://www.ethanol.org/.

39. S. Lee, J. G. Speight and S. K. Loyalka, *Handbook of Alternative Fuel Technology*. Boca Raton, FL: CRC Press, 2007.

40. Flexible fuel vehicles: Providing a renewable fuel choice. March 2010. Available at http://www.afdc.energy.gov/afdc/pdfs/47505.pdf.

41. J. W. Wulff, M. Hulett and S. Lee, "Internal combustion system using acetylene fuel," U.S. Patent 6,076,487, June 20, 2000.

42. J. W. Wulff, M. Hulett and S. Lee, "A dual fuel composition including acetylene for use with diesel and other internal combustion engines," U.S. Patent 6,287,351, September 11, 2001.

43. J. E. Wenzel, J. Picou, M. Factor and S. Lee, "Kinetics of supercritical water reformation of ethanol to hydrogen," in *Energy Materials, 2007 Proceedings of the Materials Science and Technology Conference*, Published by the Minerals, Metals, and Materials Society, Detroit, MI, 2007, pp. 1121–1132.

44. S. Lee and Y. T. Shah, *Biofuels and Bioenergy: Processes and Technologies*. Boca Raton, FL: CRC Press, 2012.

# Ethanol from Lignocellulosics

*Sunggyu Lee*

## CONTENTS

## 12.1 INTRODUCTION

### 12.1.1 LIGNOCELLULOSE

The structural materials produced by plants to form cell walls, leaves, stems, and stalks are composed primarily of three different types of biobased macromoles, which are typically classified as cellulose, hemicellulose, and lignin. These biobased macromolecules are collectively called *lignocellulose, lignocellulosic biomass*, or *lignocellulosic materials*. A generalized plant cell wall structure is like a composite material in which rigid cellulose fibers are embedded in a cross-linked matrix of lignin and hemicellulose that binds the cellulose fibers, as illustrated in Figure 10.1.

The composition of these three types of macromolecules is widely varying depending upon the types of lignocellulosic biomass. Generally speaking, the dry weight of a typical cell wall consists of approximately 35%–50% cellulose, 20%–35% hemicellulose, and 10%–25% lignin [1]. Others may claim that cellulose typically accounts for 40%–50% of woody biomass, while lignin and hemicellulose each accounts for about 20%–30%. Although lignin comprises only 20%–30% of typical lignocellulosic biomass, it provides 40%–50% of the overall heating value of the biomass, due to its higher calorific value (CV) than cellulose and hemicellulose. This explains why chemical conversion or beneficial use of lignin is very important in fuel/energy utilization of lignocellulosic resources.

Lignocellulosic biomass structures also contain a variety of plant-specific chemicals in the matrix. These include extractives (such as resins, phenolics, and other chemicals) and minerals (calcium, magnesium, potassium, and others) that will leave behind ash when the biomass is combusted. The trace minerals and major elements in lignocellulosic materials display a high degree of variability for most of the elements between different species, between different organs within a given plant, and also depending on the growing conditions including the soil characteristics [2]. In addition to their potential environmental and health effects, trace minerals can play nontrivial roles in the downstream chemical treatments, including catalytic conversion of thermochemical intermediates of lignocellulose.

Cellulose is a large polymeric molecule composed of many hundreds or thousands of monomeric sugar (glucose) molecules. Since glucose is a monosaccharide, cellulose may be considered a *polysaccharide*. The molecular linkages in cellulose form linear chains that are rigid, highly stable, and resistant to chemical attack. Due to its linear polymeric structure, cellulose exhibits crystalline properties [3]. Cellulose may be somewhat soluble in a suitable solvent, just as most crystalline polymers. However, cellulose molecules in their crystalline form are packed so tightly that even small molecules such as water cannot easily permeate the structure. Logically, it would be even more difficult for larger enzymes to permeate or diffuse into the cellulose structure. Cellulose exists within a matrix of other polymers, mainly hemicellulose and lignin, as mentioned earlier.

On the other hand, hemicellulose consists of short and highly branched chains of sugar molecules. It contains both five-carbon sugars (usually D-xylose and L-arabinose) and six-carbon sugars (such as D-galactose, D-glucose, and D-mannose) as well as uronic acid. For example, galactan, found in hemicellulose, is a polymer of the sugar galactose, whose solubility in water is 68.3 g per 100 g of water at room temperature. Uronic acid is a sugar acid that has both a carbonyl and a carboxylic function. Hemicellulose is amorphous due to its highly branched macromolecular structure [3] and is relatively easy to hydrolyze to its constituent simple sugars, both five-carbon and six-carbon sugars. When hydrolyzed, the hemicellulose from hardwoods releases sugary products high in xylose (a five-carbon sugar or pentose), whereas the hemicellulose contained in softwoods typically yields more six-carbon sugars. Even though both five-carbon sugars and six-carbon sugars, illustrated in Figure 12.1, are simple fermentable sugars, there is a discerning difference in their fermentation chemistry and process characteristics with regard to specific yeasts and enzymes involved.

Humans have had far more extensive and successful experience in the fermentation of six-carbon sugars (hexoses) than five-carbon sugars (pentoses or xyloses), as well evidenced by a long history of manufacturing alcoholic beverages throughout the world. This statement is still valid for fuel ethanol fermentation as well. Many years ago, it was believed that xylose could not be fermented by yeasts. However, in recent years, a number of yeasts have been found to be capable of fermenting xylose into ethanol [4,5]. Genetic engineering of xylose fermentation in yeasts has also been carried out with successful outcomes [6].

Lignin is a complex and highly cross-linked aromatic polymer that is covalently linked to hemicellulose. Lignin contributes to the stabilization of a mature cell wall. It is a macromolecule whose typical molecular weight exceeds 10,000. Due to its cross-linked structure, lignin is generally more difficult to process, extract, hydrolyze, or react than cellulose or hemicellulose. Therefore, degradation or biodegradation of a cross-linked structure is the first and most important step for biofuel production from the cellulosic feedstocks, thus increasing the availability of the three types of macromolecules. Therefore, efficient conversion of lignin would result in a substantial increase in fuel yield as well as an enhanced economic outlook with utilization of lignocelluloses.

## 12.1.2 Cellulose Degradation, Conversion, and Utilization

Ethanol has garnered a great deal of attention as an alternative liquid fuel source to gasoline or as a gasoline blend to reduce the consumption of conventional gasoline. The ethanol alternative fuel program has been most seriously pursued by Brazil and the United States. In Brazil, all cars are run on either gasohol (a 22 vol.% mixture of ethanol with gasoline, or E22, mandated in 1993; a 25% blend, or E25, mandated since 2007) or pure ethanol (E100). In Brazil, the National Program of Alcohol, *PROALCOOL*, started in November 1975, and was created in response to the first oil crisis of 1973. This program effectively changed the consumption pattern of transportation fuels in the country. In 1998, these ethanol-powered cars consumed about 2 billion gallons of ethanol per year, and about 1.4 billion gallons of ethanol was additionally used for producing gasohol (E22, i.e., 22% ethanol and 78% gasoline) for other cars [7]. In March 2010, a milestone of 10 million flex-fuel ethanol-powered vehicles produced in Brazil was achieved. The Brazilian program has successfully

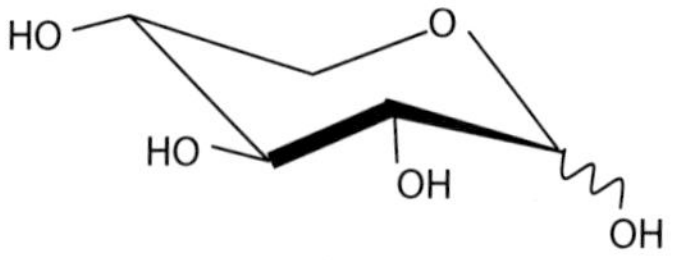

Xylose

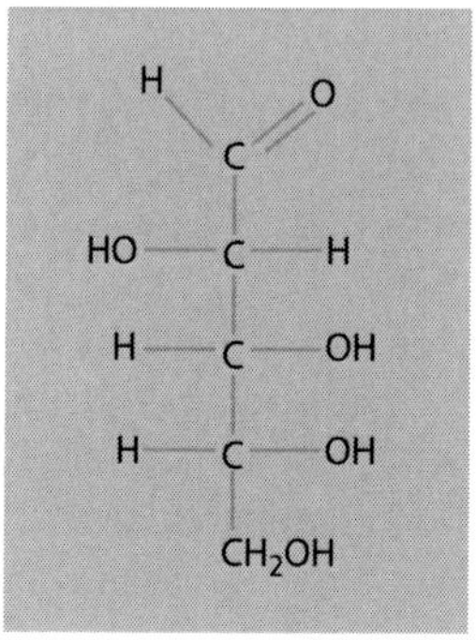

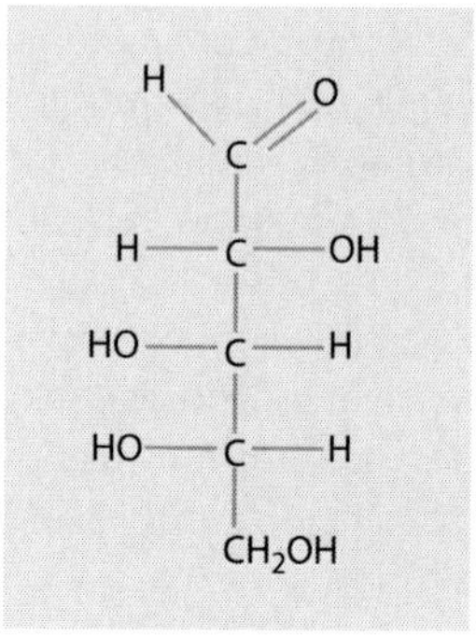

D-(–)-Arabinose

L-(+)-Arabinose

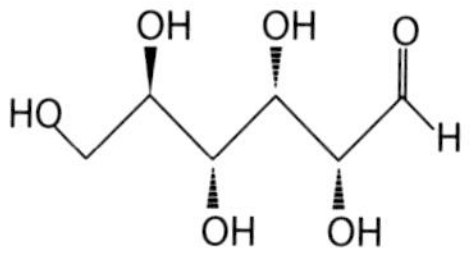

Glucose

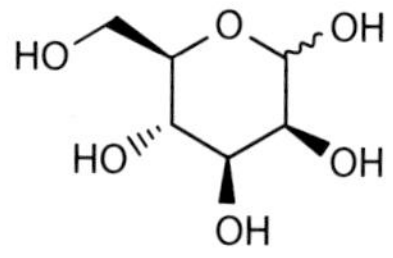

Mannose

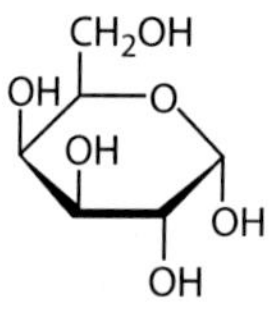

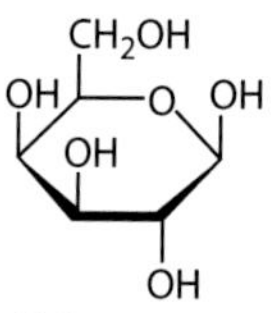

α-D-Galactopyranose

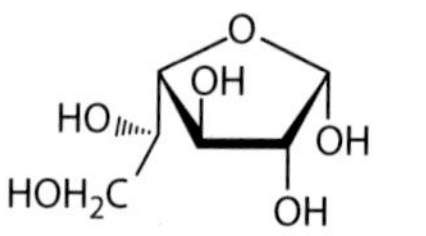

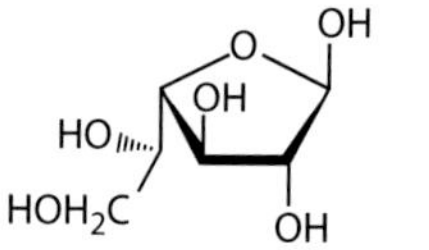

β-D-Galactopyranose

α-D-Galactofuranose

β-D-Galactofuranose

β-D-Galactose

**FIGURE 12.1**   Molecular structures of five-carbon and six-carbon sugars.

demonstrated large-scale production of ethanol from sugarcanes and established bioethanol as a sustainable motor fuel. In 2010, Brazil produced about 6.92 billion gallons [8] of ethanol.

The U.S. ethanol-based alternative fuel program has a shorter history than Brazil, and its ethanol production is mainly corn-based. The U.S. corn ethanol production has grown very rapidly in order to meet the sharply rising demand for oxygenated fuel to be blended with conventional gasoline. The U.S. production of corn ethanol in 2000, 2007, and 2010 was 1.62, 6.5, and 13.2 billion gallons, respectively. The eightfold increase in a matter of 11 years is remarkable. In 2010, the fuel ethanol production by these two countries accounted for about 90% of the world's industrial ethanol production.

Regarding the atmospheric concentrations of greenhouse gases (GHGs), the National Research Council (NRC), responding to a request from Congress and with funding from the U.S. Department of Energy, emphasizes the need for substantially more research and development on renewable energy sources, improved methods of utilizing fossil fuels, energy conservation, and energy efficient technologies. The *Energy Policy Act (EPACT) of 1992* was passed by the U.S. Congress to reduce the nation's dependence on imported petroleum by requiring certain fleets to acquire alternative fuel vehicles, which are capable of operating on nonpetroleum fuels. Alternative fuels for vehicular purposes, as defined by the Energy Policy Act, include ethanol, natural gas, propane, hydrogen, biodiesel, electricity, methanol, and p-series fuels. *P-Series fuels* are a family of renewable fuels that can substitute for gasoline. The *Energy Policy Act (EPACT) of 2005* changed the U.S. energy policy by providing tax incentives and loan guarantees for energy production of various types, which included tax reductions for alternative motor vehicles and fuels including bioethanol [9].

The market for transportation fuels has been dominated by petroleum-based fuels and that trend is expected to continue. A great many researchers have worked on the biological production of liquid fuels from biomass and coal [10]. They have found microorganisms that can produce ethanol from biomass, convert natural gas into ethanol, and convert syngas derived from coal gasification into liquid fuels. These microorganisms are found to have great potentials for being energy efficient and promising for industrial production. If successfully developed and implemented, the microbial process works at ordinary temperature and pressure and offers significant advantages over chemical processes, such as direct coal liquefaction and Fischer–Tropsch synthesis, which operate under severe conditions to produce liquid fuels from coal. Researchers have also focused on using lignin as a renewable source to derive traditional liquid fuel. Lignins are produced in large quantities in the United States as by-products of paper and pulp industry. As a consequence, the prices of some lignin products, such as lignosulfonates or sulfonated lignins, are relatively low. Lignosulfonates are used mainly as plasticizers in making concrete and also used in the production of plasterboard. Global production of lignin for various industrial applications is estimated to be quite high, even though reliable statistical data are unavailable. In China, the national lignin production has grown from 32 million tonnes in 2006 to 45 million tonnes in 2010, at a relatively fast growth rate.

## 12.2   LIGNOCELLULOSE CONVERSION

### 12.2.1   ETHANOL

#### 12.2.1.1   Manufacture of Industrial Alcohol

Industrial alcohol can be produced (1) synthetically from ethylene, (2) as a by-product of certain industrial operations, or (3) by the fermentation of sugars, starch, or cellulose. There are two principal processes for the synthesis of alcohol from ethylene. The original method (first carried out in the 1930s by Union Carbide) was the *indirect hydration process*, alternately referred to as the strong sulfuric acid–ethylene process, the ethyl sulfate process, the esterification hydrolysis process, or the sulfation hydrolysis process. The other synthetic process, designed to eliminate the use of sulfuric acid, is the *direct hydration process*. In the direct hydration process, ethanol is manufactured by directly reacting ethylene with steam. The hydration reaction is exothermic and reversible, that is, the maximum conversion is limited by chemical equilibrium.

$$CH_2 = CH_2 + H_2O_{(g)} \rightleftharpoons CH_3CH_2OH_{(g)} \quad \left(-\Delta H_{298}^{\circ}\right) = 45 \text{ kJ/mol}$$

Only about 5% of the reactant ethylene is converted into ethanol per each pass through the reactor. By selectively removing the ethanol from the equilibrium product mixture and recycling the unreacted ethylene, it is possible to achieve an overall 95% conversion. Typical reaction conditions are as follows: 300°C, 6–7 MPa, and employing phosphoric (V) acid catalyst adsorbed onto a porous support of silica gel or diatomaceous earth material. This catalytic process was first utilized in a large scale by the Shell Oil Company in 1947.

In addition to the direct hydration process, the sulfuric acid process, and fermentation routes to manufacture ethanol, several other processes have also been suggested [11–14]. However, none of these have been successfully implemented on a commercial scale.

#### 12.2.1.2   Fermentation Ethanol

Fermentation, one of the oldest chemical processes known to humans and most widely practiced by them, is used to produce a variety of useful products and chemicals. In recent years, however, many of the products that can be made by fermentation are also synthesized from petroleum feedstocks, often at lower costs or more selectively. It is also true that modern efforts of exploiting renewable biological resources rather than nonrenewable petroleum resources as well as focusing on green technologies, thereby alleviating the process involvement of harmful chemicals, are strong drivers for biological treatment processes such as fermentation. The future of the fermentation industry, therefore, depends on its ability to utilize the high efficiency and specificity of enzymatic catalysis to synthesize complex products, and also on its ability to overcome variations in the quality and availability of the raw materials.

As discussed in Chapter 11, ethanol can be quite easily derived by fermentation processes from any material that contains sugar(s) or sugar precursors. The raw materials used in the manufacture of ethanol via fermentation are classified as sugars, starches, and cellulosic materials [15,16]. Sugars can be directly converted to ethanol by simple chemistry, as fully explained in Chapter 11. Starches must first be hydrolyzed to fermentable sugars by the action of enzymes. Likewise, cellulose must first be converted to sugars, generally by the action of mineral acids (i.e., inorganic acids such as the common acids sulfuric acid, hydrochloric acid, and nitric acid). Once the simple sugars are formed, enzymes from yeasts can readily ferment them to ethanol. Therefore, an efficient process conversion of cellulosic material into simple sugars as well as a robust and efficient enzymatic fermentation of different forms of sugars to ethanol are crucially important for the development of a successful process technology for cellulosic ethanol.

### 12.2.1.3 Fermentation of Sugars

A widely utilized form of sugar for ethanol fermentation is the blackstrap molasses, which contain about 30–40 wt.% sucrose, 15–20 wt.% invert sugars such as glucose and fructose, and 28–35 wt.% of nonsugar solids. The direct fermentation of sugarcane juice, sugarbeet juice, beet molasses, fresh and dried fruits, sorghum, whey, and skim milk have been considered, but none of these could compete economically with molasses. From the viewpoint of industrial manufacture of ethanol, sucrose-based substances such as sugarcane and sugarbeet juices present many advantages, including their relative abundance and renewable nature. Molasses, the noncrystallizable residue that remains after the sucrose purification, has additional advantages; it is a relatively inexpensive raw material, readily available, and already used for industrial ethanol production. Molasses are used in dark brewed alcoholic beverages such as dark ales and also for rum. Bioethanol production in Brazil uses sugarcane as feedstock and employs first-generation technologies based on the use of the sucrose content of sugarcane. The enhancement potential for sugarcane ethanol production in Brazil was discussed by Goldemberg and Guardabass [17] in two principal areas of (1) productivity enhancement and (2) area expansion.

Park and Baratti [18] have studied the batch fermentation kinetics of sugar beet molasses by *Zymomonas mobilis*. *Z. mobilis* is a rod-shaped gram-negative bacterium that can be found in sugar-rich plant saps. It degrades sugars to pyruvate using the *Entner–Doudoroff pathway* [19]. The pyruvate is then fermented to produce ethanol and carbon dioxide as the only products. This bacterium has several advantageous properties that make it competitive with the yeasts and, in some aspects, superior to yeasts; important examples include higher ethanol yields, higher sugar uptake, higher ethanol tolerance and specific productivity, and lower biomass production. When cultivated on molasses, however, *Z. mobilis* generally shows poor growth and low ethanol production as compared to cultivation in glucose media [18]. The low ethanol yield is explained by the formation of by-products such as levan and sorbitol. Other components of molasses such as organic salts, nitrates, or the phenolic compounds could also be inhibitory for the bacterial growth [20]. Therefore, its acceptable and utilizable substrate range is restricted to simple sugars such as glucose, fructose, and sucrose. Park and Baratti [18] found that in spite of good growth and prevention of levan formation, the ethanol yield and concentration were not sufficient for the development of an industrial process. In a study by Kalnenieks et al. [21], potassium cyanide (KCN) at submillimolar concentrations (20–500 µM) inhibited the high respiration rates of aerobic cultures of *Z. mobilis* but, remarkably, stimulated culture growth. Effects of temperature and sugar concentration on ethanol production by *Z. mobilis* have been studied by scientists. Cazetta et al. [22] investigated the effects of temperature and molasses concentration on ethanol production. They used factorial design of experiments (DOEs) in order to study varied conditions concurrently; the different conditions investigated included varying combinations of temperature, molasses concentration, and culture times. They concluded that the optimal conditions found for ethanol production were 200 g/L of molasses at 30°C for 48 h, and this produced 55.8 g ethanol/L [22].

Yeasts of the *Saccharomyces* genus are mainly used in industrial processes for ethanol fermentation. One well-known example is *Saccharomyces cerevisiae*, which is most widely used in brewing beer and wine. However, *S. cerevisiae* cannot ferment D-xylose, the second most abundant sugar form of the sugars obtained from cellulosic materials. One microorganism that is naturally capable of fermenting D-xylose to ethanol is the yeast *Pichia stipitis*; however, this yeast is not as ethanol- and inhibitor-tolerant as the traditional ethanol-producing yeast, that is, *S. cerevisiae*. Therefore, its industrial application is impractical, unless significant advances are made. There have been efforts that attempt to generate *S. cerevisiae* strains that are able to ferment D-xylose by means of genetic engineering [23]. Scientists have been working actively to ferment xylose with high productivity and yield by developing variants of *Z. mobilis* that are capable of using $C_5$-sugars (pentoses or xyloses) as a carbon source [24]. Advances with promising results are being reported in the literature.

As a significant advance in metabolistic changes brought about by genetic engineering, Tao et al. altered *Escherichia coli* B strain, which is an organic acid producer, to *E. coli* strain KO11, which is an ethanol producer. The altered KO11 strain yielded 0.50 g ethanol/g xylose using 10% xylose solution at 35°C and pH of 6.5 [25]. This result provides an example of how the output of a microbe can be altered.

Utilizing a combination of metabolic engineering and systems biology techniques, two broad methods for developing more capable and more tolerant microbes and microbial communities are the recombinant industrial and native approaches [26]. The two methods differ as follows:

1. *Recombinant industrial host approach*: Insert key novel genes into known robust industrial hosts with established recombinant tools.
2. *Native host approach*: Manipulate new microbes with some complex desirable capabilities to develop traits needed for a robust industrial organism and to eliminate unneeded pathways.

The research on yeast fermentation of xylose to ethanol has been very actively studied; particular emphasis has been placed on genetically engineered *S. cerevisiae*. *S. cerevisiae* is a safe microorganism that plays a traditional and major role in modern industrial bioethanol production [27]. *S. cerevisiae* has several advantages including its high ethanol productivity as well as its high ethanol and inhibitor tolerance. Unfortunately, this yeast does not have the capability of fermenting xylose. A number of different strategies based on genetic engineering and advanced microbiology have been applied to engineer yeasts to become capable of efficiently producing ethanol from xylose. These novel strategies included (1) the introduction of initial xylose metabolism and xylose transport, (2) changing of the intracellular redox balance, and (3) overexpression of xylulokinase and pentose phosphate pathways [27]. One of the pioneering studies was that of Sedlak et al. [28], which involves the development of genetically engineered *Saccharomyces* yeasts that can coferment both glucose and xylose to ethanol. Even though their recombinant yeast *S. cerevisiae* with xylose metabolism added was found to be the most effective yeast, they still utilized glucose more efficiently than xylose. According to their experimental results, following rapid consumption of glucose in less than 10 h, xylose was metabolized more slowly and less completely. In fact, xylose was not totally consumed even after 30 h. Ideally, xylose should be consumed simultaneously [26] with glucose at a similar efficiency and speed; however, the newly added capability of cofermentation of both glucose and xylose was of groundbreaking discovery. Furthermore, they found that while ethanol was the most abundant product from glucose and xylose metabolism, small amounts of the metabolic by-products of glycerol and xylitol were also obtained [28]. The above two issues clearly suggest the areas for improvement, viz., higher efficiency for xylose fermentation and optimization and by-product control, and they are certainly the subjects of intense research investigation.

## 12.2.2 SOURCES FOR FERMENTABLE SUGARS

### 12.2.2.1 Starches

As fully discussed in Chapter 11, grains such as corn generally provide cheaper ethanol feedstocks in most regions of the world, and the industrial conversion may be kept relatively inexpensive because they can be stored more easily than most sugar crops, which often must be reduced to a form of syrup prior to storage. Furthermore, grain milling ethanol process produces a variety of by-products that can be marketed for a number of value-added end uses such as protein meal in the animal feeds [29]. Fermentation of starch from grains is inherently more complex than sugars, involving more processing steps, because the starch must first be converted to sugar and then fermented to ethanol. A simplified equation for the conversion of starch to ethanol can be written as

$$C_6H_{10}O_5 + H_2O \xrightarrow[\text{Fungal amylase}]{\text{Enzyme}} C_6H_{12}O_6 \xrightarrow{\text{Yeast}} 2C_2H_5OH + 2CO_2$$

As shown in Figure 12.2, in manufacturing grain alcohol, the distiller produces a sugar solution from feedstock, ferments the sugar to ethanol, and then separates the ethanol from water by distillation.

Among the disadvantages in the use of grains are its fluctuations in feedstock price. Critics of corn ethanol have been making remarks in relation to "fuel vs. food" and cited that the recent food price increase has something to do with the corn ethanol manufacture, while others oppose to this view offering statistical data and logical reasons [31]. Ethanol in gasoline boosts the fuel's octane rating and also helps cleaner burning. In the United States, ethanol is currently the most popular oxygenated blend fuel.

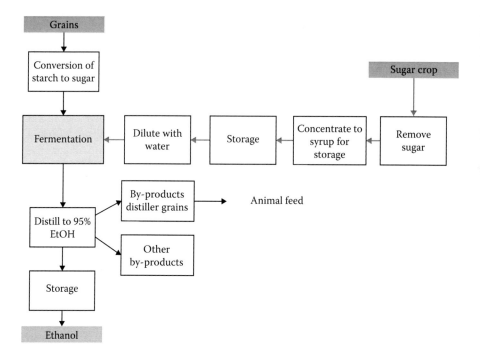

**FIGURE 12.2** Synthesis of ethanol from grains and sugar crops. (From U.S. Congress Office of Technology Assessment, "Energy from biological process," U.S. Government Printing Office, Washington, D.C., Tech. Rep. 2, pp. 142–177, 1980.)

### 12.2.2.2 Cellulosic Feedstock

Enormous amounts of carbohydrate-containing cellulosic wastes are generated every year throughout the world. Cellulose from wood, agricultural residues, and wastes from pulp and paper mills must first be converted to *sugar* before it can be fermented. Cellulosic ethanol is claimed to reduce GHG emission by more than 90% over conventional petroleum-based fuels [32]. In addition, cellulosic ethanol is free from the criticism of "food vs. fuel," since it is not derived from food crops. Based on these reasons, lignocellulosic ethanol is classified as a second-generation biofuel. New ways of reducing the cost of cellulosic ethanol production include the development of effective pretreatment methods, replacement of acidic hydrolysis with efficient enzymatic hydrolysis, commercialization of robust enzymes, and fine-tuning of enzymatic hydrolysis and fermentation times, in addition to the fermentation selectivity and effectiveness for both $C_6$- and $C_5$-sugars.

## 12.3 HISTORICAL PERSPECTIVE OF ALCOHOL FERMENTATION TECHNOLOGY

Although less heralded, cellulosic ethanol has a fairly long history. One of the first recorded attempts at commercializing a cellulosic ethanol process was made in Germany as early as 1898. The process was based on the use of dilute acid to

hydrolyze the cellulose to glucose and the subsequent fermentation of glucose to ethanol. The reported productivity was 7.6 L of ethanol per 100 kg of wood waste, equivalent to 18 gallons per short ton. As an early process, the conversion of wood waste
into ethanol was quite remarkable; the process was further enhanced in Germany
to yield about 50 gallons of ethanol per short ton of biomass. In the United States,
this process was further enhanced during World War I by adopting a single-stage
dilute sulfuric acid hydrolysis process, by which the overall ethanol yield per input
biomass was about 50% lower than the original German version, but the throughput
of the process was much higher. This American process was short-lived due to a
significant decrease in lumber production in the postwar era. However, this process
was brought to commercial operation again during World War II for production of
butadiene by ethanol conversion to ultimately produce synthetic rubber. Even though
the process achieved an ethanol yield of 50 gallons/dry ton of wood cellulose, this
level of productivity was far from profitable and the process was halted after the war.
Even though commercial production had been stopped, active research on cellulosic
ethanol continued throughout the world, even more intensifying as a result of several
rounds of petroleum crises, booming bioethanol consumption, and rapid advances in
biotechnology.

In 1978, the Gulf Oil researchers [33] designed a commercial scale plant producing $95 \times 10^6$ L/year of ethanol by simultaneous enzymatic hydrolysis of cellulose and fermentation of resulting glucose as it is formed, thereby overcoming
the problem of product inhibition. The process consisted of a unique pretreatment,
which involved the grinding and heating of the feedstock followed by hydrolysis
with a mutant bacterium, also specially developed for this purpose. Mutated strains
of the common soil mold *Trichoderma viride* were able to process 15 times more
glucose than natural strains. Simultaneous hydrolysis and fermentation reduced the
time requirement for the separate hydrolysis step, thus reducing the production cost
and increasing the yield. Also, the process did not use acids that would increase the
equipment costs. The sugar yields from the cellulose were about 80% of what was
theoretically achievable, but the small amount of hemicellulose in the sawdust was
not converted. This fact demonstrated a need for an effective pretreatment to cause
hemicellulose separation. As advances in enzyme technology have been realized, the
acid hydrolysis process has been gradually replaced by a more efficient enzymatic
hydrolysis process. In order to achieve efficient enzymatic hydrolysis, chemical and/
or biological pretreatment of the cellulosic feedstock has become necessary to prehy-
drolyze hemicelluloses in order to separate them from a lignin or lignin-held structure.
The researchers of the Forest Products Laboratory of the U.S. Forest Service (USFS)
and the University of Wisconsin-Madison developed the Sulfite Pretreatment to
Overcome Recalcitrance of Lignocellulose (SPORL) for robust enzymatic saccharification of lignocellulose [34].

*Cellulase* is a class of enzymes that catalyze cellulolysis, which breaks cellulose
chains into glucose molecules. In recent years, various enzyme companies and biotechnology industries have contributed significant technological breakthroughs in
cellulosic ethanol technology through the development of highly potent cellulase
enzymes as well as the mass production of these enzymes for enzymatic hydrolysis with economic advantages. Research, development, and demonstration (RD&D)

efforts in cellulase enzyme by many international companies such as Novozymes, Genencor, Iogen, SunOpta, Verenium, and Dyadic International, and also by the U.S. national laboratories including the National Renewable Energy Laboratory (NREL), are quite significant.

Cellulosic ethanol garnered strong endorsements and received significant support from U.S. President George W. Bush in his State of the Union address, delivered on January 31, 2006, that proposed to expand the use of cellulosic ethanol. In his address, President Bush outlined the *Advanced Energy Initiative (AEI 2006)* to help overcome America's dependence on foreign energy sources and the *American Competitiveness Initiative (ACI 2006)* to increase R&D investment and strengthen education. In the following year, President Bush announced, in his State of the Union address on January 23, 2007, a proposal mandate for 35 billion gallons of ethanol by 2017. It has been recognized by the experts that the maximum capacity of corn ethanol on the U.S. market would be about 15 billion gallons per year, and therefore, this mandate would have to be met by some 20 billion gallons per year of cellulosic ethanol. President Bush's plan also included $2 billion funding for the period of 2007–2017 for cellulosic ethanol plants with an additional $1.6 billion announced by the USDA on January 27, 2007. The *Energy Independence and Security Act (EISA) of 2007* established long-term renewable-fuel production targets through the *Second Renewable Fuel Standard (RFS2)*. The RFS2 expanded upon the initial corn–ethanol production volumes and timeline of the original RFS, under which the U.S. EPA is responsible for implementing regulations to ensure that increasing volumes of biofuels for the transportation sector are produced. The U.S. EPA released its final rule for the expanded RFS2 in February 2010, through which its statutory requirements established specific annual volumes, for the total renewable-fuel volume, from all renewable fuel sources [35]. U.S. EPA's implementation of the RFS2 would position the United States for making significant improvements in the GHG footprint due to the transportation sector. In February 2010, the White House under the leadership of President Barack Obama released "Growing America's Fuel," which is a comprehensive roadmap to advanced fuel deployment [36].

In 2004, the researchers at the NREL, in collaboration with two major industrial enzyme manufacturers (Genencor International and Novozymes Biotech), achieved a dramatic reduction in cellulase enzyme costs, which was one of the major stumbling blocks in commercialization of cellulosic ethanol. Cellulases belong to a group of enzymes known as glycosyl hydrolases, which cleave (hydrolyze) chemical bonds linking a carbohydrate to another molecule. The novel technology involves a cocktail of three types of cellulases, viz., endoglucanases, exoglucanases, and β-glucosidases. These enzymes synergistically work together to attack cellulose chains, pulling them away from the crystalline structure, breaking off cellobiose molecules (two linked glucose residues), splitting them into individual glucose molecules, and making them available for further enzymatic processing. This breakthrough work is claimed to have resulted in a 20- to 30-fold cost reduction and earned NREL and collaborators an R&D 100 Award [37]. This is certainly a milestone accomplishment in cellulosic ethanol technology development. However, there are a number of areas that still need to be enhanced and addressed for the development and implementation of cost-effective cellulosic ethanol technology, and they include

a. Further cost reductions are required in cellulase enzyme manufacture.
b. New routes need to be developed to enhance enzymatic efficiencies.
c. Development of enzymes with higher heat tolerance and improved specific activities is highly desired.
d. Better matching of enzymes with plant cell-wall polymers needs to be achieved.
e. A high-solid enzymatic hydrolysis process with enhanced efficiency needs to be developed.

The U.S. Department of Energy Workshop Report summarizes, in scientific details, the identified research needs in the area (Biofuels Joint Roadmap, June 2006) [26].

In recent years, major advances are also being made in utilizing genetic engineering and advanced microbiology in the development of robust microbe systems that are capable of efficiently cofermenting both $C_5$ and $C_6$ sugars, and that are resistant to inhibitors and tolerant against process variability.

Another effort for production of cellulosic ethanol is via catalytic conversion of gaseous intermediates produced by thermochemical conversion of cellulosic materials without the use of enzymes. Certainly, there is a trade-off between the purely chemical route and the enzymatic route in various aspects, including conversion efficiency, product selectivity, reaction speed, capital cost, overall energy efficiency, raw material flexibility, and more. A large commercialization effort launched by Range Fuels in 2007, based on catalytic conversion of thermochemical intermediates derived from biomass, was terminated in 2011 without meeting its original goals.

## 12.4   AGRICULTURAL LIGNOCELLULOSIC FEEDSTOCKS

The world has depended so heavily on natural gas and petroleum for its energy needs and the manufacture of most organic materials due to the fact that gases and liquids are easier and more convenient to handle than solids. Solid materials like wood, on the other hand, are difficult to collect, transport, and process into components, which can make desired products for energy. As such, solid materials are seriously lacking in continuous processability and render logistical problems in their utilization in large quantities.

Agricultural lignocellulose is vastly abundant, inexpensive, and renewable because it is made via photosynthesis with the aid of solar energy. Greater biomass utilization could also help ameliorate solid waste disposal problems. In 2009, 243 million tons of municipal solid wastes (MSWs) were generated in the United States, which is equivalent to about 4.3 lbs of wastes per person per day. Of these wastes, 28.2% were paper and paperboard, 13.7% yard clippings, 6.5% wood, and 14.1% food scraps [38]. Considering that some food scraps contain cellulosic materials, about 50% of the total MSWs is cellulosic and could be converted to useful chemicals and fuels [39].

Although lignocellulose is inexpensive, it still involves transformational efforts to convert to fermentable sugars. Furthermore, as shown in Figure 12.3, lignocellulose has a complex chemical structure with three major components, each of which

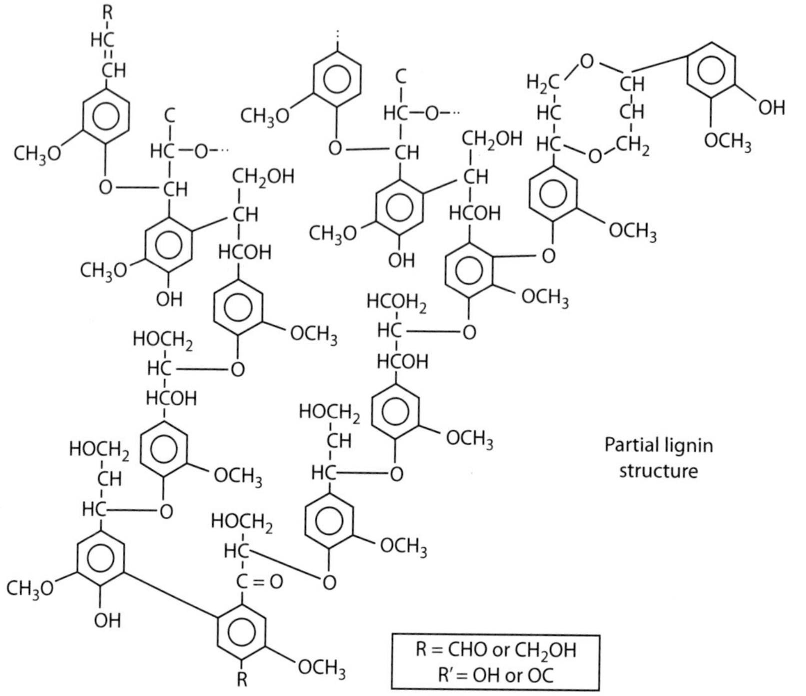

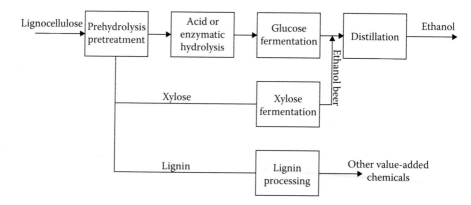

**FIGURE 12.3** Major polymeric components of plant materials. (Adapted from C&EN, "Major polymeric components of plant materials," *Chem. Eng. News*, 1990.)

must be processed separately to make the best use of high efficiencies inherent in the biological process. The three major components of lignocellulose are crystalline cellulose, hemicellulose, and lignin.

A general scheme for the conversion of lignocellulose to ethanol is shown in Figure 12.4. The lignocellulose is pretreated to separate or liberate the xylose and, sometimes, the lignin from the crystalline cellulose. This step is very important,

**FIGURE 12.4** Generalized conversion scheme of lignocellulose to ethanol.

since the efficiency of pretreatment directly affects the efficiency of the ensuing steps. The xylose can then be fermented to ethanol, whereas the lignin can be further processed to produce other liquid fuels and valuable chemicals. Crystalline cellulose, the largest (around up to 50%) and most useful fraction of lignocellulose, remains behind as a solid after the pretreatment and is sent to an enzymatic hydrolysis process that breaks the cellulose down into glucose. Enzymes, the biological catalysts, are highly specific; hence, the hydrolysis of cellulose to sugar does not further break down the sugars. Enzymatic processes are capable of achieving a yield approaching 100%. The glucose is then fermented to ethanol and combined with the ethanol from xylose fermentation. This dilute beer, that is, dilute ethanol–water solution, is then concentrated to fuel grade ethanol via distillation and further purification such as pressure swing adsorption (PSA).

The hemicellulose fraction, the second major component at around 25%, is primarily composed of xylan, which can be easily converted to the simple sugar xylose (or pentose). Xylose constitutes about 17% of woody angiosperms and accounts for a substantially higher percentage of herbaceous angiosperms. Therefore, xylose fermentation or conversion is essential for commercial bioconversion of lignocellulose into ethanol or other biochemicals. Generally speaking, xylose is more difficult than glucose to convert or ferment to ethanol, based on the current level of science and technology. From this process standpoint, it would be more beneficial to find or develop a more robust and optimal microorganism that can ferment both glucose and xylose to ethanol in a single fermenter with high yield and selectivity. Methods have been identified using new strains of, or, metabolically engineered yeasts [23], bacteria, and processes combining enzymes and yeasts. Although none of these fermentation processes are yet ready for commercial use, considerable progress has been made.

Lignin, the third major component of lignocellulose (around 25%), is a large random phenolic polymer. In lignin processing, the polymer is broken down into fragments containing one or two phenolic rings. Extra oxygen and side chains are stripped from the molecules by catalytic methods, and the resulting phenol groups are reacted with methanol to produce methyl aryl ethers. Methyl aryl ethers, or arylmethylethers, are high value octane enhancers that can be blended with gasoline.

## 12.5 CELLULOSIC ETHANOL PROCESS TECHNOLOGY

In this section, various process stages of the typical cellulose ethanol fermentation technology, as illustrated in Figure 12.4, are explained.

### 12.5.1 Acid or Chemical Hydrolysis

Acid hydrolysis of cellulosic materials has long been practiced and is relatively well understood. Among the important specific factors in acid hydrolysis are (1) surface-to-volume ratio, (2) acid concentration, (3) temperature, and (4) treatment time. The surface-to-volume ratio is especially important in that it also determines the yield of glucose. Therefore, smaller particle size that has higher surface-to-volume ratio results in better hydrolysis, in terms of the extent and rate of reaction [41]. A higher liquid-to-solid ratio also results in a faster reaction. A trade-off must be made between the optimum ratio and economic feasibility because the increase in the cost of equipment parallels the increase in the ratio of liquids to solids. For chemical hydrolysis, a liquid-to-solid ratio of *10:1* seems to be most suitable [41].

In a typical system for chemically hydrolyzing cellulosic wastes, the wastes are milled to fine particle sizes. The milled material is immersed in a weak acid (0.2% to 10%), the temperature of the suspension is elevated to 180°C to 230°C, and a moderate pressure is applied. Eventually, the hydrolyzable cellulose is transformed into sugar. However, this reaction has no effect on the lignin, which is also present. The yield of glucose varies, depending upon the nature of raw waste. For example, 84–86 wt.% of kraft paper or 38–53 wt.% of the ground refuse may be recovered as sugar. The sugar yield increases with the acid concentration as well as the elevation of temperature. A suitable concentration of acid ($H_2SO_4$) is about 0.5% of the charge.

A two-stage, low-temperature, and ambient-pressure acid hydrolysis process that utilizes separate unit operations to convert the hemicellulose and cellulose to fermentable sugars was developed [42] and tested by the Tennessee Valley Authority (TVA) and the U.S. Department of Energy (DOE). Laboratory- and bench-scale evaluations showed more than 90% recovery and conversion efficiencies of sugar from corn stover. Sugar product concentrations of more than 10% glucose and 10% xylose were achieved. The inhibitor levels in the sugar solutions never exceeded 0.02 g/100 mL, which is far below the level shown to inhibit fermentation. An experimental pilot plant was designed and built in 1984. The acid hydrolysis pilot plant provided fermentable sugars to a 38 L/h fermentation and distillation facility built in 1980. The results of their studies are summarized as follows:

- Corn stover ground to 2.5 cm was adequate for the hydrolysis of hemicellulose.
- The time required for optimum hydrolysis in 10% acid at 100°C was 2 h.
- Overall xylose yields of 86% and 93% were obtained in a bench-scale study at 1 and 3 h reaction times, respectively.
- Recycled leachate, dilute acid, and prehydrolysis acid solutions were stable during storage for several days.
- Vacuum drying was adequate in the acid concentration step.
- Cellulose hydrolysis was successfully accomplished by cooking stover containing 66% to 78% acid for 6 h at 100°C. Yields of 75% to 99% cellulose conversion to glucose were obtained in the laboratory studies.
- Fiberglass reinforced plastics of vinyl ester resin were used for construction of process vessels and piping.

### 12.5.1.1 Process Description

The process involves two-stage sulfuric acid hydrolysis, relatively low temperature, and a cellulose prehydrolysis treatment with concentrated acid. Figure 12.5 is a schematic flow diagram of the TVA process. Corn stover is ground and mixed with the

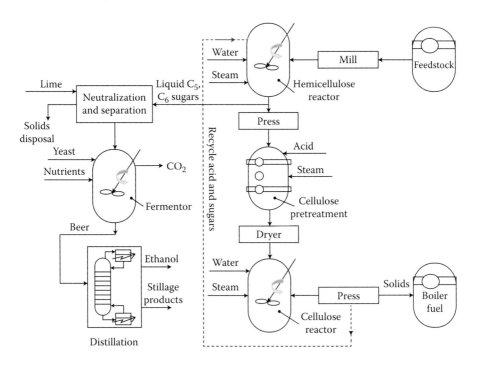

**FIGURE 12.5** Low-temperature low-pressure two-stage acid hydrolysis concept for conversion of nonwoody feedstocks to ethanol. (From G. E. Farina et al., *Energy Sources*, 10, 231–237, 1988.)

dilute sulfuric acid (about 10% by weight). The hemicellulose fraction of the stover is converted to pentose (xylose) sugars by heating the solution to 100°C for 2 h in the first hydrolysis reactor. Raw corn stover contains, on a dry basis, about 40% cellulose, 25% hemicellulose, and 25% lignin. Sulfuric acid for the hydrolysis reaction is provided by recycling the product stream from the second hydrolysis step, which contains the sulfuric acid and hexose sugars. The pentose and hexose sugars, which are primarily xylose and glucose, respectively, are leached from the reactor with warm water. The sugar-rich leachate is then neutralized with lime (calcined limestone, CaO; or calcined dolomite, CaO·MgO), filtered to remove precipitated materials, and fermented to produce ethanol [43].

Residue stover from the first hydrolysis step (hemicellulose conversion reactor) is dewatered and prepared for the second hydrolysis step (cellulose conversion reactor) by soaking (prehydrolysis treatment step) in sulfuric acid (about 20%–30% concentration) from 1 to 2 h. The residue is then screened, mechanically dewatered, and vacuum dried to increase the acid concentration to 75%–80% in the liquid phase before entering the cellulose reactor. The second hydrolysis reactor operates at 100°C and requires a time of 4 h. The reactor product is filtered to remove solids (primarily lignin and unreacted cellulose). Since the second hydrolysis reactor product stream contains about 10% acid, it is used in the first hydrolysis step to supply the acid required for hemicellulose hydrolysis. Residue from the reactor is washed to recover the remaining sulfuric acid and the sugar not removed in the filtration step.

Lignin is the unreacted fraction of the feedstock, which can be burned as a boiler fuel. It has the heating value of about 5270 kcal/kg (or 9486 Btu/lb), which is comparable to that of subbituminous coal. Other products such as surfactants, concrete plasticizers, and adhesives can also be made from lignin. Stillage can be used to produce several products, including methane. Preliminary research showed that 30 L of biogas containing 60% methane gas was produced from a liter of corn stover stillage. For each liter of ethanol produced, 10 L of stillage was produced [42].

All process piping, vessels, and reactors in contact with corrosive sulfuric acid were made of fiberglass reinforced vinyl ester resin. The dryer was made of carbon steel and lined with Kynar®, which is a trademark of Arkema Inc. (formerly Atofina Inc.) for poly(vinylidene fluoride) (PVDF). Conveyor belts were also made of acid-resistant material. Mild steel agitator shafts were coated with Kynar® or Teflon®, which is DuPont's trademark for polytetrafluoroethylene (PTFE). Heat exchangers were made with chlorinated poly(vinyl chloride) (CPVC) pipe shells and Carpenter 20 stainless-steel coils. Carpenter 20, also known as Alloy 20, is a nickel–iron–chromium austenitic alloy that was developed for maximum corrosion resistance to acid attack, in particular, sulfuric acid attack. Pumps were made with nonmetallic compound Teflon® lining or Carpenter 20 stainless steel. The two filter press units had plates made of polypropylene (PP) [42].

## 12.5.2 Enzymatic Hydrolysis

Cellulose has a polymerized structure as 1-4,β-glucosidic linkage and is insoluble in water. Each cellulose molecule is an unbranched polymer of 15 to 10,000 D-glucose units. Hydrolysis of crystalline cellulose is a rate-controlling step in the conversion of biomass to ethanol because aqueous enzyme solutions have difficulty acting on insoluble, impermeable, highly structured cellulose. Therefore, making soluble enzymes act on insoluble cellulose is one of the principal challenges in the process development of cellulosic ethanol.

Cellulose needs to be efficiently solubilized such that an entry can be made into cellular metabolic pathways. Solubilization is brought about by enzymatic hydrolysis catalyzed by a *cellulase* system of certain bacteria and fungi. Cellulase is a class of enzymes, produced primarily by fungi, bacteria, and protozoans, that catalyze the hydrolysis of cellulose, that is, cellulolysis, as described below.

### 12.5.2.1 Cellulase Enzyme System

There are several different kinds of cellulases, and they differ mechanistically and structurally. Each cellulolytic microbial group has an enzyme system unique to it. The enzyme capabilities range from those with which only soluble derivatives of cellulose can be hydrolyzed to those with which a cellulose complex can be disrupted. Although it is a usual practice to refer to a mixture of compounds that have the ability to degrade cellulose as *cellulase*, it is actually composed of a number of distinctive enzymes. Based on the specific type of reaction catalyzed, the cellulases may be characterized into five general groups, viz., endocellulase, exocellulase, β-glucosidase, oxidative cellulase, and cellulose phosphorylase.

1. *Endocellulase* cleaves internal bonds to disrupt the crystalline structure of cellulose and expose individual cellulose polysaccharide chains.
2. *Exocellulase* detaches two or four saccharide units from the ends of the exposed chains produced by endocellulase, resulting in disaccharides or tetrasaccharides, such as cellobiose. *Cellobiose* is a disaccharide with the formula of $[HOCH_2CHO(CHOH)_3]_2O$. There are two principal types of exocellulases or cellobiohydrolases (CBH): (1) *CBH-I* works processively from the reducing end, and (2) *CBH-II* works processively from the nonreducing end of cellulose. In this description, processivity (or progressivity) is the ability of an enzyme to repetitively continue its catalytic function without dissociating from its substrate. By an active enzyme being adhered onto the surface of a solid substrate, the chance for reaction is significantly enhanced.
3. *β-glucosidase or cellobiase* hydrolyzes the exocellulase products, that is, disaccharides and tetrasaccharides, into individual monosaccharides.
4. *Oxidative cellulases* depolymerize and break down crystalline cellulose molecules by radical reactions, as in the case with a cellobiose dehydrogenase (acceptor), which is an enzyme that catalyzes the chemical reaction of

cellobiose + acceptor ⇌ cellobiono-1,5-lactone + reduced acceptor

by which cellobiose is dehydrogenated while the acceptor is reduced, thereby resulting in a reduction–oxidation reaction.
5. *Cellulose phosphorylases* depolymerize cellulose using phosphates instead of water.

In most cases, the enzyme complex breaks down cellulose to β-glucose (whose hydroxyl group on $C_1$ position and $-CH_2OH$ group on $C_5$ position are on the same side of the plane, i.e., cis-arrangement). This type of cellulase enzyme is produced mainly by symbiotic bacteria. Symbiotic bacteria are bacteria living in symbiosis (close and long-term interaction) with another organism or each other.

On the other hand, enzymes that hydrolyze hemicellulose are usually referred to as *hemicellulase* and are still commonly classified under cellulases. However, enzymes that break down lignin are not classified as cellulase, strictly speaking. Along with diverse types of enzymes, it must be clearly pointed out that a principal challenge in hydrolytic degradation of biomass into fermentable sugars is how to make these different enzymes work together as a synergistic enzyme system. For example, cellulases and hemicellulases are secreted from a cell as free enzymes or extracellular cellulosomes (complexes of cellulolytic enzymes created by bacteria). The collective activity of these enzymes in a system is likely to be more active than, or at least quite different from, the individual activity of an isolated enzyme.

The enzymes described above can be classified into two types, viz., progressive (also known as processive) and nonprogressive (or nonprocessive) types. Progressive cellulase will continue to interact with a single polysaccharide strand, while nonprogressive cellulase will interact once, disengage, and then engage another polysaccharide strand.

Based on the enzymatic capability, the cellulase is characterized into two groups, namely, $C_1$ enzyme (or $C_1$ factor) and $C_X$ enzyme (or $C_X$ factor) [41]. The $C_1$ factor is regarded as an "affinity" or prehydrolysis factor that transforms highly ordered (crystalline) cellulose, that is, cotton fibers or Avicel, into linear and hydroglucose chains. The $C_1$ factor has little effect on soluble derivatives. Raw *cotton* is composed of 91% pure cellulose. As such, it serves as an essential precursor to the action of $C_X$ factor. The $C_X$ (hydrolytic) factor breaks down the linear chains into soluble carbohydrates, usually cellobiose (a disaccharide) and glucose (a monosaccharide).

Microbes rich in $C_1$ enzyme are more useful in the production of glucose from the cellulose. Moreover, since $C_1$ phase proceeds more slowly than the subsequent step, it is the rate-controlling step. Among the many microbes, *Trichoderma reesei* surpasses all others in the possession of $C_1$ complex. *T. reesei* is an industrially important cellulolytic filamentous fungus and is capable of secreting large amounts of cellulases and hemicellulases [44]. Recent advances in cellulase enzymology, cellulose hydrolysis (cellulolysis), strain enhancement, molecular cloning, and process design and engineering are bringing *T. reesei* cellulases closer to being a commercially viable option for cellulose hydrolysis [45]. The site of action of cellulolytic enzymes is important in the design of hydrolytic systems ($C_X$ factor). If the enzyme is within cell mass, the material to be reacted must diffuse into the cell mass, which would be highly undesirable. Therefore, the enzymatic hydrolysis of cellulose usually takes place extracellularly, where the enzyme is diffused from the cell mass into the external medium.

Another important factor in the enzymatic reaction is whether the enzyme is adaptive or constitutive. A *constitutive enzyme* is present in a cell at all times, whereas *adaptive enzymes* are found only in the presence of a given substance and the synthesis of the enzyme is triggered by an inducing agent. Most of the fungal cellulases are adaptive [15,41]. Cellobiose is an inducing agent with respect to *T. reesei*. In fact, depending on the circumstances, cellobiose can be either an inhibitor or an inducing agent. It is inhibitory when its concentration exceeds 0.5% to 1.0%. Cellobiose is an intermediate product and is generally present in concentrations low enough to permit it to serve as a continuous inducer [46].

A milestone achievement (2004) accomplished by the NREL in collaboration with Genencor International and Novozyme Biotech is of significance in making effective cellulase enzymes at significantly reduced costs, as mentioned in an earlier section.

## 12.5.3 ENZYMATIC PROCESSES

All enzymatic processes basically consist of four major steps that may be combined in a variety of ways: *pretreatment, enzyme production, hydrolysis,* and *fermentation,* as represented in Figure 12.6.

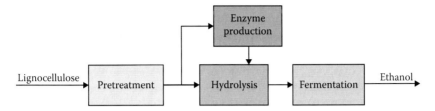

**FIGURE 12.6**   Schematic of fungal enzyme hydrolysis process. (From J. D. Wright, *Chem. Eng. Prog.*, 84, 62–74, 1988.)

#### 12.5.3.1   Pretreatment

It has long been recognized that some form of pretreatment is necessary to achieve reasonable rates and yields in the enzymatic hydrolysis of biomass. Pretreatment has generally been practiced to reduce the crystallinity of cellulose, to lessen the average degree of polymerization of the cellulose and the lignin–hemicellulose sheath that surrounds the cellulose, and to alleviate the lack of available surface area for the enzymes to attack. A typical pretreatment system consists of size reduction, pressure sealing, heating, reaction, pressure release, surface area increase, and hydrolyzate/solids separation [47].

Mechanical pretreatments such as intensive ball milling and roll milling have been investigated as a means of increasing the surface area, but they require exorbitant amounts of energy. The efficiency of the chemical process can be understood by considering the interaction between the enzymes and the substrate. The hydrolysis of cellulose into sugars and other oligomers is a solid phase reaction in which the enzymes must bind to the surface to catalyze the reaction. Cellulase enzymes are large proteins, with molecular weights ranging from 30,000 to 60,000 and are thought to be ellipsoidal with major and minor dimensions of 30 to 200 Å. The internal surface area of wood is very large; however, only about 20% of the pore volume is accessible to cellulase-sized molecules. By breaking down the tight hemicellulose–lignin matrix, hemicellulose and lignin can be separated and the accessible volume can be greatly increased. This removal of material greatly enhances the enzymatic digestibility.

The hemicellulose–lignin sheath can be disrupted by either acidic or basic catalysts. Basic catalysts simultaneously remove both lignin and hemicellulose but suffer large consumption of base through neutralization by ash and acid groups in the hemicellulose. In recent years, more attention has been focused on the acidic catalysts. They can be mineral acids or organic acids generated *in situ* by autohydrolysis of hemicellulose.

Various types of pretreatments are used for biomass conversion. The pretreatments that have been studied in recent years include steam explosion autohydrolysis, wet oxidation, organosolv, and rapid steam hydrolysis (RASH). The major objective of most pretreatments is to increase the susceptibility of cellulose and lignocellulose material to acid and enzymatic hydrolysis. Enzymatic hydrolysis is a very sensitive indicator of lignin depolymerization and cellulose accessibility. Cellulase enzyme systems react very slowly with untreated material; however, if the lignin barrier around the plant cell is partially disrupted, then the rates of enzymatic hydrolysis are increased dramatically.

Most pretreatment approaches are not intended to actually hydrolyze cellulose to soluble sugars but rather to generate a pretreated cellulosic residue that is more readily hydrolyzable by cellulase enzymes than native biomass. *Dilute acid hydrolysis* processes are currently being proposed for several near-term commercialization efforts until lower-cost commercial cellulase preparations become available. Such dilute acid hydrolysis processes typically result in no more than 60% yields of glucose from cellulose.

### 12.5.3.1.1 Autohydrolysis Steam Explosion

A typical autohydrolysis process [48] uses compressed liquid hot water at a temperature of about 200°C under a pressure that is higher than the saturation pressure, thus keeping hot water in liquid phase, to hydrolyze hemicellulose in minutes. Hemicellulose recovery is usually high, and unlike the acid-catalyzed process, no catalyst is needed. A process schematic is shown in Figure 12.7. Very high temperature processes may lead to significant pyrolysis, which produces inhibitory compounds. The ratio of the rate of hemicellulose hydrolysis to that of sugar degradation (more pyrolytic in nature) is greater at higher temperatures. Low-temperature processes have lower xylose yields and produce more degradation products than well-controlled, high-temperature processes, which use small particles. According to a study by Dekker and Wallis [49], pretreatment of bagasse by autohydrolysis at 200°C for 4 min and explosive defibration resulted in a 90% solubilization of the hemicellulose (a heteroxylan) and in the production of a pulp that was highly susceptible to hydrolysis by cellulases from *T. reesei*. Saccharification yields were 50% after 24 h at 50°C (pH 5.0) in enzymatic digests containing 10% (w/v) bagasse pulps and 20 filter paper cellulase units (FPU), while their saccharification yield could be increased to 80% at 24 h by the addition of exogenous β-glucosidase from *Aspergillus niger*.

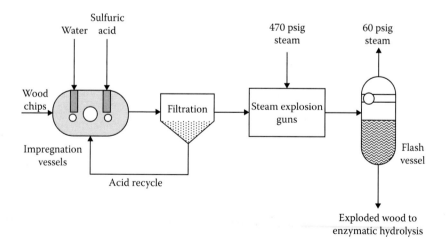

**FIGURE 12.7** Steam explosion pretreatment process flow diagram. (From J. D. Wright, *Chem. Eng. Prog.*, 84, 62–74, 1988.)

In general, xylose yields in autohydrolysis are low (30%–50%). An autohydrolysis system is used as the pretreatment in separate hydrolysis and fermentation (SHF). The reaction conditions are 200°C for 10 min, with a xylose yield of 35%.

Steam consumption in autohydrolysis is strongly dependent upon the moisture content of the starting material. Wet feedstocks require considerably more energy because of the high heat capacity of retained water. An important advantage of auto-hydrolysis is that it breaks the lignin into relatively small fragments that can be easily solubilized in either base or organic solvents.

*Steam explosion process* [50] was first developed in 1925 for hardboard production and was also applied on aspen wood in the early 1980s. In a typical steam explosion process, cellulosic material is heated using high-pressure steam (20–50 atm, 210°C–290°C) for a short period (seconds to minutes). At the increased pretreatment pressure, water molecules diffuse into the inner microporous structure of the lignocellulose [47]. In this process, some steam condenses under high pressure, thereby wetting the material. The wetted material is then driven out of a reactor (i.e., ejected from a reactor) through a small nozzle by a pressure difference. Due to a rapid decrease in the pressure, the material is ejected through the discharge valve. The term "explosion" is used due to the process characteristics of ejection driven by a sudden large pressure drop of steam.

### 12.5.3.1.2  Dilute Acid Prehydrolysis

Lower temperature operation with reduced sugar degradation is achieved by adding a small amount of mineral acid to the pretreatment process. The acid increases reaction rates at a given temperature, and the ratio of the hydrolysis rate to the degradation rate is also increased.

A compromise between the reaction temperature and the reaction time exists for acid-catalyzed reactions. As for autohydrolysis, however, conditions explored range from several hours at 100°C to 10 s at 200°C with a sulfuric acid concentration of 0.5% to 4.0%. Acid catalysts have also been used in steam explosion systems with similar results. Xylose yields generally range from 70% to 95%. However, sulfuric acid processes produce lignin that is more condensed (52% of the lignin extractable in dilute NaOH) than that produced by the autohydrolysis system. Sulfur dioxide has also been investigated as a catalyst to improve the efficiency of the pretreatments. Use of excess water increases the energy consumption and decreases the concentration of xylose in the hydrolyzate, thus decreasing the concentration of ethanol that can be produced in the xylose fermentation step. In a study by Ojumu and Ogunkunle [51], production of glucose was achieved in batch reactors from hydrolysis of lignocellulose under extremely low acid (ELA) concentration and high-temperature condition by pretreating the sawdust by autohydrolysis *ab initio*. The maximum glucose yield obtained was reported to be 70% for the pretreated sawdust at 210°C in the eighteenth minute of the experiment. This value is about 1.4 times the maximum glucose level obtained from the untreated sawdust under the nominally same condition [51].

Acid hydrolysis process has a long history of more than a 100 years. As an alternative to the dilute acid hydrolysis, concentrated acid-based hydrolysis processes are also conceivable and available. However, these types of processes are generally more expensive to operate and render handling difficulties [52]. Sulfuric acid is the most common choice of catalyst; however, other mineral acids like hydrochloric, nitric, and trifluoroacetic acids ($CF_3COOH$) have also been used.

### 12.5.3.1.3 Organosolv Pretreatment

The *Organosolv process* is a pulping technique that uses an organic solvent to solubilize lignin and hemicellulose. The process was first developed as an environmentally benign alternative to kraft pulping. Its main advantages include the production of high-quality lignin for added values and easy recovery and recycle of solvents used in the process, thereby alleviating environmental stress on the water stream.

In this type of pretreatment of lignocellulose, an organic solvent (ethanol, butanol, or methanol) is added to the pretreatment reaction to dissolve and remove the lignin fraction. In the pretreatment reactor, the internal lignin and hemicellulose bonds are broken and both fractions are solubilized, while the cellulose remains as a solid. After leaving the reactor, the organic fraction is removed by evaporation (distillation) in the liquid phase. The lignin then precipitates and can be removed by filtration or centrifugation. Thus, this process cleanly separates the feedstock into a solid cellulose residue, a solid lignin that has undergone a few condensation reactions, and a liquid stream containing xylon, as shown in Figure 12.8. The Organosolv process is usually carried out at an elevated temperature of 140°C–230°C under pressure. High temperature is somewhat dictated by the desired bond cleavage reactions involving

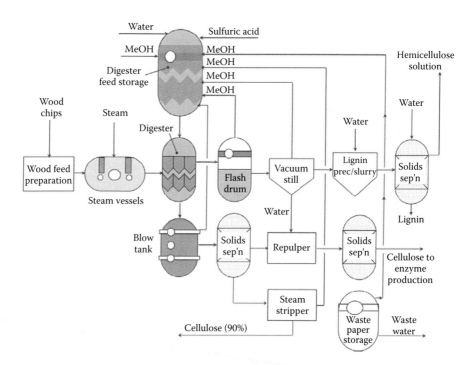

**FIGURE 12.8**  Organosolv pretreatment process. (From J. D. Wright, *Chem. Eng. Prog.*, 84, 62–74, 1988.)

the liberation of lignin, while the high pressure is needed to keep the solvent process operation in the liquid phase. Ethanol has been regarded as a preferred solvent for Organosolv due to its low price, availability, and easy solvent recovery. Butanol has also shown promise because of its superior capability of high lignin yield and immiscibility with water, which make solvent recovery simple without energy-intensive distillation. While butanol's effectiveness is quite appealing, its cost is considered to be somewhat prohibitive. As explained, a principal concern in these processes is the complete recovery of the solvent, which affects the overall process economics; therefore, process engineering and optimization become important factors in the process economics.

Results have shown that there are some reactions occurring during the Organosolv process that strongly affect the enzymatic rate [53]. These reactions could be due to the physical or chemical changes in lignin or cellulose. In general, Organosolv processes have higher xylose yields than the other processes because of the influence of the organic solvent on hydrolysis kinetics. In a recent study, Pan et al. [54] applied the ethanol Organosolv pretreatment to lodgepole pine killed by the mountain beetle and achieved 97% conversion to glucose. They recovered 79% of the lignin using the conditions of 170°C, 1.1 wt.% $H_2SO_4$, and 65 vol.% ethanol for 60 min.

### 12.5.3.1.4  Combined RASH and Organosolv Pretreatment

Attempts have been made to improve the overall process efficiency by combining the two individual pretreatments of RASH and Organosolv. Rughani and McGinnis [53] have studied the effect of a combined RASH–Organosolv process upon the rate of enzymatic hydrolysis and the yield of solubilized lignin and hemicellulose. A schematic diagram of the process is shown in Figure 12.9. For the Organosolv pretreatment, the steam generator is disconnected and the condensate valve closed. The rest of the reactor setup is similar to the typical RASH procedure.

The Organosolv processes at low temperature are generally ineffective in removing lignin, as explained earlier; however, combining the two processes leads to increased solubilization of lignin and hemicellulose. RASH temperature is the major factor in maximizing the percentage of cellulose in the final product. The maximum yield of solubilized lignin was obtained at a temperature of 240°C for RASH and 160°C for the Organosolv process.

### 12.5.3.1.5  Ionic Liquid Pretreatment

Ionic liquids are a relatively new class of solvents that have recently gained popularity as environmentally friendly alternatives to organic solvents. An ionic liquid is a salt composed of anions and cations that are poorly coordinated and has a melting point typically below 100°C. Ionic liquids are also referred to as liquid electrolytes, ionic melts, liquid salts, ionic glasses, and the like. There are thousands of substances that fall into this category. Ionic liquids have been demonstrated as very efficient solvents in the fields of hydrogenation, esterification, nanomaterial synthesis, biocatalysis, and selective extraction of aromatics [55,56].

The development of a novel biomass pretreatment technology using ionic liquids has only recently been initiated. The first demonstration of an ionic liquid as a cellulose solvent under relatively mild processing conditions was reported

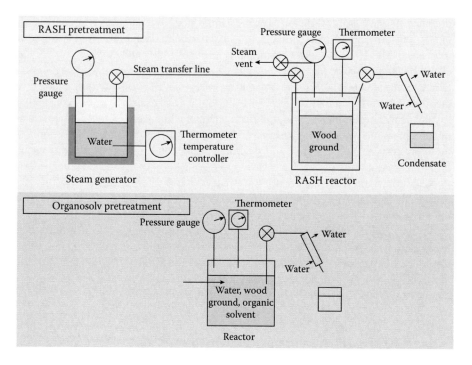

**FIGURE 12.9**   A combined RASH and Organosolv pretreatment scheme. (From L. Rughani and G. D. McGinnis, *Biotechnol. Bioeng.*, 33, 681–686, 1989.)

in 2002 by Swatloski [56]. In experiments using a range of anions and 1-butyl-3-methylimidazolium cations, some ionic liquids were able to completely dissolve microcrystalline cellulose, and the cellulose was recovered through the addition of an antisolvent such as water or ethanol. Moreover, the recovered product could be regenerated into a wide range of shapes and morphologies. The most effective cellulose solvents were the ionic liquids that contain chloride anions. An important finding associated with this novel pretreatment method is that enzymes can more efficiently hydrolyze into glucose the amorphous cellulose produced by ionic liquids than the microcrystalline cellulose found naturally in lignocellulose [55,57].

Ionic liquids are an exciting area of new scientific discovery and inherently possess many processing merits in lignocellulose pretreatment. More in-depth R&D work needs to be conducted before a commercially viable process can be fully developed and exploited.

### 12.5.3.2   Enzyme Production and Inhibition

The enzyme of commercial interest is the cellulase, needed for the hydrolysis of the cellulose, that is, cellulolysis. Cellulase is a multicomponent enzyme system consisting of endo-β-1,4-glycanases, exo-β-1,4-glucan glucohydrolases, and exo-β-1,4-glucan cellobiohydrolase. Cellobiose is the dominant product of this system but is highly inhibitory to the enzymes and is not usable by most organisms. Cellobiase hydrolyzes cellobiose to glucose, which is much less inhibitory and highly fermentable. Many fungi produce this cellobiase, and most of the work that is presently being conducted is on *T. reesei* (*viride*). The cellulase produced by *T. reesei* is much less inhibited than other cellulases that have a major advantage for industrial purposes [58].

The type of inhibition exhibited by cellulases is the subject of much debate in research. Although most of the researchers favor competitive inhibition [59–64], some cellulases are noncompetitively [46,62,65,66] or uncompetitively inhibited [60]. Uncompetitive inhibition takes place when an enzyme inhibitor binds only to the complex formed between the enzyme and the substrate, whereas noncompetitive inhibition takes place when an enzyme inhibitor and the substrate may both be bound to the enzyme at any given time. On substrates like Solka-Floc® (purified cellulose), wheat straw, and bagasse (biomass remaining after sugarcane stalks are crushed to extract their juice), *T. reesei*-produced enzyme is competitively inhibited by glucose and cellobiose. On the other hand, some enzymes are noncompetitively inhibited by cellobiose using other substrates like rice straw and Avicel®. Avicel® is a registered trade name for microcrystalline cellulose that has been partially hydrolyzed with acid and reduced to a fine powder, and it is used as a fat replacer. *T. viride* is uncompetitively inhibited by glucose in a cotton waste substrate [60].

Many mutants have been produced following *T. reesei*. The most prominent among these is the Rut C-30 [67], the first mutant with β-glucosidase production [43]. Other advantages of the strain include its hyperproducing properties and the fact that it is carbolite-repression resistant. The term "hyperproduction" means excessive production.

Cellulases from thermophilic bacteria have also been extensively examined. Among these, *Clostridium thermocellum* is perhaps the most extensively characterized organism. *C. thermocellum* is an anaerobic, thermophilic, cellulolytic, and ethanogenic bacterium capable of directly converting cellulosic substrate into ethanol. The enzymes isolated from thermophilic bacteria may have superior thermal stability and hence will have longer half-lives at high temperatures. Although this is not always the case, cellulases isolated from *C. thermocellum* have high specific activities [68], especially against crystalline forms of cellulose that have proven to be resistant to other cellulase preparations.

Enzyme production with *T. reesei* is difficult because cellulase production discontinues in the presence of easily metabolizable substrates. Thus, most production work has been carried out on insoluble carbon sources such as steam exploded biomass or Solka-Floc® [69]. Solka-Floc® is composed of β-1,4-glucan units; is white, odorless, and flavorless; and has varying particle sizes [70]. In such systems, the rate of growth and cellulase production is limited because the fungi must secrete the cellulase and carry out slow enzymatic hydrolysis of the solid to obtain the necessary carbon. Average productivities have been approximately 100 IU/L/h. (Hydrolytic activity of cellulose is generally represented in terms of an international filter unit [IU]. This is a unit defined in terms of the amount of sugar produced per unit time from a strip of Whatman filter paper.) The filter paper unit is a measure of the combined activities of all three enzymes on the substrate. High productivities have been reported with *T. reesei* mutant in a fed-batch system using lactose as a carbon source and steam exploded aspen as an inducer. Although lactose is not available in sufficient quantities to supply a large ethanol industry, this does suggest that it may be possible to develop strains that can produce cellulases with soluble carbon sources such as xylose and glucose.

Increases in productivities dramatically reduce the size and cost of the fermenters used to produce the enzyme. More rapid fermentation technologies would also decrease the risk of contamination and might allow for less expensive construction. Alternatively, using a soluble substrate may allow simplification of fermenter design or allow the design of a continuous enzyme production system. Low-cost but efficient enzymes for lignocellulosic ethanol technology must be developed in order to reduce the operational cost and improve the productivity of the process.

### 12.5.3.3 Cellulose Hydrolysis

#### 12.5.3.3.1 Cellulase Enzyme Adsorption

The enzymatic hydrolysis of cellulose proceeds by adsorption of cellulase enzyme on the lignacious residue as well as the cellulose fraction. The adsorption on the lignacious residue is also interesting from the viewpoint of the recovery of enzyme after the reaction and recycling it for use on the fresh substrate. Obviously, the recovery efficiency is reduced by the adsorption of enzyme on lignacious residue, since a large fraction of the total operating cost is due to the production of enzyme. Since the capacity of lignacious residue to adsorb the enzyme is influenced by the pretreatment conditions, the pretreatment should be evaluated, in part, by how much enzyme adsorbs on the lignacious residue at the end of hydrolysis as well as its effect on the rate and extent of the hydrolysis reaction.

The adsorption of cellulase on cellulose and lignacious residue has been investigated by Ooshima et al. [71] using cellulase from *T. reesei* and hardwood pretreated by dilute sulfuric acid with explosive decomposition. The cellulase was found to adsorb on the lignacious residue as well as on the cellulose during hydrolysis of the pretreated wood. A decrease in the enzyme recovery in the liquid phase with an increase in the substrate concentration has been reported due to the adsorption on the lignacious residue. The enzyme adsorption capacity of the lignacious residue decreases as the pretreatment temperature is increased, whereas the capacity of the cellulose increases with higher temperature. The reduction of the enzyme adsorbed on the lignacious residue as the pretreatment temperature increases is essential for improving the ultimate recovery of the enzyme as well as enhancing the enzyme hydrolysis rate and extent. Lu et al. [72] conducted an experimental investigation on cellulase adsorption and evaluated the enzyme recycle during the hydrolysis of $SO_2$-catalyzed steam-exploded Douglas-fir and posttreated steam-exploded Douglas-fir substrates [72]. After hot alkali peroxide posttreatment, the rates and yield of hydrolysis attained from the posttreated Douglas fir were significantly higher, even at lower enzyme loadings, than those obtained with the corresponding steam-exploded Douglas fir. This work suggests that enzyme recovery and reuse during the hydrolysis of posttreated softwood substrates could result in a reduced need for the addition of fresh enzyme during softwood-based bioconversion processes [72].

An enzymatic hydrolysis process involving solid lignocellulosic materials can be designed in many ways. The common denominators are that the substrates and the enzyme are fed into the process, and the product stream (sugar solution), along with a solid residue, leaves it at various points. The residue contains adsorbed enzymes, which are lost when the residue is removed from the system. Therefore, the design must be made in such a way that minimizes this loss.

In order to ensure that the enzymatic hydrolysis process is economically efficient, a certain degree of enzyme recovery is essential. Both the soluble enzymes and the enzyme adsorbed onto the substrate residue must be reutilized. It is expected that the loss of enzyme is influenced by the selection of the stages at which the enzymes in solution and adsorbed enzymes are recirculated and the point where the residue is removed from the system. Vallander and Erikkson [46] defined an enzyme loss function, $L$, assuming that no loss occurs through filtration:

$$L = \frac{\text{amount of enzyme lost through removal of residue}}{\text{amount of enzyme at the start of hydrolysis}}$$

They developed a number of theoretical models to conclude that an increased enzyme adsorption leads to an increased enzyme loss. The enzyme loss decreases if the solid residue is removed late in the process. Both the adsorbed and dissolved enzymes should be reintroduced at the starting point of the process. This is particularly important for the dissolved enzymes. Washing of the entire residue is likely to result in significantly lower recovery of adsorbed enzymes than if a major part (60% or more) of the residue with adsorbed enzymes is recirculated. An uninterrupted hydrolysis over a given time period leads to a lower degree of saccharification than when hydrolyzate is withdrawn several times. Saccharification is also favored if the residue is removed at a late stage. Experimental investigations of the theoretical hydrolysis models have recovered more than 70% of the enzymes [46].

### 12.5.3.3.2  Mechanism of Hydrolysis

The overall hydrolysis is based on the synergistic action of three distinct cellulase enzymes and is dependent on the concentration ratio and the adsorption ratio of the component enzymes: endo-β-gluconases, exo-β-gluconases, and β-glucosidases.

Endo-β-gluconases attack the interior of the cellulose polymer in a random fashion [43], exposing new chain ends. Because this enzyme catalyzes a solid phase reaction, it adsorbs strongly but reversibly to the microcrystalline cellulose (also known as *avicel*). The strength of the adsorption is greater at lower temperatures. This enzyme is necessary for the hydrolysis of crystalline substrates. The hydrolysis of cellulose results in a considerable accumulation of reducing sugars, mainly cellobiose, because the extracellular cellulase complex does not possess cellobiose activity. Sugars that contain aldehyde groups that are oxidized to carboxylic acids are classified as *reducing sugars*.

Exo-β-gluconases remove cellobiose units (which are disaccharides with the formula $[HOCH_2CHO(CHOH)_3]_2O$) from the nonreducing ends of cellulose chains. This is also a solid-phase reaction, and the exo-β-gluconases adsorb strongly on both crystalline and amorphous substrates. The mechanism of the reaction is complicated because there are two distinct forms of both endo-enzymes and exo-enzymes, each with a different type of synergism with the other members of the complex. As these enzymes continue to split off cellobiose units, the concentration of cellobiose in the solution may increase. The action of exo-β-gluconases may be severely inhibited or even stopped by the accumulation of cellobiose in the solution.

β-Glucosidases act on and hydrolyze the cellobiose to glucose. Glucosidase may be defined as any enzyme that catalyzes hydrolysis of glucoside. β-Glucosidase catalyzes the hydrolysis of terminal, nonreducing β-D-glucose residues with the release of β-D-glucose. The effect of β-glucosidase on the ability of the cellulase complex to degrade avicel has been investigated by Kadam and Demain [73]. They determined the substrate specificity of the β-glucosidase and demonstrated that its addition to the cellulase complex enhances the hydrolysis of avicel, specifically by removing the accumulated cellobiose. A thermostable β-glucosidase form, *C. thermocellum*, which is expressed in *E. coli,* was used to determine the substrate specificity of the enzyme. The hydrolysis of cellobiose to glucose is a liquid-phase reaction, and β-glucosidase adsorbs either quickly or not at all on cellulosic substrates. β-Glucosidase's action can be slowed or halted by the inhibitive action of glucose accumulated in the solution. The accumulation may also induce the entire hydrolysis to a halt as inhibition of the β-glucosidase results in a build-up of cellobiose, which in turn inhibits the action of exo-gluconases. Therefore, the hydrolysis of the cellulosic materials depends on the presence of all three enzymes in proper amounts. If any one of these enzymes is present in less than the required amount, the other enzymes will be inhibited or lack the necessary substrates to act upon.

The hydrolysis rate generally increases with increasing temperature. However, because the catalytic activity of an enzyme is also related to its shape, the deformation of the enzyme at high temperature can inactivate or destroy the enzyme. To strike a balance between increased activity and increased deactivation, it is preferable to run fungal enzymatic hydrolysis at approximately 40°C–50°C.

While enzymatic hydrolysis is preferably carried out at a low temperature of 40°C–50°C, dilute acid hydrolysis is carried out at a substantially higher temperature. Researchers at the NREL reported results for a dilute acid hydrolysis of softwoods in which the conditions of the reactors were as follows [74]:

1. Stage 1: 0.7% sulfuric acid, 190°C, and a 3 min residence time
2. Stage 2: 0.4% sulfuric acid, 215°C, and a 3 min residence time

Their bench-scale tests also confirmed the potential to achieve yields of 89% for mannose, 82% for galactose, and 50% for glucose, respectively. Fermentation with *S. cerevisiae* achieved ethanol conversion of 90% of the theoretical yield [75].

### 12.5.3.4 Fermentation

Cellulose hydrolysis and fermentation can be achieved by two different process schemes, depending upon where the stage of fermentation is actually carried out in the overall process sequence: (1) separate hydrolysis and fermentation (SHF) or (2) simultaneous saccharification and fermentation (SSF). The acronyms SHF and SSF are very commonly used in the field.

## 12.5.3.4.1  Separate Hydrolysis and Fermentation

In SHF, the hydrolysis is carried out in one vessel, and the hydrolyzate is then fermented in a second vessel. The most expensive items in the overall process cost are the cost of feedstock, enzyme production, hydrolysis, and utilities. The feedstock and utility costs are high because only about 73% of the cellulose is converted to ethanol in 48 h, while the remainders of the cellulose, hemicellulose, and lignin are burned or gasified. Enzyme production is a costly step due to the large amount of the enzyme used in an attempt to overcome the end-product inhibition as well as to its slow reaction rate. The hydrolysis step is also expensive due to the large capital and operating costs associated with large-size tanks and agitators. The most important parameters are the hydrolysis section yield, the product quality, and the required enzyme loading, all of which are interrelated. Yields are typically higher in more dilute systems where inhibition of enzymes by glucose and cellobiose is minimized. Increasing the amount of enzyme loading can help to overcome inhibition and increase the yield and concentration, although it undoubtedly increases the overall cost. Increased reaction times also make higher yields and concentrations, but significantly increases the capital and operating cost. Cellulase enzymes from different organisms can result in markedly different performances. Figure 12.10 shows the effect of yield at constant solid and enzyme loading and the performance of different enzyme loadings. Increase in enzyme loading beyond a particular point turned out to be of no use. It would be economical to operate at a minimum enzyme loading level, or the enzyme could be recycled by appropriate methods. As the cellulose is hydrolyzed, the endo- and exo-gluconase components are released back into the solution. Because of their affinity for cellulose, these enzymes can be recovered and reused by contacting the hydrolyzate with fresh feed. The amount of recovery is limited because of β-glucosidase, which does not adsorb on the feed. Some of the enzyme remains attached to the lignin and unreacted cellulose, and enzymes are thermally denatured during hydrolysis. A major difficulty in this type of process is maintaining their sterility; otherwise, the process system would be contaminated. The power consumed in agitation is also significant and does affect the economics

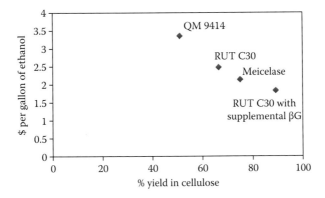

**FIGURE 12.10**   Effect of yield on selling price of ethanol. In the drawing, βG denotes β-gluconase. (From J. D. Wright, *Chem. Eng. Prog.*, 84, 62–74, 1988.)

of this process [43]. Even though the effect of yield on the selling price of ethanol in the figure was based on more classical ethanol production processes, it does explain the importance of yield on the final product cost.

Um and Hanley [69] carried out an experimental study on high-solid enzymatic hydrolysis and fermentation of Solka-Floc® into ethanol. To lower the ethanol distillation cost of fermentation broths, a high initial glucose concentration is desired. However, an increase in the glucose concentration typically reduces the ethanol yield due to the decreased mass and heat transfer rate. To overcome the incompatible temperatures between the enzymatic hydrolysis (50°C) and fermentation (30°C), saccharification, followed by fermentation (SFF), was employed with relatively high-solid concentrations (10% to 20%) using a portion loading method. Glucose and ethanol were produced from Solka-Floc®, which was first digested by enzymes at 50°C for 48 h, followed by fermentation. In this process, commercial enzymes were used in combination with a recombinant strain of *Zymomonas mobilis*. The highest ethanol yields of 83.6%, 73.4%, and 21.8%, based on the theoretical amount of glucose, were obtained with substrate concentrations of 10%, 15%, and 20%, respectively. These values also correspond to 80.5%, 68.6%, and 19.1%, based on the theoretical amount of the cell biomass and soluble glucose present after 48 h of SFF. In addition to the substrate concentration effects, they also investigated the effects of reactor configurations [69].

As a classic study of the mechanism of the enzymatic hydrolysis of cellulose, Fan et al. [76] investigated the effects of major structural features of cellulose on enzymatic hydrolysis. They found that the hydrolysis rate is mainly dependent upon the fine structural order of cellulose, which can be best represented by the crystallinity rather than the simple surface area [76].

### 12.5.3.4.2 Simultaneous Saccharification and Fermentation

The operating cost of the SSF process is generally lower than that of SHF as long as the process integration is synergistically accomplished. As the name of the process implies, both the hydrolysis and fermentation are carried out in the same vessel. In this process, yeast ferments the glucose to ethanol as soon as the glucose is produced, thus preventing the sugars from accumulating and causing end-product inhibition. Using the yeast, *Candida brassicae,* and the Genencor enzyme (by Genencor International), the yield is increased to 79%, and the ethanol concentration produced is 3.7% [43,77].

Even in SSF, cellobiose (the soluble disaccharide sugar) inhibition occurs to an appreciable extent. The enzyme loading for SSF is only 7 IU/g of cellulose, compared to 33 IU/g in SHF. The cost of energy and feedstock is somewhat reduced because of the improved yield, and the increased ethanol concentration significantly reduces the cost of distillation and utilities. The cost of the SSF process is slightly less than the combined cost of hydrolysis and fermentation in the SHF process. The decreasing factor of the reactor volume due to the higher concentration of ethanol offsets the increasing factor in the reactor size caused by the longer reaction times (7 days for SSF vs. 2 days for hydrolysis and 2 days for fermentation). Earlier studies showed that fermentation is the rate controlling step and the enzymatic hydrolysis process is not. With recent advances and developments in recombinant yeast strains that are capable of effectively fermenting both glucose and xylose, the process configurational considerations for commercial exploitation as well as the determination of a rate-limiting step for the overall process technology may have to change accordingly.

The hydrolysis is carried out at 37°C, and increasing the temperature increases the reaction rate. However, the ceiling temperature is usually limited by the yeast cell viability. The concentration of ethanol is also a limiting factor. (This was tested by connecting a flash unit to the SSF reactor and removing the ethanol periodically. This technique showed productivities up to 44% higher.) Recycling the residual solids may also increase the process yield. However, the most important limitation in enzyme recycling comes from the presence of lignin, which is inert to the enzyme. High recycling rates increase the fraction of lignin present in the reactor and cause handling difficulties.

Two major types of enzyme recycling schemes have been proposed: one in which enzymes are recovered in the liquid phase and the other in which enzymes are recovered by recycling unreacted solids [43]. Systems of the first type have been proposed for SHF processes, which operate at 50°C. These systems are favored at such a high temperature because increasing temperature increases the proportion of enzyme that remains in the liquid phase. Conversely, as the temperature is decreased, the amount of enzyme adsorbed on the solid increases. Therefore, at lower temperatures encountered in SSF processes, solid recycling becomes a more effective option.

### 12.5.3.4.3  Comparison between the SSF and SHF Processes

SSF systems offer large advantages over SHF processes thanks to their reduction in end-product inhibition of the cellulase enzyme complex. The SSF process shows a higher yield than SHF (88% vs. 73% in an earlier example) and greatly increases product concentrations (equivalent glucose concentration of 10% vs. 4.4%). The most significant advantage of the SSF process is the enzyme loading, which can be reduced from 33 to 7 IU/g cellulose; this cuts down the cost of ethanol production appreciably. With constant development of low-cost enzymes, the comparative analysis of the two processes will inevitably be changing. A comparative study between the approximate costs of two processes was reported in Wright's article [43]. The results show that, based on the estimated ethanol selling price from a production capacity of 25,000,000 gallons/year, SSF is found to be more cost-effective than SHF by a factor of 1:1.49, that is, $\$_{SHF}/\$_{SSF} = 1.49$. It should be clearly noted that the number quoted here is the ratio of the two prices, not the direct dollar value of the ethanol selling price. Furthermore, this study was also based on the enzymes and bioconversion technologies available in the mid-1980s, which are significantly different from the most advanced current technologies of the twenty-first century. However, this ethanol production cost comparison between two different process configurations provides an idea about the complexity of interrelated cost factors among the reaction rates, temperature, processing time, enzyme adsorption, enzyme loading and recoverability, product inhibition, and more.

For the very same process economic reasons, it is anticipated that a *hybrid hydrolysis and fermentation (HHF) process* configuration is going to be widely accepted as a process of choice for production of lignocellulosic fuel ethanol, which begins with a separate prehydrolysis step and ends with a simultaneous saccharification (hydrolysis) and fermentation (SSF) step. In the first stage of hydrolysis, higher-temperature enzymatic cellular saccharification takes place, whereas in the second stage of SSF, mesophilic (moderate-temperature) enzymatic hydrolysis and biomass sugar fermentation takes place simultaneously. The optimized process configurational scheme would have to change if a specific enzyme, proven to be highly efficient and cost-effective, is found to be intolerant against certain inhibitors that are associated with any of these processing steps.

### 12.5.3.5  Xylose Fermentation

Since xylose accounts for 30%–60% of the fermentable sugars in hardwood and herbaceous biomass, the fermentation of xylose to ethanol becomes an important issue. The efficient fermentation of xylose and other hemicellulose constituents is essential for the development of an economically viable process to produce ethanol from lignocellulosic biomass. Needless to say, cofermentation of both glucose and xylose with comparably high efficiency would be ideally desirable. As discussed earlier, xylose fermentation using *pentose yeasts* has proven to be difficult due to several factors including the requirement for $O_2$ during ethanol production, the acetate toxicity, and the production of xylitol as a by-product. Xylitol (or xyletol) is a naturally occurring low-calorie sugar substitute with anticariogenic (preventing production of dental caries) properties.

Other approaches to xylose fermentation include the conversion of xylose to xylulose (a pentose sugar, part of carbohydrate metabolism, that is found in the urine of individuals with the condition pentosuria [78]) using xylose isomerase prior to fermentation by *S. cerevisiae* and the development of genetically engineered strains [79].

A method for integrating xylose fermentation into the overall process is illustrated in Figure 12.11. In this example, dilute acid hydrolysis was adopted as a pretreatment step. The liquid stream is neutralized to remove any mineral acids or organic acids liberated in the pretreatment step and is then sent to xylose fermentation. Water is added before the fermentation, if necessary, so that organisms can make full use of the substrate without having the yield limited by end-product inhibition. The dilute ethanol stream from xylose fermentation is then used to provide the dilution water for the cellulose–lignin mixture entering SSF. Thus, the water that enters during the pretreatment process is used in both the xylose fermentation and the SSF process.

The conversion of xylose to ethanol by *recombinant E. coli* has been investigated in pH-controlled batch fermentations [80]. Relatively high concentrations of ethanol (56 g/L) were produced from xylose with good efficiencies. In addition to xylose, all other sugar constituents of biomass, including glucose, mannose, arabinose, and galactose, can be efficiently converted to ethanol by recombinant *E. coli*. Neither oxygen nor strict maintenance of anaerobic conditions is required for ethanol production by *E. coli*. However, the addition of base to prevent excessive acidification is essential. Although less base was needed to maintain low pH conditions, poor ethanol yields and slower fermentations were observed below the pH of 6. Also, the addition of metal ions, such as calcium, magnesium, and ferrous ions, stimulated ethanol production [80]. In general, xylose fermentation does not require precise

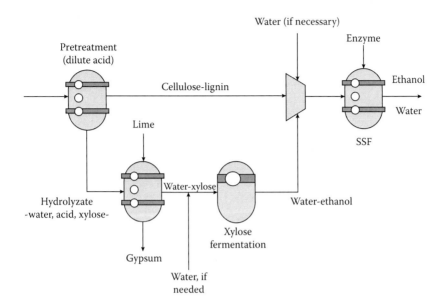

**FIGURE 12.11** Integration of xylose fermentation and SSF. (From J. D. Wright, *Chem. Eng. Prog.*, 84, 62–74, 1988.)

temperature control, provided the broth temperature is maintained between 25°C and 40°C. Xylose concentrations as high as 140 g/L have been positively tested to evaluate the extent to which this sugar inhibits growth and fermentation. Higher concentrations considerably slow down the growth and fermentation. Ingram and coworkers [80–83] demonstrated that recombinant *E. coli* expressing plasmid-borne *Z. mobilis* genes for pyruvate decarboxylase (PDC) and alcohol dehydrogenase II (ADHII; adhB) can efficiently convert both hexose and pentose sugars to ethanol. *Ethanologenic E. coli* strains require simpler fermentation conditions, produce higher concentrations of ethanol, and are more efficient than pentose-fermenting yeasts for ethanol production from xylose and arabinose [84].

A study by Sedlak et al. [28] successfully developed *genetically engineered Saccharomyces yeasts* that can ferment both glucose and xylose simultaneously to ethanol. According to their experimental results, following rapid consumption of glucose in less than 10 h, xylose was metabolized more slowly and less completely. While the xylose conversion was quite significant by this genetically engineered yeast strain, xylose was still not totally consumed even after 30 h. Ideally, xylose should be consumed simultaneously [26] with glucose at a similar efficiency and speed; however, this newly added capability of cofermentation of both glucose and xylose has given a new promise in the lignocellulosic ethanol technology leading to technological breakthroughs. They also found that while ethanol was the most abundant product from glucose and xylose metabolism, small amounts of the metabolic by-products of glycerol and xylitol were also obtained [28]. Certainly, later studies will be focused on the development and/or refinement of more efficient engineered strains, ethanol production with higher selectivity and speed, and optimized process engineering and flowsheeting.

### 12.5.3.6 Ethanol Extraction during Fermentation

In spite of the considerable efforts devoted to the fermentative alcohols, industrial applications have been delayed because of the high cost of production, which depends primarily on the energy input to the purification of dilute end products, the low productivities of cultures, and the high cost of enzyme production. These issues are directly linked to inhibition phenomena.

Along with the conventional unit operations, liquid–liquid extraction with biocompatible organic solvents, distillation under vacuum, and selective adsorption on the solids have demonstrated the technical feasibility of the extractive fermentation concept. Lately, membrane separation processes that decrease biocompatibility constraints have been proposed. These include dialysis [85] and reverse osmosis [65]. More recently, the concept of supported liquid membranes have been reported. This method minimizes the amount of organic solvents involved and permits simultaneous realization of the extraction and recovery phases. Enhanced volumetric productivity and high substrate conversion yields have been reported [86] via the use of a porous Teflon® sheet (soaked with isotridecanol) as support for the extraction of ethanol during semicontinuous fermentation of *Saccharomyces bayanus*. This selective process results in ethanol purification and combines three operations: fermentation, extraction, and re-extraction (stripping) as schematically represented in Figure 12.12. As shown and suggested, novel process ideas can further accomplish maximized alcohol production, energy savings, and reduced cost in production.

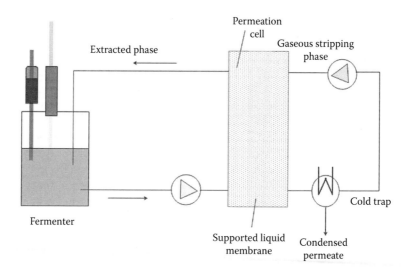

**FIGURE 12.12** Extractive fermentation system: (1) fermenter; (2) permeation cell; (3) supported liquid membrane; (4) extracted phase; (5) gaseous stripping phase; (6) cold trap; and (7) condensed permeate. (From P. Christen et al., *Biotechnol. Bioeng.*, 36, 116–123, 1990.)

## 12.5.4 LIGNIN CONVERSION

Lignin is produced in large quantities, approximately 250 billion lbs per year in the United States, as a by-product of the paper and pulp industry. Lignins are complex amorphous phenolic polymers, which are not sugar-based; hence, they cannot be fermented into ethanol. Lignin is a random polymer made up of phenyl propane units, where the phenol unit may be either a guaiacyl or syringyl unit (Figure 12.13). These units are bonded together in many ways, the most common of which are α- or β-ether linkages. A variety of C–C linkages are also present but are less common (Figure 12.14). The distribution of linkage in lignin is random because lignin formation is a free radical reaction that is not under enzymatic control. Lignin is highly resistant to chemical, enzymatic, and microbial hydrolysis due to extensive cross-linking. Therefore, lignin is frequently removed simply to gain access to cellulose.

Lignin monomer units are similar to gasoline, which has a high octane number; thus, breaking the lignin molecules into monomers and removing the oxygen (i.e., deoxygenation) make them useful as liquid transportation fuels. The process for lignin conversion is mild hydrotreating to produce a mixture of phenolic and hydrocarbon materials, followed by reaction with methanol to produce *methyl aryl ether*. The first step usually consists of two principal parts: (1) hydrodeoxygenation (removal of oxygen and oxygen-containing groups from the phenol rings) and (2) dealkylation (removal of ethyl groups or large side chains from the rings). The major role of this stage is to carry out these reactions to remove the unwanted chains without carrying the reaction too far; otherwise, this would lead to excessive consumption of hydrogen and produce saturated hydrocarbons, which are not as effective as octane

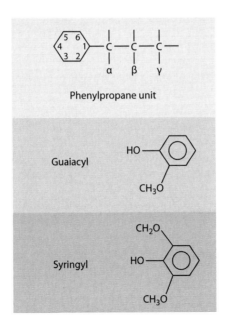

**FIGURE 12.13** Monomer units in lignin. (From J. D. Wright, *Chem. Eng. Prog.*, 84, 62–74, 1988.)

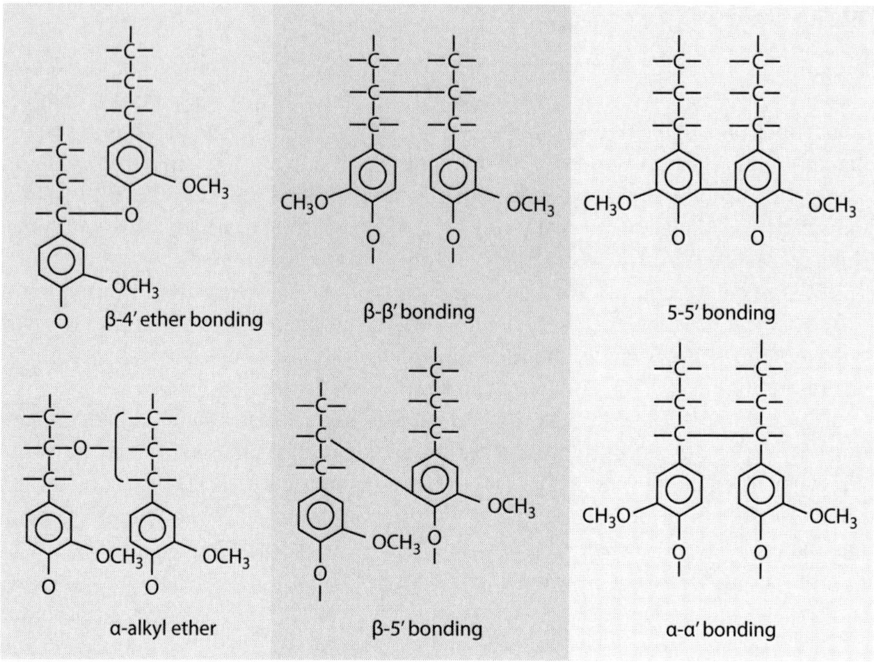

**FIGURE 12.14** Ether and C–C bonds in lignin. (From J. D. Wright, *Chem. Eng. Prog.*, 84, 62–74, 1988.)

enhancers as are aromatic compounds. Further, excessive consumption of hydrogen would represent additional cost for the conversion process. Catalysts to carry out these reactions have dual functions. Metals such as molybdenum and molybdenum/ nickel catalyze the deoxygenation, while the acidic alumina support promotes the carbon–carbon bond cleavage.

Although lignin chemicals have applications in drilling muds, as binders for animal feed, and as the base for artificial vanilla, they have not been previously used as surfactants for oil recovery. According to Naee [87], lignin chemicals can be used in two ways in chemical floods for enhanced oil recovery. In one method, *lignosulfonates* are blended with tallow amines and conventional petroleum sulfonates to form a unique mixture that costs about 40% less to use than chemicals made solely from petroleum or petroleum-based products. In the second method, lignin is reacted with hydrogen or carbon monoxide to form a new class of chemicals called *lignin phenols*. Because they are soluble in organic solvents, but not in water, these phenols are good candidates for further conversion to produce chemicals that may be useful in enhanced oil recovery (EOR).

## 12.5.5 COPRODUCTS OF CELLULOSIC ETHANOL TECHNOLOGY

In order to reduce the ethanol production cost from lignocellulosic materials, it is imperative to expand or develop the market for the process coproducts, by-products, or derivatives. Unlike the mature corn ethanol industry, the by-product (or coproduct) industry for the lignocellulosic ethanol industry is neither very well defined nor established yet. Potential coproducts include hemicellulose hydrolyzate (xylose), cellulose hydrolyzate (glucose of mixed sugars), cell mass, enzymes, soluble and insoluble lignins, lignin-derived chemicals and fuels, solid residues, etc. Other valuable products include xylitol, which is a sugar alcohol sweetener and is produced by hydrogenation of xylose (an aldehyde) into a primary alcohol.

## 12.6 ENERGY BALANCE FOR ETHANOL PRODUCTION FROM BIOMASS

Biomass process development depends upon the economics of the conversion process, be it chemical, enzymatic, or a combination of both. A number of estimates have been computed based upon existing or potential technologies. One obvious factor is that, regardless of the process, transportation of the biomass material from its source to the site of conversion must be kept to an absolute minimum. Approximately 35% of the expected energy is consumed by transporting the substrate a distance of 15 mi. [68]. This considerable expenditure of energy, simply to transport the starting material, dictates that any conversion plant be of moderate size and in close proximity to the production source of the starting material.

There are some objections to the production and use of ethanol as a fuel. Most important is the criticism that producing ethanol can consume more energy than the finished ethanol contains. The European analysis takes wheat as the feedstock and includes estimates of the energy involved to grow the wheat, transport it to the distillery, make the alcohol, and transport it to a refinery for blending it with petrol. It allows credit for by-products, such as animal feed from wheat, for savings on petrol that comes from replacing 5% with alcohol, and from the energy gained from the increase of 1.25 octane points. As fully explained in Chapter 11, more recent and very extensive assessment on the net energy value (NEV) of corn ethanol technology [88] using the advanced process technology as well as more realistic industrial data [89,90] decisively showed that the prevailing corn ethanol process in the United States generates a significantly positive NEV. As for the cellulosic ethanol, a number of factors and issues including the feedstock diversity and availability, the use of nonfood crops, minimal or no use of fertilizers, nonuse of arable land, and more complex but still evolving conversion process technology make such an assessment far more difficult and less meaningful. Yet, to confine debates on biomass fuels solely to the process energy balance would be misleading. Based on the merits of cellulosic biofuels as well as regional strengths, a number of cellulosic biofuel plants based on diverse process technologies are being operated and under construction throughout the world [91].

Energy requirements to produce ethanol from different crops were evaluated by Da Silva et al. [92]. The industrial phase is always more energy intensive, consuming from 60% to 75% of the total energy. The energy expended in crop production includes all the forms of energy used in agricultural and industrial processing, except the solar energy that plants use for growth. The industrial stage, including extraction and hydrolysis, alcohol fermentation, and distillation, requires about 6.5 kg of steam per liter of alcohol. It is possible to furnish the total industrial energy requirements from the by-products of some of the crops. Thus, it is also informative to consider a simplified energy balance in which only agricultural energy is taken as input and only ethanol is taken as the output, the bagasse supplying energy for the industrial stage, for example. Furthermore, technological data are often very difficult or nearly impossible to compare between different options due to the wide variety of feedstock crops as starting lignocellulose. Therefore, the U.S. Department of Energy–sponsored projects choose corn stover as the model feedstock [93]. This selection is based on the fact that corn stover is the most abundant and concentrated biomass resource in the United States, and its collection can leverage the existing corn ethanol infrastructure, including corn harvesting and ethanol production [93].

Unlike the cellulosic ethanol technology, sugarcane ethanol technology is far more straightforward and as such the energy balance evaluation is relatively straightforward. The energy balance results for ethanol production from sugarcane in Zimbabwe have shown that the energy ratio is 1:52 if all the major outputs are considered and 1:15 if ethanol is considered as the only output. The reported value of the net energy ratio for ethanol production from sugarcane in Brazil [92] is 2:41 and in Louisiana, U.S.A. [94], it is 1:85. The low ratio in Zimbabwe is due to (a) the large energy input in the agricultural phase, arising from a large fertilizer need, and (b) the large fossil-based fuel consumption in the sugarcane processing. As shown in this comparative example, the energy balance results are dependent upon a large number of factors including process conversion technology, agricultural technology, climate and soil quality, logistical issues, and much more.

NEV of cellulose ethanol from switchgrass was analyzed by Schmer et al. [95]. In this study, perennial herbaceous plants such as switchgrass were evaluated as cellulosic bioenergy crops. Two major concerns of their investigation were the net energy efficiency and economic feasibility of switchgrass and similar crops. This was a baseline study that represented the genetic material and agronomic technology available for switchgrass production in 2000 and 2001. Their study reported the following:

1. The annual biomass yields of established fields averaged 5.2–11.1 Mg·ha$^{-1}$·year$^{-1}$ with a resulting average estimated net energy yield (NEY) of 60 GJ·ha$^{-1}$·year$^{-1}$.
2. Switchgrass produced 540% more renewable than nonrenewable energy consumed.
3. Switchgrass monocultures managed for high yield produced 93% more biomass yield and an equivalent estimated NEY than previous estimates from human-made prairies that received low agricultural inputs.
4. Estimated average GHG emissions from cellulosic ethanol derived from switchgrass were 94% lower than estimated GHG emissions from gasoline.

Generally speaking, the cost of production of ethanol decreases with an increase in capacity of the production facility, as is the case with most petrochemical industries. However, the minimum total cost corresponds to a point of inflection, at which point an increase in the production cost for every increase in the plant capacity is seen [42]. The possibility of the existence of an empirical relationship between the plant size or output and the production cost has also been examined using various production functions and the computed $F$ values at a 5% level of significance [96]. It is also imaginable that if the average distance of raw material transportation and acquisition becomes excessively long due to the increased plant capacity, then the production cost can be adversely affected by the plant size.

Xylose fermentation is being carried by bacteria, fungi, yeast, enzyme–yeast systems, or genetically engineered microorganisms. The advanced fermentation technology would reduce the cost by 25% or more in the case of herbaceous-type materials, as shown in the study by Schemer et al. [95]. Efforts are being made to achieve the yield of 100% and an increased ethanol concentration.

Lignin is another major component of biomass and accounts for its large energy content because it has a much higher energy content per pound than carbohydrates. Since it is a phenolic polymer, it cannot be fermented to sugar and is instead converted to materials like methyl aryl ethers, which are compatible with gasoline as an efficient octane enhancer. The combination of the above processes has the potential to produce transportation fuels at a competitive price.

## 12.7   PROCESS ECONOMICS AND STRATEGIC DIRECTION

McAloon et al. [97] studied the cost of ethanol production from lignocellulosic materials in comparison to that from corn starch. As properly pointed out in their study, the cost comparison was made between the mature corn–ethanol industry and the emerging lignocellulosic ethanol industry. Based on the fixed price of the year 2000, the cost of fuel ethanol production from lignocellulose processes was determined to be $1.50/gal, whereas that from corn processes was $0.88/gal [97]. Needless to say, the cost values determined in 2000 cannot be considered valid for the current year due to significant changes during the period in infrastructural and raw material costs as well as variable operating costs.

In order to make the lignocellulosic biorefinery technology a success, the following must be resolved:

1. The lignocellulose feedstock collection and delivery system has to be established on an economically sound basis. Feedstock preparation also becomes an issue.
2. Each step of the process technology needs to be separately investigated for various options, and the interactions and connectivity between the steps must be completely evaluated. Interactive effects between stages become very important, since one stage's product and by-products may function as the next stage's inhibitors.
3. A thorough data base for a variety of different feedstocks must be established. A different feedstock can be chosen as a model feedstock for different countries and regions, depending upon the local availability, logistical constraints, and infrastructural benefits. Further, conversion technologies should be readily adaptable to other lignocellulosic feedstocks and agricultural residues [93].

4. Large-scale demonstration is crucially important for commercial operational experience as well as to minimize the risk involved in scale-up efforts. Further, such an operation on a large scale helps demonstrate environmental life cycle analysis whose results are more meaningful.

5. Low-cost but highly efficient enzymes for the technology must be developed in order to reduce the operational cost and improve the productivity. Current efforts by NREL, Genencor International, and Novozymes Biotech are very significant and noteworthy in this regard. Further advances will make the full-scale commercialization of cellulosic ethanol economically more feasible and profitable.

## REFERENCES

1. B. C. Saha, "Lignocellulose biodegradation and applications in biotechnology," in *Lignocellulose Biodegradation*, B. C. Saha and K. Hayashi, Eds. ACS Symposium Series, vol. 889, Washington, DC: Am. Chem. Soc., 2004, Chapter 1, pp. 2–34.
2. D. Cohen, "Form and distribution of trace elements in biomass for power generation," QCat Technology Transfer Center, Australia, Tech. Rep. 48, July 2004.
3. G. Odian, *Principles of Polymerization*. Hoboken, NJ: Wiley-Interscience, 2004.
4. M. Bettiga, O. Bengtsson, B. Hahn-Hagerdal and M. F. Gorwa-Grauslund, "Arabinose and xylose fermentation by recombinant *Saccharomyces cerevisiae* expressing a fungal pentose utilization pathway," *Microb. Cell Fact.*, vol. 8, pp. 1–12, 2009.
5. Y. Lin, P. He, Q. Wang, D. Lu, Z. Li, C. Wu and N. Jiang, "The alcohol dehydrogenase system in the xylose-fermenting yeast *Candida maltosa*," *PloS One*, vol. 5, pp. 1–9, 2010.
6. T. W. Jeffries and N. Q. Shi, "Genetic engineering for improved xylose fermentation by yeasts," *Adv. Biochem. Eng. Biotechnol.*, vol. 65, pp. 117–161, 1999.
7. E. L. La Rovere, The Brazilian ethanol program: Biofuels for transport. Presented at International Conference for Renewable Energies, 2004. Available at http://www.renewables2004.de/ppt/Presentation4-SessionIVB(11-12.30h)-LaRovere.pdf.
8. F. O. Lichts, Industry statistics: 2010 world fuel ethanol production. Renewable Fuels Association, 2011. Available at http://www.ethanolrfa.org/pages/statistics#E.
9. Federal Energy Regulatory Commission. *Energy Policy Act (EPAct) of 2005*. Available at http://www.gpo.gov/fdsys/pkg/PLAW-109publ58/pdf/PLAW-109publ58.pdf.
10. K. T. Klasson, M. D. Ackerson, E. C. Clausen and J. L. Gaddy, "Bioconversion of synthesis gas into liquid or gaseous fuels," *Enzyme Microb. Technol.*, vol. 14, pp. 602–608, 1992.
11. G. Ellis, *Chemistry of Petroleum Derivatives*. New York: Reinhold Publishing Corp., 1937.
12. C. A. Judice and L. E. Pirkle, U.S. Patent 3,095,458, 1963.
13. W. K. Lewis, U.S. Patent 2,045,785, June 30, 1936.
14. S. A. Miller, *Ethylene and its Industrial Derivatives*. London: Ernest Benn Ltd., 1969.
15. J. E. Bailey and D. F. Ollis, *Biochemical Engineering Fundamentals*, 2nd Ed. New York: McGraw-Hill, 1986.
16. A. Demirbas, "Political, economic and environmental impacts of biofuels: A review," *Applied Energy*, vol. 86, pp. S108–S117, 2009.
17. J. Goldemberg and P. Guardabassi, "The potential for first-generation ethanol production from sugarcane," *Sustain. Chem. Green Chem.*, vol. 4, pp. 17–24, 2010.
18. S. C. Park and J. Baratti, "Batch fermentation kinetics of sugar beet molasses by *Zymomonas mobilis*," *Biotechnol. Bioeng.*, vol. 38, pp. 304–313, 1991.

19. J. M. Willey, L. M. Sherwood and C. Woolverton, *Prescott's Microbiology*, 8th Ed. New York: McGraw-Hill, 2011.

20. L. Viikari, "Carbohydrate metabolism in Zymomonas," *Crit. Rev. Biotechnol.*, vol. 7, pp. 237–261, 1988.

21. U. Kalnenieks, N. Galinina, M. M. Toma and R. K. Poole, "Cyanide inhibits respiration yet stimulates aerobic growth of *Zymomonas mobilis*," *Microbiology*, vol. 146, pp. 1259–1266, 2000.

22. M. L. Cazetta, M. A. P. C. Celligoi, J. B. Buzato and I. S. Scarmino, "Fermentation of molasses by *Zymomonas mobilis*: Effects of temperature and sugar concentration on ethanol production," *Bioresour. Technol.*, vol. 98, pp. 2824–2828, 2007.

23. A. Eliasson, C. Christensson, C. F. Wahlbom and B. Hahn-Hägerdal, "Anaerobic xylose fermentation by recombinant *Saccharomyces cerevisiae* carrying XYL1, XYL2, and XKS1 in mineral medium chemostat cultures," *Appl. Environ. Microbiol.*, vol. 66, pp. 3381–3386, 2000.

24. M. Zhang, *Zymomonas mobilis*, special topics session, microbial pentose metabolism. Presented at 25th Symposium on Biotechnology for Fuels and Chemicals. May 5, 2003. Available at http://www1.eere.energy.gov/biomass/pdfs/34264.pdf (accessed March 25, 2014).

25. H. Tao, "Engineering a homo-ethanol pathway in *Escherichia coli*: Increased glycolytic flux and levels of expression of glycolytic genes during xylose fermentation," *J. Bacteriol.*, vol. 183, pp. 2979–2988, 2001.

26. U.S. DOE Office of Science and Office of Energy Efficiency and Renewable Energy, "Breaking the biological barriers to cellulosic ethanol: A joint research agenda. A research roadmap resulting from the biomass to biofuels workshop, December 7–9, 2005, Rockville, MD," Tech. Rep. DOE/SC-0095, June 2006.

27. A. Matsushika, H. Inoue, T. Kodaki and S. Sawayama, "Ethanol production from xylose in engineered *Saccharomyces cerevisiae* strains: Current state and perspectives," *Appl. Microbiol. Biotechnol.*, vol. 84, pp. 37–53, 2009.

28. M. Sedlak, H. J. Edenberg and N. W. Y. Ho, "DNA microarray analysis of the expression of the genes encoding the major enzymes in ethanol production during glucose and xylose co-fermentation by metabolically engineered *Saccharomyces* yeast," *Enzyme Microb. Technol.*, vol. 33, pp. 19–28, 2003.

29. J. Bonnardeaux, "Potential uses for distillers grains," Department of Agriculture and Food, Government of Western Australia, South Perth, WA 6151, Australia, 2007.

30. U.S. Congress Office of Technology Assessment, "Energy from biological process," U.S. Government Printing Office, Washington, D.C., Tech. Rep. 2, pp. 142–177, 1980.

31. Ethanol Across America, The impact of ethanol production on food, feed and fuel. 2008. Available at http://www.cleanfuelsdc.org/pubs/documents/FoodFeedandFuel08.pdf.

32. Novozymes. Cellulosic biofuel greatly benefits society. 2013. Available at http://bioenergy.novozymes.com/en/cellulosic-ethanol/advantages/cellic/Pages/default.aspx.

33. H. Szamant, "Big push for a biomass bonanza," *Chem. Week*, vol. 122, pp. 40, 1978.

34. J. Y. Zhu, X. J. Pan, G. S. Wang and R. Gleisner, "Sulfite pretreatment (SPORL) for robust enzymatic saccharification of spruce and red pine," *Bioresour. Technol.*, vol. 100, pp. 2411–2418, 2009.

35. T. Eggeman and C. Atiyeh, "The role for biofuels," *Chem. Eng. Prog.*, vol. 106, pp. 36–38, 2010.

36. The White House. *Growing America's Fuel: An Innovation to Achieving the Biofuels Target*. Available at http://www.whitehouse.gov/sites/default/files/rss_viewer/growing_americas_fuels.PDF (accessed September 2013).

37. National Renewable Energy Laboratory. Reducing enzyme costs increases market potential of biofuels. June 2010. Available at http://www.nrel.gov/docs/fy10osti/47572.pdf.

38. U.S. EPA. Wastes—non-hazardous wastes—municipal solid wastes. Available at http://www.epa.gov/osw/basic-solid.htm (accessed March 23, 2014).
39. I. S. Goldstein, "Department of Wood and Paper Science, North Carolina University," *C&EN*, pp. 68, September 10, 1990.
40. C&EN, "Major polymeric components of plant materials," *Chem. Eng. News.* September 10, 1990.
41. L. F. Diaz, G. M. Savage and C. G. Golueke, "Critical review of energy recovery from solid wastes," *CRC Critic. Rev. Environ. Control*, vol. 14, pp. 285–288, 1984.
42. G. E. Farina, J. W. Barrier and M. L. Forsythe, "Fuel alcohol production from agricultural lignocellulosic feedstocks," *Energy Sources*, vol. 10, pp. 231–237, 1988.
43. J. D. Wright, "Ethanol from biomass by enzymatic hydrolysis," *Chem. Eng. Prog.*, vol. 84, pp. 62–74, 1988.
44. R. Kumar, S. Singh and O. V. Singh, "Bioconversion of lignocellulosic biomass: Biochemical and molecular perspectives," *J. Ind. Microbiol. Biotechnol.*, vol. 35, pp. 377–391, 2008.
45. L. Viikari, M. Alapuranen, T. Puranen, J. Vehmaanperä and M. Siika-Aho, "Thermostable enzymes in lignocellulose hydrolysis," *Adv. Biochem. Eng. Biotechnol.*, vol. 108, pp. 121–145, 2007.
46. L. Vallander and K. Erikkson, "Enzymatic hydrolysis of lignocellulosic materials: I. Models for the hydrolysis process—a theoretical study," *Biotechnol. Bioeng.*, vol. 38, pp. 135–138, 1991.
47. J. B. Cort, P. Pschorn and B. Stromberg, "Minimize scale-up risk," *Chem. Eng. Prog.*, vol. 106, pp. 39–49, 2010.
48. F. Carvalheiro, L. C. Duarte and F. M. Girio, "Hemicellulose biorefineries: A review on biomass pretreatments," *J. Sci. Ind. Res.*, vol. 67, pp. 849–864, 2008.
49. R. F. H. Dekker and A. F. A. Wallis, "Enzymic saccharification of sugarcane bagasse pretreated by autohydrolysis–steam explosion," *Biotechnol. Bioeng.*, vol. 25, pp. 3027–3048, 1983.
50. W. H. Mason, "Process and apparatus for disintegrating wood and the like," U.S. Patent 1,578,609, March 30, 1926.
51. T. V. Ojumu and O. A. Ogunkunle, "Production of glucose from lignocellulosics under extremely low acid and high temperature in batch process—autohydrolysis approach," *J. Appl. Sci.*, vol. 5, pp. 15–17, 2005.
52. M. Galbe and G. Zacchi, "A review of the production of ethanol from softwood," *Appl. Microbiol. Biotechnol.*, vol. 59, pp. 618–628, 2002.
53. L. Rughani and G. D. McGinnis, "Combined rapid steam hydrolysis and organosolv pretreatment of mixed southern hardwoods," *Biotechnol. Bioeng.*, vol. 33, pp. 681–686, 1989.
54. X. J. Pan, D. Xie, R. W. Yu, D. Lam and J. N. Saddler, "Pretreatment of lodgepole pine killed by mountain pine beetle using the ethanol organosolv process: Fractionation and process optimization," *Ind. Eng. Chem. Res.*, vol. 46, pp. 2609–2617, 2007.
55. B. A. Simmons, S. Singh, B. M. Holmes and H. W. Blanch, "Ionic liquid pretreatment," *Chem. Eng. Prog.*, vol. 106, pp. 50–55, 2010.
56. R. P. Swatloski, "Dissolution of cellulose with ionic liquids," *J. Am. Chem. Soc.*, vol. 124, pp. 4974–4975, 2002.
57. A. P. Dadi, "Mitigation of cellulose recalcitrance to enzymatic hydrolysis by ionic liquid pretreatment," *Appl. Biochem. Biotechnol.*, vol. 137, pp. 407–421, 2007.
58. M. T. Holtzapple, M. Cognata, Y. Shu and C. Hendrickson, "Inhibition of *Trichoderma reesei* cellulase by sugars and solvents," *Biotechnol. Bioeng.*, vol. 38, pp. 296–303, 1991.
59. P. J. Blotkamp, M. Takagi, M. S. Pemberton and G. H. Emert, "Biochemical engineering: Renewable sources of energy and chemical feedstocks," in *AIChE Symposium Series*, J. M. Nystrom and S. M. Barnett, Eds. New York: AIChE, 1978.
60. P. L. Beltrame, P. Carniti, B. Focher, A. Marzetti and V. Sarto, "Enzymatic hydrolysis of cellulosic materials: A kinetic study," *Biotechnol. Bioeng.*, vol. 26, pp. 1233–1238, 1984.

61. K. Ohmine, H. Ooshima and Y. Harano, "Kinetic study on enzymatic hydrolysis of cellulose by cellulase from *Trichoderma viride*," *Biotechnol. Bioeng.*, vol. 25, pp. 2041–2053, 1983.

62. M. Okazaki and M. Young, "Kinetics of enzymatic hydrolysis of cellulose: Analytic description of mechanistic model," *Biotechnol. Bioeng.*, vol. 20, pp. 637–663, 1078.

63. D. Y. Ryu and S. B. Lee, "Enzymatic hydrolysis of cellulose: Determination of kinetic parameters," *Chem. Eng. Commun.*, vol. 45, pp. 119–134, 1986.

64. G. Gonzales, G. Caminal, C. de Mas and J. L. Santin, "A kinetic model for pretreated wheat straw saccharification by cellulose," *J. Chem. Technol. Biotechnol.*, vol. 44, pp. 275, 1989.

65. A. Garcia, E. L. Lannotti and J. L. Fischer, "Butanol fermentation liquor production and separation by reverse osmosis," *Biotechnol. Bioeng.*, vol. 28, pp. 785–791, 1986.

66. L. Vallander and K. Erikkson, "Enzymatic hydrolysis of lignocellulosic materials: II. experimental investigation of theoretical hydrolysis process models for an increased enzyme recovery," *Biotechnol. Bioeng.*, vol. 38, pp. 139–144, 1991.

67. Z. Szengyel, G. Zacchi, A. Varga and K. Réczey, "Cellulase production of *Trichoderma reesei* Rut C 30 using steam-pretreated spruce. Hydrolytic potential of cellulases on different substrates," *Appl. Biochem. Biotechnol.*, vol. 84–86, pp. 679–691, 2000.

68. V. Moses, D. G. Springham and R. E. Cape, *Biotechnology—the Science and the Business*. Newark, NJ: Harwood Academic, 1991.

69. B. H. Um and T. R. Hanley, "High-solid enzymatic hydrolysis and fermentation of Solka Floc into ethanol," *J. Microbiol. Biotechnol.*, vol. 18, pp. 1257–1265, 2008.

70. International Fiber Corporation. *Solka Floc*. 2014. Available at http://www.ifcfiber.com/products/solkafloc.php.

71. H. Ooshima, D. S. Burns and A. O. Converse, "Adsorption of cellulase from *Trichoderma reesei* on cellulose and lignacious residue in wood pretreated by dilute sulfuric acid with explosive decompression," *Biotechnol. Bioeng.*, vol. 36, pp. 446–452, 1990.

72. Y. Lu, B. Yang, D. Gregg, J. N. Saddler and S. D. Mansfield, "Cellulase adsorption and an evaluation of enzyme recycle during hydrolysis of steam-exploded softwood residues," *Appl. Biochem. Biotechnol.*, vol. 98–100, pp. 641–654, 2002.

73. S. Kadam and A. Demain, "Addition of cloned beta-glucosidase enhances the degradation of crystalline cellulose by the *Clostridium thermocellum* cellulase complex," *Biochem. Biophys. Res. Commun.*, vol. 161, pp. 706–711, 1989.

74. R. Torget, "Milestone completion report: Process economic evaluation of the total hydrolysis option for producing monomeric sugars using hardwood sawdust for the NREL bioconversion process for ethanol production," National Renewable Research Laboratory, Golden, CO, 1996.

75. Q. Nguyen, "Milestone completion report: Evaluation of a two-stage dilute sulfuric acid hydrolysis process," National Renewable Research Laboratory, Golden, CO, 1998.

76. L. T. Fan, Y. Lee and D. H. Beardmore, "Mechanism of the enzymatic hydrolysis of cellulose: Effects of major structural features of cellulose on enzymatic hydrolysis," *Biotechnol. Bioeng.*, vol. 22, pp. 177–199, 1980.

77. D. D. Spindler, C. E. Wyman, A. Mohagheghi and K. Grohmann, "Thermotolerant yeast for simultaneous saccharification and fermentation of cellulose to ethanol," *Appl. Biochem. Biotechnol.*, vol. 17, pp. 279–293, 1988.

78. "Pentosuria," *The American Heritage Stedman's Medical Dictionary*. New York: Houghton Mifflin Co., 2004.

79. A. Sarthy, L. McConaughy, Z. Lobo, A. Sundstorm, E. Furlong and B. Hall, "Expression of the *Escherichia coli* xylose isomerase gene in *Saccharomyces cerevisiae*," *Appl. Environ. Microbiol.*, vol. 53, pp. 1996–2000, 1987.

80. D. S. Beall, K. Ohta and L. O. Ingram, "Parametric studies of ethanol production from xylose and other sugars by recombinant *Escherichia coli*," *Biotechnol. Bioeng.*, vol. 38, pp. 296–303, 1991.

81. K. Ohta, D. S. Beall, J. P. Mejia, K. T. Shanmugam and L. O. Ingram, "Genetic improvement of *Escherichia coli* for ethanol production: Chromosomal integration of *Zymomonas mobilis* genes encoding pyruvate decarboxylase and alcohol dehydrogenase II," *Appl. Environ. Microbiol.*, vol. 57, pp. 893–900, 1991.

82. L. O. Ingram and T. Conway, "Expression of different levels of ethanologenic enzymes from *Zymomonas mobilis* in recombinant strains of *Escherichia coli*," *Appl. Environ. Microbiol.*, vol. 54, p. 404, 1988.

83. F. Alterthum and L. O. Ingram, "Efficient ethanol production from glucose, lactose, and xylose by recombinant *Escherichia coli*," *Appl. Environ. Microbiol.*, vol. 55, pp. 1943–1948, 1989.

84. K. Skoog and B. Hahn-Hagerdal, "Xylose fermentation," *Enzyme Microbiol. Technol.*, vol. 10, pp. 66–80, 1988.

85. K. H. Kyung and P. Gerhardt, "Continuous production of ethanol by yeast immobilized in membrane-contained fermenter," *Biotechnol. Bioeng.*, vol. 26, pp. 252, 1984.

86. P. Christen, M. Minier and H. Renon, "Enhanced extraction by supported liquid membrane during fermentation," *Biotechnol. Bioeng.*, vol. 36, pp. 116–123, 1990.

87. D. G. Naee, "ACS Press Conference, 200th National Meeting of the ACS, August 1990," p. 17, September 10, 1990.

88. H. Shapouri, J. A. Duffield and M. Wang, "The energy balance of corn ethanol: An update," Washington, DC: U.S. Department of Agriculture, October 2004.

89. S. Mueller and K. Copenhaver, "News from corn ethanol: Energy use, co-products, and land use," in *Near-Term Opportunities for Biorefineries Symposium*, Champaign, IL, October 11–12, 2010.

90. S. Mueller, Detailed report: 2008 national dry mill corn ethanol industry. University of Illinois-Chicago. May 4, 2010. Available at http://ethanolrfa.3cdn.net/2e04acb7ed88d0 8d21_99m6idfc1.pdf (accessed March 21, 2014).

91. P. de Groot and D. Hall, "Power from the farmers," *New Scientist*, pp. 50–55, 1986.

92. J. G. Da Silva, G. E. Serra, J. R. Moreira, J. C. Concalves and J. Goldenberg, "Energy balance for ethyl alcohol production from crops," *Science*, vol. 201, pp. 903–906, 1978.

93. Bioethanol process based on enzymatic cellulose hydrolysis. March 2002. Available at http://www1.eere.energy.gov/biomass/pdfs/sugar_enzyme.pdf (accessed February 2014).

94. C. S. Hopkinson and J. W. Davy, "Net energy analysis of alcohol production from sugarcane," *Science*, vol. 207, pp. 302–304, 1980.

95. M. R. Schmer, K. P. Vogel, R. B. Mitchell and R. K. Perrin, "Net energy of cellulosic ethanol from switchgrass," *Proc. Natl. Acad. Sci. USA*, vol. 105, pp. 464–469, 2008.

96. L. Gladius, "Some aspects of the production of ethanol from sugar cane residues in Zimbabwe," *Solar Energy*, vol. 33, pp. 379–382, 1984.

97. A. McAloon, F. Taylor, W. Yee, K. Ibsen and R. Wooley, "Determining the cost of producing ethanol from corn starch and lignocellulosic feedstocks," National Renewable Energy Laboratory, Golden, CO, Tech. Rep. NREL/TP-580-28893, 2000.

# 13 Biodiesel

*Sunggyu Lee*

## CONTENTS

## 13.1 WHAT IS BIODIESEL?

### 13.1.1 DEFINITION OF BIODIESEL

Biodiesel is an alternative fuel that is similar to conventional petroleum-based diesel fuel, usually derived from vegetable oil or animal fat, and exclusively formulated for diesel engines. Biodiesel is typically manufactured by chemically reacting the lipids derived from plant oil or animal fat with a simple alcohol such as methanol or ethanol, which is known as a transesterification reaction. Alternately, biodiesel may be defined as mono-alkyl esters of long-chain fatty acids derived from vegetable oils or animal fats that meet ASTM D6751 [1].

Biodiesel is meant to be used in standard diesel engines without modification and can be used in pure form or in any mixture combination with petroleum-based diesel fuel. Therefore, biodiesel should be clearly distinguished from straight vegetable oils (SVOs) or waste oils that are used for converted diesel engines. The latter do not meet the specification standards of ASTM D6751.

**441**

The name or word *biodiesel* comes from the fact that the fuel is derived from biological sources and the fuel is, at least originally, meant to be used in diesel engines. Biodiesel is considered a renewable fuel as it is produced from plant oils and animal fats [2]. However, it is a reality that "biodiesel" as a name may potentially imply and inadvertently encompass more than what is technologically defined as biodiesel. In other words, biodiesel, as technologically defined, is far more specific than simply "a biologically derived fuel that can be used on diesel engines."

Biodiesel is a renewable alternative fuel for diesel engines comprised of a long-chain mono-alkyl ester and is the product of the *transesterification* reaction of triglycerides with low-molecular-weight alcohols such as methanol and ethanol. According to the *National Biodiesel Board (NBB)*, which is the national trade association representing the biodiesel industry in the United States, *biodiesel* is defined as "a domestic, renewable fuel for diesel engines derived from natural oils like soybean oil, and which meets the specifications of ASTM D6751" [3]. While ASTM D 6751 provides the original specifications for 100% pure biodiesel (B100), there are other objective fuel standards and specifications for biodiesel fuel blends:

- Biodiesel blends up to 5% (B5) to be used for on- and off-road diesel applications (ASTM D975-08a)
- Biodiesel fuel blends from 6% to 20% (B6–B20) (ASTM D7467-09)
- Residential heating and boiler applications (ASTM D396-08b)

### 13.1.2 MARKET ADOPTION OF BIODIESEL

There is a major global push enacted by the *Kyoto protocol* (adopted on December 11, 1997; entered into force on February 15, 2005) to reduce emissions of greenhouse gases (GHGs), more particularly, carbon dioxide [4]. The current world is seriously concerned with energy sustainability and affordability, as many industrialized and developing nations are economically hurting from escalating costs of energy and fuels, in particular, petroleum-based transportation fuels. Biodiesel is one of the alternative fuels that can help the world address these issues. Biodiesel is considered a mostly carbon-neutral fuel and is completely biodegradable. Since the enactment of the Energy Policy Act of 2005 (EPAct 2005), which created the Renewable Fuel Standard (RFS) program and under the Energy Independence and Security Act of 2007 (EISA 2007) that expanded the RFS program to include diesel, the use of biodiesel in the United States has been increasing. In the United Kingdom, the Renewable Transport Fuels Obligation (RFTO) requires suppliers to include 5% renewable biofuel in all transport fuels marketed in the United Kingdom, which effectively mandates B5 (5% biodiesel + 95% petrodiesel) as road diesel [5].

## 13.2 HISTORICAL BACKGROUND

Biodiesel has been around for quite some time, and therefore, it is not a newly invented product. However, it has never been considered a viable fuel until recently. The transesterification of triglycerides was discovered by E. Duffy and J. Patrick as early as 1853; in fact, this happened several decades before the first functional diesel engine was invented. It was not until 1893 that *Rudolf Diesel* invented the first diesel engine and designed it to run on peanut oil. Later, in the 1920s, the diesel engine was redesigned to run on petrodiesel, a fossil fuel derived from petroleum crude [6]. Since petrodiesel had been much cheaper to produce compared to any biofuel, there has not been many active developments in the biodiesel infrastructure. It was not until 1977 that the first industrial biodiesel process using ethanol was patented. Later, in 1979, South Africa started research on the transesterification of sunflower oil. After 4 years, South African agricultural engineers published a process for fuel-quality, engine-tested biodiesel. An Austrian company, Gaskoks, used this process to build the first biodiesel pilot plant in 1987 and later built an industrial-scale plant in 1989. The industrial-scale plant was capable of processing 30,000 tons of rapeseed per year [6]. As a matter of fact, rapeseed oil has also become the primary feedstock for biodiesel in Europe (estimates for 2006: more than 4.0 million tons of rapeseed oil went into biodiesel) [7]. During the 1990s, many European countries such as Germany and Sweden started building their own biodiesel plants. By 1998, approximately 21 countries had some sort of commercial biodiesel production. On April 23, 2009, the European Union (EU) adopted the *Renewable Energy Directive (RED)*, which included a 10% target for the use of renewable energy in road transport fuels by 2020. It also established the environmental sustainability criteria that biofuels consumed in the EU have to comply with, covering a minimum rate of direct GHG emission savings as well as restrictions on the types of land that may be converted to production of biofuel feedstock crops [8,9]. In September of 2005, the state of Minnesota became the first U.S. state to mandate that all diesel fuel sold in the state contain a certain part biodiesel, requiring a content of at least 2% biodiesel (B2 and up). This established for the first time in the United States that biodiesel blend fuel was no longer a choice but a standard and mandate. Since the passage of EPAct 2005 and the enactment of EISA 2007, which effectively mandated the *RFS* program including diesel fuel, biodiesel use in the United States has been sharply increasing. The biodiesel industry in the United States reached a key milestone by producing over 1 billion gallons in 2011. Biodiesel production in the EU exceeded 9.5 million metric tons in 2010 [10], which is roughly equivalent to 2.85 billion U.S. gallons.

## 13.3 FEEDSTOCK RENEWABILITY

Plant oils or vegetable oils, which are renewable in nature and can be used as feedstock for biodiesel, may or may not be edible. Examples of edible oils include almond oil, coconut oil, corn oil, olive oil, peanut oil, rapeseed oil, sesame oil, soybean oil, walnut oil, and many more. Examples of inedible vegetable oils include linseed oil, tung oil, and castor oil used in the manufacture of lubricants, paints, cosmetics, pharmaceuticals, and other industrial applications. Even though most current efforts in producing biodiesel have been using edible oils, biofuel and, in particular, biodiesel can also be produced using inedible oils [11].

The most frequently used method of characterizing the chemical structure of a vegetable oil is the fatty acid composition, which shows the distribution of different kinds of fatty acids, the carbon numbers of ingredient fatty acids, location of double bonds in the fatty acid molecular structure, ratio of saturated versus unsaturated fatty acids, and more. Since vegetable oils contain more unsaturated bonds, that is, double bonds, in their aliphatic fractions than animal fats, the occurrences of these unsaturated bonds in the chemical structures of vegetable oils are quite important. It is also important to note that the resultant biodiesel properties as well as the applicable process technology are strongly dependent upon the chemical structures of the feedstock vegetable oils. Since the main ingredients of vegetable oils are triglycerides or triacylglycerides (TAGs), which are esters derived from glycerin and three fatty acids, we may call these "characteristic fatty acids" for their parent vegetable oils. Table 13.1 shows fatty acid compositions of common edible oils in terms of percentage by weight of total fatty acids.

In Table 13.1, C18:0 as a notation means that it is made of an 18-carbon chain with 0 double bond, whereas C18:2 denotes an 18-carbon chain with 2 double bonds in the molecular structure. The specific gravity of most edible vegetable oil ranges between 0.91 and 0.93, depending upon the kind of vegetable oil, specific composition of the oil, purity, and other factors. For comparison, the specific gravity of castor oil, an inedible oil, is approximately 0.957–0.961. This does not necessarily mean that the density of an inedible oil is higher than that of an edible oil. Needless to say, the cost of inedible vegetable oil is significantly lower than that of edible vegetable oil. Furthermore, the use of inedible vegetable oils in biodiesel manufacture would not cause a "fuel versus food" dilemma.

## 13.4 TRANSESTERIFICATION PROCESS FOR BIODIESEL MANUFACTURE

Biodiesel can be produced in a number of different ways. The process can be operated either as a batch process or as a fully continuous process. It is usually performed catalytically, using a strong base or acid as the catalyst. Alternately, it can be operated noncatalytically, using supercritical methanol. The most popular process of biodiesel manufacture in industry currently uses methanol as the alcohol and *sodium hydroxide (NaOH)* as the base catalyst and is a continuous process.

Biodiesel is most popularly produced by the *transesterification* reaction of triglycerides. Triglyceride is also referred to as triacylglycerol (TG) or TAG. Triglycerides are found as main constituents in plant oils and animal fats, and the molecular structure of a triglyceride is shown in Figure 13.1. Triglycerides found in plant oil are typically more unsaturated than those in animal fats. As shown, typical triglycerides found in these feedstocks are fairly large molecules, that is, substantially larger than those in petroleum-based transportation fuels. Furthermore, the molecular structure involves a significant amount of oxygen. While some oxygen content is considered good, excessively high oxygen content has detrimental effects on fuel properties. Transesterification effectively handles the molecular weight issues, while it does not address the oxygen content issue.

Transesterification occurs when the triglycerides are mixed with an alcohol, typically either methanol or ethanol. As the hybridized terminology of *trans- + esterification* implies, transesterification is a chemical reaction in which the aliphatic organic group (R-) of an ester is exchanged with another aliphatic organic group (R'-) of an alcohol, thereby producing a different ester and a different alcohol. In

**TABLE 13.1**

**Fatty Acid Compositions of Common Edible Oils (by Weight Percentage of Total Fatty Acids)**

| Oil | Unsat/ Sat Ratio | Saturated | | | | | Monounsaturated | Polyunsaturated | |
| --- | --- | --- | --- | --- | --- | --- | --- | --- | --- |
| | | Capric Acid C10:0 | Lauric Acid C12:0 | Myristic Acid C14:0 | Palmitic Acid C16:0 | Stearic Acid C18:0 | Oleic Acid C18:1 | Linoleic Acid (ω6) C18:2 | Alpha Linolenic Acid (ω3) C18:3 |
| Almond oil | 9.7 | – | – | – | 7 | 2 | 69 | 17 | – |
| Canola oil | 15.7 | – | – | – | 4 | 2 | 62 | 22 | 10 |
| Cocoa butter | 0.6 | – | – | – | 25 | 38 | 32 | 3 | – |
| Cod liver oil | 2.9 | – | – | 8 | 17 | – | 22 | 5 | – |
| Coconut oil | 0.1 | 6 | 47 | 18 | 9 | 3 | 6 | 2 | – |
| Corn oil (maize oil) | 6.7 | – | – | – | 11 | 2 | 28 | 58 | 1 |
| Cottonseed oil | 2.8 | – | – | 1 | 22 | 3 | 19 | 54 | 1 |
| Flaxseed oil | 9.0 | – | – | – | 3 | 7 | 21 | 16 | 53 |
| Grapeseed oil | 7.3 | – | – | – | 8 | 4 | 15 | 73 | – |
| Olive oil | 4.6 | – | – | – | 13 | 3 | 71 | 10 | 1 |
| Palm oil | 1.0 | – | – | 1 | 45 | 4 | 40 | 10 | – |
| Palm olein | 1.3 | – | – | 1 | 37 | 4 | 46 | 11 | – |

*(continued)*

**TABLE 13.1 (Continued)**
**Fatty Acid Compositions of Common Edible Oils (by Weight Percentage of Total Fatty Acids)**

| Oil | Unsat/Sat Ratio | Saturated | | | | | Monounsaturated | Polyunsaturated | |
| | | Capric Acid C10:0 | Lauric Acid C12:0 | Myristic Acid C14:0 | Palmitic Acid C16:0 | Stearic Acid C18:0 | Oleic Acid C18:1 | Linoleic Acid (ω6) C18:2 | Alpha Linolenic Acid (ω3) C18:3 |
|---|---|---|---|---|---|---|---|---|---|
| Palm kernel oil | 0.2 | 4 | 48 | 16 | 8 | 3 | 15 | 2 | – |
| Peanut oil | 4.0 | – | – | – | 11 | 2 | 48 | 32 | – |
| Safflower oil[a] | 10.1 | – | – | – | 7 | 2 | 13 | 78 | – |
| Sesame oil | 6.6 | – | – | – | 9 | 4 | 41 | 45 | – |
| Shea nut | 1.1 | – | 1 | – | 4 | 39 | 44 | 5 | – |
| Soybean oil | 5.7 | – | – | – | 11 | 4 | 24 | 54 | 7 |
| Sunflower oil[a] | 7.3 | – | – | – | 7 | 5 | 19 | 68 | 1 |
| Walnut oil | 5.3 | – | – | – | 11 | 5 | 28 | 51 | 5 |

*Source:* http://www.scientificpsychic.com/fitness/fattyacids1.html; http://www.connectworld.net/whc/images/chart.pdf; http://curezone.com/foods/fatspercent.asp.

*Note:* Percentages may not add up to a total of 100% due to other constituents not listed in the table. Where percentages vary, average values are used.
a Not high-oleic variety.

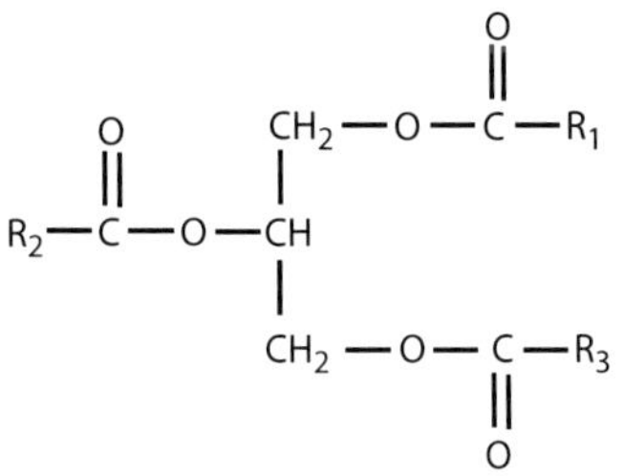

**FIGURE 13.1** Molecular structure of triglyceride.

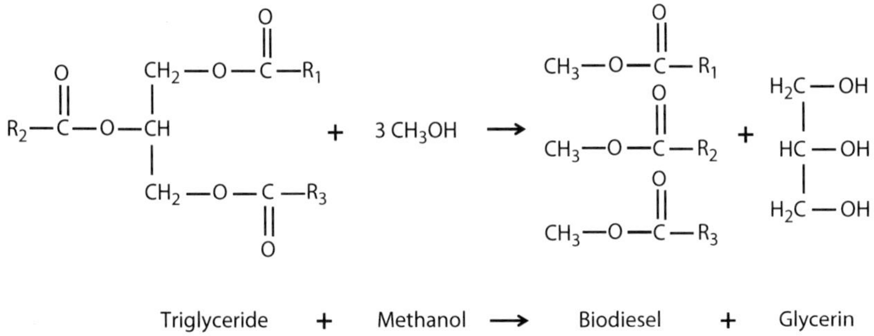

Triglyceride + Methanol ⟶ Biodiesel + Glycerin

**FIGURE 13.2** Stoichiometric reaction chemistry of transesterification.

other words, the starting ester (triglyceride) and monohydric alcohol (methanol) are converted by the transesterification reaction into a simpler form of esters (mono-alkyl ester or biodiesel) and a more complex form of alcohol, that is, trihydric alcohol (glycerin). In this reaction of transesterification of triglycerides, three alcohol molecules liberate the long-chain fatty acids from the glycerin backbone by bonding (i.e., esterification) with the carboxyl group carbons in the triglyceride molecule, as shown in Figure 13.2. The products of the transesterification reaction are a glycerin molecule and three long-chain mono-alkyl ester molecules, otherwise commonly known as biodiesel. Therefore, some of the chemical structural properties of the original oil are retained in the aliphatic organic groups (R-, or $R_1$-, $R_2$-, or $R_3$-) of the product biodiesel. More specifically, if a biodiesel is produced by transesterification reaction between soy triglycerides and methanol, the resultant biodiesel ester is often referred to as methyl soyate or soy biodiesel. Similarly, rapeseed oil-based biodiesel is called rapeseed biodiesel. The transesterification reaction is usually catalyzed using a strong base such as NaOH or KOH. The base helps to catalyze this reaction by removing the hydrogen in the hydroxyl group on the alcohol molecules.

## 13.4.1 Pretreatment of Oil

Before oils and fats can react to form biodiesel, they must go through a *pretreatment* process. The first stage of the pretreatment process involves filtering to remove dirt and other particulate matter (PM) from the oil. Next, water must be removed from the oil because it will hydrolyze the triglycerides to form fatty acid and glycerin. Fatty acids are undesirable products of the biodiesel industry, since free fatty acids (FFAs) can directly react with the base catalyst to form soap, which is certainly unwanted for biodiesel manufacture. If soap formation is active, the process would require an additional amount of base catalyst to compensate for the reactive depletion due to undesirable chemical reactions. Finally, the pretreated oil must be tested for *FFA* content. Typically, less than 1% FFA in oil is acceptable for processing without further provisional treatments. FFAs are long-chain carboxylic acids that have broken free from the triglycerides, typically resulting from thermal degradation of triglycerides due to a prolonged exposure to heat. These acids can increase soap formation in the reactor, as mentioned earlier. Too much soap in the reactor causes substantial operational and technological difficulties such as the following: (1) soap formation becomes a reason for an additional amount of base catalyst usage to overcome its wasteful depletion by the soap formation reaction; (2) additional problems arise in product separation; and (3) as an extreme case, the formed soaps mix with water from the fuel wash stage to create an emulsion that can seriously slow down or even prevent settling of the wash water layer from the product biodiesel layer.

There are two ways to deal with the FFAs in the oil. An acid can be added to the oil to convert the FFAs into biodiesel; this is the case with an acid-catalyzed esterification reaction. Alternatively, they can be neutralized, turned into soap, and removed from the oil.

### 13.4.2 Transesterification Reaction in a Biodiesel Reactor

After the oil pretreatment step, the oil is then sent to the *reactor*. The feed methanol that reacts with the oil in the reactor also has to go through some pretreatment or premixing. Before the methanol is sent to the reactor, it goes through a mixer, where it is combined with the sodium hydroxide (NaOH) catalyst. The oil and methanol/catalyst mixture are then fed into the reactor to undergo the transesterification reaction. Methanol is usually fed in excess of around 1.6 times the stoichiometric amount, and the reactor is kept at around 60°C. With the aid of the base catalyst, the reaction is able to proceed up to *98% conversion.*

### 13.4.3 Product and By-Product Separation

The effluent stream from the reactor is fed into a *separator.* The glycerin by-product has a much greater density (glycerin specific gravity at 25°C = 1.263) than the biodiesel (specific gravity at 60°C = 0.880) and is therefore easily removed via gravitational separation. In other words, the glycerin by-product forms the lower layer, while the biodiesel product ends up in the top layer.

### 13.4.4 Purification

After the biodiesel is separated from the glycerin by-product, it goes through a *purification process.* The first step is to neutralize the remaining catalyst by adding an acid to the biodiesel. Then, the biodiesel is sent through a stripper to remove any methanol left from the reactor. This unconsumed methanol is then recovered and recycled back to the methanol/catalyst mixer. After the methanol removal, the biodiesel goes through a water wash to remove all the soaps and salts (e.g., neutralized salt of NaOH catalyst) generated during transesterification and neutralization. The biodiesel is then dried and stored as the final product.

The glycerin by-product, also known as crude glycerin, also goes through some levels of purification. The crude glycerin contains a considerable amount of methanol, which comes out unreacted due to its excess amount of feed to the reactor. The glycerin goes through a *distillation* separation that recovers a great deal of the methanol for recovery and recycling. The recycled methanol collects most of the water that entered the process, and therefore, it must go through a separate distillation column to be purified. As mentioned earlier, water is an unwelcome species in the reactor stage. The glycerin that comes out of the distillation process is pure glycerin that can be marketed for other industries including pharmaceutical and cosmetics industries. A schematic of the biodiesel manufacturing process via transesterification is shown in Figure 13.3.

As mentioned above, crude glycerin is a mixture of glycerin, methanol, and salts. Crude glycerin can be sold as is at a lower price or further purified into pharmaceutical grade glycerin. A marketable grade of crude glycerin is generally at least 80% glycerin with less than 1% methanol. Crude glycerin that has lower levels of glycerin or higher levels of methanol often has little or no value; this is especially true in the current era of surplus supply of glycerin on the market. While glycerin is overly abundant on the world marketplace due to rapidly increased biodiesel production as well as no new

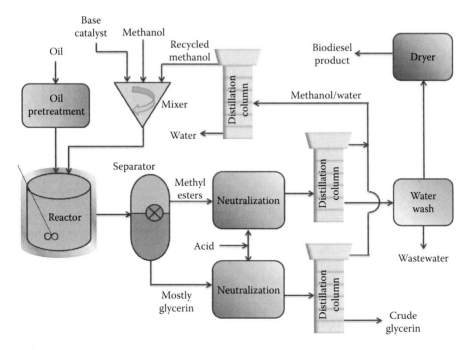

**FIGURE 13.3**    A schematic of the biodiesel manufacturing process via transesterification.

significant end uses for glycerin, the purification of crude glycerin into pure glycerin is quite energy intensive and costly. Efficient chemical conversion of glycerin or crude glycerin into other value-added chemicals and petrochemicals, besides the conventional end uses established in the food, pharmaceutical, and cosmetics sectors, would help stabilize the market price of glycerin and provide additional income to the biodiesel industry, thus ultimately improving the industry's gross margin and profitability.

## 13.5 PROPERTIES OF BIODIESEL

### 13.5.1 CETANE RATING OR CETANE NUMBER

The two most beneficial properties of biodiesel are its higher cetane ratings (CRs) and better lubricating properties than the ultra-low-sulfur diesel (ULSD). The CR is a measure of a fuel's ignition delay, which may be defined as the time period between the beginning of the fuel injection and the first identifiable pressure increase during the fuel combustion. Conventional diesel engines run well with a CR of 40 to 55. The CR of biodiesel ranges between 50 and 60, whereas that of ULSD ranges between 45 and 50. On the other hand, the CR of vegetable oil ranges between 35 and 45. Addition of biodiesel to the petrodiesel such as B5 and B10 would increase the fuel's CR. Also, addition (or blending) of biodiesel in petrodiesel in a low concentration is reported to reduce fuel system wear, resulting in a beneficial effect to diesel engines.

### 13.5.2 CALORIFIC VALUE OR HEATING VALUE

The lower calorific value (LCV) or lower heating value (LHV) of biodiesel is 37.37 MJ/kg, whereas that for low-sulfur diesel is 42.612 MJ/kg. The higher calorific value (HCV) or higher heating value (HHV) of biodiesel is 40.168 MJ/kg, whereas that for low-sulfur diesel is 45.575 MJ/kg. Both heating values (HVs) of biodiesel are approximately 12% lower than those of low-sulfur diesel. Variations in the HVs for biodiesel are mainly from the variability of the biodiesel feedstocks, that is, the source of triglycerides. A downside of biodiesel in comparison to petrodiesel is its lower combustion energy content, even though the difference may not be viewed as very substantial. However, it has been claimed by fuel engineers that the ultimate fuel efficiency of biodiesel is comparable to that of petrodiesel, despite its lower energy density, thanks to several compensating factors including more complete combustion and better lubricity. The lower combustion energy content of biodiesel is attributable to its oxygenated molecular structure, in contrast to the nonoxygenated hydrocarbon structures of petrodiesel. Highly oxygenated structures of biodiesel adversely affect the cold flow properties of biodiesel, which are represented by the fuel's cloud point (CP), pour point, and cold filter plugging point (CFPP) [12]. For these reasons, an economically prudent method of deoxygenation of biodiesel is an area that needs to be further developed.

### 13.5.3 GENERAL PHYSICAL PROPERTIES OF BIODIESEL

Biodiesel has a color varying from golden or light brown to dark brown. The color depends upon the originating feedstock of biodiesel. Biodiesel is immiscible with water. Biodiesel has a high boiling point and a low vapor pressure. The boiling point data of biodiesel are very scarcely reported in the literature [13], since a smoke point is usually reached before a boiling point during its measurement, thus making the measurement itself quite difficult. At a smoke point, ingredients of biodiesel are degrading, that is, going through pyrolytic decomposition. Based on the published literature values, the boiling points of pure methyl esters of C18:0, C18:1, and C18:2 are 625K, 622K, and 639K, respectively [14]. Yuan et al. [13] developed models, based on the Antoine equation and a group contribution method, for predicting vapor pressure and normal boiling point of pure methyl esters and biodiesel fuels. The flash point of biodiesel (typically >130°C) is significantly higher than that of petroleum diesel (64°C) or gasoline (−45°C), which makes biodiesel substantially easier to handle and less flammable. Biodiesel has a density of approximately 0.880 g/cm$^3$, which is slightly higher than that of petrodiesel (0.85 g/cm$^3$) and substantially higher than that of gasoline (0.71–0.77 g/cm$^3$).

### 13.5.4   COLD FLOW PROPERTIES

Pure biodiesel (B100) has poor cold-temperature properties as straight diesel fuel. Biodiesel's *CP* and *CFPP* are both high, thus making pure biodiesel (B100) unsuitable as a cold-climate fuel without blending or additives. When biodiesel is cooled below a certain temperature, some ingredient molecules of biodiesel start to aggregate and form crystals. This temperature varies depending on the feedstock of biodiesel but is consistently quite high. As the biodiesel is further cooled and the crystals become larger than one-quarter of the wavelength of visible light, the fuel system starts to look cloudy. This point is known scientifically as the CP. The CP measurement follows the ASTM 2500.

The lowest temperature at which biodiesel can pass through a 45 μm filter is called the CFPP. A high CFPP tends to clog up vehicle engines more easily at cold temperatures. As biodiesel is further cooled below the CFPP, it will gel and eventually solidify. This point is called the *gel point*.

Another important cold flow property is the *pour point*, which is defined as the lowest temperature where the fuel is observed to flow. As all these temperatures are generally higher for pure biodiesel (B100) than petrodiesel, biodiesel (B100) freezes substantially faster than most petrodiesels. Commercial additives developed for diesel to improve its cold flow properties are mostly applicable to biodiesel and biodiesel blends [12]. However, development of more potent and effective additives for biodiesel is certainly desirable and would stimulate the wider utilization of biodiesel or crop oil-based fuels.

### 13.5.5   MATERIAL COMPATIBILITY

As for the material compatibility, biodiesel is quite different from petrodiesel. Biodiesel is compatible with *high-density polyethylene (HDPE)*, while it is incompatible with polyvinylchloride (PVC) and polystyrene (PS), as PS is readily soluble in biodiesel and PVC is slowly dissolved in biodiesel. Polypropylene (PP) is affected by biodiesel, showing swell increase (by 8%–5%) and reduced hardness (by about 10%).

Polyurethane (PUR) is also affected by biodiesel, showing some swell increase (by 6%). Biodiesel affects some natural rubber and all nitrile rubber products, while biodiesel is compatible with commonly used Viton®-type synthetic rubbers in modern vehicles. Studies indicate that Viton types B and F (FKM-GBL-S and FKM-GF-S) are more resistant to acidic biodiesel [15]. Biodiesel affects and is affected by many metals including copper, zinc, tin, lead, cast iron, and brass, whereas biodiesel does not affect stainless steel (304 and 316), carbon steel, and aluminum. Brass, bronze, copper, lead, tin, and zinc may accelerate, by catalytic activities, the oxidation process of biodiesel, resulting in generation of fuel insolubles or gels and salts. Therefore, these metals should be avoided as materials of construction for piping, regulators, and fittings in applications where biodiesel is expected to be in contact with them.

Neat biodiesel (B100) may degrade some hoses, gaskets, seals, elastomers, glues, and plastics with prolonged exposure. Acceptable storage tank materials for biodiesel include aluminum, steel, fluorinated polyethylene, fluorinated PP, Teflon, and most fiber glasses [16].

## 13.6  PROSPECTS AND ECONOMICS

Based on the current level of process economics associated with the transesterification of vegetable oil in the United States, biodiesel requires a subsidy from the government in order to compete with petrodiesel and other fossil fuels. The scheduled expiration and delayed but retroactive reinstatement of biodiesel subsidy in the United States ($1 a gallon credit for diesel fuel created from biomass and a 10-cent credit available to small agriprocessors who make biodiesel) has lately been intensely contested and debated by Congress. Biodiesel plant owners have to sell their crude glycerin by-product for a decent value to stay profitable. Or, they need to convert the glycerin by-product and coproduct into other value-added chemicals and/or products. Successful development of profitable markets for by-products and coproducts may be a key determinant of the overall success of biodiesel industries. However, as fossil fuels become more expensive, biodiesel becomes more of a feasible fuel alternative.

In the United States, soybean oil is the largest biodiesel feedstock, followed by corn oil, yellow grease, and tallow. As of 2013, roughly one-quarter of soybean oil produced in the United States is being used for biodiesel manufacture. In 2012, the United States produced 969 million gallons of biodiesel (B100) [17]. However, the current level of United States production is substantially below the industry capacity, which means that biodiesel manufacturing facilities are being grossly underutilized.

The future of the biodiesel industry depends strongly on the cost of feedstocks. The raw-material cost is a substantial portion of the biodiesel manufacturing cost. The current biodiesel industry's gross margin is very poor without taking into account governmental subsidy. Food cost has increased substantially in recent years, which also escalated the raw-material cost of vegetable oils for the conventional biodiesel industry, even though some blame the biofuel industry for the food price hike. In this regard, the algae biofuel option appears to be promising, since it does not compete for food or require arable land for algae growth.

Biodiesel offers many benefits to the environment that are worthy of note. It is considered a mostly *carbon-neutral fuel* because all the carbon dioxide emitted from burning biodiesel originally came from plants and animals that removed it from the air. However, it is not completely carbon neutral because to process biodiesel all the way from plant oil and animal fat, some nonrenewable energy is inevitably required. It is, nonetheless, a huge improvement. It reduces carbon dioxide emissions by 78% compared to petrodiesel. Biodiesel even offers a reduction in carbon monoxide and sulfur emissions. It does, however, have slightly higher nitrous oxide ($N_2O$) emissions. Biodiesel is also biodegradable and nontoxic. Biodiesel is available today as a pure (or neat) fuel, known as B100, and also as a blended fuel with petrodiesel, such as B2, B5, B10, and B20. Blends such as B20 (20% biodiesel and 80% petrodiesel) burn cleaner than petrodiesel alone, thus reducing emissions of harmful air pollutants such as carbon monoxide, volatile organic compounds (VOCs), soot, and PM. Biodiesel has good potential to play a large role in the future fuel economy as well as in the transportation fuel sector as more sustainable and environmentally cleaner fuel.

## REFERENCES

1. National Biodiesel Board (NBB), Biodiesel basics, 2013. Available at http://www. biodiesel.org/what-is-biodiesel/biodiesel-basics.
2. J. V. Gerpen, Biodiesel processing and production, *Fuel Process. Technol.*, vol. 86, pp. 1097–1107, 2005.
3. National Biodiesel Board (NBB), Biodiesel: America's advanced biofuel, June 2013. Available at http://www.biodiesel.org/docs/default-source/ffs-engine_manufacturers/ biodiesel-benefits-and-oem-positions.pdf?sfvrsn=6.
4. United Nations Framework Convention on Climate Change, Kyoto protocol, 2013. Available at http://unfccc.int/kyoto_protocol/items/2830.php.
5. UK Department for Transport, Renewable transport fuels obligation, November 5, 2012. Available at https://www.gov.uk/renewable-transport-fuels-obligation.
6. Progressive Fuel Limited, Biodiesel history, 2012. Available at http://www.progressive fuelslimited.com/biodiesel.asp.
7. Soyatech, Rapeseed facts, 2013. Available at http://www.soyatech.com/rapeseed_facts. htm.
8. P. Al-Riffai, B. Dimaranan and D. Laborde, Global trade and environmental impact study of the EU biofuels mandate; final report. ATLASS Consortium, 2010. Available at http://trade.ec.europa.eu/doclib/docs/2010/march/tradoc_145954.pdf.
9. P. Al-Riffai, B. Dimaranan and D. Laborde, European Union and United States biofuel mandates: Impacts on world markets. Inter-American Development Bank, 2010. Available at http://idbdocs.iadb.org/wsdocs/getdocument.aspx?docnum=35529623.
10. European Biodiesel Board (EBB), 2010–2011: EU biodiesel industry production forcasts show first decrease in 2011 since data is gathered, October 18, 2011. Available at http://www.ebb-eu.org/EBBpressreleases/EBB%20press%20release%202010%20 prod%202011_capacity%20FINAL.pdf.
11. M. Mathiyazhagan, A. Ganapathi, B. Jaganath, N. Renganayaki and N. Sasireka, Production of biodiesel from non-edible plant oils having high FFA content, *Int. J. Chem. Env. Eng.*, vol. 2, pp. 119–122, 2011.
12. National Biodiesel Board (NBB), Biodiesel cold flow basics, a set of PowerPoint slides prepared by NBB, 2014. Available at http://www.biodiesel.org/extpages/search-results? q=cold%20flow%20basic.

13. W. Yuan, A. C. Hansen and Q. Zhang, Vapor pressure and normal boiling point predictions for pure methyl esters and biodiesel fuels, *Fuel*, vol. 84, pp. 943–950, 2005.

14. M. S. Graboski and R. L. McCormick, Combustion of fat and vegetable oil derived fuels in diesel engines, *Prog. Energy Combust. Sci.*, vol. 24, pp. 125–164, 1998.

15. E. W. Thomas, R. E. Fuller and K. Terauchi, Fluoroelastomer compatibility with biodiesel fuels, DuPont Performance Elastomers, LLC, 2007.

16. National Biodiesel Board (NBB), Materials compatibility, 2014. Available at http://www.biodiesel.org/docs/ffs-performace_usage/materials-compatibility.pdf.

17. U.S. DOE Energy Information Administration (EIA), Monthly biodiesel production report: U.S. Biodiesel capacity and production, September 27, 2013. Available at http://www.eia.gov/biofuels/biodiesel/production/table1.pdf.

# 14 Algae Fuel

*Sunggyu Lee*

## CONTENTS

## 14.1 TERMINOLOGY AND DEFINITIONS

Algae are a widely diverse group of simple organisms that are typically autotropic and range from unicellular to multicellular forms. Algae in Latin means "seaweed." Algae are often classified into two types, namely, microalgae and macroalgae. Microalgae are unicellular, while macroalgae are multicellular and possess some plantlike characteristics. Typically, microalgae require the aid of a microscope to be seen, while macroalgae can be readily seen without the aid of a microscope. Most algae are photosynthetic but lacking the many distinct cell and organ types that are typically found in land plants. The largest and most complex marine forms of macroalgae are called seaweeds. While both microalgae and macroalgae are potent biofuel feedstocks, more development efforts for biofuel are currently focused on microalgae due to a variety of reasons that include (1) their superfast growth characteristics and (2) higher lipid content than macroalage. However, even though macroalgae have very low lipid content, they have some other advantages that include (1) simple harvest method due to their larger physical dimensions and (2) higher content of carbohydrates, which can also be converted into liquid and/or gas. In this chapter, more emphasis is placed on the technology for deriving liquid transportation fuel from microalgae.

Algal biofuel is often classified as a *third-generation biofuel* and is, to some extent, linked to utilization of carbon dioxide as its feedstock. For comparison, *first-generation biofuels* refer to the fuels that have been derived from biological sources like starch, sugar, animal fats, and vegetable oils, whereas *second-generation biofuels* are derived from lignocellulosic sources. Therefore, first-generation biofuels are generally related to biomasses that are edible, whereas second-generation biofuels are derived from a variety of nonedible feedstocks such as lignocellulosic materials and municipal solid wastes (MSWs). The oils under the category of first-generation biofuels are obtained using the conventional techniques of production. Some of the most typical types of first-generation biofuels include vegetable oil, conventional biodiesel, corn ethanol, and so forth. Examples of second-generation biofuels include cellulosic ethanol and biomass derived Fischer–Tropsch fuels.

## 14.2  INTRODUCTION

Despite the ever-escalating price of petroleum products and rapidly growing concerns regarding carbon dioxide emissions, the world still remains heavily dependent upon fossil fuels. The 2011 International Energy Outlook (IEO2011) predicts by its *reference case scenario* that the total world consumption of marketed energy will increase by roughly 42% by 2035 from 2010, with an increase in liquid fuel consumption of 30% by 2035 from 2010 in the transportation sector [1]. This increase in demand, however, cannot be met by petroleum alone. Global efforts to satisfy the liquid fuel demand by greater use of renewable and sustainable biofuels have been intensified in recent years. They include sugar ethanol, corn ethanol, crop oil biodiesel, cellulosic ethanol, algal biodiesel, and more. Algae oil has been proposed as both a sustainable and an economically feasible solution to alternative liquid transportation fuels.

Many commodities, such as biodiesel and bioplastics, are synthesized using the triglycerides found in vegetable oils, while other petrochemicals can also be derived or synthesized using processing by-products such as glycerin. Algae, specifically *microalgae*, are a promising source of oil because, compared to other crops, they have fast growth rates, potential for higher yield rates, and the ability to grow in a wide range of conditions [2]. The yield of oil per unit area of algae is at least seven times greater than that of palm oil, which is the second-highest-yielding crop [3]. Another benefit of using algae is that it overcomes the "food versus fuel" issue of other vegetables and grains since algae is not a food crop and it does not take arable land from other crops. Algae grown on 9.5 million acres (out of a total of 2.3 billion acres of U.S. land), compared to the 450 million acres used for other crops, could provide enough algal biofuel to replace all petroleum transportation fuels in the United States [3]. However, based on the currently available technologies, harvesting of algae and the extraction of oil is technologically challenging and energy intensive [2]. There are several different methods proposed and developed for extracting oil, an important step in making algal biofuels as well as algal bioplastics. Unlike conventional straight vegetable oils (SVOs) and conventional biodiesels based on crop oils, which are first-generation biofuels, algae fuels are classified by many as third-generation biofuels.

The lipid containing oil from algae must be separated from the proteins, carbohydrates, and nucleic acids. The steps for extracting algae oil involve breaking the cell wall, separating the oil from the remaining biomass, and purifying the oil [4].

Oil can be extracted from algae by either mechanical or chemical methods. The three well-known methods for extraction of oil from algae are *expeller pressing* (or oil pressing), *subcritical solvent extraction*, and *supercritical fluid extraction* [2]. Other methods include *enzymatic extraction, osmotic shock,* and *ultrasonic-assisted extraction* [3]. While expeller pressing and ultrasonic-assisted extraction are mechanical processes, subcritical and supercritical extraction methods are chemical processes. Each of the methods has its advantages and drawbacks. Extraction of oil by expeller pressing is simple and straightforward, but requires the algae to be fully dried, which is energy intensive. Further, the extraction efficiency is not very high, and a substantial amount of unextracted oil is left behind. The benefit of solvent extraction is that the algae do not need to be fully dried, but common subcritical solvents, such as hexane, pose environmental, health, and safety concerns [4]. Further, while most solvent is recovered and reused, its associated cost is also burdensome. Supercritical fluid extraction (SFE) may be the most efficient method as an optimally designed process could extract almost all the oil and provide the highest purity since supercritical fluids are selective [2]. Furthermore, extraction with supercritical $CO_2$ (sc-$CO_2$) eliminates the use of harmful solvents. However, its high-pressure operation and required high-pressure equipment increase the overall process cost.

Algae oil extracted contains a high concentration of triglycerides, which can be converted into biodiesel using the conventional transesterification process. Carbohydrate content of algae oil can be fermented into bioethanol or biobutanol. Algae biomass can be converted into biogasoline or algal jet fuel by a series of process treatments. The transesterification product of algae oil is usually called algae biodiesel. Specific conversion technologies, which could be employed or involved for each type of algal biofuel, are not repeated in this chapter, since the background technological information is presented in the relevant chapters throughout this book.

## 14.3 MICROALGAE AND GROWTH

*Microalgae* or microphytes are a division of algal organisms that encompass diatoms, the green algae, and the golden algae. These microscopic organisms are incredibly efficient solar energy converters that perform photosynthesis and are capable of very fast growth in either freshwater or saline environments. Although this value varies from species to species, and depends upon the cultivation conditions, roughly 50% of the weight of algae is lipid oil. Algae are typically cultivated in either open or closed ponds, photobioreactors, or hybrid systems of both. Once the algae have matured, they are harvested and processed to extract the algae's oil.

## 14.4 ALGAE HARVESTING

Collecting, concentrating, and processing algae consists of separating algae from the growth medium, drying, and processing it to obtain the desired product. Separating algae from its growth medium is generally referred to as *algae harvesting*. The term *algae harvesting* technologically refers to the activity of concentration of fairly diluted (ca. 0.02%–0.06% total suspended solids [TSSs]) algae suspension until a slurry or paste containing 5%–25% TSS or greater is obtained. This process step involves a major effort of water removal or dewatering, which represents a significant operating cost. Specific harvesting methods depend primarily on the type of algae and growth media. The high water content of algae must be removed to enable further processing. The most common harvesting processes include (1) *microscreening*, (2) *flocculation*, and (3) *centrifugation* [5,6]. The three methods represent different unit operations of filtration, flotation, and centrifugation, respectively. Therefore, these harvesting steps must be energy efficient and relatively inexpensive; as such, selecting easy-to-harvest algae strains becomes quite important. Macroalgae harvesting, even though it may be simple, requires substantial manpower, whereas microalgae can be harvested more easily using microscreens, centrifugation, flocculation, or froth flotation. As mentioned earlier, this chapter is focused more on the microalgae biofuel technology.

### 14.4.1 Microscreening Harvesting of Algae

*Membrane filtration* is one of the algae harvesting methods and is usually aided by a vacuum pump. Membrane filtration provides well-defined pore openings to separate algal cells from the culture. An advantage of the membrane filtration harvesting method is that it is capable of collecting and concentrating microalgae or cells of very low initial density (concentration). However, concentration by membrane filtration is somewhat limited to small volumes and leads to the clogging and fouling of the filter (membranes) by the packed cells when a vacuum is applied. Fouling and clogging of the membrane surface due to increased concentration of algal cells results in sharp declines in flux and requires frequent maintenance.

A modified filtration method involves the use of a reverse-flow vacuum in which the pressure operates from above rather than below, making the process more gentle and avoiding or alleviating the packing of cells on the membrane. This method itself has been modified to allow a relatively large volume of water to be concentrated in a short period of time (20 L to 300 mL in 3 h or concentrating of nearly 70 times in 3 h) [6].

*Cross-flow filtration* is a purification separation technique, typically employed for submicron-sized materials, where the majority of the feed (algae–water suspension) flow travels tangentially across the surface of the filter rather than perpendicularly into the filter. It is advantageous over standard filtration, since the filter cake is being constantly washed away during the filtration process, thereby increasing the service time during which the filtration device can be continuously operated without maintenance stoppage.

*Cross-flow microfiltration (MF)* was investigated by Hung and Liu [7] for separation of green algae, *Chlorella* sp., from freshwater under several different transmembrane pressures (TMPs), and also with both laminar and turbulent flows by varying the cross-flow velocity. The study examined the hydrodynamic conditions and interfacial phenomena of MF of green algae and revealed the interrelations among the cross-flow velocity, MF flux decline, and TMP [7].

*Forward osmosis (FO)* is an emerging membrane separation process, and it has recently been explored for microalgae separation. It is claimed that FO membranes use relatively small amounts of external energy compared to the conventional methods of algae harvesting. The driving force through a semipermeable membrane for FO separation is an osmotic pressure gradient, such that a "draw" solution of high concentration (relative to that of the feed solution of dilute algae suspension) is used to induce a net flow of water through the membrane into the draw solution, thus effectively separating the feed water from its solutes (microalgae). Zou et al. [8] studied FO algae separation by comparing two different draw solutions of NaCl and $MgCl_2$, and also examining the efficacy as well as their membrane fouling characteristics.

Sometimes, concentrated algae may be collected with a *microstrainer*. When a microstrainer is used to collect algae, the processed algae–water suspension may look faintly green, indicating that it could be further concentrated. However, due to its eventual clogging, a microscreen alone is usually insufficient for long-term continuous or large-scale operation; substantial energy and labor input are required to remove the clogging and reopen the flow channels in the microscreen.

A novel process for *harvesting, dewatering, and drying (HDD)* of algae from an algae–water suspension has been developed by Algaeventure Systems, LLC. In this process, a superabsorbent polymer (SAP) fabric belt is put in contact with the bottom of the screen (water meniscus), thereby enabling the movement of a vast amount of water without moving the algae and achieving dewatering. This is based on the fact that water–water hydrogen bonding is stronger than the weak intermolecular forces between water and algae. As such, reduced surface tension, enhanced capillary effect, and modified adhesion effect can be built in, and the system can be designed to be continuous. In a prototype testing, an exceptional rate of HDD was achieved with a very low power input. A schematic of the Algaeventure Systems harvester is shown in Figure 14.1 [9].

*SAPs* can absorb and retain very large amounts of a liquid (such as an aqueous solution) in comparison to their own mass and have been widely used in baby diapers, personal hygiene and care products, chlorine dioxide gel, and water-soluble or hydrophilic polymer applications. Some examples of SAPs include sodium salt of poly(acrylic acid), potassium salt of poly(acrylic acid), polyvinyl alcohol copolymers, polyacrylamide copolymer, ethylene maleic anhydride copolymer, cross-linked polyethylene oxide, and cross-linked carboxymethylcellulose.

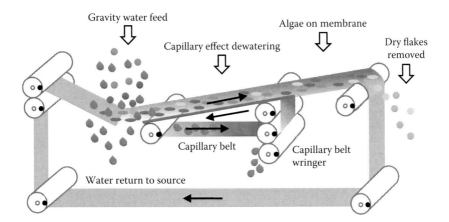

**FIGURE 14.1** A schematic of the Algaeventure Systems harvester. (From Algae Venture Systems, *Harvesting, Dewatering and Drying*, http://14w.algaevs.com/harvesting-dewatering-and-drying.)

### 14.4.2 Algae Harvesting by Flocculation

*Algae flocculation* is a method of separating algae from its medium by using chemicals to force the algae to form lumps or aggregates. The main disadvantage of this separation method is that the additional chemicals are difficult to remove or recover from the separated algae, thus making it inefficient and uneconomical for commercial use. The cost to remove or recover these chemicals may be too expensive to be commercially viable, unless a technological breakthrough is achieved.

Flocculating agents, or *flocculants*, are chemicals that promote flocculation by causing colloids and other suspended particles in liquids to aggregate, forming a floc. In general, there are two types of flocculants commonly used: (1) inorganic flocculants and (2) organic polymer/polymer electrolyte flocculants. Alum (hydrated potassium aluminum sulfate [$KAl(SO_4)_2 \cdot 12H_2O$]) and ferric chloride ($FeCl_3$) are the chemical flocculants most frequently used to harvest algae [5]. A commercial product called "chitosan," popularly used for water purification, can also be used as a flocculant but is far more expensive than other flocculants. The shells of crustaceans, such as shrimp, lobster, crabs, and crayfish, are ground into powder and processed to obtain chitin, a polysaccharide found in the shells, from which chitosan is derived via deacetylation. Water that is more brackish or saline requires additional chemical flocculant to induce flocculation [5]. High-molecular-weight organic polymers are considered good flocculants, since several segments of a polymer can attach themselves to the surface of a colloidal particle while the remaining segments are extended into the solution [10].

Harvesting via chemical flocculation alone, by the current technoeconomic standards, is a method that may be too expensive for large operations. Besides the chemical flocculation discussed above, there are different methods of flocculation of algae, including autoflocculation [11], bioflocculation [12], and electroflocculation. Bioflocculation has been studied and practiced in wastewater and sewage treatment. The efficacy of algae flocculation depends upon a large number of factors that are usually poorly understood or ignored, including cell size, cell shape, cell wall thickness, cell surface and interfacial properties, and so forth. By considering these properties as well as appropriate combinations of flocculation with other concentrating techniques, there is room for substantial enhancement in algae flocculation practice.

### 14.4.3 Algae Harvesting by Centrifugation

The choice of a good harvesting method for algae is crucial to the efficiency of the overall process in terms of capital investments as well as operating costs. The key factors for comparison include high cake dryness in the separated algae and low specific energy demand during the process, which may be defined as the energy demand per unit mass of algae harvested.

In a *single-stage harvesting process* using disc stack centrifuges, the algae–water suspension is directly fed into the centrifuge. Inside the centrifuge, the suspension is separated into a mostly clear water phase and an algae concentrate. The algae concentrate is drawn out periodically and has a fluid/creamy consistency. Since the whole suspension has to be put into rotation to create a centrifugal force up to 10,000 *g*, the specific energy demand is relatively high. Therefore, this single-stage harvesting process is suitable, especially for small and middle-sized facilities.

*Disc stack centrifuges* have been successfully operated for separation of two different liquid phases and solids from each other in a continuous process [13]. The operational principle of a disc stack centrifuge is described below. A good example of separation of two different liquid phases is separation of biodiesel methyl ester and by-product glycerin, whereas a good example of solid–liquid separation is dewatering algae from an algae–water suspension. *Alfa Laval* offers a variety of sizes and types of disc stack centrifuges for such industrial applications [14]. In a disc stack centrifuge used for liquid–solid separation, the denser solids are pushed outward by centrifugal forces against the rotating bowl wall, while the less dense liquid phases form inner concentric layers. By inserting specially designed disc stacks where the liquid phases meet, very high separation efficiency is achieved. The solids, such as algae cakes, can be removed manually, intermittently, or continuously, depending upon the specific process design and application. The separated liquid phase overflows in the outlet area on top of the bowl into recovery vessels, which are sealed off from each other to prevent potential cross-contamination [14].

The *Flottweg enalgy process* [15] is a two-stage algae harvesting process consisting of (1) preconcentration via static settling, filtration, flocculation, or dissolved air flotation (DAF) and (2) bulk harvesting using the Flottweg Sedicanter in order to dewater the algae suspension to concentrate. In contrast to the aforementioned single-stage process, only a small, predewatered, part of the algae suspension is separated by centrifugation in this two-stage process, thus reducing the energy demand drastically. While the preconcentrator provides a clear water phase as an initial step, the Flottweg Sedicanter dewaters the algae concentrate to obtain a solids cake with 22%–25% dry substance [15]. A schematic of the Flottweg two-stage harvesting process is shown in Figure 14.2.

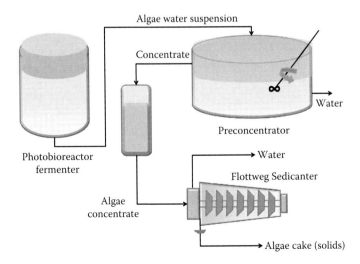

**FIGURE 14.2** Flottweg two-stage enalgy process of algae harvesting.

## 14.5 ALGAE OIL EXTRACTION

Three major methods of extracting oil from algae are (1) solvent extraction with hexane, (2) expeller/press, and (3) SFE. The extracted oil can then either be used as SVO or further processed by a transesterification reaction to produce biodiesel. Algal oil is considered a "balanced" carbon-neutral fuel, since any $CO_2$ taken out of the atmosphere by the algae is returned when the algae biofuels are burned. It has been estimated by Weyer et al. [16] that the theoretical maximum limit of unrefined oil that could be produced from algae is 354,000 $L \cdot ha^{-1} \cdot year^{-1}$ or 38,000 $gal \cdot ac^{-1} \cdot year^{-1}$, while limits for the practical cases examined in their study range from 46,300 to 60,500 $L \cdot ha^{-1} \cdot year^{-1}$.

### 14.5.1 EXPELLER PRESSING EXTRACTION OF ALGAE OIL

*Expeller pressing* methods have a long history of extracting vegetable oil from oil seeds; the methods involve some means of mechanically squeezing oil from crushed oil-containing seeds under applied pressure. Most cooking oils are produced via expeller pressing of a variety of feedstocks including maize, sunflower, soya, sesame, coconut, mustard seed, and groundnuts. Even though this mechanical extraction method of oil from oil seeds is technologically very simple, the process has been significantly enhanced with development of energy-efficient and mechanically superior machineries. Further, large processing plants have advantages over small plants in terms of reduced processing cost and extraction efficiency; most countries have such centralized large plants for cooking oil manufacture. These plants are also known as "oil refineries," more precisely, vegetable oil refineries. However, logistical burdens and transportation costs of raw materials and finished products also make small-scale highly efficient plants relevant and viable options. This is especially true for algae oil and biodiesel processing. With development of diverse and alternative fuels, in particular, liquid fuels, the word *refinery* is used in many areas other than the conventional crude oil refinery. They include corn, coal, tire oil, and vegetable oil refineries, and more.

*Oil presses* are typically used for vegetable oil and biodiesel processing on both large and small scales. There are two different kinds of large-scale processing methods involving oil presses, one being *hot* processing and the other being *cold* processing. In hot processing, the system includes a steam cooker and oil press. The steam cooker is mainly for pretreatment of oil seeds, which achieves shell softening and seed swelling. In cold processing, the machine operates at a low temperature, for example, 80°C, when it presses the seeds. An advantage of cold processing is that the extraction environment is not destructive to the nutrients in the oil. Dry algae can also be processed, quite similarly to oil seeds, using an oil press for extraction of algae oil, by mechanically rupturing the cell walls and collecting the extracted oil.

### 14.5.2 ULTRASONICALLY ASSISTED EXTRACTION

Ultrasonic extraction or *ultrasonication* can enhance and accelerate the algae oil extraction processes. Ultrasonication uses ultrasonic frequencies that are greater than 20 kHz. In an ultrasonic reactor, intense sonication of liquids generates ultrasonic waves that propagate into the liquid medium. During the low-pressure cycle, high-intensity small vacuum bubbles are created in the liquid due to the pressure imbalance. When these bubbles attain a certain critical dimension (cavity size), they collapse violently during a high-pressure cycle. As these bubbles collapse vigorously near the algae cell walls, that is, implode, they create shock waves, locally high pressure, and high-speed liquid jets. The resultant shear forces cause algae cell walls to mechanically rupture and release or help release their contents (algae lipids) into the solvent medium. This process of bubble formation and subsequent collapse is mechanistically called *cavitation* [17]. The advantages of this process include the following:

1. Dry cake is not required for oil extraction.
2. No caustic chemical is involved. The system is less corrosive.
3. The ultrasonication process can be used in conjunction with enzymatic extraction or other extraction methods.
4. Environmental impact is minimal.

However, the process is not yet proven on a large scale, and the energy cost needs to be lowered for it to be economically competitive. While ultrasonication can be employed in a solvent extraction process, the process is usually classified as mechanical extraction, mainly based on its cell wall rupture mechanism, that is mechanically induced.

### 14.5.3 Single Step Extraction Process by OriginOil, Inc.

While many new process technologies have been developed to extract the SVO from algae, this section will specifically describe the Single Step Extraction process developed by OriginOil, Inc.

The Single Step Extraction process begins with the mature algae entering the system as an algae-and-water suspension. Before entering the extraction tank, the stream is subjected to pulsed electromagnetic fields and pH modification in a process known as *quantum fracturing*. As the terminology implies, quantum fracturing creates the fluid fracturing effect, thereby mechanically distressing algae cells [18]. The electromagnetic fields are generated using a low-voltage power input, and the pH is modified using carbon dioxide, which helps optimize electromagnetic delivery and assists in cell degradation. The electromagnetic field created causes algae cells to release internal lipids. After quantum fracturing, the processed culture passes into a gravity clarifier while a return culture stream recycles into the inlet stream. The gravity clarifier separates the processed culture into layers of oil, water, and biomass. The lipid layer exit stream produces SVO, while the water layer exits via a recycle stream to the bioreactor. The biomass can then be harvested for a number of purposes including livestock feed, ethanol processing, and biomass gasification. A schematic of the OriginOil Single Step Extraction process [19] is shown in Figure 14.3.

This innovative method can extract approximately 97% of the lipid oil contained in algae cells. Typical extraction values for expeller press extraction and SFE are approximately 75% and 100%, respectively, suggesting that this method would

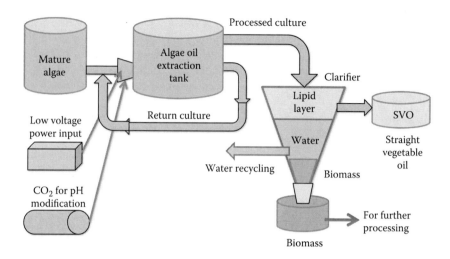

**FIGURE 14.3**  A schematic of the OriginOil single-stage algae oil extraction process.

be competitive, with industrial significance. This process does not require heavy machinery, chemicals, or dewatering of the feedstock and is thus claimed to use much less energy than traditional extraction processes.

The OriginOil Single Step Extraction process is an enhancement to the traditional algae oil extraction methods. The unique quantum fracturing method of extraction utilized by the Single Step Extraction process reduces the processing cost by eliminating the need for chemical input or energy-intensive machinery. As shown in the flow sheet, the OriginOil process uses a quite-different approach in oil extraction, which is based on a sequence of oil extraction → solids separation → dewatering basically in a single integrated step, while the conventional approach follows a sequential order of solids separation → dewatering → oil extraction [18]. As results of this novel process alignment, no initial dewatering is required, substantial energy savings can be expected, and the capital expenditure becomes reduced. With the biodiesel and bio-oil industries' growth, the Single Step Extraction technology will prove to be an invaluable tool for keeping algae-derived fuels competitive with petroleum.

## 14.5.4  Solvent Extraction of Algae Oil

Algal oil can be extracted from microalgae using an effective chemical solvent [20]. Hexane, cyclohexane, benzene, ether, acetone, and chloroform have proven to be effective in oil extraction of microalgae paste [21]. Among these, hexane has long been used as an oil extraction solvent in the food industry and is relatively inexpensive. A chemical method is usually faster in terms of the extraction speed and requires a lower energy input for the extraction process itself. One of the drawbacks of using chemical solvents for algae oil extraction is the safety issues involved in working with the chemicals. Care must be taken to avoid exposure to chemical vapors and direct contact with the chemical, either of which can cause serious personal injury. Benzene is classified as a carcinogen, and most ethers are highly flammable. Another disadvantage of using a chemical is the additional cost of recovery of the chemical for reuse in the process.

In a broader classification of algae oil extraction technology, chemical extraction methods include (1) the hexane solvent method, (2) Soxhlet extraction, and (3) SFE.

1. *Hexane solvent method*

   *Hexane solvent extraction* has long been used effectively for vegetable oil extraction. A very good example is in the production of soybean oil, for which hexane solvent extraction is predominantly used in industrial production due to its lower energy consumption and higher extraction efficiency (oil yield) in comparison to hydraulic presses, that is, the expeller method. Furthermore, the hexane extraction technology can be used as a stand-alone process for algae oil extraction, or it can also be used in conjunction with the physical extraction technology of the oil press/expeller method.

   If a chemical extraction process utilizing cyclohexane as an extracting solvent is employed for algae oil extraction in conjunction with an expeller method, the envisioned process scheme is as follows.

   The algae oil and lipids are first extracted using an expeller. The remaining pulp and biomass is then mixed with cyclohexane to extract the residual oil content remaining in the residue. The algae oil dissolves in cyclohexane, whereas the pulp and residues do not. The biomass is filtered out from the solution. The biomass rejected here can be used for other energy generation processes such as biomass gasification. As the last stage, the algae oil and cyclohexane are separated by distillation. Cyclohexane is recycled back to the solvent extraction stage. This two-stage extraction process of combined cold expeller press and hexane solvent extraction is capable of achieving an extraction efficiency of higher than 95% of the total oil present in the algae [20].

2. *Soxhlet extraction*

   *Soxhlet extraction* is a distillative extraction method that uses a chemical solvent, and the extraction unit is equipped with a heated solvent reflux mechanism. These days, Soxhlet extraction is very widely used in chemical laboratories for extraction of a variety of materials (chemical, biological, and polymeric) where the desired compound has a limited solubility in a chosen solvent. Interestingly, the original Soxhlet extraction process was invented in 1879 by Franz von Soxhlet, a German agricultural chemist, for the extraction of a lipid from a solid material [22].

   Oils are extracted from the algae via repeated washing, or percolation (trickling through), with an organic solvent, such as hexane or petroleum ether, under hot reflux in a specially designed glassware setup equipped with a condenser, a distillation path, a siphon arm, a thimble, and a distillation pot [20]. Soxhlet extraction is meant to be a laboratory process for a small-scale operation, but it is very useful for initial technology development. The extraction process is usually slow, taking an hour to several days, depending upon the specifics of the extraction process conditions as well as the specific solvent chosen for the extraction. The energy efficiency per unit mass of product yield is inevitably very low.

3. *SFE of algae oil*

Supercritical fluids have synergistic properties of both liquid-like and gas-like properties, such as low viscosity, high material diffusivity, and high solvent density, which make them good media for selective extraction [23]. The properties of low viscosity and high molecular diffusivity are gas-like properties, whereas high solvent density is more of a liquid-like property. A fluid is supercritical when both the temperature and pressure are above its critical point values. For example, the critical pressure and temperature of $CO_2$ are 72.8 atm and 31.16°C, respectively [24], while those for $H_2O$ are 218 atm and 371°C. sc-$CO_2$ has received a great deal of attention in many chemical process applications, due to its low critical temperature and inexpensive abundance. The low critical temperature bears extra significance when a supercritical fluid process is applied to biological or temperature-sensitive materials. As most other supercritical solvents, the solvent power of sc-$CO_2$ increases as the solvent density of carbon dioxide is increased.

The process for SFE of oil from algae is similar to that of any vegetable oil. The sc-$CO_2$ acts as a selective solvent to extract the oil. The oil is soluble in sc-$CO_2$, in particular, very-high-pressure $CO_2$, but the proteins and other solids are not [24]. To further increase the yield, a cosolvent, such as methanol or ethanol, can be used to increase the solubility of the more polar components of the oil [23]. In a cosolvent mode of supercritical fluid processing, the second solvent's concentration is usually kept on the low side, mainly due to the resultant critical temperature of a mixture fluid and the solvent recovery and replenishment cost. Generally speaking, SFE has the capability of yielding high-quality oil and high biomass, and sc-$CO_2$ is considered an environmentally benign solvent that gives low environmental impact.

A semi-batch supercritical extraction process of vegetable oil is described in a book by McHugh and Krukonis [24]. The term "semi-batch" means that the fluid flows continuously through the reactor while the solids do not. In this process, the extraction vessel is filled with crushed algae. The algae must be crushed in order for the oil to be accessible to the sc-$CO_2$ because the extraction rate is limited by the mass transfer rate through the cell wall of a whole cell [23]. sc-$CO_2$ is passed through the algae, extracting the oil and leaving the solid residues in the vessel. The pressure of the mixture of sc-$CO_2$ and oil is reduced to a subcritical level so that the oil can precipitate and be separated out. The $CO_2$ is then repressurized and recycled back into the extractor vessel. Several vessels can be used to optimize the system efficiency so that while some are being charged, extraction is happening in others, in an alternating sequence. A schematic of the process flow diagram is shown in Figure 14.4. Certainly, a fully continuous SFE process is conceivable with implementation of a mechanism (or system) for continuous feed of dry algae into a supercritical reactor and removal of solid biomass from the reactor.

The solubility of all vegetable triglycerides is approximately the same and depends largely on the temperature and pressure conditions of a supercritical fluid. At 70°C and 800 atm, clearly above the critical temperature

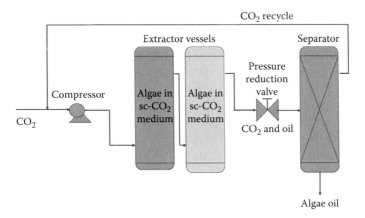

**FIGURE 14.4**   A schematic of the semi-batch supercritical algae oil extraction process. (From McHugh, M.A., and Krukonis, V.J., *Supercritical Fluid Extraction: Principles and Practice,* Butterworths, Boston, 1986.)

and pressure of $CO_2$, triglycerides and $CO_2$ become miscible, and decreasing the pressure to 200 atm will cause the oil will separate from the $CO_2$ [24]. This fairly low operating temperature allows for extraction of highly unsaturated triglycerides without degradation [23]. While the process shown in Figure 14.4 was an early attempt of SFE of algae oil, the pressures of 200 and 800 atm are excessively high for most chemical processing operations. Obviously, the high pressure of 800 atm was necessary to increase the solvent power of sc-$CO_2$. In order to reduce the operating pressure, different solvent and cosolvent combinations can be developed for process synergism and implemented accordingly. Supercritical extraction of algae oil is still in the research stage and not practiced on an industrial scale. Research on making biodiesel from algae and SFE of algae oil is being carried out in many laboratories throughout the world, including the Sandia National Lab [3].

For the production of goods from algae oil to be feasible, the cost must be competitive with that of petroleum-based or other biomass-based products. The cost of the oil extraction process is a critical component to the overall cost of production [4]. The biggest obstacle facing the economical use of supercritical extraction technology is the lack of the capability of continuously feeding and removing the solids [24], as well as the ability to extract oil at a manageably low pressure by using an appropriate solvent combination. One possible option to overcome this obstacle is to have multiple vessels whose stages of operation can be alternated during the process cycles and stages. Or, a continuous solid feeding and removal system needs to be implemented in the reactor system without disrupting the supercriticality of the mixture system. Another possibility for process enhancement is the use of a synergistically effective cosolvent supercritical fluid system, which has an enhanced solubility toward algae oil without requiring an excessively high-pressure condition. Another economical obstacle currently preventing production of biofuels based on SFE from being competitive is the capital cost associated with use of expensive and energy-intensive high-pressure equipment. Recent advances achieved in supercritical fluid technology in other chemical, biological, and petrochemical areas—in particular, pressure vessel designs, advanced reactor materials, solids handling capability, binary and ternary solvent systems for maximum process synergism, ability to tune the fluid properties, ingenious energy and process integration, and development of high-value by-products—could offer substantially enhanced process options for algae oil extraction based on the supercritical fluid technology.

### 14.5.5 ENZYMATIC EXTRACTION

The enzymatic process uses select enzymes to degrade algae cell walls, and in the process, system water acts as a solvent medium for enzyme action. Since no additional solvent is involved, this process facilitates an easier downstream fractionation. The advantages of the enzymatic extraction process include the following:

1. The process does not require dry cakes for oil extraction.
2. No caustic chemicals are required or involved.
3. Mild process conditions are used.
4. The process can be synergistically integrated with other processes such as ultrasonification.
5. Environmental impact is minimal.

The process is also in an early stage of development, and finding efficient and robust enzymes for the process is a challenge. Unless a cost-effective enzyme is developed and proven for the process, this process would be more expensive than hexane solvent extraction. However, there are ample opportunities with this approach, since the process can be utilized in conjunction with many other mechanical extraction technologies; as well, a drastically different product portfolio may be developed for highly value-added coproducts and by-products.

## 14.6 BY-PRODUCT UTILIZATION

The technical report from the National Renewable Energy Laboratory (NREL) discusses further uses for the by-products of the extraction process [4]. The high-value products such as pigments and omega-3 fatty acids could be separated and sold. Omega-3 fatty acids are a group of three fats including ALA (alpha-linolenic acid, found in plant oils), EPA (eicosapentaenoic acid, typically found in marine oils), and DHA (docosahexaenoic acid, also commonly found in marine oils). Chemically, omega-3 fatty acids are polyunsaturated fatty acids with a double bond (C=C) at the third carbon atom from the end of the carbon chain. Omega-3s are essential fatty acids and vital for metabolism. Since they cannot be synthesized by the human body, they are sold as dietary supplements at a high price.

Other high-value by-products could be developed as coproducts and exploited for commercial potential. The remaining high-protein solids could be used for animal feed. The residual biomass, which would be a very significant amount from a large-scale algae oil refinery, could also be used for further energy generation such as biomass gasification followed by methanol and dimethylether (DME) synthesis or feedstock for a combined heat and power (CHP) process. As very successfully demonstrated by the corn ethanol refinery, profitable utilization of by-products is very important for the overall process economics of the algae-based biofuel process.

Development of specialty end uses of algae oil can provide another valuable option for the algae industry. Such specialty end uses include utilization of algae oil as raw material or feedstock for manufacture of bio-based solvents, bio-based lubricants, bio-based polymers, biodegradable plastics, bio-based functional fillers, and more. While the cost to manufacture products from algae oil is not currently competitive with that of petroleum-based products, further research in the area could make the process and its products economically feasible, while contributing to the sustainability of society.

## REFERENCES

1. U.S. Energy Information Administration (EIA), "International energy outlook 2011," U.S. Department of Energy, Washington, DC, Tech. Rep. DOE/EIA-0484, September 19, 2011.
2. A. Demirbas and F. M. Demirbas, "Importance of algae oil as a source of biodiesel," *Energy Conversion and Management*, vol. 52, pp. 163–170, 2011.
3. M. S. Kent and K. M. Andrews, "Biological research survey for the efficient conversion of biomass to biofuels," Sandia National Laboratories, Albuquerque, New Mexico, Tech. Rep. SAND2006-7221, 2007.
4. A. Milbrandt and E. Jarvis, "Resource evaluation and site selection for microalgae production in India," National Renewable Energy Laboratory, Golden, CO, Tech. Rep. NREL/TP-6A2-48380, 2010.
5. Oilgae, Algae harvesting—Flocculation, 2013. Available at http://14w.oilgae.com/algae/har/flc/flc.html.
6. Oilgae, Algae harvesting—Filtration, 2013. Available at http://14w.oilgae.com/algae/har/fil/fil.html.
7. M. T. Hung and J. C. Liu, "Microfiltration for separation of green algae from water," *Colloids and Surfaces B: Biointerfaces*, vol. 51, pp. 157–164, 2006.
8. S. Zou, Y. Gu, D. Xiao and C. Y. Tang, "The role of physical and chemical parameters on forward osmosis membrane fouling during algae separation," *Journal of Membrane Science*, vol. 368, pp. 356–362, 2011.
9. Algae Venture Systems, *Harvesting, Dewatering and Drying*. 2013. Available at http://14w.algaevs.com/harvesting-dewatering-and-drying.
10. G. Shelef, A. Sukenik and M. Green, "Microalgae harvesting and processing: A literature review," Solar Energy Research Institute, Golden, CO, Tech. Rep. SERI/STR-231-2396, 1984.
11. A. Sukenik and G. Shelef, "Algal autoflocculation—Verification and proposed mechanism," *Biotechnology and Bioengineering*, vol. 26, pp. 142–147, 1984.
12. Oilgae, Bio-flocculation of algae research project at Newcastle University. *Oilgae Blog*, 2008. Available at http://14w.oilgae.com/blog/2008/12/bio-flocculation-of-algae-research-project-newcastle-university.html.

13. J. J. Milledge and S. Heaven, "Disc stack centrifugation separation and cell disruption of microalgae: A technical note," *Environment and Natural Resources Research*, vol. 1, pp. 17–24, 2011.

14. Alfa Laval, Alfa laval—Disc stack centrifuge technology. 2013. Available at http://local. alfalaval.com/en-us/key-technologies/separation/separators/dafrecovery/Documents/ Alfa_Laval_disc_stack_centrifuge_techonology.pdf.

15. Flottweg Separation Technology, Flottweg centrifuges for efficient algae harvesting. Flottweg AG. Vilsbiburg, Germany, 2010. Available at http://14w.flottweg.de/cms/ upload/downloads/old/algen_Internetversion_englisch.pdf.

16. K. M. Weyer, D. R. Bush, A. Darzins and B. D. Willson, "Theoretical maximum algal oil production," *Bioenergy Research*, vol. 3, pp. 204–213, 2010.

17. Hielscher Ultrasound Technology, Biodiesel from algae using ultrasonication, 2013. Available at http://14w.hielscher.com/ultrasonics/algae_extraction_01.htm.

18. R. Eckelberry, "OriginOil—Algae harvesting, dewatering and extraction," in *World Biofuels Markets Congress*, Amsterdam, The Netherlands, March 16, 2010.

19. OriginOil, Single-step extraction, March 2012. Available at http://14w.originoil.com/ technology-old/single-step-extraction.html.

20. Oilgae, Extraction of algal oil by chemical methods, 2013. Available at http://14w. oilgae.com/algae/oil/extract/che/che.html.

21. P. Mercer and R. E. Armenta, "Developments in oil extraction from microalgae," *European Journal of Lipid Science and Technology*, vol. 113, pp. 539–547, 2011.

22. F. Soxhlet, "Die gewichtsanalytische Bestimmung des Milchfettes," *Polytechnisches Journal (Dingler's)*, vol. 232, pp. 461–465, 1879.

23. Z. Cohen and C. Ratledge, "Supercritical fluid extraction of lipids and other materials from algae," in *Single Cell Oils*, edited by Z. Cohen and C. Ratledge, AOCS Publishing, Urbana, IL, 2010. Chapter 14.

24. M. A. McHugh and V. J. Krukonis, *Supercritical Fluid Extraction: Principles and Practice*. Butterworths, Boston, 1986.

# 15 Thermochemical Conversion of Biomass

*Sunggyu Lee*

## CONTENTS

## 15.1 BIOMASS AND ITS UTILIZATION

### 15.1.1 DEFINITION OF THE TERM "BIOMASS"

The term "biomass" has been an important part of legislation enacted by Congress for many decades and has evolved over time, resulting in a variety of differing and sometimes conflicting definitions [1]. These definitions are critical to all parties engaged in the research, development, finance and application of biomass to produce energy. The term biomass is more generally defined as "different materials of biological origin that can be used as a primary source of energy" [2–5]. Alternately, biomass is defined as "plant materials and animal wastes used especially as a source of fuel" [6]. While these biomass definitions contain the generalized statements for the origins of the materials and/or their intended uses and applications, the definitions are not meant to provide the sufficient and necessary conditions for certain specific material to be classified, or qualified, as biomass.

Riedy and Stone (2010) explained the evolving nature of biomass definitions and analyzed its trend in the biomass-related legislation [1]. Based on the common definition that biomass is a biologically originated matter that can be converted into energy, more readily conceivable and common examples of biomass include food crops, nonfood crops for energy generation, crop residues, woody materials and by-products, animal wastes, and residues of biological fuel processing operation. Over the past decades, however, the term *biomass* has grown to encompass algae and algae processing residues, municipal solid wastes (MSWs), yard wastes, and food waste. The term still remains highly flexible and open to divergent interpretations, including specific inclusions and special exceptions, often based on a number of factors involving technoeconomic considerations, technological advances and new scientific findings, renewability and sustainability issues, environmental and climate change concerns, ecological issues, strategic directions of local and federal governments, regional economic strengths and weaknesses, and more. Simply put, biomass is a very broad term and encompasses a wide variety of matters. This is why the term *biomass* itself has been a part of modern legislation that is promulgated by the Congress. Legislation can have many purposes: to regulate, to authorize, to provide funds and incentives, to sanction, to grant, to proscribe, to declare, or to restrict. Using a globally generic definition of biomass in specific legislation would be not only grossly insufficient and inappropriate, but also potentially conflicting and controversial [1].

This book deals with conversion of biomass into biofuels and bioenergy, generally speaking. For the same reason described above, the individual chapters of this book are subdivided, based on the specific types of biomass and their associated transformation technologies, into the technologically categorized topics of corn ethanol, cellulosic ethanol, biodiesel, algae biodiesel, waste-to-energy, biomass pyrolysis and gasification, and so forth.

The discussion of biomass definitions lately has centered around the issues involving: (1) the types of forestry products considered eligible biomass sources, (2) the lands where biomass removal can occur, specifically Federal and Indian lands, and (3) the kinds and types of wastes that qualify as biomass, specifically, MSW and construction and demolition (C&D) debris [1].

### 15.1.2 Renewability and Sustainability of Biomass Feedstock

An aspect that is quite attractive in biomass utilization is its renewability that ultimately guarantees nondepletion of the resource. Considering all plants and plant-derived materials, all energy is originally captured, transformed, and stored via a natural process of photosynthesis. Strictly following the aforementioned definitions of biomass, it can be safely said that energy from biomass has been exploited by humans for a very long time in all geographical regions of the world. The combustion, or incineration, of biological substances such as woody materials and plant oils has long been exploited to provide warmth, illumination, and energy for cooking. It has been estimated that, in the late 1700s, approximately two-thirds of the volume of wood removed from the American forest was for energy generation [3]. Since wood was one of the only renewable energy sources readily exploitable at the time, its use continued to grow till the mid- to late-1800s, when petroleum was discovered and town gas infrastructure based on coal gasification was introduced. It was reported that during the 1800s, single households consumed an average of 70 to 145 m³ of wood annually for heating and cooking [7,8]. A small percentage of the rural communities in the United States still use biomass for these purposes. Countries like Finland use the direct combustion of wood for a nontrivial percentage of their total energy consumption [9]. Furthermore, Finland has spent significant research and development (R&D) efforts in biomass utilization programs and successfully developed a number of advanced biomass conversion technologies. Finland and the United States are not the only countries that use biomass consumption for supplementing their total energy usage. In fact, the percentage of biomass energy of the total energy consumption for a country is far greater in African nations and many other developing countries. An assessment by the World Energy Council (WEC) [10] reported that the 1990 biomass usage in all forms accounted for 1070 MTOE, which is approximately 12% of the global energy consumption of 8811 MTOE assessed for the same year. MTOE stands for metric tons of oil equivalent. In 2010, about 16% of global energy consumption came from renewables, of which about 10% was contributed from traditional biomass, which was mainly used for heating; 3.4% from hydroelectricity; and 2.8% from so-called new renewables which included small hydro, modern biomass, wind, solar, geothermal, and biofuels [11]. The last category of new renewables has been very rapidly growing based on the development of advanced technologies as well as the global fear of depletion of conventional petroleum fuel.

On a larger scale, biomass is currently the primary fuel in the residential sector in many developing countries. Their biomass resources may be in the form of wood, charcoal, crop waste, or animal waste. For these countries, its most critical function of biomass fuel is for cooking, with the other principal use being lighting and heating. The dependence on biomass for critical energy supply for these countries is generally decreasing, whereas that for industrialized countries is more strategically targeted for the new generation of biomass energy. According to REN21 [11], the top five nations in terms of the existing biomass power capacity in 2011 are the United States, Brazil, Germany, China, and Sweden, in the order of one to five. The top two nations of this list are also the top two nations in bioethanol production, not coincidentally. In other words, the biomass power category has, so far, been propelled and dominated mostly by the bioethanol transportation fuel sector.

### 15.1.3  WOODY BIOMASS AND ITS UTILIZATION

Of diverse biomass resources, woody biomass is of particular interest for biomass energy. Woody biomass is used to produce bioenergy and a variety of bio-based products including lumber, composites, pulp and paper, wooden furniture, building components, round wood, ethanol, methanol and chemicals, and energy feedstocks including firewood.

The U.S. national *Energy Policy Act of 2005 (EPAct 2005)* recognized the importance of a diverse portfolio of domestic energy. This policy outlined 13 recommendations that are designed to increase America's use of renewable and alternative energy. One of these recommendations directed the Secretaries of the Interior and Energy to reevaluate access limitations to federal lands in order to increase renewable energy production, such as biomass, wind, geothermal, and solar. The Departments of Agriculture and Interior are jointly implementing the *National Fire Plan (NFP)*, the President's Healthy Forests Initiative, the Healthy Forest Restoration Act, and the Tribal Forest Protection Act of 2004 to address the risk of catastrophic wildland fires, reduce their impacts on communities, assure firefighting capabilities for the future, and improve forest and rangeland health on federal lands by thinning biomass density. The NFP includes five key points: (1) firefighting preparedness, (2) rehabilitation and restoration of burned areas, (3) reduction of hazardous fuels, (4) community assistance, and (5) accountability [12].

On June 18, 2003, the U.S. Departments of Energy, Interior, and Agriculture jointly announced an initiative to encourage the use of woody biomass from forest and rangeland restoration and hazardous fuels treatment projects. The three Departments signed a Memorandum of Understanding (MOU) on Policy Principles for *Woody Biomass Utilization for Restoration and Fuel Treatment on Forests, Woodlands, and Rangelands*, supporting woody biomass utilization as a recommended option to help reduce or offset the cost and increase the quality of the restoration of hazardous fuel reduction treatments [12,13].

One of the gateway process technologies for bioenergy generation from woody biomass is gasification, whose resultant product is biomass synthesis gas, also known as biomass syngas or biomass gas. The biomass syngas is similar in nature and composition to coal-based or natural gas (NG)-based syngas, while differences largely originated from the source-specific properties. Similarly to the syngas generated via coal gasification or NG reformation, biomass syngas is also rich in hydrogen, carbon monoxide, carbon dioxide, and methane, and as such this syngas can be used as building-block chemicals [14] for a variety of synthetic fuels and petrochemicals, and also as a feedstock for electric power generation.

### 15.1.4 Thermal and Thermochemical Conversion of Biomass

There are five thermal or thermochemical approaches that are commonly used to convert biomass into an alternative fuel/energy, namely, direct combustion, gasification, liquefaction, pyrolysis, and partial oxidation. These five modes of conversion are also applicable to the conventional utilization of coal. When biomass is heated under an oxygen-deficient environment, it generates bio-oil and biogas that consists primarily of carbon dioxide, methane, and hydrogen. When biomass is gasified at a higher temperature with an appropriate gasifying medium, synthesis gas is produced, which has similar compositions as coal syngas. This syngas can be directly

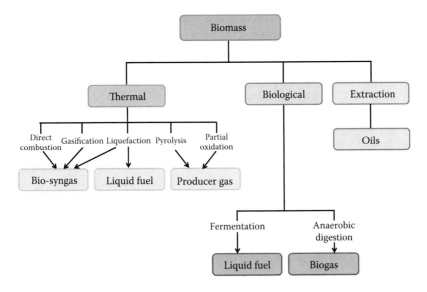

**FIGURE 15.1**  Conversion of biomass to alternative fuels.

burned or further processed for other gaseous or liquid fuel products. In this sense, thermal or thermochemical conversion of biomass is very similar to that of coal [5]. Figure 15.1 shows a variety of process options of biomass treatment and utilization as well as their resultant alternative fuel products.

Of a variety of thermochemical conversion options of biomass, this chapter is mainly focused on fast pyrolysis and gasification of biomass due to their technological significance, while other options for thermochemical conversion are discussed whenever deemed relevant.

## 15.2  ANALYSIS AND COMPOSITION OF BIOMASS

### 15.2.1  SIMILARITIES AND DIFFERENCES BETWEEN BIOMASS AND COAL AS FEEDSTOCK

Even though biomass gasification has long been practiced on a variety of scales with and without modern scientific understanding, the subject itself has greatly benefitted from coal science and technology, which has been researched far more in-depth [15]. Most of the scientific tools developed for and/or applied to the coal technology are more or less relevant to the biomass utilization technology, and they include the analytical methods, solids handling technology, chemical reaction pathways, reactor designs and configurations, process integration, waste heat recovery and energy integration, gas cleanup, product separation, safety practice and measures, and much more. However, the compositional differences between coal and biomass feedstocks as well as their impacts on the processing technologies must be clearly understood for full and beneficial exploitation of the advances and innovations made in the coal processing technology. The differences between biomass and coal feedstocks, which are quite significant, are summarized below.

1. The hydrogen content in biomass is significantly higher than that of coal. Coal is a very mature product of a lengthy and slow coalification process whose principal chemical reaction is carbonization, while biomass is not. The coal rank basically indicates the degree of carbonization that has progressed, and a higher-rank coal is a petrologically older coal than a lower-rank coal. Therefore, the H/C ratio of coal is much lower than that of biomass. The H/C ratio of a higher-rank coal is also lower than that of a lower-rank coal.

2. A higher H/C ratio of biomass feedstock makes it generally more reactive than coal, for the conventional transformational process treatments. Roughly speaking, it is easier to liquefy biomass than coal or easier to obtain liquid hydrocarbon products from biomass than from coal.

3. Biomass typically has much higher moisture content than coal. This statement is applicable to both forms of moisture, that is, equilibrium moisture and chemically bound moisture. Among various ranks of coal, lignite, the lowest-rank coal, which is also the youngest, has the highest moisture content. Therefore, among all ranks of coal, lignite may be considered the closest to biomass in terms of both proximate analysis and ultimate analysis. Due to its higher moisture content, the heating value of biomass is inherently lower than that of coal.

4. Biomass contains a significantly higher oxygen content than coal due to its highly oxygenated molecular structures of carbohydrates (or polysaccharides), cellulose, glycerides, fatty acids, and so forth. Weathered coals, that is, coal exposed to the atmosphere after mining, shows an increased level of oxygen content compared with freshly mined coal. However, the oxygen content of weathered coal is still far lower than that of typical biomass. Due to the high oxygen content of biomass fuel, its heating value is substantially lower than that of coal. It is also imperative that deoxygenation of biomass is an important technological option for biomass-to-fuel conversion.

5. Coal contains 0.5–8 wt.% sulfur (S), whereas biomass has little or no sulfur content. Coal with lower than 1 wt.% sulfur may be roughly classified as low-sulfur coal, while coal with higher than 3 wt.% sulfur may be considered high-sulfur coal. No such designation or classification is necessary for typical biomass. Sulfurous compounds, if not removed, in coal syngas affect the downstream processing severely by poisoning the catalysts and also by triggering corrosion on metallic parts and equipment. In this regard, biomass is considered a *sulfur-free* raw material, which serves as a great advantage. Sulfurous compounds in coal syngas typically include $H_2S$, carbonyl sulfide (COS), and mercaptans (R-SH), whose prevailing abundance depends largely on the gasifying environment as well as the feed coal composition. Furthermore, coal sulfur is subdivided largely into three different forms of sulfur, that is, *pyritic sulfur, organic sulfur, and sulfatic sulfur* [14]. However, such a subcategorization for forms of sulfur is unnecessary for biomass. Owing to its low sulfur content, biomass can be used as a blending feedstock for coal-fired power plants.

6. Alkali metals such as sodium (Na) and potassium (K) as well as low-boiling heavy metals such as lead (Pb) and cadmium (Cd) are typically present in raw biomass syngas [16]. This is not as severe for coal syngas, and the trace element problems with coal syngas are more source specific. Due to the trace mineral components in the biomass syngas, downstream processing of biomass syngas, in particular, catalytic processing, requires rather comprehensive purification pretreatment of feed syngas or use of robust and poison- and fouling-resistant catalysts. Therefore, full and thorough analysis of trace minerals in biomass syngas or bioliquid must be obtained before a process technology is designed.

7. The fuel analysis of both coal and biomass is represented by proximate analysis and ultimate analysis. Proximate analyses and ultimate analyses of a variety of biomass samples found in the literature are presented in Tables 15.1 and 15.2, respectively.

8. Due to the high abundance of moisture, high oxygen content, and noncombustible impurities in biomass, the heating value of biomass is typically much lower than that of coal. The energy density of biomass feedstock on a volume basis is therefore substantially inferior to that of coal.

9. Biomass has substantially higher volatile matter (VM) content than coal, while it has much lower fixed carbon (FC) content than coal. Therefore, a large amount of hydrocarbon species can be extracted/obtained from biomass simply via devolatilization or pyrolysis, whereas devolatilization or pyrolysis of coal generates a high amount of char.

## TABLE 15.1
### Proximate Analysis of Biomass Species (Dry Basis)

| | Fixed Carbon, % | Volatile Matter, % | Ash, % | HHV, kJ/kg (Measured) | Data Source |
|---|---|---|---|---|---|
| Cotton shell briquette | 17.1 | 77.8 | 5.1 | 19,055 | [17] |
| Blockwood | 14.6 | 83.3 | 2.1 | 18,261 | [17] |
| Plywood | 21.8 | 74.2 | 4.0 | 19,720 | [17] |
| Corn stover | 17.6 | 78.7 | 3.7 | 17,800 | [18] |
| Corn cob | 18.5 | 80.1 | 1.4 | 18,770 | [18] |
| Wood chips | 23.5 | 76.4 | 0.1 | 19,916 | [19] |
| Hazelnut shell | 28.3 | 69.3 | 1.4 | 19,300 | [18] |
| Redwood | 19.9 | 79.7 | 0.4 | 20,720 | [20] |
| Softwood | 28.1 | 70.0 | 1.7 | ~20,000 | [18] |
| Eucalyptus | 21.3 | 75.4 | 3.3 | 18,640 | [17] |
| Douglas fir | 25.8 | 73.0 | 1.2 | 22,098 | [21] |
| Walnut | 20.8 | 78.5 | 0.7 | 19,967 | [17] |
| Wheat straw | 23.5 | 63.0 | 13.5 | ~17,000 | [17] |
| Rice straw (ground) | 16.2 | 68.3 | 15.5 | 15,614 | [17] |

*Note:* All percentages are in dry weight percent. HHV is higher heating value in kJ/kg.

**TABLE 15.2**

**Ultimate Analysis of Biomass Species (Dry Basis)**

| | Elemental Analysis (Dry wt.%) | | | | | Data Source |
|---|---|---|---|---|---|---|
| | **C** | **H** | **O** | **N** | **S** | |
| Cotton shell briquette | – | – | – | – | – | [17] |
| Blockwood | 46.9 | 6.07 | 43.99 | 0.95 | 0 | [17] |
| Plywood | – | – | – | – | | [17] |
| Corn stover | – | – | – | – | – | [18] |
| Corn cob | 46.58 | 5.87 | 45.46 | 0.47 | 0.01 | [18] |
| Wood chips | 48.1 | 5.99 | 45.74 | 0.08 | 0 | [19] |
| Hazelnut shell | 52.9 | 5.6 | 42.7 | 1.4 | - | [18] |
| Redwood | 50.64 | 5.98 | 42.88 | 0.05 | 0.03 | [20] |
| Softwood | 52.1 | 6.1 | 41 | 0.2 | - | [18] |
| Eucalyptus | 46.04 | 5.82 | 44.49 | 0.3 | 0 | [17] |
| Douglas fir | 56.2 | 5.9 | 36.7 | 0 | 0 | [21] |
| Walnut | 48.2 | 6.25 | 43.24 | 1.61 | – | [17] |
| Wheat straw | 45.5 | 5.1 | 34.1 | 1.8 | – | [17] |
| Rice straw (ground) | – | – | – | – | – | [17] |

*Note:* All percentages are in dry weight percent.

10. Biomass is generally composed of softer organic materials, and its grindability or pulverizability is poor using common size-reduction equipment. Considering the irregular shapes and nonuniform compositions of untreated biomass components, cost-effective size reduction for manageable transportation as well as continuous reactor feeding often becomes a technological challenge. Pretreatment of biomass feedstock is usually required for industrialized utilization.

11. While both biomass gasification and coal gasification encounter varying degrees of tar formation during thermochemical transformation, the severity of tar formation is typically more significant with biomass gasification. While tar is collectively a carcinogenic species and environmentally hazardous if not contained, it condenses at reduced temperatures, thereby blocking and clogging pipelines and valves as well as fouling process equipments and parts. Therefore, tar removal and/or conversion becomes an important part of biomass processing technology.

## 15.2.2 Analysis of Biomass

Similar to coal, the *proximate analysis* and *ultimate analysis* of specific biomass material provide very valuable information about the biomass feedstock. This compositional information provides the science and engineering information needed to identify or determine the fuel heating value, ash amount projected, maximum achievable gasification and liquefaction efficiency, moisture content of feedstock, VM content, predicted behavior of the feedstock in a processing environment, contents of trace minerals, and much more. The proximate analysis is a procedure for determination, by prescribed methods, of moisture (MO), VM, FC, and ash. The amount of FC is determined by the difference after taking MO, VM, and ash from the total. The term *proximate analysis* involves neither determination of quantitative amounts of chemical elements nor determination other than those categorically named or prescribed. The group of analyses involved in proximate analysis is defined in ASTM D3172. On the other hand, the ultimate analysis is a procedure of the determination of the elemental composition of the organic portion of carbonaceous materials, as well as the total ash and moisture. The ultimate analysis is also referred to as *elemental analysis*. And this analysis is also determined by prescribed testing methods.

An extensive tabulation of both proximate and ultimate analysis data on over 200 biomass species was presented in Channiwala's PhD dissertation [22]. Some representative values of proximate analyses and ultimate analyses of a variety of biomass species, as found from the literature sources, are presented in Tables 15.1 and 15.2, respectively. For comparison purposes, the classification of coal and typical analysis is also tabulated in Table 15.3.

Comparing between the analyses of coal and biomass, the following generalized statements can be made:

## TABLE 15.3
## Coal Classification and Analysis

| | Average Analysis—Dry and Ash-Free (daf) Basis | | | | | | |
|---|---|---|---|---|---|---|---|
| | Volatile Matter (wt.%) | Hydrogen (wt.%) | Carbon (wt.%) | Oxygen (wt.%) | Heating Value (kJ/kg) | $\dfrac{C}{H}$ | $\dfrac{C+H}{O}$ |
| Anthracite | | | | | | | |
| Meta- | 1.8 | 2.0 | 94.4 | 2.0 | 34,425 | 46.0 | 50.8 |
| Anthracite | 5.2 | 2.9 | 91.0 | 2.3 | 35,000 | 33.6 | 42.4 |
| Semi | 9.9 | 3.9 | 91.0 | 2.8 | 35,725 | 23.4 | 31.3 |
| Bituminous | | | | | | | |
| Low volatile | 19.1 | 4.7 | 89.9 | 2.6 | 36,260 | 19.2 | 37.5 |
| Med volatile | 26.9 | 5.2 | 88.4 | 4.2 | 35,925 | 16.9 | 25.1 |
| High volat. A | 38.8 | 5.5 | 83.0 | 7.3 | 34,655 | 15.0 | 13.8 |
| High volat. B | 43.6 | 5.6 | 80.7 | 10.8 | 33,330 | 14.4 | 8.1 |
| High volat. C | 44.6 | 4.4 | 77.7 | 13.5 | 31,910 | 14.2 | 6.2 |
| Subbituminous | | | | | | | |
| Subbitum. A | 44.7 | 5.3 | 76.0 | 16.4 | 30,680 | 14.3 | 5.0 |
| Subbitum. B | 42.7 | 5.2 | 76.1 | 16.6 | 30,400 | 14.7 | 5.0 |
| Subbitum. C | 44.2 | 5.1 | 73.9 | 19.2 | 29,050 | 14.6 | 4.2 |
| Lignite | | | | | | | |
| Lignite A | 46.7 | 4.9 | 71.2 | 21.9 | 28,305 | 14.5 | 3.6 |

*Source:* Lee, S., *Alternative Fuels,* Taylor & Francis, Philadelphia, PA, 1996.

**TABLE 15.4**

**Standardized Testing Procedure for Biomass Fuels**

| Test | Standardized Procedure | Desired Units |
|------|------------------------|---------------|
| C | ASTM D5373 | Weight% |
| H | ASTM D5373 | Weight% |
| N | ASTM D5373 | Weight% |
| Cl | ASTM D3761 | mg/kg |
| S | ASTM D4239 | Weight% |
| Proximate | ASTM D3172 | Weight% |
| Moisture | ASTM D2013 | Weight% |
| Ash | ASTM D5142 | Weight% |
| Heat of combustion | ASTM D5865 | kJ/kg or Btu/lb |

1. Biomass has a very high oxygen (O) content, which is the second most abundant atomic species present in biomass and is nearly as much as the carbon (C) content. However, the oxygen content in coal is much lower than the carbon content, and this trend is even more noticeable with higher-rank coals. The higher the rank of a coal, the lower its oxygen content. It may be said that deoxygenation, or oxygen rejection, is an important part of the petrological process of coalification.
2. Due to the high oxygen content in biomass, the heating value of biomass is much lower than that of coal. Bio-oil derived from biomass also has a high oxygen content, which makes the oil more corrosive to metallic parts and piping. Therefore, efficient use of biomass as fuel or a fuel precursor will involve a certain level of oxygen rejection, that is, deoxygenation, in its process scheme.
3. The H/C ratio of biomass is substantially higher than that of coal. Owing to the high H/C ratio, the reactivity of biomass is generally higher than that of coal, and its processability is also better than coal's.
4. Among various ranks of coal, lignite is the closest to biomass in a number of properties, including its high moisture content, high oxygen content, low carbon content, and low heating value. As such, lignite has often been considered as a cofed companion fuel with biomass.

The standardized analysis of biomass fuel is conducted following the ASTM Standards, and Table 15.4 shows the list of these codes for specific analyses.

## 15.2.3 THERMOCHEMICAL CONVERSION OF BIOMASS

Thermochemical treatment of biomass can convert biomass into solid, liquid, and gaseous fuel products whose compositional distribution is governed by the imposed process treatment conditions. The solid product is usually char (or bio-char), the liquid product is bio-oil, and the gaseous product is bio-syngas. The process also involves formation of unwanted by-products of tar, which must be removed or converted.

**TABLE 15.5**
**Biomass Treatment Processes and Their Product Distribution**

| | Process Treatment | | | Typical Product Compositions | | |
| --- | --- | --- | --- | --- | --- | --- |
| | Temperature (L, M, H) | Residence Time | Air or O$_2$ (Y, N) | Solid Char | Liquid Bio-Oil | Gas or Syngas |
| Carbonization | Low | Long | N | 35% | 30% | 35% |
| Fast pyrolysis | Medium | Short | N | 12% | 75% | 13% |
| Gasification | High | Long | Y | 10% | 5% | 85% |

The thermochemical conversion process involves heating of the biomass feedstock, which triggers a very large number of parallel and consecutive reactions including devolatilization of VM, pyrolytic decomposition of hydrocarbons and other carbonaceous matter, gas–solid gasification reactions, coke and char formation, tar and its precursor formation, and more. Simply speaking, depending upon the processing temperature, type of reactor vessel, and reactor residence time, thermochemical treatment of biomass can be regrouped into three basic types of process treatment, namely, *carbonization*, *fast pyrolysis*, and *gasification*. As shown in Table 15.5, the principal or intended product of fast pyrolysis is a liquid fuel, whereas the desired product of gasification is a gaseous fuel [23]. The primary product of carbonization is solid char, while other types of products, both oil and gas, are also produced in significant amounts.

Even though it is not listed in Table 15.5, indirect liquefaction via bio-syngas route is also a viable option for liquid fuel production, as well demonstrated in the fields of coal and NG syngas [5,14,24,25]. As explained in the chapter on Coal Liquefaction, indirect liquefaction goes through two stages of process treatment, namely, gasification followed by liquefaction, by which liquid hydrocarbon fuels such as methanol, dimethyl ether (DME), higher alcohols, gasoline, diesel, and jet fuel are synthesized using the syngas produced during the gasification stage. In this technology, biomass syngas is a thermochemical intermediate for the next-stage synthesis of liquid fuel.

### 15.2.4   ANALYSIS OF BIOMASS FEEDSTOCK AND PRODUCT COMPOSITIONS

Fast pyrolysis of biomass generates a wide variety of organic and inorganic chemical compounds, and the product compositions vary significantly depending upon the types of feedstocks as well as the process treatment conditions the biomass is subjected to. Therefore, studies of process modeling and technoeconomic analysis are often carried out using model substances carefully chosen for the specific process and typical feedstock [26]. The analysis of corn stover samples used by Mullen et al. [27] for their fast pyrolysis study is presented in Table 15.6, as an example of the compositional analysis of biomass feedstock. They carried out the fast pyrolysis in a bubbling fluidized bed (BFB) of quartz sand at a temperature of 500°C.

Inorganic elemental composition of corn stover used for the aforementioned pyrolysis determined by x-ray fluorescence (XRF) is given in Table 15.7. Also, compared in the same table are the XRF analysis data for corn cobs, which were also tested for fast

**TABLE 15.6**

**Ultimate and Proximate Analysis of Corn Stover by Mullen et al.**

| Ultimate Analysis (Dry Basis) | | Proximate Analysis (Wet Basis) | |
|---|---|---|---|
| Element | Mass% | Ingredient | Mass% |
| Carbon (C) | 46.60 | Moisture | 25.0 |
| Hydrogen (H) | 4.99 | Volatile matter | 52.8 |
| Oxygen (O) | 40.05 | Fixed carbon | 17.7 |
| Nitrogen (N) | 0.79 | Ash | 4.5 |
| Sulfur (S) | 0.22 | | |
| Ash | 6 | | |

*Source:* Mullen, C.A. et al., *Biomass Bioenergy*, 34, 2010.

**TABLE 15.7**

**Inorganic Elemental Compositions of Corn Cobs and Corn Stovers by XRF**

| Inorganic Element | Corn Cobs (g/kg or 1000 ppm) | Corn Stover (g/kg or 1000 ppm) |
|---|---|---|
| Si | 5.33 | 27.9 |
| Al | 0.18 | 5.09 |
| Fe | 0.08 | 2.35 |
| Ca | 0.23 | 3.25 |
| Mg | 0.55 | 2.34 |
| Na | 0.10 | 0.23 |
| K | 10.38 | 4.44 |
| Ti | 0.003 | 0.37 |
| Mn | 0.01 | 0.98 |
| P | 1.11 | 2.15 |
| Ba | 0.11 | 0.02 |
| Sr | 0.002 | 0.005 |
| S (inorganic) | 0.14 | 0.05 |

*Source:* Mullen, C.A. et al., *Biomass Bioenergy*, 34, 2010.

pyrolysis by Mullen et al. [27]. As can be seen, the elemental compositions of corn cob and corn stover are quite different. K was the most abundant element in corn cob, while Si was the most abundant element in corn stover. High levels of K and P in both samples are expected, while the high levels of Ca, Mg, Al, Fe, and Mn in corn stover are noteworthy. Mineral matter in the biomass feedstocks can reappear as contaminants or trace elements in bio-oils and bio-syngas, which can potentially affect the catalytic activity of the downstream processing by fouling and/or poisoning.

The yield data of the U.S. Department of Agriculture (USDA) fast pyrolysis of corn stover by Mullen et al. [27] are shown in Table 15.8. The pyrolysis product distribution in terms of the product phases was bio-oil 61.7%, bio-char 17.0%, and noncondensable gas (NCG) 21.9%. As explained earlier and also summarized in

## TABLE 15.8
## Product Analysis of Fast Pyrolysis of Corn Stover

| Gaseous Compounds | Vol.% |
|---|---|
| $CO_2$ | 40.3 |
| CO | 51.6 |
| $H_2$ | 2.0 |
| $CH_4$ | 6.0 |
| HHV (MJ/kg) | 6.0 |
| **Bio-Oil Compounds** | **Mass%** |
| C | 53.97 |
| H | 6.92 |
| N | 1.18 |
| S | <0.05 |
| O | 37.94 |
| Ash | <0.09 |
| HHV (MJ/kg) | 24.3 (dry) |
| **Bio-Char** | **Mass%** |
| C | 57.29 |
| H | 2.86 |
| N | 1.47 |
| S | 0.15 |
| O | 5.45 |
| Ash | 32.78 |
| HHV (MJ/kg) | 21.0 |

Table 15.5, the principal product of fast pyrolysis of biomass, that is, corn stover in this example, is bio-oil, while the by-products are NCG and bio-char.

From the product compositions, the following observations are of significance:

1. The gaseous effluent of fast pyrolysis has a heating value of only 6.0 MJ/kg. The gas composition is dominated by carbon oxides (CO and $CO_2$), followed by methane and hydrogen.
2. High levels of oxygen in the effluent gas show that the gaseous effluent served at least as an outlet for deoxygenation of biomass.
3. Bio-oil also showed a very high level of oxygen, which is the reason for its low heating value, compared to other liquid hydrocarbons. In order to enhance the fuel quality of bio-oil as well as to enhance the fast pyrolysis process, a systematic and efficient rejection of oxygen from the products' molecular structures, that is, deoxygenation, would become crucially important.
4. Bio-char showed a heating value nearly as high as that of bio-oil, even though it contained a high level of ash.
5. Bio-char showed a high C/H ratio, which is indicative of its lack of volatile hydrocarbons. Thus, bio-char is a useful by-product of the fast pyrolysis process of biomass.

## 15.3   CHEMISTRY OF BIOMASS GASIFICATION

*Gasification* is a thermochemical conversion process that transforms macromolecular carbonaceous matter contained in fossil fuels and biological substances into simpler gaseous molecular products. The gaseous products from the gasification reaction are called synthesis gas, syngas, or producer's gas. Depending upon the original carbonaceous feedstocks, the syngas may be further labeled as coal syngas, NG syngas, or *biomass syngas.* Gasification of biomass usually takes place at an elevated temperature with the aid of a gasifying medium, which may be regarded as a gaseous reactant. Since gasification involves both heat and chemical(s) that induce concurrent thermal decomposition and chemical reactions, gasification process is classified as a *thermochemical conversion* process.

### 15.3.1   Chemical Reactions Taking Place during Biomass Gasification

Typical biomass gasification takes place in the presence of injected air (or oxygen) and steam under high pressure at an elevated temperature, typically T > 850°C. In this regard, typical biomass gasification is very similar to advanced coal gasification process technologies [5,28,29]. The chemical reactions taking place in a biomass gasifier are very complex, and they include (1) pyrolytic decomposition of hydrocarbons and oxygenated organics such as carbohydrates (or saccharides) and cellulose, (2) further decomposition of fragmented hydrocarbons (of reduced molecular weights) into lighter hydrocarbons and oxygenates such as aldehydes, (3) recombination reactions involving methylene and methyl radicals as well as unsaturated hydrocarbons, (4) partial oxidation of hydrocarbons and oxygenates, (5) steam gasification of hydrocarbons and oxygenates, (6) water gas shift (WGS) reaction, (7) formation of polycyclic aromatic hydrocarbons (PAHs) and potential coking precursors, (8) formation of tar, and (9) carbon dioxide gasification of carbonaceous materials, and more. Some of the representative reactions are described below.

$$C_xH_y \rightarrow C_aH_b + C_cH_d + eH_2$$

$$C_xH_y \rightarrow C_fH_g + hCH_4 + jH_2$$

$$C_uH_vO_w \rightarrow C_kH_lO_{w1} + C_mH_nO_{w2} + pH_2O + qH_2 + rCO_2 + sCO$$

$$C_{x1}H_{y1} \rightarrow C_{f1}H_{g1} + k(CH_2)$$

$$(CH_2) + H_2 \rightarrow CH_4$$

$$(CH_2) + (CH_2) \rightarrow C_2H_4$$

$$C_xH_y + \left(\frac{x}{2} + \frac{y}{4}\right)O_2 \rightarrow xCO + \frac{y}{2}H_2O$$

$$C_uH_vO_w + \left(\frac{u}{2} + \frac{v}{4} - \frac{w}{2}\right)O_2 \rightarrow uCO + \frac{v}{2}H_2O$$

$$C_xH_y + xH_2O \rightarrow xCO + \left(x + \frac{y}{2}\right)H_2$$

$$C_uH_vO_w + (u-w)H_2O \rightarrow uCO + \left(\frac{v}{2} + u - w\right)H_2$$

$$CO + H_2O \leftrightarrow H_2 + CO_2$$

$$C_xH_y + xCO_2 \rightarrow 2xCO + \frac{y}{2}H_2$$

$$C_uH_vO_w + uCO_2 \rightarrow 2uCO + \left(\frac{v}{2} - w\right)H_2 + wH_2O$$

The first five reactions represent pyrolytic decomposition reactions of hydrocarbons and oxygenates, which provide some explanation for the formation of methane and lighter hydrocarbon species. The last five reactions explain the formation of carbon oxides and hydrogen, principal ingredients of biomass syngas. One can also notice that the above reactions are analogous to the four classical gasification reactions of carbon and concurrent WGS reaction, as shown below [14].

$$C_s + H_2O \rightarrow CO + H_2 \text{ [steam gasification]}$$

$$C_s + CO_2 \rightarrow 2CO \text{ [Boudouard reaction]}$$

$$C_s + 2H_2 \rightarrow CH_4 \text{ [hydrogasification]}$$

$$C_s + O_2 \rightarrow CO/CO_2 \text{ [partial oxidation]}$$

$$CO + H_2O \leftrightarrow H_2 + CO_2 \text{ [WGS reaction]}$$

where $C_s$ denotes carbon on the solid surface. All other species in the stoichiometric reactions are gaseous species, and the first four reactions are basically gas–solid reactions. The WGS reaction takes place both homogeneously in the gas phase as well as heterogeneously on solid surfaces.

The last three reactions, as written, are exothermic, whereas the first two are endothermic at their typical operating conditions. The WGS reaction can proceed in either the forward or reverse direction, depending upon the temperature and imposed/developed reaction environment. The forward WGS reaction is exothermic, whereas the reverse WGS (RWGS) reaction is endothermic. The forward WGS reaction is thermodynamically favored, that is, $K_p > 1$, for $T < 814°C$.

Chemical equilibrium constants for a wide range of temperatures for selected chemical reactions that are of significance to the gasification of carbon are listed in Table 15.9. While the values are for reactions of carbon, not biomass or coal char, they still provide useful information for reactions involving carbonaceous matter. As the C/H ratio of solid materials increases or the number of carbon atoms in a hydrocarbon molecule increases, their thermodynamic equilibrium values become closer to those of carbon reactions. Furthermore, if biomass is pretreated before any gasification reactions, then the equilibrium values listed for carbon reactions in the table would be more relevant and closer to the actual values.

The temperature where $K_p = 1$, that is, $\ln K_p = 0$, has some extra significance, by indicating the general location of the chemical equilibrium shift. The temperatures where $K_p = 1$ for steam gasification, Boudouard reaction, hydrogasification, and WGS reaction are 947K (674°C), 970K (697°C), 823K (550°C), and 1087K (814°C), respectively. For example, it may be said that for the steam gasification of carbon to proceed in the forward direction, the gasification temperature must be higher than 674°C. It must be noted that all practical gasification reactions are always carried out at $T > 850°C$ due to the kinetic reasons based on the reaction rates. Among the reactions listed in Table 15.9, the temperature-dependent variation of the equilibrium constant is by far the weakest for the WGS reaction, thus exhibiting an easily reversible nature of the chemical equilibrium for a very wide range of temperatures. This is the reason why the WGS reaction equilibrium becomes a player in nearly all syngas reaction systems under widely varying process conditions.

When the coal gasification reaction is explained or modeled, most technologists denote and simplify coal more or less as carbon, that is, $C_{(s)}$, based on the fact that the hydrogen content of coal is much lower than that for most hydrocarbons. However, such practice in the case of biomass gasification would become an oversimplification, since the oxygen and hydrogen contents in biomass feedstock are much higher than those of high-rank coal. The abundance of oxygenated functional groups such as hydroxyl (–OH) groups in biomass makes most decomposition and transformation reactions proceed much more easily, compared to coal.

### 15.3.1.1 Pyrolysis or Thermal Decomposition

Pyrolysis or thermal decomposition is the molecular breakdown of organic materials such as hydrocarbons via cleavages of chemical bonds at elevated temperatures without the involvement of oxygen or air. Typical chemical bond cleavages during pyrolysis involve the C–C and C–H bonds at most operating temperatures, while it is substantially more difficult to break double bonds of C=C and C=O at most practical conditions. As can be imagined by the chemical bonds that are typically broken during pyrolysis, the pyrolysis reaction starts at a temperature as low as 150°C–200°C, where the intrinsic reaction rate is very slow and the extent of reaction is far from

**TABLE 15.9**
**Chemical Equilibrium Constants for Carbon Reactions**

| | | | | $\ln K_p$ | | | |
|---|---|---|---|---|---|---|---|
| T, K | 1/T, K$^{-1}$ | C + 1/2 O$_2$ = CO | C + O$_2$ = CO$_2$ | C + H$_2$O = CO + H$_2$ | C + CO$_2$ = 2 CO | CO + H$_2$O = CO$_2$ + H$_2$ | C + 2H$_2$ = CH$_4$ |
| 300 | 0.003333 | 23.93 | 68.67 | −15.86 | −20.81 | 4.95 | 8.82 |
| 400 | 0.0025 | 19.13 | 51.54 | −10.11 | −13.28 | 3.17 | 5.49 |
| 500 | 0.002 | 16.26 | 41.26 | −6.63 | −8.74 | 2.11 | 3.43 |
| 600 | 0.001667 | 14.34 | 34.40 | −4.29 | −5.72 | 1.43 | 2.00 |
| 700 | 0.001429 | 12.96 | 29.50 | −2.62 | −3.58 | 0.96 | 0.95 |
| 800 | 0.00125 | 11.93 | 25.83 | −1.36 | −1.97 | 0.61 | 0.15 |
| 900 | 0.001111 | 11.13 | 22.97 | −0.37 | −0.71 | 0.34 | −0.49 |
| 1000 | 0.001 | 10.48 | 20.68 | 0.42 | 0.28 | 0.14 | −1.01 |
| 1100 | 0.000909 | 9.94 | 18.80 | 1.06 | 1.08 | −0.02 | −1.43 |
| 1200 | 0.000833 | 9.50 | 17.24 | 1.60 | 1.76 | −0.16 | −1.79 |
| 1300 | 0.000769 | 9.12 | 15.92 | 2.06 | 2.32 | −0.26 | −2.1 |
| 1400 | 0.000714 | 8.79 | 14.78 | 2.44 | 2.80 | −0.36 | −2.36 |

*Source:* Walker, P.L. et al., "Gas reactions in carbon," in *Advances in Catalysis*, Eley, D.D. et al., eds., Academic Press, New York, 1959.

completion in any reasonable time. Most practical pyrolysis of hydrocarbons without using any catalyst is conducted at a temperature higher than 400°C. Pyrolysis involves the concurrent change of chemical compositions and physical phases, and the reaction process is irreversible.

If high-molecular-weight hydrocarbons are pyrolyzed in an oxygen-deprived environment at a temperature of 400°C–650°C, lighter hydrocarbons (with reduced carbon numbers and lower molecular weights), hydrogen, and solid char typically would be formed. Lighter hydrocarbons nearly always involve methane ($CH_4$), ethylene ($C_2H_4$), ethane ($C_2H_6$), and other fragmented hydrocarbons, of which methane is most dominant. Depending upon the pyrolysis conditions, liquid-range hydrocarbons, $C_4$–$C_{15}$, are also obtained. Char formation is believed to be via a route similar to the formation of PAHs, which involves polymerization of highly reactive free radicals of fragmented hydrocarbons and unsaturated hydrocarbon species. Hydrogen formation during pyrolysis is via cleavage reactions of C–H chemical bonds of the original and intermediate hydrocarbon molecules, and some of the hydrogen molecules formed in such a manner are recombined with methyl radicals and ethylene, thereby producing methane and ethane. As illustrated, pyrolysis of hydrocarbons can yield materials of all three different phases (i.e., solid, liquid, and gas) as its end products depending upon the treatment conditions. Further, the actual number of chemical reactions involved and the number of final and intermediate chemical species are countless. Therefore, pyrolysis collectively represents a class of chemical reactions taking place as thermochemical decomposition and concurrent side reactions. A more generalized chemical reaction equation for hydrocarbon pyrolysis may be written as

$$C_aH_b \rightarrow cCH_4 + C_dH_e + C_fH_g + hH_2$$

$$a = c + d + f$$

$$b = 4c + e + g + 2h$$

$$C_uH_vO_w \rightarrow C_kH_lO_{w1} + C_mH_nO_{w2} + pH_2O + qH_2 + rCO_2 + sCO$$

$$u = k + m + r + s$$

$$v = l + n + 2p + 2q$$

$$w = w1 + w2 + p + 2r + s$$

In the first reaction expression, methane is explicitly written on the product side, since methane is always a major hydrocarbon product species of hydrocarbon pyrolysis.

Biomass chemical compounds are heavily oxygenated, and biomass also contains a high level of moisture. Therefore, thermal decomposition or pyrolysis of biomass also generates carbon oxides, in addition to the aforementioned pyrolysis products of condensable hydrocarbons, oxygenates, methane, and hydrogen. If biomass is microbially degraded in anaerobic conditions, it generates a product gas rich in methane and carbon dioxide. This product gas is called *biogas*, or *land-fill gas*. A process system developed for exploiting this biogas is called *anaerobic digester*, which can produce methane-rich gas from waste materials on a small-scale to medium-scale unit.

The scientific definition of pyrolysis presented in this section does preclude the oxygen involvement in its mechanistic reaction steps. This was necessary in defining the pyrolysis as a thermochemical reaction by itself. However, it should be clearly noted that the actual reaction of pyrolysis can also occur as a component reaction of many reactions simultaneously taking place in widely different chemical process environments, including both oxidative environments as well as reducing environments. In such environments, pyrolytic decomposition reactions compete with other chemical reactions also occurring in the system, and as such, the reaction environment becomes much more complex in terms of both the nature and the total number of simultaneous reactions taking place in the system.

It is also true that pyrolysis alone in the absence of oxygen can be targeted in certain process environments, as is the case with *fast pyrolysis of biomass*. Typical fast pyrolysis processes are operated at a temperature that is substantially lower than typical gasification temperatures of steam gasification, Boudouard reaction, hydrogasification and partial oxidation. Hence, fast pyrolysis as a transformation process is more or less strictly a combination of devolatilization and pyrolytic decomposition reactions in an oxygen-deprived environment. Therefore, the reaction speed of fast pyrolysis at an elevated temperature is very fast, nearly instantaneous. Furthermore, the principally targeted product of fast pyrolysis of biomass is liquid-phase *bio-oil*, not gaseous synthesis gas. Relative ease in production of liquid fuel (bio-oil) from biomass via fast pyrolysis is one of the main advantages of biomass utilization.

Since biomass has high moisture content and is also rich in oxygen species, biomass can be gasified without any gasifying agent additionally introduced into the reactor. This type of gasification is called *pyrogasification or pyrolytic gasification*. Pyrogasification of biomass takes advantage of both pyrolysis and gasification, and it can be carried out either catalytically [30,31] or noncatalytically. In pyrogasification, no separate gasifying medium or oxygen (or air) is introduced; it is expected that a gasifying reactant such as steam has to be *in situ* provided from biomass pyrolysis. In pyrogasification, biomass pyrolysis also produces bio-char, and this bio-char reacts with steam via steam gasification to generate product gas.

Pyrolytic gasification of wood using a stoichiometric nickel aluminate catalyst was carried out by Arauzo and coworkers [31] in a fluidized-bed reactor, and near-equilibrium yields of products were obtained above 650°C. While they obtained 85%–90% gas yields, tar production was not detected. They further tested the process using a modified nickel–magnesium aluminate stoichiometric catalyst and also an addition of potassium as a promoter. They found that addition of magnesium to the catalyst crystalline lattice enhanced attrition resistance of the catalyst with a minor loss of gasification activity and an increased production of coke. However, they found little effect from addition of a potassium component. Catalyst fouling by carbon deposition, that is, surface coverage by coke, was shown, and regeneration of magnesium-containing catalyst by carbon burn-off was also demonstrated.

Asadullah et al. [30] comparatively evaluated the catalytic performances of Rh/ $CeO_2/SiO_2$, steam reforming catalyst G-91, and dolomite for a number of different biomass gasification modes, including pyrolytic, $CO_2$, $O_2$, and steam. With respect to the biomass conversion to product gas and selectivity of useful gaseous species, $Rh/CeO_2/SiO_2$ has shown superior results in all gasification modes. In the pyrogasification case, about 79% of the carbon in biomass was converted to the product gas at 650°C. There was no tar detected in the effluent gas stream. The gasifier used for their experiment was a lab-scale continuously fed fluidized-bed reactor.

### 15.3.1.2 Partial Oxidation

In chemical processes that generate synthesis gas from fossil or biomass feedstocks, partial oxidation has been proven to be an important pivotal gasification reaction. The partial oxidation has several inherent merits, namely,

1. The reaction rate is very fast, thus contributing to the reduction of reactor volume.
2. The reaction irreversibly proceeds over a very wide range of temperatures.
3. The reaction generates exothermic heat of reaction, which helps sustain the system's energy balance as well as enhances its heat transfer process.
4. The reaction is universally and nearly equally efficient on all hydrocarbon molecules of widely different carbon numbers.
5. Partial oxidation of hydrocarbons generates hydrogen and carbon monoxide as principal product species, which are major components of synthesis gas product.
6. The partial oxidation reaction is an excellent companion reaction with many other chemical reactions, including steam gasification, steam reformation, Boudouard reaction, and so forth.
7. If partial oxidation is properly used in conjunction with other gasification reactions, synergistic effects can result in (1) efficient process energy management including an autothermal operation, (2) higher gas yield and/or higher gasification efficiency, (3) higher conversion of carbon, (4) tailor-made gas composition or control of $H_2/CO$ ratio in syngas product, (5) reduction of char formation or resistance against coking, (6) reduced formation of tar, and more.

Most advanced coal gasification processes such as the Texaco gasifier and Shell gasifier utilize partial oxidation of coal as a principal reaction [5]. The reaction is usually carried out in the copresence of steam, which gets involved in steam gasification as well as WGS reaction as

$$C_{(s)} + \frac{1}{2}O_2 \rightarrow CO$$

$$C_{(s)} + H_2O \rightarrow CO + H_2$$

$$CO + H_2O \leftrightarrow CO_2 + H_2$$

If partial oxidation is poorly managed or improperly designed, an unnecessarily high extent of complete combustion of hydrocarbons can take place, resulting in a large amount of carbon dioxide, thereby wasting the useful heating value of feedstock hydrocarbons as well as increasing the greenhouse gas (GHG) formation even without subjecting the fuel to useful end uses.

### 15.3.1.3   Steam Gasification

Steam reacts with carbonaceous materials including hydrocarbons, carbohydrates, oxygenates, NG, and even graphite at elevated temperatures and generates carbon monoxide and hydrogen. The stoichiometric chemical reactions in this class of reaction include

$$C_{(s)} + H_2O_{(g)} = CO_{(g)} + H_{2(g)}$$

$$Coal + Steam = CO + H_2$$

$$CH_4 + H_2O = CO + 3H_2$$

$$C_aH_b + aH_2O = aCO + \left( a + \frac{b}{2} \right) H_2$$

The first reaction represents steam *gasification* of carbon, whereas the second reaction is steam gasification of coal. The third reaction is steam reformation of methane (or, *methane steam reformation [MSR]*), whereas the fourth reaction is known as reformation of hydrocarbon fuels. The chemical equilibrium favors the forward reaction of steam gasification of carbon, if the temperature of reaction exceeds 674°C, as explained in Section 15.3.1 and Table 15.9. This threshold temperature (and its vicinity) for the forward reaction progress is nearly universally applicable to all hydrocarbon species including coal, irrespective of their carbon numbers. As clearly shown, product hydrogen in these reactions at least partially originates from water (steam) molecules. Without separately going through a water-splitting reaction, this reaction efficiently extracts hydrogen out of water molecules while carbon atoms in the hydrocarbon molecules react with oxygen atoms from water molecules. As expected, the forward reactions, that is, steam gasification reactions as written, are highly endothermic at practical operating temperatures, requiring high energy input.

As mentioned earlier in the pyrolysis section, even when hydrocarbons are reformed or gasified by steam at elevated temperatures, thermochemical conversion due to pyrolysis is also taking place as a competing and parallel reaction to the steam gasification reaction. If a hydrocarbon feedstock, such as coal and biomass, is introduced into a reactor where steam gasification is desired at an elevated temperature, the resultant reaction usually proceeds as an *apparent two-stage reaction*, namely, appearing to be the pyrolysis reaction followed by steam gasification. Mathematically, this apparent two-stage reaction process can be explained by the result of superposition of two parallel reactions between one very fast reaction (pyrolysis) and one slow reaction (gasification). In this explanation, easily pyrolyzable components are rapidly broken down in the early period (e.g., in a matter of a few seconds), whereas much slower gasification takes place more steadily over a much longer period of time (e.g., in a matter of 0.5–3 h). In coal gasification studies, some researchers interpreted this early-stage pyrolysis result as an initial conversion [32].

Since biomass generally contains a high level of moisture, the steam gasification reaction is nearly always present with or without a separate feed of steam into the reactor, except in the case of fast pyrolysis. In the fast pyrolysis of biomass, the typical temperature of operation is around 500°C, which is substantially lower than the steam gasification temperature, and therefore, the biomass moisture is not involved in a steam gasification reaction. On the other hand, a high moisture content of biomass represents an unnecessarily high heat duty for thermochemical conversion of biomass, thereby lowering the energy efficiency of the process.

### 15.3.1.4 Boudouard Reaction or Carbon Dioxide Gasification Reaction

Among the endothermic gasification reactions of hydrocarbons, the speed of the carbon dioxide gasification reaction is the slowest at practical operating temperatures. Most advanced gasification technologies produce carbon dioxide as a component in their syngas products. However, gasification using $CO_2$ has not been popularly attempted, due to its poorer thermal efficiency and inferior energetics compared to steam gasification. However, due to the growing concerns of GHG emissions as well as the roles of carbon dioxide as a major GHG, various technologies including the capture of $CO_2$, the reduction of $CO_2$, the utilization of $CO_2$ in carbon gasification, and the conversion of $CO_2$ into other petrochemicals are actively pursued and developed. Gasification of biomass and/or coal coupled with $CO_2$ management is also an environmentally prudent option.

Complete combustion of biomass or fossil fuels generates carbon dioxide. Since carbon dioxide is chemically very stable, its reactivity is limited. Therefore, the conversion of carbon dioxide into far more reactive carbon monoxide is one of the technological options, whereas the direct conversion of carbon dioxide into hydrocarbons is another. The two types of reactions are categorized under the *reduction of carbon dioxide*, and finding energetically prudent pathways for $CO_2$ reduction is a challenge in the modern fuel chemistry. The chemical reactions leading to the reduction of $CO_2$ include the Boudouard reaction and the RWGS reaction:

$$C_{(s)} + CO_{2(g)} = 2CO_{(g)}$$

$$CO_{2(g)} + H_{2(g)} = CO_{(g)} + H_2O_{(g)}$$

As can be seen from Table 15.9, the temperature for $K_p > 1$ for the forward reactions as written to proceed for the Boudouard reaction and the RWGS reaction are 697°C and 814°C, respectively. Furthermore, the RWGS reaction requires hydrogen as a reactant, which generally makes the process conversion costly.

Lee et al. [33] studied the kinetics of carbon dioxide gasification of various coal char samples for a temperature range between 800°C and 1050°C using a unified intrinsic kinetic model and compared with the literature values obtained for various

**TABLE 15.10**

**Activation Energy for $CO_2$ Gasification Reaction of Coal/Char/Graphite**

| Sample | Arrhenius Activation Energy, E | | Investigators |
|---|---|---|---|
| | kcal/mol | kJ/mol | |
| Carbon | 59–88 | 247–368 | Walker et al. [34] |
| Anthracite, coke | 49–54 | 205–226 | von Frederdorff [35] |
| Coke | 68 | 285 | Hottel et al. [36] |
| Graphite | 87 | 364 | Strange and Walker [37] |
| Montana Rosebud char | 60 | 251 | Lee et al. [33] |
| Illinois no. 6 char | 58 | 243 | Lee et al. [33] |
| Hydrane no. 49 char | 65 | 272 | Lee et al. [33] |

carbon, coal, and char samples. The Arrhenius activation energy values obtained for the carbon dioxide gasification for these samples are shown in Table 15.10 [33].

Obtained from independent investigations by various investigators on diverse carbonaceous materials, the activation energy values for the kinetic rate equations for carbon dioxide gasification are around 60 kcal/mol or 250 kJ/mol. This high activation energy is also indicative of the nature of chemical reaction, which requires a high-temperature reaction to attain a practically significant reaction rate. If we check the E/RT value for the carbon dioxide gasification at 1000°C, then the value becomes

$$\frac{E}{RT} = \frac{60,000}{1.987*1273} = 23.7$$

This E/RT value is within the range of the values for most industrially practiced petrochemical reactions, as often used as a rule of thumb. The study [33] also established that the kinetic rate of the noncatalytic carbon dioxide gasification of coal char at practical operating conditions, such as 900°C and 250 psi, is substantial. The noncatalytic reaction rate was found to be about two to four times slower than that of the steam gasification at the nominally same T and P conditions.

### 15.3.1.5 Hydrogasification

The term "hydrogasification" is the reaction between carbonaceous material and hydrogen, strictly speaking. However, an augmented definition of "hydrogasification" involves the reaction of carbonaceous material in a hydrogen-rich environment to generate methane as a principal product. The gasification reaction in certain steam environments, such as in the copresence of steam and hydrogen, often qualifies for this hydrogen-rich environment for methane generation. The latter is called "steam hydrogasification" [38]. However, the steam gasification whose principal goal is to produce syngas should be still referred to as "steam gasification," not simply as "hydrogasification." *Hydro-* as a prefix is used for "of water" or "of hydrogen," depending upon the situation. As far as the gasification of coal and biomass is concerned, the prefix *hydro-* means "of hydrogen," as in the case of "hydrocracking."

Carbonaceous materials undergo *hydrocracking* under high pressures of hydrogen at elevated temperatures. Hydrocracking generates lighter hydrocarbons as cleavage products from larger hydrocarbons. While the hydrocracking reaction is chemically distinct from hydrogasification of carbon, the difference between the two becomes little when it is applied to coal or coal char whose molecular structure is deficient in hydrogen.

Unlike other gasification reactions involving steam and carbon dioxide, this gasification reaction is *exothermic*, that is, generating exothermic heat of reaction, as

$$C_{(s)} + 2H_{2(g)} = CH_4 \quad \left(-\Delta H^0_{298}\right) = 74.8 \text{ kJ/mol}$$

Coal char hydrogasification can be regarded as two simultaneous reactions differing considerably in their reaction rates [32], as also mentioned in the earlier section on steam gasification. This statement of apparent two-stage reactions of pyrolysis and gasification is valid for the biomass gasification as well. Due to the high moisture content in raw untreated biomass, biomass hydrogasification always involves steam gasification inevitably, where all three principal modes of gasification—including pyrolysis, steam gasification, and hydrogasification—take place simultaneously. Of these reactions, the pyrolysis is by far the fastest chemical reaction at typical operating conditions.

Hydrogasification of carbons and biomass can be catalyzed for faster and more efficient reactions [39]. Many metallic ingredients have been shown to have catalytic effects on hydrogasification of coal char and carbon, and these catalysts include aluminum chloride [40], iron-based catalysts [41], nickel-based catalysts [42], and calcium salt-promoted iron group catalysts [43].

An interesting study was carried out by Porada [44], in which hydrogasification and pyrolysis of basket willow (*Salix viminalis*), bituminous coal, and a 1:1 mixture of the two were compared. Their study employed nonisothermal kinetics approach, in which the temperature of reaction was increased at a constant rate of 3K/min from ambient temperature to 1200K under a hydrogen pressure of 2.5 MPa. Of the test samples, the highest gas yields were obtained during hydrogasification of coal, and the lowest yields were observed in basket willow processing. It was also established that the conversion ratio to $C_1$–$C_3$ hydrocarbons from these samples under a relatively low $H_2$ pressure was approximately five times higher than the pyrolysis conducted in an inert atmosphere. This clearly explains that the beneficial role of hydrogen gas is very significant in gasification of biomass as well as coal.

### 15.3.1.6 Water Gas Shift Reaction

The WGS reaction plays an important role in manufacturing hydrogen, ammonia, methanol and other chemicals. Nearly all synthesis gas reactions involve the WGS reaction in some manner. The WGS reaction is of a reversible kind, whose net proceeding direction can be relatively easily reversed by changing the gaseous compositions as well as varying the temperature of the reaction. As shown in Table 15.9, the temperature dependence of the chemical reaction equilibrium constant ($K_p$) is the mildest of the important syngas reactions considered in the table. Furthermore, the temperature when $K_p$ becomes unity is around 814°C, where most carbon gasification reactions begin to be kinetically active.

The WGS reaction has dual significance. The forward WGS reaction converts carbon monoxide and water into additional hydrogen and carbon dioxide. This reaction is utilized to enhance the hydrogen yield from raw or intermediate syngas, such as the raw product of steam reformation of hydrocarbons. If the WGS reaction is exploited in its reverse direction, that is, as the RWGS reaction, carbon dioxide can be reduced to carbon monoxide that is far more reactive than carbon dioxide. This enables further chemical conversion of carbon monoxide into other useful petrochemicals, instead of direct conversion of carbon dioxide, which is a much more difficult task. The catalytic RWGS reaction could be very useful as a reduction method of carbon dioxide.

In many industrial reactions, WGS reaction is a companion reaction to the principally desired reaction in the main stage, as evidenced in methanol synthesis and in SMR. Whenever deemed appropriate, WGS reaction is also carried out as a secondary-stage reaction to result in additional conversion of water gas into hydrogen, as desired by a fuel reformer to generate hydrogen for proton exchange membrane (PEM) fuel cell application. Since the forward WGS is an exothermic reaction, low temperature thermodynamically favors higher CO conversion. However, its intrinsic kinetic rate without an aid of an effective catalyst is inevitably low at low temperatures. Therefore, most WGS reactions are carried out catalytically at a low temperature, such as 180°C–240°C. This type of catalyst is called a *low-temperature shift* (LTS) catalyst, which has long been used industrially. One of the LTS catalyst formulations is coprecipitated $Cu/ZnO/Al_2O_3$ catalyst, whose formulation is also known for the low-pressure methanol synthesis catalyst [24].

In biomass gasification for generation of bio-syngas, WGS reaction also plays an important role, since it can produce additional hydrogen and also be used to control the ratio of $H_2/CO$ in the synthesis gas composition. The WGS reaction not only enhances the targeted gas composition with a higher selectivity but also prepares a syngas more suitable for the next-stage conversion by adjusting it for an optimal syngas composition.

## 15.3.2 Bio-Syngas

Depending upon the compositions of resultant gaseous products, syngas may be classified as (1) balanced syngas whose $H_2$:CO molar ratio is close to 2:1, (2) unbalanced syngas whose $H_2$:CO molar ratio is substantially lower than 2:1, (3) $CO_2$-rich syngas whose $CO_2$ molar concentration exceeds 10%, and so forth. The balanced syngas is also referred to as hydrogen-rich syngas, while unbalanced syngas is called CO-rich syngas. The gaseous product can also be classified based on its heating value (HHV) as (1) high-Btu gas, >700 Btu/scf; (2) medium-Btu gas, 300–700 Btu/scf; and (3) low-Btu gas, <300 Btu/scf. Alternately, syngas may be classified based on its origin as (1) NG-derived syngas, (2) coal-derived syngas, (3) biomass syngas, and (4) coke oven gas. These descriptions are commonly used for all types of syngas derived from a variety of feedstocks including NG, shale gas, coal, and biomass. The first category of classification has been used widely in the synthesis of clean liquid fuels such as methanol and DME synthesis, whereas the second category has come from the classical design of coal gasifiers [5].

For further clarification of the terminology, *biogas* usually stands for a gas produced by anaerobic digestion of organic materials and is largely comprised of methane (about 50% or higher) and carbon dioxide. Therefore, the term *biogas* is not interchangeable with the term *biomass syngas*. The methane-rich biogas is a medium-Btu gas and also called "marsh gas," "landfill gas," or "swamp gas." As the name implies, swamp gas is produced by the same anaerobic processes (where oxygen is absent and unavailable for biological conversion) that take place during the underwater decomposition of organic matters in wetlands. *Anaerobic digestion* is a series of biochemical processes in which microorganisms break down biodegradable material in the absence or serious deficiency of oxygen. The biochemical processes carried out by microorganisms consist of four principal stages, namely, (1) hydrolysis, (2) acidogenesis, (3) acetogenesis, and (4) methanogenesis. Therefore, the biogas obtained from anaerobic digesters is not classified as a thermochemical intermediate of biomass conversion.

### 15.3.3 Tar Formation

*Tar* is neither a chemical name for certain molecular species nor a clearly defined terminology in materials. Tar has been operationally defined in gasification work as the material in the product stream that is condensable in the gasifier or in downstream processing steps and/or subsequent conversion devices and parts [45]. This physical definition is inevitably dependent upon the types of processes, the nature of feedstocks, and specifically applicable treatment conditions. Producer gases from both biomass gasification and coal gasification contain tars. The generalized composition of tars is mostly aromatic, and the average molecular weight is fairly high. Even with the very same biomass feed, the amount and nature of tars formed are different depending upon the type of gasifiers used and process conditions employed. Similarly, the same gasifier would generate different amounts and types of tars depending upon the feedstock properties and compositions. Therefore, successful implementation of efficient gasification technology depends on the effective control of tar formation reactions as well as the efficient removal and/or conversion of tar from the produced gas.

A number of investigators studied various aspects of tar in terms of its formation, maturation scheme, properties, molecular species, and relationship between the tar yield and the reaction temperature. Elliott (1988) extensively reviewed the composition of biomass pyrolysis and gasifier tars from various gasification processes and proposed a tar maturation scheme [46], as shown in Figure 15.2.

Nickel-based catalysts are known to be effective in biomass gasification for tar reduction to produce synthesis gases, because of their relatively lower cost and good catalytic effects. Several different types of nickel-based catalysts for biomass gasification were reviewed with respect to tar reduction efficiency by Wu and Williams [47].

Several methods for the sampling and analysis of tar have been developed. Most of these methods are based on the condensation of tar in a liquid phase or adsorption

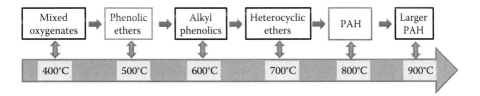

**FIGURE 15.2** Tar maturation scheme proposed by Elliott (1988). (From Milne, T.A., and Evans, R.J., Biomass gasifier "tars"; their nature, formation, and conversion, US Department of Energy, National Energy Technology Laboratory, 1998. Elliott, D.C., ed., *Relation of Reaction Time and Temperature to Chemical Composition of Pyrolysis Oils*, American Chemical Society, Washington, DC, 1988.)

of tar on a solid material. The collected samples are subsequently analyzed gravimetrically or by using a gas chromatograph (GC). The Solid Phase Absorption (SPA) method was originally developed by KTH, Sweden, and according to the SPA method, a gas sample is passed through an amino-sorbent, which collects all tar compounds [48]. The ensuing step is to use different solvents to collect aromatic and phenolic compounds separately. These compounds are then analyzed on a GC, and positive identification of the condensed material is achieved by a GC-mass spectrometer (GC-MS). The tar amount determined by the GC analysis is called *GC-detectable tar.* Tar can also be analyzed gravimetrically, and the gravimetrically determined tar is called *gravimetric tar.* Gravimetric tar is evaporation/distillation residue from particle-free solution(s) determined by gravimetric analysis. Both GC-detectable tar and gravimetric tar are reported in $mg/m^3$. Both chromatographic and gravimetric determination of tar are based on the European Technical Specification, TC BT/TF 143 WI CSC 03002.4: 2004 (E), developed by Technical Committee CEN/BT/TF 143 "Measurement of Organic Contaminants (Tar) in Biomass Producer Gas" [49]. This Technical Specification is applicable to sampling and analysis of tars and particles in the concentration range between $1 mg/m^3$ and $300 g/m^3$ at all relevant sampling port conditions (0°C–900°C and 0.6–60 bars). The application of the Technical Specification allows determination of four different analytical values [49], namely,

- The concentration of gravimetric tars in $mg/m^3$
- The sum of the concentrations GC-detectable tars in $mg/m^3$
- The concentration of individual organic compounds in $mg/m^3$
- The concentration of particles in $mg/m^3$

## 15.4 FAST PYROLYSIS OF BIOMASS

*Pyrolysis of biomass* is an important process option either as a pretreatment for gasification (i.e., the first stage of two-stage gasification) or as an independent process such as *fast pyrolysis.* The former is usually aimed at producing bio-syngas, whereas the latter is intended to produce a liquid fuel product. Typical biomass pyrolysis takes place actively at around 500°C and produces a liquid product via fast cooling (shorter than 2 s) of volatile pyrolytic products. The liquid product produced is called *bio-oil* or *pyrolysis oil*. Bio-oil is considered GHG neutral, since it only puts back into the atmosphere what was initially removed by the plant during its lifetime [50]. Bio-oil is nearly sulfur-free. As can be seen from the prevailing pyrolysis temperature of ~500°C, pyrolysis of biomass as a process treatment is quite similar to oil shale pyrolysis [51] and coal pyrolysis [14], wherein hydrocarbon species are devolatilized and thermally cracked. Bio-oil production via biomass pyrolysis is typically carried out via *flash-pyrolysis* or *fast pyrolysis*. The *biomass fast pyrolysis* process [23,52,53] is a thermochemical conversion process that converts biomass feedstocks into gaseous, solid, and liquid products via heating of biomass in the absence of oxygen or air. The principal product of typical fast pyrolysis of biomass is a liquid product of bio-oil. As the name of the process implies, the process is intended to take place very fast, in a matter of a couple of seconds or shorter, that is, $\tau < 2$ s. As such, heat and mass transfer conditions in the reactor become crucially important in both design and operation of fast pyrolysis of biomass. A variety of reactor designs have been proposed and tested on pilot scales, and they include bubbling fluidized-bed (BFB) reactors, circulating fluidized-bed (CFB) reactors, rotating cone reactors, vacuum reactors, ablative tubes, and more. Table 15.11 shows a partial list of operational pilot-scale biomass pyrolysis units.

The typical biomass fast pyrolysis process involves several stages of operation, including *biomass feed drying, comminuting, fast pyrolysis, separation of char, and liquid recovery*.

Fast pyrolysis of biomass has several distinct merits as an alternative fuel process technology, and they are as follows:

- The principal product of fast pyrolysis biomass is bio-oil, that is, a liquid product. As such, storage and transportation of the product is easy.
- The process takes place very fast, in a matter of 0.5–2 s, and as such, the reactor residence time is very short.

## TABLE 15.11
### Pilot-Scale Fast Pyrolysis of Biomass

| Process | Type of Reactor | Capacity | References |
|---|---|---|---|
| Dynamotive Energy Systems (Canada) | Bubbling fluidized bed | 400 kg/h | [50,54] |
| Union Fenosa (Spain) | Bubbling fluidized bed | 200 kg/h | [55] |
| Wellman Process Eng. Ltd. (United Kingdom) | Fluidized bed | 250 kg/h | [56] |
| Resource Technology (RTI) | Fluidized bed | 20 kg/h | [57] |
| Red Arrow/Ensyn | Circulating fluid bed | 1000 kg/h | [53] |
| VTT/Ensyn (Finland) | Circulating fluid bed | 20 kg/h | [58] |
| ENEL/Ensyn (Italy) | Circulating transported bed | 625 kg/h | [59] |
| BTG/KARA (Netherlands) | Rotating cone | 200 kg/h | [60] |
| Pyrovac | Vacuum, stirred bed | 3500 kg/h | [61] |
| Enervision (Norway) | Ablative tube | | [62] |
| Fortum Oy (Finland) | Own | 350 kg/h | [63] |

- The process technology is very widely and universally applicable to a variety of biomass feedstocks.
- The process chemistry is simple and straightforward.
- The process equipment needed is relatively simple and not complex.
- Due to the high reactor throughput and simple chemistry, the capital cost is not high.
- Small-scale process feasibility has been demonstrated. However, process economics for a small-scale power generation (<5 $MW_e$) based on fast pyrolysis of biomass is substantially less favorable than that of a larger-scale (>10 $MW_e$) system.

However, the drawbacks of fast pyrolysis of biomass include the following:

- The moisture level of the feed biomass needs to be controlled below 10% or even lower. Otherwise, the feed water and reaction-produced water will end up in the final liquid oil product.
- The biomass feedstock needs to go through size reduction and preconditioning. The feed material has to be in particulate form in order to minimize the heat and mass transfer resistance, preferably ~2 mm for bubbling bed and ~6 mm for CFB.
- The quality of bio-oil product is generally poor. Bio-oil has a high oxygen content, which makes the oil more corrosive and less stable, besides possessing a lower heating value. Bio-oil also contains metallic compounds and nitrogen species, which can foul and deactivate most of the fuel-upgrading catalysts. The upgrading process would require a large amount of hydrogen and could become costly.
- The process requires a very fast heating rate, which can be costly in both operation and capital investment.
- The overall energy efficiency of the process is not high, by itself.
- Large-scale process operation may be subjected to significant logistical burdens of feedstock collection, storage, and pretreatment.

The first biocrude in the United States was produced at a 30 kg/h scale [64]. The operating conditions employed were a the reactor temperature of approximately 500°C and a residence time of 1 s, that is, at a typical fast pyrolysis condition. Since then, Canadian researchers have converted woody biomass into fuel via pyrolysis in a 200 kg/h pilot plant [65]. The fuel oil substitute was produced (on 1000 ton/day dry basis) at approximately $3.4/GJ, based on the 1990 fixed price. At the time, the cost for light fuel oil was $4.0–4.6/GJ [66], thus indicating that the pyrolyzed biomass fuel was a more economical alternative. However, the skepticism of high transportation costs of the biomass to the pyrolyzer outweighing potential profits had limited the R&D funding for the following years. This was when the petroleum-based liquid fuel price was much lower than that in the twenty-first century. To circumvent this biomass transportation and logistical problem, the Energy Resources Company (ERCO) in Massachusetts developed a *mobile pyrolysis prototype* for the US EPA.

*ERCO*'s unit was designed to accept biomass with 10% moisture content at a rate of 100 tons/day. At this rate, the system had a minimal net energy efficiency of 70% and produced gaseous, liquid, and char end products. The process, which was initially started using an outside fuel source, became completely self-sufficient shortly after its start-up. This was achieved by implementing a cogeneration system to convert the pyrolysis gas into the electricity required for operation. A small fraction of the pyrolysis gas is also used to dry the entering feedstock to the required 10% moisture. A simplified version of *ERCO's mobile unit* is shown in Figure 15.3 [67].

The end products are pyrolysis oil and pyrolytic char, both of which are more economical to transport than the original biomass feedstock. The average heating values for the pyrolysis oil and char are 10,000 and 12,000 Btu/lb, respectively [67]. The pyrolysis gas, which has a nominal heating value of 150 Btu/scf, is not considered an end product since it is directly used in the cogeneration system. If classified based on the coal syngas criterion, typical pyrolysis gas of biomass would be classified as a low-Btu gas whose usual criterion for the heating value is less than 300 Btu/scf [5]. The mobility, self-sufficiency, and profitability of the system lifted some of the hesitancy of funding research on the pyrolysis of biomass. In addition, ERCO's success led to additional investigation of "dual" or cogeneration systems, which produce both useful heat and electric power, that is, combined heat and power (CHP).

In a conventional *fluidized-bed fast pyrolysis reactor*, fine particles (2–3 mm size) of biomass are introduced into a bed fluidized by a gas, which is usually a recirculated product gas [68]. High heat and mass transfer rates result in rapid heating of biomass particles via convective heat transfer. Char attrition and bio-oil contamination may take

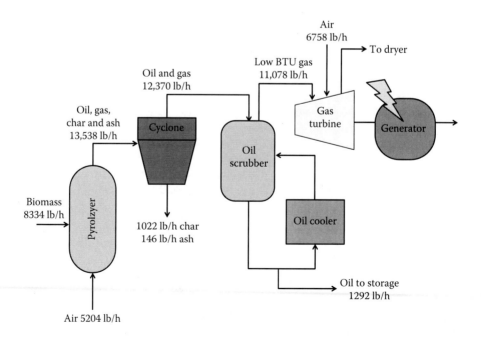

**FIGURE 15.3** A schematic and material balance for ERCO's mobile pyrolysis unit.

place, and char carbon could appear in the bio-oil product. Heat can be supplied externally to the bed or by a hot fluidizing gas. In most designs, reactor heat is usually provided by heat exchanger tubes through which hot gas from char combustion flows. The reactor effluent leaves from the top of the reactor as noncondensable gas (NCG), char, bio-oil vapor, and aerosol. The fluid bed may contain, besides biomass particles, fluidizing media such as *hot sand particles* for enhanced heat transfer. *Aerosol* is generally defined as a suspension of solid and liquid particles in a gas. The term *aerosol* includes both the particles, and the suspending gas and the particle size may range from 0.002 to larger than 100 µm [49]. There is some dilution of the products due to the use of fluidizing gas, which makes it more difficult to condense and then separate the bio-oil vapor from the gas exiting the condensers. The general operational principle is quite similar to that of the traditional gas–solid fluidized-bed reactor [69]. The fluidized-bed fast pyrolysis process has been pilot tested by several companies including Dynamotive [54], Wellman [56] and Agri-Therm [70], as listed in Table 15.11. Special versions of fluidized-bed reactors popularly used in biomass fast pyrolysis are BFB and CFB reactors. A principal difference between the two types is that the former has a freeboard space above the fluid bed, whereas the latter achieves a full entrainment of the particle–fluid mixture in the reactor (Figure 15.4).

*BFB reactors* have long been used in chemical and petroleum processing. An earlier version of the bubbling fluidized sand bed reactor was utilized by the *Waterloo Fast Pyrolysis Process* (WFPP) [71,72]. Larger units based on the BFB are a 200 kg/h system by Union Fenosa (Spain) and a 400 kg/h system by Dynamotive (Canada). Both systems were based on the WFPP developed at the University of Waterloo (Canada) and designed by its spin-off company Resource Transforms International (RTI) in Canada [55]. BFBs occur when the incoming carrier gas velocity is sufficiently above the minimum fluidization velocity to cause the formation of *bubble-like* structures within the particulate bed. In such a condition, the bed appears more or less bubbling. A variety of different designs have also been introduced. The particulate or granular bed in the BFB may be composed of biomass particles only

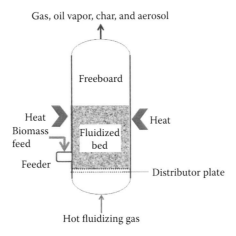

**FIGURE 15.4** Fluidized-bed reactor for fast pyrolysis of biomass.

without any inert media, and all the process heat can be supplied by hot fluidizing gas. Alternately, the granular bed can contain hot heat-transfer solid medium such as indirectly heated sand, which enhances the heat transfer efficiency and generally allows for a larger throughput for the process. An earlier version of the Dynamotive system used NG to heat their pilot-scale reactor, but most modern designs use the exothermic heat of char combustion to supply the necessary heat to the pyrolysis reactor. The BFB is, in principle, self-cleaning, as by-product char is carried out of the reactor by the product gases and oil vapors [55]. For this process feature, the density of char will have to be less than that of fluidizing media so that the by-product char will literally "float" on top of the bed. The allowable particle size range for this type of reactor is quite narrow and should be carefully managed. Further, the bio-oil produced has a carbon contamination possibility, since the oil vapor has to pass through the char-rich layer on its way out of the bed [55]. The gas flow rate for this reactor will have to be determined based on the dimensions of fluidizing media and the desired residence time of gas and oil vapor in the freeboard section of the reactor, which is above the bed. The residence time is generally between 0.5 and 2.0 s. The BFB is currently most popularly used for both fast pyrolysis and gasification processes of biomass. In order to achieve a short residence time for volatiles, a shallow bed depth, a high gas flow rate, or both are usually utilized [57,73]. A high gas-to-biomass feed ratio is adopted for necessary fluidization and a short residence time, which, in turn, results in product dilution and lowers the thermal efficiency of the process.

*CFB reactors* and derivative types of reactor design are frequently utilized for fast pyrolysis of biomass. The basic concept of circulating fluid-bed reactors involves efficient and rapid heat transfer in a convective mode and short residence times for both biomass particles and product vapors. Fine biomass particles are introduced into a CFB of hot sand. Hot sand and biomass particles move together with the transport gas, which is usually a recirculated product gas. In reactor engineering, this transport gas is usually referred to as a fluidizing gas. In practice, the residence times for solid biomass particles are not uniform and only a little longer than the volatiles. Therefore, solid recycling of partially reacted feed would become necessary, or a very fine particle size of biomass would have to be used. Most CFB reactors are dilute-phase units, and their heat transfer rates are high but not as high as particularly desired, since the mode of heat transfer is gas–solid convective heat transfer [57,74]. If a twin-bed reactor system is used, that is, the first for fast pyrolysis and the second as a char combustor to reheat the circulating solids, as shown in Figure 15.5, there is a strong possibility for ash carryover to the pyrolysis reactor and ash buildup in circulating solids [57]. The ash attrition and char carryover problem could also be high, and if not controlled properly, some level of contamination of bio-oil products is also possible. One of the main advantages of the CFB is the possibility to achieve a short and controllable residence time for char [75]. The alkaline compounds in biomass ash are known to possess a catalytic effect for cracking organic molecules contained in volatile vapors, thereby potentially lowering the volatile bio-oil yield. *Red Arrow and VTT processes* are based on circulating fluid-bed reactors.

A *rotating cone fast pyrolysis system* operates based on the idea of intensive mixing between biomass particles and hot sand particles, thereby providing good heat

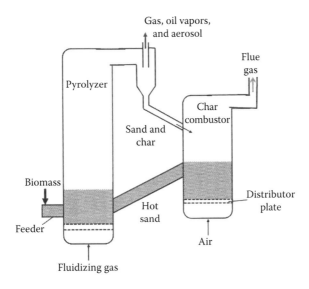

**FIGURE 15.5** Circulating fluidized-bed reactor for fast pyrolysis of biomass.

and mass transfer. This type of reactor requires very fine to fine biomass particle size. This process does not require any carrier gas for its operation, and therefore, the reactor size can be made compact. The *BTG's fast pyrolysis* process developed by Biomass Technology Group (The Netherlands) is based on a *modified rotating cone reactor* [60]. In this reactor, efficient heat transfer between the hot sand particles and biomass particles is accomplished, while a good portion of process heat is retained in the hot sand particles. A wide variety of different biomass feedstocks can be processed in the pyrolysis process, which is operated at 450°C–600°C. Before entering the reactor, the biomass feedstock must be reduced in size to a particulate form finer than 6 mm, and its moisture content to below 10 wt.%. Sufficient excess heat is normally available from the pyrolysis plant to dry the feed biomass from 40–50 wt.% moisture to below 10 wt.%. A schematic of a rotating cone reactor is shown in Figure 15.6.

From the BTG process, up to 75 wt.% pyrolysis oil and 25 wt.% char and gas are produced as primary products [60]. Since no "inert" carrier gas is used in this process, no additional gas heating is required, and the pyrolysis products are undiluted vapor. This undiluted vapor flow allows the downstream equipment to be of a minimum size. In a condenser, the oil vapor product is rapidly cooled, yielding the oil product and some permanent (noncondensable) gases. In only a few seconds of process treatment, the biomass is transformed into pyrolysis oil. Biomass char and hot sand used in the reactor are recycled to a combustor, where char is combusted to reheat the sand. After reheating the sand by char combustion heat, the sand is recirculated back to the reactor. The permanent gases (or NCG) can be utilized in a gas engine to generate electricity or simply flared off. In principle, no external utilities are required for operation of the process, that is, energy-wise, it is self-sufficient. A schematic of the BTG process for fast flash of biomass is shown in Figure 15.7.

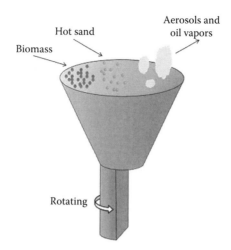

**FIGURE 15.6** A schematic of a rotating cone reactor for fast pyrolysis of biomass.

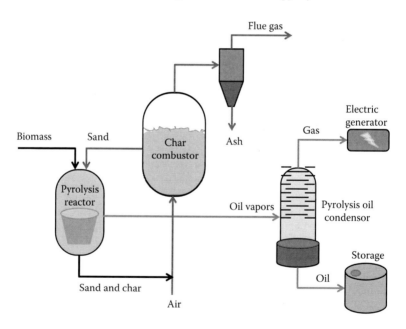

**FIGURE 15.7** A schematic of BTG process for fast pyrolysis of biomass.

The *vacuum pyrolysis process* operates under low pressure (vacuum or atmospheric) and has principal process merits in its processability of larger biomass particles as well as its short residence time for volatiles. These biomass particles are taken out using a vacuum pump from the reactor regardless of the particles' residence time. Due to the lack of convective gas flow inside the reactor, however, the heat and mass transfer rates are slower than those for the fluidized-bed reactors, hence requiring a longer residence time for biomass particles in the vacuum reactor. A longer biomass residence time, in turn, makes the reactor and equipment size inevitably larger. Biomass in a vacuum reactor moves downward by gravity and rotating scrapers through the multiple hearth pyrolyzer, with the temperature increasing from about 200°C to 400°C, as shown in Figure 15.8. *Pyrovac's pyrocycling process* is based on the vacuum pyrolysis technology [61]. According to Pyrovac, their process technology has the following features:

- There is no need to pulverize or grind the feed biomass materials. Particles of up to 20 mm can be fed without any difficulty.
- The process system can be operated under torrefaction mode (<300°C) or pyrolysis mode (>450°C). The first mode generates bio-char as a principal product, whereas the latter mode produces bio-oil.
- The process can be operated under vacuum or atmospheric conditions to enhance the production of either bio-oil or bio-char. The process operation as well as its product portfolio are more versatile compared to competing process technologies.
- The process adopts a moving and stirred-bed reactor.
- Molten salt heat carrier at 575°C is in indirect contact with biomass feedstock, thus aiding in efficient heat transfer.
- There are two heating plates with internal raking systems.
- The process uses two condensing towers. The first tower mainly collects heavy bio-oil and contains little water and acids. The second tower mainly recovers the aqueous acidic phase.

The *Ablative pyrolysis process* is based on the heat transfer taking place when a biomass particle slides over a hot surface, as shown in Figure 15.9. High pressure applied to biomass particles on a hot reactor wall or surface to provide good contact is achieved by centrifugal or mechanical motion [68]. This type of reactor does

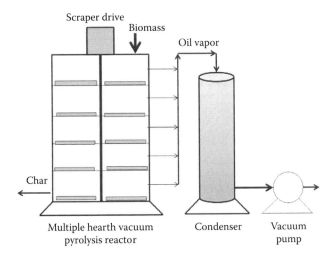

**FIGURE 15.8**  A schematic of vacuum pyrolysis reactor for biomass fast flash.

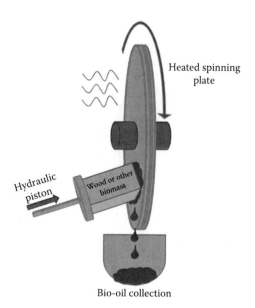

Heated spinning
plate

Hydraulic
piston

Wood or other
biomass

Bio-oil collection

**FIGURE 15.9**   A schematic of an ablative pyrolysis reactor for biomass fast flash.

not require a small particle size and can handle large particles without difficulty. The reactor does not require any carrier gas or sweep gas. The reactors of cyclonic type have difficulty in achieving sufficiently long residence times for the biomass particles, which are required to allow a high degree of conversion. Therefore, it is usually necessary to recycle partially reacted solids back to the reactor, as is the case with a CFB reactor. A high degree of char attrition also takes place and tends to contaminate the product bio-oil with a high level of carryover carbon [57]. Some variations of the ablative pyrolysis process include the cone type and plate type for hot surfaces. These processes are mechanically more complex and difficult to scale up, since the moving parts are subjected to high temperatures required for pyrolysis. Further, the loss of thermal energy from the ablative process in general could be high, since the hot surface needs to be at a substantially higher temperature than the desired pyrolysis temperature.

Fast pyrolysis of biomass can be achieved in an *auger-type reactor* or *auger reactor*, in which an auger or an advancing screw assembly drives the biomass and hot sand through the reactor barrel. The operational principle is very similar to that of a polymer twin-screw extruder, as shown in Figure 15.10. This type of reactor achieves a very good mixing of materials in the reactor, thereby enhancing the heat and mass transfer efficiency. In an auger-type reactor for biomass fast pyrolysis, the usual mechanism of heat transfer is via direct heat transfer between hot sand and biomass particles.

At the outlet of the reactor, solids including hot sands, char, and ash are separately recovered from the oil vapors and aerosol. Hot sand is reheated using char combustion heat and recirculated back to the reactor, while the oil vapor and aerosol are sent to the condenser for bio-oil recovery. The process does not require a carrier gas, but

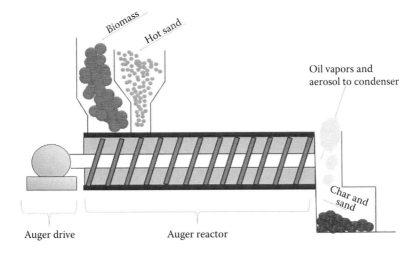

**FIGURE 15.10** An auger-type reactor for biomass fast pyrolysis.

the biomass particle size is preferentially small for smooth operation. The design of auger reactors has received extensive benefits from the industrial practice and design experience of polymer and twin-screw extruder industries [76]. Therefore, this type of process is relatively straightforward to design and fabricate and is deemed to be suitable for a small-scale production.

## 15.5 BIOMASS GASIFICATION PROCESSES

The first biomass gasification system investigated at the pilot scale was a fluidized bed, which incorporated dry ash-free (daf) corn stover as the gasifier feed. Corn stover has been selected as the feed to the gasifier since 1977 [77], even when the U.S. corn production was less than half of the 2010 production of 312 million metric tons [78]. The amount of corn stover that could be sustainably collected and made available in 2003 was estimated to be 80–100 million dry metric tons/year [79]. Of this total, potential long-term demand for corn stover by nonfermentative applications such as biomass gasification in the United States was estimated to be about 20 million dry tons/year. An early pilot-scale system designed and operated at Kansas State University, shown in Figure 15.11, has a 45.5 kg bed capacity [80]. Fluidizing gas and heat for the biomass gasification were supplied by the combustion of propane in the presence of air. The particulates and char were removed using a high-temperature cyclone. A venturi scrubber was then used to separate the VM into NCG, a tar-oil fraction, and an aqueous waste fraction. Raman et al. [80] conducted a series of tests with temperatures ranging from 840K to 1020K. The optimal gas production was obtained using a feed rate of 27 kg/h and a temperature of 930K. At these conditions, $0.25 \times 10^6$ Btu/h of gas was produced. This was enough to operate a 25 hp internal combustion engine operating at 25% efficiency [80].

Another one of the extensively studied gasification systems for biomass conversion is Sweden's *VEGA gasification system*. Skydkraft AB, a Swedish power

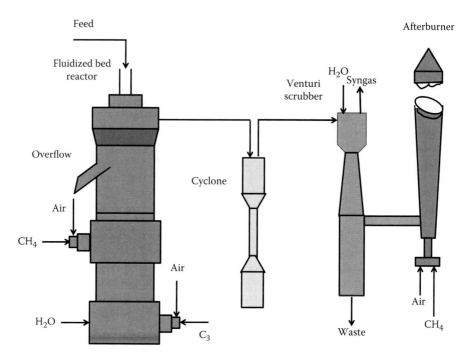

**FIGURE 15.11.**    KSU's pilot-scale fluidized-bed gasifier of corn stover.

company, decided in June 1991 to build a cogeneration power plant in Värnamo, Sweden, to demonstrate the *integrated gasification combined-cycle (IGCC)* technology. Bioflow, Ltd., was formed in 1992 as a joint venture between Skydkraft and Alstrom to develop the pressurized air-blown CFB gasifier technology for biomass [81]. The biomass IGCC (*BIGCC*) was commissioned in 1993 and fully completed in 1995. VEGA is a biomass-fuel based IGCC system that combines heat and power (CHP) for a district heating system. It generated 6.0 MW$_e$ and 9.0 MW$_{th}$ for district heating of the city of Värnamo, Sweden. This was the first complete BIGCC for CHP from biomass feedstock. As shown in Figure 15.12, the moisture of the entering biomass feedstock is removed via a "biofuel dryer" to decrease gaseous emissions [81]. The dried biomass is then converted into a "biofuel" in a combined-cycle gasifier. The resulting gas is cooled before entering the heat recovery boiler and distribution to the district heating. The gasifier is known as the *Bioflow Gasifier* or *Bioflow Pressurized Circulating Fluidized-Bed Gasifier.*

Biomass gasification and power generation technology has long been developed with significant technological advances in Finland, where about 20% of total energy consumption derives from biomass. This high percentage of biomass energy utilization is mainly due to the recycling of biowaste produced as a by-product of the forest industry. VTT Energy and Condens Oy of Finland developed a new type of fixed-bed biomass gasifier [58], whose configuration is based on forced feed flow that allows the use of low bulk-density fibrous biomass feedstock. This gasifier is a combination of updraft and cocurrent gasification technologies, where gasifying

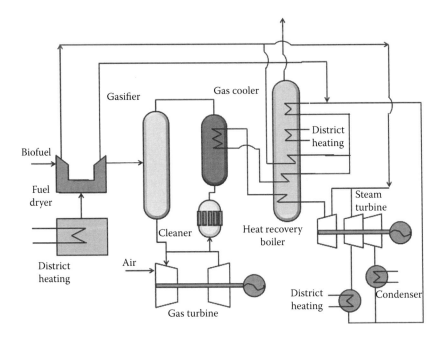

**FIGURE 15.12** A schematic of VEGA process for biomass gasification.

medium and solid feed move upward through the gasifying section of the reactor. In 1999 to 2001, a 500 kW$_{th}$ pilot plant was operated in a test facility and very positive test results were obtained. Some of the principal features of the technology include the following [58,82]:

- Fuel feeding is not based on natural gravity alone.
- The process is suitable for various biomass residues and waste-derived fuels.
- The process achieves high carbon conversion and generates low tar content.
- The process can be scaled up to above 8 MW, unlike its predecessor technology of Bioneer gasifiers.
- There was no problem with leaking feeding systems or blocking gas lines.
- The VTT successfully demonstrated a variety of feedstocks including
  - Forest wood residue chips (moisture level of 10–55 wt.%)
  - Sawdust and wood shavings
  - Crushed barks of maximum moisture of 58%
  - Demolition wood
  - Residue from plywood and furniture industry
  - Recycled fuel manufactured from household waste
  - Sewage sludge in conjunction with other fuels

Condens Oy is offering this technology for a wide range of fuel feedstocks. The Kokemäki (Finland) CHP plant of 1.8 MW$_e$/3.9 MWt$_h$ based on this technology was started up in 2005 [82].

The most common method of gasifying biomass is using an *air-blown CFB gasifier with a catalytic reformer*, even though there are many different variations. Most fluidized-bed gasification processes use close-coupled combustion with very little or no intermediate gas cleaning [83]. This type of process is typically operated at around 900°C, and the product gas from the gasifier contains $H_2$, CO, $CO_2$, $H_2O$, $CH_4$, $C_2H_4$, benzene, and tars. Gasification uses oxygen (or air) and steam to help the process conversion, just as in the advanced gasification of coal [5]. While the effluent gas from the fluidized-bed gasifier contains a decent amount of syngas composition, the hydrocarbon content is also quite substantial. Therefore, the gasifier effluent gas cannot be directly used as syngas for further processing for other liquid fuels or chemicals without major purification steps. This is the reason why the gasifier is coupled with a catalytic reformer, where hydrocarbons are further reformed to synthesis gas. In this stage, the hydrocarbon content including methane is reduced by 95% or better. A very successful example is Chrisgas, a European Union-funded project, which operates an 18 $MW_{th}$ circulating fluidized gasification reactor at Värnamo, Sweden. They use a pressurized CFB gasifier operating on oxygen/steam, a catalytic reformer, and a WGS conversion reactor that enriches the hydrogen content of the product gas. The process also uses a high-temperature filter. The project has been carried out by the Växjö Värnamo Biomass Gasification Centre (VVBGC). The use of oxygen instead of air is to avoid a nitrogen dilution effect, which, if not avoided, adds an additional burden of nitrogen removal during downstream processing [83].

Another CFB gasification process by Termiska Processor AB (TPS) in Nyköping, Sweden, developed for small- to medium-scale electric power generation is using biomass and refuse-derived fuel (RDF) as their feedstocks [84]. The process is based on an air-blown low-pressure CFB gasifier that operates at 850°C–900°C and 1.8 bar. The raw product gas has a tar content of 0.5%–2.0% of dry gas with a heating value of 107–188 Btu/scf. As such, the raw product gas is a low-Btu gas, if the coal syngas classification scheme is followed. The process has merits of good fuel flexibility, good process controllability and low-load operation characteristics, uniform gasifier temperature due to highly turbulent movement of biomass solids, high gasification yield, additional features of catalytic *tar cracking*, and fines recycling from a secondary solids separation. The tar in the syngas is catalytically cracked by *dolomite (CaCO₃·MgCO₃)* in a separate reactor vessel at 900°C immediately following the gasifier. The full calcination of dolomite is active at this temperature, as the chemical equilibrium constant ($K_p$) for calcite ($CaCO_3$) decomposition becomes unity at approximately 885°C [5]. A schematic of a pilot-scale TPS process with a tar cracker is shown in Figure 15.13. A waste-fueled gasification plant was constructed based on the TPS CFB process by Ansaldo Aerimpianti SpA in Grève-in-Chianti, Italy [85].

*Indirect gasification* is another gasification process technology that takes advantage of the unique properties associated with biomass feedstock. As such, indirect gasification of biomass is substantially different from most coal-based gasification process technologies. For example, biomass is low in sulfur, low in ash, highly reactive, and highly volatile. In an indirect gasification process, biomass is heated indirectly using an external means such as heated sands as in the *Battelle process* [86]. A typical gaseous product from an indirect gasifier is close to medium-Btu gas. Battelle began this process R&D in 1980, and has continued till now, accumulating a

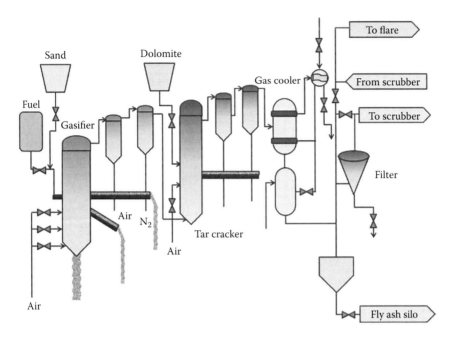

**FIGURE 15.13**   A schematic of TPS biomass gasification system with a tar cracking unit.

substantial amount of valuable data regarding biomass gasification and utilization through demonstration plant operation. Battelle's process is known as the *FERCO SilvaGas process*, which is commercialized by FERCO Enterprise. The principal gasifying medium for the process is steam. A commercial-scale demonstration plant of the SilvaGas process was constructed in 1997 at Burlington, Vermont, at the Burlington Electric Department (BED) McNeil station. The design capacity of this plant is 200 tons/day of biomass feed (dry basis). The McNeil station uses conventional biomass combustion technology, a stoker gate, conventional steam power cycle, and an electrostatic precipitator (ESP)-based particulate matter (PM) removal system. The gas produced by the SilvaGas gasifier is used as a cofired fuel in the existing McNeil power boilers [86]. The product gas has a heating value of about 450–500 Btu/scf, which is in the medium-Btu gas range. A schematic of the FERCO SilvaGas process is shown in Figure 15.14.

CUTEC-Institut Gmb of Germany recently constructed and operated an oxygen-blown CFB gasifier of 0.4 $MW_{th}$ capacity coupled with a catalytic reformer [87]. Part of their product gas is, after compression, directly sent to a *Fischer–Tropsch synthesis (FTS) reactor* for liquid hydrocarbon synthesis. The CFB gasifier of *the CUTEC process* was operated at 870°C, whereas the fixed-bed Fischer–Tropsch synthesis reactor was operated at 150°C–350°C and 0.5–4.0 MPa using a *fused iron catalyst* (Fe/Al/Ca/K/Mg = 100/1.7/2.5/0.7). Their pilot-scale process system was successfully demonstrated with a variety of biomass feedstocks with wide ranges of particle sizes and moisture levels, including sawdust (~3 mm), wood pellets (6–18 mm), wood chips (~10 mm), and chipboard residues (~30 mm). This process, once fully

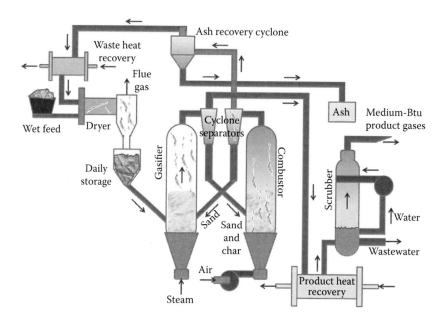

**FIGURE 15.14** FERCO SilvaGas biomass gasification process. (From Paisley, M.A. et al., "A promising power option—the FERCO SilvaGas biomass gasification process—operating experience at the Burlington gasifier," in *Proceedings of ASME Turbo Expo 2001 June 4–7, 2001,* New Orleans, Louisiana, USA, 2001.)

developed for large-scale operation, has a good potential for a single-train biomass-to-liquid (BtL) fuels conversion process [87]. The CUTEC's idea of directly linking between the biomass gasification and the FTS is very similar to that adopted by indirect coal liquefaction based on the FTS.

Another process option for biomass gasification for syngas production involves the use of an *entrained flow reactor.* This type of process is operated at a very high temperature, around 1300°C, and without use of catalyst. The high temperature is necessary due to the fast reaction rate required for an entrained flow reactor whose reactor residence time is inherently very short. If a specific biomass feed has a high ash content, which is not very typical for biomass, slag can be formed at such a high temperature. Learning from the research developments in coal gasification [14], a slagging entrained flow gasifier may be adopted for high-ash biomass conversion. Another important process requirement besides high temperature and short residence time is the particle size of the solid feed; it must be very fine for efficient entrainment as well as for better conversion without mass transfer limitations. However, pulverization or milling of biomass to very fine particle sizes is energy intensive and costly, in general. To facilitate an efficient size reduction of biomass feed, two options are most commonly adopted, namely, torrefaction and pyrolysis. *Torrefaction* is a mild thermal treatment at a temperature of 250°C–300°C, which converts solid biomass into a more brittle and easily pulverizable material that can be treated and handled just like coal [88]. This torrefied product is often called *"bio-coal."* Thus, pulverized torrefied biomass can be treated just as coal, and most entrained flow gasifiers designed for coal can be smoothly converted for torrefied bio-coal without much adaptation. Torrefaction as a process has long been utilized in many applications including the coffee industry. Torrefaction of biomass can alleviate some of the logistical problems involved with biomass feedstock collection and transportation. However, more study is needed for the biomass industry to make it more tuned for biomass and optimized as an efficient pretreatment technique. Gases produced during the torrefaction process may be used as an energy source for torrefaction, thus accomplishing a self-energy supply cycle.

An example of entrained flow biomass gasification can be found from the *Buggenum IGCC* plant, whose capacity is 250 MW$_e$ [89]. NUON has operated this process and their demonstration test program using, from 2001 through 2004, 6000 M/T of sewage sludge, 1200 M/T of chicken litter, 1200 M/T of wood, 3200 M/T of paper pulp, 50 M/T of coffee, and 40 M/T of carbon black as cofeeds with coal. A typical particle size of biomass feed was smaller than 1 mm, and pulverization of wood was more difficult than that of chicken litter and sewage [89]. In their test program, they also mentioned torrefaction as a pretreatment option. Their experience with a variety of biomass feedstocks provides valuable operation data for future development in this area.

As explained in the fast pyrolysis section of this chapter, biomass pyrolysis takes place actively at around 500°C and produces a liquid product via fast cooling (shorter than 2 s) of volatile pyrolytic products. As also mentioned earlier, the liquid product produced is called *bio-oil*. The produced bio-oil can be mixed with char (biomass char or bio-char) to produce a *bio-slurry*. Bio-slurry can be more easily fed, as a pumpable slurry, to the gasifier for efficient conversion. Bio-slurry is somewhat analogous to coal-oil slurry (COM) [5]. A successful example of using bio-slurry is found from the *FZK process* [90,91]. Forschungszentrum Karlsruhe (FZK) developed a process that produces syngas from agricultural waste feeds like straws. They developed a flash pyrolysis process that is based on twin screws for pyrolysis, as explained earlier as an auger pyrolyzer. The process concept is based on the Lurgi-Ruhrgas coal gasification process [5,15]. A 5–10 kg/h process development unit (PDU) is available at the FZK company site. In this process, straw is flash-pyrolyzed into a liquid that is subsequently mixed with char to form a bio-oil/bio-char slurry. The slurry is pumpable and alleviates technical difficulties involved in solid biomass feeding and handling. This slurry is transported and added to a pressurized oxygen-blown entrained gasifier. The operating conditions of the gasifier at Freiberg, Germany, involve a slurry throughput of 0.35–0.6 tons/day, 26 bars, and 1200°C–1600°C. The current FZK process concepts involve gasification of flash-pyrolyzed wood products, slow-pyrolyzed straw char slurry (with water condensate), and slow-pyrolyzed straw char slurry (with pyrolysis bio-oil) [90]. Slurries from straws have been efficiently converted into syngas with high conversion and near-zero methane content [83].

Their ultimate objective is the development of an efficient *BtL* plant. A simplified block diagram of the FZK process concept leading to BtL is shown in Figure 15.15.

Canadian developments in biomass gasification for the production of medium- and high-Btu gases have also received worldwide technological attention. The *BIOSYN gasification* process was developed by Biosyn Inc., a subsidiary of Nouveler Inc., a division of Hydro-Quebec. The process is based on a BFB gasifier containing a bed

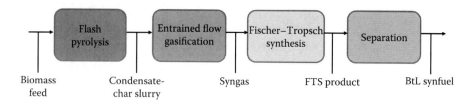

**FIGURE 15.15** FZK process concept for BtL synfuel.

of silica (or alumina) and can be operated at a pressure as high as 1.6 MPa. They tested the process extensively during 1984 till 1988 on a 10 ton/h demonstration plant that was comprised of a pressurized air- or oxygen-fed fluidized-bed gasifier [92]. The system has the ability to utilize a diversified array of feedstocks including whole biomass, fractionated biomass, peat, and MSW. The primary end use for the biogas is replacing the oil currently used in industrial boilers. It also has the added capability for producing synthesis gas for methanol or low-energy gas production. Later, they used a 50 kg/h BIOSYN gasification PDU, and the test program also proved the feasibility of gasifying a variety of other feedstocks, such as primary sludges, RDF, rubber residues containing 5%–15% Kevlar, granulated polyethylene, and polypropylene [93].

Another emerging process option for biomass gasification involves *supercritical water gasification of biomass*. Supercritical water (SCW) is water existing under a condition where both temperature and pressure are above its critical temperature and pressure, that is, $T > 374°C$ and $P > 218$ atm. At supercritical conditions, water exhibits extraordinary properties that are quite different and distinct from those of ambient water, as compared qualitatively in Table 15.12 [94].

## TABLE 15.12
### Physicochemical and Transport Properties of Water at Supercritical and Ambient Conditions

| Ambient Water | Properties Compared | Supercritical Water |
|---|---|---|
| Negligible to low | Organic solubility | Very high |
| Very high to high | Inorganic solubility | Negligible to very low |
| Higher | Density | Medium to high |
| Higher | Viscosity | Lower |
| Lower | Diffusivity | Higher |
| ~80 | Dielectric constant | 5.7 at critical point |
| High | Polarity | Low |
| Not | Corrosivity | Somewhat |
| Lower | Energetics | Highly energized |
| Fire extinguishing | Oxidation | Ideal combustion medium |
| 9.2 mg/L | Oxygen solubility | In any proportion |
| Low | $H_2$ solubility | Very low |

Biomass feedstocks can be gasified in a SCW medium at a temperature higher than about 650°C and pressure higher than 22.1 MPa [95,96]. While common gasification technology requires wet biomass to be sufficiently dried before gasification treatment, the technology based on gasification in SCW can handle wet biomass as is, without energy- and cost-intensive drying of the feed material. Boukis et al. [96] studied biomass gasification in near-critical and supercritical conditions using a pilot-scale process system called VERNA (a German acronym for "experimental facility for the energetic exploitation of agricultural matter") that had a throughput capacity of 100 kg/h and a maximum reaction temperature of 660°C at 28 MPa. The process system was capable of preparing large biomass particles into about <1 mm particle size using a cutting mill followed by a colloidal mill. The reactor was a downflow type, and the reactor system could handle the separation of brines and solids from the bottom of the reactor [96].

The principal gaseous products of SCW gasification are hydrogen, carbon monoxide, carbon dioxide, methane, and ethane. A generalized reaction scheme involves a *reformation* (steam gasification) reaction of hydrocarbons and oxygenates as well as *pyrolytic decomposition* reactions involving the cleavages of both C–C and C–H bonds. Carbon dioxide concentration usually increases with the temperature of reaction, while carbon monoxide decreases. The increase of carbon dioxide in the product stream is due to the results of the forward WGS reaction that converts carbon monoxide and water into carbon dioxide and hydrogen [97]. The methane formation for all ranges of gasification temperature is believed to have originated from pyrolytic decomposition of hydrocarbons and their intermediates, not by the methanation reaction of syngas, that is, $CO + 3H_2 = CH_4 + H_2O$ [98].

From the kinetics standpoint, the reformation reaction is more active than pyrolytic decomposition at higher temperatures, whereas the pyrolysis reaction is faster and more active than the reformation reaction at lower temperatures [99]. As such, the two pseudo-first reaction rates meet and cross each other at some point in the temperature domain, when the rates are plotted against the temperature or a reciprocal of temperature. This is a crossover point in the rates between the two representative chemical reactions, where the two reactions have an identical pseudo–first-order reaction rate. The higher the crossover temperature is, the more difficult the gasification (or reformation). The location of this kinetic crossover point is different from chemical to chemical and from one biomass type to another. The extent of the gasification or gasification efficiency, which is defined as the total carbon appeared in gas phase products divided by the total carbon in biomass feedstock entered into the reactor, depends very strongly upon the imposed reaction conditions such as the reaction temperature; space time; feed biomass/water ratio; molecular structures of biomass pertaining to the H/C ratio, O content, and –OH content; isothermality/nonisothermality of the reactor; monolithic catalytic effect from the reactor wall materials, and so forth. As the gasification temperature increases, the gasification efficiency generally increases to a certain point. However, the gasification efficiency does not change much with the pressure, as long as the water is in its supercritical fluid region. Picou et al. [100] demonstrated that the hydrocarbon reformation in SCW can be operated in an *autothermal mode* with numerous advantages including (1) 100% gasification efficiency, (2) energy-wise self-sustainable operation, and (3) comparable or enhanced product gas composition and yield. They demonstrated their technology using jet fuel in SCW at temperatures up to 775°C on a high-nickel Haynes Alloy 282 reactor system. *Autothermal reformation (ATR)* has been successfully practiced in the field of SMR [101], wherein a substoichiometric amount of air (or oxygen) is fed to the reformer for sacrificial oxidation reaction of hydrocarbons, thus generating the exothermic heat and promoting a reformation reaction. The conventional ATR has been operated catalytically on an industrial-scale subcritical reformation process at much higher temperatures, unlike the process Picou et al. [100] demonstrated using noncatalytic supercritical reformation technology. As shown in Table 15.12, oxygen fully dissolves in SCW, thereby easily facilitating an autothermal mode of operation in a SCW reactor.

Bouquet et al. [98] showed direct experimental evidence that hydroxyl functional groups present in the molecular structure of a feed chemical play an important role by making the reformation (or gasification) reaction proceed more efficiently via a mechanistically simpler pathway leading to syngas, that is, hydrogen and carbon oxides. They experimentally compared the kinetic results of the SCW gasification reactions of three different kinds of $C_3$ alcohols, namely, isopropyl alcohol, propylene glycol, and glycerin. The three alcohols used in the experiment represent monohydric, dihydric, and trihydric alcohols of $C_3$ hydrocarbons, respectively. The results bear significance in biomass gasification in SCW, since biomass is rich in hydroxyl groups due to its abundant cellulosic ingredients. In particular, biomass feedstock pretreatment should be carried out in such a way that hydroxyl groups in the molecular structure should not be prematurely destroyed or extracted if the biomass is to be gasified in SCW.

## 15.6 UTILIZATION OF BIOMASS SYNTHESIS GAS

Synthesis gas obtained by biomass gasification can be utilized in a variety of ways and end uses. While all conventional syngas utilization methods are conceivable and applicable, the process economics, based on the current fuel market and the current level of available technology, may not be favorable for the manufacture of bulk petrochemicals that have traditionally relied on the syngas derived from either NG or coal. However, in certain niche markets, biomass synthesis gas can favorably compete against the NG-based syngas or coal syngas. Furthermore, biomass syngas is $CO_2$ neutral and renewable as well as may be better suited for small- to medium-scale operations.

Syngas obtained from biomass gasification can be used for *indirect liquefaction* processes by which syngas is converted to liquid transportation fuels such as methanol, DME, ethanol, higher alcohols, gasoline, diesel, and jet fuels, as shown in Figure 15.16.

The following will have to be accomplished in order to convert the biomass syngas into clean liquid fuels that can be used in the conventional energy/fuel infrastructure:

1. Innovative process integration schemes need to be devised.
2. Highly efficient energy integration between the intraprocess and interprocess steps needs to be devised and achieved.

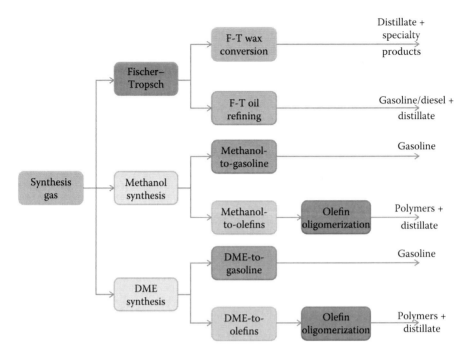

**FIGURE 15.16**  Transformation of syngas into liquid transportation fuels.

3. Effects of trace minerals in the biomass syngas on the process catalysts need to be fully understood and managed.
4. Robust and highly effective catalyst systems need to be developed and demonstrated in the long term.
5. Efficient conversion technology of $CO_2$-rich syngas needs to be developed and refined.
6. Biomass pretreatment technology needs to be enhanced.
7. Conversion technologies using mixed feedstocks need to be developed and refined.
8. Gas cleanup technology needs to be sufficiently enhanced in terms of the efficacy and cost.
9. Gas separation and purification technology needs to be enhanced.

## REFERENCES

1. M. J. Riedy and T. C. Stone, Defining biomass—A comparison of definitions in legislation, Mintz, Levin, Cohn, Ferris, Glovsky, and Popeo, PC, http://www.mintz.com/newsletter/2009/Special/Defining_Biomass.pdf, 2010.
2. R. N. Sampson, "Biomass management and energy," *Water, Air and Soil Pollution*, vol. 70, pp. 139–159, 1993.
3. D. W. MacCleery, "American forests: A history of resiliency and recovery," U.S. Dept. of Agriculture Forest Service, Tech. Rep., p. 59, 1993.
4. G. Trebbi, "Power-production options from biomass: The vision of a Southern European utility," *Bioresource Technology*, vol. 46, pp. 23–29, 1993.

5. S. Lee, J. G. Speight and S. K. Loyalka, *Handbook of Alternative Fuel Technology*. Boca Raton, FL: CRC Press, 2007.

6. "Biomass," Merriam-Webster.com, Merriam-Webster, 2014. Web. March 24, 2014.

7. R. N. Sampson, "Forest management and biomass in the USA." *Water, Air and Soil Pollution*, vol. 70, nos. 1–4, pp. 519–532, 1993.

8. R. A. Reese, "Herbaceous biomass feedstock production," *Energy Policy*, vol. 21, pp. 726–734, 1993.

9. J. Nurmi, "Heating values of the above ground biomass of small-sized trees," *ACTA Forestalia Fennica*, vol. 236, pp. 2–30, 1993.

10. World Energy Council (WEC), *Energy for Tomorrow's World*. New York: St. Martin's Press, 1993.

11. REN21 (Renewable Energy Policy Network for the 21st Century), "Renewables 2011: Global status report," 2011. http://www.ren21.net/ (accessed March 2013).

12. U.S. Forest Service, Woody biomass utilization, http://www.fs.fed.us/woodybiomass/whatis.shtml, 2011.

13. United States Department of Agriculture, United States Department of Energy and United States Department of the Interior, Memorandum of understanding on policy principles for woody biomass utilization for restoration and fuel treatments on forests, woodlands, and rangelands, http://www.fs.fed.us/woodybiomass/documents/BiomassMOU_060303_final_web.pdf, 2003.

14. S. Lee, *Alternative Fuels*. Philadelphia, PA: Taylor & Francis, 1996.

15. J. G. Speight, *The Chemistry and Technology of Coal*. New York: Marcel Dekker, 1994.

16. U.S. Department of Energy, Energy Efficiency and Renewable Energy, Biomass Program, Biomass program: Trace metal scavenging from biomass syngas using novel sorbents, http://www1.eere.energy.gov/biomass/pdfs/trace_metal.pdf, 2011.

17. J. Parikh, S. A. Channiwala and G. K. Ghosal, "A correlation for calculating HHV from proximate analysis of solid fuels," *FUEL*, vol. 84, pp. 487–494, 2005.

18. A. Demirbas, "Calculation of higher heating values of biomass fuels," *FUEL*, vol. 76, pp. 431–434, 1997.

19. B. M. Jenkins, "Downdraft gasification characteristics of major California residue derived fuels," Ph.D. Dissertation, University of California-Davis, Davis, CA, 1980.

20. B. M. Jenkins and J. M. Ebeling, "Correlation of physical and chemical properties of terrestrial biomass with conversion," in *Symposium Energy from Biomass and Waste IX*, Institute of Gas Technology, 1985, p. 371.

21. D. A. Tillman, *Wood as an Energy Resource*, Academic Press, New York, 1978.

22. S. A. Channiwala, "On biomass gasification process and technology development— Some analytical and experimental investigation," Ph.D. Thesis, Mechanical Engineering Dept., IIT, Mumbay, 1992.

23. A. V. Bridgwater, "Biomass fast pyrolysis," *Thermal Science*, vol. 8, pp. 21–49, 2004.

24. S. Lee, *Methanol Synthesis Technology*. Boca Raton, FL: CRC Press, 1990.

25. S. Lee, *Methane and its Derivatives*. New York: Marcel Dekker, 1997.

26. M. W. Wright, J. A. Satrio, R. C. Brown, D. E. Daugaard and D. D. Hsu, "Technoeconomic analysis of biomass fast pyrolysis to transportation fuels," National Renewable Energy Laboratory, Golden, CO, Tech. Rep. NREL/TP-6A20-146586, November 2010.

27. C. A. Mullen, A. A. Boateng, N. Goldberg, I. M. Lima and K. B. Hicks, "Bio-oil and bio-char production from corn cobs and stover by fast pyrolysis," *Biomass and Bioenergy*, vol. 34, pp. 67–74, 2010.

28. KBR Homepage, Coal gasification, http://www.kbr.com/Technologies/Coal-Gasification/, August 2011.

29. U.S. Department of Energy, National Energy Technology Laboratory (NETL), Gasification systems: Key area—Gasifier optimization, http://www.netl.doe.gov/technologies/coalpower/gasification/adv-gas/index.html, 2011.

30. M. Asadullah, T. Miyazawa, S. Ito, K. Kunimori and K. Tomishige, "Catalyst performance of Rh/CeO$_2$/SiO$_2$ in the pyrogasification of biomass," *Energy and Fuels*, vol. 17, pp. 842–849, 2003.

31. J. Arauzo, D. Radlein, J. Piskorz and D. S. Scott, "Catalytic pyrogasification of biomass. Evaluation of modified nickel catalyst," *Industrial and Engineering Chemistry Research*, vol. 36, pp. 67–75, 1997.

32. C. Y. Wen and J. Huebler, "Kinetic study of coal char hydrogasification. Rapid initial reaction," *Industrial and Engineering Chemistry Process Design and Development*, vol. 4, pp. 142–147, 1965.

33. S. Lee, J. C. Angus, R. V. Edwards and N. C. Gardner, "Non-Catalytic coal char gasification," *AIChE Journal*, vol. 30, pp. 583–593, 1984.

34. P. L. Walker, F. Rusinko and L. G. Austin, "Gas reactions in carbon," in *Advances in Catalysis*, D. D. Eley, P. W. Selwood and P. B. Weisz, Eds. New York: Academic Press, 1959, p. 133.

35. C. G. von Frederdorff and M. A. Elliott, "Coal gasification," in *Chemistry of Coal Utilization, Supplementary Volume*, H. H. Lowry, Ed. New York: Wiley, 1963, p. 892.

36. H. C. Hottel, G. C. Williams and P. C. Wu, "The reaction of coke with carbon dioxide," *American Chemical Society, Division of Fuel Chemistry*, vol. 10, no. 3, pp. 58–71, 1966.

37. J. F. Strange and P. L. Walker, "Carbon-carbon dioxide reaction—Langmuir-Hinshelwood kinetics at intermediate pressures," *Carbon*, vol. 14, pp. 345–350, 1976.

38. National Energy Technology Laboratory (NETL), U.S. Department of Energy, Gasifipedia, http://www.netl.doe.gov/technologies/coalpower/gasification/gasifipedia/4-gasifiers/4-1-4-3_hydro.html, 2011.

39. D. S. Scott, *Catalytic Hydrogasification of Wood*. Canada: Renewable Energy in Canada, 1986.

40. K. Walter, S. Friedman, L. V. Frank and R. W. Hiteshu, "Coal hydrogasification catalyzed by aluminum chloride," *ACS Fuel Preprint*, vol. 12, pp. 43–47, 1968.

41. K. Asami and Y. Ohtsuka, "Hydrogasification of brown coal with active iron catalysis," *ACS Fuel Preprint*, vol. 37, pp. 1951–1956, 1992.

42. Y. Nishyama and T. Haga, "Low temperature hydrogasification of carbons using nickel based catalyst," in *New Horizons in Catalysis: Part 7B. Proceedings of the 7th International Congress on Catalysis*, T. Seiyama and K. Tanabe, Eds. Tokyo, Japan: Elsevier, 1980, pp. 1434–1438.

43. T. Haga and Y. Nishyama, "Promotion of iron-group catalysts by a calcium salt in hydrogasification of carbons at elevated pressures," *Industrial and Engineering Chemistry Research*, vol. 26, pp. 1202–1206, 1987.

44. S. Porada, "A comparison of basket willow and coal hydrogasification and pyrolysis," *Fuel Processing Technology*, vol. 90, pp. 717–721, 2009.

45. T. A. Milne and R. J. Evans, Biomass gasifier "tars"; their nature, formation, and conversion. U.S. Department of Energy, National Energy Technology Laboratory, Pittsburg, PA, 1998.

46. D. C. Elliott, Ed., *Relation of Reaction Time and Temperature to Chemical Composition of Pyrolysis Oils*. Washington, D.C.: American Chemical Society, 1988.

47. C. Wu and P. T. Williams, "Nickel-based catalysts for tar reduction in biomass gasification," *Biofuels*, vol. 2, pp. 451–464, 2011.

48. C. Brage, Q. Yu, G. Chen and K. Sjöström, "Use of amino phase adsorbent for biomass tar sampling and separation," *Fuel*, vol. 76, pp. 137–142, 1997.

49. Technical committee CEN/BT/TF 143, Measurement of organic contaminants (tar) in biomass producer gas. Biomass gasification—Tar and particles in product gases—Sampling and analysis, http://www.eeci.net/results/pdf/CEN-Tar-Standard-draft-version-2_1-new-template-version-05-11-04.pdf, 2004.

50. Dynamotive Energy Systems, BioOil, http://www.dynamotive.com/industrialfuels/biooil/, 2011.

51. S. Lee, *Oil Shale Technology*. Boca Raton, FL: CRC Press, 1991.
52. S. Czernik, Reviews of fast pyrolysis of biomass. National Renewable Energy Laboratory, http://www.nh.gov/oep/programs/energy/documents/biooil-nrel.pdf, 2002.
53. S. Czernik and A. V. Bridgwater, "Overview of applications of biomass fast pyrolysis," *Energy and Fuels*, vol. 18, pp. 590–598, 2004.
54. R. L. Bain, An introduction to biomass thermochemical conversion. Presented at DOE/ NASLUGC Biomass and Solar Energy Workshops, http://www.nrel.gov/docs/gen/ fy04/36831e.pdf, 2004.
55. M. Ringer, V. Putsche and J. Scahill, "Large-scale pyrolysis oil production: A technology assessment and economic analysis," U.S. Department of Energy, National Energy Research Laboratory, Golden, CO, Tech. Rep. NREL/TP-510-37779, November 2006.
56. Conversion and Resource Evolution (CARE), Ltd. 250 kg/h biomass fast pyrolysis plant for power generation. Wellman Process Engineering Ltd., Oldbury, England, http://www.care. demon.co.uk/projectprofile07.pdf, 1998–2002.
57. D. S. Scott, P. Majerskib, J. Piskorz and D. Radlein, "A second look at fast pyrolysis of biomass—The RTI process," *Journal of Analytical and Applied Pyrolysis*, vol. 51, pp. 23–37, 1999.
58. VTT Technical Research Centre of Finland, "Review of Finnish biomass gasification technologies," Tech. Rep. Report No. 4, 2002.
59. A. T. Gradassi, "Fast pyrolysis of biomass: Work in progress at ENEL produzione," in *IEA—Clean Coal Sciences Agreement 23rd Meeting*, Pisa, Italy, 2002.
60. Biomass Technology Group (BTG), Pyrolysis oil, http://www.btgworld.com/index.php? id=22&rid=8&r=rd, November 2011.
61. C. Roy, "Vacuum pyrolysis breakthrough," *Pyrolysis Network*, no. 10, p. 4, 2000.
62. A. V. Bridgwater, "Principles and practice of biomass fast pyrolysis processes for liquids," *Journal of Analytical and Applied Pyrolysis*, vol. 51, pp. 3–22, 1999.
63. U.S. Department of Energy—Energy Efficiency and Renewable Energy—Biomass Program, Pyrolysis and other thermal processing, http://www1.eere.energy.gov/biomass/ printable_versions/pyrolysis.html, 2011.
64. R. L. Bain, "Electricity from biomass in the United States; Status and future direction," *Bioresource Technology*, vol. 46, pp. 86–93, 1993.
65. Y. Solantausta, "Wood-pyrolysis oil as fuel in diesel-power plant," *Bioresource Technology*, vol. 46, pp. 177–188, 1993.
66. F. Rick and U. Vix, "Product standards for pyrolysis products for use a s a fuel," in *Biomass Pyrolysis Liquids Upgrading and Utilization*, A. V. Bridgwater and G. Grassi, Eds. London: Elsevier Pub., 1991, pp. 177–218.
67. W. W. Skelley, J. W. Chrostowski and R. S. Davis, "The energy resources fluidized bed process for converting biomass to electricity," in *Symposium on Energy from Biomass and Wastes VI*, January 25–29, 1982, Lake Buena Vista, FL, pp. 665–705.
68. R. C. Brown and J. Holmgren, Fast pyrolysis and bio-oil upgrading, http://www.ascension- publishing.com/BIZ/HD50.pdf, 2011.
69. O. Levenspiel, *Chemical Reaction Engineering*. Hoboken, NJ: Wiley, 1999.
70. Agri-Therm, Agri-therm pyrolysis systems, http://agri-therm.com/, 2011.
71. J. Piskorz, P. Majerski, D. Radlein, D. S. Scott and A. V. Bridgwater, "Fast pyrolysis of sweet sorghum and sweet sorghum bagasse," *Journal of Analytical and Applied Pyrolysis*, vol. 46, pp. 15–29, 1998.
72. J. Piskorz and D. S. Scott, "The composition of oils obtained by the fast pyrolysis of different woods," *ACS Fuel Preprint*, vol. 32, pp. 215–222, 1987.
73. P. Kaushal, S. A. Mirhidi and J. Abedi, "Fast pyrolysis of biomass in bubbling fluidized bed: A model study," *Chemical Product and Process Modeling*, vol. 6, Article 24, pp. 1–36, 2011.

74. A. V. Bridgwater, S. Czernik and J. Piskorz, "An overview of fast pyrolysis," in *Progress in Thermochemical Biomass Conversion*, A. V. Bridgewater, Ed. Oxford, U.K., Blackwell Science, 2001, pp. 977–997.

75. M. van der Velden, X. Fan, A. Ingramz and J. Baeyens, "Fast pyrolysis of biomass in a circulating fluidized bed," in *2007 ECI Conference on the 12th International Conference on Fluidization—New Horizons in Fluidization Engineering*, Harrison Hot Springs, Canada, May 13–18, 2007.

76. P. G. Anderson, "Twin screw extrusion," in *Encyclopedia of Chemical Processing*, S. Lee, Ed. New York: Taylor & Francis, 2007.

77. W. R. Benson, "Biomass potential from agricultural production," in *Proceedings: Biomass-a Cash Crop for the Future? Conference on the Production of Biomass from Grains, Crop Residues, Forages and Grasses for Conversion to Fuels and Chemicals*, Kansas City, MO, 1977.

78. The Guardian Data Blog, U.S. corn production and use for fuel ethanol, http://www.guardian.co.uk/environment/datablog/2010/jan/22/us-corn-production-biofuel-ethanol, 2010.

79. K. L. Kadam and J. D. McMillan, "Availability of corn stover as a sustainable feedstocks for bioethanol production," *Bioresource Technology*, vol. 88, pp. 17–25, 2003.

80. K. P. Raman, W. P. Walawender, Y. Shimizu and L. T. Fan, "Gasification of corn stover in a fluidized bed," in *Bio-Energy 80*, Atlanta, GA, 1980.

81. B. Bodland and J. Bergman, "Bioenergy in Sweden: Potential, technology, and application," *Bioresource Technology*, vol. 46, nos. 1–2, pp. 31–36, 1993.

82. Finnish Ministry of the Environment, Communications Unit, Gasified biomass for efficient power and heat generation—FACTS on environmental protection, http://www.ymparisto.fi/download.asp?contentid=68164&lan=fi, 2007.

83. A. van der Drift and H. Boerrigter, "Synthesis gas from biomass for fuels and chemicals," in *Report of Workshop on "Hydrogen and Synthesis Gas for Fuels and Chemicals," Organized IEA Bioenergy Task 33, SYNBIOS Conference*, Stockholm, Sweden, 2006.

84. K. R. Craig and M. K. Mann, Cost and performance analysis of biomass-based integrated gasification combined-cycle (BIGCC) power systems. National Renewable Energy Laboratory, U.S. Department of Energy, http://www.nrel.gov/docs/legosti/fy97/21657.pdf, 1996.

85. J. Yan, P. Alvfors, L. Eidensten and G. Svedberg, "A future for biomass," *ASME Magazine*, October 1997. http://www.memagazine.org/backissues/membersonly/october97/features/biomass/biomass.html (accessed in 2011).

86. M. A. Paisley, J. M. Irving and R. P. Overend, "A promising power option—The FERCO SilvaGas biomass gasification process—operating experience at the burlington gasifier," in *Proceedings of ASME Turbo Expo 2001, June 4–7, 2001*, New Orleans, LA, 2001, pp. 1–7.

87. M. Claussen and S. Vodegel, "The CUTEC concept to produce BtL-fuels for advanced power trains," in *International Freiberg Conference on IGCC and XtL TECHNOLOGIES*, Freiberg, Germany, 2005.

88. Y. T. Shah, J. Ontko, T. Gardner, D. Barry and W. Summers. "Torrefaction," in *Encyclopedia of Chemical Processing*, S. Lee, Ed., Taylor & Francis, New York, 2012.

89. C. Wolters, M. Canaar and J. Kiel, Co-generation of biomass in 250 MWe IGCC plant "Willem-Alexander-Centrale." Presented at Biomass Gasification Workshop at Rome, Italy, http://www.gastechnology.org/webroot/downloads/en/IEA/IEARomeWSKiel.pdf, 2004.

90. E. Dinjus, German developments in biomass gasification. Presented at IEA Renewable Working Party Bioenergy Task 33: Thermal Gasification, Innsbruck, Austria, http://www.gastechnology.org/webroot/downloads/en/IEA/Fall05AustriaTaskMeeting/GermanyGasificationActivities.pdf, 2005.

91. E. Henrich, The status of the FZK concept of biomass gasification. Presented at 2nd European Summer School on Renewable Motor Fuels, http://www.baumgroup.de/Renew/download/5%20-%20Henrich%20-%20slides.pdf, 2007.

92. R. D. Hayes, "Overview of thermochemical conversion of biomass in canada," in *Biomass Pyrolysis Liquids: Upgrading and Utilization*, A. V. Bridgwater and G. Grassi, Eds. Amsterdam, The Netherlands: Elsevier Science Publ., 1991.

93. S. Babu, "Biomass gasification for hydrogen production—process description and research needs," http://ieahia.org/pdfs/Tech_Report,_Babu,_IEA_Bioenergy_Thermal_Gas_Task.pdf (accessed March 21, 2014).

94. S. Lee, H. B. Lanterman, J. E. Wenzel and J. Picou, "Noncatalytic reformation of JP-8 fuel in supercritical water for production of hydrogen," *Energy Sources: Part A: Recovery, Utilization, and Environmental Effects*, vol. 31, pp. 1750–1758, 2009.

95. M. J. Antal Jr., S. G. Allen, D. Schulman and X. Xu, "Biomass gasification in super-critical water," *Industrial and Engineering Chemistry Research*, vol. 39, pp. 4040–4053, 2000.

96. N. Boukis, U. Galla, H. Muller and E. Dinjus, "Biomass gasification in supercritical water experimental process achieved with the VERENA pilot plant," in *15th European Biomass Conference and Exhibition*, Berlin, Germany, pp. 1013–1016, 2007.

97. J. Picou, M. S. Stever, J. S. Bouquet, J. E. Wenzel and S. Lee, "Kinetics of the noncata-lytic water gas shift reaction in supercritical water production," *Energy Sources, Part A: Recovery, Utilization, and Environmental Effects*, in press, 2014.

98. J. S. Bouquet, R. E. Tschannen, A. C. Gonzales and S. Lee, "The effects of carbon-to-oxygen ratio upon supercritical water reformation for hydrogen production," in *2011 AIChE Annual Meeting, Symposium on Alternative Fuels I*, Minneapolis, MN, October 16–20, 2011.

99. S. Lee, H. B. Lanterman, J. Picou and J. E. Wenzel, "Kinetic modeling of JP-8 reforming by supercritical water," *Energy Sources, Part A: Recovery, Utilization, and Environmental Effects*, vol. 31, pp. 1813–1821, 2009.

100. J. Picou, J. E. Wenzel, H. B. Lanterman and S. Lee, "Hydrogen production by non-catalytic autothermal reformation of aviation fuel using supercritical water," *Energy and Fuels*, vol. 23, pp. 6089–6094, 2009.

101. K. Aasberg-Petersen, "Synthesis gas," in *Encyclopedia of Chemical Processing*, S. Lee, Ed. London: Taylor & Francis, pp. 2933–2946, 2005.

# 16 Energy Generation from Waste Sources

*Sunggyu Lee*

## CONTENTS

## 16.1   INTRODUCTION

Solid wastes are, by definition, any wastes other than liquids or gases that are no longer deemed valuable and therefore discarded.[1] Such wastes typically originate from either the residential community (i.e., municipal solid waste [MSW]) or commercial and light-industrial communities. Wastes generated from manufacturing activities of heavy-industrial and chemical industries are typically classified as hazardous wastes. As regulations continue to get stricter with decreasing land availability, alternative uses for the waste must be found in order to recover the residual heating values as well as to alleviate landfill-overburdening problems. As indicated by the heating values listed in Table 16.1, the generation of waste-derived fuels appears to be very promising from both the environmental and energy points of view.[2,3]

In a landfill, the biodegradable components of the MSW decompose over time and produce methane. Methane is a very potent *greenhouse gas* (*GHG*) that is known to be 23 times more potent than carbon dioxide. Therefore, the biodegradable components in MSW, such as food and paper wastes that end up in landfills, must be reduced in a good waste management plan. Many nations have implemented regulations and specific efforts to significantly reduce this. A good example is the European Union's (EU's) *landfill directive of 1999*, which targets reducing the amount of biodegradable materials going to landfills by 65% of the 1995 level by 2016. From this perspective, energy generation from MSW is regarded as one of the very few viable options to efficiently cope with the GHG emission problem from landfills.

Although *refuse-derived fuel* (*RDF*) is attractive from the standpoints of resource conservation as well as waste reduction, there are serious concerns that waste treatment for energy generation may cause new environmental problems. Most of the processing difficulties arise from the heterogeneous and nonuniform nature of the waste feed itself, which, in turn, generates a very widely varying spectrum of treated intermediates and by-products. Often, these intermediates and by-products are environmentally hazardous, if not isolated or managed improperly.

### TABLE 16.1
### Comparison of Heating Values of Various Waste-Derived Fuels

| Fuel Source | Btu/lb |
|---|---|
| Yard wastes | 3000 |
| Municipal solid waste | 6000 |
| Combustible paper products | 8500 |
| Textiles and plastics | 8000 |
| Bituminous coal (average) | 11,300 |
| Anthracite coal (average) | 12,000 |
| Spent-tire rubber | 13,000–16,000 |
| Crude oil (average) | 17,000 |
| Natural gas (425 ft³) | 13,500 |

This chapter will address the development of alternative energy sources from various solid waste classifications, except biological and agricultural wastes, which are covered separately in Chapter 15.

## 16.2  ENERGY RECOVERY FROM MSW

### 16.2.1  INTRODUCTION

The recovery of energy from MSW has been practiced for centuries. The burning, or incineration, of wastes such as wooden planks and miscellaneous household products was first used to produce warmth. This idea has become the basis for energy generation from today's MSW. For instance, each year, Sweden burns 1.5 million tons of MSW to meet approximately 15% of its *district heating* requirements.[4] The heating value of this MSW incineration is approximately one-third the heating value of coal combustion. Besides incineration, "gaseous fuels" can also be obtained by *anaerobic digestion* in conjunction with landfill gas (LFG) recovery. Anaerobic digestion is discussed in detail in Chapter 12.

### 16.2.2  GASIFICATION OF MSW

One method of recovering usable energy from MSW is gasification. The US Environmental Protection Agency (EPA) is currently investigating the use of the *Texaco gasification process* for generating a medium-Btu gas from MSW.[5] A simplified Texaco process (Figure 16.1) gasifies the MSW under high pressure by the injection of air and steam with concurrent gas/solid flow. After separation of the noncombustible waste, water or oil is added to the combustible MSW to form a pumpable slurry. This is then pumped under pressure to the gasifier. In the gasifier, the slurry is reacted with air at high temperatures. The resultant gaseous product is then sent to a scrubber system to remove any pollutants and impurities.

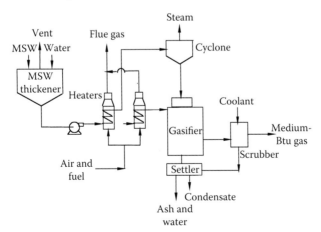

**FIGURE 16.1**  Simplified Texaco gasification process for the conversion of MSW to a medium-Btu gas.

The Texaco gasification process was originally developed for gasification of coal (refer to Chapter 2). Recently, it has been modified to treat soils contaminated with hydrocarbons, as well as to recover usable energy from MSW and polymeric wastes. Although the process has demonstrated 85% remediation efficiency for contaminated soils, it is still in the preliminary experimental stages for the MSW-to-energy application.[5] A demonstration of the MSW-to-energy process using a pilot-scale Texaco gasifier was successfully conducted in 1995 at Texaco's Montebello Research Laboratory in South El Monte, CA.[5] It was successfully demonstrated that the gasification technology offers an environmentally acceptable option for waste treatment.

### 16.2.3　ANAEROBIC DIGESTION OF MSW

Anaerobic digestion of solid wastes is a process very similar to that used at *wastewater treatment* facilities and also to that used in *biogas production*. Anaerobic bacteria, in the absence of oxygen, are used to break down the organic matter of the waste. Frequently, the MSW is mixed with sewage sludge from the treatment plant to enhance the efficiency of the digestion. During the "conversion," a mixture of methane and carbon dioxide gases is produced. The product gas is called "landfill gas" (LFG). The typical ratio of the gas mixture is approximately 50% methane and 50% carbon dioxide. Even without further treatment, the off-gas has a heating value of 450 to 600 Btu/ft$^3$. It should be noted that the principal ingredients (both methane and $CO_2$) of the off-gas are major GHGs. Therefore, its efficient capture and recovery is very important from the environmental standpoint.

With rising energy costs and diminishing landfill space, the use of anaerobic digestion to generate a potential fuel source from MSW is an attractive alternative. A relatively new technology of LFG recovery has been developed to aid in the collection of gases generated from the anaerobic digestion of solid wastes. In 1980, 23 landfills were used as a source of methane production.[6] It is a reliable source of energy, since it is generated 24 h a day and 365 days a year. By properly using the LFG in energy generation, landfills can significantly reduce their emission of methane, a major GHG, and reduce the need for generation of energy from other fossil fuels, thereby reducing emissions of carbon dioxide, sulfur dioxide, $NO_x$, and other harmful combustion by-products. The use of LFG for combined heat and power (CHP) cogeneration in Europe is very active, with a large number of plants currently in operation. In the United States, several hundred projects of LFG utilization in creative areas are currently being operated. These creative LFG utilization projects include CHP generation, heating greenhouses, firing brick kilns, producing high-Btu gas, fueling garbage trucks, and more.[7]

The vast majority of current research and development (R&D) focuses on the generation of liquid fuels instead of the gaseous fuels from anaerobic digestion owing to the high capital cost associated with methane collection and purification as well as carbon dioxide sequestration.

Production of liquid fuels from LFG has several advantages. First, low-sulfur, low-ash fuels can be made for commercial use.[8] Second, liquid fuels are traditionally much easier to store, handle, and transport than their gaseous counterparts. Finally, the production of the fuel aids in the battle against pollution by municipal wastes.[9,10]

By utilizing the waste as an alternative source to generate fuel, less MSW will have to be disposed of in landfills, and GHG emission due to unrecovered LFG will also be reduced. Although this statement holds true for the production of gaseous fuel, the generation of liquid fuels utilizes more MSW. In fact, processes have been developed that are capable of producing over a barrel of pyrolytic fuel oil from a ton of MSW.[7]

## 16.2.4 Pyrolysis of MSW

Figure 16.2 contains typical material distribution data for MSW generation in the United States in the early 1990s. Figure 16.3 shows more recent data for the early 2000s on MSW distribution in the United States by relative percentages. Because raw MSW contains both noncombustible and combustible components, the first step in producing a liquid fuel is to concentrate the combustible components. This is usually achieved with a rotating screen to remove noncombustible materials such as glass and dirt. An air classifier is used to remove the "light ends" such as plastics, wood, and small metals. Heavier components, ceramics, heavy metals, and aluminum are routed for disposal in the landfill. By removing these noncombustible materials, the heating value of the raw MSW becomes approximately 7000 Btu/lb on a wet basis. The combustible components are sent to a shredder to reduce their size, and then to a pyrolysis unit to generate the fuel by *pyrolysis reactions*.

In the past, pyrolysis of MSW was mainly used to generate a gaseous fuel. However, recent research has found that pyrolyzing a cellulose-based waste at 116°C and atmospheric pressure will generate a liquid fuel. In 1988, approximately 80% of the 180 million tons of waste generated in the United States had a cellulose base.[11] The estimated percentages of cellulosic components of MSW did not change much in the 1990s and 2000s, as shown in Figures 16.2 and 16.3. Pober and Bauer[8] were able to utilize this particular type of waste to obtain a fuel that contained 77% of the

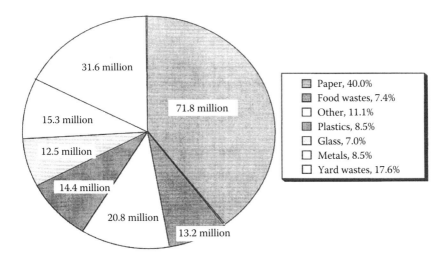

**FIGURE 16.2** Material distribution of MSW collected in the United States by weight (tons) in the early 1990s.

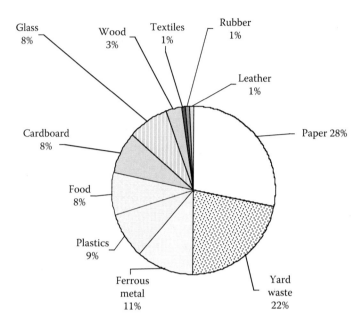

**FIGURE 16.3** Breakdown of material distribution of MSW collected in the United States in 2002.

heating value of typical petroleum fuels. The cellulosic components of the refuse comprise papers, newsprints, packing materials, wood clippings, and yard wastes.[12]

A typical pyrolytic reaction for the cellulose component of MSW may be written as

$$3C_8H_{10}O_5 \xrightarrow{\Delta} 8H_2O + 2CO + 2CO_2 + CH_4 + H_2 + 13C + C_6H_8O \quad (16.1)$$

where $C_6H_8O$ represents a "family" of liquid products. The exact composition of $C_6H_8O$ is dependent on feedstock composition and reaction temperature. Research conducted by the US Bureau of Mines has further demonstrated the successful pyrolysis of 1 ton of MSW at temperatures ranging from 500°C to 900°C.[13] At these conditions, the end product composition is similar to that given in Table 16.2.

The light oil primarily comprises benzene, whereas the liquor contains dissolved organics in water. The gaseous product resembles that of a typical *town gas*. The heating value of the gas is 447 Btu/ft³, which translates into a heat recovery of 82%.[14] Town gas is also known as *manufactured gas*, which typically contains hydrogen, carbon monoxide, methane, and volatile hydrocarbons. Prior to natural gas supplies and transmission in the United States, town gas played a major role in providing gaseous fuel for lighting and heating in both residential and industrial sectors.

When the pyrolysis temperature is higher than 350°C, small quantities of polyethylene chips can be added to the cellulose feedstock. Operating temperatures ranging from 100°C to 400°C decrease the amount of gaseous product and increase the formation of the liquid product (i.e., $C_6H_8O$ family).

**TABLE 16.2**

**Final Product Composition from the Pyrolysis of 1 Ton of MSW**

| Component | Mass or Volume |
|---|---|
| Char | 154–424 lb |
| Tar | 0.5–6 gal |
| Light oil | 1–4 gal |
| Liquor | 97–133 gal |
| Gas | 7.38–18 scf |

*Source:* From Bell, P.R., and Varjavandi, J.J., *Waste Management, Control, Recovery and Reuse*, Ann Arbor Science, MI, 1974.

In addition to the liquid fuel or oil, pyrolysis generates a medium-BTU gas stream that, after purification, can be recycled as a supplemental fuel within the plant. Process water and char are also generated. The process water may have to be treated before discharge or can be used in the plant as heat exchanger water. All the pyrolysis products have the potential of being useful fuels or intermediates for producing other valuable products for use in the petrochemical industry. A simplified schematic of the pyrolysis process such as the one developed by Pober and Bauer[8] and the full-scale plant in Ames, IA, is shown in Figure 16.4. A schematic of the pilot plant used to demonstrate the pyrolysis of MSW, which generated the results presented in Table 16.2, is shown in Figure 16.5.[13]

An example of a successful commercial-scale MSW pyrolysis plants in full operation is the Müllpyrolyseanlage (MPA) MSW pyrolysis plant, which is located outside of the City of Burgau, Germany.[15] The plant is on a 3-acre lot adjacent to a closed landfill, surrounded by farmlands. The plant was commissioned in mid-1984 and is currently in full operation, serving 120,000 residents and processing about 38,580 tons per year of MSW, which include both residential and industrial wastes as well as sewage sludge. The process utilizes a thermal pyrolysis process designed by WasteGen U.K. Ltd. The process is typically operated at 400°C–900°C in the complete absence of oxygen. The produced syngas mainly contains hydrogen,

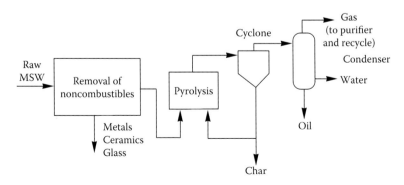

**FIGURE 16.4**  A simplified process schematic for the pyrolysis of MSW.

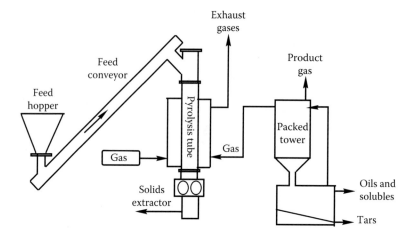

**FIGURE 16.5** A solid waste pyrolysis system. (From Bell, P.R., and Varjavandi, J.J., *Waste Management, Control, Recovery and Reuse*, Ann Arbor Science, MI, 1974.)

carbon monoxide, carbon dioxide, and methane, and the air emission control system removes most of the air pollutants. The syngas is used in boilers, gas turbines, and internal combustion engines to generate electricity, or it is used for the manufacture of chemicals. The solid char material is used as an absorbent, and the inorganic ash is sent for disposal. The process does not require any special presegregation or pretreatment of MSW feedstock. However, MSW must be shredded to a maximum size of 12 in. The average heating value of MSW feedstock is about 3660 Btu/lb, ranging between 2150 and about 6000 Btu/lb. A schematic of the *MPA MSW Pyrolysis Process* is shown in Figure 16.6.[15]

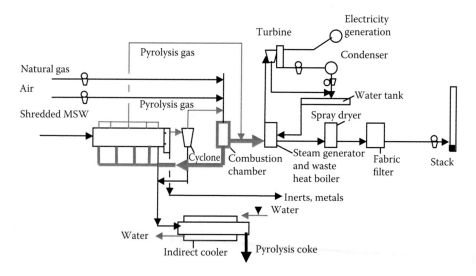

**FIGURE 16.6** A schematic of the Müllpyrolyseanlage (MPA) MSW pyrolysis process.

Although there are a number of MSW-derived fuel systems in full operation or being started up throughout the world, there are still developmental issues in regard to process design and engineering, pollution control and monitoring, equipment design and selection, and application-related information.[16] The cost–benefit analysis along with the environmental impact analysis must be carefully conducted for any chosen process and site.

## 16.3  ENERGY GENERATION FROM POLYMERIC WASTES

### 16.3.1  INTRODUCTION

A significantly important component of MSW is polymeric materials. Although polymer waste only accounts for 8.5% by mass of the total MSW disposed of in the United States, plastics represent over 28% by volume.[17] Furthermore, most plastic wastes are not biodegradable and constitute a long-term waste management problem. The high volume-to-mass ratio and the inertness to biological reactions make plastics and polymeric wastes the most important target for recycling, not for disposal. Polymeric wastes range from packaging materials used in the food industry to various parts in automobiles to high-density polyethylene (HDPE) and polyethylene terephthalate (PET) containers such as pop bottles, laundry detergent bottles, milk jugs, and so forth. In 1993, over 50% of all the food packaged in Europe for distribution utilized plastics.[4] It is estimated that over 65% of the food packaging in the United States comes from plastics. As of 1978, the Ford Motor Company estimated that the average junked car contained 80 kg of plastic and nontire rubber.[18] The Ford Motor Company also stated that the number was expected to rise at the rate of 4% per year. Metallic parts were replaced by plastic ones to reduce the vehicle weight as well as the component cost. About 7.5% by weight of an automobile came from plastic components in 1998, which is a substantial increase from 4.6% in 1977.[19] The same trend is also observed in the computer and electronics industry, which depends very heavily on plastics for their fabrication materials. Moreover, the expected usable life of electronic goods and computer components is typically quite short, thus generating an enormous amount of scrap polymeric wastes.

Over the next few decades, the use of polymeric materials will continue to increase, owing to their versatility, functional values, and low energy requirements for production. Current technology enables over 200 million kg of plastic and rubber materials (excluding tire rubber) to be recovered from shredded automobiles.[18] The impetus to reuse polymeric wastes instead of disposing of them in landfills is primarily owing to the inability of polymers to rapidly degrade once dumped or buried at landfills.

### 16.3.2  MECHANICAL RECYCLING

The manufacture of bottle containers from HDPE and PET is perhaps the largest use of polymeric materials. Some of the more common items manufactured from HDPE and PET are soft drink bottles, juice containers, milk jugs, laundry detergent bottles, spring water bottles, and motor oil cans. Austria introduced a regulation in October 1993 stating that over 90% of all HDPE containers must be recycled instead of placed in landfills.[20] Germany has been the most aggressive in its demand for plastics recycling. In 1993, over 12% of all German municipalities were active in collecting 12,000 tons of polymeric containers.[21,22] Officials have projected that this number will increase to 80,000 tons of containers from 62% of the municipalities by 1996, and nearly 100% by the early 2000s.

Most other countries have also jumped on the recycling bandwagon. The Netherlands plans to recycle 35% of all plastics and to recover energy by incinerating another 45%.[23] Italy currently recycles over 40% of its containers. Several states within the United States have also enforced mandatory recycling of HDPE. In order to ensure consumer involvement, states such as Michigan impose a deposit "tax" on all their HDPE bottles. The consumer is able to recover the deposit only if the item is returned to designated stores or distributors. Other states have developed a recycling "lottery."[1] Sanitary officials randomly pick an area and check for proper recycling of domestic waste. The homeowners complying with all recycling guidelines receive $200. In 2013, nearly all municipalities in the Organization for Economic Co-operation and Development (OECD) nations adopted rigorous recycle programs for plastics.

There are a great many different types of plastics. The American Society of Plastics Industry (SPI) has developed a standard code system to help consumers identify and sort the main types of plastics. There are seven designated types, and each type is represented by a number: "1" for PET, "2" for HDPE, "3" for polyvinyl chloride (PVC), "4" for low-density polyethylene (LDPE), "5" for polypropylene (PP), "6" for polystyrene (PS), and "7" for all others or miscellaneous, such as melamine and polycarbonate (PC).

Most recycled plastics are used to produce, by melt processing, polymeric objects and articles usually of inferior quality or for rugged applications. However, recycling alone is not the solution to polymeric waste. Research conducted at DOW Chemical Co. has shown that more than 52% of all HDPE bottles would have to be recovered before mechanical recycling could save more energy than employing a waste-to-energy (WTE) process such as incineration.[24] For this reason, scientists are striving to develop processes to generate fuel from polymeric wastes.

### 16.3.3 Waste-to-Energy Processes

There are a number of technologies available for the generation of fuel energy, either in gaseous or liquid form, from polymeric waste. These technologies include pyrolysis, thermal cracking, catalytic cracking, and degradative extrusion followed by partial oxidation. Brief descriptions of these technologies follow. However, it is important to note that these are not the only processes available. As the technology advances, so do the processes used for the recovery of energy from wastes.

#### 16.3.3.1 Pyrolysis

Pyrolysis of polymers is, by definition, the thermal decomposition of plastic waste back into oil and/or gas using heating processes in oxygen-starved or oxygen-deprived environments. This is the process typically used for WTE generation because it has several advantages. Besides being a proven technology, pyrolysis has relatively good adaptability to fluctuations in the quality of the feedstock, is simple to operate, and is also economical. In addition, refineries, which will utilize the chemically recycled polymers, already have pyrolysis units in operation. SPI's Council for Solid Waste Solutions originally comprised three of the leading petroleum companies, Amoco, Mobil, and Chevron, even though all these companies went through corporate mergers with other leading companies. The council investigated the use of oil refineries equipped with pyrolysis units for the conversion of mixed plastics into hydrocarbons.[25] Ideally, these hydrocarbons would be identical to those split from petroleum oils.

Research conducted by Chambers et al.[18] implements the pyrolysis of polymeric wastes in the presence of molten salts. The molten salts are used to enhance the production of a particular desired oil–gas mix based on their excellent heat transfer properties. For instance, when a mixture of LiCl–KCl and 10% CuCl was used as the pyrolysis medium at 520°C, over 35% of the shredded polymer was converted to fuel oil.[18]

Molten salts have also been successful at slightly lower operating temperatures. When the salts are used during a standard pyrolysis operation at 420°C, the gaseous fraction is minimized. This enables higher liquid and solid fractions to be recovered. Although the nature of chemical reactions between the molten salts and polymeric waste is not yet completely understood, it does appear that production of a particular product mix is optimized by the presence of the salts. Typically, the recovered fractions consist of light oils, aromatics, paraffin waxes, and monomers.[26] When pyrolysis utilizes molten salts, care must be taken to decrease the amount of corrosion and contamination of the pyrolysis chamber due to the salts.

In 1980, researchers at Germany's Federal Ministry of Research and Technology (FMRT) demonstrated the feasibility of pyrolyzing polymeric wastes in a 10 kg/h pilot plant. Figure 16.7 shows a simplified flow diagram of the laboratory test

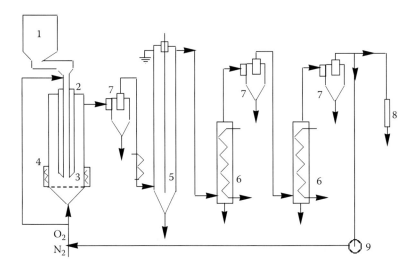

**FIGURE 16.7**    A simplified flow diagram of Germany's Federal Ministry of Research and Technology laboratory test plant—1: feed hopper, 2: downpipe and cooling jacket, 3: fluidized bed reactor, 4: heater, 5: electrostatic precipitator, 6: intensive cooler, 7: cyclone, 8: gas sampler, and 9: compressor. (From Kaminsky, W., *Resour. Recovery Conserv.*, 5, 1980.)

plant used in the development of the full-scale pilot plant. The main body of the FMRT pilot plant is a fluidized bed reactor with a space–time ratio of 0.4 kg/h/L. Preliminary experiments with the pilot plant enabled 40%–60% of polymeric feed to be recovered as a usable liquid product.[27] The major components of this liquid product were benzene, toluene, styrene, and $C_3$–$C_4$ hydrocarbons. Table 16.3 presents the data for all of the end products for two pyrolysis experiments. These experiments utilized waste polyethylene and spent syringes as the feed at pyrolysis temperatures of 810°C and 720°C, respectively.

As of 1981, the pilot plant had accumulated over 600 h of successful operation. The researchers at FMRT have also demonstrated that the plant was self-sufficient in regard to energy needs. In other words, FMRT's facility was able to utilize one-half of the pyrolysis gas produced.[27]

**TABLE 16.3**
**Composition of Pyrolysis Products from Preliminary Pilot Plant Studies**

| Identified Products | Polyethylene (wt%) | Spent Syringes (wt%) |
|---|---|---|
| Hydrogen | 1.2 | 0.49 |
| Methane | 18.8 | 18.82 |
| Ethane | 6.2 | 7.75 |
| Ethylene | 17.9 | 13.73 |
| Propane | 0.2 | 0.08 |
| Propene | 7.2 | 10.67 |
| Butene | 1.0 | 3.32 |
| Butadiene | 1.5 | 1.39 |
| Cyclopentadiene | 0.8 | 2.79 |
| Other aliphatics | 1.3 | 3.46 |
| Benzene | 21.6 | 13.62 |
| Toluene | 3.8 | 3.84 |
| Xylene, ethylbenzene | 0.2 | Trace |
| Styrene | 0.4 | 0.43 |
| Indane, indene | 0.6 | 0.46 |
| Naphthalene | 3.7 | 2.46 |
| Methylnaphthalene | 0.6 | 0.92 |
| Diphenyl | 0.3 | 0.33 |
| Fluorene | 0.1 | 0.14 |
| Phenanthrene | 0.6 | 0.33 |
| Other aromatics | 0.7 | 1.15 |
| Carbon dioxide | 0.0 | Trace |
| Carbon monoxide | 0.0 | Trace |
| Water | 0.0 | Trace |
| Acetonitrile | 0.0 | Trace |
| Waxes, tars | 9.3 | 5.07 |
| Carbon residue, fillers | 1.8 | 5.80 |
| Balance | 99.8 | 97.05 |

### 16.3.3.2  Thermal Cracking

Thermal cracking is similar to pyrolysis in that it is a high-temperature process. When polymeric wastes are cracked in an oxygen-free environment at temperatures above 480°C, a mixture of gas–liquid hydrocarbon is produced. At higher temperatures (650°C–760°C), more of the gaseous product is generated. Conversely, at lower temperatures, up to 85% of the product is a liquid hydrocarbon.[28] Both the gas and liquid forms of the converted mixed-polymeric waste can be utilized as a feed stream by petroleum facilities.

### 16.3.3.3  Catalytic Cracking

Scientifically speaking, catalytic cracking is an extrapolation of thermal cracking. The same operating principles apply; the primary difference is the addition of a catalyst to enhance the cracking process. A typical catalytic cracker consists of two large reactor vessels, one to react the feed over the hot catalyst and the other to regenerate the spent catalyst by burning off carbon (fouled carbon on the catalytic surface) with air. Using two catalytic reactors enables the process to be run on a continuous basis. Figure 16.8 is a simplified schematic of *Mobil Oil's* process for the generation of gasoline from polymeric wastes via catalytic cracking.[29] Because the theory behind catalytic cracking is not new, the main focus of research is to determine the optimum catalyst for the cracking of the waste polymers.

Studies using *organotin* compounds have shown promise for generation of fuel from polyurethanes (PURs). However, the highest activity was exhibited when used for glycolysis (oxidation of glucose), not cracking.[30] The use of chromium compounds has also been investigated. Scheirs and associates[31] have demonstrated the feasibility of the use of chromium to aid in the cracking and pyrolysis of HDPE. Although chromium compounds appear to be relatively effective, more research needs to be

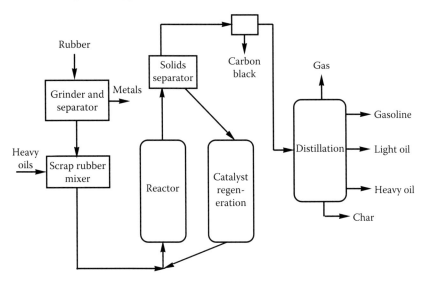

**FIGURE 16.8**  A simplified schematic of the Mobil Oil Corp. process for gasoline production from polymeric waste.

done to determine the optimum solubility of the catalyst for each polymeric compound. The addition of platinum and iron over activated carbon has also been investigated. Specifically, the activity of the catalysts for the degradation of PP waste into aromatic hydrocarbons was studied. The addition of these metals increased the yield of the aromatics from PP. It has been speculated that the increase in activity is influenced by the methyl branching of the PP. However, the exact mechanism is still not clearly understood. Especially, aromatization from aliphatic polymeric materials needs to be investigated mechanistically.

Perhaps the most widely studied classification of catalysts for cracking operations is that of solid acid catalysts.[32–34] Specifically, HZSM-5, HY, and rare-earth metal-exchanged Y-type (REY) zeolites and silica–alumina have been investigated.[34] Only HZSM-5 was found to be unsuitable for cracking polymeric wastes, while HY, REY, and silica–alumina were all capable of producing at least 30% gasoline (or gasoline-range hydrocarbons) and 20% heavy oil. The differences between the catalysts arise in the production of the coke and gas fractions. Formation of coke is not desirable, because it causes catalytic deactivation by fouling and plugging the pore paths. Songip et al.[34] found that the incorporation of rare-earth metals in HY zeolite increased the gasoline yield and decreased coke formation. Regardless of the catalyst used, catalytic cracking, owing to lower-temperature operation, appears to be more technologically sound than pyrolysis. However, catalytic processes are often costlier than noncatalytic counterparts because of the costs involved in catalysts, catalytic reactors, and catalyst regeneration steps.

### 16.3.3.4   Degradative Extrusion

Degradative extrusion is, in principle, not a new technology; however, its use as a process for energy generation is a new technology. The basis of the technology is that at high temperatures, under the simultaneous effect of shearing, it is possible to break down complex mixtures of plastics into homogeneous low-molecular-weighted polymer melts.[35] These polymer melts could replace the heavy oils used in the production of synthesis gas. The polymer-derived heavy oil would be fed directly into a partial-oxidation fluidized bed chamber. The end product from the extruded waste at 800°C would include methanol, one of the primary feedstocks for the chemical industry.[36] Depending on the operating conditions, the methanol could be sent directly to a chemical plant as feedstock without undergoing any subsequent treatment. However, if the gas undergoes low-temperature decomposition, then carbon dioxide and hydrogen would be the end products. Figure 16.9 contains a possible process schematic for the generation of methanol from polymeric wastes. Degradative extrusion is the reverse of reactive extrusion, in which a specially desired chemical reaction is carried out on the polymeric backbone, such as graft copolymerization.

### 16.3.3.5   Depolymerization via Supercritical Water Partial Oxidation

A depolymerization process based on supercritical water partial oxidation was developed by Lee and coworkers.[37,38] The process takes advantage of the extraordinary physicochemical and transport properties of supercritical water to create a unique reactive environment for selective cleavage of C–C bonds of the polymer molecules. While the depolymerization reaction takes place in a pseudohomogeneous phase of a supercritical water mixture without an aid of a catalyst, both the depolymerization

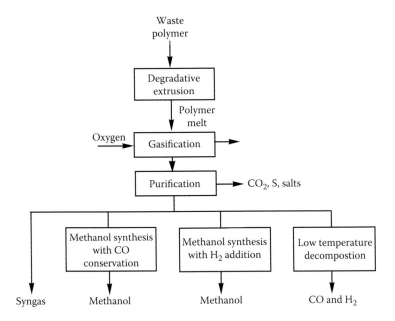

**FIGURE 16.9** A process scheme for the degradative extrusion and partial oxidation of polymeric wastes for methanol generation.

efficiency and the rate of reaction are found to be enhanced by the presence of the partial oxidation of feed polymers. A substoichiometric, small amount of oxygen input promotes oxidative cleavage of C–C bonds and generates an exothermic heat-of-oxidation reaction that helps maintain the system temperature. In supercritical water, whose temperature and pressure conditions both exceed the critical temperature of 371°C and the critical pressure of 218 atmospheres, oxygen dissolves in any proportion, thus creating a very effective homogeneous oxidation environment. The supercritical water depolymerization process has been successfully applied to a variety of polymers on a laboratory minipilot system based on a Hastelloy C-276 alloy reactor. The unique feature of this process is in its selective depolymerization essentially leading to "monomerization" of polymer molecules, that is, depolymerizing polymer molecules back to their monomer molecules with high selectivity. For example, propylene monomer is a dominant product species of PP depolymerization, styrene monomer is a dominant product species from PS depolymerization, and so on.[39] A variety of polymer materials have been tested with high efficiency and selectivity, and they include HDPE, LDPE, PP, PS, polyisoprene (PI), and ground scrap tire.

## 16.4 FUEL PRODUCTION FROM SPENT TIRES

### 16.4.1 INTRODUCTION

Scrap tire disposal has become a global problem of epidemic proportions. In 1977, the number of tires scrapped in Japan was 47 million a year.[40] More alarming is the fact that this number has doubled over the past 5 years. In 1992, the United Kingdom scrapped over 25 million tires per year.[41] Add the annual scrapping of 250 million passenger tires in the United States, and the outlook becomes even grimmer.[42]

Traditional handling of the scrap tires was by landfills; however, the acute shortage of viable landfills has all but eliminated this as a means of disposal. In fact, several of the Midwestern US states have issued laws that close landfills to tires.[43] State-registered private collectors must dispose of the tires at approved, legal dumps and recyclers. But even the number of legal landfills is dwindling. This has forced researchers to find an economical and efficient alternative for the spent tires.

Currently there are four key areas for "marketing" the spent tires. The first uses shredded tires as a "clean dirt" for road embankments and landfill liners.[44] The second use is as a rubber-modified asphalt. Tire rubber asphalt, or tire crumb, can be used for playgrounds, for running tracks, or as an ingredient of highway-paving material. The third use of spent tires is for electric power generation by combusting ground tire chips together with coal. The fourth, and most important area, is for tire-derived fuels (TDFs).[45] Under proper conditions, spent tires are a clean fuel with a 15% higher Btu value than coal with a low sulfur content.[46] In fact, it was estimated that by 1997, over 150 million scrap tires would be used for the generation of TDFs.[47]

TDFs can be obtained by several methods. The first is by incineration.[45,48] Britain's tire incinerator burns approximately 90,000 tons of rubber a year. With this amount, the Wolverhampton facility will generate 25 MW of energy, which is enough to power a small town.[49] In the United States, Illinois Power incinerates shredded tire chips to supplement their soft coal. This co-combustion of tire chips and coal is practiced widely in the United States. The direct incineration of the chips will utilize approximately 15.6 million tires per year and will make up 2% of the total fuel consumed at the plant. Other processes, such as thermal cracking and depolymerization, recover the oil, char, and gases from the scrap tires as separate "product" streams.[50,51] However, the most well-known method for the generation of TDFs is from direct pyrolysis.[42,44,52–55]

## 16.4.2 PYROLYSIS OF SPENT TIRES

The recovery of energy from spent tires is not a new process. In 1974, the U.K. Department of Trade and Industry's Warren Spring Laboratory conducted the first tests to recover energy from spent passenger tires.[53] These initial experiments demonstrated that it was possible to break the tires down into oil and gas by heating it in a closed retort, followed by distillation of the gaseous products. As research continued, it was found that the final bottoms product, or char, could be further treated for the manufacture of activated carbon (as shown in Figure 16.10).[56] During the late 1980s, the use of pyrolysis for making TDFs became common. A number of U.S. and European patents have been granted for the pyrolysis of scrap tires.[57–60] Although each patent is somewhat different, the main operating principles and conditions are quite similar.

In each instance, the scrap tires are first reduced in size, by grinding, chipping, or pelletizing, then sent to a clarifier for the removal of the scrap metal. The method presented by Williams et al.[61] represents the basic process utilized as the "starting block" by most researchers. The temperature is initially at 100°C while the rubber is loaded

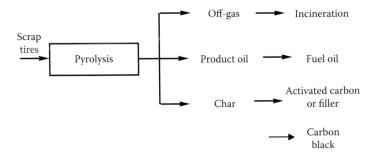

**FIGURE 16.10**  Possible end products from the pyrolysis of scrap tires.

into the reactor. After 1 h, the temperature is ramped up to 300°C–500°C, depending on the process. The reactor is held at the final temperature for a minimum of 2 h. The gas fraction is sent to a distillation apparatus for subsequent purification and analysis. The oil and char are separated using a second column. The slight variations on the pyrolysis process and its end products are presented in the following sections.

### 16.4.2.1  Occidental Flash Pyrolysis

Occidental Chemical's flash pyrolysis system was first demonstrated in late 1971. The process was able to produce a high-quality fuel oil at a moderate temperature and pressure. The advantage of the process was that the pyrolysis reaction was achieved without having to introduce hydrogen or using a catalyst. The process was divided into three main sections: feed preparation, flash pyrolysis, and product collection. A simplified schematic of the overall process is shown in Figure 16.11.[61,62]

The most time-consuming aspect of Occidental's process is the feed preparation. During this stage, the tires are debeaded and shredded to approximately 3 in. in size. A magnet is then used to remove all the metal components. The remaining material is further shredded to 1 in. before being ground to −24 mesh. The grinding of the

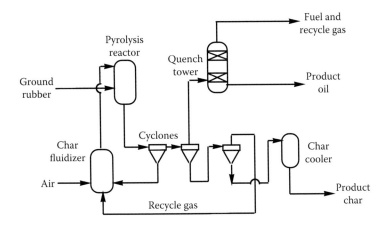

**FIGURE 16.11**  A simplified schematic of Occidental's flash pyrolysis system.

tires to such fine particles enabled the flash pyrolysis to occur at a quicker rate. The quick devulcanization of the ground tire rubber enabled a shorter residence time, which, in turn, decreased product cracking.[63]

After leaving the pyrolysis reactor, the gaseous stream was sent to a quench tower to separate the two end products. The product oil was collected and sent to a storage facility. The recovered gas was recycled to the char fluidizer and pyrolysis reactor as a supplemental process fuel. The solid components that remained in the bottom of the pyrolysis reactor were sent through three cyclones. The cyclones were used to separate the solid particles by size, as well as to cool the material back down to room temperature. At the end of the process, a 35 wt% carbon black was obtained. Analysis conducted on the carbon black showed that it had high-enough quality for direct reuse in the rubber industry.[61]

### 16.4.2.2  Fluidized Thermal Cracking

The first commercial use for *fluidized thermal cracking* (*FTC*) dates back to the 1930s for coal gasification. Similar to other pyrolysis processes, the FTC process burns waste with high combustion efficiency; however, it has a few other added advantages. Because it is a fluidized bed process, there is a rapid mixing of solid particles, which enables uniform temperature distribution; thus, the operation can be simply controlled. Fluidization also enhances the heat and mass transfer rates, which, in turn, decreases the amount of CO emitted. Researchers have been able to adjust the FTC operating conditions to decrease the $SO_x$ and $NO_x$ emissions.[64]

The Nippon Zeon Company has conducted extensive studies on TDFs via thermal cracking. The precommercial process feeds crushed tire chips to the fluidized bed using a screw feeder. Air heated by a preheating furnace is fed to the reactor bottom to elevate the reactor temperature near cracking conditions, 400°C–600°C, before the chips are introduced. This enables the tire chips themselves to maintain the high cracking temperature. A continuous cyclone is used to remove the char. The cracked gases are brought into contact with the recovered oil from the quench tower. Part of the recovered oil is recycled to the quench tower, and the remainder is sent to an oil storage tank for further purification. The uncondensed gas or noncondensable gas (NCG) is sent to a treatment process to remove the hydrogen sulfide. A simplified flowchart of the Nippon Zeon plant in Tokuyama is shown in Figure 16.12. The estimated break-even cost of the pyrolysis plant is $0.25 per tire.[65]

The preliminary studies used for the development of the Tokuyama plant had very promising results. All the end products produced could either be used directly as a supplemental fuel source at the plant, or sent off-site for petroleum and chemical industries. A typical end product distribution of the FTC process is given in Table 16.4.[66]

### 16.4.2.3  Carbonization

*Carbonization* is another form of pyrolysis, which can convert over one-half of a scrap tire into usable products. Carbonization processes operate at much higher temperatures than typical pyrolysis processes. At this higher-temperature condition, the main product is char. The char is purified into carbon black, one of the main components in the manufacture of tires. In 1974, the cost of carbonization was

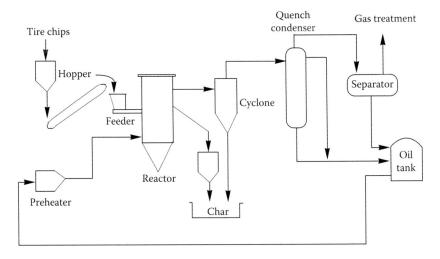

**FIGURE 16.12**  Nippon Zeon scrap tire fluidized cracking process. (From Saeki, Y., and Suzuki, G., *Rubber Age*, 108, 1976.)

**TABLE 16.4**

**End Product Distribution for Tires Cracked at 450°C**

| Product | Amount Produced (kg) | Percent |
|---------|----------------------|---------|
| Oil     | 257                  | 52.0    |
| Char    | 166                  | 33.6    |
| Gas     | 72                   | 14.4    |
| Total   | 495                  | 100.0   |

approximately three times the cost of making carbon black from standard petroleum operations.[67] Although this number decreased by only 15% in the next decade, the incentive of utilizing a waste material offsets the remaining cost.[68]

## 16.4.3  CO-COMBUSTION OF SCRAP TIRES AND TDFS

Tire rubber can be combusted together with other fossil fuels as well as biomass without major alteration of an existing combustion facility. Energy generation using scrap tires and TDFs as a supplemental fuel to coal is of particular interest and is being practiced throughout the world. The rationale behind the use of tire rubber as a supplemental fuel to conventional coal in electric power generation is based on the following[69–71]:

1. Excellent combustion characteristics and high heating value
2. Lower cost and good availability of raw material feedstock
3. Relatively low sulfur content, especially on a Btu basis
4. Reducing the environmental burden and health effects of tire stockpiles

Typical heating values of tire rubber are 15,000 to 16,000 Btu/lb, whereas typical bituminous coal (washed) has a heating value of 11,000 to 13,000 Btu/lb. Further, both carbon and hydrogen contents[70] of tire rubber are excellent, namely, 89 to 90 wt% for C and 7.5 wt% for H. These values are substantially better than those for coals. Typically, the sulfur content of scrap tires and TDFs is 1.2%–1.8%. Considering the sulfur dioxide ($SO_2$) emission potential per fuel's generated energy value (pound $SO_2$ produced per million Btu generated), the result is closer to that for low-sulfur-containing coals.

However, co-combustion of tire rubber and TDF may alter the emission characteristics of the combustion facility.[69–71] Some studies showed that using TDF resulted in slight increases in pollutant emissions, whereas others showed slight decreases in the mass emission rates of these pollutants.[70] There is no consistent trend. Based on the Purdue University Wade Utility Plant Facility test runs[69] comparing pure coal combustion and combustion of coal (95%) + TDF (5%), the atmospheric emissions of most trace metals increase when TDF is co-combusted with coal. In particular, zinc and cadmium emissions were drastically higher with TDF-blended fuel feed, whereas mercury emission was about the same. A comprehensive recent study of the effect of TDF supplement to fuel on air emissions was performed at Riverside Cement Inc. in Oro Grande, CA.[70] Based on the testing of 41 specific air toxins potentially emitted from a cement kiln, the average of three runs with and without TDF supplement (about 4.5% TDF by weight) showed that the mass emission rates of 22 air toxins were lower with TDF supplement, whereas the mass emission rates of only two compounds, zinc and anthracene, were higher with TDF-supplemented coal feed.[70] The two compounds are not classified as hazardous air pollutants (HAPs).

As can be seen, results of the effects on pollutant emission with TDF-supplemented fuel can be quite conflicting and point toward different directions.[70] Therefore, results of one plant study should not be generalized to another, because the prevailing conditions may differ very widely. Environmental impacts and health effects of tire co-combustion need to be carefully assessed for each plant with all relevant factors taken into account, namely, the combustion process itself, combustion conditions, plant design, control equipment and its efficiency with respect to specific trace elements, feed material properties, blend ratios, analysis techniques, and more.

## 16.4.4 IFP Spent-Tire Depolymerization Process

A relatively new approach to the decomposition of used tires is the IFP process developed by the French Institute of Petroleum (in French, Institut Français du Pétrole).[41] Unlike the other process, IFP does not require pretreatment of the scrap tire. The whole tires are placed in a basket and lowered into a 600 L reactor. Hot oil at 380°C is sprinkled onto the tire's surface. A chemical reaction between the hot oil and tires causes the depolymerization of the rubber. After the tires have been completely depolymerized, the reactor is cooled to 100°C. The off-gases are sent to a distillation column. The column separates the gas and light hydrocarbon fractions. The resultant gas is sent for further purification into a $C_4$–$C_6$ fraction and a gasoline ($C_8$–$C_{10}$) fraction, and the light hydrocarbons are recycled to the reactor. The recycled hydrocarbon fraction is used to dilute the viscous fuel oil generated during the depolymerization process. Upon evacuating the reactor, the scrap metal is removed and sent off-site for disposal.

The IFP process is different in that it is a batch process.[50] Because it is operated batchwise, the concern over having enough tires for continuous feed is eliminated. Depolymerizing 4000 tons per year of waste tires has been estimated to generate over 15,200 tons of fuel oil, 360 tons of gasoline, 560 tons of gas, and 600 tons of metallic waste.

## 16.4.5 DRY DISTILLATION OF SPENT TIRES

Researchers have investigated the use of a specialty dry distillation apparatus for producing a gaseous fuel from spent tires. The process can be operated with either shredded or whole tires without removing metallic wire meshes. The tires enter the top of the distillation tower via a conveyor belt. Once filled, the tower is sealed off to prevent the gaseous product from escaping. Combustion is initiated by burners located at the bottom of the tower and then sustained by the introduction of process air. As combustion continues, the hot gas rises to the top of the tower. The rising gas has a dual purpose. The first is to assist in the combustion of the tires located at the top of the tower. Second, the rising of the gas acts in the same manner as a traditional distillation column for the separation of the end products. The gas is cooled down by the gas cooler. Part of the gas is liquefied and recovered in the form of oil. Recovered oil is equivalent to class B heavy oil.[72] The residue (char and tire cord) is removed from the bottom of the tower by a conveyor. Once cooled, the residue is separated into individual components.

Hiroshi and Haruhiko[73] operated a pilot plant scale of the tower continuously for 250 h. At the end of the process time, it was found that approximately 40 wt% of the tires was converted into usable gaseous fuel. The remaining 60% was composed of char and tire cord. Research is being conducted on improving the amount and purity of the recovered gas. In addition, Hiroshi and Haruhiko are also investigating the potential uses for the recovered char.

A process schematic of the *Direct Dry Distillation of Tires* by Fujikasui Engineering Co. is shown in Figure 16.13.[72]

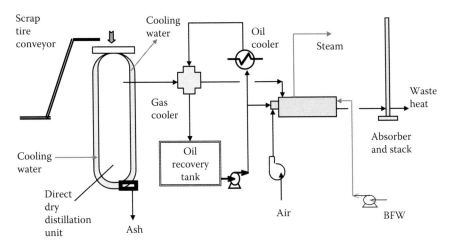

**FIGURE 16.13** A process schematic of the direct dry distillation of tires.

### 16.4.6 Goodyear's Devulcanization Process

Goodyear Rubber and Tire Co. has developed a unique process for devulcanizing scrap tire rubber using a novel supercritical fluid technology.[74,75] The technology uses sec-butanol as a supercritical fluid, which functions both as a pyrolysis medium and as a chain cleavage agent.[74] The solvent is unique in the sense that it not only facilitates C–S and C–C bond cleavage reaction in a near-homogeneous state, but also dissolves the reacted fragments and takes them away from other reactive intermediates. The solvent, sec-butanol, is fully recoverable and reusable in the process, thus minimizing the solvent cost. Furthermore, the objective of the process is quite different from that of other processes in the sense that the product of Goodyear's process treatment is devulcanized rubber of high quality and intended for use in manufacturing tires as a tread-compounding ingredient. A laboratory test of 20 phr (parts per hundred parts of rubber) of the process reclaimed materials has shown equal or superior mechanical properties and cure characteristics, when compared with fresh tread-compounding formulations (i.e., without any reclaimed material).[75] The process achieves devulcanization in the truest sense without altering basic polymeric properties of rubber. They have tested an exhaustive list of other solvents that are intuitively promising and structurally attractive for similar process applications. Their results showed that sec-butanol in its supercritical condition was at least severalfold better than all other fluids. It is noteworthy that even the closest isomer such as iso-butanol did not yield any comparable results. This strong supercritical behavior of sec-butanol may be attributed to its unique molecular structure, which has the center carbon connected to all four different functional groups ($CH_3$–, $C_2H_5$–, H–, and OH–), thereby maximizing the quadruple moments of the molecule. They further enhanced their process by employing a cosolvent process, by which the amount of sec-butanol needed can be significantly reduced for the process without loss of efficiency. As a cosolvent to this process, carbon dioxide ($CO_2$) was found to be very effective in attaining process synergism.[74]

### 16.4.7 Hydrogenation of Spent-Tire Rubber

Hydrogenation, unlike pyrolysis or similar processes, is a chemical synthesis process. In simple terms, it entails the addition of hydrogen, the element that is removed from oil, to make synthetic rubber. By adding the appropriate amount of hydrogen to the waste tires, the rubber can be returned to its original form. By adding hydrogen, devulcanization, saturation, and carbon–carbon bond cleavage reactions are induced. In a sense, the process is quite similar to the hydrotreating process of heavy oil and resids.

### REFERENCES

1. Corbitt, R.A., Ed., *Standard Handbook of Environmental Engineering*, McGraw-Hill, New York, 1990.
2. Fisher, P.M. and Evans, L.R., Jr., Whole tyre recycling: The elm energy approach, *Rubber Europe Conference Proceedings*, No. 6.012, June 1993, pp. 1–6.
3. Frederick, W.J., Lisa, K., Lundy, J.R., O'Connor, W.K., Reis, K., Scott, A.T., Sinquefield, S.A., Sricharoenchaikul, V., and Van Vooren, C.A., Energy and materials from recycled paper sludge, AIChE Summer National Meeting, Denver Co., August 14–17, 1994.

4. Association of Plastics Manufacturers in Europe, Plastics: A Vital Ingredient for the Food Industry, RAPRA Technology, Ltd., Telford, Shropshire, U.K.,1993, pp. 12–30.
5. Richards, M.K., USEPA's evaluation of a Texaco gasification technology, *Proceedings: ACS Emerging Technologies in Hazardous Waste Management*, VI, Atlanta, GA, September 1994, pp. 19–21.
6. Stearns, R.P., Landfill methane: 23 sites are developing recovery systems, *Solid Waste Management Magazine*, June 1980, pp. 56–59.
7. U.S. EPA Landfill Gas Outreach Program, Energy projects and candidate landfills, http://www.epa.gov/lmop/projects-candidates/index.html, accessed October 2013.
8. Pober, K. and Bauer, H., From garbage—Oil, *CHEMTECH*, 1(3): 164–169, 1977.
9. Kagayama, M., Igarashi, M., Hasegawa, M. and Fukuda, J., Gasification of solid waste in dual fluidized-bed reactors, *ACS Symposium*, Ser. No. 130, 38, 1980, pp. 525–540.
10. Henry, J.G. and Heinke, G.W., *Environmental Science and Engineering*, Prentice Hall, NJ, 1989, p. 560.
11. Green, A.E.S., Overview of fuel conversion, *Fuel Combust. Technol.*, 12: 3–15, 1991.
12. Helt, J.E. and Mallya, N., Pyrolysis experiments with municipal solid waste components, *Proceedings of 23rd Intersociety Energy Conversion Engineering*, IEEE Service Society, NJ, 1988.
13. Bell, P.R. and Varjavandi, J.J., Pyrolysis—Resource from solid waste, proceeds, *Waste Management, Control, Recovery and Reuse*, Ann Arbor Science, MI, 1974, pp. 207–210.
14. Schlesinger, M.D. et al., Pyrolysis of waste materials from urban and rural sources, *Proceedings of Third Mineral Waste Utilization Symposium*, Chicago, IL, March 14–16, 1972, pp. 423–428.
15. Official web site of the City of Los Angeles, http://www.lacity.org/council/cd12/pdf/ Landfilling_Resources_MPA_Pyrolysis_Facility.pdf, accessed May 2006.
16. Davis, M.L. and Cornwell, D.A., *Introduction to Environmental Engineering*, 2nd ed., McGraw-Hill, New York, 1991.
17. Scott, D.S., Czernik, S.R., Piskorz, J. and Radlein, A.G., Fast pyrolysis of plastic wastes, *Energy Fuels*, 4: 407–411, 1990.
18. Chambers, C., Larsen, J.W., Li, W. and Wiesen, B., Polymer waste reclamation by pyrolysis in molten salts, *Ind. Eng. Chem. Process Des. Dev.*, 23: 648–654, 1984.
19. Web site of Automotive Learning Center, http://www.drivinginnovation.com/glossary/ faq.html, accessed May 2006.
20. Producer responsibility for packaging waste in Austria, *ENDS Rep.*, 224: 18, 1993.
21. Baker, J., Unravelling the recycling targets, *Eur. Chem. News*, 60(1596): 16–17, 1993.
22. Topfer extends plastics recycling deadlines, *Eur. Chem. News*, 60(1595): 25, 1993.
23. McMahon, P., Plastics reborn, *Chem. Eng.*, 99: 37–43, 1992.
24. Hunt, J., LCA endorses use of HDPE waste, *Pack. Week*, 9(22): 1, 1993.
25. Leaversuch, R.D., Chemical recycling brings real versatility to solid-waste management, *Mod. Plast.*, 68: 40–43, 1991.
26. Bertolini, G.E. and Fontaine, J., Value recovery from plastics wastes by pyrolysis in molten salts, *Conserv. Recycling*, 10(4): 331–343, 1987.
27. Kaminsky, W., Pyrolysis of plastic waste and scrap tires in a fluid bed reactor, *Resour. Recovery Conserv.*, 5: 205–216, 1980.
28. Romanow-Garcia, S., Plastics-planning for the future, *Hydrocarbon Process.*, 72(10): 15, 1993.
29. Sittig, M., *Organic and Polymer Waste Reclaiming Encyclopedia*, Noyce Data Corp., NJ, 1981, p. 178.
30. Vohwinkel, F., Approach to recycling polyurethane waste, *Tin Uses*, 149: 7–10, 1986.
31. Scheirs, J., Bigger, S.W. and Billingham, N.C., Effect of chromium on the oxidative pyrolysis of gas-phase HDPE as determined by dynamic thermogravimetry, *Polym. Degrad. Stab.*, 38(2): 139–145, 1992.

32. Venuto, P.B. and Habib, E.T., Jr., *Fluid Catalytic Cracking with Zeolite Catalysts*, Marcel Dekker, New York, 1979.

33. Hashimoto, K. et al., *New Developments in Zeolite Science and Technology*, Elsevier, Amsterdam, 1986, pp. 505–510.

34. Songip, A.R., Masuda, T., Kuwahara, H. and Hashimoto, K., Test to screen catalysts for reforming heavy oil from plastic wastes, *Appl. Catal. B: Environ.*, 2: 153–164, 1993.

35. Menges, G., Basis and technology for plastics recycling, *Int. Polym. Sci. Technol.*, 20(8):10–15, 1993.

36. Semel, G., Study of gasification of plastic waste, *UAPG Symposium*, Ser. No. 19, 9, 1991.

37. Lee, S., Gencer, M., Azzam, F.O. and Fullerton, K.L., Depolymerization process, U.S. Patent 5,386,055, January 31, 1995.

38. Lee, S. and Azzam, F.O., Oxidative decoupling of scrap rubber, U.S. Patent 5,516,952, May 14, 1996.

39. Lilac, W.D. and Lee, S., Kinetics and mechanisms of styrene monomer recovery from waste polystyrene by supercritical water partial oxidation, *Adv. Environ. Res.*, 6: 9–16, 2001.

40. Kawakami, S., Inoue, K., Tanaka, H. and Sakai, T., Pyrolysis process for scrap tires, *ACS Symposium* Ser. No. 130, 40, 1980, pp. 557–572.

41. Used Tyres: A crumb of comfort, *Recycl. Resour. Manage. Mag.*, 25–26, 1992.

42. English, D., Scrap tire problem could be gone by 2003, *Tire Bus.*, 11(7): 34–35, 1993.

43. Greenhut, S., Dealers, retreaders confront growing scrap tire problem, *NTDRA Dealer News*, 49(1): 36–40, 1986.

44. Clark, T., Scrap tyres: Energy for the asking, *Brit. Plast. Rubber*, 35–37, 1985.

45. McCarron, K., Maker of scrap-tire boilers sees bright future, *Tire Bus.*, 11(17): 14, 1993.

46. Sikora, M.C., Ed., Whitewall cement fires kiln with scrap tires: Tires-to-fuel process becomes a reality at LaFarge Cement, *Scrap Tire News*, 7(11): 1–10, 1993.

47. Kokish, B., Organization seeks scrap tire solutions, *Rubber Plast. News*, 23(6): 44–46, 1993.

48. SSI: Lucas furnaces—Tyres disposal by incineration, *Tyres Accessories Mag.*, 11: 68–69, 1992.

49. Pearce, F., Scrap tyres: A burning issue, *New Scientist*, 140(1900): 13–14, 1993.

50. Audibert, F. and Beaufils, J.P., Thermal depolymerization of waste tires by heavy oils: Conversion into fuels, Final Report EUR 8907 EN, Commission of European Communities: Energy, 1984.

51. Saeki, Y. and Suzuki, G., Fluidized thermal cracking processes for waste tires, *Rubber Age*, 108(2): 33–40, 1976.

52. Braslaw, J., Gealar, R.L. and Wingfield, R.C., Jr., Hydrocarbon generation during the inert gas pyrolysis of automobile shredder waste, *Polym. Prep.*, 24(2): 434–435, 1983.

53. Reed, D., Tyre pyrolysis comes on stream, *Eur. Rubber J.*, 166(6): 29–33, 1984.

54. Earle, B.A., Dallas investors purchase tire pyrolysis plant, *Rubber Plast. News*, 23(2): 3, 1993.

55. Wyman, V., Turning a profit from old tyres, *The Engineer*, 273(7066/7): 36, 1991.

56. Jackson, D.V., Resource recovery, Warren Spring Laboratory Report No. C95/85, 1985.

57. Apffel, F., Recovery process, U.S. Patent No. 4,647,443, 1987.

58. Reu, R.A., Pyrolyric conversion system, European Patent No. 446,930,A1, 1991.

59. Grispin, C.W., Jr., Pyrolytic process and apparatus, European Patent No. 162,802, 1984.

60. Roy, C., Vacuum pyrolysis of scrap tires, U.S. Patent No. 4,740,270, 1988.

61. Williams, P.T., Besler, S. and Taylor, D.T., The pyrolysis of scrap automotive tires: Influence of temperature and heating rate on product composition, *Fuel*, 69: 1474–1481, 1990.

62. Che, S.C., Deslate, W.D. and Duraiswamy, K., The occidental flash pyrolysis process for recovering carbon black and oil from scrap rubber tires, *ASME*, 76-ENAs-42, 1976.
63. Nag, D.P., Nath, K.C., Mitra, D.C. and Raja, K., A laboratory study on the utilization of waste tire for the production of fuel oil and gas of high calorific value, *J. Mines Met. Fuels*, 473–476, 1983.
64. Chang, Y.M. and Chen, M.Y., Industrial waste to energy by circulating fluidized bed combustion, *Resour. Conserv. Recycling*, 9(4): 281–294, 1993.
65. Kroschwitz, J.I., Ed., *Encyclopedia of Polymer Science and Engineering*, Vol. 14, Wiley & Sons, New York, 1988, pp. 787–904.
66. Saeki, Y. and Suzuki, G., Fluidized thermal cracking process for waste tires, *Rubber Age*, 108(2): 33–40, 1976.
67. Kiefer, I., U.S. EPA Report No. SW-32c.1, 1974.
68. Jarrell, J., International Patent, WO 93/12198, 1993.
69. Carleton, L.E., Giere, R., Lafree, S.T. and Tishmack, J.K., Investigation of atmospheric emissions from co-combustion of tire and coal, paper presented at the *Annual Meeting of the Geological Society of America*, Seattle, WA, November 2–5, 2003.
70. Karell, M., Regulation impacts on scrap tire combustion: Part II, 2006. MALCOM PIRNIE web site, http://www.pirnie.com/resources_pubs_air_feb00_3.html.
71. Alvarez, R., Callén, M.S., Clemente, C., Gómez-Limón, D., López, J.M., Mastral, A.M. and Murillo, R., Soil, water, and air environmental impact from tire rubber/coal fluidized-bed cocombustion, *Energy Fuels*, 18(6): 1633–1639, 2004.
72. Web site by Japanese Advanced Environment Equipment, Waste Tires Direct Dry Distillation System, http://nett21.gec.jp/JSIM_DATA/WASTE/WASTE_6/html/Doc_541.html, accessed May 2006.
73. Hiroshi, K. and Haruhiko, A., Process and apparatus for dry distillation of discarded rubber tires, European Patent No. 0,072,387, 1982.
74. Benko, D.A., Beers, R.N., Lee, S. and Clark, K., Devulcanization of cured rubber, U.S. Patent No. 6,992,116, January 31, 2006.
75. Beers, R.N. and Benko, D.A., Recycling of spent tires, in *Encyclopedia of Chemical Processing*, Lee, S., Ed., Taylor & Francis, New York, 2005, pp. 2613–2623.

# 17 Geothermal Energy

*Sunggyu Lee and H. Bryan Lanterman*

## CONTENTS

## 17.1 INTRODUCTION

The natural heat of the earth is called geothermal energy. *Geothermal* is a hybrid word, combining *geo* (earth) and *thermal* (of heat). The term *geothermal energy* denotes the total thermal energy contained beneath the relatively thin and comparatively cool outer surface of the earth. This represents about 260 billion cubic miles of rocks and metallic alloys at or near their melting temperatures. Geothermal resources range from shallow ground to hot rock and water several miles below the earth's outer surface, and even farther down toward the earth's core, in the region of extremely high temperatures of molten rock called *magma*.

Geothermal energy, the second most abundant source of heat on earth after solar energy, is accessible using current technology and is concentrated in underground reservoirs, usually in the forms of steam, hot water, and hot rocks. The three applicable technology categories are geothermal heat pumps (GHPs), direct-use applications, and electric power plants. GHPs use the earth's surface as a heat sink and heat source for both heating and cooling. Direct-use applications utilize the naturally occurring geothermally heated water for heating. Electric power plants use electric turbines fed by geysers to generate electricity. As in solar energy, the utility of geothermal energy is hampered by the extent of its distribution over the earth's surface in amounts that are often too small or too dispersed.[1] This is especially serious for the generation of electricity.

The most obvious forms of geothermal energy are *geysers*, boiling pools of mud, fumaroles, and hot springs. However, a greater potential does exist in regions not yet recognized for their energy possibilities—they are *hot dry rocks* (HDRs).

Besides the vast availability and the unique distribution pattern of these resources, geothermal energy is very clean and environmentally friendly. Geothermal energy generates no (or minimal) greenhouse gases because the conversion or utilization process does not involve any chemical reaction, in particular, combustion. Geothermal fields produce only about one-sixth of the carbon dioxide that a natural-gas-fueled power plant produces and very little, if any, of the nitrous oxide or sulfur-bearing gases. Furthermore, geothermal energy is available 24 h a day, and 365 days a year, whatever the external weather conditions may be. This is in sharp contrast to other green energy technologies such as wind and solar. In fact, geothermal power plants typically have average availabilities of 95% or higher, much higher than most coal and nuclear plants. Even this high availability can be further enhanced to a level that is practically near 100%, with advances and enhancements in the process technology.

## 17.2 GEOTHERMAL ENERGY AS RENEWABLE ENERGY

### 17.2.1 NEED FOR GEOTHERMAL ENERGY

The development and importance of new, clean energy sources such as geothermal energy gather pace because of not only the depletion of petroleum resources but also the environmental problems associated with conventional energy processes. Environmental problems associated with the utilization of fossil fuel sources involve (1) emission of greenhouse gases such as $CO_2$, $CH_4$, and $N_2O$; (2) emission of $SO_x$ and $H_2S$; (3) discharge of nitrogen oxides; (4) potential emission of mercury and selenium; (5) emission of volatile and semivolatile organic compounds (*VOCs* and *SVOCs*)[2]; (6) emission of particulate matter (*PM*); and (7) contamination of soils and groundwater resources with hazardous wastes.

At a depth of about 6 mi. from the earth's surface, the temperature is higher than 100°C; thus, the total amount of geothermal energy in storage far exceeds, by several orders of magnitude, the total thermal energy accountable in all forms of nuclear and fossil fuel resources of this planet. Solar energy is the only comparable resource in terms of such vast quantities. Therefore, it is very logical, if not imperative, that our energy priorities incorporate a vital resource such as geothermal energy.

### 17.2.2 Renewability and Sustainability of Geothermal Energy

The US Department of Energy (DOE) classifies geothermal energy as *renewable*. Its source is the continuously emanating thermal energy generated by the earth's core. Each year, rainfall and snowmelt maintain the supply of requisite water to geothermal reservoirs, and production from individual geothermal fields can be sustained for decades and perhaps centuries. An accurate prediction of the sustainable service life of each field is, however, very difficult.

### 17.2.3 Occurrence of Geothermal Energy

The occurrence of geothermal heat (also known as geoheat) can be explained by one of the following theories:[3]

1. The first theory is that about 6 billion years ago, the earth was a hot molten mass of rock, and this mass has been cooling through the epochs of time, with the outer crust formed as a result of a faster cooling rate.
2. The second theory presupposes that the earth is like a giant furnace. The decaying of radioactive material within the earth provides a constant heat source.
3. The third theory is based on the presumption that geothermal heat originates from the earth's fiery consolidation of dust and gas over 4 billion years ago.

Even though a generally agreed-on explanation for the natural occurrence of geothermal energy is unavailable, a combination of the aforementioned theories is widely offered.

The interior of the earth consists of a molten fluid of extremely high-temperature rocks called *magma*, which cools and expels heat to the earth's surface according to the second law of thermodynamics. The flow of heat is from the hot source (earth core) toward the cold sink (earth surface). The cold sink (i.e., heat sink in thermodynamics) consists of the earth's crust, surface, and atmosphere. This may be considered a very slow process of heat transfer.

Figure 17.1 shows a typical geological setting of a geothermal energy source. Thermal energy from the earth's core continuously flows outward. The heat transfer from the core to the surrounding layers of rock, the *mantle*, is principally via conduction. As the temperature and pressure of the system become high enough, some mantle rocks melt and form *magma*. Because the magma as a liquid phase is less dense and more fluidlike than the surrounding rock, it slowly rises and moves toward the earth's crust, thus convecting the heat from the core. This is why a slow convective heat transfer often represents the overall heat transfer process. Sometimes, the

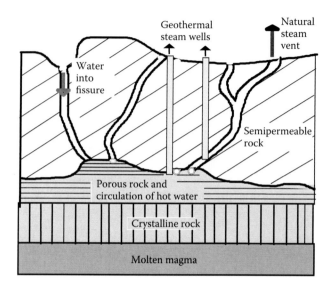

**FIGURE 17.1**   Geothermal energy source.

hot molten magma reaches all the way to the earth's surface, where it is known as *lava*. However, in most cases, the magma remains well below the earth's crust, heating neighboring rocks and water, which originates from rainwater seepage deep into the earth. The temperature of the water can be as hot as 380°C, which is even higher than the critical temperature of water, 374°C.

The water at this depth is subjected to high pressure and temperature. Depending on the imposed conditions, it may exist as a supercritical fluid. It then rises to the surface through fissures as a result of its density change, and in effect, it vents from the system, thereby reducing the pressure of the system. When the pressure decreases, the water boils, turns into steam, and rises to the surface through fissures and wells. Some of the best-known examples of hot geothermal water are hot springs or *geysers*. However, most of the hot geothermal water remains deep underground, typically trapped in cracks and porous rocks. This natural collection of hot underground water is called a *geothermal reservoir*.

The gross quantity of geothermal energy cannot be properly determined. The total energy content of the rocks down to a depth of 10 km has been estimated to be $3 \times 10^{26}$ cal. At the earth's core, that is, 6400 km deep, the temperature may reach over 5000°C. However, based on current technology, only a fraction of this heat is available as a recoverable resource. A mere 0.03%, or $10^9$ cal, of this energy is considered hot enough and near enough to the earth's surface to be recoverable; the rest of the energy is too widely dispersed over the crust or too deep to be practical.

Geothermal resources may be classified into several categories, principally based on their phases and forms, as shown in Table 17.1.

Although not all resources shown in the table can be used to produce electricity, they can still be utilized in many industrial, agricultural, and domestic areas. Research conducted by the US Geological Survey[4] has identified the significant research problems that need to be solved for full utilization of geothermal resources.

**TABLE 17.1**

**Classification of Geothermal Resources**

| Types of Geothermal Resources | Temperature (Approximate, °C) |
|---|---|
| 1. Convective hydrothermal resource | |
| Vapor dominated | 240 |
| Hot water dominated | 30–350 |
| 2. Other hydrothermal resources | |
| Sedimentary basins/regional aquifers (hot fluid in sedimentary rocks) | 30–150 |
| Geopressured (hot fluid under pressure that is greater than hydrostatic) | 90–200 |
| Radiogenic (heat generated by radioactive decay) | 30–150 |
| 3. Hot rock resources | |
| Part still molten (magma) | >600 |
| Solidified (hot, dry rock) | 90–650 |

## 17.2.4 ADVANTAGES OF GEOTHERMAL ENERGY

Geothermal energy resources are continuous, reliable, sustainable, and clean and can be cost-competitive in meeting baseload capacity needs. Specific advantages of geothermal systems include the following:

1. Indigenous energy—Geothermal energy helps reduce dependence on fossil or nuclear fuels and, as such, helps boost the economic benefits in the region.
2. Clean energy—Use of geothermal energy helps reduce combustion-related emissions.
3. Diversity of use—Geothermal energy has three common economic uses: electricity generation, direct use of heat, and GHPs.
4. Long-term resource potential—With optimum development strategies, geothermal energy can provide a significant portion of a nation's long-term energy needs.
5. Flexible system sizing—Current power generation projects range in capacity from a 200 kW system in China to 1200 MW at The Geysers in California. Additional units can be installed in increments depending on the growth of the electricity demand.
6. Power plant longevity—Geothermal power plants are designed for a life span of 20 to 30 years. With proper resource management strategies, life spans can exceed design periods.
7. High availability and reliability—*Availability* is defined as the percentage of the time that a system is capable of producing electricity. Availability of 95%–99% is typical for modern geothermal plants, compared to a maximum of 80%–85% for coal and nuclear plants.[5]
8. Combined use—Geothermal energy can be simultaneously used for both power generation and direct-use applications.

9. Low operating and maintenance costs—The annual operation and maintenance costs of a geothermal electric system are typically 5%–8% of the capital cost.

10. Land area requirement—The land area required for geothermal power plants is smaller per megawatt than for almost every other type of power generation plant.

11. Enhanced standard of living—Geothermal systems can be installed at remote locations without requiring other industrial infrastructure. The region can prosper without pollution.

## 17.2.5 GLOBAL GEOTHERMAL ENERGY

The current total installed capacity of geothermal power stations throughout the world is over 8200 MW.[6] The United States remains the biggest producer of electricity from geothermal energy, as shown in Table 17.2. The developing countries

**TABLE 17.2**
**Installed Geothermal Electricity Generation Capacity**

| Country | 1990 | 1995 | 1998 |
|---|---|---|---|
| Argentina[a] | 0.67 | 0.67 | 0 |
| Australia | 0 | 0.17 | 0.4 |
| China | 19.2 | 28.78 | 32 |
| Costa Rica | 0 | 55 | 120 |
| El Salvador | 95 | 105 | 105 |
| France (Guadeloupe) | 4.2 | 4.2 | 4.2 |
| Greece[a] | 0 | 0 | 0 |
| Guatemala | 0 | 0 | 5 |
| Iceland | 44.6 | 49.4 | 140 |
| Indonesia | 144.75 | 309.75 | 589.5 |
| Italy | 545 | 631.7 | 768.5 |
| Japan | 214.6 | 413.7 | 530 |
| Kenya | 45 | 45 | 45 |
| Mexico | 700 | 753 | 743 |
| New Zealand | 283.2 | 286 | 345 |
| Nicaragua | 70 | 70 | 70 |
| Philippines | 891 | 1191 | 1848 |
| Portugal (Azores) | 3 | 5 | 11 |
| Russia | 11 | 11 | 11 |
| Thailand | 0.3 | 0.3 | 0.3 |
| Turkey | 20.4 | 20.4 | 20.4 |
| United States | 2774.6 | 2816.7 | 2850 |
| Total | 5866.72 | 6796.98 | 8240 |

[a] Argentina and Greece closed their pilot plants.

accounted for 35% of the total for 1995 and 46% for 1999. About 2850 MW of electricity generation capacity is available from geothermal power plants in the western United States. The major geothermal fields along with their capacities are shown in Table 17.3.[7] Direct-use geothermal technologies utilize naturally hot geothermal water for commercial greenhouses, crop dehydration, fish farming, bathing, and district community heating. *GHPs* use the constant temperature of the top 50 ft of the earth's surface to heat buildings in the winter and cool them in the summer. Table 17.4 shows the direct use of geothermal heat in various categories in the United States.[5,8]

Fossil fuels, namely oil, coal, and gas, provide 86% of all the energy used in the United States. In 2004, renewable energy sources supplied just 6.1% of the total US energy consumption. Most of the renewable energy consumption was provided by hydroelectric power (44.5%) and biomass utilization (46.5%), whereas only 5.6% came from geothermal sources. Figure 17.2 shows the history and projections for US energy consumption from fuel sources during the period between 1980 and 2030.[9]

As shown in the figure, the projected future growth in the area of nonhydro and nonnuclear renewable energies is substantial. When the relative cost of electricity generation from geothermal sources decreases, the popularity of geothermal power

**TABLE 17.3**
**Major Geothermal Power Plants in the United States**

| Location | Capacity Installed (MW) |
|---|---|
| The Geysers, CA | 2115 |
| East Mesa, CA | 119 |
| Salton Sea | 198 |
| Heber, CA | 94 |
| Mammoth, CA | 7 |
| Coso, CA | 225 |
| Amadee, CA | 2 |
| Wendel, CA | 0.6 |
| Puna, HI | 18 |
| Steamboat, NV | 31 |
| Beowave, NV | 17 |
| Brady, NV | 6 |
| Desert Peak, NV | 9 |
| Wabuska, NV | 1.2 |
| Soda Lake, NV | 3.6 |
| Stillwater, NV | 14 |
| Empire Farms, NV | 4.8 |
| Roosevelt, UT | 20 |
| Cove Fort, UT | 4.2 |
| Total | 2889.4 |

*Source:* Wright, P.M., *The American Association of Petroleum Geologists Bulletin*, Vol. 23, No. 12, 366, October 1989.

**TABLE 17.4**
**US Geothermal Direct-Use Projects**

| Applications | Number of Sites | Thermal Capacity (MW) | Annual Energy (GWh) |
|---|---|---|---|
| Geothermal heat pumps | Most states | 2072 | 2402 |
| Space and district heating | 126 | 188 | 433 |
| Greenhouses | 39 | 66 | 166 |
| Aquaculture | 21 | 66 | 346 |
| Resorts/pools | 115 | 68 | 426 |
| Industrial processes | 13 | 43 | 216 |
| Total | | 2503 | 3989 |

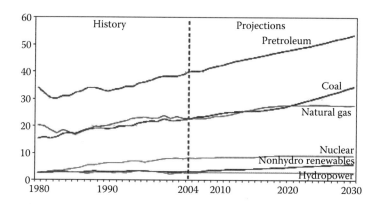

**FIGURE 17.2**　US energy consumption by fuel sources: past, current, and future forecast. Unit used is quadrillion Btu. (From US Department of Energy, 2006.)

generation will undoubtedly increase. Figure 17.3 shows a cost comparison among various modes of power generation.

## 17.3　HISTORY OF GEOTHERMAL ENERGY DEVELOPMENTS

Ancient people regarded the depths of the earth with horror as hell, the seat of malignant gods, who were responsible for natural phenomena like earthquakes and volcanic eruptions. Nevertheless, in ancient times, the Romans, and in modern times, the Icelanders, Japanese, Turks, Koreans, and others, have used its potential for baths and space heating.

The Larderello field in Tuscany, Italy, first began to produce electricity in 1904, which developed over the next 10 years to a capacity of 250 KW. In Japan, Beppu was the first site for experimental geothermal work in 1919, and these experiments led to a pilot plant in 1924 producing 1 KW of electricity. Somewhat earlier than this, the Japanese had begun to use geoheat to warm their greenhouses. In Iceland, municipal heating was provided using hot thermal waters in the 1930s, and it is still the major source of heating today.

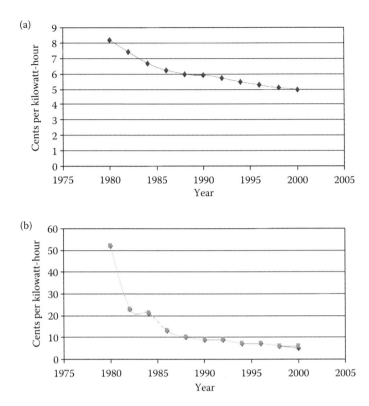

**FIGURE 17.3** Cost comparison of electricity generation between (a) geothermal energy and (b) solar thermal energy.

It was not until the early 1920s, that the United States examined the possibilities of commercial usage of geothermal steam. However, the competition from hydroelectric power was too keen to promote further development at that time. Today, the largest geothermal power plant in the world is located in California at The Geysers, which is probably the largest reservoir of geothermal steam in the world.[10] In recent years, the US DOE's *GeoPowering the West* (*GPW*) program has been working to further geothermal energy efforts.[11]

Growing concerns about the environmental effects of increasing $CO_2$ and methane in the atmosphere are working to enhance the role of geothermal resources worldwide. Hence, the utilization of geothermal energy to generate electric power dominates all other applications.

A number of factors that have boosted the production of geothermal energy are the following:[12]

1. The economics of geothermal energy became more favorable, owing to the increase in petroleum and natural gas prices.
2. The cost of producing geothermal energy decreased during 1980 and 2000.

3. Legislative actions and measures encouraging geothermal developments have been in place for many countries. Examples in the United States include the Energy Policy Act and the National Geologic Mapping Act in the early 1990s.[13]
4. The implementation of the Clean Air Act Amendments of 1990 also provides an economic benefit because of the well-developed technology for control of gas emissions from geothermal power plants.
5. The Amendment of the Public Utilities Regulatory Act removed the 80 MW limit from independent power plants selling electricity to utilities and is expected to improve the competitiveness of geothermal energy.

## 17.4 GEOTHERMAL PROCESSES AND APPLICATIONS

### 17.4.1 GEOTHERMAL POWER PLANTS

Geothermal resources may be described as hydrothermal, HDR, or geopressured. Hydrothermal resources contain hot water, steam, or a mixture of water and steam. Although research into ways of efficiently extracting and using the energy contained in HDR and geopressured resources continues, virtually all current geothermal power plants operate on hydrothermal resources.

The characteristics of the hydrothermal resource determine the power cycle of the geothermal power plant. A resource that produces dry steam uses a direct steam cycle. A power plant for a liquid-dominated resource with a temperature above 165°C typically uses a flash steam cycle. For liquid-dominated resources with temperatures below 165°C, a binary cycle is the best choice for power generation. Power plants on liquid-dominated resources often benefit from combined cycles, using both flash and binary energy-conversion cycles.

#### 17.4.1.1 Direct Steam Cycle

Direct steam[14] is also referred to as *dry steam*. As the term implies, steam is routed directly to the turbines, thus eliminating the need for the boilers used by conventional natural gas and coal power plants. Figure 17.4 shows a schematic of a direct steam cycle power generation process. In a direct steam cycle power plant, a geothermal turbine can operate with steam that is far from pure. Chemicals and compounds in solid, liquid, and gaseous phases are transported with the steam to the power plant. At the power plant, the steam passes through a separator that removes water droplets and particulate before it is delivered to the steam turbine. The turbines are of conventional design with special materials, such as 12Cr steel and precipitation-hardened stainless steel, to improve reliability in geothermal services.

The other components present in the direct steam geothermal cycle include the following:

1. A condenser used to condense turbine exhaust steam. Both direct-contact and surface condensers are used in direct steam geothermal power plants.
2. Noncondensable gas-removal system, to remove and compress the noncondensable gases. A typical system uses two stages of compression. The first

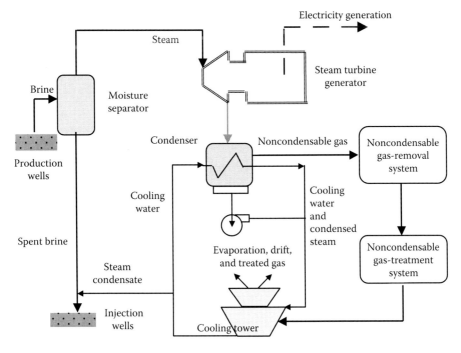

**FIGURE 17.4** Direct steam cycle geothermal power plant.

stage is a steam jet ejector. The second stage is another steam jet ejector, a liquid ring vacuum pump, or a centrifugal compressor.

3. The cooling tower employs a multicell wet mechanical draft design. Cooling is accomplished primarily by evaporation. Water that is lost from the cooling system to evaporation and drift is replaced by steam condensate from the condenser.

4. An injection well. Excess water is returned to the geothermal resource in an injection well.

The direct steam cycle is typical of power plants at The Geysers in northern California, the largest geothermal field in the world. The primary operator of The Geysers is Calpine Corporation.

### 17.4.1.2 Flash Steam Cycle

Flash steam[14] is the steam produced when the pressure on a geothermal fluid is reduced. A flash steam cycle for a high-temperature liquid-dominated resource is shown in Figure 17.5. This dual-flash cycle is typical of most larger flash steam geothermal power plants. Single-flash cycles are frequently selected for smaller facilities.

Geothermal brine, or a mixture of brine and steam, is delivered to a flash vessel at the power plant by either natural circulation or pumps in the production wells. At the entrance to the flash vessel, the pressure is reduced to produce flash steam. The steam is delivered to the high-pressure inlet to the turbine. The remaining brine

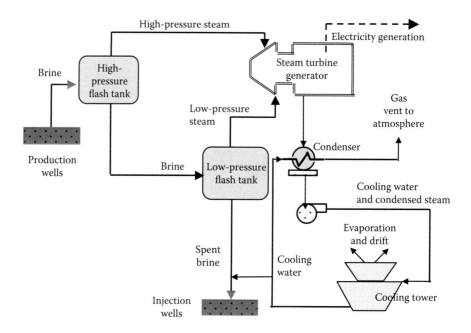

**FIGURE 17.5**   Double-flash steam cycle geothermal power plant.

drains to another flash vessel, in which the pressure is again reduced to produce low-pressure flash steam.

The other components present in the double-flash steam geothermal cycle include the following:

1. Direct-contact condenser: As hydrogen sulfide is not produced in large quantities.
2. A cooling tower: This cycle also uses a multicell wet mechanical draft cooling tower. The water lost because of evaporation and drift is replaced by steam condensate.
3. The excess water and spent brine from the flash vessels are injected back into the geothermal resource in an injection well.

### 17.4.1.3   Binary Cycle

A binary-cycle[6] geothermal power plant employs a closed-loop heat exchange system in which the heat of geothermal fluid (*primary fluid*) is transferred to a lower-boiling heat transfer fluid (*secondary fluid*) that is thereby vaporized and used to drive a turbine or generator set. In other words, a binary cycle uses a secondary heat transfer fluid instead of steam in the power generation equipment. Binary geothermal plants have been in service since the late 1980s. A binary cycle is the economic choice for hydrothermal resources, with temperatures below approximately 165°C. A typical binary cycle is shown in Figure 17.6.

The binary cycle shown in Figure 17.6 uses isobutane ($i\text{-}C_4H_{10}$) as the binary heat transfer fluid. Heat from geothermal brine vaporizes the binary fluid in the brine heat

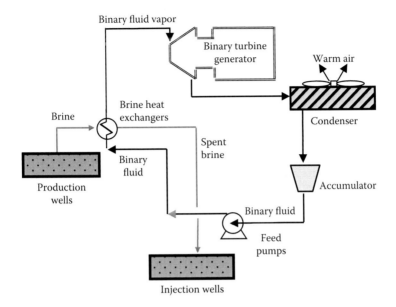

**FIGURE 17.6** Binary-cycle geothermal power plant.

exchanger. Spent brine is returned to the resource in injection wells, and the binary fluid vapor drives a turbine generator. The turbine exhaust vapor is delivered to an air-cooled condenser, in which the vapor is condensed. Liquid binary fluid drains to an accumulator vessel before being pumped back to the brine heat exchangers to repeat the cycle. The brine heat exchangers are typically shell-and-tube units fabricated from carbon steel.

### 17.4.1.4  Hot Dry Rock (Dry Geothermal Sources) Systems

Because the vast majority of the geothermal heat resources of the world exist as HDR sources rather than water (hydrothermal) systems, it is only natural for this energy source to receive more attention from geothermalists. The more accessible HDR resources in the United States alone would provide an estimated 650,000 quads of heat, 1 quad (1 quadrillion Btu) being equivalent to the amount of energy contained in 171.5 million barrels of oil. Because annual US energy consumption is approximately 84 quads, whoever figures out how to economically tap even a fraction of the potential in HDR could earn a place in history.[3,15,16]

HDR is a deeply buried crystal rock at a usefully high temperature. Current engineering designs plan to tap its heat by drilling a wellbore, fracturing or stimulating preexisting joints around the wellbore, and directionally drilling another wellbore through the fracture network. Cold water then flows down one wellbore, pushes through the fractured rock, warms, returns up the other wellbore, and drives a power plant. The major technical uncertainty is establishing the fracture network between the two wellbores. If adequate connectivity can be established and a sufficiently large fracture surface area can be exposed between the two wellbores, HDR can be a very competitive source of energy.

Figure 17.7 is a schematic diagram of the experimental Los Alamos System in New Mexico. Water at 65°C and 1000 psia is pumped into the hydraulic fracture

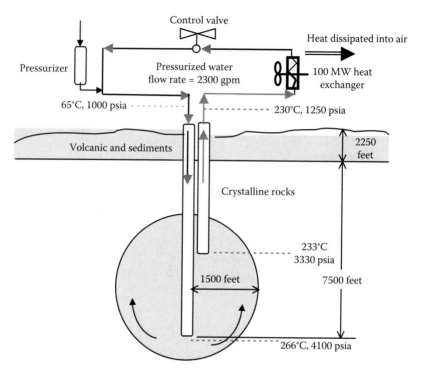

**FIGURE 17.7** Experimental configuration and operating conditions in the Los Alamos hydraulic fracture network.

network, approximately 3000 ft in diameter, and circulated at 7500 ft, where temperatures range between 260°C and 320°C. The pressure is around 4100 psia. The water is then pumped out of the ground, and when it reaches the surface, its temperature is 230°C at 1250 psia.

In this experimental system, the hot water is circulated through an air-cooled heat exchanger with the extracted heat dissipated to the atmosphere.

### 17.4.1.5 Freshwater Production

Less than 2% of the earth's retained water supply is available for drinking. The oceans, atmosphere, rocks or rock formations, and polluted resources contain the remaining 98%. From the standpoint of water shortage, all the systems recognized to date (desalination, recycling, and transportation over long distances) consume enormous amounts of energy and have also proved to be uneconomical. Geothermal resources, on the other hand, contain vast reservoirs of hot water and steam, and some of these produce electricity and freshwater as by-products. The geothermal resource satisfies two main criteria for alleviating water shortages, namely:

1. An energy source for distillation processes such as multistage flash (MSF) and vertical tube evaporator (VTE)
2. An ample supply of water

## 17.4.2 DIRECT USE OF GEOTHERMAL HEAT

Direct-heat use is one of the oldest, most versatile, and also the most common form of utilization of geothermal energy. The term *direct use* means that geothermal heat is used directly without first converting it to electricity. The warm water or steam exiting the ground will be piped into the dwelling or structure to provide warmth. Direct heating is obviously an older technology than geothermal power generation and is widely practiced.

### 17.4.2.1 Space and District Heating

Direct heating can be applied in what are known as either district or space heating systems, the distinction being that space heating systems serve only one building, whereas district heating systems serve many structures from a common set of wells. Direct heating has made the greatest progress and development in Iceland, where the total capacity of the operating geothermal district heating system is 800 MW.[17] Figure 17.8 shows an example of a district heating system.

Each system has to be adapted to the local situation, depending on the type of geothermal resource available, the population density of the area and the predicted population growth, the type of buildings requiring heating or cooling, and above all, the local climate. Geothermal district heating pumps are capital intensive in the early stages. The principal costs are initial investments for production and injection wells, downhole and circulation pumps, heat exchangers, and pipelines, as well as the distribution network. A high load density usually makes district heating economically more feasible because the cost of the distribution network transporting hot water to consumers is shared. Importantly, operating costs are comparatively low, thus making the long-term cost much more favorable. Geothermal district heating

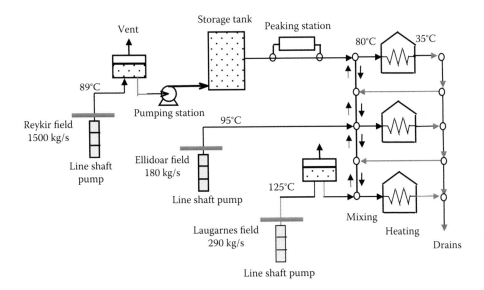

**FIGURE 17.8**   A schematic of Hitaveita Reykjavikur (Reykjavik district heating system).

systems offer significant life-cycle cost savings to consumers, as much as 30%–50% of the cost of using natural gas or oil.

At present, a very successful district heating system exists in San Bernardino, CA. The water production system, consisting of two wells, yields an average flow of 5200 L/min at 54°C water. The system currently serves 33 buildings, including government centers, a prison, a new blood bank facility, and other private buildings.

### 17.4.2.2   Agricultural Applications

One specific application of direct heating is greenhouse heating. This is one of the most common worldwide applications of geothermal energy. Fruits, vegetables, flowers, and ornamental plants are successfully grown year-round, in geothermally heated greenhouses using low-temperature sources (<38°C). Geothermal energy can extend short growing seasons and significantly reduce fuel costs. One example is a 650 m² greenhouse in California, utilizing a geothermal well 150 m deep that supplies 67°C water. The well is capable of supplying heat for an additional 1800–3700 m² of greenhouse. It is noteworthy that the energy crisis experienced in California in 2001 posed a very severe threat to greenhouse farmers who relied on electricity or natural gas heating. As of 2003, there were at least 37 greenhouse operations based on the geothermal energy in the United States.[18]

Another direct-heating application involves aquaculture, which is the raising of freshwater or marine organisms in a controlled environment. Geothermally heated water produces excellent yields of high-quality fish and crustaceans under accelerated growth conditions. Furthermore, geothermal aquaculture permits breeding in the winter, allowing fish farmers to harvest their products when product availability is low and market prices are high. As of 2003, there were at least 58 aquaculture sites using geothermal energy in the United States.[18] About 15 aquaculture operations were clustered in the southern California area.

### 17.4.2.3   Balneology

Balneology involves the use of geologically heated water/brine/mud sources for bathing purposes and is also said to possess healing and prophylactic properties. Balneology is centuries old and has been practiced by Etruscans, Romans, Greeks, Turks, Mexicans, Japanese, Koreans, Americans, and undoubtedly, others.

### 17.4.2.4   Industrial Process Heat

Industrial processes can be heat intensive and commonly use either steam or superheated water with temperatures of 150°C or higher. This makes industrial processes the highest-temperature users of geothermal direct-heat applications. However, lower temperatures can suffice in some cases, especially for some drying applications. Two of the largest industrial users of geothermal heat are a diatomaceous-earth drying plant in Iceland and a paper and pulp processing plant in New Zealand.

### 17.4.3   Geothermal Heat Pumps

The *GHP*, also referred to as the *ground source heat pump* (*GSHP*), uses the earth as a heat source for heating and as a heat sink for cooling. The GSHP uses a reversible refrigeration cycle combined with a circulating ground loop to efficiently provide either heating or cooling from electricity. The basic mechanism is the same as that of an air-source heat pump but operates more efficiently because the temperature of the ground is more favorable than that of the air, that is, the ground is warmer in the winter and cooler in the summer than the air. Additionally, the ground temperature is fairly constant throughout the year, even at depths of as little as 5 to 10 ft.

The typical components of a residential GSHP during heating and cooling cycles are shown in Figures 17.9 and 17.10. The major components of the system are the ground loop and a refrigeration unit composed of the compressor, primary heat exchanger, expansion valve, and secondary heat exchanger. The refrigeration cycle utilizes the same unit operation steps as an air-source heat pump or typical home air conditioning unit. In what is referred to as a closed-loop system, a water-and-antifreeze mixture circulates through a pipe buried in the ground and transfers thermal energy between the ground and the primary heat exchanger in the heat pump. Depending on the mode of operation, either heating or cooling is provided based on the reversible valve that allows the refrigerant to reverse the order of the operations of the cycle. Therefore, the primary heat exchanger, which consists of a water-to-refrigerant loop, can act as an evaporator or condenser. Also, included in this system, is a heat exchanger following the compressor that provides heat to a hot water heater; this is often referred to as a desuperheater. The Geothermal Heat Pump Consortium Inc. offers technical, educational, and promotional support for geoexchange systems.[19]

GHPs offer a distinct advantage over the use of air as a source or sink because the ground is at a more favorable temperature. Compared to atmospheric air, the ground is the warmer in winter and cooler in the summer. Therefore, GHPs demonstrate

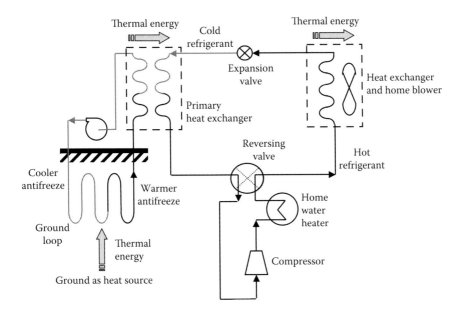

**FIGURE 17.9**   GSHP during heating cycle.

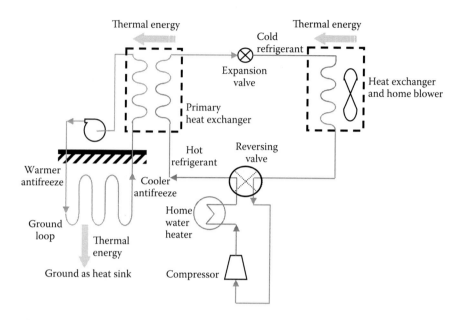

**FIGURE 17.10** GSHP during cooling cycle.

better performance over air-source heat pumps. GHPs also reduce electricity consumption by approximately 30% compared to air-source heat pumps. Aided by utility-sponsored programs, GHPs are becoming increasingly popular throughout the world. In the United States, the GHP industry is expanding at a growth rate of 10%–20% annually. As of 2004, more than 200,000 GHPs are being operated in US homes, schools, and commercial buildings.[20]

The industrial and other potential applications of geothermal energy suggest that great economic advantages could be gained from dual-purpose or multipurpose plants combining power production with one or more other applications. Such plants would enable the costs of exploration drilling and certain other items to be shared among two or more end users.

## 17.5  SCIENTIFIC AND TECHNOLOGICAL DEVELOPMENTS

### 17.5.1  MAJOR RESEARCH EFFORTS

The following major activities are examples of US government-funded research being conducted in accordance with its research and development (R&D) strategy:[5,11]

1. Advanced techniques to detect and delineate hidden geothermal resources are being developed, including remote-sensing techniques and improvements of various electric and acoustic methods.
2. Slim-hole drilling and coring, a cost-effective option for exploratory drilling, needs to be improved. This research includes developing slim-hole reservoir engineering techniques and logging tools.

3. Improved materials that are capable of withstanding the high temperature and corrosive nature of geothermal brines are being developed.
4. Methods to increase the net brine effectiveness of geothermal power plants are being pursued, as are ways to reduce power plant costs.

Significant research activities in progress at national laboratories and universities in the United States and the world include those of Sandia National Laboratory,[11,21] Lawrence Berkeley National Laboratory,[11] Brookhaven National Laboratory,[6,11] National Renewable Energy Laboratory,[20] Los Alamos National Laboratory,[22–24] The Geysers,[12] Camborne School of Mines,[6] European Hot Dry Rock Industries,[6] Stanford University and Leningrad's Mining Institute,[6] Electric Power Research Institute (EPRI), Geo-Heat Center of the Oregon Institute of Technology, Southern Methodist University Geothermal Laboratory, and Virginia Polytechnic Institute and State University.

## 17.5.2  TECHNOLOGY UPDATES

The development of a successful geothermal energy project relies on a variety of specialized technologies as well as their cost-effectiveness.

### 17.5.2.1  Exploration Technology

Exploration is key to the discovery of new geothermal resources. It identifies geothermal resources, estimates resource potential, and establishes resource size, depth, and potential production. It relies on surface measurements of subsurface geological, geochemical, and geophysical conditions to develop a conceptual model of the system. Geothermal exploration of unmapped regions typically proceeds in two basic phases, reconnaissance and detailed exploration. During the reconnaissance phase, regional geology and fracture systems are studied, such as young volcanic features, tectonically active fault zones (as deduced from seismic information), and overt or subtle geothermal manifestations. If the reconnaissance phase confirms that the province has geothermal potential and that specific sites in the province should be explored further, the second phase focuses on one or more individual prospects covered in the reconnaissance phase.

### 17.5.2.2  Brine-Handling Technology

Brine is a geothermal solution containing appreciable concentrations of sodium chloride or other salts. The chemical composition, including the salinity of geothermal fluids, varies greatly from one reservoir to another. Variations in chemistry and salinity affect the design, maintenance, and longevity of wells and surface equipments. Recent advances in this area include the following:[5]

1. Use of scale-inhibiting chemicals to reduce carbonate scaling of flashing wells
2. Development of pH modification to control silica scaling in power plants
3. Development of highly effective computer programs to estimate and predict chemistry effects in geothermal systems
4. Continued development of polymeric cement coating to reduce corrosion in heat exchangers and process piping

### 17.5.2.3   Environmental Issues of Geothermal Energy Utilization

Even though geothermal energy is one of the cleanest and safest means of generating electric power, its effects on water resources, air quality, and noise during geothermal development and operation must be understood and mitigated. Among these are emissions to air (particularly of hydrogen sulfide), land use, and disposal of solid wastes. Effects can vary greatly from site to site.

Steam and flash plants emit mostly water vapor (steam). Binary power plants run on a closed-loop system; therefore, zero discharge of gases is accomplished. Geothermal industries have developed advanced technologies to recycle minerals in geothermal fluid so that little or no disposal or emissions occur. The examples are found from The Geysers power plants in northern California that separate and use sulfur for sulfuric acid production, and also from the Salton Sea power plants in southern California recycling salts from geothermal brine, recovering silica from mineralized brine for use as fillers in concrete, and extracting zinc for additional plant profitability.

## 17.6   CONCLUSION

Geothermal resources are continuously renewable sources of energy regardless of climate or weather conditions, unlike wind or solar energy. Reliability, sustainability, and cleanness make geothermal energy especially attractive as a source of baseload electricity generation or for direct-use applications that need constant heat or energy. Geothermal power plants compete economically with coal, oil, and nuclear plants in meeting baseload capacity needs, with significant environmental advantages. The next generation of geothermal power plants will be designed using long-term projections for resource production as the basis for cycle selection, optimization, and system design.

Current HDR technology is competitive with modern coal-fired plants in regions with geothermal gradients exceeding 60°C/km.[15] Reasonable improvements in reservoir performance or reductions in drilling and completion costs may substantially lower the effective cost of HDR power. In areas with steep geothermal gradients, the use of HDR may demonstrate a substantial cost advantage over coal. This advantage may increase over time, allowing the use of HDR to produce a significant portion of the future electricity of the world.

Owing to its practicality and low operating costs, direct application of geothermal energy is expected to grow in popularity, especially in geothermally favored regions. Diverse applications are expected to be developed in this field, and more advances in GHP technology are also expected. Advances in materials, process integration and design, resource management, instrumentation, and drilling technology will undoubtedly enhance the global utilization of geothermal energy.

## REFERENCES

1. Geothermal Resource Group, *Geothermal Resource and Technology in the United States*, National Academy of Sciences, Washington, DC, 1, 1979.
2. Speight, J.G. and Lee, S., *Handbook of Environmental Technology*, 2nd ed., Taylor & Francis, New York, 2000.

3. Chermisinoff, P.N. and Morresi, A.C., *Geothermal Energy Technology Assessment*, Technomic Publishing Co., CT, 1970.
4. Griffin, R.D., *CQ Researcher*, Vol. 2, No. 25, 1992, pp. 575–588.
5. U.S. Department of Energy, United States Geothermal Energy—Equipment and Services for Worldwide Application, 1994, DOE/EE—0044.
6. Lynn, M. and Reed, M.J., *Energy Sources*, Vol. 14, 1992, p. 443.
7. Wright, P.M., *The American Association of Petroleum Geologists Bulletin*, Vol. 23, No. 12, 1989, p. 366.
8. U.S. Department of Energy, Geothermal Energy: 1992 Program Overview, 1992.
9. *Annual Energy Outlook 2006*, Energy Information Administration (EIA), U.S. Department of Energy, Washington, DC, 2006.
10. Hadfield, P., *New Scientist*, 1990, Issue No. 1703, p. 58.
11. Web site of Geothermal Technologies Program, Energy Efficiency and Renewable Energy, U.S. Department of Energy, 2004, http://www.eere.energy.gov/geothermal/deployment_gpw.html.
12. Reed, M.J., *Geotimes*, Vol. 36, No. 2, 1991, p. 16.
13. Reed, M.J., *Geotimes*, Vol. 38, No. 2, 1993, p. 12.
14. Phair, K.A., *Mechanical Engineering*, 1994, Vol. 116, p. 76.
15. Harden, J., *Energy*, Vol. 17, No. 8, 1992, p. 777.
16. Tenenbaum, D., *Technology Review*, Vol. 98, 1995, p. 38.
17. Dickson, M.H. and Fanelli, M., *Energy Sources*, Vol. 6, 1994, p. 349.
18. Boyd, T.L. and Lund, J.W., Geothermal heating of greenhouses and aquaculture facilities, paper presented at International Geothermal Conference, Session #14, Reykjavik, Iceland, September 2003, http://jardhitafelag.is/papers/PDF_Session_14/S14Paper029.pdf.
19. Web site of Geothermal Heat Pump Consortium, Inc., 2004, http://www.geoexchange.org/about/how.html.
20. Web site of National Renewable Energy Laboratory, 2004, http://www.nrel.gov/documents/geothermal_energy.html.
21. Feature Article, *Power Engineering*, Vol. 93, 1989, p. 50.
22. Joyce, C., *New Scientist*, 1989, Issue No. 1652, p. 58.
23. Anderson, I., *New Scientist*, Vol. 111, 1986, p. 22.
24. Feature Article, *Mechanical Engineering,* Vol. 113, 1991, p. 10.

# 18 Nuclear Energy

*Sudarshan K. Loyalka*

## CONTENTS

## 18.1 NUCLEAR FISSION AND NUCLEAR REACTOR PHYSICS

The neutron was discovered in 1932. The following years witnessed intense study of its properties and interactions with matter. This neutral particle is about 2000 times the mass of an electron and is scattered and absorbed by different materials. The nature and rate of its reaction are determined by the nuclei of the host material and the energy of the neutron. Moreover, the nuclei that absorb neutrons can become radioactive and be transmuted to other types of nuclei, through radioactive decays. Neutrons can also split (fission) some nuclei (the fissile isotopes such as U-233, U-235, and Pu-239). Such fission is a complex process that produces new nuclei, beta and gamma radiation, and a few neutrons themselves. The products are energetic (the kinetic energy of fission products, energy of radiation), deriving their energy from the binding energy of the nucleus. Consequently, the new neutrons (two to three on average) released in fission provide the basis for a chain reaction. This chain reaction can be sustained (each successive generation has the same number of neutrons) or multiplied (each successive generation has more neutrons), and it can be used for a controlled and a sustained as well as an explosive release of energy.[1-21]

Fission of a single nucleus releases about 200 MeV ($3.2 \times 10^{-11}$ J) of energy. This energy is distributed, in a power reactor of modern design, approximately as shown in Table 18.1 and Figure 18.1.

**TABLE 18.1**

**Distribution of Fission Generated Energy in Time and Position**

| Type | Process | Percentage of Total Released Energy | Principal Position of Energy Deposition |
|---|---|---|---|
| | **Fission** | | |
| I: Instantaneous energy | Kinetic energy of fission fragments | 80.5 | Fuel material |
| | Kinetic energy of newly born fast neutrons | 2.5 | Moderator |
| | $\gamma$ energy released at time of fission | 2.5 | Fuel and structures |
| II: Delayed energy | Kinetic energy of delayed neutrons | 0.02 | Moderator |
| | $\beta$-decay energy of fission products | 3.0 | Fuel materials |
| | Neutrinos associated with $\beta$ decay | 5.0 | Nonrecoverable |
| | $\gamma$-decay energy of fission products | 3.0 | Fuel and structures |
| | **Neutron Capture** | | |
| III: Instantaneous and delayed energy | Nonfission reactions due to excess neutrons plus $\beta$- and $\gamma$-decay energy of $(n, \gamma)$ products | 3.5 | Fuel and structures |
| Total | | 100 | |

*Source:* El-Wakil, M.M., *Nuclear Heat Transport*, International Textbook Company, now available from the American Nuclear Society, 1971. With permission.

Of this 200 MeV, about 190 MeV, or 95%, is recoverable energy, as the neutrinos do not interact with matter, and escape from the system without depositing energy. One kilogram of U-235 contains $2.563 \times 10^{24}$ U-235 nuclei, and its fission would release energy of about $78 \times 10^6$ MJ (which is the same as $21.6 \times 10^6$ kWh or 2.47 MWy).

The fission products and beta and alpha particles are mostly deposited within a fraction of a centimeter from the point of birth (in a solid or a liquid), whereas neutrons and gammas travel a greater distance depending on their energy and the material. The prompt radiation is emitted within $10^{-17}$ to $10^{-6}$ s from an interaction, whereas the delayed radiation can be emitted within a few milliseconds to thousands of years (e.g., the long-lived isotopes).

The U-235 and neutron fission reaction can thus be described as

$$U^{235} + n \rightarrow A + B + \nu n \tag{18.1}$$

in which $A$ and $B$ can be, for example, nuclei of cesium and strontium, and $\nu$ is the number of neutrons produced. The reaction is in accordance with generalized laws of conservation, but $A$, $B$, and $\nu$ are not necessarily the same for each fission. In fact, a number of species are produced, and both $\nu$ and the energy (kinetic) of

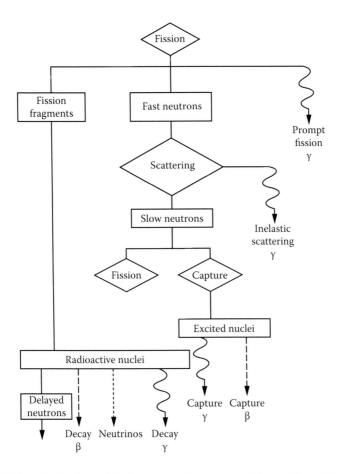

**FIGURE 18.1** Distribution of fission energy in energy and time. (From Ott, K.O., and Neuhold, R.J., *Introductory Nuclear Reactor Dynamics*, American Nuclear Society, 1985. With permission.)

product neutrons vary. Furthermore, the fission reaction rate is strongly dependent on the energy (kinetic) of the reacting neutron (and also on the kinetic energy of the reacting U-235 in certain energy ranges). Figures 18.2 and 18.3 show, respectively, the distribution of fission products and the distribution in energy of the neutrons produced in fission.

It is useful to note here that in a nuclear reactor, the overall neutron density (#/cm$^3$) is smaller than the nuclei density (#/cm$^3$) by about 10 orders of magnitude. Thus, the nuclei distribution is determined by nuclei–nuclei interactions, and the neutron distribution in space, direction, energy, and time is determined by neutron–nuclei interaction. In the long term, the fission products, actinides, and lattice suffer damage because the energetic products and neutrons affect the nuclei–nuclei interactions. Incidentally, both microscopic and macroscopic experiments have been used to study neutron–nuclei interactions, and a wealth of information on the nature of

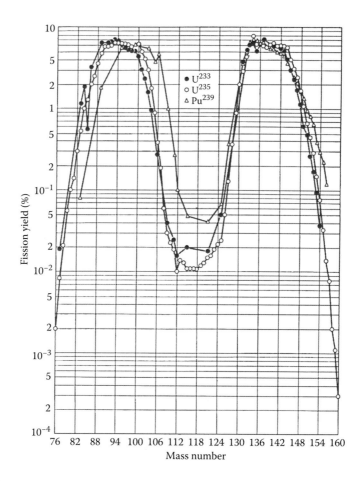

**FIGURE 18.2** Fission product distribution. (From Weinberg, A.M., and Wigner, E.P., *The Physical Theory of Neutron Chain Reactors*, University of Chicago Press, 1958. With permission.)

these interactions and their rates is now available. The rate $R_i$ (#/s) for a reaction of type "$i$" (absorption, fission, scattering) is expressed as

$$R_i(r,E,\Omega,t)drd\Omega dE = \sum_i (r,E,t)\phi(r,E,\Omega,t)drd\Omega dE \qquad (18.2)$$

in which

$$\sum_i (r,E,t) = \text{Macroscopic cross section (1/length) at } r \text{ at } E \text{ at time } t,$$

$$\phi(r,E,\Omega,t)dEd\Omega = \text{"Neutron Flux," (\# /area time) at } r \text{ at } E \text{ in } dE \text{ at } \Omega \text{ in } d\Omega \text{ at time } t,$$

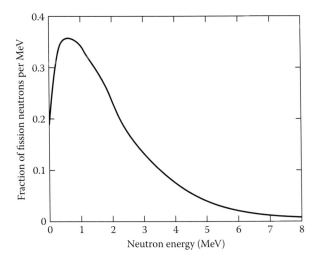

**FIGURE 18.3**   Energy distribution of neutrons produced in fission (the fission spectrum).

$d\mathbf{r}$ is a volume element at location $\mathbf{r}$, and $d\Omega$ is an elemental solid angle in direction $\Omega$. $E$ indicates energy, and $t$ is the time. A neutron balance equation can thus be constructed as

$$\frac{1}{v}\frac{\partial \phi(r,E,\Omega,t)}{\partial t} = -\Omega\nabla\phi(r,E,\Omega,t) - \sum_t (r,E,t)\phi(r,E,\Omega,t)$$

$$+ \int dE' \int d\Omega' \sum_s (r,E' \to E,\Omega' \to \Omega,t)\phi(r,E',\Omega',t)$$

$$+ S(r,E,\Omega,t)$$

(18.3)

This is often known as the *linear Boltzmann equation* for neutron transport, or just the *transport equation*, as it follows directly from the *nonlinear Boltzmann equation* for molecular distribution in the kinetic theory of gases. Note that $v$ is the speed of neutrons, the integral includes scattering as well as neutrons born in fission, and $S$ is a source term. Subscript $t$ indicates "total." The gradient is with respect to $\mathbf{r}$, and the gradient term indicates free streaming or drift. The boundary conditions for this equation are usually those of no inward neutron flow for a convex surface for a body situated in a vacuum. The initial conditions just prescribe the initial flux.

We should note that the neutron flux is related to the neutron density $n$ by

$$\phi(r,E,\Omega,t) = vn(r,E,\Omega,t)$$

and that the flux in the nuclear nomenclature is a scalar quantity. Thus,

$$\phi(r,E,\Omega,t)drdEd\Omega$$

should be understood as the path length traveled per unit time by neutrons in the elemental phase space volume $drdEd\Omega$. Thus, the inverse of the macroscopic cross section is the neutron mean free path $\lambda$ for that particular reaction, and we have $= 1/\sum$.

Progress over the last few years enables us now to compute the neutron flux for complicated geometries and reactor configurations by using combinations of analytical, deterministic, and Monte Carlo methods,[11–18] and this task has been greatly aided by advances in computational hardware. In a simplified picture, we note that for any given mass, the neutron multiplication factor (the ratio of neutrons in a generation to neutrons in the previous generation) can be written as

$$k = \frac{\text{neutrons produced}}{\text{neutrons absorbed} + \text{neutrons lost due to leakage}} \tag{18.4}$$

and is a measure of the criticality of the mass ($k > 1$, supercritical; $k = 1$, critical; $k < 1$, subcritical; $k \geq 1$ is needed to sustain a chain reaction). The associated rate equation can be written as

$$\frac{dn(t)}{dt} = \frac{k-1}{l} n(t) + s(t) \tag{18.5}$$

where $n(t)$ is the number density of neutrons (#/cm³), l is known as the neutron lifetime ($\sim 10^{-3}$ to $10^{-6}$ s), and s (#/cm³ s) is a source of neutrons. For a non-reentrant mass (surface), the factor $k$ is approximately expressed as

$$k = \frac{\int dr \int dE \int d\Omega v(E) \sum_f (r,E)\phi(r,E,\Omega)}{\int dr \int dE \int d\Omega \sum_a (r,E)\phi(r,E,\Omega) + \int dr_s \int dE \int_{\Omega.n(r_s)>0} d\Omega\Omega.n(r_s)\phi(r,E,\Omega)} \tag{18.6}$$

Here, the second term in the denominator relates to the leakage from the system, with $r_s$ being a point on the surface, and $n(r_s)$ a unit normal to the surface directed outward to the vacuum. This term is more important for small assemblies (with respect to the neutron mean free path) and less so for larger assemblies. Small assemblies generally correspond to weapons and research reactors, and the larger assemblies to cores of nuclear power plants. The macroscopic cross section is represented (we suppress the position dependence) as

$$\sum(E) = N\sigma(E) \tag{18.7}$$

where $N$, the number density of the nuclei (#/cm³) in the mass, is expressed as

$$N = \frac{0.6023 \times 10^{24} \rho}{M} \tag{18.8}$$

in which $\rho$ is the density of the mass (g/cm³), and $M$ is the molecular weight (g/gmol). $\sigma$ (cm²) is known as the microscopic cross section for interaction with neutrons. It is different for different processes (fission, absorption, scattering) and generally has a complex dependence on the material and the energy. We show a typical cross section in Figure 18.4 (note, a barn = $10^{-24}$ cm²).

For large reactors and design purposes, $k$ can be approximately expressed as

$$k_\infty = \frac{\int\limits_{all} dr \int\limits_0^\infty dE v(E) \sum\limits_f (r,E)\phi(r,E)}{\int\limits_{all} dr \int\limits_0^\infty dE \sum\limits_a (r,E)\phi(r,E)} \tag{18.9}$$

where an integral on the solid angle is understood. Further, it is conveniently expressed as

$$k_\infty = \varepsilon\eta fp$$

where the four factors are defined in Table 18.2.

We also define $E_c$ as some cutoff energy (about 1 eV), below which neutrons are regarded as "thermal" in that they have kinetic energy comparable to those of the nuclei and both gain and lose energy while interacting with nuclei (above the cutoff, the analysis can be simplified by assuming that the neutrons lose energy in collisions, scattering, with nuclei, but do not gain energy). Each of these four factors (known as the fast fission factor, reproduction factor, thermal utilization factor, and resonance escape probability, respectively) can be experimentally measured or estimated (computed). They aid greatly in understanding the role of various nuclear and material

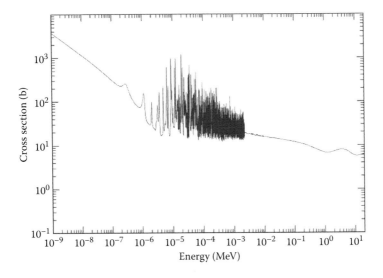

**FIGURE 18.4** U-235 total cross section. (ENDFB-VI cross-section files, obtained from www.nndc.bnl.gov.)

**TABLE 18.2**
**The Four Factors**

| Factor | Approximate Calculation or Measurement | Values (Typical of Natural Uranium and Water) | |
|---|---|---|---|
| | | Homogeneous Assembly | Heterogeneous Lattice |
| $$\varepsilon = \frac{\int_{fuel} dr \int_0^\infty dE\, \nu(E) \sum_f (r,E)\phi(r,E)}{\int_{fuel} dr \int_0^{E_c} dE\, \nu(E) \sum_f (r,E)\phi(r,E)}$$ | Insensitive to geometry. Can be estimated using approximate shapes of the neutron spectrum and cross sections. | 1.03 | 1.03 |
| $$\eta = \frac{\int_{fuel} dr \int_0^{E_c} dE\, \nu(E) \sum_f (r,E)\phi(r,E)}{\int_{fuel} dr \int_0^{E_c} dE \sum_a (r,E)\phi(r,E)}$$ | Insensitive to geometry. Can be estimated using $\phi(r,E) \sim \psi_M(E)$, thermal cross sections, and a thermal neutron beam incident on a foil in a manganese bath. | 1.34 | 1.34 |
| $$f = \frac{\int_{fuel} dr \int_0^{E_c} dE \sum_a (r,E)\phi(r,E)}{\int_{all} dr \int_0^{E_c} dE \sum_a (r,E)\phi(r,E)}$$ | Assuming $\phi(r,E) \sim R(r)\psi_M(E)$, $R$ can be measured through use of bare and cadmium-covered gold foils embedded at different points in a typical cell. | 0.9 | 0.8 |
| $$p = \frac{\int_{all} dr \int_0^{E_c} dE \sum_a (r,E)\phi(r,E)}{\int_{all} dr \int_0^\infty dE \sum_a (r,E)\phi(r,E)}$$ | Neutron absorption in thermal and resonance regions can be measured through use of bare and cadmium-covered U-238 foils. | 0.7 | 0.9 |

properties, thermal conditions, and geometrical arrangements of fuel ($UO_2$, etc.) and moderators, absorbers, and coolants in influencing $k$. Use of these four factors was quite important in early design of heterogeneous reactors, and it is still useful today. For example, it was found that arrangement of fuel in lumps or lattices leads to a higher value of $k$ over a homogeneous distribution. Also, whereas a critical reactor cannot be constructed with just natural uranium (of enrichment currently available) and light water even in the most favorable geometry, it is possible to construct critical reactors with natural uranium and graphite or heavy water as moderators, and with a gas, heavy water, or light water as coolants. Indeed, the earliest reactors were constructed with just natural uranium. We show typical values of the four factors in Table 18.2 and note how each of these can be measured or calculated approximately.

Clearly, knowledge of the neutron flux is crucial to the design of a reactor as the criticality and heat generation are directly dependent on it. The flux can be calculated if the geometry and material distribution are defined and the relevant neutron cross sections are known (from experiments or theory). The nuclear enterprise has paid detailed and careful attention to the cross sections from the beginning of the nuclear age and extensive and carefully assessed values are available for almost all materials of interest in the open literature and through government-sponsored research centers (for example, the National Nuclear Data Center at the Brookhaven National Laboratory). The geometry and material information can be used to create an input file for a Monte Carlo computer program such as MCNP[22] that has the cross-section libraries integral to it, and one can obtain the flux distribution in the reactor as well as compute the reactor's multiplication factor, power distribution, and other needed quantities. Computer programs are also available for computation of space–time variation of the fission products that accumulate in the reactor core and for devising fuel management strategies to obtain optimum power from the fuel consistent with applicable safety standards and regulations.

## 18.2   ELECTRICITY GENERATION FROM NUCLEAR REACTORS

As we have noted earlier, the energy released in a nuclear reactor is that associated with fission of fissile nuclei by neutrons, and also by emissions of beta, gamma, alpha, and neutrons by radioactive or unstable nuclei that are created by fission or absorption of neutrons by nuclei. Not all neutrons are released at the moment of fission; some are released later from the fission products, and some are also emitted by the actinides or because of photon or alpha particle reactions. The rate of this energy release (that is, the power, $P$) can be expressed as

$$P(r,t) = G_{f,p} \int_0^\infty dE \sum_f (r,E,t)\phi(r,E,t) + \sum_i G_i \lambda_i \int dr' \int_0^t dt' K_i(r,r';t,t')N_i(r',t')$$

$$(18.10)$$

in which we have made the simplifying assumption that a part of the energy is deposited locally (the first term) and is directly related to the local and instantaneous value of the neutron flux, and the rest is from the decays of the radioactive isotopes. The kernel $K$ is a measure of the contribution to power at **r** at time $t$ from a decay at **r** and $t$. In practice, computer programs are used to calculate space–time distributions of all important isotopes and the power production. For our purposes, it is sufficient to note that in an operating reactor, 95% or more of the energy is deposited in nuclear fuel, and the rest in reactor coolant and structural material. In a steady state, this energy is continuously removed to produce electricity and maintain the reactor at the desired neutronic, thermophysical, and structural conditions. We should also note that for a reactor that has operated for some time, not all the power production will stop if the neutron flux is reduced to a zero value (reactor shutdown) at the end of the operation, in that the radioactive isotopes would have accumulated and these will continue to produce power (the decay heat). This accumulation of isotopes depends on the reactor power history. The decay power after reactor shutdown for a reactor that has been operated for a long time, at some steady-state power $P_0$, can be approximately represented as

$$P(t)/P_0 = 0.066\ t^{-0.2} \tag{18.11}$$

in which $t$ is the time in seconds after the shutdown. We show a plot of the decay power in Figure 18.5.

Thus, the decay power is a significant factor in that adequate cooling must be maintained even after the reactor is shut down, as otherwise, the reactor fuel could melt, and fission products may be released.

Nuclear reactor power plants typically have three essential parts:

1. Nuclear reactor core, where the nuclear fuel is fissioned and energy and associated radiation is produced. The core must be designed to contain the radiation and fission products.
2. A coolant system (the primary), which removes energy from the core.

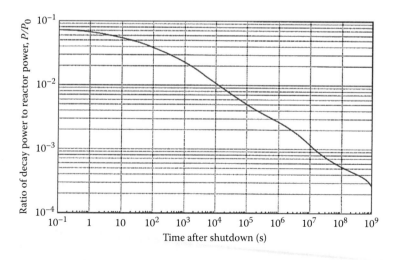

**FIGURE 18.5** Decay heat after shutdown for a reactor that had been operated for a very long time. (From American Nuclear Society Standard, ANS 5.1.)

3. A coolant system (the secondary), which transfers the energy (removed by the primary from the core) to an electricity generator through a turbine, and to the environment through a condenser and associated cooling tower or system. Some designs do not use a secondary coolant system at all and transfer energy from the primary to the turbine directly, whereas other designs require use of a tertiary system. Generally, about one-third of the energy generated in the core is converted to electricity, and about two-thirds is dissipated (lost) to the environment.

We show a typical nuclear power plant schematic in Figure 18.6.

A considerable portion of "2" and all of "3" are similar to those in coal-fired steam plants and do not require any special elaboration here. We will hence focus more on part "1," and we will also discuss parts of "2."

The reactor core of a modern 1000 MW$_e$ reactor is generally composed of about 250 reactor fuel assemblies ("bundle"), with each assembly consisting of about 200 fuel rods. Thus, there may be about 50,000 fuel rods in the reactor core. Generally, the fuel is $UO_2$, in which the fresh fuel, the uranium, has been enriched to 2%–4% (by weight) in U-235. The pencil-thin fuel rods are each about 4 m long and are contained in individual Zircaloy cans. There is some small spacing (filled with helium at the beginning) between a rod and its Zircaloy can ("cladding") to hold xenon, krypton, iodine, and other gases that are released during fission in the fuel. The core is placed in a thick pressure vessel, and coolant (water) is pumped through the core at high velocities. Each fuel rod can thus be envisaged as central to a cooling channel, in which the colder water (~573 K) enters the bottom, is heated by the fuel, and exits

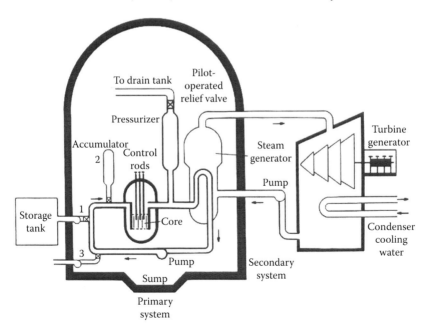

**FIGURE 18.6**   Schematic of a pressurized water reactor (PWR) plant. (From LANL report.)

relatively hot (~610 K) at the top. The high coolant temperature at exit is required by 2T Carnot cycle thermal efficiency considerations, and this in turn dictates the choice of coolants, fuel material, operating pressure (which can be as high as 15 MPa to prevent boiling), structural materials, and so forth. Once the coolant and the fuel are chosen, other design aspects depend on these choices.

During steady operation, there are typical drops in temperature of about 500, 200, 30, and 30 K (radial), respectively, across the fuel (from its centerline to surface), the gap, the clad, and the coolant. Obviously, these drops depend on thermophysical properties (thermal conductivity) of the materials, the structural conditions of the fuel and clad, the hydrodynamic conditions (the Reynolds and Prandtl numbers), and the heat generation rate in the fuel, and they are different at different locations in the core. A useful quantity here is the linear heat generation rate, which is typically about 15 to 20 kW/m of a fuel rod. Thus, each fuel rod generates 60 to 80 kW, and 50,000 or so fuel rods generate 3000 to 4000 $MW_{th}$ (megawatt thermal) power. With an overall thermal efficiency of 33% or so, this corresponds to 1000 to 1300 $MW_e$ (megawatt electric) of power generation, with the rest rejected to the atmosphere via cooling towers and so forth.

We show a view of a pressurized water reactor (PWR) core and pressure vessel in Figure 18.7 and schematics of a fuel assembly and rod in Figure 18.8. Typical fuel rod arrangement and the temperature drop across a fuel rod are shown in Figure 18.9.

As a nuclear reactor generates power, U-233, U-235, Pu-239, and other fissile isotopes are fissioned (consumed), and fertile isotopes such as Th-232 and U-238 are converted to fissile isotopes U-233 and Pu-239. The space–time concentrations of all the isotopes are controlled through reactor designs and fuel management strategies. Typically, because of neutronic and structural considerations, one-third of the fuel assemblies are replaced with fresh fuel assemblies each year, requiring reactor shutdowns for the refueling period. In a typical PWR that has been operating for a long time, most of the power is generated from U-235, whereas the rest derives from Pu-239 that is continuously generated from U-238. Some reactor designs also permit online refueling of reactors, and shutdowns for refueling are then not necessary. There is considerable interest at this time to design and employ high- and ultrahigh-burnup fuels that have refueling times of 10 to 15 years.

## 18.2.1 Reactor Control and a Toy Model

The reactor control requires careful attention to detail. Equation 18.5 does not tell the entire story in that not all neutrons are emitted at the same time, and some neutrons are born "delayed" through some fission products. Also, some fission products (such as xenon) are strong neutron absorbents, and they have both local and global effects on the reactor criticality, as they are both born (through fission products and their transmutation) and destroyed (through decay and neutron capture) continuously during the reactor operation. The multiplication factor $k$ is further dependent on the reactor temperature (for example, there is an increase in neutron absorption in U-238 resonances as the temperature goes up, leading to a decrease in $k$) and coolant conditions (an increase in temperature leads to lower density or phase change from liquid to vapor, and these changes, known as the density and void effects, respectively, both lead to a decrease in $k$). A simple model (a point kinetics or toy model) that captures effects of some of these phenomena could be written as

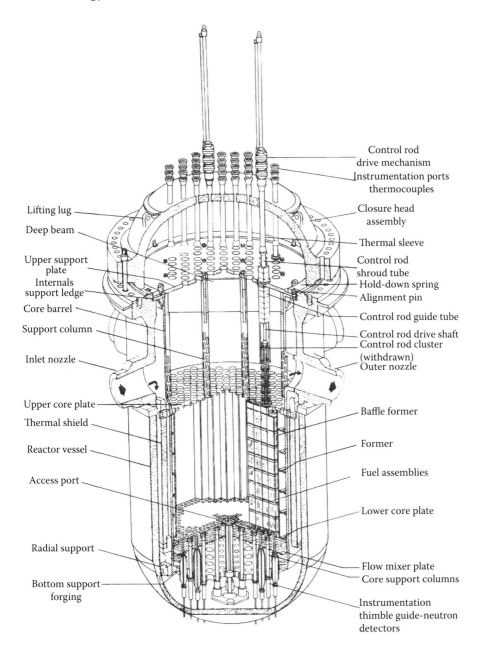

**FIGURE 18.7** A view of a pressurized water reactor (PWR) core and pressure vessel. (Westinghouse, from Connolly, T.J., *Foundations of Nuclear Engineering*, John Wiley & Sons, 1978. With permission.)

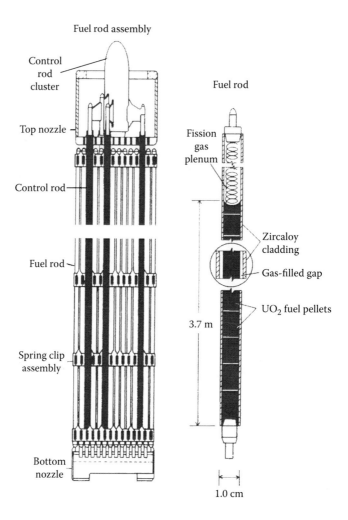

**FIGURE 18.8** Schematic of a fuel assembly and rod. (From LANL report.)

$$\frac{dn(t)}{dt} = \frac{(k(t,U(t))-1)-\beta}{\ell} n(t) + \lambda^c c(t) - \sigma_a^{Xe} Xe(t) vn(t) + s(t)$$

$$\frac{dc(t)}{dt} = \frac{\beta^c}{\ell} n(t) - \lambda^c c(t)$$

$$\frac{dl(t)}{dt} = \frac{\beta^I}{\ell} n(t) - \lambda^I l(t) \tag{18.12}$$

$$\frac{dXe(t)}{dt} = \lambda^I I(t) - \lambda^{Xe} Xe(t) - \sigma_a^{Xe} Xe(t) vn(t)$$

$$\frac{dU(t)}{dt} = P(t) - \dot{m}(t)(h_{out}(t) - h_{in}(t))$$

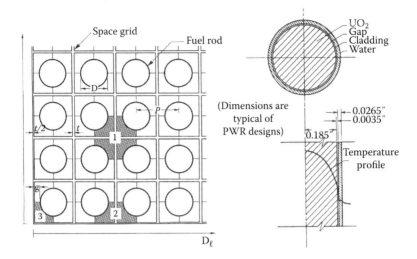

**FIGURE 18.9** Fuel rod arrangement with coolant channels in a PWR and typical temperature drop across a fuel rod.

with appropriate initial conditions. Here, $U$ is the internal energy of the core, $\dot{m}$ is the coolant mass flow rate through the core, $h$ is the enthalpy of the coolant, and $P$ is the power generation in the core. We have

$$P(t) = VG_f \sum_f vn(t) \tag{18.13}$$

in which $V$ is the total volume of the core (assuming that the cross sections are for the homogenized core in some sense). Also, $\beta$s are fractional coefficients that indicate generation due to fissions of the delayed neutron precursor "c" (actually, there are several, but we have shown only one), iodine-135 (I), and xenon-135 (Xe). The delayed neutron effects are most important for the routine short-term control of the reactor, whereas other effects are important for both short-term and long-term control of the reactor.

The multiplication factor $k$ has a complex dependence on the internal energy of the core, but this dependence can be expressed in some simple ways through the use of a summation of contributions from separate effects via coefficients that provide a measure of the change in $k$ with respect to these effects. For example, one constructs the equation

$$\frac{dk(t,U(t))}{dt} = \left( \frac{\partial k}{\partial U} \right) \frac{dU}{dt} + \dots \tag{18.14}$$

and develops it as dictated by insights and measurements or calculations. The nonlinear system of ordinary differential equations then can be numerically solved, and considerable insights can be gained in the overall working of the system through simulations. In fact, simulators based on similar principles have played a very significant role in reactor operator training, just as they have been crucial in the aircraft industry.

## 18.3  NUCLEAR FUEL CYCLE

Natural uranium and thorium, and their compounds and transmuted products (Pu-239 and U-233, respectively), are the fuel resources for nuclear reactors. At present, natural uranium contains 99.3% U-238 and 0.7% U-235 in its compounds. Natural thorium occurs mainly as Th-232 in its compounds. Th-232 is not fissile, but through neutron absorption and subsequent decays of products, it can be converted to U-233, which is fissile. As we have noted earlier, through similar processes, U-238 is converted to Pu-239, which is also fissile. Thus, fertile materials such as U-238 and Th-232 can be converted or "bred" into fissile materials. Because both U-238 and Th-232 are plentiful in the earth's crust, the nuclear fuel resource can be expanded 100-fold or more through breeding over the ones that are naturally available in the fissile form. Although all nuclear reactors convert fertile materials into fissile materials, special designs can enhance the breeding ratio (BR), which is defined as

$$BR = \frac{\text{Average rate of production of fissile isotopes}}{\text{Average rate of loss of fissile isotopes}} \qquad (18.15)$$

And it is possible to achieve BR > 1, guaranteeing long-term nuclear fuel supply (thousands of years). BR depends on the neutron spectrum (thermal, epithermal, or fast) in the reactor and the fuel (mainly the fertile content of uranium and thorium, that is, U-238 or Th-232 content in the fuel). Note that, fast neutrons produce a larger number of neutrons per fission as compared to thermal neutrons, as shown in Figure 18.10.

The effect is, however, dependent on the isotope. Significantly, for a fast-neutron spectrum, there is a greater availability of fission-produced neutrons for capture by the fertile isotopes, which results in a greater production of fissile isotopes. Thus, fast reactors (these are reactors in which the neutron spectrum is rich in fast-neutron content through avoidance of light moderating materials) can lead to a BR > 1, though certain thermal reactors based on U-233 (and thorium) can also lead to a BR > 1. Generally, however, for all light-water reactors, BR < 1, and such reactors are known as converters rather than breeders (for which, by definition, BR > 1).

Indeed, as we have noted earlier, nuclear power plants can be built with reactor cores that are either natural uranium based or use enriched uranium. One can also use mixed fuels, which are based on various combinations of thorium, uranium, and plutonium. The fissile materials produced in reactors can thus be recycled, either through special reactors or through most reactors of present designs, with appropriate adjustments.

Because nuclear reactors initially developed in the same time frame as nuclear weapons, the development of nuclear reactor power plants in various countries largely followed the expertise and resources that were developed in conjunction with nuclear weapons programs. Almost all present-day power plant designs derive from the initial work in the United States, Canada, United Kingdom, Russia (the Soviet Union), France, and Sweden. The countries that developed or used uranium technologies for weapons work preferred slightly enriched uranium (as the technology was already available to them and the enriched uranium leads to higher power densities, smaller cores, and longer fuel replacement times), whereas the other countries

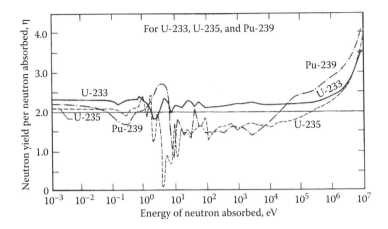

**FIGURE 18.10** Net neutrons produced per absorption in fissile isotopes, as a function of the energy of the neutron absorbed. (From ERDA-1541, June 1976.)

preferred (or had no other realistic choice than) to use natural uranium with graphite or heavy water, with water or a gas as a coolant. The fuel resource needs for the former are more complicated, as the enrichment at the scales needed is an expensive process and is available now only in a few nations.

The nuclear fuel cycle initially involves exploration and mining of uranium and thorium ores. These ores are widely available, and the principal resources are in the United States, Russia, China, Australia, South Africa, Gabon, Congo, Niger, India, and so forth. After mining (through open-pit mining or *in situ* leaching from the ground or rocks), the uranium ore (~0.1% uranium oxide content) is milled (and leached and precipitated out in an acid solution) to an oxide powder. In this form, the powder is known as *yellowcake* (~80% uranium oxide content). The remainder of the ore is known as the *tailings*; it is slightly radioactive and toxic and is a waste that must be properly disposed of.

At a conversion facility, the yellowcake is first refined to uranium dioxide, which can be used as the fuel in reactors that use natural uranium. Further processing, however, is needed if this uranium were to be enriched in U-235 for the reactors that require such uranium. For enrichment, the yellowcake is converted into uranium hexafluoride gas through use of hydrogen fluoride and is shipped in containers to an enrichment facility where either a diffusion or a centrifuge process is used (other methods such as laser isotope separation are still not in commercial use). The first process relies on a slight difference in the diffusion of the U-235 hexafluoride molecule from that of the U-238 hexafluoride molecule through a membrane (the diffusion coefficient is inversely proportional to the square root of the molecular mass), whereas the second method relies on the mass difference (inertia) between the two molecules.

In practice, thousands of stages are used. The diffusion process is highly energy intensive because of the pumping requirements, whereas the centrifuge process requires special rotor materials, motors, bearings, and so forth (the practical details are not in the public domain but are reported to have been clandestinely disseminated to a number of countries in the last few years). The enriched uranium hexafluoride

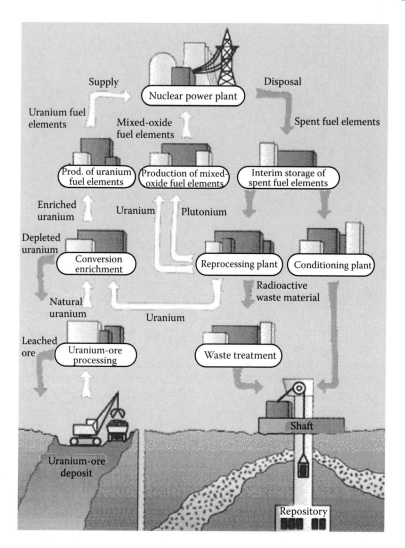

**FIGURE 18.11** Nuclear fuel cycle. (From http://www.infokreis-kernenergie.org/e/brenn stoffkreislauf.cfm.)

is next reconverted to produce enriched uranium oxide. This oxide then is sintered (baked) at a high temperature (over 1673 K) to produce pellets, which are then used in fuel rods and assemblies. In all this processing, great care is taken to ensure quality control with respect to content of fuel and its size and shape. Extensive efforts are also made to avoid accidental criticality.

As the reactor fuel is used in a reactor, it undergoes enormous transformations. The fissile isotopes are fissioned and lead to fission products, neutrons, other radiation, and heat section. The neutrons transmute the fission products and also produce many actinide species through absorption in the fertile isotopes (some of which will fission again). The fuel cracks because of heat and stresses, the fission products

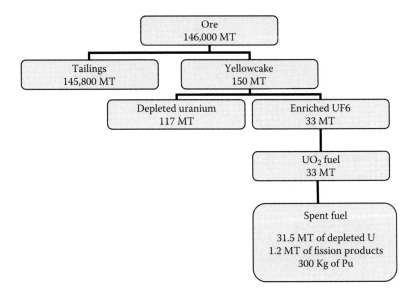

**FIGURE 18.12** Uranium ore needed for annual operation of a 1000 MW$_{th}$ reactor and the associated material balance and waste.

migrate within the fuel, and gases such as xenon and iodine accumulate in the gap between a fuel rod and its cladding. The degradation in the fuel thermophysical properties requires that the fuel be taken out of the reactor after a certain time period and that fresh fuel replace it. This generally is a batch process in that in most reactors, this "refueling" is done once a year when a reactor is shut down for a couple of weeks, and approximately one-third of the reactor core is replaced. The fuel that is taken out of the reactor would have been generally used for a period of about 3 years and is known as the *spent fuel*. There are certain reactors in operation (for instance, the CANDU reactor, which we will discuss later) in which the refueling is done online, and a shutdown is not required.

The spent fuel contains fission products, actinides (both minor and major), original fissile isotopes, and other material. It is both radioactive and hot and is stored at the reactor site in concrete-lined and cooled water pools for 6 months or longer. This storage is temporary, as eventually, the fuel must be reprocessed for extraction of the fissile isotopes (chiefly U-235 and Pu-239) and shipped for longer-term storage and disposal at remote sites. Figure 18.11 is a schematic of the fuel cycle.

The reprocessing permits recycling of unused U-235, enables use of Pu in a reactor, and reduces the volume of the waste that needs to be disposed. Although these actions are very desirable from an economic viewpoint and the technology has been demonstrated, both the recycling of Pu and waste disposal issues have been very controversial. It has been contended that the availability of Pu can lead to weapons proliferation and that it is very difficult to demonstrate long-term safe storage of the nuclear waste (even in geological sites such as the Yucca mountain site in the United States).

We show a typical material balance for the fuel needed for the annual operation of a 1000 MW$_{th}$ reactor in Figure 18.12.

## 18.4   TYPES OF REACTORS

We have discussed the PWR design earlier. Other types of reactors in present use are as follows:

> *The boiling water reactor (BWR):* This type of reactor differs from a PWR in that water is allowed to boil fully in the upper portions of the reactor core and that no heat exchanger loop (the secondary) is used. The water is maintained at about 6–7 MPa (instead of 15.5 MPa as in a PWR). The steam passes through a separator at the top of the core and then goes on to drive turbines; the condensed water is pumped back to the core. The steam does contain some radioactive material, so a modest shielding on the turbine side is required. Lower pressures and avoidance of steam generators simplify the plant design, but then there is the additional complication of a two phase flow in the core. Also, the pressure vessel is larger (it needs to accommodate the separator), and the control blades are inserted from the bottom, thereby precluding gravitational insertion of control rods in an emergency. Several different designs of BWRs and their containments are presently in existence, but in the United States, these reactors are designed only by the General Electric Co. We show a schematic of a BWR in Figure 18.13.
>
> *The pressurized heavy-water reactor (PHWR):* These reactors, sometimes also known as CANDU because of their primary development in Canada, employ heavy water as a moderator and either the heavy water or light water or both as a coolant. These reactors use natural uranium or slightly enriched uranium and pressurized tubes for the coolant, thereby avoiding large pressure vessels. Because of the use of the tubes, fuel can be replaced "online" without a reactor shutdown for refueling.
>
> *The graphite-moderated gas-cooled reactors (Magnox and HTGR):* These reactors use graphite as a moderator and a gas ($CO_2$ or helium) as a coolant. Because, like heavy water, graphite is a good moderator and a poor neutron absorber, graphite-moderated reactors can use natural uranium as fuel. Note that a gas, because of its low density, is quite transparent to neutrons.

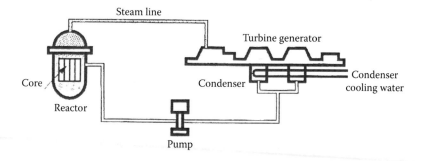

**FIGURE 18.13**   Schematic of a boiling water reactor. (Courtesy of US Department of Energy.)

Gases are, however, poor coolants as compared to liquids, and these reactors have low specific heat generation and comparatively large surface areas and, hence, large volumes. But one can achieve very high temperatures (hence the name *high-temperature gas reactor*) as phase changes and dissociations or surface reactions can be avoided, particularly with the use of a noble gas such as helium. However, helium does pose a difficulty both in terms of its cost (in countries other than the United States) and the ease of its leakage from the system.

*The graphite-moderated water-cooled reactors* (*RBMK types*): These reactors use graphite as a moderator and light water as a coolant. Natural uranium or slightly enriched uranium can be used as fuel. The Chernobyl reactor was this type. These reactors are more compact as compared to the gas-cooled reactors and also put less demand on coolant pumping power. These reactors, however, can have a positive temperature coefficient in that under certain circumstances, an increase in reactor power can lead to an additional automatic increase in power (a positive feedback). Thus, if other means of automatic or manual control of the reactor are disabled, an increase in reactor power could potentially occur that would terminate only after an excursion and reactor core disassembly, resulting in serious damage to the plant and, possibly, its surroundings.

*Liquid metal fast-breeder reactors* (*LMFBR*): The BR depends on average neutron energy as the cross sections for absorption and fission are functions of neutron energy. For uranium-fueled reactors, a fast spectrum (that is, the neutrons are mostly at energies higher than 100 keV or so) can lead to a BR higher than unity. The need for a fast spectrum precludes use of low-atomic-mass materials in the reactor core (excepting gases, which, because of their low density, have large mean free paths for neutrons at all energies). This requires consideration of liquid metals or their alloys. Sodium, lead–bismuth, and mercury have all merited consideration, and several power reactors have been built, tested, and extensively used with sodium as a coolant. Generally, the cores are designed with zones of different fuel compositions (seed and blanket) to maximize breeding and facilitate fuel reprocessing. Sodium does pose challenges because of its corrosiveness and high melting point (it is solid at room temperature), but it is an excellent coolant. One of the earliest reactors to produce electric power was the Experimental Breeder Reactor-I. The Experimental Breeder Reactor-II (EBR-II) operated for a long time. The Fast Fuel Test Facility (FFTF) was a sodium-cooled test reactor, and so is the Kalpakkam reactor in India. The French Phoenix and Super Phoenix reactors both have been LMFBRs and have produced power. In the United States, the experience has been mixed. The Fermi-I, the first commercial LMFBR built near Detroit, was eventually shut down because of a flow blockage and consequent temperature rise that damaged the core. The Clinch River reactor was aborted because of economic and political considerations. We show a schematic of a LMFBR in Figure 18.14.

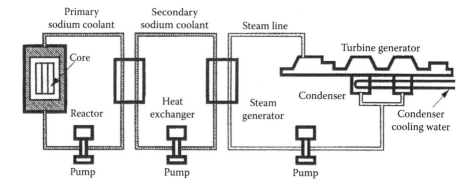

**FIGURE 18.14**   Schematic of a liquid metal fast-breeder reactor. (Courtesy of US Department of Energy.)

**TABLE 18.3**

**Basic Features of Major Power Reactor Types**

| | | | | Fuel | |
| Reactor Type | Neutron Spectrum | Moderator | Coolant | Chemical Form | Approximate Fissile Content (All $^{235}$U Except LMFBR) |
|---|---|---|---|---|---|
| Water cooled | Thermal | | | | |
| PWR | | $H_2O$ | $H_2O$ | $UO_2$ | ~3% enrichment |
| BWR | | $H_2O$ | $H_2O$ | $UO_2$ | ~3% enrichment |
| PHWR (CANDU) | | $D_2O$ | $D_2O$ | $UO_2$ | Natural |
| SGHWR | | $D_2O$ | $H_2O$ | $UO_2$ | ~3% enrichment |
| Gas cooled | Thermal | Graphite | | | |
| Magnox | | | $CO_2$ | U metal | Natural |
| AGR | | | $CO_2$ | $UO_2$ | ~3% enrichment |
| HTGR | | | Helium | UC $ThO_2$ | ~7%–20% enrichment[a] |
| Liquid metal cooled | Fast | None | Sodium | | |
| LMR | | | | U/Pu metal; $UO_2/PuO_2$ | ~15%–20% Pu |
| LMFBR | | | | $UO_2/PuO_2$ | ~15%–20% Pu |

*Source:*   Ott, K.O., and Neuhold, R.J., *Introductory Nuclear Reactor Dynamics*, American Nuclear Society, 1985. With permission.

[a]   Older operating plants have enrichments of more than 90%.

We summarize the basic features of these reactor types and their thermodynamic parameters in Tables 18.3 and 18.4, respectively.

The nationwide distribution of nuclear power plants is shown in Table 18.5. Table 18.6 shows the distribution of reactors by their types. These data are current as of September 2013 according to the International Atomic Energy Agency (IAEA).

## 18.4.1 ADVANCED REACTORS AND CONCEPTS

Over the past 60 years, many nuclear reactor designs have been considered. Many have been utilized for commercial and routine production of electric power. Considering that France produces almost 70% of its electric power needs through existing nuclear reactor designs, it should be clear that given the political will and public acceptance, power plants based on the existing designs can provide electric power to most countries in the immediate future.

However, from a long-range technical point of view, as well as current and anticipated socioeconomic and political viewpoints, it is important to consider new reactor designs. The historical development of nuclear power plants is well depicted by Figure 18.15.[23]

A major consideration in the new developments is the desire to extend the utilization of fuel resources through breeding of fertile material (which are plentiful but do not directly produce energy) to fissile material (which are not plentiful but produce energy) and minimization of nuclear waste material. Figure 18.16 shows how the advancements would lead to less waste and more sustainable energy production.

Two types of initiatives have received broad industrial and government support in this regard:

1. The first initiative consists of some incremental but important modifications of the existing designs. These modifications are based on the criteria of credible plans for regulatory acceptance, existence of industrial infrastructure, commercialization, cost sharing between industry and government, demonstration of economic competitiveness, and reliance on existing fuel cycle industrial structure. The reactor designs that have emerged are the Advanced Boiling Water Reactor (ABWR-1000, ESBWR), the Advanced Pressurized Water Reactor (AP600, AP1000, SWR-1000), Pebble Bed Modular Reactor (PBMR), International Reactor and Innovative and Secure (IRIS), and Gas-Turbine Modular Helium Reactor (GT-MHR). These reactors, and the demonstration plants, could come into operation by the year 2010.

2. The second initiative,[24] known as the Generation IV Initiative (the GEN-IV Initiative), has the goals of *sustainability* (the ability to meet the needs of present generations while enhancing and not jeopardizing the ability of future generations to meet society's needs indefinitely), *safety and reliability, economics*, and *proliferation resistance, and physical protection*. The emphasis here is on reactor concepts and associated nuclear energy systems that will satisfy the following criteria:
   a. Meet clean air objectives
   b. Promote long-term availability of systems and effective fuel utilization for worldwide energy production
   c. Excel in safety and reliability
   d. Have very low likelihood and degree of reactor core damage
   e. Eliminate the need for off-site emergency response
   f. Have a clear life-cycle cost advantage over other energy sources
   g. Have a level of financial risk comparable to other energy sources

**TABLE 18.4**

**Typical Characteristics of the Thermodynamic Cycles for Six Reference Power Reactor Types**

| Characteristic | BWR | PWR(W) | PHWR | HTGR | AGR | LMFBR |
|---|---|---|---|---|---|---|
| Reference design | | | | | | |
|   Manufacturer | General Electric | Westinghouse | Atomic Energy of Canada, Ltd. | General Atomic | National Nuclear Corp. | Novatome |
|   System (reactor station) | BWR/6 | (Sequoyah) | CANDU-600 | (Fulton) | HEYSHAM 2 | (Superphenix) |
| Steam cycle | | | | | | |
|   No. of coolant systems | 1 | 2 | 2 | 2 | 2 | 3 |
|   Primary coolant | $H_2O$ | $H_2O$ | $D_2O$ | He | $CO_2$ | Liq. Na |
|   Secondary coolant | — | $H_2O$ | $H_2O$ | $H_2O$ | $H_2O$ | Liq. Na/$H_2O$ |
| Energy conversion | | | | | | |
|   Gross thermal power, $MW_{th}$ | 3579 | 3411 | 2180 | 3000 | 1550 | 3000 |
|   Net electric power, $MW_e$ | 1178 | 1148 | 638 | 1160 | 618 | 1200 |
|   Efficiency (%) | 32.9 | 33.5 | 29.3 | 38.7 | 40.0 | 40.0 |
| Heat transport system | | | | | | |
|   No. of primary loops and pumps | 2 | 4 | 2 | 6 | 8 | 4 |

| | | | | | |
|---|---|---|---|---|---|
| No. of intermediate loops | — | 4 | — | — | 8 |
| No. of steam generators | — | 4 | 4 | 6 | 4 | 8 |
| Steam generator type | — | U tube | U tube | Helical coil | Helical coil | Helical coil |
| Thermal hydraulics | | | | | | |
| Primary coolant | | | | | | |
| Pressure (MPa) | 7.17 | 15.5 | 10.0 | 4.90 | 4.30 | ~0.1 |
| Inlet temp. (°C) | 278 | 286 | 267 | 318 | 334 | 395 |
| Average outlet temp. (°C) | 288 | 324 | 310 | 741 | 635 | 545 |
| Core flow rate (Mg/s) | 13.1 | 17.4 | 7.6 | 1.42 | 3.91 | 16.4 |
| Volume (L) or mass (kg) | — | $3.06 \times 10^5$ | $1.20 \times 10^5$ | (9550 kg) | $5.3 \times 10^6$ | $(3.20 \times 10^6 \text{ kg})$ |
| Secondary coolant | | | | | | |
| Pressure (MPa) | — | 5.7 | 4.7 | 17.2 | 16.0 | ~0.1/17.7 |
| Inlet temp. (°C) | — | 224 | 187 | 188 | 156.0 | 345/235 |
| Outlet temp. (°C) | — | 273 | 260 | 513 | 541.0 | 525/487 |

**TABLE 18.5**
**Nuclear Power Units by Nation**

| Nation | In Operation | | Total | |
|---|---|---|---|---|
| | # Units | Net MW | # Units | Net MW |
| Argentina | 2 | 935 | 3 | 1627 |
| Armenia | 1 | 375 | 1 | 375 |
| Belgium | 7 | 5927 | 7 | 5927 |
| Brazil | 2 | 1884 | 3 | 3129 |
| Bulgaria | 2 | 1906 | 2 | 1906 |
| Canada | 19 | 13,500 | 19 | 13,500 |
| China | 18 | 13,860 | 46 | 41,644 |
| China (Taiwan) | 6 | 5028 | 8 | 7628 |
| Czech Republic | 6 | 3804 | 6 | 3804 |
| Finland | 4 | 2752 | 5 | 4352 |
| France | 58 | 63,130 | 59 | 64,730 |
| Germany | 9 | 12,068 | 9 | 12,068 |
| Hungary | 4 | 1889 | 4 | 1889 |
| India | 20 | 4391 | 27 | 9215 |
| Iran | 1 | 915 | 1 | 915 |
| Japan | 50 | 44,215 | 52 | 46,865 |
| Korea, Republic | 23 | 20,739 | 28 | 27,059 |
| Mexico | 2 | 1530 | 2 | 1530 |
| Netherlands | 1 | 482 | 1 | 482 |
| Pakistan | 3 | 725 | 5 | 1355 |
| Romania | 2 | 1300 | 2 | 1300 |
| Russia | 33 | 23,643 | 43 | 32,025 |
| Slovakia | 4 | 1816 | 6 | 2696 |
| Slovenia | 1 | 688 | 1 | 688 |
| South Africa | 2 | 1860 | 2 | 1860 |
| Spain | 8 | 7567 | 8 | 7567 |
| Sweden | 10 | 9408 | 10 | 9408 |
| Switzerland | 5 | 3308 | 5 | 3308 |
| Ukraine | 15 | 13,107 | 17 | 15,007 |
| United Arab Emirates | 0 | 0 | 2 | 2690 |
| United Kingdom | 16 | 9231 | 16 | 9231 |
| United States | 100 | 98,560 | 103 | 101,959 |
| Totals | 434 | 370,543 | 503 | 437,739 |

*Source:* http://www.iaea.org/pris/.

　　h. Provide assurance against diversion or theft of weapons-usable materials
　　i. Provide physical security of the reactor and facilities

Nearly 100 concepts that can be generally classified in the broad categories of water cooled, gas cooled, liquid metal cooled, and nonclassical (molten salt, gas-core, heat pipe, and direct energy conversion) have been proposed and considered.

## TABLE 18.6
### Nuclear Power Units by Reactor Type

| Reactor Type | In Operation | | Total | |
|---|---|---|---|---|
| | # Units | Net MW$_e$ | # Units | Net MW$_e$ |
| Pressurized light-water reactors (PWR) | 270 | 249,621 | 327 | 306,896 |
| Boiling light-water reactors (BWR) | 84 | 78,122 | 88 | 83,372 |
| Gas-cooled reactors, all types | 15 | 8040 | 16 | 8240 |
| Heavy-water reactors, all types | 48 | 23,961 | 53 | 27,173 |
| Graphite-moderated light-water reactors (LGR) | 15 | 10,219 | 15 | 10,219 |
| Liquid metal-cooled fast-breeder reactors (LMFBR) | 2 | 580 | 4 | 1839 |
| Totals | 434 | 370,543 | 503 | 437,739 |

*Source:* http://www.iaea.org/pris/.

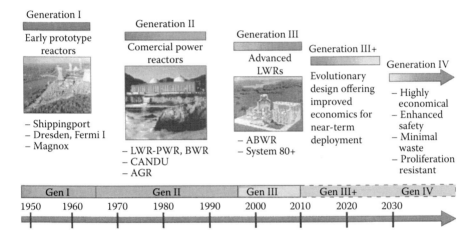

**FIGURE 18.15** Historical progression of nuclear reactor development. (From *A Technology Road Map for Generation IV Nuclear Energy Systems*, USDOE, 2002.)

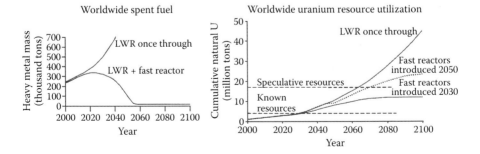

**FIGURE 18.16** Gain in sustainable nuclear resources with Pu recycling. (From *A Technology Road Map for Generation IV Nuclear Energy Systems*, USDOE, 2002.)

Six leading candidates that have been identified for research funding by the US Department of Energy are as follows:

- Gas-cooled fast reactor (GFR)
- Lead-cooled fast reactor (LFR)
- Molten salt reactor (MSR)
- Sodium-cooled fast reactor (SFR)
- Supercritical water-cooled reactor (SWCR)
- Very-high-temperature reactor (VHTR)

We show their schematics in Figure 18.17 and summarize important parameters in Table 18.7.

We also note that, while there are about 200 small power reactors operating worldwide, there has been much discussion of small modular reactors (SMR), particularly in the United States, and significant research programs have been launched by private industry with financing from the government. These reactors will be less than 250 $MW_e$ each, generally underground, and will be mostly PWR types. A large pressure vessel will house most main components (core, steam generator, pressurizer), and the number of valves and so forth will be minimized. The anticipated start date for the first such reactor is about 2025. These reactors will have the advantages of lower costs, rapid construction, improved safety, lower source terms, and smaller evacuation zones. The challenges include protection against floods, accident management, cost and nuclear proliferation, and so forth.

Interesting as these concepts are, they do not constitute a radical departure from the reactor designs of the past. The research and development issues with these GEN-IV reactors are mostly related to long-term operations at high temperatures and thus concern materials compatibility, corrosion and damage, safety (particularly for fast spectra), and fuel processing and recycling. But basically, the GEN-IV reactors are recycles of old designs, and no new physics is involved.

An interesting design concept under exploration is the accelerator-driven reactor.[25] In this design, a proton beam is used to strike a lead or other heavy target, creating high-energy (fast) neutrons (known as *spallation neutrons*). These neutrons drive an otherwise subcritical reactor, and a steady state of neutron population is maintained in a subcritical ($k < 1$) reactor (see Equation 18.5, where $dn/dt = 0$, but $s \neq 0$, leading to $n = (s)/(k - 1)$). These reactors avoid any chances of large reactivity insertions and, hence, large reactor accidents. These reactors can use any fissile material and have BRs because of fast spallation neutrons. These designs are receiving some attention in the international community.

We should also note that all the preceding designs rely on conversion of fission heat through a thermodynamic steam or gas cycle to electricity and thus have low conversion efficiencies (maximum of about 50%). There are research efforts underway to explore direct conversion schemes in which the fission energy can be first converted to photonic energy through formation of excimers and photonic emissions by these excimers, and then the conversion of this photon energy to electric energy through the use of photoelectronic devices. In principle, one might then be able to get higher conversion efficiencies as a 2T Carnot cycle is avoided.[26]

(a)

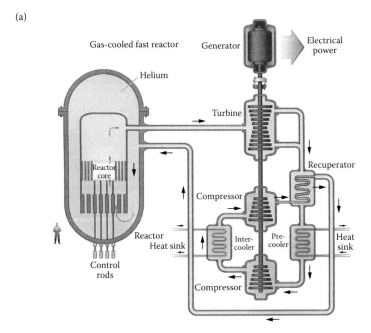

(b)

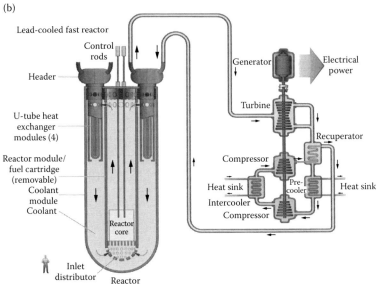

**FIGURE 18.17** Schematics of GEN-IV reactors. (a) Gas-cooled fast reactor (GFR). (From *A Technology Road Map for Generation IV Nuclear Energy Systems*, USDOE, 2002.) (b) Lead-cooled fast reactor (LFR). (c) Molten salt reactor (MSR). (d) Sodium-cooled fast reactor (SFR). (e) Supercritical water-cooled reactor (SWCR). (From Lapeyre, B. et al., *Introduction to Monte-Carlo Methods for Transport and Diffusion Equations*, Oxford, 2003, original in French, 1998. With permission.) (f) Very-high-temperature reactor. (From *A Technology Road Map for Generation IV Nuclear Energy Systems*, USDOE, 2002.)

(c)

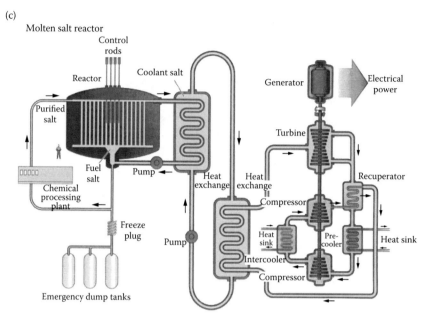

(d)

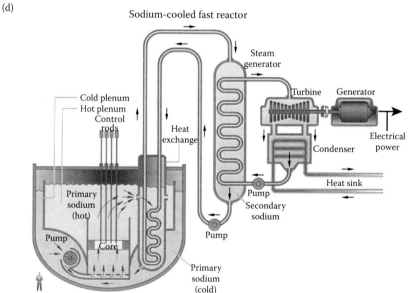

**FIGURE 18.17** (Continued) Schematics of GEN-IV reactors. (a) Gas-cooled fast reactor (GFR). (From *A Technology Road Map for Generation IV Nuclear Energy Systems*, USDOE, 2002.) (b) Lead-cooled fast reactor (LFR). (c) Molten salt reactor (MSR). (d) Sodium-cooled fast reactor (SFR). (e) Supercritical water-cooled reactor (SWCR). (From Lapeyre, B. et al., *Introduction to Monte-Carlo Methods for Transport and Diffusion Equations*, Oxford, 2003, original in French, 1998. With permission.) (f) Very-high-temperature reactor. (From *A Technology Road Map for Generation IV Nuclear Energy Systems*, USDOE, 2002.)

(e)

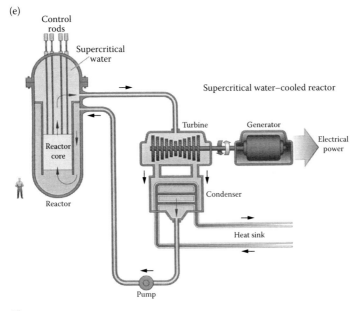

(f)

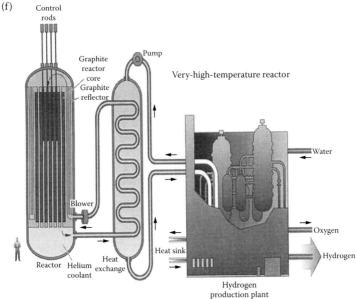

**FIGURE 18.17** (Continued) Schematics of GEN-IV reactors. (a) Gas-cooled fast reactor (GFR). (From *A Technology Road Map for Generation IV Nuclear Energy Systems*, USDOE, 2002.) (b) Lead-cooled fast reactor (LFR). (c) Molten salt reactor (MSR). (d) Sodium-cooled fast reactor (SFR). (e) Supercritical water-cooled reactor (SWCR). (From Lapeyre, B. et al., *Introduction to Monte-Carlo Methods for Transport and Diffusion Equations*, Oxford, 2003, original in French, 1998. With permission.) (f) Very-high-temperature reactor. (From *A Technology Road Map for Generation IV Nuclear Energy Systems*, USDOE, 2002.)

**TABLE 18.7**
**Some Technical Specifications of GEN-IV Reactors**

| Reactor Parameter | Reactor Type | | | | | |
|---|---|---|---|---|---|---|
| | GFR | LFR | MSR | SFR | SCWR | VHTR |
| Power | 600 MW$_{th}$ | 125–400 MW$_{th}$ | 1000 MW$_e$ | 1000–5000 MW$_{th}$ | 1700 MW$_e$ | 600 MW$_{th}$ |
| Net plant efficiency (%) | 48 | | 44 to 50 | | 44 | >50 |
| Reference fuel compound | UPuC/SiC with about 20% Pu | Metal alloy or nitride | U or Pu fluorides dissolved in Na/ZR fluorides | Oxide or metal alloy | UO$_2$ with stainless-steel or Ni-alloy cladding | ZrC-coated particles in blocks, pins or pebbles |
| Moderator | N/A | N/A | Graphite | N/A | Water | Graphite |
| Coolant | Helium | Lead-eutectic | | Sodium | Water | Helium |
| Coolant inlet/outlet Temperature (°C) | 490/850 | NA/550 | 565/700(850) | NA/550 | 280/510 | 640/1000 |
| Coolant pressure (bar) | 90 | 1 | | 1 | 250 | Variable |
| Neutron spectrum | Fast | Fast | Thermal | Fast | Thermal/fast | Thermal |
| Conversion (breeding) ratio | Self-sufficient | 1.0 | Burner | 0.5–1.30 | | |
| Average power density (MW$_{th}$/m$^3$) | 100 | | 22 | 350 | 100 | 6–10 |
| Burnup damage | 5% FIMA; 60 dpa | 100 GWD/MTHM | | 150–200 GWD/MTHM | 45 GWD/MTH; 10–30 dpa | |
| Earliest deployment | 2025 | 2025 | 2025 | 2015 | 2025 | 2020 |

*Source:*  *A Technology Road Map for Generation IV Nuclear Energy Systems,* USDOE, 2002.

*Note:*  dpa, displacements per atom; FIMA, fissions of initial metal atoms; GWD, gigawatt days; MTHM, metric ton heavy metal; NA, not available; N/A, not applicable.

## 18.4.2 HYDROGEN PRODUCTION

Fuel cells for automotive transport will require hydrogen. It has been suggested that nuclear energy (reactors) may provide a very effective means of generating hydrogen. The schemes under consideration include the following:[27]

- Steam methane reforming, using nuclear energy for the endothermic heat of reaction
- Conventional electrolysis, using nuclear-generated electricity
- Thermochemical cycles for water splitting
- Hybrid cycles combining thermochemical and electrolytic steps
- High-temperature electrolysis using nuclear electricity and heat

The steam methane reformation is based on the reaction

$$CH_4 + 2H_2O + 185 \text{ kJ} \rightarrow CO_2 + 4H_2 \tag{18.16}$$

The advantage of using nuclear energy to produce steam for this above process is that the process itself has been extensively studied. The use of nuclear energy to produce steam avoids the need for methane combustion to produce steam, and the $CO_2$ so produced is easier to sequester than $CO_2$ resulting from methane burning. The disadvantages are that the $CO_2$ is nevertheless produced and needs to be sequestered. Also, a large amount of methane (natural gas) is used.

In the thermochemical cycle under consideration (see Figure 18.18), the nuclear-generated heat is used to split hydrogen from water through use of sulfur dioxide and iodine reactions. The process has the advantage that the raw stocks (iodine and sulfur dioxide) are not consumed and are recycled. Its disadvantages are that a high-temperature reactor must be built to test this idea.

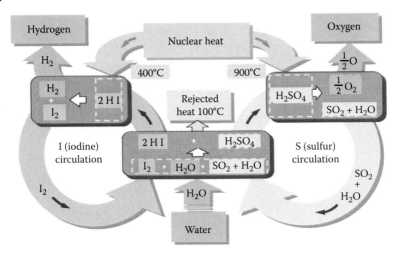

**FIGURE 18.18**   The thermochemical cycle for producing hydrogen from water and nuclear energy. (Courtesy of INEL.)

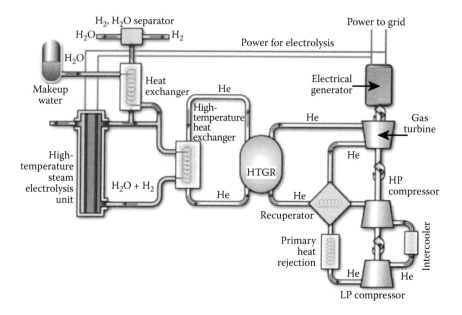

**FIGURE 18.19** High-temperature electrolysis using heat and electricity generated from a high-temperature gas reactor. (From *A Technology Road Map for Generation IV Nuclear Energy Systems*, USDOE, 2002.)

The high-temperature electrolysis process is quite straightforward, as shown in Figure 18.19, but it has a lower conversion efficiency than that of the thermochemical cycle.

Note that both the VHTR and GFR of the GEN-IV designs would also be quite suitable for hydrogen production because of the high temperature of the exiting helium.

## 18.5   PUBLIC CONCERNS OF SAFETY AND HEALTH

Public concerns regarding safety and health issues associated with the operation of nuclear reactors have their genesis in the dread of nuclear weapons and the fact that the developments of the weapons and the nuclear reactors have overlapped and proceeded in a coincidental time frame. Fission weapons are, however, characterized by small size; the explosive part is only a few centimeters in diameter. Power density is extremely high, and explosion time is a few microseconds. The design is specific to achieving the explosion in that the fissile material is compressed and brought to supercriticality and kept supercritical for a short time frame in a specific fashion, and the neutrons are introduced at an appropriate time to get the desired explosive energy release. Commercial nuclear reactors, on the other hand, are large, with cores that are approximately 12 ft in diameter. Their power density is low. Also, energy is produced over a long period of time. Design safeguards protect against power excursions of the type the weapons are designed for.

Generally, in a nuclear explosion, 50% of the damage comes from thermal radiation, 35% from the blast, and only the remaining 15% from the short-term and delayed radiation (beta, gamma, and alpha radiation associated with decay of radioisotopes). The radiation can cause cellular, glandular, and DNA damage and induce cancer through inhalation or other means of exposure (digestion or skin exposure). The thermal radiation and the blast are not the causes for concern with respect to nuclear reactors, as the reactors cannot explode like weapons. But all reactors can encounter circumstances in which the rate of heat generation in the reactor core exceeds the heat removal rate. Let us consider the first law of thermodynamics as applied to the entire reactor core (we neglect conduction, radiation losses, and so forth, to simplify the arguments):

$$\frac{dU(t)}{dt} = -\dot{m}(t)(h_{out}(t) - h_{in}(t)) + P(t) \tag{18.17}$$

where $U$ is the internal energy of the core, $\dot{m}$ is the coolant mass flow rate through the core, $h$ is the enthalpy, and $P$ is the power generation in the core. The steady state corresponds to the two terms on the right balancing out, but obviously, if the second term exceeds the first, the core will heat up. For example, if the coolant flow slows down or stops owing to a breakage of piping or loss of power to a pump, and $P$ is not reduced correspondingly, the core will heat up. Similarly, if $P$ is increased, but $\dot{m}$ is not (or if $h_{out}$ is not, which relates to conditions at the exit and, hence, the convective heat transfer coefficient between the fuel and the coolant), the core can again heat up.

Under normal operations, these imbalances are adjusted through passive and active controls. Most reactors are also designed to be somewhat inherently safe in that an increase in fuel or coolant temperature leads to an automatic reduction in $k$ and, hence, the power generation $P$. Also, expansion of the core leads to a reduction in $k$ and, hence, $P$. The rate constants are affected by delayed emission of neutrons, and thermal reactors have most neutrons at relatively slow speeds, thus reducing the overall rate of increase of neutron populations should $k$ be increased accidentally. Thus, generally, with most reactors, concerns are focused more on accidents that start with loss of coolant flow or the coolant itself and rapid increase in power $P$ that, for example, could result from sudden withdrawal or lack of insertion of a control rod.

Under normal conditions, radioactive isotopes are contained in the fuel rods. The cladding, water coolant system, piping and pressure vessel, containment, and engineered safety features (sprays, ice condensers, suppression pools, etc.) are designed to limit the release of radioactive isotopes during accidents (see Figure 18.20[28]). Natural processes (physicochemical reactions, deposition, settling, coagulation, fragmentation, aerosol growth, resuspension, etc.) may act to reduce or enhance the release fractions.

Overall, there are several aspects of the nuclear fuel cycle that have been a cause for public concern:

- Release of radioactive material to the environment during the mining, processing, and transport (shipping) of fresh or used nuclear fuel
- Releases of radioactive material during normal operation or accidents at nuclear power plants

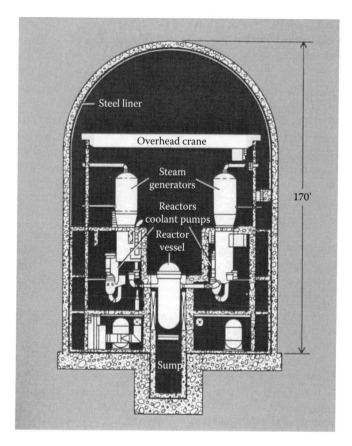

**FIGURE 18.20** Cross section of a typical containment building for a pressurized water reactor (PWR). The concrete building houses the entire primary system, the pressure control system, ventilation equipment, and part of the emergency core-cooling system. The various components are encased in concrete and surrounded by a 0.63-cm-thick steel liner. (From *Los Alamos Science*, Vol. 2, No. 2, Los Alamos Scientific Laboratory, 1981. With permission.)

- Short-term and long-term storage of nuclear waste and releases from the storage sites
- Proliferation of technology and the attendant risk of terrorism

Nuclear industry, utility groups, national governments, and the IAEA have all addressed these issues in depth, and stringent regulations at national levels, as well as guidance at international levels, have been formulated. These are enforced or followed to reduce risks to the public from postulated or real accidents. Yet, it must be realized that accidents have occurred, and will occur, and neither accidents nor releases of radionuclides can be completely prevented. The best one can do is reduce the probabilities of accidents, and then, when the accidents do occur, reduce the consequences associated with them. But risk management is expensive, and good risk–benefit and cost–benefit analysis are needed to arrive at regulatory requirements that would find public and institutional support.[29–36] This support has varied greatly in different countries and at different times.

Table 18.8 gives the half-lives and radioactive inventories of some important isotopes that are produced in a nuclear power plant. Health hazards are largely associated with the longer-lived, volatile isotopes of I, Cs, Sr, Pu, Ru, and Te, which emit beta and gamma radiation, and Pu and other actinides, which emit alpha particles and neutrons also.

Accident sequences that can cause vaporization of reactor inventory are those initiated by a loss of coolant, accidents, and severe transients (primary coolant pipe break, main steam line break in a PWR, control rod ejection, pressure vessel failure, etc.) Details of such sequences are discussed in the Reactor Safety Study (WASH-1400).

The amount and timing of the release of radioactive substances from a reactor plant to the environment is referred to as a *nuclear source term*. More broadly, source terms are characterized by the radionuclides that are released to the environment as

## TABLE 18.8
## Important Radioactive Nuclides (in a 3412 $MW_{th}$ PWR Operated for 3 Years, as Predicted by Computations)

| Radionuclide | Half-Life (Days) | Inventory (Ci × 10⁻⁸) | Radionuclide | Half-Life (Days) | Inventory (Ci × 10⁻⁸) |
|---|---|---|---|---|---|
| | | Iodine Isotopes | | | |
| I-131 | 8.05 | 0.87 | I-133 | 0.875 | 1.8 |
| I-132 | 0.0958 | 1.3 | I-135 | 0.280 | 1.7 |
| | | Noble Gases | | | |
| Kr-85 | 3.950 | 0.0066 | Kr-88 | 0.117 | 0.77 |
| Kr-85m | 0.183 | 0.32 | Xe-133 | 5.28 | 1.8 |
| Kr-87 | 0.0528 | 0.57 | Xe-135 | 0.384 | 0.38 |
| | | Cesium Isotopes | | | |
| Cs-134 | $7.5 \times 10^2$ | 0.13 | Cs-137 | $1.1 \times 10^4$ | 0.065 |
| | | Other Fission Products | | | |
| Sr-90 | $1.103 \times 10^4$ | 0.048 | Ba-140 | $1.28 \times 10^1$ | 1.7 |
| Ru-106 | $3.66 \times 10^2$ | 0.29 | Ce-144 | $2.84 \times 10^2$ | 0.92 |
| Te-132 | 3.25 | 1.3 | | | |
| | | Actinide Isotopes | | | |
| Pu-238 | $3.25 \times 10^4$ | 0.0012 | Pu-241 | $5.35 \times 10^3$ | 0.052 |
| Pu-239 | $8.9 \times 10^6$ | 0.00026 | Cm-242 | $1.63 \times 10^2$ | 0.014 |
| Pu-240 | $2.4 \times 10^6$ | 0.00028 | Cm-244 | $6.63 \times 10^3$ | 0.0084 |

*Source:* Williams, M.M.R., and Loyalka, S.K., *Aerosol Science: Theory and Practice with Special Applications to the Nuclear Industry*, Pergamon, Oxford, 1991. With permission.

*Note:* A Curie, Ci, signifies $3.7 \times 10^{10}$ emissions of a radioactive particle per second, and is often used in describing radioactivity.

well as the time dependence of the release, the size distributions of the aerosols released, the location (elevation) of the release, the time of containment failure, the warning time, and the energy and momentum released with the radioactive material. The definition of the source term is slightly loose as different computer programs may require different inputs. Still, it is clear that source terms will be closely related to the vapors, gases, and particles in suspension in the reactor containment (or building) at a given time, and the states of this suspension and the containment. If a containment does not fail (and is not bypassed), then regardless of the complicated phenomenology that takes place inside the containment during the accident, the source term would be zero, and no direct harmful effects to the public would result.

The determination of source terms within well-defined bounds is not simple. First, a range of severe accident scenarios, with corresponding initiating events, must be studied. Using probabilistic methods (fault and event trees), a probability of occurrence can then be assigned to the given accident scenario. Next, an integrated analysis of all that occurs in the plant needs to be carried out. Detailed physicochemical, neutronic, and thermal hydraulic models with an extensive database (separate effects) and integrated computer programs (as verified against a range of integral experiments) are required. This task can be quite overwhelming as the number of molecular species involved is large, temperatures and pressures can be high, and the associated flows can be quite complex. High radiation fields are also present, and depending on the specific type of accident, the situation can be very dynamic.

Note that, in 1957, the WASH-740 reports recommended an exclusion zone of radius $R$ (miles) around a nuclear plant of power $P$ ($MW_{th}$), based on the formula

$$R = \left(\frac{P}{10}\right)^{1/2} \sim 17\,\text{mi for a 3000 MW}_{th}\ \text{plant}$$

This formula is not based on realistic estimates. Rather, all material from the plant is assumed to disperse without any mitigating mechanisms. In 1957, the Windscale accident occurred, in which 100% of the noble gases, 12% of the I inventory, and 10% of the Cs inventory of the core were released in to the environment. This accident was the basis for the TID-14844 criteria for licensing (1962, regulatory guides 1.3 and 1.4), which stipulates that release from the core to the environment will consist of 100% of noble gas, 50% of I (in gaseous form), and 10% of the nonvolatile (solids) inventory. It was also specified that the containment would retain half of the I (of that released) and all of the solids. Further retentions could occur because of particular containment designs and engineered safety features. In these guidelines, containment is assumed not to fail but to leak.

The year 1975 was significant in the history of reactor safety analysis and the source term. The WASH-1400 report provided estimates on the frequency of accidents and related consequences. The analysis was specific to a PWR and a BWR. The report showed that reactors are very safe and pose only an extremely small risk.

The March 28, 1979, accident at the Three Mile Island-2 (TMI-2) plant near Harrisburg, PA, was rather serious. In the WASH-1400 nomenclature, the accident sequence was TMLQ (transient with loss of flow), initiated by a pressure relief valve stuck in an open position and later exacerbated by operator actions that shut down emergency cooling water. The source terms, however, were a matter of great surprise. Whereas all the noble gases were released (no surprise), the releases of volatiles (I, Cs) were three to four orders of magnitude smaller than predicted by a TID-14844 type of analysis. These observations and their implications for safety analysis were noted in a series of papers. It was argued that in wet reducing environments, iodine and cesium are quite reactive and do not stay in the vapor phase but either plate out or react with aerosols that settle or deposit on the walls. Thus, the chemistry of the environment and the aerosol dynamics clearly play a vital role in the estimation of source terms.

The Chernobyl accident (April 26, 1986) was characterized by a large source term (large releases). At Chernobyl, however, the situation was quite different in that the reactor design was not inherently safe (there could be a rapid increase in power), and the building was not designed to be a containment. Also, chemical explosions rendered any containment possibilities ineffective.

In the past, several nuclear reactors (SL-1, SNAPTRAN, Crystal River-3, Windscale-1, KIWI-TNT, HTRE-3, SPERT-1, TMI-2) have, unintentionally or otherwise, experienced core damage accidents. A common observation is that large releases to the atmosphere occurred only when "dry" situations prevailed (Windscale-1, a graphite-moderated reactor, and HTRE-3, a zirconium-hybrid-moderated reactor). In all light-water reactor accidents, releases to the atmosphere were rather small.

The Chernobyl accident is in contrast to the other accidents mentioned earlier. The RBMK reactor #4 was destroyed, lives were lost, and sizeable releases of radioactivity occurred, extending over a period of several days. The Chernobyl reactor started operating in December 1983 and, by April 1986, had an average fuel burnup of 10.3 MWd/kgU. The accident on April 26, 1986, was caused by operator errors that led to a prompt critical excursion exacerbated by positive void coefficients. It is estimated that within a second (1:23:44 a.m. to 1:23:45 a.m. local time), the fuel went from about 603 K to 2273 K, and during the next second, a substantial part of the fuel melted and vaporized (above 3033 K). The reactor core then disassembled violently, and the excursion was terminated. In the following hours, the heat from the excursion and the radioactive decay was redistributed in the fuel and the relatively cooler graphite. Eventually, the graphite began to burn and continued to burn for several days. This exothermic reaction heated the core further. Sand and other materials (boron carbide, dolomite, Pb, etc.) were dropped on the core, and eventually, by May 6, the fire was effectively extinguished and the release terminated. The violent disassembly of the core and the graphite fire led to some 30 additional fires in and around the reactor. Firefighting efforts were carried out in a radioactive environment, leading to two immediate and 29 subsequent deaths of site personnel. The radioactive releases occurred over a 10-day period. The releases contained noble gases, fuel and core debris including fuel fragments and large chunks of metal and graphite, a large number of large-sized particles (tens to hundreds of micrometers), and an even larger number of submicron aerosol particles. The latter were transported through the air over large distances within the northern hemisphere.

In the previous edition of this book (2007), we noted that since the Chernobyl accident, there had been no significant reactor accident worldwide. We stated, "There was, however, an accident in Japan at the Tokaimura fuel processing facility, where some fissile materials in solution were poured in excess in a tank. The tank became critical, and a small nuclear excursion and radiation release took place. A worker lost his life because of radiation exposure, and a few other workers were also affected. Release of radiation external to the plant was small. There have been additional instances, for example, in the United States, which have shown that reactor pressure vessels and their components can suffer greater structural damage during operation than previously envisaged, but so far it has been possible to detect major problems in time and to take corrective actions."

The circumstances have changed dramatically since our writing then. On March 11, 2011, the great Tohuku Earthquake in Japan and the accompanying tsunami caused extensive damage to the Fukushima Daiichi nuclear power complex in Japan (a detailed and very informative account of the accident and the aftermath is given in an American Nuclear Society Report, http://fukushima.ans.org/). This complex consists of six nuclear plants of various vintages of BWRs dating back to 1971 and power ratings of 460 to 1100 MWe. Units 4 to 6 were already in some stages of shutdown and outage, and the remaining units 1 to 3 were also all quickly shut down following the earthquake (through sensing of earthquakes). The tsunami caused extensive damage to property and life (about 15,700 people were killed, 4650 went missing, 5300 injured, 131,000 displaced) in a wide area around the plant, and the ~15 m waves also overflowed the 5 m sea walls around the plant and flooded the reactor support systems of units 1 to 4 such as emergency diesel generators. Since the backup battery systems have supply times of about 8 h, and since the off-site power was lost because of damage to power lines, it was a case of station blackout (SBO), where the reactor core could not be cooled except by natural means. There were many other difficulties also associated; as with pump failures and loss of electricity, the rejection of heat to the outside could not be effectively carried out. The core quickly heated up, and there was hydrogen generation in several units because of zirconium clad–steam interaction. There were also concerns of spent-fuel water interactions (in these plants, the spent fuel is stored inside the reactor containments around the top of the reactor vessel). The hydrogen was not vented from the containments, it accumulated, and there were hydrogen explosions in three of the reactor containments. Overall, there was significant damage in units 1 to 4, and some nuclear fuel in units 1 to 3 melted, posing challenges to both short-term and long-term cooling. Large amounts of water were sprayed to maintain cooling, which resulted in this water being contaminated with radioactive substances such as cesium and strontium, and it has continued to require storage in hundreds of tanks on-site. One or more of these tanks have leaked and are requiring additional attention.

The total release of airborne radioactivity from the accidents, however, has not been as large as one might have feared. There were no radiation-related reported deaths among the public or the plant staff. The long-term effects are, at present, not known but are assumed to be small.

The other consequences of the accident, however, have been large. First, there has been economic damage exceeding hundreds of billions of dollars due to long-term public evacuation from a ~12 mile zone surrounding the plants, contamination of the soil, and long-term loss of electric power. Second, there was erosion of public and political confidence in nuclear power, resulting in a virtual shutdown of all nuclear power plants in Japan and several in Germany. Third, there have been questions about the soundness of the construction of nuclear power plants in seismically challenged and/or coastal areas and the nuclear industry's ability to manage large-scale accidents of this type.

Obviously, all accidents are also opportunities for learning lessons and improving the technology so that such accidents are avoided in the future. These lessons have included many suggestions, including the following:

1. Improved protections against earthquakes, floods, and other natural events
2. Inclusion of property damage analysis among consequences of severe accidents
3. Creations of centralized stations for safety support (pumps, generators, filtration devices)
4. Improvements in emergency power supplies independent of damage to surrounding infrastructure (long-term batteries, power generators)
5. Installation and strengthening of vents with filters
6. Greater scrutiny of problems related to construction of multiple units at one site, and in particular, sharing of safety systems among them
7. Minimization of spent-fuel storage at plant sites
8. Improvements in nuclear fuel designs to avoid hydrogen generation
9. Improvements in long-term heat removal
10. Improved source term analysis and dynamic real-time predictive capabilities
11. Improved instrumentation for accident diagnostics

Many of these suggestions are being followed in earnest, and changes are being implemented, with more to follow.

### 18.5.1 Nuclear Weapons Proliferation

Fission of 1 kg of U-235 or Pu-239 releases an energy equivalent to that obtained in explosion of about 20 kT of TNT. World War II imperatives led to the Manhattan Project in the United States and construction, testing, and use of the first nuclear weapon in 1945. There has been no other combat use of nuclear weapons, but there have been many other detonations, both aboveground and underground. The fission bombs have been surpassed with vastly more powerful (~50 MT, 1 MT = 1000 kT) hydrogen or thermonuclear bombs, where a fission bomb is used to create a fusion reaction. The United States, Russia (Soviet Union), United Kingdom, France, China, India, and Pakistan have all tested nuclear weapons. It is also widely accepted that Israel has produced and stockpiled nuclear weapons and that South Africa had also produced nuclear weapons and perhaps detonated one. Many other nations (Argentina, Brazil, Iraq, Iran, Libya, and North Korea) have pursued nuclear weapons technology clandestinely at one time or another, and several of them are currently pursuing it (it detonated one in October 2006). It is also widely accepted that North Korea has a few nuclear weapons and that it is working on more. Nuclear weapons are comparatively compact, and these can be delivered by airplanes, missiles, ships, barges, or even trucks. There is speculation that suitcase-size nuclear weapons exist.

Nuclear weapons technology was born in wartime, and many of its practical aspects have since been well guarded (classified), not only by the United States but by other nations also. Many Manhattan Project documents, and the subsequent nuclear literature, however, provide considerable insights into the basic technology. Aspects of the technology acquired with respect to power production, including uranium enrichment, fuel processing for plutonium production, experience with radiation and its detection, neutronics, metallurgy, specialized electronics, and theoretical understanding and computational capabilities have direct applications in the weapons area. Although it is far-fetched to imagine homemade nuclear bombs, a determined nation or a well-financed group can hide a weapons program under the guise of a power program, and indeed, some nations have already done so. Concerns remain that nuclear material and weapons can be stolen also, or nuclear shipments can be hijacked. These are all legitimate concerns, and similar concerns now surround biotechnology and aspects of chemical technology. The most effective means of alleviating concerns here can only be international understanding and control, and arms reduction. There are, however, no simple answers to the issue of proliferation, as the political parts of it overwhelm any technical fixes one might attempt. Also, the knowledge (for example, with respect to the centrifuge technology) has diffused to an extent that the proliferation is occurring from secondary and tertiary sources, over which control is very difficult.

### 18.5.2 Nuclear Waste Disposal

As challenging as the problems of reactor safety and nuclear weapons proliferation are, nothing has caught the public attention more than the problems of nuclear waste disposal, whether it be the uranium mill tailings or the spent reactor fuel. This is a bit unfortunate as good technical solutions to the waste disposal problems either are available or can be formulated. Uranium (together with several of its decay products) is naturally radioactive, and it is possible to reduce much of the fuel cycle products to a radioactive state that has a life span of about 1000 years only. This, for example, can be done by separating major actinides from the spent fuel and recycling them for fission (power production) in the standard or specialized reactors. The integrated fast reactor (IFR) is an example of such a reactor concept, wherein pyroprocessing (electrolytic separation at high temperatures) can be used to process spent fuel on-site. Mixed oxide fuel reactors are another example, where Pu-based fuel is used together with U-based fuel.

The problem, however, is presently being addressed through two stages:

1. On-site storage, for several years, of the spent fuel
2. Relocation to temporary or permanent storage sites in the future

These permanent sites are to be geologically stable and safe from water percolation and other climactic factors that might lead to radioactive material dispersal. In the United States, a site is being constructed at Yucca Mountains in the State of Nevada, and spent reactor fuel (assemblies) will be transported via trucks from nuclear power plant fuel storage pools to this site contingent on the results of a drawn-out and on-again-off-again regulatory and public or political process. There are also other options under consideration, including monitored retrievable storage.

Other nations have made progress, with respect to both reprocessing and recycling of the spent fuel in reactors, as well as long-term storage and disposal. For example, in France, the fuel is reprocessed and recycled, as well as stored after reprocessing (wherein the waste is encapsulated in vitreous material that is impervious to fluids and is environmentally stable) in salt formations.

### 18.5.3 TERRORISM

Most recently, there has been considerable concern regarding "dirty bombs," or radio-active material dispersion, in which, for example, a reactor fuel assembly is exploded with dynamite and the material is dispersed in a dense population. Sabotage of nuclear reactor power plants is also feared. These aspects have required, and will require, enhanced security at nuclear power plants and improved safeguards on nuclear materials.[37]

In conclusion, we should note that at least one natural reactor existed millions of years ago in Africa. As Lamarsh[2] has stated,

> Oklo is the name of a uranium mine in the African nation of Gabon, where France obtains much of the uranium for her nuclear program. When uranium from this mine was introduced into a French gaseous diffusion plant, it was discovered that the feed uranium was already depleted below the 0.711 wt% of ordinary natural uranium. It was as if the uranium had already been used to fuel some unknown reactor.
>
> And so it had. French scientists found traces of fission products and Trans Uranic (TRU) waste at various locations within the mine. These observations were puzzling at first, because it is not possible to make a reactor go critical with natural uranium, except under very special circumstances with a graphite or heavy water moderator, neither of which could reasonably be expected to have ever been present in the vicinity of Oklo. The explanation of the phenomenon is to be found in the fact that the half-life of U-235, 7.13E8 years, is considerably shorter than the half-life of U-238, 4.51E9 years. Since the original formation of the Earth, more U-235 has therefore decayed than U-238. This, in turn, means that the enrichment of natural uranium was greater years ago than it is today. Indeed about 3 billion years ago this enrichment was in the neighborhood of 3 wt%, sufficiently high to form a critical assembly with ordinary water, which is known to have been present near Oklo at that time.

Thus, nuclear energy from fission is really not that new! Our challenge is to recognize that we have an amazing source of energy that, with good institutional safeguards, can provide, in environmentally safe ways, abundant energy to humankind for thousands of years to come.

## 18.6 NUCLEAR FUSION

Fusion is the inverse of fission in that certain light nuclei can combine together to create a pair of new nuclei that have a greater stability than the reactants. Fusion reactions are the primary source of power in stars, including the sun.[38–41] To understand conditions under which fusion can occur, let us look at a much-studied[1,38–49] reaction

$$D + T \rightarrow He^4 + n + 17.6 \text{ MeV} \tag{18.18}$$

where D indicates a deuteron or deuterium nucleus ($H_1^2$), T indicates a triton or tritium nucleus ($H_1^3$), and "n" indicates a neutron. Because 1 kg of D contains $3.0115 \times 10^{26}$ D nuclei, its fusion with an equal number of T nuclei (1.5 kg) would release about $8.5 \times 10^8$ MJ (or expressed differently, $2.36 \times 10^8$ kWh or 26.9 MWy). This is about 10 times the energy released from the fission of 1 kg U-235.

Note that both D and T are positively charged particles; their interaction is governed by repulsive coulombic forces. Large gravitational or inertial (acceleration) forces or high temperatures are needed to overcome this repulsion for any significant fusion reaction to occur. This is a crucial difference from fission, where such special conditions for the reaction are not needed. The 17.6 MeV that is released in this reaction is associated with the kinetic energy of the neutron (14.1 MeV) and the $He^4$ (3.5 MeV). The rate equations for this reaction and the associated energy balance (the first law of thermodynamics) can be approximately described by a point kinetics model as

$$\frac{dN_D}{dt} = -R_{DT}(T)N_D N_T + \dot{N}_{D,netgain}$$

$$\frac{dN_T}{dt} = -R_{DT}(T)N_D N_T + \dot{N}_{T,netgain}$$

$$\frac{dN_{He^4}}{dt} = R_{DT}(T)N_D N_T - \dot{N}_{He^4,netloss} \tag{18.19}$$

$$\frac{dN_n}{dt} = R_{DT}(T)N_D N_T - \dot{N}_{n,netloss}$$

$$\frac{dU}{dt} = V\left( f_n G_n \frac{dN_n}{dt} + f_{He^4} G_{He^4} \frac{dN_{He^4}}{dt} \right) - \dot{Q}_{loss}$$

Here, $N$ indicates the concentration of the subscripted species, $U$ the internal energy of the system, $V$ its volume, $f$ the fraction of energy $G$ (14.1 MeV or 3.5 MeV) that is deposited in the system, and $Q_{loss}$ the power removed from the system through losses or that is removed through cooling to drive a turbine and generate electricity. $R$ ($m^3/s$) is a measure of the reaction rate (here $T$ indicates the temperature, and it should not cause any confusion with the symbol for triton, as they are used in different contexts), and it is approximately given by[1,44]

$$R_{DT}(T) = 3.7 \times 10^{-18} \frac{\exp(-20/(k_B T)^{1/3})}{(k_B T)^{2/3}} \tag{18.20}$$

in which kT has been expressed in units of keV (note that the Boltzmann constant $k_B = 0.861735 \times 10^{-7}$ keV/K). We show in Figure 18.21 a plot of this expression.

Note that $R$ is significant only at very high temperatures that are not realizable. But, because at the high temperatures that are needed, the material will be in a state of plasma (mixture of ions, electrons, and neutrals) with very low densities, the perfect gas equation of state applies, and we have

$$n_D \sim \frac{1}{T}, \quad n_T \sim \frac{1}{T} \tag{18.21}$$

A plot of the expression $R_{DT}n_D n_T$ (Figure 18.22) shows that the D + T reaction has a maximum at about $45 \times 10^6$ K.

It also turns out that for a D + D reaction, the reaction rate is about 1/25 of that for the D + T reaction for a given temperature. Overall, then, the focus has been on the D + T reaction. What is clear is that we need temperatures in the millions of degrees to get reaction rates that would enable significant energy generation. The deuterium needed for the system can be obtained from heavy water, whereas the tritium can be generated from a reaction of the neutrons with lithium.

The physics and engineering challenges of achieving such high temperatures and still confining the plasma for any significant time in a fusion reactor are enormous (the ions that move at very high speeds, are light, and leak). Note that the plasma (because of its very low density) is essentially transparent to neutrons, and only the alpha energy is available for the plasma heating and maintenance of a high temperature in the plasma. A schematic of the fusion power plant is given in Figure 18.23, and it shows some of the engineering concepts involved.

Although there have been efforts to investigate engineering designs of the fusion plants, much of the focus so far has been on achieving plasma confinement and toward investigations of conditions that could lead to breaking even (that is, $dU/dt = 0$,

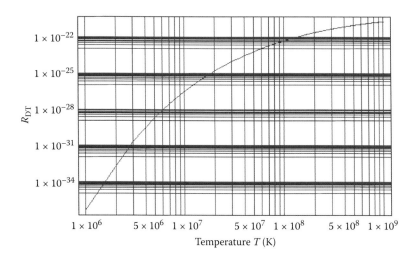

**FIGURE 18.21** The reaction parameter as a function of temperature.

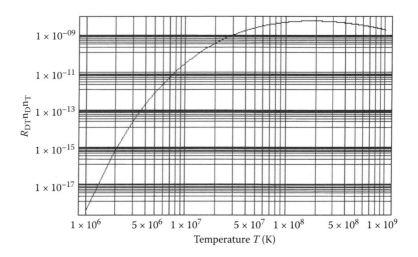

**FIGURE 18.22**   The normalized reaction rate as a function of temperature.

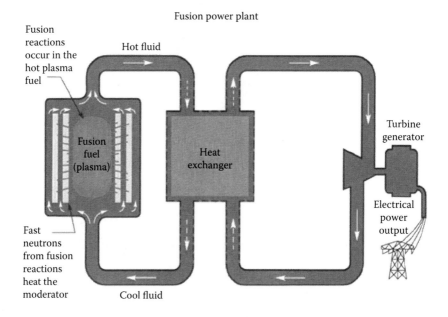

**FIGURE 18.23**   A schematic of a fusion power plant.

at a significant temperature with external energy input for plasma heating). There are two concepts that play a major role in current research:

1. Magnetic confinement (mirror and tokamak)
2. Laser-driven inertial fusion

In magnetic confinement, the charged particles are sought to be confined through the use of high magnetic fields. The fields alter ion trajectories and, in a mirror (an open system), reflect the particles back toward the plasma at the two ends. In the tokamak (which comprises a toroidal geometry and is a closed system), the ions move in loops within the plasma. Because of the high temperature of the plasma, it is essential to keep the plasma away from surfaces. Schematics of the tokamak concept are shown in Figure 18.24. At present, there is a substantial international effort (known as the ITER program) on tokamaks.

Laser-driven inertial confinement uses laser ablation and resulting compression of a small hollow glass sphere to create very high densities for the reactants (D + T, where the sphere is filled with D and some T, and most of the T is obtained from Li + n reaction; Li is coated on the inside surface of the hollow sphere). Very-high-energy lasers and precise focusing of the laser beams (from several directions) on the sphere are needed to achieve uniform compression.

The challenges of controlled fusion for electricity generation are enormous: plasma confinement, high temperatures, very-high-powered lasers, precise and economic target fabrication, damage from high-energy neutrons, and so forth. But the

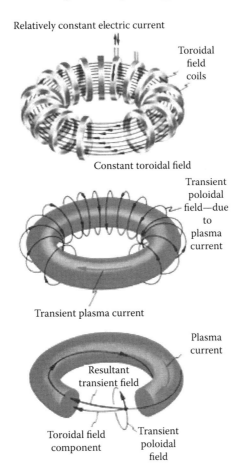

**FIGURE 18.24** Toroidal confinement.

promises are also huge: unlimited raw material supply, very little stored energy in the system, negligible possibilities of catastrophic accidents, lower radioactive material and waste, and so forth, as compared to fission-based reactors. We hope that eventually, the challenges will be met, as every year, there is some progress in our understanding of the physics and engineering of fusion. There has been, however, a significant reduction in government research funding in this area as the realizable benefits are distant.

## REFERENCES

1. Connolly, T.J., *Foundations of Nuclear Engineering*, John Wiley & Sons, New York, 1978.
2. Lamarsh, J., *Introduction to Nuclear Engineering*, Addison-Wesley, Reading, MA, 1983, p. 181.
3. Glasstone, S. and Sesonske, A., *Nuclear Reactor Engineering*, Van Nostrand Reinhold, New York, 1981.
4. Lamarsh, J., *Nuclear Reactor Theory*, Addison-Wesley, Reading, MA, 1966.
5. Henry, A., *Nuclear Reactor Analysis*, MIT Press, Cambridge, MA, 1975.
6. Duderstadt, J., *Nuclear Reactor Theory*, John Wiley & Sons, New York, 1976.
7. Ott, K.O. and Bezella, W.A., *Introductory Nuclear Statistics*, American Nuclear Society, La Grange Park, IL, 1989.
8. Ott, K.O. and Neuhold, R.J., *Introductory Nuclear Reactor Dynamics*, American Nuclear Society, La Grange Park, IL, 1985.
9. Stacey, W.M., *Nuclear Reactor Physics*, John Wiley & Sons, New York, 2001.
10. El-Wakil, M.M., *Nuclear Heat Transport*, International Textbook Company, now available from the American Nuclear Society, La Grange Park, IL, 1971.
11. Todreas, N.E. and Kazimi, M.S., *Nuclear Systems*, Vol. I, Hemisphere, New York, 1990.
12. Lahey, R.T. and Moody, F.J., *The Thermal Hydraulics of a Boiling Nuclear Reactor*, American Nuclear Society, La Grange Park, IL, 1977.
13. Weinberg, A.M. and Wigner, E.P., *The Physical Theory of Neutron Chain Reactors*, University of Chicago Press, Chicago, 1958.
14. *Reactor Physics Constants*, ANL-5800, U.S. Govt., Washington, DC, 1963.
15. Davison, B., *Neutron Transport Theory*, Oxford University Press, London, 1957.
16. Bell, G.I. and Glasstone, S., *Nuclear Reactor Theory*, Van Nostrand, New York, 1974.
17. Williams, M.M.R., *The Slowing Down and Thermalization of Neutrons*, John Wiley & Sons, New York, 1966.
18. Williams, M.M.R., *Mathematical Methods in Particle Transport Theory*, Butterworths, London, 1971.
19. Case, K.M. and Zweifel, P.F., *Linear Transport Theory*, Addison-Wesley, Reading, MA, 1967.
20. Duderstadt, J. and Martin, W., *Transport Theory*, John Wiley & Sons, New York, 1982.
21. Lewis, E.E. and Miller, W.F., *Computational Methods of Neutron Transport*, American Nuclear Society, La Grange Park, IL, 1993.
22. MCNP—A General Monte Carlo N-Particle Transport Code, Version 5, Los Alamos National Laboratory, Los Alamos, NM, 2003.
23. A Technology Roadmap for Generation IV Nuclear Energy Systems, U.S. DOE Nuclear Energy Research Advisory Committee and the Generation IV International Forum, December, 2002.
24. Lapeyre, B., Pardoux, E. and Sentis, R., *Introduction to Monte-Carlo Methods for Transport and Diffusion Equations*, Oxford University Press, London, 2003, original in French, 1998.

25. Garwin, R.L. and Charpak, G., *Megawatts and Megatons*, Knopf, New York, 2001.
26. Steinfelds, E.V., Ghosh, T.K., Prelas, M.A., Tompson, R.V. and Loyalka, S.K., Development of radioisotope energy conversion systems, in *ICAPP Conference Proceeding*, Space Nuclear Power, 2003.
27. Lake, J.A., Nuclear Hydrogen Production, Idaho National Engineering and Environmental Laboratory, paper available at http://nuclear.inel.gov, 2003.
28. *Los Alamos Science*, Vol. 2, No. 2, Los Alamos Scientific Laboratory, Los Alamos, NM, 1981.
29. Cochran, R. and Tsoulfanidis, N., *The Nuclear Fuel Cycle: Analysis and Management*, American Nuclear Society, La Grange Park, IL, 1999.
30. Faw, R.E. and Shultis, J.K., *Radiological Assessment: Sources and Doses*, American Nuclear Society, La Grange Park, IL, 1999.
31. Shultis, J.K. and Faw, R.E., *Radiation Shielding*, American Nuclear Society, La Grange Park, IL, 2000.
32. U.S. Nuclear Regulatory Commission, Reactor safety study: An assessment of accident risks in U.S. Power Plants, Report No. WASH-1400, NUREG-75/014, 1975.
33. McCormick, N.J., *Reliability and Risk Analysis: Methods and Nuclear Power Applications*, Academic Press, New York, 1981.
34. Wilson, R., Araj, A., Allen, O., Auer, P., Boulware, D.G., Finlayson, F., Goren, S., et al. Report to the American Physical Society of the study group on radionuclide release from severe accidents at nuclear power plants, *Rev. Mod. Phys.*, 57(3), Part II, S1–S154, 1985.
35. U.S. Nuclear Regulatory Commission, Severe accident risks: An assessment for five U.S. nuclear power plants, (Final Summary Report) NUREG-1150, Vol. 1–2, 1990.
36. Williams, M.M.R. and Loyalka, S.K., *Aerosol Science: Theory and Practice with Special Applications to the Nuclear Industry*, Pergamon, Oxford, UK, 1991.
37. Ghosh, T.K., Prelas, M.A., Viswanath, D.S. and Loyalka, S.K., *Science and Technology of Terrorism and Counter-Terrorism*, Marcel Dekker, New York, 2002.
38. von Weizsacker, C.F., *Phys. Z.*, 38, 176, 1937.
39. Gamow, G. and Teller, E., *Phys. Rev.*, 53, 608, 1938.
40. Bethe, H.A. and Critchfield, C.H., *Phys. Rev.*, 54, 248, 1938.
41. Bethe, H.A., Energy production in stars, *Phys. Rev.*, 55, 434, 1939.
42. Rose, D.J. and Clarke, M., *Plasma and Controlled Fusion*, MIT Press, Cambridge, MA, 1961.
43. Glasstone, S., *Controlled Nuclear Fusion*, U.S. Atomic Energy Commission, Washington, DC, 1965.
44. Kamash, T., *Fusion Reactor Physics*, Ann Arbor Science, Ann Arbor, MI, 1973.
45. Miley, G.H., *Fusion Energy Conversion*, American Nuclear Society, La Grange Park, IL, 1976.
46. Dolan, T.J., *Fusion Research: Principles, Experiments and Technology*, Pergamon, Oxford, UK, 1982.
47. Stacey, W.M., *Fusion: An Introduction to the Physics and Technology of Magnetic Confinement Fusion*, John Wiley & Sons, New York, 1984.
48. Duderstadt, J.E. and Moses, G.R., *Inertial Confinement Fusion*, John Wiley & Sons, New York, 1982.
49. Fowler, T.K., Nuclear power: Fusion, in *More Things in Heaven and Earth: A Celebration of Physics at the Millennium*, Bederson, B., Ed., Springer, New York, 1999.
50. Loyalka, S.K., Transport theory, in *Encyclopedia of Physics*, Vol. 2, 3rd ed., Trigg, L. and Lerner, R., Eds., Wiley-VCH, Weinheim, Germany, 2005.
51. Wirtz, K., *Lectures on Fast Reactors*, American Nuclear Society, La Grange Park, IL, 1982.
52. *Controlled Nuclear Chain Reaction: The First Fifty Years*, American Nuclear Society, La Grange Park, IL, 1992.

53. Glasstone, S., *The Effects of Nuclear Weapons*, U.S. Atomic Energy Commission, Washington, DC, first published 1950, 1964.
54. Till, C.E., Nuclear fission reactors, in *More Things in Heaven and Earth: A Celebration of Physics at the Millennium*, Bederson, B., Ed., Springer, New York, 1999.
55. Knief, R., *Nuclear Energy Technology: Theory and Practice of Commercial Nuclear Power*, McGraw-Hill, New York, 1981.
56. Melese, G. and Katz, R., *Thermal and Flow Design of Helium Cooled Reactors*, American Nuclear Society, La Grange Park, IL, 1984.
57. Simpson, J.W., *Nuclear Power from Underseas to Outer Space*, American Nuclear Society, La Grange Park, IL, 1994.
58. Serber, R., *The Los Alamos Primer*, U.S. Govt., first published as LA-1, April 1943, declassified 1965, annotated book, University of California Press, Berkeley, CA, 1992.

*Mihaela F. Ion and Sudarshan K. Loyalka*

## CONTENTS

## 19.1  INTRODUCTION

Fuel cells are devices that produce electrical energy through electrochemical processes, without combusting fuel and generating pollution of the environment. They represent a potential source of energy for a wide variety of applications.

Sir William Robert Grove invented the fuel cell in 1839, and further improvements have been added over the years by many investigators, a significant contribution being made by Francis Bacon in the 1930s. By the 1960s, fuel cells were already being used in the National Aeronautics and Space Administration's (NASA's) space exploration missions. Over the next few decades, the interest in fuel cells declined considerably due to their extremely high associated costs and relatively poor performance for daily applications. More recently, the increasing demand for cleaner energy sources has brought the fuel cells back into the public attention. At present, there is a major push for commercialization of fuel cells, with a potential for applications from portable to automotive and even power-generating stations.

## 19.2  BASIC CONCEPTS

### 19.2.1  Design Characteristics

The core structure of a generic fuel cell includes two thin *electrodes* (anode and cathode), located on the opposite sides of an *electrolyte* layer (Figure 19.1).

The electrochemical reactions occur at the *electrodes*. For most fuel cells, the catalytic fuel decomposition occurs at the anode, where ions and electrons split. They recombine at the cathode, where by-products of the reaction are created (i.e., water or $CO_2$). As high rates of reaction are desired in functioning of the fuel cell, thin layers of catalysts are applied to the electrodes. Furthermore, materials used for electrodes require specific properties such as high conductivity and high ionization and deionization properties, and they must have sufficient permeability to the fuel/oxidant and electrolyte (i.e., they have to be made of porous materials).

The electrolyte layer is made of either a solute or a solid material and plays an important role during operation, as a conductor for the ions between the two electrodes. Characterized by strong insulating properties, the electrolyte does not allow the transfer of electrons. These are conducted through one of the electrodes via an external pathway to the other electrode, closing the electric circuit of the cell.

### 19.2.2  Operation

Fuel cells are a type of galvanic cell, and they operate similarly to conventional batteries, converting the chemical energy of the reactants directly into electrical energy. However, there is an important difference between them, which lies in the way the chemical energy is transformed into electrical energy. A battery uses the chemical energy *stored* within the reactants inside the battery, whereas a fuel cell *converts* the chemical energy provided by an external fuel/oxidant mixture into electrical energy. Thus, batteries use chemical energy until the reactants are completely depleted, and at the end of their lifetime, they can be either recharged or just thrown away. Fuel

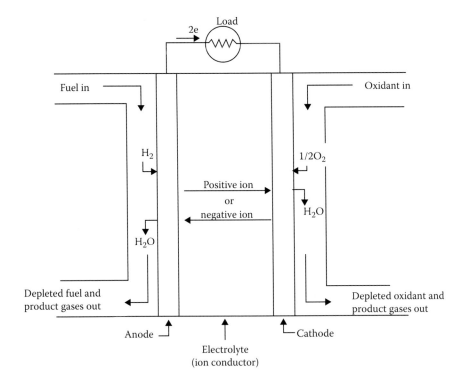

**FIGURE 19.1** Design of a generic fuel cell. (From EG&G Services, Parsons Inc. and Science Applications International Corporation, *Fuel Cell Handbook*, fifth edition, contract no. DE–AM26–99FT40575, for US Department of Energy, October 2000.)

cells, on the other hand, can provide electrical output as long as the supply of fuel and oxidant is maintained.

Typically, hydrogen-rich fuels are used for operation, and the most common fuels include gases (i.e., hydrogen, natural gas, or ammonia), liquids (i.e., methanol, hydrocarbons, hydrazine), or coal. A preliminary conversion (*reforming*) process is required for all fuels, except for direct hydrogen. The oxidant used at the cathode is usually oxygen or air.[1,2]

During operation, both hydrogen-rich fuel and oxygen/air are supplied to the electrodes. Hydrogen undergoes catalytic oxidation at one of the electrodes and splits into ions and electrons. Oxygen undergoes a reduction reaction at the other electrode. Both ions and electrons travel from one electrode to the other, using different pathways. Ions travel through the electrolyte, and electrons are forced through a separate pathway via the current collectors to the other electrode, where they combine with oxygen to create water or other by-products, such as $CO_2$.

### 19.2.3 THERMAL EFFICIENCY

Although the input (chemical energy, $E_{ch}$) and output (electrical energy, $E_e$) of the operation are the same for fuel cells and heat engines, the conversion process is

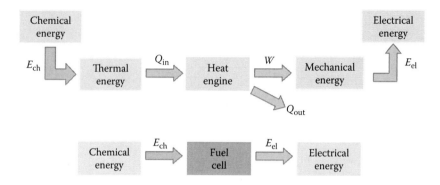

**FIGURE 19.2** Energy conversion processes for heat engine and fuel cell.

different. Heat engines use chemical energy to produce intermediate heat, which is subsequently transformed into mechanical energy, which in turn leads to electrical energy. Fuel cells use a direct conversion process, transforming the chemical energy directly into electrical energy (see Figure 19.2). Thus, when comparing fuel cells with heat engines, two aspects of the second law of thermodynamics have to be considered: heat absorption from a reservoir to use for operation and energy losses to the surroundings.

### 19.2.3.1 Heat Absorption from a Reservoir to Use for Operation

Fuel cell operation does not require two different temperature reservoirs, and thus, any temperature restrictions associated with the Carnot cycle are eliminated. In comparison, heat engines operate between a hot source and a cold sink. Their thermal efficiency ($\eta$) is calculated as the amount of net work ($W$) done for the heat ($Q_{in}$) absorbed by the engine. The amount of net work is determined as the difference between the absorbed ($Q_{in}$) and the rejected ($Q_{out}$) heat. In case of an ideal, reversible heat engine, the entropy remains constant, and

$$\frac{Q_{out}}{Q_{in}} = \frac{T_{out}}{T_{in}}.$$

The maximum efficiency of such an engine is given by

$$\eta_{HE,max} = 1 - \frac{T_{out}}{T_{in}}.$$

A high value is therefore obtained using either a very low $T_{out}$ (for an ideal machine, this can be zero) or a very high combustion temperature, $T_{in}$. In reality, $T_{out}$ is approximately 300 K, which is the ambient temperature. At the other end, the combustion temperature cannot be too high (i.e., 2000–3000 K), owing to temperature restrictions of the materials.

### 19.2.3.2 Energy Losses to the Surroundings

Processes in real heat engines or fuel cells are irreversible, and losses occur. The real efficiency is always less than the theoretical efficiency. To properly compare the efficiency of heat engines and fuel cells, the theoretical values have to be considered in each case. For fuel cells, this is calculated as the maximum electrical work ($W_{el}$) done for the total thermal energy available or the enthalpy of the fuel ($\Delta H_0$). $W_{el}$ is given by the change in the chemical energy, or the Gibbs free energy of the electrochemical process, and it is calculated as the difference between the total heat available and the heat produced during operation. The maximum efficiency can be written as

$$\eta_{FC,max} = 1 - \frac{T\Delta S_0}{\Delta H_0}$$

where $\Delta S_0$ represents the entropy change of the system.

### 19.2.4 CELL VOLTAGE

Under ideal conditions, operation of fuel cells is performed without any losses. This can be seen in Figure 19.3, which presents the cell voltage and current characteristics for ideal and real situations. The most important losses that occur during normal operation are

- Activation losses, which are directly dependent on the reaction rates
- Ohmic losses, caused by resistance to flows of ions and electrons through media
- Concentration losses, due to changes in the concentration of reactants

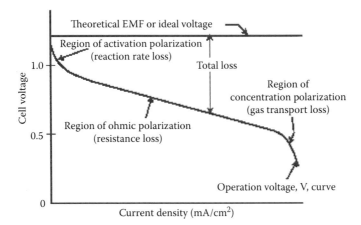

**FIGURE 19.3** Fuel cell voltage/current characteristic. (From EG&G Services, Parsons Inc. and Science Applications International Corporation, *Fuel Cell Handbook*, fifth edition, contract no. DE–AM26–99FT40575, for US Department of Energy, October 2000.)

## 19.3  FUEL CELL SYSTEM

### 19.3.1  General Description

Major applications require power input that is far above the output produced by a single fuel cell unit (i.e., approximately 1 $W/cm^2$ electrode area for hydrogen fuel cells). Multiple fuel cells are therefore arranged in stacks to produce the desired power output. Preliminary conversion processes are required to supply the fuel cells with the appropriate fuel for operation, and after the electrochemical reactions occur, further conversion processes are required to transform the electrical output from the fuel cells into a form accessible to various applications. The whole ensemble comprising the fuel processor, stack of fuel cells, and power conditioner represents the *fuel cell system*. A schematic representation of a typical fuel cell system configuration is shown in Figure 19.4.

The component playing the major role in the whole system is the *fuel cell stack* (or *power section*). A fuel cell stack is usually made of at least 50 fuel cell units of various configurations. Here, the electrochemical reactions that transform chemical energy directly into electrical energy occur (see Figure 19.5). This structure, which resembles a sandwich, is further placed between two *bipolar separator plates*. The bipolar plates have two important operating functions, as current collectors and separator plates: as *current collectors*, they conduct the electrons produced by the oxidation of hydrogen, and *as separator plates*, they provide the necessary physical separation of the flows for adjacent cells as well as the required electrical connections.[3]

Typically, hydrogen-rich gas is used for fuel cell operation. Because only the hydrogen component of these fuels reacts at the electrodes, a reforming process is performed in the *fuel processor* for all the fuels, except for the direct hydrogen, before the gas enters the fuel cell.

The *power conditioner* is the section of the fuel cell system where the direct current (DC) obtained from the fuel cell stack is converted into alternating current (AC), when required. It is also designed to adjust the current and voltage of the stack to produce the desired power output.

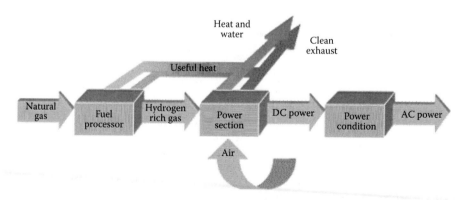

**FIGURE 19.4**   Fuel cell system. (From US Department of Defense Fuel Cell Demonstration Program, Fuel Cell Descriptions, www.dodfuelcell.com/fcdescriptions.html.)

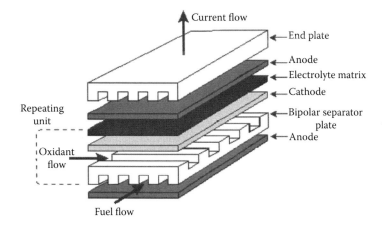

Current flow

End plate
Anode
Electrolyte matrix
Cathode

Repeating unit

Bipolar separator plate
Anode

Oxidant flow

Fuel flow

**FIGURE 19.5** Fuel cell stack components. (From Energy Center of Wisconsin, Fuel Cells for Distributed Generation, A Technology and Marketing Summary Report, 193-1, 2000.)

### 19.3.2 FUEL CELL CLASSIFICATION

Fuel cells can be classified according to various criteria, based on electrolyte, operating temperature, fuel or oxidant used, reforming process, and so forth. The most common criterion, also used to name these devices, is the type of the electrolyte used for operation. Five major categories can be identified, based on the type of the electrolyte:

- Proton exchange membrane fuel cells (PEMFCs)
- Alkaline fuel cells (AFCs)
- Phosphoric acid fuel cells (PAFCs)
- Molten carbonate fuel cells (MCFCs)
- Solid oxide fuel cells (SOFCs), which can be further divided into two sub-categories, namely, the intermediate temperature solid oxide fuel cells (ITSOFCs), with operating temperatures less than 800°C, and the tubular solid oxide fuel cells (TSOFCs), with temperatures over 800°C[4]

A summary of the characteristics of fuel cells is presented in Table 19.1.[4–7] The PEMFCs and AFCs operate at lower temperatures and are mainly developed for transportation and small utilities. The PAFCs operate at higher temperatures, being designed for medium-scale power applications. The MCFCs and SOFCs operate at high temperatures and are intended for large power utilities.

## 19.4 LOW-TEMPERATURE FUEL CELLS

### 19.4.1 PROTON EXCHANGE MEMBRANE FUEL CELLS

The PEMFCs were developed originally by General Electric in the 1960s for NASA's space explorations. Over the years, these fuel cells have been known under various names, such as *ion exchange membrane, solid polymer electrolyte, proton exchange*

**TABLE 19.1**
**Fuel Cell Characteristics**

| Characteristics | PEMFC | AFC | PAFC | MCFC | SOFC — ITSOFC | SOFC — TSOFC |
|---|---|---|---|---|---|---|
| **Operating Parameters** | | | | | | |
| Temperature[a] (°C) | 80 | 65–220 | 150–220 | ~650 | 600–800 | 800–1000 |
| Pressure[b] (atm) | 1–5 | | 1–8 | 1–3 | 1–15 | |
| Efficiency[c] (%) | 40–50 | 40–50 | 40–50 | 50–60 | 45–55 | |
| Power density[d] (kW/kg) | 0.1–1.5 | 0.1–1.5 | 0.12 | — | 1–8 | |
| **Cell Components[e]** | | | | | | |
| Electrolyte | Proton exchange membrane (solid) | Potassium hydroxide (liquid) | Phosphoric acid (liquid) | Molten carbonate salt (liquid) | Ceramic (solid) | |
| Electrodes | Carbon based | Carbon based | Graphite based | Nickel and stainless steel based | Ceramic | |
| Catalyst | Platinum | Platinum | Platinum | Nickel | Perovskites | |
| **Reactants[f]** | | | | | | |
| Charge carrier | H$^+$ | OH$^-$ | H$^+$ | CO$_3^{2-}$ | O$^{2-}$ | |
| Fuel | H$_2$ (reformate) | H$_2$ (pure) | H$_2$ (reformate) | H$_2$/CO/CH$_4$ (reformate) | H$_2$/CO/CH$_4$ (reformate) | |
| Reforming process | External | — | External | External/internal | External/internal | |
| Oxidant | O$_2$/air | O$_2$ | O$_2$/air | CO$_2$/O$_2$/air | O$_2$/air | |
| **Operation** | | | | | | |
| Water management[g] | Evaporative | Evaporative | Evaporative | Gaseous product | Gaseous product | |
| Heat management[g] | Process gas / Independent cooling medium | Process gas / Electrolyte calculation | Process gas / Independent cooling medium | Internal reforming / Process gas | Internal reforming / Process gas | |

**Advantages/Disadvantages[h]**

| | | | | |
|---|---|---|---|---|
| Advantages | High current and power density; long operating life | High current and power density; high efficiency | Advanced technology | High efficiency; internal fuel processing; high-grade waste heat | Internal fuel processing; high-grade waste heat; potentially inexpensive |
| Disadvantages | CO intolerance; water management; noble metal catalyst | $CO_2$ intolerance | Efficiency; lifetime; noble metal catalyst | Electrolyte instability; lifetime | High temperature; efficiency; low ionic conductivity |

**Applications[i]**

| | | | | |
|---|---|---|---|---|
| Type | Motive/small utility | Aerospace | Small utility | Utility | Utility |
| Scale | 0.1 kW–10 MW | 0.1–20 kW | 200 kW–10 MW | >100 MW | >100 MW |

*Source:* [a]EG&G Services, Parsons Inc. and Science Applications International Corporation, *Fuel Cell Handbook*, fifth edition, contract no. DE–AM26–99FT40575, for US Department of Energy, October 2000. [b]Penner, S.S., Ed., *Int. J.*, 20, 1995; Energy Center of Wisconsin, Fuel Cells for Distributed Generation. A Technology and Marketing Summary Report, 193-1, 2000. [c]European Commission, A Fuel Cell Research, Development and Demonstration Strategy for Europe up to 2005, 1998 ed., Belgium, 1998. With permission. [d]Norbeck, J.M., *Hydrogen Fuel for Surface Transportation*, Society of Automotive Engineers, Warrendale, PA, 1996. [e]EG&G Services, Parsons Inc. and Science Applications International Corporation, *Fuel Cell Handbook*, fifth edition, contract no. DE–AM26–99FT40575, for US Department of Energy, October 2000; Mehta, S.K., and Bose, T.K., Ed., *Proceedings of the Workshop on Hydrogen Energy Systems*, presented at the Official Opening of the Hydrogen Research Institute, in collaboration with Canadian Hydrogen Association, Canada, 1996; Energy Center of Wisconsin, Fuel Cells for Distributed Generation. A Technology and Marketing Summary Report, 193-1, 2000. [f]EG&G Services, Parsons Inc. and Science Applications International Corporation, *Fuel Cell Handbook*, fifth edition, contract no. DE–AM26–99FT40575, for US Department of Energy, October 2000; Mehta, S.K., and Bose, T.K., Ed., *Proceedings of the Workshop on Hydrogen Energy Systems*, presented at the Official Opening of the Hydrogen Research Institute, in collaboration with Canadian Hydrogen Association, Canada, 1996. With permission. [g]EG&G Services, Parsons Inc. and Science Applications International Corporation, *Fuel Cell Handbook*, fifth edition, contract no. DE–AM26–99FT40575, for US Department of Energy, October 2000. [h]Decher, R., *Direct Energy Conversion—Fundamentals of Electric Power Production*, Oxford University Press, 1997. With permission. [i]Mehta, S.K., and Bose, T.K., Ed., *Proceedings of the Workshop on Hydrogen Energy Systems*, presented at the Official Opening of the Hydrogen Research Institute, in collaboration with Canadian Hydrogen Association, Canada, 1996. With permission.

*membrane*, or simply, *polymer electrolyte* fuel cells. They use hydrogen as fuel, oxygen or air as oxidant, and a solid polymer membrane as electrolyte.

### 19.4.1.1 Design Characteristics

The core of the PEMFC design consists of a proton conducting membrane (the electrolyte), located between two platinum-impregnated porous electrodes. Teflon gaskets and current collectors are added to these components to complete a single fuel cell unit. The core of the fuel cell is usually less than a millimeter thick and is referred to as the *membrane-electrode assembly* (MEA). Depending on the mode of fabrication, the MEA can include either the membrane along with the catalyst layers only or the whole ensemble of the previously mentioned components plus the carbon electrodes. Figure 19.6 shows a schematic representation of the PEMFC manufactured by Ballard Power Systems. A general view of the cell hardware and its cross section are presented in Figure 19.7.

#### 19.4.1.1.1 Electrolyte

Various electrolyte materials have been developed over the years for use in PEMFCs, and there is still extensive ongoing research focused on improving the materials currently used or finding new solutions. Currently, the thickness of the membrane is approximately 50–175 μm, and recent developments show that stable operation conditions can be obtained with membranes only 10–25 μm thick.[8]

Most of the membranes used to date in PEMFC have a fluorocarbon polymer-based structure to which sulfonic acid groups are attached. The key characteristic of these materials is that, although the acid molecules are fixed to the polymer, the protons on these acid groups are free to travel through the membrane. The most well known are the Nafion membranes, which have been developed by DuPont over more than three decades. These types of membranes are thin, nonreinforced films based

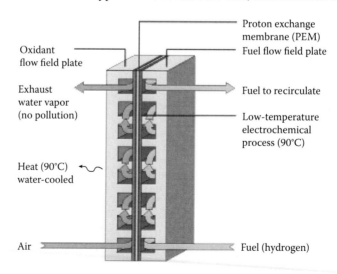

**FIGURE 19.6** PEMFC, Ballard Power Systems. (From Ballard Power Systems, How the Ballard® fuel cell works, www.ballard.com.)

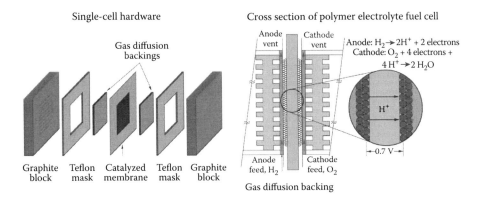

Single-cell hardware                    Cross section of polymer electrolyte fuel cell

Gas diffusion backings

Graphite   Teflon   Catalyzed   Teflon   Graphite
block      mask     membrane    mask     block

Anode: $H_2 \rightarrow 2H^+ + 2$ electrons
Cathode: $O_2 + 4$ electrons + $4H^+ \rightarrow 2H_2O$

Anode vent      Cathode vent

$H^+$

$\leftarrow 0.7\,V \rightarrow$

Anode feed, $H_2$      Cathode feed, $O_2$

Gas diffusion backing

**FIGURE 19.7**    PEMFC design. (From Gottesfeld, S., "The Polymer Electrolyte Fuel Cell, Materials Issues in a Hydrogen Fueled Power Source," Los Alamos National Laboratory, Materials Science and Technology Division, White Paper on LANL Hydrogen Education, Web site: http://education.lanl.gov/RESOURCES/h2/gottesfeld/education.html.)

on the Nafion resin, a perfluorinated polymer.[9] The structure of the Nafion membranes is given in Figure 19.8.

This type of membrane is usually prepared by modifying a basic polymer (polyethylene) through a process called *perfluorination*, where the hydrogen is substituted with fluorine. The modified polymer is polytetrafluoroethylene (PTFE) or Teflon. A side chain of sulfonic acid $HSO_3$ is then added to PTFE through the "sulfonation" process. The end of the chain is an $SO_3-$ ion, and the $HSO_3$ group is ionically bonded. The resulting structure combines strong hydrophobic properties of the fluorocarbon polymer backbone with strong hydrophilic properties of the terminal sulfonic acid function. It is an excellent proton conductor, is durable (owing to strong bonds between the fluorine and carbon), and shows good chemical resistance.[2,10]

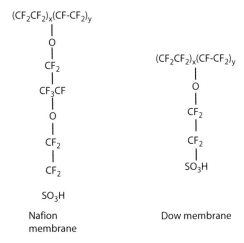

$(CF_2CF_2)_x(CF\text{-}CF_2)_y$
|
O
|
$CF_2$
|
$CF_3CF$
|
O
|
$CF_2$
|
$CF_2$

$SO_3H$

Nafion membrane

$(CF_2CF_2)_x(CF\text{-}CF_2)_y$
|
O
|
$CF_2$
|
$CF_2$
|
$SO_3H$

Dow membrane

**FIGURE 19.8**    Structural characteristics of PEMFC membranes. (From Bloemen, L.J., and Mugerwa, M.N., (Eds.), *Fuel Cell Systems*, Plenum Press, New York, 1994.)

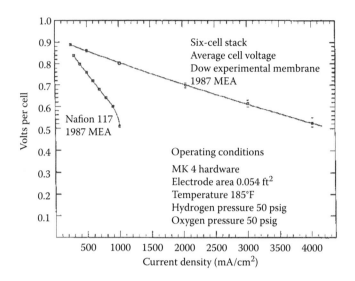

**FIGURE 19.9**   PEMFC performance using Nafion and Dow membranes. (From EG&G Services, Parsons Inc. and Science Applications International Corporation, *Fuel Cell Handbook*, fifth edition, contract no. DE–AM26–99FT40575, for US Department of Energy, October 2000.)

Tests showed that the PEMFC performance levels improved with the membrane developed by Dow Chemical.[11] Although this membrane remains a perfluorosulfonic acid membrane, its structure is characterized by a shorter side chain and, thus, a lower equivalent weight compared to Nafion. Conductivity and hydrophilic properties are slightly enhanced, and durability is still maintained. The Dow membrane was tested between 1987 and 1988 at Ballard Power Systems, and the results showed significant increases in PEMFC performance levels (Figure 19.9).

### 19.4.1.1.2 Electrodes

The anode and cathode have identical structures, consisting of two layers in close contact with each other and with the membrane. The roles of the electrodes in operation of the fuel cell are summarized in Table 19.2.

The layer situated adjacent to the membrane is the *catalyst layer*, which provides the area where the electrochemical reactions occur (Figure 19.7). It is a platinum carbon composite film, about 5–10 μm thick, used to increase the reaction rates. Because of the high cost of platinum, sustained technological efforts were focused on reduction of the platinum load, originally about 4 mg Pt/cm² but currently almost 0.2 mg Pt/cm² and even lower values, with high performance levels.[12,13]

Typically, carbon paper is used when a compact design of fuel cells is desired; the most used brand is Toray paper. However, if only a simple assembly is preferred, carbon cloth is sufficient. Thickness of the backing layer is typically between 100 and 300 μm. Figure 19.10 shows details of the MEA structure with the catalyst and backing layers.[14]

**TABLE 19.2**

**Roles of Electrodes in PEMFC Operation**

| Electrode | Layer | Role |
|---|---|---|
| Anode | Catalyst | Catalysis of anode reaction |
| | | Proton conduction into membrane |
| | | Electron conduction into gas diffusion layer |
| | | Water transport |
| | | Heat transport |
| | Gas diffusion | Fuel supply and distribution (hydrogen/fuel gas) |
| | | Electron conduction |
| | | Heat removal from reaction zone |
| | | Water supply (vapor) into electrocatalyst |
| Cathode | Catalyst | Catalysis of cathode reaction |
| | | Oxygen transport to reaction sites |
| | | Proton conduction from membrane to reaction sites |
| | | Electron conduction from gas diffusion layer to reaction zone |
| | | Water removal from reactive zone into gas diffusion layer |
| | | Heat generation/removal |
| | Gas diffusion | Oxidant supply and distribution (air/oxygen) |
| | | Electron conduction toward reaction zone |
| | | Heat removal |
| | | Water transport (liquid/vapor) |

*Source:* Hoogers, G., Ed., *Fuel Cell Technology Handbook*, CRC Press, Boca Raton, FL, 2003. With permission.

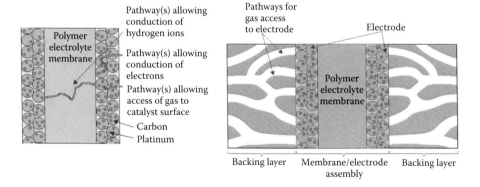

**FIGURE 19.10** MEA structure design details. (From Thomas, S., and Zalbowitz, M., Fuel Cells—Green Power, LA–UR–99–3231, Los Alamos National Laboratory, 1999. With permission.)

### 19.4.1.1.3 Teflon Masks and Current Collectors

The single-cell structure is completed by two Teflon masks and two high-density graphite plates. The Teflon masks are gaskets that confine the gas flow to the active area, providing an effective seal along the periphery of the membrane. The graphite plates are current collectors, and they also contain gas flow fields at the same time. In a fuel cell stack, the current collector plates contain gas flow fields on both sides, and they become *bipolar plates*.

## 19.4.1.2 Operation Characteristics

Hydrogen gas is supplied to the anode, where it dissociates into hydrogen atoms in the presence of the platinum catalyst. The atoms further split into protons and electrons, which travel separate ways from the anode to the cathode. Protons are conducted through the electrolyte membrane, and electrons are forced to go via an external circuit to the cathode, producing electricity. Oxygen is supplied to the cathode, where a reduction process occurs and water and heat are created as by-products. Figure 19.11 shows an illustration of the PEMFC principle of operation. The basic reactions for the PEMFC are

$$\text{Anode: } 2H_2 \rightarrow 4H^+ + 4e^-$$

$$\text{Cathode: } O_2 + 4H^+ + 4e^- \rightarrow 2H_2O$$

$$\text{Cell reaction: } 2H_2 + O_2 \rightarrow 2H_2O$$

Continuous research efforts over the years have led to significant improvements in the performance levels of the PEMFC (Figure 19.12).

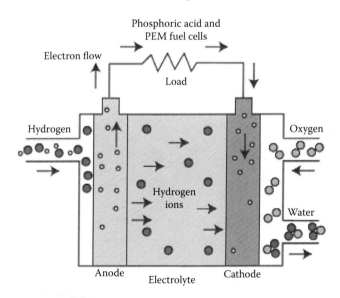

**FIGURE 19.11** PEMFC and PAFC operation principle. (From US Department of Defense Fuel Cell Test and Evaluation Center, "Fuel Cell Basics," www.fctec.com/index.html.)

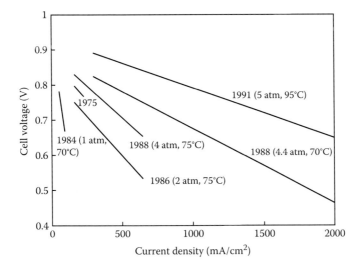

Current density (mA/cm$^2$)

**FIGURE 19.12** PEMFC performances. (From EG&G Services, Parsons Inc. and Science Applications International Corporation, *Fuel Cell Handbook*, fifth edition, contract no. DE–AM26–99FT40575, for US Department of Energy, October 2000.)

The typical output is approximately 0.7 V per cell unit, and the power density is usually higher compared to other fuel cells, which translates into a smaller size of the fuel cell stack. For transport applications, Asia Pacific Fuel Technologies produces 3 kW 64-cell stacks, which are 25 cm high and have an active area of 150 cm$^2$. The Mark 902 fuel cell module produced by Ballard Power Systems has the dimensions of 80.5 × 37.5 × 25.0 cm and yields an 85 kW rated net output.[15] The Nexa™ power module, Ballard's first volume-produced PEMFC designed to be integrated into stationary and portable applications, is 56 × 25 × 33 cm, with a rated net output of 1200 W (see Figure 19.13).[15]

PEMFCs are intended to be used also in small applications, such as portable devices (i.e., laptops; see Figure 19.14), electronics, and so forth.

### 19.4.2 ALKALINE FUEL CELLS

The development of the AFC started almost seven decades ago, when researchers started to realize that hydrogen fuel cells with alkaline electrolytes can be used in commercial applications. The first notable solution was the high-power-density AFC developed by Sir Francis Bacon, with an output of 0.6 V at 1.11 A/cm$^2$ current density and 240°C operating temperature.[16] AFCs were used by NASA on their space explorations during the 1960s and 1970s.

The AFCs utilize potassium hydroxide (KOH) as an electrolyte of variable concentration, either in aqueous solution or in stabilized matrix form. The KOH concentration varies with the operating temperature, increasing from 35 wt% for low temperatures to about 85 wt% for high temperatures. The electrolyte is contained in a porous asbestos matrix, and the catalysts are typically made of nickel (Ni) and

**FIGURE 19.13** Nexa power module. (From Ballard Power Systems, "Ballard® Fuel Cell Power Module Nexa™," www.ballard.com, 2002.)

**FIGURE 19.14** Ballard fuel cell. (Fuel Cells 2000, "Transportation Fuel Cells—Technical Info," www.fuelcells.org.)

silver (Ag). Noble metals, metal oxides, or spinels are also considered among the materials used to fabricate the catalysts.[17] Further details of the AFC components used for the space applications are given in EG&G Services, Parsons Inc., and Science Applications International Corporation's *Fuel Cell Handbook*, fifth edition.[4]

Hydrogen and oxygen are supplied to the electrodes similarly to PEMFCs. The KOH electrolyte is extremely sensitive to potential poisoning with CO or reaction with $CO_2$, and thus, only pure hydrogen and oxygen can be used as reactants for the electrochemical processes. The carrier in this case is the hydroxyl ion (OH), which travels from the cathode to the anode, where it combines with $H_2$ and creates water and electrons (Figure 19.15). If the electrolyte is in a solution form, it mixes up with the water created at the anode. To ensure proper operation of the fuel cell unit, it is required that water be continuously removed from the electrolyte. Electrons formed at the anode are conducted to the external circuit to create the electrical output and then forced to the cathode, closing the circuit. The basic electrochemical reactions for the AFC are

$$\text{Anode: } 2H_2 + 4OH^- \rightarrow 4H_2O + 4e^-$$

$$\text{Cathode: } O_2 + 2H_2O + 4e^- \rightarrow 4OH^-$$

$$\text{Cell reaction: } 2H_2 + O_2 \rightarrow 2H_2O$$

Owing to the fact that only pure fuel or oxidant can be used in operation, the AFC is used for specialized applications. Space explorations, military use, and research are among the few areas in which these fuel cells are utilized. Efforts are being made to broaden the spectrum of terrestrial daily applications. Tests and demonstrations have shown that AFC hybrid vehicles are potential technological solutions for the near future in transportation. In the 1970s, an AFC-based hybrid vehicle was tested for 3 years, using liquid hydrogen and oxygen, and KOH liquid electrolyte.[18] The improved version of this vehicle was tested again in 1998, using a system consisting of AFC and rechargeable alkaline manganese dioxide–zinc (RAM) batteries. The

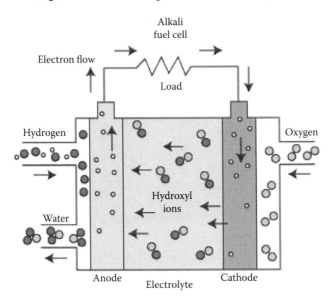

**FIGURE 19.15** AFC operation principle. (US Department of Defense Fuel Cell Test and Evaluation Center, "Fuel Cell Basics," www.fctec.com/index.html.)

operating lifetime of these AFCs is anticipated to be about 4000 h, and the mass production cost is expected to become comparable to the cost of the currently used heat engines ($50 to $100 per kW).[19]

### 19.4.3 PHOSPHORIC ACID FUEL CELLS

These cells have been developed for medium-scale stationary applications and are the only commercialized type of fuel cells. The technology employed for the PAFC is the most well-known technology developed to date for fuel cells.

The PAFC has a similar design with the PEMFC. The electrolyte used for the PAFC is concentrated phosphoric acid ($H_3PO_4$), allowing operation at temperatures higher than the PEMFC (i.e., over 100°C). This electrolyte is contained in a silicon carbide matrix, and catalysts are typically made of Pt. Technological advances of the components of this type of fuel cells have been extensively documented over the last 40 years, and a brief summary is presented in Table 19.3.

Operation of PAFCs is similar to PEMFCs, as can be seen in Figure 19.11. Hydrogen-rich fuel is supplied to the anode, where protons and electrons split and start traveling to the cathode, following different pathways through the membrane layer (protons) and via an external circuit, producing electricity (electrons). At the cathode, they will combine with oxygen, and water and heat are obtained as by-products. The basic reactions for PAFCs are the same as for PEMFCs.

A 200 kW PureCell™ 200 power system (formerly known as the PC25 system) has been commercially available. Each unit of this provides more than 900,000 Btu

---

**TABLE 19.3**
**PAFC Component Characteristics**

| Component | ca. 1965 | ca. 1975 | Current Status |
|---|---|---|---|
| Anode | PTFE-bonded Pt black | PTFE-bonded Pt/C Vulcan XC-72[a] | PTFE-bonded Pt/C Vulcan XC-72[a] |
| | 9 mg/cm$^2$ | 0.25 mg Pt/cm$^2$ | 0.1 mg Pt/cm$^2$ |
| Cathode | PTFE-bonded Pt black | PTFE-bonded Pt/C Vulcan XC-72[a] | PTFE-bonded Pt/C Vulcan XC-72[a] |
| | 9 mg/cm$^2$ | 0.5 mg Pt/cm$^2$ | 0.5 mg Pt/cm$^2$ |
| Electrode support | Ta mesh screen | Carbon paper | Carbon paper |
| Electrolyte support | Glass fiber paper | PTFE-bonded SiC | PTFE-bonded SiC |
| Electrolyte | 85% $H_3PO_4$ | 95% $H_3PO_4$ | 100% $H_3PO_4$ |

*Source:* EG&G Services, Parsons Inc. and Science Applications International Corporation, *Fuel Cell Handbook* fifth edition, contract no. DE–AM26–99FT40575, for US Department of Energy, October 2000.

[a] Conductive oil furnace black, product of Cabot Corp. (Typical properties: 002 d-spacing of 3.6 Å by x-ray diffusion, surface area of 220 m$^2$/g by nitrogen adsorption, and average particle size of 30 μm by electron microscopy.)

**TABLE 19.4**
**PureCell™ 200 Power System Performance Data**

| Feature | Characteristics |
|---|---|
| Rated electrical capacity | 200 kW/235 kVA |
| Voltage and frequency | 480/277 V, 60 Hz, 3 phase |
| | 400/230 V, 50 Hz, 3 phase |
| Fuel consumption | Natural gas: 2100 cft/h @ 4–14″ water pressure |
| | Anaerobic digester gas: 3200 cft/h at 60% $CH_4$ |
| Efficiency (lower heating value [LHV] basis) | 87% total: 37% electrical, 50% thermal |
| Emissions | <2 ppmv CO, <1 ppmv $NO_x$, and negligible $SO_x$ (on 15% $O_2$, dry basis) |
| Thermal energy available | |
| Standard | 900,000 Btu/h @ 140 F |
| High grade heat recovery options | 450,000 Btu/h @ 140 F and 450,000 Btu/h @ 250 F |
| Sound profile | 60 dBA @ 30 ft |
| Power module | |
| Dimensions and weight | 121″ (H) × 114″ (W) × 212″ (L); 40,000 lb |
| Cooling module | |
| Dimensions and weight | 50″ (H) × 49″ (W) × 162″ (L); 1700 lb |

per hour of heat, and hundreds are already being used throughout the world. Table 19.4 presents performance data of this system.[20]

## 19.5 HIGH-TEMPERATURE FUEL CELLS

### 19.5.1 Molten Carbonate Fuel Cells

The electrolyte is a mixture of lithium carbonate (~68%) and potassium carbonate (~32%), contained in a lithium aluminum oxide ($LiAlO_2$) matrix. Hydrogen and CO are used for the electrochemical reactions, and water and $CO_2$ result as by-products. Catalysts are typically made of nickel.

Operation of MCFC is shown in Figure 19.16. $CO_2$ and $O_2$ are supplied at the cathode, and they react with the available electrons. The resulting carbonate ions travel to the anode, where they combine with the hydrogen to produce water, $CO_2$, and electrons. These electrons are then forced to go back to the cathode through the external pathway to create electricity. The basic reactions for MCFCs are

$$\text{Anode: } CO_3^{2-} + H_2 \rightarrow H_2O + CO_2 + 2e^-$$

$$\text{Cathode: } CO_2 + 1/2\,O_2 + 2e^- \rightarrow CO_3^{2-}$$

$$\text{Cell reaction: } H_2 + 1/2\,O_2 + CO_2 \text{ (cathode)} \rightarrow H_2O + CO_2 \text{ (anode)}$$

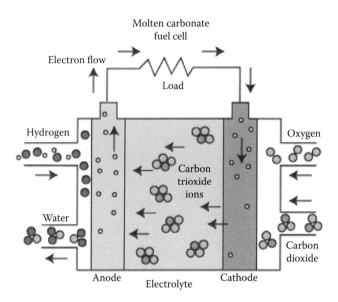

**FIGURE 19.16** MCFC operation principle. (From US Department of Defense Fuel Cell Test and Evaluation Center, "Fuel Cell Basics," www.fctec.com/index.html.)

The MCFC operates at much higher temperatures (about 650°C) compared to PEMFCs and PAFCs, which makes it possible to process the fuel internally, thus increasing the overall efficiency of the fuel cell and minimizing emissions. They are intended to be used for power plant applications. Over the years, various attempts have been made to commercially develop plants based on the MCFC. In the 1990s, companies such as MC Power and FuelCell Energy (formerly known as Energy Research Corp.) installed MCFC power plants for testing and development purposes. For example, a 2 MW unit was installed in 1996 in California by FuelCell Energy to test the design of commercial units. Several MCFC systems in the United States, Europe, and Japan are now operational.

## 19.5.2 SOLID OXIDE FUEL CELLS

The least mature technology of the fuel cells, SOFCs are characterized by extremely high operating temperatures. The fuel/oxidant mixture is less restricted, compared to all the other fuel cells, owing to the high operating temperature of the cell, which allows for more combinations. Fuel can be hydrogen, CO, or $CH_4$, and the oxidant can be $CO_2$, $O_2$, or air; catalysts are made of perovskite materials. A solid coated zirconia oxide ceramic ($Y_2O_3$-stabilized $ZrO_2$) is used as electrolyte.

Operation of SOFC is shown in Figure 19.17. In this case, oxygen ions formed at the cathode from reaction of oxygen and electrons travel through the electrolyte to the anode. There, they combine with fuel, creating by-products (i.e., water) and

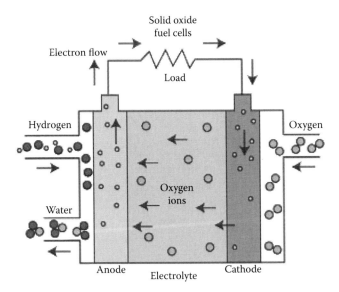

**FIGURE 19.17**  SOFC operation principle. (From US Department of Defense Fuel Cell Test and Evaluation Center, "Fuel Cell Basics," www.fctec.com/index.html.)

electrons. These electrons travel to the cathode through an external circuit, producing electricity. The basic reactions are

$$\text{Anode: } 2H_2 + 2O^{2-} \rightarrow 2H_2O + 4e^-$$

$$\text{Cathode: } O_2 + 4e^- \rightarrow 2O^{2-}$$

$$\text{Cell reaction: } 2H_2 + O_2 \rightarrow 2H_2O$$

SOFCs are intended to be used for large power and cogeneration utilities. A number of 25 kW SOFC-based systems are already in use, and a 100 kW SOFC has been in use for some time in Germany.

## 19.6  HYDROGEN PRODUCTION AND STORAGE

### 19.6.1  HYDROGEN PRODUCTION

Fuel cells operate with hydrogen-rich fuels, and either direct hydrogen or reformed fuels are typically used. Currently, industrial production of hydrogen is designed to accommodate the required supply for producing ammonia, which is largely used in agriculture as fertilizer and in oil refineries to produce automotive fuels. A number of methods can be used to obtain hydrogen, as illustrated in Figure 19.18; the most notable example is extraction from fossil fuels. Other technologies, such as water

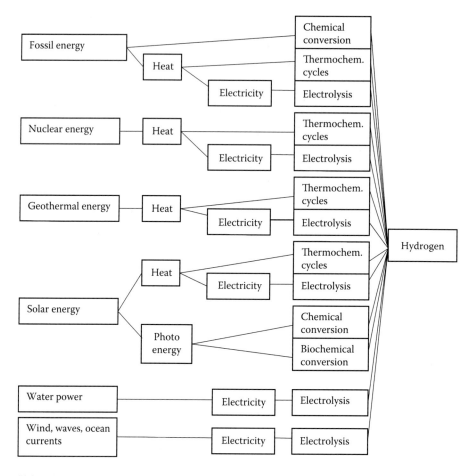

**FIGURE 19.18** Hydrogen production sources.

electrolysis, are employed only on a much smaller scale because of their high costs and so forth.

### 19.6.1.1 Fossil Fuels

Hydrogen is extracted from fossil fuels through various techniques, such as steam reforming, partial oxidation (or a combination of them), and gasification.

*Steam reforming* is a well-established technology, which uses natural gas as feedstock. This process takes place at temperatures between 750°C and 1000°C. Methane reacts with water over a catalyst (usually nickel, supported by alumina) and produces the hydrogen-rich gas that is further used by fuel cells. The overall process takes place in two steps:

- Steam reforming: $CH_4 + HO_2 \rightarrow CO + 3H_2$
- Shift reaction: $CO + H_2O \rightarrow CO_2 + H_2$

Methanol is also used for producing hydrogen, and the reaction takes place at temperatures between 200°C and 300°C, the catalyst being made of copper, supported by zinc oxide:

- Steam reforming: $CH_3OH + H_2O \rightarrow CO_2 + 3H_2$
- Shift reaction: $CO + H_2O \rightarrow CO_2 + H_2$

*Partial oxidation* is typically used to process heavy oil fractions; the exothermic reaction in this case does not require the presence of a catalyst. If applied to natural gas or methane, presence of a catalyst becomes necessary. The following reactions for methane are given as an example:

- Partial oxidation: $CH_4 + \dfrac{1}{2}O_2 \rightarrow CO + H_2$
- Shift reaction: $CO + H_2O \rightarrow CO_2 + H_2$

The advantage of steam reforming technology is that its output has the highest hydrogen concentration compared to other technologies based on fossil fuel. However, it does not offer fast start-up and dynamic response. Partial oxidation, on the other hand, produces only low concentrations of hydrogen combined with a fast start-up and dynamic response. A natural question is, thus, what would happen if we try to combine both these technologies, using the advantages of each of them? This combination is known as *autothermal reforming*, and efforts have been made to develop various reformers, such as the HotSpot fuel processor.[21] Table 19.5 shows a comparison of the gas compositions obtained after using different options for the reforming process, using methanol as fuel.

The third technology, *coal gasification*, is achieved through coal reaction with oxygen and steam at high temperatures, and it uses all types of coals for the process.

A disadvantage common to all these technologies is that one of the by-products of the reforming reactions is $CO_2$, a significant contributor to the environment pollution. To eliminate this, various other technologies are being developed as potential solutions for "$CO_2$-free" hydrogen production from fossil fuels, such as the *pyrolytic cracking* of natural gas:[22]

$$CH_4 \rightarrow C + 2H_2$$

**TABLE 19.5**

**Gas Composition of Reformer Outputs**

| Composition (Dry gas, %) | Steam Reforming | Partial Oxidation | Autothermal Reforming |
|---|---|---|---|
| $H_2$ | 67 | 45 | 55 |
| $CO_2$ | 22 | 20 | 22 |
| $N_2$ | — | 22 | 21 |
| $CO$ | — | — | 2 |

*Source:* Hoogers, G., Ed., *Fuel Cell Technology Handbook*, CRC Press, Boca Raton, FL, 2003. With permission.

### 19.6.1.2 Water Electrolysis

Another way of obtaining hydrogen is through water electrolysis. This technology is based on decomposition of water into hydrogen and oxygen with the help of electricity. Although its development began in the nineteenth century, water electrolysis has never reached the level of large-scale production, because it uses electricity as input, and this has a direct impact on the overall cost of producing hydrogen. Costs associated with this technology are considerably higher than for obtaining hydrogen directly from the fossil fuels.[23] The contribution of water electrolysis technology to the total production of hydrogen represents only about 0.5%.

### 19.6.1.3 Other Sources

Nuclear or renewable energies are also considered potentially "$CO_2$-free" sources of hydrogen production.

### 19.6.2 HYDROGEN STORAGE

Hydrogen storage represents one of the difficult issues associated with operation of the fuel cell systems. This is because hydrogen is characterized by low energy density and high specific energy. There are a number of technologies currently used for storage, such as liquefaction or compression. Other storage methods based on carbon nanofibers, metal hydrides, or glass microspheres are currently under investigation.

Hydrogen liquefaction can be done through several techniques. For small-scale systems, the Stirling process is considered an important solution. An alternative to the Stirling refrigerator is the magnetocaloric refrigeration process, based on the isentropic demagnetization of a ferromagnetic material near its Curie point temperature. For large-scale applications, the Claude process represents a viable economic solution.

Different types of pressure cylinders or tanks are currently used for storing compressed hydrogen with a typical maximum pressure up to 30 MPa. Cylinders are made of lightweight composite materials, which reduce the overall weight of the storage. It is also proposed to design pressure cylinders or tanks with better space filling of hydrogen compared to the current solutions. A "conformable technology" has been developed by Sleegers and Thiokol Propulsion's Group, which allows storing more fuel in pressure tanks while reducing the overall tank weight. This technology uses a combination of multicell conformable tanks located such that maximum fuel storage is obtained.

Other technologies, such as hydrogen storage in solid forms, are being researched, and recent results have been reported using carbon (nanotubes, activated carbon, fullerenes), metal hydrides, and glass microspheres. Carbon storage is a very attractive technology, which provides the overall system with high energy density, reliability, and safety. This technology is based on gas-on-solids adsorption of hydrogen, and it is being developed for small applications, in which safety and weight of the device are key characteristics.[24] Metal hydrides are another interesting option currently explored for hydrogen storage, where hydrogen reacts chemically with a metal. Table 19.6 shows the characteristics of a number of metal

**TABLE 19.6**
**Hydrogen Storage Properties of Metal Hydrides**

| Metal Hydride System | Mg/ MgH$_2$ | Ti/ TiH$_2$ | V/VH$_2$ | Mg$_2$Ni/ Mg$_2$NiH$_4$ | FeTi/ FeTiH$_{1.95}$ | LaNi$_5$/ LaNi$_5$H$_{5.9}$ | LH$_2$ |
|---|---|---|---|---|---|---|---|
| Hydrogen content as mass fraction (%) | 7.7 | 4.0 | 2.1 | 3.2 | 1.8 | 1.4 | 100.0 |
| Hydrogen content by volume (kg/dm$^3$) | 0.101 | 0.15 | 0.09 | 0.08 | 0.096 | 0.09 | 0.077 |
| Energy content based on higher heating value (HHV) (MJ/kg) | 9.9 | 5.7 | 3.0 | 4.5 | 2.5 | 1.95 | 143.0 |
| Energy content based on LHV (MJ/kg) | 8.4 | 4.8 | 2.5 | 3.8 | 2.1 | 1.6 | 120.0 |
| Heat of reaction (kJ/Nm$^3$) | 3360 | 5600 | — | 2800 | 1330 | 1340 | — |
| Heat of reaction (kJ/mol) | 76.3 | 127.2 | — | 63.6 | 30.2 | 30.4 | — |
| Heat of reaction as fraction of HHV (%) | 26.7 | 44.5 | — | 22.2 | 10.6 | 10.6 | — |
| Heat of reaction as fraction of LHV (%) | 31.6 | 52.6 | — | 26.3 | 12.5 | 12.6 | — |

*Source:* Hoogers, G., Ed., *Fuel Cell Technology Handbook*, CRC Press, Boca Raton, FL, 2003. With permission.

hydride systems. The third option, glass microspheres, takes advantage of the variation of glass permeability with temperature and fills the microspheres with hydrogen to trap it inside.

## 19.7 CURRENT PERFORMANCES

### 19.7.1 OPERATIONAL ISSUES

#### 19.7.1.1 Water and Heat Management

Although good results have been obtained using solid polymers as membranes, there are a number of issues that require special attention during PEMFC operation, such as water and heat management. Conductivity properties are extremely sensitive to the level of hydration of membranes, and maintaining adequate humidity conditions is a challenging task. A fine balance of equilibrium has to be maintained to avoid either flooding or dehydration of the membrane. Water management depends on several factors, such as operating parameters (temperature and pressure), water content, and the presence of the impurity ions in the membranes. PEMFCs typically operate efficiently at approximately 80°C at atmospheric pressure; if increased above 100°C, dehydration of the membrane occurs, and conductivity of the membrane decreases significantly. The water content depends on the water transport, which is a complex phenomenon still not very well understood. Diffusion and electroosmosis are considered to be the processes responsible for the water transport, and various models have been developed describing the mechanism and the factors influencing it.[25–28] Impurities in membranes are due to the impurities present in the fuel or oxidant or to the corrosion of materials, and water management can be seriously affected by their presence.[29,30]

### 19.7.1.2 CO Poisoning

Hydrogen-rich fuels used for fuel cells are either pure hydrogen or reformed fuel. As briefly described in Section 19.6.1, during the reforming process, CO is produced, and trace amounts of carbon monoxide remain present in the flow fed to the electrodes. For low-temperature fuel cells using platinum catalysts, the presence of carbon monoxide even at trace levels is detrimental, with CO having an affinity for platinum and thus poisoning the catalysts. The overall performance of fuel cells deteriorates as hydrogen is blocked from reaching the catalysts. Although further fuel processing with complete removal of CO is critical for fuel cell operation, these additional processes require design modifications of the system, which, in the end, translate into higher overall costs.

### 19.7.1.3 Hydrogen Safety

Handling of hydrogen, similar to any other flammable fuel, entails a number of hazards. Although since the Hindenburg incident in 1937 the public has considered hydrogen to be extremely dangerous, it has been demonstrated that risks associated with it are manageable and are similar to those with other gaseous fuels. Hydrogen is indeed characterized by high volatility and flammability, but it also has very low density, which means that it disperses extremely rapidly, and the ignition and detonation levels are not easily reached.

### 19.7.2 Cost

Despite numerous improvements over the years (i.e., reducing of the membrane thickness), the cost of the PEMFC membranes remains high, the price of Nafion membranes, for example, being $500 to $1000 per $m^2$ based on quantity acquired, or approximately $100 per kW electric power.[21] It is estimated, however, that the cost of the membranes will decrease with the increase of the production volume; their relative trends are shown in Figure 19.19. To achieve lower costs and improve performance of the membranes, different technologies have been investigated during

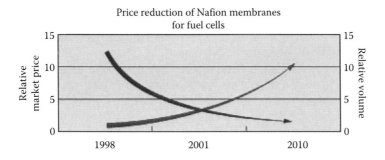

**FIGURE 19.19** Price/volume trends for Nafion membranes. (From DuPont Fuel Cells, DuPont™ Nafion® Membranes and Dispersions, http://www.dupont.com/fuelcells. With permission.)

the last few years. Summary descriptions of these investigations are presented in the work of Hoogers[21] and Paddison.[31]

### 19.7.3 Environmental Impact

The most attractive feature of fuel cells is their minimum impact on the environment during operation. An interesting comparison is shown in Figure 19.20, where drastic reductions in greenhouse gas emissions can be observed when comparing fuel cells with internal combustion engines. If direct hydrogen is used as primary fuel for fuel cell operation, greenhouse gas emissions are practically zero.

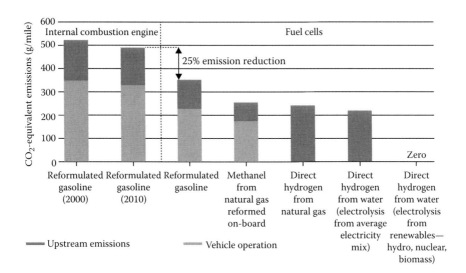

**FIGURE 19.20** Greenhouse gas emissions. (From Industry Canada, "Canadian Fuel Cell Commercialization Roadmap," March 2003.)

## 19.8 RESEARCH AND DEVELOPMENT ISSUES

Extensive efforts are currently underway to improve fuel cell performance along with a reduction of the overall cost. To achieve this, issues like theoretical modeling, finding material and design alternatives for components or the entire fuel cell, and finding viable solutions for hydrogen storage are under intense scrutiny worldwide.

The diffusion mechanism of protons through water or the membrane of PEMFCs is one of the theoretical aspects that have been under evaluation over the years, numerous attempts being made to identify the best model describing the phenomenon. These models are either deterministic[32,33] or statistical approaches.[31,34] The pore structure of the membranes is assumed to have different geometries. For example, a cylindrical pore structure is used in the model described in the work of Pintauro and Yang,[33] with either a uniform or variable pore radius and pore-wall distribution of fixed charges. This model is an ion/solvent model that predicts the multicomponent salt separation by ion exchange membranes. Another example of the theoretical modeling developments is the mechanism describing the transport of the proton in water (*Grotthuss mechanism*), which considers that transport is done through consecutive proton migration steps between adjacent water ($H_2O$) molecules.[35] Other models assume that the diffusion process of the proton is facilitated through formation of complex structures, such as $H_5O_2^+$ (two $H_2O$ molecules sharing the same proton) and $H_9O_4^+$ (three $H_2O$ molecules strongly bonded to the $H_3O^+$ core).[36,37]

New types of membranes for PEMFCs that would allow operation of the fuel cells at higher temperatures are also being investigated. Improved cathodes, advanced catalysts, and optimized gas diffusion layers are among the objectives of present contracts awarded in the United States and elsewhere.

**FIGURE 19.21** Direct methanol fuel cell stack (30 units). (From Department of Energy, "Direct Methanol Fuel Cell," http://www.ott.doe.gov.)

New types of cells are being developed, such as direct methanol fuel cells (DMFCs). The DMFCs (see Figure 19.21) are basically PEMFCs with a slightly modified anode catalyst made of platinum–ruthenium and using methanol as fuel. These fuel cells eliminate the reforming process of the hydrogen-rich fuel of PEMFCs but do not eliminate the $CO_2$ emissions produced through chemical reactions, which are greenhouse gas contributors.

A recent account of research and development that provides pathways to commercialization is given in the UD Department of Energy's Pathways to Commercial Success.[38]

## REFERENCES

1. Appleby, A.J. and Foulkes, F.R., *Fuel Cell Handbook*, Van Nostrand Reinhold, New York, 1989.
2. Larminie, J. and Dicks, A., *Fuel Cell Systems Explained*, John Wiley & Sons, Chichester, UK, 2000.
3. Stobart, R., Ed., *Fuel Cell Technology for Vehicles*, PT-84, Society of Automotive Engineers, Warrendale, PA, 2001.
4. EG&G Services, Parsons, Inc. and Science Applications International Corporation, *Fuel Cell Handbook* (Fifth Edition), Contract No. DE–AM26–99FT40575, for U.S. Department of Energy, West Virginia, October 2000.
5. Penner, S.S., Ed., Commercialization of fuel cells, *Energy: Int. J.*, 20(5), Pergamon Press, New York, 1995.
6. Mehta, S.K. and Bose, T.K., Ed., *Proceedings of the Workshop on Hydrogen Energy Systems*, presented at the Official Opening of the Hydrogen Research Institute, in collaboration with Canadian Hydrogen Association, Canada, 1996.
7. European Commission, A Fuel Cell Research, Development and Demonstration Strategy for Europe up to 2005, 1998 ed., Belgium, 1998.
8. Gottesfeld, S. and Wilson, M.S., Polymer electrolyte fuel cells as potential power sources for portable electronic devices, in *Energy Storage Systems for Electronics*, Osaka, T. and Datta, M., Eds., Gordon and Breach Science Publishers, The Netherlands, 2000.
9. DuPont Fluoroproducts, DuPont™ Nafion® PSFA Membranes N-112, N-1135, N-115, N-117, N-1110 Perfluorosulfonic Acid Polymer, Product Information NAE101, November 2002.
10. Decher, R., *Direct Energy Conversion—Fundamentals of Electric Power Production*, Oxford University Press, New York, 1997.
11. Dow Liquid Separations, DOWEX ion exchange resins—Fundamentals of ion exchange, based on a paper by R.M. Wheaton, L.J. Lefevre—Dow Chemical U.S.A., June 2000.
12. Wazikoe, M., Velev, O.A. and Srinivasan, S., Analysis of proton exchange membrane fuel cell performance with alternate membranes, *Electrochim. Acta*, 40(3), 335, 1995.
13. Wilson, M.S., Valerio, J.A. and Gottesfeld, S., Low platinum loading electrodes for polymer electrolyte fuel cells fabricated using thermoplastic ionomers, *Electrochim. Acta*, 40(3), 355, 1995.
14. Thomas, S. and Zalbowitz, M., *Fuel Cells—Green Power*, LA–UR–99–3231, Los Alamos National Laboratory, New Mexico, 1999.
15. Ballard Power Systems Inc., Ballard® Fuel Cell Power Module NEXA™, Ballard® Power to Change the World®, http://www.ballard.com, 2002.
16. Perry, M.L. and Fuller, T.F., A historical perspective of fuel cell technology in the 20th century, *J. Electrochem. Soc.*, 149(7), S59, 2002.

17. Cabot, P.L., Guezala, E., Calpe, J.C., García, M.T. and Casado, J., Application of Pd-based electrodes as hydrogen diffusion anodes in alkaline fuel cells, *J. Electrochem. Soc.*, 147(1), 43, 2000.
18. Kordesch, K. and Simader, G., *Fuel Cells and Their Applications*, John Wiley & Sons, New York, March 1996.
19. Kordesch, K., Gsellmann, J., Cifrain, M., Voss, S., Hacker, V., Aronson, R.R., Fabjan, C., Hejze, T. and Daniel-Ivad, J., Intermittent use of a low-cost alkaline fuel cell-hybrid system for electric vehicles, *J. Power Sources*, 80, 190, 1999.
20. UTC Power, PureCell™ 200 Commercial Fuel Cell Power System, http://www.golden stateenergy.com/videojunio2004/UTC%20Fuel%20Cell%20Brochure.pdf, 2003.
21. Hoogers, G., Ed., *Fuel Cell Technology Handbook*, CRC Press, Boca Raton, FL, 2003.
22. Pohl, H.W., Ed., *Hydrogen and Other Alternative Fuels for Air and Ground Transportation*, John Wiley & Sons, Germany, 1995.
23. Norbeck, J.M., *Hydrogen Fuel for Surface Transportation*, Society of Automotive Engineers, Warrendale, PA, 1996.
24. Workshop on Storage of Hydrogen on Carbon Nanostructured Materials: Held at Institut de Recherche sur l'Hydrogene Trois-Rivieres, Canada, 2000.
25. Janssen, G.J.M., A phenomenological model of water transport in a proton exchange membrane fuel cell, *J. Electrochem. Soc.*, 148(12), A1313, 2001.
26. Motupally, S., Becker, A.J. and Weidner, J.W., Diffusion of water in Nafion 115 membranes, *J. Electrochem. Soc.*, 147(9), 3171, 2000.
27. Zawodzinski, T.A., Jr., Springer, T.E., Davey, J., Lopez, R.C., Valerio, J. and Gottesfeld, S., A comparative study of water uptake by and transport through ionomeric fuel cell membranes, *J. Electrochem. Soc.*, 140(7), 1981, 1993.
28. Nguyen, T.V. and White, R.E., A water and heat management model for proton-exchange-membrane fuel cells, *J. Electrochem. Soc.*, 140(8), 2178, 1993.
29. Okada, T., Theory for water management in membranes for polymer electrolyte fuel cells—Part 1: The effect of impurity ions at the anode side on the membrane performances, *J. Electroanal. Chem.*, 465, 1, 1999.
30. Okada, T., Theory for water management in membranes for polymer electrolyte fuel cells—Part 2: The effect of impurity ions at the cathode side on the membrane performances, *J. Electroanal. Chem.*, 465, 18, 1999.
31. Paddison, S.J., The modeling of molecular structure and ion transport in sulfonic acid based ionomer membranes, *J. New. Mater. Electrochem. Syst.*, 4, 197, 2001.
32. Thampan, T., Malhotra, S., Tang, H. and Datta, R., Modeling of conductive transport in proton–exchange membranes for fuel cells, *J. Electrochem. Soc.*, 147(9), 3242, 2000.
33. Pintauro, P.N. and Yang, Y., Mathematical analysis of transport in ion–exchange membranes, in *Tutorials in Electrochemical Engineering—Mathematical Modeling*, Electrochemical Society Proceedings Vol. 99–14, Savinell, R.F., Fenton, J.M., West, A., Scanlon, S.L. and Weidner, J.W., Eds., The Electrochemical Society, Inc., New Jersey, 1999.
34. Paddison, S.J., Paul, R. and Zawodzinski, T.A., Jr., A statistical mechanical model of proton and water transport in a proton exchange membrane, *J. Electrochem. Soc.*, 147, 617, 2000.
35. de Grotthuss, C.J.T., Sur la Décomposition de l'Eau et des Corps qu'Elle Tient en Dissolution à l'Aide de l'Électricité Galvanique, *Ann. Chim.*, LVIII, 54, 1806.
36. Eigen, M., Proton transfer, acid-base catalysis and enzymatic hydrolysis, *Angew. Chem. Int. Ed. Engl.*, 3, 1964.
37. Marx, D., Tuckerman, M.E., Hutter, J. and Parrinello, M., The nature of the hydrated excess proton in water, *Nature*, 397, 601, 1999.
38. Pathways to Commercial Success, Fuel Cell Technologies Program, prepared by Pacific Northwest National Laboratory for the US Department of Energy Fuel Cell Technologies Program, September 2012.

# Index

Page numbers followed by f and t indicate figures and tables, respectively.